HYDRA 2000

Proceedings of the XXVIth Congress of the International Association for Hydraulic Research, London, 11–15 September 1995.

1. Integration of research approaches and applications
Edited by D. A. Ervine

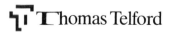

Thomas Telford

International conference organized by the International Association for Hydraulic Research and the Institution of Civil Engineers

Published by Thomas Telford Services Ltd, Thomas Telford House, 1 Heron Quay, London E14 4JD

First published 1995

Distributors for Thomas Telford books are
USA: American Society of Civil Engineers, Publications Sales Department, 345 East 47th Street, New York, NY 10017-2398
Japan: Maruzen Co. Ltd, Book Department, 3–10 Nihonbashi 2-chome, Chuo-ku, Tokyo 103
Australia: DA Books and Journals, 648 Whitehorse Road, Mitcham 3132, Victoria

A CIP catalogue record for this book is available from the British Library.

Classification
Availability: unrestricted
Content: collected papers
Status: refereed papers
User: civil and hydraulic engineers

ISBN: 0 7277 2056 2

Printed and bound in Great Britain by Redwood Books, Trowbridge, Wiltshire.

Preface

The International Association of Hydraulic Research (IAHR), founded in 1935, is a world wide organisation of engineers and scientists interested in hydraulics (including hydrology and fluid mechanics) and its practical application, including industrial and environmental aspects. IAHR stimulates and promotes both basic and applied research and strives to have science and technology contribute to the optimisation of world water management and industrial processes. Through its members, IAHR accomplishes its goal by a wide variety of activities, offering a common infrastructure for the hydraulic research community, one of the most important of these activities being the biennial congresses.

HYDraulics Research and its Application next century—HYDRA 2000—was the main theme of the twenty-sixth IAHR Congress held at the Institution of Civil Engineers in London from 11 to 15 September 1995. This congress theme was chosen so as to bring together researchers, hydraulics users and practitioners to expose and discuss matters of common interest.

The main theme was broken down into four topics:

1. **Integration of research approaches and applications.**
2. **Industrial hydraulics and multi-phase flows.**
3. **Future paths in maritime hydraulics.**
4. **The hydraulics of water resources and their development.**

The papers for each of these topics are produced in separate volumes of the proceedings. Additionally there are communications (short papers or technical notes), and student papers which were submitted for the John F. Kennedy Student competition.

At the Congress an invited lecture was given on each of the four topics by leading authorities. The extended papers on which these four lectures were based are also included in the proceedings, which thus provide an up-to-date account of the role of hydraulics research and the part it can play in the optimisation of future developments.

Not included in the proceedings are papers and discussion at five special seminars and two forums which formed an integral part of the Congress.

W. R. White
Chairman of the Local Organising Committee

Acknowledgements

In recent years the average attendance at IAHR congresses has exceeded 500, with worldwide participation. The organisation of an event of this size is a major task and involves many people. In the case of the London Congress I would particularly like to thank:

1. The Local Organising Committee:

> P. Ackers (Editorial Board), *Consultant*
> P. Avery, *British Hydromechanics Research Group*
> S. J. Darby (Finance), *National Rivers Authority*
> Dr D. A. Ervine, *Glasgow University*
> Professor A. J. Grass, *University College London*
> Professor P. Novak (Technical Programme), *Consultant*
> R. G. Purnell, *Ministry of Agriculture, Fisheries and Food*
> Professor R. H. J. Sellin (Social Programme), *University of Bristol*
> R. L. Soulsby, *HR Wallingford*
> D. G. Wardle, *Consultant*
> Professor B. B. Willetts, *University of Aberdeen*
> [Dr W. R. White (Chairman), *HR Wallingford*]

2. The Conference Office of the Institution of Civil Engineers:
   C. Chin, R. Coninx, B. O'Donoghue, S. Frye, N. Kerwood, J. Morris, S. Walton

3. The four editors of the Congress proceedings for their hard work in administering the review procedures and collating each volume.

4. The distinguished IAHR lecturer, Sir Geoffrey Palmer of the University of Iowa, USA.

5. The convenors of the seminars, the forums and the student competition.

6. All those 'behind-the-scenes' helpers including those who reviewed papers and the partners of most of the LOC members whose talents and energy were used extensively.

I would also like to thank the Institution of Civil Engineers for hosting the Congress and providing a degree of financial security to the conference organisers.

In addition I also gratefully acknowledge the support and sponsorship of the following:

> British Hydromechanics Research Group
> HR Wallingford Group Ltd
> Ministry of Agriculture, Fisheries and Food
> EMAP Business Communications
> Participants in the Exhibition
> Advertisers in the Exhibition Brochure

W. R. White
Chairman of the Local Organising Committee

Reviewing and editing

The number of papers and short communications submitted totalled over 350, an excellent response to the request for papers for which the Organising Committee is very grateful. However, this was many more than could be accommodated within the Congress timetable. It had in any case been the intention to review all submissions in terms of their relevance to the chosen topics and their quality, and so each paper and communication was reviewed by two experts in the field, may of them from overseas. In bringing the number of accepted papers for the Technical Sessions down to a manageable total of about 250, the Editors were able to set a high standard, but were faced with some tough decisions. The acceptance of a paper for presentation is therefore a stamp of very good quality, but we regret that inevitably a proportion of carefully prepared submissions had to be rejected: we realise the disappointment of authors if their hard work seems to have been marginalised.

The Editorial Board is very appreciative of the expertise and ready co-operation of our band of reviewers. They had a difficult task to complete in a very short time, but their skills have ensured a high technical standard for this XXVIth Congress. May I also add my thanks to the Members of the Editorial Board and especially to the four Theme Editors for their expertise and enthusiasm.

P. Ackers
Editor-in-Chief

Editor's preface

The papers presented in the volume are concerned with how we can marry the often diverse aspects of hydraulic engineering, namely research and practice. Researchers have been sometimes accused of having no concept of the needs of practising engineers and end users, whereas practitioners have been sometimes accused of sticking with well established methods even though better techniques have evolved. The final choice of papers therefore reflects a desire to make research practically relevant, and in some way, more accessible to the practising community.

With considerable input from Technical Division 1 (Methods in Hydraulics) and in particular, the Fluid Mechanics Section, it was agreed to have the following sub-themes:

A. Turbulent flow and exchange processes: linking small and large scales.
B. Interaction of physical and numerical modelling and its support by informatic tools.
C. Advanced experimentation and instrumentation in complex or extreme flow conditions.
D. Dynamic and stochastic approaches in hydraulic and hydrological engineering.

If you imagine the four sub-themes above to be horizontal layers of a cake, then when the cake is cut vertically, you have a much better idea of the rich diversity of papers in each sub-theme.

The initial papers deal with the problem of turbulence, which is perhaps one of our most fundamental problems. Its random, unsteady and three-dimensional motion over a wide range of eddy sizes is difficult to measure and even more difficult to model or simulate. What is encouraging about the papers on turbulence is the desire clearly demonstrated to apply turbulence measurements and models to real situations, such as outlet plumes, compound channels, sediment transport problems, estuary behaviour and hydraulic structures.

The sub-theme B emphasises the symbiotic nature of physical and numerical modelling in hydraulics. Results from physical models can be used directly in the refinement of numerical models, which in turn can feed-back into more realistic and refined experiments, setting-up a process of mutual advancement of both areas. This is helped not only by more powerful and less expensive computers, but just as importantly, advances in instrumentation, which is the focus of sub-theme C. The final sub-theme, D, on dynamic and stochastic approaches also includes some papers on hydrology.

In all these sub-themes, there are clearly recognised research themes covering areas of current practical interest. These include estuary behaviour, open channels of complex cross section, dispersion of pollutants, flow resistance in rivers, sediment deposition and scour problems, spillway behaviour and hydraulic structures. There are also some fascinating papers on new modelling techniques of genetic modelling and the use of artificial neural networks as well as parameter optimisation.

I hope you enjoy reading the papers contained in this volume. They represent a significant volume of human endeavour, and also reveal the concerns and interests of hydraulic engineers as we approach the new millennium.

Finally, I would like to thank the fifty reviewers who gave of their time and efforts to ensure the quality of the presentations in this volume was maximised.

D. A. Ervine
Editor, Theme 1

Contents

B. Interaction of physical and numerical modelling and its support by informatic tools

Papers

C. Advanced experimentation and instrumentation in complex or extreme flow conditions

D. Dynamic and stochastic approaches in hydraulic and hydrological engineering

INTEGRATION OF
RESEARCH APPROACHES AND APPLICATIONS

Jean J. Peters
Consultant, River Specialist
44 Rue Ph. de Champagne, B-1000 Brussels
Part-time Professor Vrije Universiteit Brussel
and Université Catholique de Louvain-la-Neuve

SUMMARY

The integration (or the lack of integration) of research approaches and applications is discussed through examples from the daily practice of fluvial hydraulics. This issue is quite important in a period in which hydraulic research is conducted mainly through informatics and when researchers would tend to loose contact with the real environment. It was not the purpose to provide solutions, rather to open a debate that is urgently needed, especially regarding organization and funding of hydraulic research and about education in hydraulics.

INTRODUCTION

Research in hydraulics has changed quickly over the past 30 years. Scale models are progressively replaced by numerical models. Both technologies are complementary and are used together for problem solving. Expert systems are introduced, allowing to include the fuzzy aspects of our knowledge: the experience in hydraulics can not all be translated into mathematical formulations.

Today, less funds are made available for fundamental research. Research Institutes are more than before involved in engineering projects, in which research and studies are to be conducted under sometimes stringent time constraints: there is often not time enough for more in-depth analysis.

New observation technologies open new perspectives for the research.

Modern instruments allow to make detailed measurements of all kind of hydraulic phenomena. This can allow our profession to understand better the physics of water and sediment movement.

Our education system produces excellent specialists, but often specialised only in one aspect of hydraulics. The use of hydro-informatics and the poor appeal of field investigations are possible reasons for the young hydraulic engineer or researcher to lose contact with our environment, with physics.

ABOUT THE RESEARCH APPROACHES AND APPLICATIONS

HISTORICAL AND GENERAL CONSIDERATIONS

First were the hydraulic applications, then came research ... Garbrecht (1983) gave a comprehensive historic overview of hydraulic research. For thousands of years, experimentation has furthered the learning process; skills were developed through failures, through the learning-by-doing mechanism. The need for hydraulic works, vital for survival, has originated the experimentation and the research.

Little is known about the first water specialists; historians and anthropologists try to find clues for the scientific knowledge acquired by ancient civilisations. Best documented are the historical achievements in North Africa, the middle East or Asia. Less well known are those in Central and South America.

Pre-Inca populations had gained extensive experience in design of water reservoirs and distribution systems, special irrigation devices for high altitude - with underground water release devices in pottery resulting in reduced ground hardening through evaporation -, water collection and drainage schemes (called "waru-waru's") - giving high crop yields despite the harsh climatic conditions in the elevated plateaux. In these ancient times, hydraulic works were conceived without models, from experience: the observation tools were the eye and the brain, the model was conceptual, obtained by unconscious or deliberate experimentation.

For the last two or three centuries, hydraulic research matured rapidly and powerful tools were generated, mainly from the nineteenth century on: numerical and scale modelling allowed hydraulic engineers to design the engineering works needed for the economic development.

From the late twentieth century on, hydraulic laboratories started using these tools with growing success and hydraulic research became a prerequisite to design.

The development of hydraulic research benefited from the discoveries in fluid dynamics. Different laws governing fluid movement were translated in mathematical terms. As pointed out by Cunge (1987), the efficiency of hydrodynamics lies in the fact that one can obtain, from the basic equations, conclusions concerning all phenomena belonging to the field, even those which are not yet known or have not yet been experimentally investigated. Hydraulics requires experimentation, and hydraulic equations give only an approximate description of phenomena that were observed, and only for the experimentally investigated range. Use in different environments or outside the range may not be acceptable or yield doubtful results.

As mentioned by Cunge (1987), the "golden age" of experimental hydraulic research (1930s to the early 1960s) - the era of the laboratories - benefited from the experience of their leaders who often were simultaneously engineers involved in engineering practices, fluid dynamics specialists and experimentalists. In the sixties, scale modelling reached its culmination point, also with the development of better laboratory instrumentation. It was however progressively replaced by mathematical models. The advent of the micro-computer produced a revolution in hydraulic research, making the tools more widely distributed and user-friendly. Today, hydro-informatics is the new fashion. Scale modelling became too costly in Europe and North America and has been progressively replaced by numerical modelling, leaving only special domains for physical modelling.

Both tools still shared the market in the seventies, but the latter has today outstripped the former in many applications. When scale modelling is required, it is often conducted in countries under development, because of the relative labour costs. This is, besides the main goal of development, one of the reasons why several hydraulic laboratories and institutes were created with foreign aid in these countries during the two or three last decades.

The fate of the field work - and research - was quite different. Prototype measurements were always considered as tedious, hazardous, costly and difficult to interpret. Often, the collection of hydraulic field data was organised merely in view of their use as input for the models or for calibration of the experimental coefficients, not so much for gathering knowledge and understanding of processes. It came even to the point that some declared without fear that field work was not necessary any more because the mathematical modelling would produce the requested data.

Obviously, the described trend must be linked to the evolution of our societies, in which field work is considered as less valuable then laboratory work. In present times, most younger scientists prefer using their computer than going out in the field.

Today, research in hydrodynamics and in hydraulics is mostly conducted by different scientists. Moreover, hydraulic research and hydraulic engineering are less and less integrated. Plate (1987) mentions the virtual impossibility for having hydraulic engineering and hydraulic research conducted by the same specialist. He further underlines the complexity of the present hydraulic engineering in which hydraulics is only part of the design process. Underestimation of the importance of hydraulic studies for design of hydraulic works has lead to failures or less efficient design, as will be shown further in this paper.

GENERAL APPRAISAL OF PRESENT INTEGRATION OF THE RESEARCH APPROACHES AND APPLICATIONS

Before discussing the present integration of research approaches and applications, it is necessary to have these terms better specified.

Fundamental research in hydraulics aims to find fundamental relationships based on scientific principles. We have to admit that little can be discovered that would revolutionise the field of hydraulics (Cunge 1987). There are however still smaller findings that make us progress in particular fields, e.g. in turbulence or secondary flow phenomena.

Applied research in hydraulics makes use of the basic principles and the mathematical relationships established through the fundamental research, e.g. the use of new theories about turbulence or secondary flow phenomena in the field of sediment transport. The relative domains of fundamental and applied research are not always easy to make.

"Hydraulic studies" is sometimes used instead of "applied research", although the former is closer to hydraulic engineering. Research aims to produce the tools for design engineers, who will use it for solving problems through hydraulic studies.

The success of the integration of research approaches and applications is quite diverse, depending on the domain considered in hydraulics. In the industrial hydraulics, when boundaries of the considered systems are mostly well known and/or controlled, research is better integrated than in other domains dealing with natural systems. The further discussions will mainly focus on domains for which research is hardly needed for achieving better applications through an improved, basic understanding of the hydraulic processes involved; fluvial hydraulics is such a challenging domain.

NEED FOR BETTER INTEGRATION?

GENERAL
The need for integration can be illustrated with many examples of engineering works that failed because of the lack of integration:
- water treatment stations having deficient functioning or unable to restore the river water quality;
- deficient access to harbours on alluvial rivers because of unpredicted sedimentation;
- water retention reservoirs silted up much quicker than foreseen;
- river training works such as retards or groynes working inefficiently or collapsed;
and so on.

The reasons for these failures may be many, among which obviously wrong decision-making or improper funding. But quite frequently were the hydraulic studies considered as of secondary importance. The client is often not aware of the limitations of the study tools and expects the designer to provide solutions in the same manner as in structural design.

Another reason is the lack of planning, of anticipation, so that research has to be conducted under pressure of time. This is particularly true with engineering related to natural hazards. After a flood, the political world suddenly becomes very sensitive for the subject and funds are made available for studies and works (not necessarily for data acquisition). However, these studies have then to be performed too quickly, not allowing for an in-depth assessment of the causes for the flood events and of the mechanisms that produced the damages. The situation is worse in Third World countries when problems are much more complex, difficult and of larger scale. Extrapolation of the existing experience may be quite hazardous.

Too strong believe or confidence in the research- and study tools may also be incriminated, for the client as well as for researchers.

In the following sections, the needs for better integration will be illustrated with examples and referring when possible to the sub-themes that will be discussed during the Congress:

I.A.   Turbulent flow and exchange processes: linking small and large (engineering) scales
I.B.   Interaction of physical and numerical modelling and its support by informatic tools
I.C.   Advanced experimentation and instrumentation in complex or extreme flow conditions
I.D.   Dynamic and stochastic approaches in hydraulic and hydrological engineering

The examples are taken from our own experience, often in still ongoing projects, for which it is not always possible to refer to literature.

TRANSPORT OF POLLUTANTS IN RIVERS

This topic has links with all four sub-themes I.A, I.B, I.C and I.D. It covers a domain in which the demand for research is obvious and for which the performance of the engineering tools is generally poor.

Numerical Modelling

A comprehensive state-of-the-art review for transport concepts in rivers was given by Holley (1987). The author refers to the Fickian-type models and exposes how they can cope with processes other than turbulent "diffusion", such as the spreading by differential advection or the exchange with dead zones.

As mentioned by Holley, Taylor (1921) showed that even for a stationary, homogeneous turbulence, although there is a relationship between the turbulence characteristics and the turbulent diffusion coefficients, the latter increases with time and becomes constant only after an elapsed time related to the integral time scale. During this initial period (called the "Fischer period"), application of the Fickian-type models for mixing is not possible, only after it (in the so-called "Taylor period") and the initial variation in mixing coefficient needs to be considered in modelling transport problems.

In the section about the exchanges with dead zones, Holley discusses the potential impact of a given size and type of temporal zones. These can be large (harbour inlets or bays) or small (at the bed, behind cobbles or dunes). The traditional approach had been different for a single dead zone or for those distributed along the bed of the river. For the latter, a term has to be added to the transport equation; for the former also an equation has to be added to account for the exchange between dead zone and main flow (Hays et al. 1966)

In 1973, a student in physics performed flume tests at the Hydraulic Research Laboratory in Borgerhout (Antwerp, Belgium) in the frame of his bachelor thesis report. The purpose was to analyse the effect of channel shape on the mixing coefficient in a one-dimensional Fickian model, all parameters in the equation being kept constant. To adjust the roughness, small concrete blocks were disposed on the flume bed. The tracer (Rhodamine Wt) was injected with a sophisticated system so that it would be immediately homogeneously mixed throughout the section. Measurements in lower sections displayed a skewed record of tracer concentration with time and the effect of differential convection.

This student decided to abandon the deterministic approach and to build a stochastic model (Michel 1974). It required obviously two parameters, one for the probability for capture of the tracer in a dead zone (e.g. behind a roughness element) and one for the probability for release back to the main flow. The equation was solved and the computations fitted very well the experimental results. Evidently, the calibration of two coefficients is less easy than the single one of the Fick equation. However, no distinction is needed any more between the Fischer period and the Taylor period (for constant mixing coefficient). Initially, the concentration curve is skewed, but with time it becomes symmetrical, similar to the one obtained with a Fick model Figure 1).

The model was used in several studies, e.g. about the contamination of the Semois river by the city sewers of Arlon, in Belgium. A few days before defending his thesis, we discovered a paper by Sayre & Conover (1967) in which the same stochastic approach was followed for modelling the movement of traced sediments: a marked sediment particle could be covered by a progressing bed form and re-exposed later on to entrainment by the flow.

Interesting is the fact that differential advection can be treated with

7

the same approach as the dead zones - the similarity is obvious - what explains the good results obtained in the application of Michel's model for which coefficients remain fairly constant in rivers in which differential advection and dead zones are the main causes for mixing. This stochastic approach could thus be an alternative for application in rivers for which the mixing coefficients of Fickian models vary too widely. In many natural rivers, turbulence has only little contribution to the overall mixing, differential advection contributing most to it.

Field Experiments and Scale Modelling
Field experiments and scale modelling remain necessary for improving our understanding of the mixing processes in rivers, especially the contribution of secondary flow phenomena such as turbulence, spiral flow, bursting etc. Scale effects are however limiting the use of physical models. Field investigations were till recently hampered by the difficulty to measure the complex flow structure in a natural environment. New instrumentation is now being tested that will provide the researcher with a wealth of information, namely the Acoustic Doppler Current Profiler (ADCP).

The ADCP was originally developed for ocean research but became operational in rivers some four years ago. It is composed of an array of four sonar's oriented to the bed, each beam at 90° of its neighbour and with a small angle to the vertical (Figure 2). The equipment is highly sophisticated, but its functioning may be summarized as follows: the ultrasonic waves are reflected by the bottom as in a traditional echo-sounder but at each depth, part of the energy is back-scattered by the sediments suspended in the water. Each beam is electronically divided in cells (gauging depth windows) in which the refracted energy is measured. For each depth, both the intensity and the Doppler frequency shift are measured in all beams. The scattered echo of the river bed allows to measure the velocity of the vessel relatively to the bottom. The transmitted intensity yields a measure of the amount of sediments in suspension. The Doppler frequency shift of three beams is sufficient to determine the velocity in magnitude and in direction.

The application of the ADCP is discussed more in detail in the section on fluvial hydraulics, but it is evident that the instrument is a powerful tool for describing in detail the flow structure in the prototype. The river becomes the model on scale 1:1. The numerical modelling would most benefit of the advances in field investigations and scale modelling could become of minor importance, except for systematic research on particular mechanisms such as the spiral flows in meandering channels. Indeed, the geometric characteristics of the channel in a scale model can be adjusted as wanted.

ENGINEERING ASPECTS OF FLUVIAL MORPHOLOGY
White (1987) presented at the XXIIth I.A.H.R. Congress in Lausanne a review of engineering aspects of fluvial morphology. In the last section on "research needs" he states that "... Today we have computers which are capable of solving equations at an incredible rate but in some areas of fluvial hydraulics we have yet to find those equations."

In his presentation at the same Congress, Ackers (1987) deals for a large part with physical modelling of sediment transport. He insists on the importance of understanding the phenomena themselves, as well as on the need to combine physical with numerical modelling and other calculations. He also states that "... well documented measurements of prototype behaviour are the only proof of success."

The following examples are given for identifying areas for which integration of research approaches and applications are urgently needed. But there is urgent need to get quickly a better insight in some phenomena with comprehensive and detailed field observations.

Controlling the Morphology of the Pirai River
The Pirai river drains a rather small catchment at the Andes foothills in Bolivia. During its 400 Km long journey to the Río Grande, one of Amazon's tributaries, it passes quickly from high altitudes with steep slopes to mild slopes, creating a sizable alluvial fan in which the river course avulsed repeatedly (Figure 3).

The particular geomorphology of the catchment and the rain patterns produce flash floods. In its middle course, in the vicinity of Santa Cruz de la Sierra (a major city with booming economic activities), peak flows may reach more than 3,000 m³/s although the maximum monthly discharge does not exceed 40 m³/s. The river bed width amounts to several hundreds of metres. Sediments eroded from the upper catchments are transported during flood events only and deposited in the alluvial fan, creating rich agricultural land. The river bed is composed of fine sands (about 0.300 mm), the result of deposition of finer suspended load at the end of the flood. During the flood, selective erosion and transport makes most likely that the bed material is quite coarse, as evidenced by the cobbles and gravels found just beneath the top soil. Local companies are active in river sand and gravel mining.

A major flood occurred in August 1983, lasting about one day only, but with a peak discharge exceeding 2,300 m³/s that destroyed infrastructure (e.g. all bridges on the Pirai except one) and damaged severely the city of Santa Cruz and rural areas. International aid was provided for rebuilding infrastructure, but also to erect a flood protection scheme for the city of Santa Cruz and to set up a Master Plan for the Pirai catchment. After completion of this programme, new severe floods occurred in 1991 and 1992, damaging the flood protection scheme of Santa Cruz and inducing a river course avulsion in the alluvial fan, close to the town of Montero (Figures 3 & 4). The analysis of the damages are interesting to assess how research approaches and applications can be integrated.

Design of bank protection at Santa Cruz de la Sierra

Design of bank protection can be based on scour depths predicted with scale model tests or with computations. However, the river behaviour is not always taken into account properly. In the case of Santa Cruz, a first project for flood protection included training over the entire stretch of the river along the city. In a later phase, only the right bank (city side) was protected. The lay-out followed the actual course, "freezing" its morphology, with only some minor bank adjustments.

The Pirai river is usually described as a braided river. Most of the time, the flow distributes through an intricate system of minor channels, each one only some decimeters deep (Figure 5). During flood events, the stream pattern becomes more meandering or even straight.

The importance of geomorphology was overlooked during the studies, especially the presence of higher river banks along the left margin, in the Santa Cruz stretch, belonging to tertiary deposits (see Figure 3). The plan shape of the low discharge channels is determined by the shape of the eroding banks; during the next upcoming flood, the flow will be guided and the more or less meandering flow leaving one bank may take a course not aligned with the overall shape of the opposite bank.

The flood protection scheme of Santa Cruz de la Sierra was conceived along the right river margin only, with a retreated levee and a bank revetment along the existing bankline. This river margin has a complex overall planform, due to the local attack by flow "rebouncing" from the left margin, which is more resistant to erosion (see Figure 5).

At present, another strategy is being tested: to alleviate the bank attack by re-aligning the opposite margin. Bank excavation is used here as complementary (as an active recurrent measure) to bank protection (as a passive training work). The methodology was successfully applied in two sites.

Figures 6 & 7 show the repair to a bank revetment at a place where the flow attacked straight on the river margin. The revetment was a gabion-type with a falling apron. Computation of bed scour with various formulas gave diverse results, so that the length of the falling apron was chosen arbitrary, shorter than the most pessimistic calculations. During a severe flood, no damage was observed all along the river bank as long as the flood flow was oriented alongside the margin. However, at a place where the flow attacked straight on the bank, the whole bank profile collapsed due to very deep, local scour.

River avulsion in the alluvial fan at Montero
The Pirai river experiences regular avulsion of its course in the region of the alluvial fan, close to the city of Montero, where it is joined by a major confluent, the Güenda river. The place is an important agricultural centre because of the fertility of the soils.

Two major dykes, built for supporting road and railway, cross the floodplain and interrupt overland flow during extreme flood events (see Figure 4). They have induced significant changes in the morphology of the river system that are too complex to be explained here. However, the analysis made (Peters 1994) shows that human intervention in the natural avulsion process had contributed to the first avulsion some 15 years ago and triggered the more recent one in 1992.

After the first avulsion, a new river channel was excavated almost along the former river track, because the avulsing channel posed a too great threat to the city of Montero. A flood levee followed the new channel all along its right margin (city side). The channel cross-section was designed for a limited flood discharge and had a small sinuosity: the idea was that consecutive floods would widen it up to normal size. Instead, the channel began to develop a strong meandering pattern, that started from upstream and extended progressively in downstream direction. Splays occurred regularly, inundating the floodplain and destroying crops, but no new avulsion occurred (Figures 8, 9, 10 and 11).

In February 1992, a new avulsion developed at the same place as in 1980. The original structure closing the inlet of the avulsing channel was composed of a dyke, with jetties as retard structures. These had functioned well for a long period. However, the change in river morphology at that place had finally resulted in a flow attacking straight on the closure. Again, as in the case of the bank revetment in Santa Cruz, a structure failed because of an attack for which it was not designed (Figure 12).

Conclusions for the Pirai case
The experience at Santa Cruz let us conclude:
- design of river training works must be based on an in-depth assessment of the river behaviour, e.g. planform evolution, possible flow patterns, influence of stage on these;
- formulas for calculation of scour depths are only valid for usual flow patterns, e.g. parallel to the river margin; there is a need for research on scour under other flow conditions, particular for the considered river;
- constructions in floodplain must be designed taking into account their potential impact on overland flow and sediment transport
- in dynamic alluvial rivers, planform control should be made not only with training works; recurrent measures must be considered as effective tools for managing the river behaviour.

Sediment Transport and Morphology in Large Alluvial Rivers
Several large projects are currently ongoing or planned on large alluvial rivers. The Flood Action Plan in Bangladesh is likely the most challenging. Impact of river works and of constructions in the floodplain must be assessed. Models are used, but some specialists question their validity, especially because of peculiar values found for model calibration coefficients, or because of the difficulty to simulate the complex flow and sediment phenomena. One of the Flood Action Plan Project, the River Survey Project (FAP 24) was set up to collect all-weather reliable flow and sediment data in the major rivers of Bangladesh and to study their hydraulic and morphological behaviour.

A similar study of hydraulic and morphological behaviour for improving navigation conditions was conducted on the Zaïre inner delta from 1968 to 1988 (Peters 1978; 1981; Peters & Wens 1991).

In his paper in Lausanne, White (1987) discusses aspects such as sediment grading, resistance of alluvial channels, channel shape, size and slope, channel patterns. In his list of research needs appear, among other:
- movement of graded sediments
- impact of unsteady flow on sediment transport and channel form
- prediction of channel plan form and its change through time

Coleman (1969), in his paper about the Jamuna (Brahmaputra) river suggested to set up a classification of large delta, stating that a better understanding of river behaviour in one large delta could help solving problems in another.

Advanced instrumentation is used in the River Survey Project in Bangladesh, besides more traditional equipment. Quite interesting results were obtained sofar with the Acoustic Doppler Current Profiler, with various sediment sampling devices, with remote sensing and so on. Surprisingly (or may be not) these data seem to confirm conclusions reached over the past 27 years in the Zaïre delta with very robust, conventional instruments. Some are discussed in the following sections:
- sediment transport and sorting processes
- bed forms and roughness
- formation of bars and islands

<u>Sediment Transport and Sorting Processes</u>
Sediment rating is fundamental for assessing the impact of river works on the future behaviour of rivers, their channels and bars. The data are usually presented in double-log diagrams because of the huge scatter observed. Analysis of sediment gauging data with formulas yield ratio's between computed and measured that are often quite large. These ratio's vary from formula to formula. The poor reliability of field data is mostly given as an explanation for the poor results.

The sediment transport formula compute bed load, or suspended load or total load, some all of these. The concepts of bed load and suspended load were discussed by many authors and bed load is accepted as that part of the sediment that rolls, slides on the bed or that saltates close to the bed, i.e. in a layer which thickness amounts only to a few diameters of bed particles, possibly up to a few centimetres. Particle sizes to be used in bed load or in suspended load formula are different, but they are taken equal at all levels from the bed in each mode of transport. Only a few formula allow to account for graded sediments (e.g. Einstein 1971, Ackers & White 1973, Belleudy et al. 1987). No formula allows to account for differential movement in suspension. Rouse's law is the standard for the distribution of concentration in a vertical profile.

Measurements performed since 1968 in the Zaïre river demonstrated that sorting was very important, certainly in relation with morphology. In a single cross-section, the size of bed material may range from 150 μm on a submerged bar at 3 m depth to 1,000 μm in the 30 m deep channel. The two main channels bifurcate and take each half of the river flow, but carry very different sediment loads, with sizes that are significantly different.

Precise measurements performed on many verticals above diverse bed form types show that coarser bed particles move above the bed in layers which thickness may reach half a metre. The analysis of these direct sediment transport measurements with the detailed bathymetry and surface flow patterns lead to the conclusion that a new approach was needed for defining sediment load in relation with morphological studies. A close analysis of the vertical profiles of both sediment concentration and size revealed two domains, with distinct vertical distribution laws (Peters & Goldberg 1989). Bed material load may move as a so-called "near-bottom bed-material load", i.e. that particles that would normally stay in contact with or very close to the bed may move in a kind of suspension up to a certain level.

This point was discussed with Bagnold in 1971 when his formula (1966) was tried for 2-D computations in relation with morphological changes (Peters 1977). The sediment transport measured in the lowest 0.50 metre was taken as the load affecting the changes in bed morphology (could be called the 'morphological load').

The same phenomena were observed with measurements performed in the Jamuna river with the same sampling instrument as used in the Zaïre river (See Figure 13). Several studies are presently conducted in a restricted area. By doing so it would be possible to assess the relative contribution on sediment transport of other flow mechanisms, such as large-scale secondary flow phenomena and flow cells; also the influence of bed forms could be investigated. In this case, the ADCP technology yields a wealth of information on all kind of flow phenomena. It is too early to draw definite conclusions, but the observations must at least be taken into account for the further morphological modelling.

Bed forms and Roughness

Bed forms were studied extensively in laboratory flumes. The distinction is made between lower and higher regime bed forms. The existence of two flow regimes makes that scale models for large alluvial rivers can not simulate properly the overall morphological changes, but can be utilized for local studies (Peters 1990).

Bed form characteristics are used to determine channel roughness and some functions provide a simultaneous solution of the resistance and the transport function. In the observations made in the Zaïre river, different types of dunes were identified: they have different sizes and aspects and the smallest ones may appear over a flat bed or above the large scaled ones (Peters 1977). Detailed measurements and observations during flood events made clear that the small scaled dunes occur at high stages (lower flow regime?) and that their appearance could be related to a kind of sediment overloading. Part of these data were used in a study of Delft Hydraulics about the flood levels in the Rhine river in the Netherlands. Figure 14 gives the relationships between observed and computed values of roughness k and Chezy coefficients given for 3 types of bed forms by the formulas of Van Rijn and Engelund: large scale dunes, small scale dunes with more rounded shape and small scale dunes with steep lee side (DELFT HYDRAULICS 1986).

The analysis of the bed forms is part of the experimental method developed for predicting the morphological changes of the Zaïre river, mainly the movement of channels, bars and islands. The appearance of one type instead of another is interpreted, apart of and together with other indicators, in terms of sediment overloading at that particular place. Local deviations of the water slope are associated with bed form changes. The analysis of bed forms helps steering the dredging operations, dredging then being used as a morphological agent.

Bed form data are presently analyzed in the Bangladesh rivers within the River Survey Project. Side scan sonar is used for determining the planform of the dunes. On the basis of the preliminary results it appears that some doubts may be expressed about a simple relationship between bed forms and roughness. There is clearly a need for further research in prototype to understand how flow resistance is build up in large rivers, and not only because of bed form geometry.

<u>Formation of Bars and Islands</u>
Bathymetric observations in large rivers are tedious, time consuming and costly. Therefore, use of remote sensing is being promoted widely for planform studies. For some applications, and especially in numerical modelling, river geometry is given with cross-sectional data. Cunge (1987) refers also to "an 'engineering type' intervention with the use of an 'equivalent bathymetry', which is not that actually measured but that which, with the available computer code and budget, will give acceptable results." He adds "Such solution is acceptable only if the 'adaptation' of the bathymetry is made by the hydraulic engineer, who has the field experience, knows exactly the purpose of the study and has intimate knowledge of and experience in the use of the simulation code. If not, essential features may well be lost in the process."

We can only support this viewpoint, but add to it that the analysis of river bathymetry (not only of the part submerged at the moment of the survey but also the dry part that will be inundated at high flow!) is needed for identifying the movement of the morphological load, as demonstrated by the studies on the Zaïre river. This identification is part of the method to assess and predict the river behaviour. Satellite images does not give sofar the details that are needed.

The drawback of satellite images in visible light is that they cannot give relief, only an impression of it for mountainous regions. Radar satellite images combined with hydraulic models were developed recently to reveal the bathymetry in some areas of the North Sea (Hesselmans et al. 1994). A possible application of these techniques in rivers would need some more research. The present conjunctive use of remote sensing and bathymetric charting in the River Survey Project in Bangladesh yields promising results.

Measurements with ADCP of flow structure around bars and islands will allow to assess the relative contribution of various mechanisms, among which spiral flow, near-bottom bed material load associated with erosion-deposition patterns, confluences and bifurcations and so on. The studies show that only the integration of the various approaches would give better insight in the mechanisms that make bars and islands move.

RESEARCH FUNDING THROUGH APPLICATIONS

Research is more often conducted at the occasion of engineering projects. Most of the above mentioned examples are related to internationally funded projects (the European Commission for the Pirai river and Bangladesh). These projects have a study component that can not be used for in-depth scientific research, because they are development projects which money should be entirely devoted to the development. Some applied research can possibly be allowed, but no fundamental research, for which other sources of funding exist. However, the available budgets do not permit these comprehensive and intensive efforts as needed today in fluvial hydraulics.

Natural disasters, such as the floods in Asia, in the Americas and in Europe in 1992, 1993 and 1994 can give the impetus (and funds) for new research activities. But experience learns that memory is short and funds will soon be diverted to more visible needs. Concerted action should be promoted, but giving the smaller institutes the same chances as the large ones. The North must take up its duties and make sure that fundamental and applied research would be possible for answering some of the basic questions in view of solving acute problems such as those related to floods and inundations.

Field work should be promoted (and funded) because many phenomena still lack appropriate mathematical formulation, or, more basically, a sufficient understanding.

EDUCATIONAL ASPECTS

Education is a key issue debated during the last IAHR Congresses and valuable recommendations have already been made. However, we must realize that the young engineers and researchers in hydraulics are usually well educated, with a good scientific background, but that they often lack skills and broadmindedness that are needed for solving the hydraulic problems, today more than ever in a multidisciplinary manner.

As stated by Cunge (1987), the objective for research in numerical modelling is the development of hydraulic simulations to be used by non-specialists for solving everyday problems. So there will be a need for a small number of knowledge centres, where the "non-trivial problems" could be dealt with. This statement is valid not only for numerical modelling. This would however not solve the problem of the gap existing usually between the desk engineer or researcher and the field specialist. Particularly in fluvial hydraulics, the challenges are now to better understand our physical systems. Should education have to adapt to be able to cope with this challenge or do we have to develop skills in the everyday practice? Don't we have to organise our education system in such a way that courses could be shared by the specialist-researcher and the specialist-engineer? We come to realize that many theories have their limits; we must therefore train the students to remain critical.

A better integration of research approaches and applications should be achieved also through adapting our education in hydraulics, confronting engineers and researchers with the various and diverse facets of our profession.

REFERENCES

Ackers P. & W.R. White 1973, "Sediment transport: new approach and analysis," Proc. ASCE, Nov

Ackers P. 1987, "Scale Models, Examples of how, why and when - with some if'" Proceedings Technical Session ?, XXII IAHR Congress, Lausanne

Bagnold R.A.. 1966, "An approach to the sediment transport problem from general physics," U.S.-Geol. Surv. Professionnal paper 422-I

Belleudy Ph, le Rahuel & T. Yang, " ," Proceedings Technical Session ?, XXII IAHR Congress, Lausanne

Coleman J.M. 1969, "Brahmaputra river: channels processes and sedimentation," Sedimentary Geology, Vol. 3, Nos 2-3, pp. 129-239

Cunge J.A. 1987, "Numerical Hydraulics Modelling; Late '80s context and cross-roads," Proceedings Technical Session ?, XXII IAHR Congress, Lausanne

DELFT HYDRAULICS 1986, "Dimensies van beddingsvormen onder permanente stromingsomstandigheden by hoog sediment-transport (bed forms dimensions under stationary flow conditions by high sediment transport rates)" Delft Hydraulics Laboratory Research Report M 2130/Q 232

Einstein A.B. 1971, "Sediment transport measurements in a gravel river," Journal of Hydraulics Division, ASCE, HY11, Nov. 1971

Garbrecht G. 1987, "Hydraulics and Hydraulic Research: Historical Review," A.A. Balkema, Rotterdam

Hesselmans G.H.F.M., G.J. Wesink, C.J. Calkoen & H. Sidhu 1994, "Application of ERS-1 SAR data to support the routing of offshore pipelines. Delft Hydraulics Report H1900

Holley E.R. 1987, "Transport of Pollutants in Rivers," Proceedings Technical Session ?, XXII IAHR Congress, Lausanne

Hays J.R., Krenkel P.A. and Schnelle, K.B. 1966, "Mass transport mechanisms in open-channel flow," Vanderbilt University, San. and Wat. Res. Engrg., Dept. of Civ. Engrg., Technical Rp. No. 8

Michel D. 1974, "Modèle stochastique de dispersion turbulente dans les cours d'eau," Thesis Report for obtaining a Bachelor of Science Degree in Physics, Université Libre de Bruxeles

Peters J.J. 1977, "Sediment transport phenomena in the Zaïre river," in "Bottom Turbulence". Proceedings of the 8th International Liege Colloquium on Ocean Hydrodynamics, Elsevier Oceanography Series, Vol. 19

Peters J.J. 1978, "Discharge and sand transport in the braided zone of the Zaïre estuary," Netherlands Journal of Sea Research, Nr 12

Peters J.J. 1981, "Water and sediment gauging of the Zaïre (Congo)." Proceedings IXXth IAHR Congress, New Delhi: Vol. 2

Peters J.J. & A. Goldberg 1989, "Flow data in large alluvial channels," Intern. Conf. on Interaction of Computational Methods and Measurements in Hydraulics and Hydrology - HYDROCOMP '89, Dubrovnik, Yugoslavia, Elsevier's Applied Sciences

Peters J.J. 1990, "Scaling of sediment transport phenomena in large alluvial rivers with very low slopes," Proceedings NATO Workshop on Movable Bed Physical Models, Delft (De Voorst) August 18-21, 1987, edited by H. W. Shen, Kluwer Academic Publishers, NATO ASI Series, Series C: Mathematical and Physical Sciences - Vol. 312: 149-158

Peters J.J. - F. Wens 1991, "Maintenance dredging in the navigation channels in the Zaïre inner delta," COPEDEC III Conference, Mombasa

Peters J.J. 1994, "Manejo de los Ríos en la Cuenca del Piraí," own publication, funded by the European Commission (in Spanish)

Plate E.J. 1987, "The role of the Research in Hydraulic Engineering," IAHR Bulletin 1987, Vol. 4, Nr 1

Jean J. Peters

Sayre W.W. & W.J. Conover1967, "General two-dimensional stochastic model for the transport and dispersion of bed-material sediment particles; IAHR XIIth Congress, Fort Collins, Colorado, Vol. 2

Sayre W.W. 1973, "Natural Mixing Processes in Rivers," Chapter 6 in "River Mechanics III: Environmental Impact on Rivers," Ed. H.W. Shen, Fort Collins, Colorado

Taylor G.I. 1921, "Diffusion by Continuous Movements," Proc., London Math. Soc., 20A

White W.R. 1987, "Engineering aspects of fluvial morphology," Proceedings Technical Session A, XXII IAHR Congress, Lausanne

FIGURES

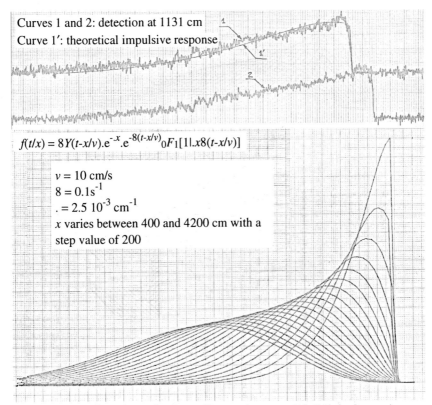

Curves 1 and 2: detection at 1131 cm
Curve 1': theoretical impulsive response

$$f(t/x) = 8Y(t\text{-}x/v).e^{-x}.e^{-8(t\text{-}x/v)}{}_0F_1[1|.x8(t\text{-}x/v)]$$

$v = 10$ cm/s
$8 = 0.1$s^{-1}
$. = 2.5 \ 10^{-3}$ cm^{-1}
x varies between 400 and 4200 cm with a step value of 200

Figure 1: Stochastic turbulent dispersion model (Michel 1974)

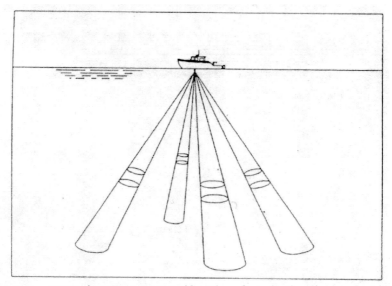

Figure 2: ADPCM beam geometry and location of one depth cell

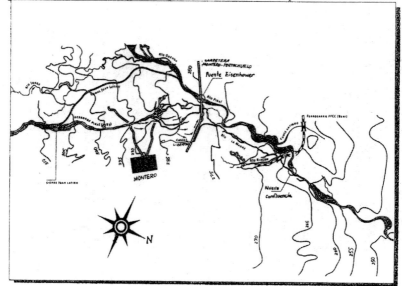

Figure 4: Area of Montero with the avulsion channel (Desborde Pirai), the man-
made channel (canal Juan Latino), the city of Montero, the main road
built on a dyke (carretera Montero-Portachuelo), the railway
(Ferrocarril FFCC Beni) and depth contours in metres above sea level.

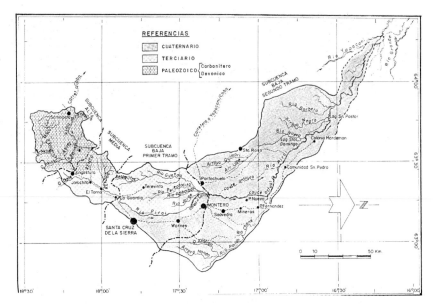

Figure 3: Geomorphology of the Pirai river catchment (Bolivia).

Figure 5:   The Pirai river at Santa Cruz de la Sierra (on the right, looking
downstream). Note the braided channel pattern (low flow conditions,
between 10 and 20 m³/s), but the general river course swinging back
and forth.
At the centre of the picture (right bank upstream of the city) bank
excavations—as a recurrent measure—and bank protection
(revetment) to re-align the general river course.

Figure 6: Repair work at the bank revetment (looking to the right bank). Note the falling apron.

Figure 7: Same repair work at the bank revetment (this time looking to the left bank). Note the left channel oriented to the right bank, previously to the place where the revetment collapsed..

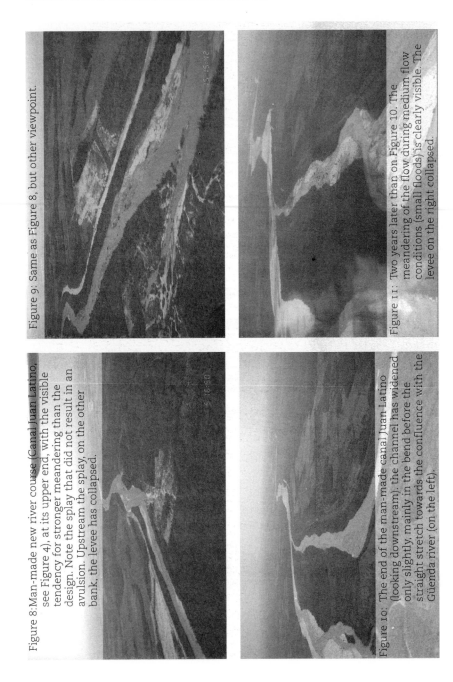

Figure 9: Same as Figure 8, but other viewpoint.

Figure 11: Two years later than on Figure 10. The meandering of the flow during medium flow conditions (small floods) is clearly visible. The levee on the right collapsed.

Figure 8: Man-made new river course (Canal Juan Latino, see Figure 4), at its upper end, with the visible tendency for stronger meandering than the design. Note the splay that did not result in an avulsion. Upstream the splay, on the other bank, the levee has collapsed.

Figure 10: The end of the man-made canal Juan Latino (looking downstream), the channel has widened only slightly, mainly in the bend before the straight stretch towards the confluence with the Güenda river (on the left).

Figure 12: The Pirai river at Montero, looking upstream (see Figure 4). Closure of the avulsion channel opened during the flood of February 1992.

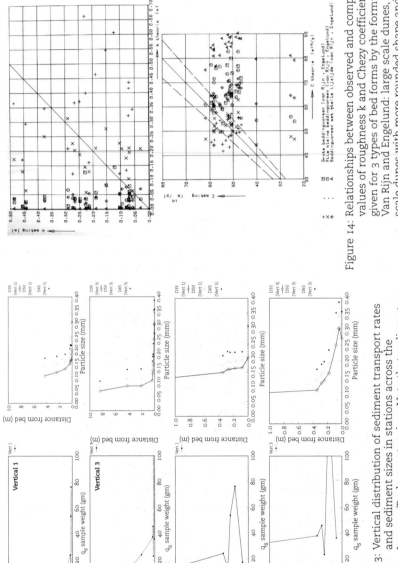

Figure 14: Relationships between observed and computed values of roughness k and Chezy coefficients given for 3 types of bed forms by the formulas of Van Rijn and Engelund: large scale dunes, small scale dunes with more rounded shape and small scale dunes with steep lee side.

Figure 13: Vertical distribution of sediment transport rates and sediment sizes in stations across the Jamuna/Brahmaputra river. Note that sediment size changes in the near-bed zone.

1A1

TURBULENT FLOW STRUCTURES AND VELOCITY DISTRIBUTION IN A TWO-STAGE CHANNEL BEND

Robert SELLIN
Department of Civil Engineering
University of Bristol
UK

INTRODUCTION

The UK Flood Channel Facility (FCF), funded originally by the Science and Engineering Research Council (now the EPSRC), was constructed at HR Wallingford in 1985-86. It was intended to provide a large scale indoor experimental facility for research teams both from the universities and from industry to study the mechanics of two-stage channel flow. The ultimate aim was to provide river engineers with a reliable design manual for two-stage channels.

The FCF, shown in Figure 1, consists of a tank in which channel models are constructed. The usable model area measures 56m x 10m, making it very suitable both for wide floodplain geometry and also for meandering channels. The channel gradient is fixed when the model, which is given a cement mortar finish, is constructed. Water can be recirculated with a maximum discharge of 1 m^3s^{-1}. Full details of the FCF are given in Knight and Sellin[1], while references to previous work on this topic will be found in Sellin, Ervine and Willetts[2]

DESCRIPTION OF TESTS

Two sets of experiments were completed, each employing an inner or main channel consisting of a series of uniform meanders set in a flat floodplain having a "valley gradient" close to 0.001. In the first set the sinuosity of the inner channel is 1.37 and in the second 2.04. The principal plan dimensions are shown in Figure 1. The idealised "natural" geometry of the inner channel can be appreciated from Figures 3 and 4.

The programme included systematic observations of stage, discharge, the velocity field and water surface levels. The stage-discharge data were presented and discussed by Sellin, Ervine and

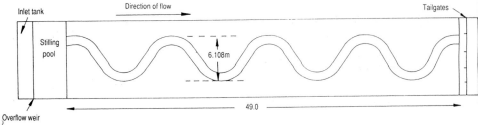

Figure 1   Plan form of the meandering two-stage channel in the FCF at HR Wallingford

Willetts[2], and the velocity distribution in the cross-over regions of the inner channel by Ervine, Sellin and Willetts[3]. The present paper reports the velocity fields and secondary circulation in the bend regions of this channel.

Detailed measurements were made for each of three flow depths, 140mm, 165mm and 200mm; the first of these represents inbank flow, which is excluded from this discussion. These depths are measured relative to the local floodplain level which is defined as +150mm above local datum at each section. The FCF discharges for the 165 and 200mm uniform depth flows were 0.058 and 0.223m³s⁻¹ respectively.

The stage-discharge curves for these tests are presented and discussed by Sellin, Ervine and Willetts[2]

In order to determine the flow structure and momentum exchange mechanisms in this compound meandering channel the following objectives and procedures were adopted. While the stage-discharge relationship, and its derivatives, reveal the global effect of this

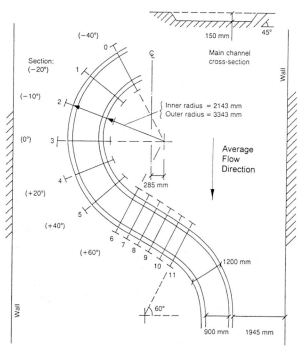

Figure 2   Details of channel study reach showing dimensions and location of measuring sections

28

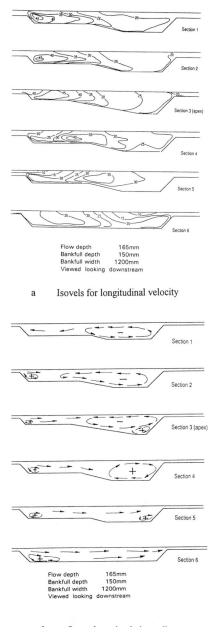

Flow depth   165mm
Bankfull depth   150mm
Bankfull width   1200mm
Viewed looking downstream

a   Isovels for longitudinal velocity

Flow depth   165mm
Bankfull depth   150mm
Bankfull width   1200mm
Viewed looking downstream

b   Secondary circulation cells

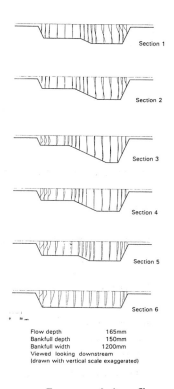

Flow depth   165mm
Bankfull depth   150mm
Bankfull width   1200mm
Viewed looking downstream
(drawn with vertical scale exaggerated)

c   Transverse velocity profiles

Figure 3   Distribution of velocity components
and inferred secondary circulation at a bend in a
two-stage channel - low overbank flow.

complex internal flow structure on the channel
conveyance it provides no direct information on
the underlying flow structures. To provide this
information point velocity measurements were
made at selected cross-sections on a notional
rectangular grid. The measurements system
adopted gave the horizontal direction as well as
the magnitude of each point velocity vector but
no information as to the vertical component. The
selective use of dye releases, recorded on video
(Ervine, Sellin and Willetts[3]), has provided a
valuable supplement to these velocity

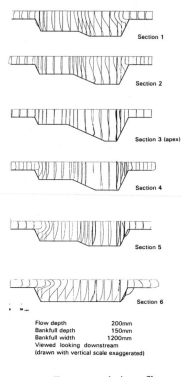

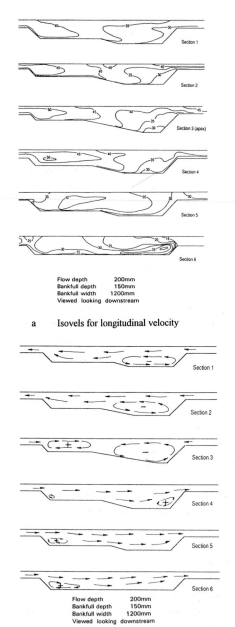

Flow depth   200mm
Bankfull depth   150mm
Bankfull width   1200mm
Viewed looking downstream

a   Isovels for longitudinal velocity

Flow depth   200mm
Bankfull depth   150mm
Bankfull width   1200mm
Viewed looking downstream
(drawn with vertical scale exaggerated)

c   Transverse velocity profiles

Figure 4   Distribution of velocity components and inferred secondary circulation at a bend in a two-stage channel - high overbank flow.

measurements. Full details of the techniques used are given in Sellin, Ervine and Willetts[2]

Figure 2 shows the location of the measuring sections, numbered 0 -11 over a representative length of the 1.37 sinuosity ($60°$ crossover angle) channel. Sections 1 - 6 are considered here as relating to the bends.

Figures 3a and 4a show the longitudinal velocity components in isovel or velocity contour form.

Flow depth   200mm
Bankfull depth   150mm
Bankfull width   1200mm
Viewed looking downstream

b   Secondary circulation cells

Figure 3 results represent shallow overbank flow (for a natural river floodplain), Figure 4 shows results for deeper flow conditions over the floodplain (for an artificial two-stage channel).

Figures 3c and 4c show the transverse *horizontal* velocity component and are drawn on the verticals on which the point measurements are made. Postulated secondary circulation cells for the bend region are shown in Figures 3b and 4b

DISCUSSION OF RESULTS.

Figure 3a, for shallow overbank flow, shows that the maximum longitudinal velocity zone stays close to the inside bank of the lower channel until section 4 is reached. From this point it weakens and moves progressively to the other side of the channel. There is evidence of a new high velocity area forming near the inside bank by the time section 6 is reached.

Figure 4a shows the situation for deeper overbank flow. However the much stronger floodplain flow associated with this relative water depth appears to dominate the longitudinal velocity distribution compared with that shown in Figure 3a. The more diffuse filament of maximum velocity is still located close to the inside channel at section 3 but it can be seen that a very concentrated high velocity zone has appeared under the outside bank at section 6. This is related to the strong flow out of the inner channel and onto the floodplain in this region, as demonstrated by the transverse velocity profiles for section 6 for this flow depth in Figure 3c. The combination of these two strong and spatially superimposed streams, flowing in different directions, suggests a zone of high shear and resulting high turbulence in this area. This is confirmed by the measured high values of wall shear stress on the floodplain here.

The transverse velocity components, Figure 3c, show how the surface cross-flow current, well established at section 1, switches direction, relative to the channel thalweg, as the sections move downstream and around the bend. The rotational sense of the principal secondary flow cell (Figure 3b) is reversed at around section 4 under the influence of the changing cross currents.

For deeper overbank flow Figure 4c shows the transverse velocity profiles and Figure 4b the secondary flow cell structure. It is clear that the surface currents, aligned more exactly now with the "valley " direction, are now dominant.

CONCLUSIONS

The information contained on the velocity and circulation diagrams considered in this section so far can be used to build up an overall picture of the principal flow mechanisms involved. Figure 5 shows such a synthesis for the 1.37 sinuosity channel although this does not include the finer distinctions noted above between the shallow and deep flow on the floodplain. Looking again at Figures 3b and

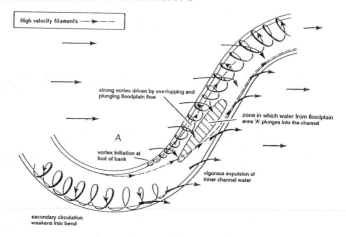

Figure 5   Representation of principle flow structures within the flooded
meander channel

4b it can be seen that the rotational cell proposed for the bend region is present at both depths studied. However for the deep flow it is growing very weak by section 4 and has vanished by section 5. The next step is to quantify the effect of these flow structures in relation to the overall hydraulic resistance of the channel. Only after this has been successfully achieved will the systematic modelling of these flows be possible.

ACKNOWLEDGEMENTS
Acknowledgements are due to fellow investigators in this project, Dr D A Ervine and Professor B B Willetts, research assistants Rosemary Greenhill, Manuel Lorena and Richard Hardwick and staff of HR Wallingford. The project was principally financed by the UK SERC.

REFERENCES
1   Knight D W and Sellin R H J, "The SERC Flood Channel Facility". J Inst of Water and Environ. Mgnt. 1987, **1**, 2 , Oct., 198-204.

2   Sellin R H J, Ervine D A and Willetts B B, "Behaviour of meandering two-stage channels". Proc. ICE, Water, Maritime and Energy, 1993, **101**, June, 99-111.

3   Ervine D A, Sellin R H J and Willetts B B, "Large flow structures in meandering compound channels". 2nd Int. Conf. on River Flood Hydraulics, York, England, publ. H R Wallingford, March 1994

1A2

Unsteady Recirculations In Shallow-Water Flows

PETER M. LLOYD and PETER K. STANSBY
Hydrodynamics Research Group, The Manchester School of Engineering,
University of Manchester, Manchester M13 9PL, United Kingdom.

ABSTRACT
Shallow-water flows around surface-piercing and submerged islands have been investigated in the laboratory. Dye visualisation and whole-field PTV vector plots are used to describe the various flow phenomena and comparisons are made with a depth-averaged computational model.

INTRODUCTION
Recirculating flows, often resulting in vortex street wakes, have been observed behind elevated topography in the atmosphere (Chopra and Hubert 1965) and in the oceans (Barkley 1972). An important category of such flows is the shallow-water case occurring in coastal and estuarial regions when the horizontal length scale of the flow is an order of magnitude or more greater than the vertical. Such recirculation zones have been studied in harbours (Falconer 1980), behind islands and headlands (Davies and Mofor 1990) and around reefs on the continental shelf (Wolanski et al. 1986). The relevance of these flows lies in their ability to trap suspended matter in the recirculation area, resulting in important pollution implications (Hamner and Hauri 1981).

A relatively small body of work has been published concerning laboratory experiments of shallow recirculating flows behind obstacles. In this paper we present some results of tests conducted using conical model islands with mildly sloping sides. As well as providing fundamental information on the effect of flow parameters and model geometry on the wake zone, the results also provide benchmark tests for the evaluation of shallow-water computational models.

EXPERIMENTAL ARRANGEMENT
The experiments were performed in a purpose-built shallow-water flume, 8.4m long by 1.5m wide which allows flows with Reynolds numbers Re_h of $O(10^4)$ or below to be studied, where $Re_h = Uh/\nu$; U = free-stream velocity, h = water depth and ν = kinematic viscosity. Results are presented using a model island

with a side slope angle of 8°, height h_i of 0.049m, bottom diameter of 0.75m and a flattened apex with a diameter of 0.05m. Dye was used in the experiments to delineate the wake flow. A certain volume of dye was introduced upstream of the model and images were produced when the dye had become mixed within the wake region.

Measurements of velocity are performed using a Particle Tracking Velocimetry (PTV) system, which yields instantaneous whole-field surface velocity maps of the model wakes. Figure 1 illustrates the PTV set-up. The test section is illuminated and several hundred 5mm solid black polypropylene beads are released onto the water surface upstream of the model. The particles in the flow are filmed from above using a standard CCD camera connected to a VCR and digital frame-grabber, providing a test section of 1.5m x 1.5m. Two video frames, separated by a short time interval(≈ 0.24s), are grabbed and digitised. In an automatic procedure they are then analysed using Fortran programs running on a 486 PC. Particles detected in the first digital image are matched to their correct partner in the second image. The individual particles leave velocity vectors in an unstructured distribution across the image plane (figure 2a) requiring an interpolation program to regularise the vector field (figure 2b). The accuracy of the resulting velocity time histories has been assessed by comparison with simultaneous LDA measurements at two points downstream of an island with a 22° side slope. This case with $U=0.088\text{ms}^{-1}$ and $h=0.080$m produced strong vortices from the island. Figure 3 shows the comparison between PTV and LDA to be good. The average PTV error was less than 10% for the more severe test with rapidly varying flow on the wake centre-line. Full details of the PTV system are presented in Lloyd and Stansby(1994).

RESULTS AND DISCUSSION
SURFACE-PIERCING CASE
When the water depth in a channel is sufficiently small, the influence of bed friction on the flow can act to stabilize large scale transverse wake oscillations. This stabilizing influence has been studied for mixing layers (Chu and Babarutsi 1988), recirculating flows (Babarutsi et al. 1989) and island wake flows (Ingram and Chu 1987). Chu et al.(1983) introduced a bottom friction stability parameter S as a measure of the stabilizing effect of bed friction relative to the de-stabilizing effect of transverse shear. Applying this parameter to model islands represented by circular cylinders, Chen and Jirka(1991) defined the bottom friction parameter as $S=c_f D/h$, in which c_f is a bottom friction coefficient and D is the cylinder diameter. For island wake flows Babarutsi et al.(1989) define a critical value of $S_c=0.40$ based on full-scale observations, such that if $S<S_c$ vortex shedding occurs and if $S>S_c$ the wake takes the form of a quasi-steady bubble.

Results are shown in figure 4a for flow around the model island of mild slope with $S_a=0.27$ (h=0.019m, $Re_h=1900$, S_a calculated using island diameter at mid depth). Figure 4b shows results with an increased influence of bottom friction, $S_a=0.40$ (h=0.0145, $Re_h=1500$), where the horizontal shear layers created by separation from the sides of the island no longer interact and a quasi-steady wake bubble is formed. These results are broadly consistent with $S_c=0.40$, with other tests showing the transition from vortex shedding to a quasi-steady bubble occurring for values of S_a between 0.35 and 0.40.

A semi-implicit Lagrangian finite difference model has been developed to solve the depth-averaged shallow-water equations, with an eddy viscosity for horizontal mixing defined locally as $\epsilon=0.10u_*h$, where u_* is the shear velocity. When the bed-friction influence is very small (S<0.20) the model is in excellent agreement with experimental measurements. When bed-friction becomes more prominent however (S>0.20), figure 5 shows how the model tends to overestimate the velocities near the island producing a significantly wider downstream wake than measured with the PTV system. The shedding period predicted by the model is nevertheless in good agreement with that measured.

SUBMERGED CASE
The similarities between a shallow-water flow and an atmospheric air flow beneath a strong inversion have been noted in the literature. Studying submerged islands in the laboratory therefore provides information important to both the flow around submerged shoals and reefs and to the flow around isolated hills.

Figure 6 shows images of dye tests conducted using the submerged model island. Even with such a mildly sloping model, figure 6a, with $h/h_i=1.10$ and $Re_h=6300$, shows a wake composed of a regular vortex street. Other measurements indicate the shed vortices to be almost 2D in nature. Although the unsteady wake is characterised by vortices with a vertical axis, the wake origin is the flow separation from a nearly horizontal surface in the narrow region of accelerated flow near the apex of the island. This separation causes strong vertical mixing in a region just downstream of the apex, resulting in a low velocity region. The horizontal shear layers between this region and the high velocity regions across from the apex produce the unsteady wake. When the water depth is increased to $h/h_i=1.20$ (figure 6b) the separation from the island no longer mixes the flow so effectively across the depth and a steady, narrow wake region is formed.

Although the unsteady wake is caused by a 3D flow region, the depth-averaged model fortuitously produces a good simulation, as shown by figures 7a and 7b. Rapid deceleration just downstream of the apex results in a wake similar to that measured using the PTV system.

CONCLUSIONS

Preliminary results for shallow-water flows around surface-piercing and submerged conical islands of mild slope have been presented. Wake behaviour for surface-piercing models is consistent with the value of the critical bottom friction parameter presented by Babarutsi et al.(1989). For the submerged island case 2D unsteady wake flow can be caused by a 3D mixing region near the apex. The criteria for the occurrence of this phenomenon have not yet been fully established, but it has been shown to be strongly dependent on water depth above the island.

REFERENCES

Babarutsi,S., Ganoulis,J. and Chu,V.H. 1989. Experimental investigation of shallow recirculating flows. *J. Hydr. Engrg.*, ASCE, 115(7), 906-924.

Barkley,R.A. 1972. Johnston Atoll's Wake. *J. Mar. Res.*, 30(2), 201-216.

Chen,D. and Jirka,G.H. 1991. Pollutant mixing in wake flows behind islands in shallow water. *Int. Symp. on Env. Hyd.*, Baalkema, 371-377.

Chopra,K.P. and Hubert,L.F. 1965. Mesoscale eddies in wake of islands. *J. Atmos. Sci.*, 22, 652-657.

Chu,V.H. and Babarutsi,S. 1988. Confinement and bed-friction effects in shallow turbulent mixing layers. *J. Hydr. Engrg.*, ASCE, 114(10), 1257-1274.

Chu,V.H., Wu,J.-H. and Khayat,R.E. 1983. Stability of transverse shear flows in a shallow channel. *Proc. 20th Cong. IAHR,* Moscow, Vol.3, 128-133.

Davies,P.A. and Mofor,L.A. 1990. Observations of flow separation by an isolated island. *Int. J. Rem. Sens.*, 11(5), 767-782.

Falconer,R.A. 1980. Numerical modelling of tidal circulation in harbours. *J. Wat. Port. Cstl. and Ocn. Div.*, ASCE, 106(WW1), 31-48.

Hamner,W.M. and Hauri,I.R. 1981. Effects of island mass: Water flow and plankton pattern around a reef in the Great Barrier Reef lagoon, Australia. *Limnol. Oceanogr.*, 26(6), 1084-1102.

Ingram,R.G. and Chu,V.H. 1987. Flow around islands in Rupert Bay: An investigation of the bottom friction effect. *J. Geo. Res.*, 92(C13), 14521-14533

Lloyd,P.M. and Stansby,P.K. 1994. Unsteady surface-velocity field measurement by video analysis. *IAHR Symp: Waves - Physical. and Num. Modelling.*, Vancouver, Canada, Vol 1, 49-59. (Extended version to appear in *IAHR J. Hyd. Res.*)

Wolanski,E., Jupp,L.B. and Pickard,G.L. 1986. Currents and coral reefs. *Oceanus,* 29(2), 83-89.

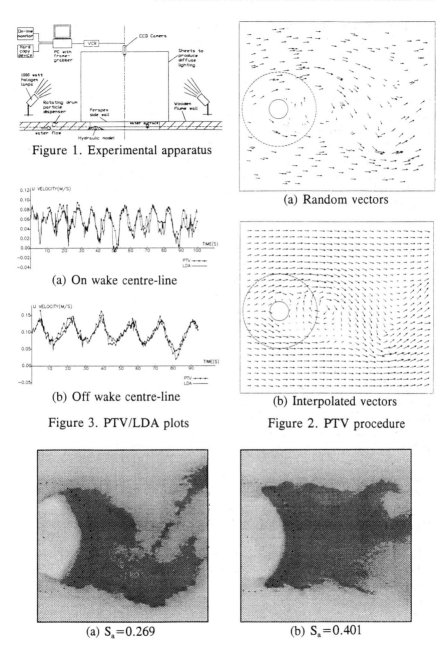

Figure 1. Experimental apparatus

(a) On wake centre-line

(b) Off wake centre-line

Figure 3. PTV/LDA plots

(a) Random vectors

(b) Interpolated vectors

Figure 2. PTV procedure

(a) $S_a = 0.269$

(b) $S_a = 0.401$

Figure 4. Dye distribution in wake of surface-piercing islands

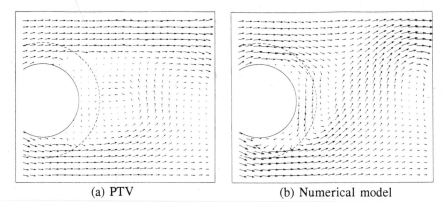

(a) PTV   (b) Numerical model

Figure 5. Velocity vector plots for $S_a = 0.269$

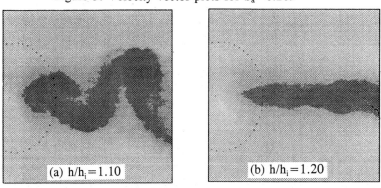

(a) $h/h_i = 1.10$ (b) $h/h_i = 1.20$

Figure 6. Dye distribution in wake of submerged islands

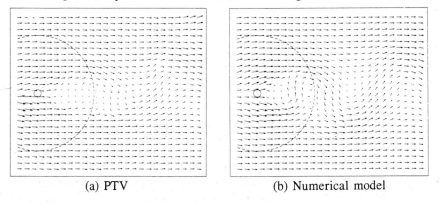

(a) PTV   (b) Numerical model

Figure 7. Velocity vector plots for $h/h_i = 1.10$

1A3

LARGE EDDY SIMULATION OF A SYMMETRIC TRAPEZOIDAL CHANNEL AND FLOODPLAIN

T. G. Thomas and J. J. R. Williams
Dept Civil Engineering,
Queen Mary and Westfield College,
London E1 4NS

ABSTRACT

This paper desribes a Large Eddy Simulation of turbulent flow in a symmetric compound channel of trapezoidal cross-section. The simulation, which was carried out in order to reproduce the SERC-Flood Channel Facility experiment 200501, captures the complex interaction between the main channel and the flood plains and predicts the bed stress distribution, velocity distribution, and the secondary circulation.

INTRODUCTION

The modelling of an open channel with a floodplain is difficult because of the complex three dimentional interaction of flow moving relatively fast in the mainchannel and slow in the floodplain. The effect of this interaction is that a large secondary flow cell is seen to develop for the whole width of the floodplain and that the bed shear stress there is increased above its infinitely wide channel value.

A series of detailed measurements of various channel and floodplain geometries has been carried out at the SERC Flood Channel Facility (SERC-FCF) in Wallingford, UK, with the aim of establishing an experimental database for validating predictive models. The full experimental data is published by Knight [1].

This paper describes the numerical calculation of one of the tests - experiment number 200501 with a Reynolds number of 430,000 - using the Large Eddy Simulation (LES) technique. LES is now well established (see the review article, Rogallo & Moin [2]) and has been used recently by Thomas & Williams [3] to predict turbulent flow and secondary circulation in a rectangular compound channel at a Reynolds number of 42,000.

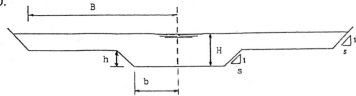

Figure 1: Channel geometry

NUMERICAL METHOD

The compound open channel is shown in figure1 where H=0.2m is the depth of the main channel, b=0.75m is the bottom width and s=2 is the sidewall slope. The floodplains are elevated a distance h=0.15m above the main channel bed and extend on either side up to a distance B where they are bounded by sloping sidewalls. The depth ratio $D_r=(1-h/H)$ is equal to 0.25, the width ratio B/b=2.2 and the aspect ratio B/H=8.25. The mean flow is assumed uniform in the streamwise direction, so that the bed slope S_0 and the energy slope l_e are identical and equal to 1.027×10^{-3}. The overall Reynolds and Froude numbers are calculated to be 430,000 and 0.627 respectively.

We chose a box length L=6H, used periodic inflow and outflow boundary conditions and imposed a rigid stress free lid at the free surface. The simulation was run with 72 cells vertically, 511 cells laterally, and 64 along the channel. A greater streamwise resolution would have been desirable but was limited by the computational resources available to us.

MEAN VELOCITY

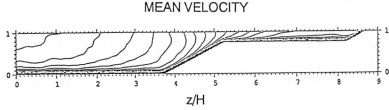

z/H

Figure 2: Mean velocity distribution: contours at $32u_\tau$, $31u_\tau$, etc.

The mean streamwise velocity profile, obtained assuming the channel surface to be perfectly smooth and made non-dimensional using u_τ, is shown in figure 2. The maximum value U_m of $32.3u_\tau$ compares with an experimentally measured value of $29.6u_\tau$ and the bulk velocity U_b equals

$27.4u_\tau$ which is also higher than expected and may also be due to insufficient streamwise resolution. A second run was carried out which allowed for the channel surface to be slightly rough (see Thomas & Williams [4] for full details) and this resulted in the bulk velocity U_b being reduced to $25.3u_\tau$.

DEPTH AVERAGED VELOCITY

The depth average streamwise velocity U_d normalised using u_τ is plotted in figure 3 for the smooth and slightly rough simulations, and for the experimental measurements.

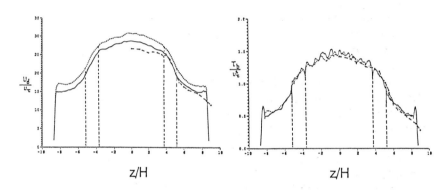

z/H   z/H

Figure 3   Figure 4
Depth averaged velocity profile   Distribution of bed stress
: - - -,smooth channel simulation; —, slightly rough simulation; •- - -•, experimental data from SERC FCF.

The smooth profile is approximately 7% higher than the slightly rough profile in the main channel and on the floodplains, and 6% higher on the sloping intersection. The slightly rough profile is in almost exact agreement with the experimental data on the sloping intersection but underpredicts at most by approximately 4% on the floodplain and overpredicts at most by approximately 8% in the middle of the main channel.

SECONDARY CIRCULATION

The secondary circulation stream function Ψ for the right half of the channel is shown in figure 5 normalised using $u_\tau H$. The dominant feature is the large clockwise circulation cell starting in the main channel near to the bottom of the sloping sidewall and extending right across the floodplain and is in close agreement with the experimental measurements of Shiono & Knight [5] taken at a $D_r=0.15$. The experimentally measured secondary circulation is shown in figure 6 in which the maximum velocity of $0.71u_\tau$ (25 mm s^{-1}) occurs at the top of the sloping interface.

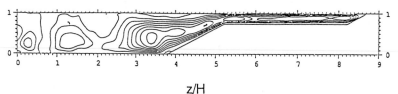

z/H

Figure 5: Secondary circulation stream function $\Psi/u_\tau H$: contours at 0.0, ± 0.01,±0.02, etc.

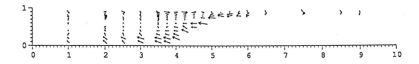

Figure 6: Secondary flow data from SERC-FCF experimental facility

BED STRESS DISTRIBUTION

Figure 4 shows the computed (smooth) and experimentally measured stress profile normalised using u_τ^2. The profiles differ by less than 2% over most of the channel, interface region and floodplains.

The computed bed stresses from the slightly rough simulation are also shown in figure 4; apart from the slightly greater statistical variation due to the shorter averaging time the result is almost identical to the smooth case.

LATERAL MOMENTUM TRANSFER BY SECONDARY CIRCULATION

The bed stress τ_b in an open channel differs from the value gdS_0 appropriate for a wide channel, where d denotes the local depth, because of the lateral gradient in the apparent shear stress τ_a. This acts on vertical planes aligned with the flow direction and is averaged over the depth. It is usual to split τ_a into two components T+J, where T represents turbulent diffusive transport (or Reynolds stress), and J represents the mean convective transport due to secondary circulation. The depth averged equation for streamwise momentum and T and J are given by

$$g d S_o - \tau_b \sqrt{1 + \frac{1}{s^2}} + \frac{\partial}{\partial z}(Td + Jd) = 0,$$

$$T = \frac{1}{d} \int -\overline{u'w'}\, dy + \frac{1}{d} \int (\nu + \nu_s)\left(\frac{\partial \bar{u}}{\partial z} + \frac{\partial \bar{w}}{\partial x}\right) dy,$$

$$J = \frac{1}{d} \int -\bar{u}\,\bar{w}\, dy.$$

The lateral distribution of τ_a, T, and J, normalised using u_τ^2 are plotted in figure 7.

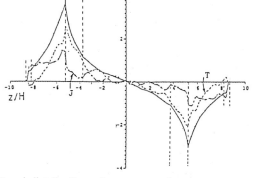

Figure 7: Lateral distribution of apparent shear stress: •- - -•, Reynolds stress component T; ◦- - -◦, secondary current component J; •—•, combined components T+J.

CONCLUSIONS

An LES simulation of a trapezoidal compound channel with extensive floodplains has been carried out at a Reynolds number of approximately 430,000. The results have been compared with experimental data from the SERC-FCF floodplain facility and the overall agreement is good.

The simulation was carried out initially using hydraulically smooth boundary conditions but these were later adjusted to allow for the fact that the channel was slightly rough. Both simulations predicted the measured bed stress distribution with very good accuracy; it follows from this that the combined lateral transport of momentum due the Reynolds stress and the secondary circulation must also have been well predicted.

REFERENCES

[1] Knight, D.W., (Editor) SERC Flood Channel Facility, reort SR 314 (May 1992) Dept. Civil Engng., Univ. Birmingham, UK.

[2] Rogallo, P.S. and Moin, P., Numerical simulation of turbulent flows. Ann. Rev. Fluid Mech. 16 (1984) 99-137.

[3] Thomas, T.G. and Williams, J.J.R., Large eddy simulation of turbulent flow in an asymmetric compound channel. Journal for Hydraulic Research. In press.

[4] Thomas, T.G. and Williams, J.J.R., Large Eddy Simulation of a Symmetric Trapezoidal Channel at a Reynolds Number of 430000. Paper submitted to the Journal for Hydraulic Research.

[5] Shiono, K. and Knight, D.W., Transverse and vertical Reynolds stress measurements in a shear layer region of a compound channel. Proc. 7th Symp. Turb. Shear Flows (1989) Stanford, Calif., 28.1.1-28.1.6.

1A4

SECONDARY CURRENTS AND EXCHANGE PROCESSES IN COMPOUND OPEN-CHANNEL FLOWS

Iehisa NEZU, Hiroji NAKAGAWA and Takashi ABE

Department of Civil and Global Environment Engineering,
Kyoto University, Kyoto 606, Japan

INTRODUCTION

Many investigations on flow resistance and the associated water-depth *vs.* discharge curves, i.e., *H-Q* curves, in compound open-channel flows have been conducted intensively for practical purposes of flood control planning. The compound flow section is divided into the two: the main channel and the flood plains. The resulting discharge calculated by 1-D hydraulic methods is, however, overestimated because a strong interaction between the high-speed main-channel flow and the low-speed flood-plain flow is ignored. This flow interaction transports longitudinal momentum from the main channel to the flood plain regions, and thus it promotes an energy loss and a substantial decrease of total discharge capacity in compound channels.

Turbulent structures in such straight compound open-channel flow indicate inherently three-dimensional (3-D) behaviors. The most important feature of compound channel flows is the strong interaction and exchange processes between the main-channel and flood-plain flows, which combines with complicated secondary currents in the cross section, as pointed out by Shiono & Knight (1991), Nezu (1994) and others.

Anisotropic turbulence near the edge between the main channel and flood plain causes intermittent upward secondary currents and therefore generates three-dimensional coherent vortices. Many researchers have pointed out significant importance about the lateral transports of momentum and suspended sediment between the main channel and flood plains by these secondary currents. This time-averaged structure of secondary currents in compound open-channel flows has first been revealed using accurate measurements with a fiber-optic laser Doppler anemometer (LDA) by Tominaga & Nezu (1991). They conducted three typical cases of experiments for asymmetrical compound open channel flows. Naot, Nezu & Nakagawa (1993) have developed computer-simulations of compound open-channel flows with secondary currents using 3-D algebraic stress model (3-D ASM); this simulation is called the "3N model". The 3N model coincided well with the LDA data of Tominaga & Nezu(1991).

In the present study, extended experiments of compound open channel flows including variable-depth flood plains were conducted with a more innovative two-components fiber-optic LDA system (Dantec-made). Secondary currents and exchange processes between the main channel and flood plain are revealed and discussed on the basis of anisotropy of turbulence. These LDA databases are also compared with the predictions calculated from the 3N model.

EXPERIMENTAL APPARATUS AND PROCEDURES

The experimental apparatus consisted of a glass flume with 8.0 m length and 30cm x 30cm cross section, as shown in Fig. 1. Acrylic boxes were placed on the bed of one side of the flume for compound channel. The slope of side wall between the main channel and the flood plain was changed into three types; 1:0 (rectangular), 1:1 and 1:2 (trapezoidal cross section). Six different cases of experiments were conducted, as indicated in Table 1.

The streamwise and vertical components of instantaneous velocities, $u(t)$ and $v(t)$, were measured accurately from the side-wall setting of the fiber probe of LDA, whereas the streamwise and spanwise components $u(t)$ and $w(t)$ were measured from the setting of the LDA probe above the free surface, as shown in Fig. 1. Therefore, all three components of mean velocities, U, V, and W, the turbulence intensities u', v' and w' and also the Reynolds stresses $-\overline{uv}$ and $-\overline{uw}$ were obtained. About 500 measuring points were traversed for each case.

Table 1 Hydraulic Conditions for Compound Open Channel Flows.

CASE	Ib	H (cm)	B_m/B	$D:B_S$	H/D	Q (ℓ/s)	$\bar{u}$ (cm/s)	U_{max} (cm/s)	Re ($\times 10^3$)	Fr
A	1/3000	6.0	0.333	1 : 0	1.2	0.92	11.5	19.9	2.4	0.15
B	1/5000	7.5	0.333	1 : 0	1.5	1.06	8.5	17.9	2.4	0.10
C	1/10000	10.0	0.333	1 : 0	2.0	1.29	6.4	16.0	2.7	0.07
D	1/10000	10.0	0.667	1 : 0	2.0	2.45	9.8	16.0	5.0	0.10
E	1/10000	10.0	0.500	1 : 1	2.0	2.64	11.1	15.9	5.7	0.11
F	1/10000	10.0	0.333	1 : 2	2.0	3.00	13.2	16.7	6.6	0.13

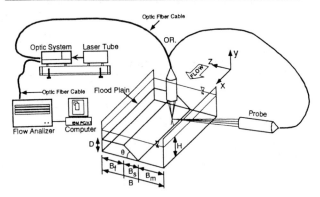

Fig. 1 Experimental Flume and Fiber-optic Laser Doppler Anemometer(LDA).

NUMERICAL CALCULATIONS

3-D numerical simulation was conducted in rectangular compound open-channels using 3-D ASM developed by Naot, Nezu & Nakagawa(1993). The model constants of this 3N model were determined from the LDA data of Tominaga & Nezu(1991), and these were also used without any modification in the present study. So, one can say that the present calculation is _real_ prediction.

EXPERIMENTAL RESULTS AND DISCUSSIONS ✓

Isovel Lines of Mean Velocity

Fig. 2 shows an example of isovel lines for case C. In this case, the flood plain is two times as large as main channel; this may represent urban rivers. The measured data coincide well with the calculated data from 3N model. It should be noted that the maximum velocity does not appear in the main channel, but on the free surface of flood plain. The second peak velocity zone occurs below the free surface in the main channel, i.e., the dip-velocity phenomena. These dip-velocity phenomena of main channel and flood plain are explained on the aspect criterion of Nezu & Nakagawa (1993) .

Of particular significance is an upward bulge from the junction edge between the main channel and flood plain. This corresponds to the upflow of secondary currents, and implies exchange processes between the two.

Secondary Currents

Fig. 3 shows velocity vector description of secondary currents (V, W) that correspond to Fig. 2. The numerical prediction coincides fairly well with the measured LDA data, although some differences between the two are seen. The angle of the measured upflow from the junction is slightly steeper than that of the calculation.

Bed Shear Stress

Fig. 4 shows the bed shear stress τ_b normalized by its averaged value $\overline{\tau_b}$. The observed value of τ_b was evaluated from the log-law of mean velocity. The configuration of the flood plain was fixed, and the depth ratio H/D was changed as 1.2, 1.5 and 2.0. As the depth of flood plain

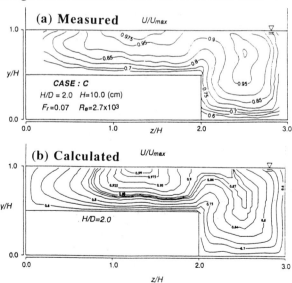

Fig.2 Isovel Lines of Primary Velocity $U(y,z)$.

increases, τ_b increases for flood plain, but it decreases for main channel, due to the very narrow main channel, i.e., B_m / B =1/3. The spanwise variation of bed shear stress $\tau_b(z)$ is explained by the secondary currents, as pointed out by Nezu & Nakagawa (1993). The value of τ_b increases at the downflow($V<0$), whereas it decreases at the upflow($V>0$). The bed shear stress on the flood plain increases toward the junction due to the interaction between the main channel and flood plain. An agreement between the measured and predicted values is not necessarily good for wide flood plain.

Reynolds Stresses

Fig. 5 shows the contour lines of Reynolds stresses $-\overline{uv}$ and $-\overline{uw}$, and the difference of normal stresses, $(\overline{w^2} - \overline{v^2})$.

Of particular significance is the sharp changes of $-\overline{uw}$ at the boundary of upflow from the junction. The value of $-\overline{uw}$ is negative on the side of flood plain, whereas it is

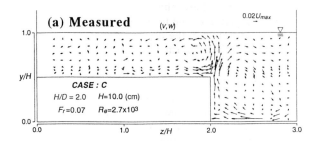

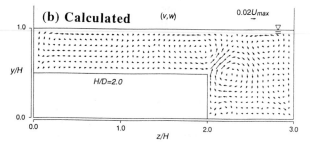

Fig.3 Vector Description of Secondary Currents (V, W).

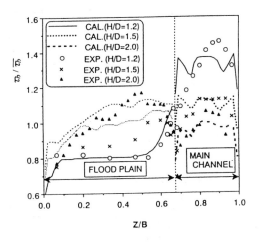

Fig.4 Spanwise Variations of Bed Shear Stress $\tau_b / \overline{\tau_b}$ for Cases A, B and C.

positive on the side of main channel. This sharp change near the junction is due to the interaction between the main channel and flood plain, and vice versa.

The value of $\left(\overline{w^2} - \overline{v^2}\right)$ expresses anisotropy of turbulence, and generates secondary currents. This value is large near the corner of channels and the junction between the two, although the accuracy of data may not be so high because of no simultaneous measurements of $v(t)$ and w(t).

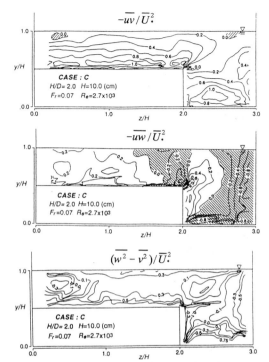

Effect of Junction Slope on Exchange

For practical purposes, it is very important to investigate the effect of junction slope between the main channel and flood plain on the exchange processes between the two.

Fig. 6 shows the secondary currents for cases E and F, in which the flow depth varies linearly from the flood plain to main channel. In the case E, the strength of secondary currents is nearly same on both sides of the slope junction. The angle of upflow is nearly vertical, and it attains up to the free surface.

The interaction

Fig.5 Contour Lines of Reynolds Stresses.

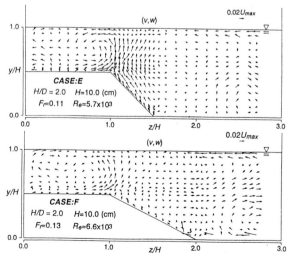

Fig.6 Secondary Currents for Slope Junction.

between the main channel and flood plain is the nearly same as the case C. On the other hand, for the case F which has milder slope, the vertical upflow of secondary currents becomes weaker.

Fig. 7 shows the effect of slope on bed shear stress τ_b. The maximum value of τ_b on the flood plain is slightly smaller as the slope is milder. However, the value of τ_b on the slope is larger than that on the bottom of main channel.

Fig. 8 shows the contours of $-\overline{uw}$ by varying the slope of junction. The interaction for the slope 1:1 is the nearly same as that for the slope 1:0.

CONCLUSIONS

The narrow compound open channel flows were measured accurately with an innovative LDA system. The databases of secondary currents and interaction were obtained and analyzed.

REFERENCES

1) Naot, D., Nezu, I. and Nakagawa, H.(1993): *J. Hydraulic Eng.*, ASCE, vol.119, pp.390-408.
2) Nezu, I.(1994): Key Address, *APD-IAHR*, Singapore, pp.1-24.
3) Nezu, I. and Nakagawa, H. (1993): *IAHR-Monograph*, Balkema, Rotterdam.
4) Shiono, K. and Knight, D.W. (1991), *J. Fluid Mech.*, vol.222, pp.617-646.
5) Tominaga, A. and Nezu, I. (1991): , *J. Hydraulic Eng.*, ASCE, vol.117, pp.21-41.

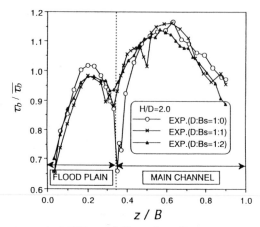

Fig.7 Effect of Slope Junction on Bed Shear Stress $\tau_b / \overline{\tau_b}$.

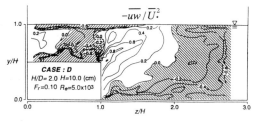

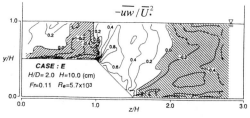

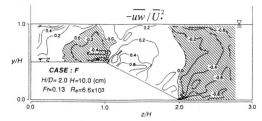

Fig.8 Effect of Slope Junction on the Reynolds Stress $-\overline{uw}$.

1A5

LATERAL SHEAR IN A COMPOUND DUCT

DAVID G. RHODES
SMMCE
Cranfield University
Shrivenham SN6 8LA
UK

DONALD W. KNIGHT
School of Civil Engineering
The University of Birmingham
Birmingham B15 2TT
UK

INTRODUCTION

Complementary to experimental work in compound open channels, a comparable zone of interaction between main channel and flood plain flows may also be engineered in a closed duct of the cross-sectional geometry illustrated in Figure 1. The horizontal plane of symmetry having zero shear stress is analogous to the free

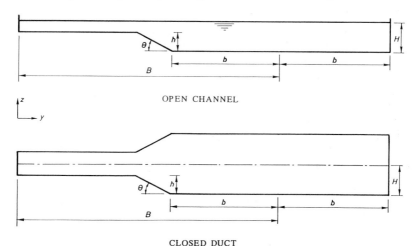

OPEN CHANNEL

CLOSED DUCT

Figure 1: Geometry of equivalent compound open channel and duct

surface, so that the model of a two-stage channel is provided by either the upper or lower half of the duct. The open circuit wind tunnel is a comparatively cheap and simple arrangement, in which the cross-sectional geometry is more easily controlled than the corresponding normal depth in a compound channel. Whereas in an open channel it is not possible to measure the velocity or turbulence fields right up to the free surface, the corresponding measurements on the horizontal plane of symmetry of a duct are routine.

The fact that the horizontal plane of symmetry does not replicate the free surface, with its effect upon the turbulence and secondary flow regimes, does make the compound duct less directly applicable to two-stage flows in the field. However, the simplified boundary condition provides a popular "halfway house" in testing out turbulence models in CFD as evidenced by a number of papers in which this choice of boundary condition has been made. It is because of the scarcity of compound duct measurements, that such CFD comparisons are often inappropriately made with compound open channel measurements.

Rhodes and Knight[3] (1994) have reported time averaged measurements of boundary shear stress and velocity distributions in compound duct flows, and the present paper applies these results to the lateral shear layer mechanism, focussing upon the effect of secondary flow. Although no measurements of turbulence or secondary flow were carried out, because of the strong similarities observed in the primary flow fields of compound ducts and open channels it is possible to infer the presence of secondary flow and its effect upon the primary flow structure.

EXPERIMENTAL PROCEDURE
Details of the experimental procedure are given by Rhodes and Knight[3] (1994). Here we provide a brief summary of the experimental parameters.

Boundary shear stress and velocity distributions were measured by Preston tube and Pitot static tube respectively in a compound duct 1231.5 mm wide with a development length of 13.1 m. Twenty seven different cross-sectional geometries were formed with two different step heights, $h = 20$ mm and $h = 40$ mm (Fig. 1), the depth H being varied to give 8 relative depths in the range $(H-h)/H = 0.15$–0.67 at each of three different wall angles at the main channel–flood plain interface, $\theta = 90°$, $45°$ and $26.6°$ respectively. The relative depth $(H - h)/H = 0.25$ was used at both step heights h. Channel aspect ratio was varied in the range $11.58 \leq b/h \leq 20.76$ and relative width in the range $1.41 \leq B/b \leq 2.16$. The Reynolds number range was $5.8 \times 10^4 \leq R \leq 10.2 \times 10^4$. Each boundary shear stress distribution was factored so as to agree with the cross-section mean boundary shear stress calculated from the measured pressure gradient by the following relationship

$$\tau_0 = -H_r \frac{\partial p}{\partial x} \tag{1}$$

DISTRIBUTION OF APPARENT SHEAR STRESS
If the Reynolds equation for incompressible flow in the streamwise direction x is depth averaged, then

$$h\frac{\partial p}{\partial x} = \frac{\mathrm{d}}{\mathrm{d}y}h\left(\overline{\tau}_{yx} - \rho\overline{UV}\right) - \tau_b\sqrt{1 + \frac{1}{s^2}} \tag{2}$$

where the overbar denotes depth averaging. Reynolds stress and secondary flow terms may be combined in what is hereafter referred to as the "depth averaged apparent shear stress" $\overline{\tau}_{a_{yx}}$.

$$\overline{\tau}_{a_{yx}} = \overline{\tau}_{yx} - \rho\overline{UV} \tag{3}$$

Equation (2) may therefore be alternatively written as

$$\text{h}\frac{\partial p}{\partial x} = \frac{\text{d}}{\text{dy}}\left(\text{h}\overline{\tau}_{a_{yx}}\right) - \tau_b\sqrt{1 + \frac{1}{s^2}} \tag{4}$$

At $y = 0$, the flood plain side wall, $\text{h}\overline{\tau}_{a_{yx}} = SF_w$. Integrating (4) with respect to y, rearranging and noting that $\partial p/\partial x$ is constant across the section

$$\overline{\tau}_{a_{yx}} = \frac{1}{\text{h}}SF_w + \frac{1}{\text{h}}\int_0^y \tau_b\sqrt{1 + \frac{1}{s^2}}\text{dy} + \frac{1}{\text{h}}\frac{\partial p}{\partial x}\int_0^y \text{hdy} \tag{5}$$

$\overline{\tau}_{a_{yx}}$ was calculated by means of (5) from the measured boundary shear stress distribution (factored as in Experimental Procedure) and measured pressure gradient. As expected, the maximum apparent shear stress always occurred at the main channel–floodplain interface, the position at which channel sub-division is usually carried out in the divided channel methods of flow calculation.

Figs. 2(a) and (b) illustrate a notable difference in the form of distribution on the flood plain, that is dependent on relative depth. For all distributions in the range $0.25 \le (H - h)/H \le 0.67$, for example at $(H - h)/H = 0.67$ shown in Fig. 2(a), there is a minimum point in the apparent shear stress distribution on the flood plain. It is most conspicuous at the higher relative depths. Below $(H - h)/H = 0.25$ the minimum can no longer be observed, and the gradient of apparent shear stress is monotonic from the side wall to the main channel–flood plain interface, for example at $(H - h)/H = 0.2$ as illustrated in Fig. 2(b).

Similar effects have been observed in the results of other workers. For example, Wright and Carstens[7] (1970), at relative depths of 0.6, 0.5 and 0.3, show distinct minima in their apparent shear stress distributions on the flood plain of a compound duct. Tominaga and Nezu[6] (1991) have a minimum point in their compound open channel measurements at a relative depth of 0.5, whereas it is no longer evident at a relative depth of 0.4. Myers[2] (1978) presents apparent shear stress distributions in compound open channel flow for relative depths of 0.202, 0.140 and 0.088, and in none of them is there a minimum. Shiono and Knight[4] (1991), for relative depths in the range 0.093–0.479, demonstrate both forms of distribution on the flood plain.

The occurrence of a minimum at high relative depths, at least in open channel flow, may be explained by reference to (3) and Tominaga and Nezu's[6] (1991) distributions of $\overline{\tau}_{yx}$ and $-\rho\overline{UV}$ (in current notation) for a relative depth of $(H - h)/H = 0.5$. The sum of the two distributions, which gives the distribution of $\overline{\tau}_{a_{yx}}$, has a minimum on the flood plain because of the dominant influence of a minimum in the distribution of $-\rho\overline{UV}$. This probably corresponds to a region in which the transverse velocity component $V(y, z)$ is relatively small, occurring where the cross-flow issuing from the salient corner region expires and the effect of the corner cells near the outer flood plain wall begins.

Considering the case where the term $\partial\overline{\tau}_{a_{yx}}/\partial y$ is positive from the flood plain side wall to the main channel–flood plain interface, ie with no minimum point,

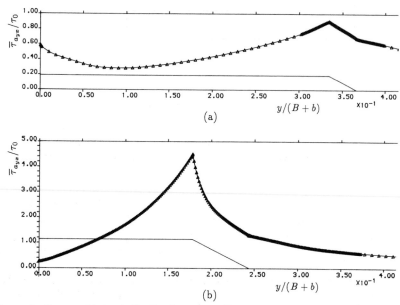

Figure 2: Forms of lateral distribution of nondimensional depth averaged apparent shear stress on the flood plain: (a) $(H - h)/H = 0.67, (b)$H-h$)/$H$=0.2

there are examples in the present data where mean velocity and bed shear stress distributions exhibit regions of near zero gradient in the y direction. In fact it is at the lowest relative depths, with the largest regions of apparently two-dimensional flow and least lateral shear due to Reynolds stresses, that this form of the distribution is observed. It appears that in such cases the secondary flow term in (3) is dominant.

FLOOD PLAIN BED SHEAR STRESS

If (2) is rearranged then for constant depth of flow $(H - h)$ the bed shear stress on the flood plain of a compound duct is given by

$$\tau_b = -(H - h)\frac{\partial p}{\partial x} + (H - h)\frac{d}{dy}\left(\overline{\tau}_{yx} - \rho\overline{UV}\right) \tag{6}$$

The equivalent expression for an open channel has the term $\rho g(H - h)S_f$ instead of $-(H - h)\partial p/\partial x$.

In Figure 3, illustrating measurements in open channel flow, it can be seen that on the flood plain the secondary flow vectors near the free surface are almost horizontal and in the negative y direction. They increase in size becoming more negative with increasing y, ie $\partial V/\partial y < 0$. Near the bed the vectors, which are again virtually horizontal, act in the positive y direction and it can be seen that

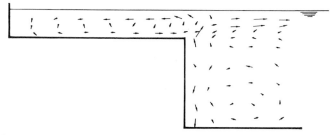

Figure 3: Secondary flow vectors in compound open channel (adapted from Tominaga *et al* 1989 and published by permission of International Association for Hydraulic Research)

$\partial V/\partial y > 0$. Given that U near the free surface is greater than U near the bed, $\mathrm{d}\left(\rho\overline{UV}\right)/\mathrm{d}y < 0$ and therefore $\mathrm{d}\left(-\rho\overline{UV}\right)/\mathrm{d}y > 0$. That is why the secondary flow term in (6) increases the bed shear stress, an effect which is observed in compound open channel flows even when the primary flow field appears to be two-dimensional (Knight and Shiono[1] 1995).

The same effect is evident in the present compound duct measurements. Table 1 shows for all three wall angles, and for selected relative depths, measurements of bed shear stress (M) on the flood plain in what appears to be two-dimensional flow from visual inspection of the lateral distribution of bed shear stress and depth mean velocity. The expected values (E) are calculated from

$$\tau_{b\infty} = -(H - h)\frac{\partial p}{\partial x} \qquad (7)$$

assuming a real two-dimensional flow. The relative depths chosen were those at which regions of near constant bed shear stress were observed on the flood plain, and the two-dimensional value was estimated by visual inspection of the distribution. Table 1 shows that, for all of the relative depths chosen, the measured bed shear stress (M) was greater than the calculated value (E) assuming real two-dimensional flow.

CONCLUSIONS

Time averaged measurements of the primary flow field in a compound duct indicate that the lateral shear layer mechanism is significantly influenced by the presence of secondary flows. Only by this effect is it possible to account for the strong lateral shear in a flow field which in other respects appears to be two-dimensional. The enhancement of bed shear stress on the flood plain is also to be explained by the presence of lateral gradients in the secondary flow field.

The importance of secondary flow has implications for the assessment of turbulence models in CFD. The present results and those of other workers show that it is necessary to judge a computational model's ability to reproduce secondary

$(H-h)/H$	$\theta = 90°$ $\tau_{b\infty}$			$\theta = 45°$ $\tau_{b\infty}$			$\theta = 26.6°$ $\tau_{b\infty}$		
	M	E	M/E	M	E	M/E	M	E	M/E
0.5	66.4	63.8	1.04	68.2	63.6	1.07	67.9	63.2	1.07
0.4	53.7	52.6	1.02	54.4	52.5	1.04	56.5	52.3	1.08
0.3	41.3	40.7	1.01	45.1	40.9	1.10	49.4	40.6	1.22
0.25	37.5	34.5	1.09	40.3	34.7	1.16	46.4	34.8	1.33
0.15	29.7	18.8	1.58	29.7	18.9	1.57	32.7	19.2	1.70

Bed shear stress values are nondimensionalised and expressed as a percentage of cross-section mean boundary shear stress

Table 1: Comparison of two-dimensional bed shear stress values on the flood plain derived from measurement (M) and expected values (E)

flow more strictly than by the usual cursory check on whether the cells exist, where they are located, whether the rotation is in the right direction and what effect they have upon the appearance of the primary isovels. The criteria must be quantitative and not merely qualitative.

References

REFERENCES

[1] Knight, D.W. and Shiono, K. River channel and flood plain hydraulics. In *Flood Plain Processes*, eds Anderson, Walling and Bates, J. Wiley, 1995, Chapter 5.

[2] Myers, W.R.C. Momentum transfer in a compound channel. *J. Hydr. Res.*, IAHR, 1978, 16(2), 139–150.

[3] Rhodes, D.G. and Knight, D.W. Velocity and boundary shear in a wide compound duct. *J. Hydr. Res.*, IAHR, 1994, 32(5), 743–764.

[4] Shiono, K. and Knight, D.W. Turbulent open-channel flows with variable depth across the channel. *J. Fluid Mech.*, 1991, 222, 617–646. (& 231, 693.)

[5] Tominaga, A., Nezu, I. and Ezaki, K. Experimental study on secondary currents in compound open-channel flow. *Proc. XXIII Congr.*, IAHR, Ottawa, Canada, 1989, A15–A22.

[6] Tominaga, A. and Nezu, I. Turbulent structure in compound open channel flow. *J. Hydr. Eng.*, ASCE, 1991, 117(1), 21–41.

[7] Wright, R.R. and Carstens, M.R. Linear momentum flux to overbank sections. *J. Hydr. Div.*, ASCE, 1970, 96(9), 1781–1793.

1A6

MIXING OF POLLUTANT CONSTITUENT IN A COMPOUND CHANNEL

DANILO T. JAQUE
DR. JAMES E. BALL
Department of Water Engineering
University of New South Wales
New South Wales, Australia

SYNOPSIS

Liquid wastes containing chemical, biological and physical constituents are discharged into flowing streams which dilute and disperse the pollutant constituents. The most common and economical form of disposal is through a side discharge over the full water depth. Experimental investigation has been conducted on the flow hydrodynamics and the mixing of a neutrally buoyant pollutant constituent in the nearfield region of a compound open channel. The investigation was restricted to the case of a side channel discharging perpendicularly into a compound open channel. The measured data were correlated using suitable characteristic length scales and relationships were obtained for both the jet trajectory and the dilution of pollutant constituents along the centerline trajectory.

INTRODUCTION

The most common and economical form of disposing liquid wastes into natural streams is through a side discharge. This type of discharge has been the focus of many recent investigations. The flow hydrodynamics in the vicinity of the discharge outlet were investigated by Rouse (1957) who measured the velocity field in the recirculating region while Mikhail et al. (1975) measured the width and length of recirculation zone using a flow visualisation technique. Wright (1977) investigated the mean behavior of buoyant jets in a crossflow. In addition, McGuirk and Rodi (1978) developed a two–dimensional depth averaged mathematical model to predict the velocity distribution and dilution of constituent mass in the nearfield region.

Previous studies on the initial dilution of pollutant constituents near the side discharge outlet have included those by Agg and Wakeford (1972), Bennett (1981), Anestis and Chu (1985), Abdel–Gawad et al. (1985), and Lee and Neville–Jones (1987). They used length scales and mathematical models to correlate all the available field and laboratory data. Results of these invetigations provided valuable information relevant to the initial dilution processes. However, investigation on the flow distribution and the dilution of constituent mass in the nearfield region of a compound open channel has not been attempted.

Presented in this paper is the result of an experimental investigation on the flow characteristic and mixing of a neutrally buoyant pollutant constituent in the nearfield region of a compound open channel flow.

DIMENSIONAL ANALYSIS AND LENGTH SCALE

Due to the complex theoretical analysis on the mixing of pollutant constituent mass issuing from a side channel into a compound channel, the use of dimensional analysis is employed. The pollutant concentration along the centerline jet trajectory downstream of the discharge outlet may be described as

$$C = f(C_o , M_o, U_o, U_m, B_o, B_m, d_o, d_m , x, y) \tag{1}$$

where C_O is the cross–sectional averaged concentration at the side channel outlet, M_o is the specific momentum flux of the side channel discharge, U_o and U_m are the cross–sectional averaged velocity and the subscripts o and m represent overbank and deeper main channel section, B is the channel width , d is the flow depth, and x and y are the spatial coordinates. The specific momentum flux M_o, can be expressed as

$$M_o = A_s V_s^2 \tag{2}$$

in which A_S and V_S are the side channel wetted cross–sectional area and cross–sectional averaged velocity. Fischer(1979) defined a characteristic length scale for the jet in terms of the cross flow velocity, U and the momentum flux, M_o which is expressed as $L_M = M_o^{1/2}/U$. The general application of this length scale is limited to a rectangular open channels, where U is a reasonable estimate of the velocity at any point.

A similar characteristic length scale appropriate for a compound channel is expressed as

$$L_M = \frac{M_o^{1/2}}{\alpha U_{av}} \tag{3}$$

where α is the ratio of the cross–section average velocity in the flood plain section to the cross–section average velocity in the deeper main channel. The parameter, U_{av} is the total cross sectional averaged velocity in the main channel.

Using the length scale presented in Eq. 3, assymptotic solutions are expected at the extremes. For a short distance from the side discharge outlet; $y \ll L_M$, the side channel momentum, M_o dominates the cross flow velocity, U. For $y \gg L_M$, the cross flow velocity, U dominates the momentum flux, M_o. Furthermore, for $y/B \ll L_M$, the influence of the far bank becomes insignificant in the solution.

By applying dimensional analysis of Eq. 1, the concentration and the jet trajectory relationships may be expressed, after appropriate simplification, as

$$C/Co = f\left(y/L_M , x/L_M , d_o/d_m \right) \quad and \quad y/l_M = f\left(x/L_M , d_o/d_m\right) \tag{4}$$

A similar function can be derived using the parameter η, to replace the x and y coordinate to give

$$C/Co = f\left(\eta/L_M , d_o/d_m \right) \quad and \quad y/L_M = f\left(\eta/L_M , d_o/d_m\right) \tag{5}$$

where η is defined as the distance along the trajectory line. Experimental data were correlated and relationships were obtained for the dilution of pollutant constituent and the jet trajectory for various side discharge momentum.

RESULTS AND DISCUSSIONS

Velocity Field and the Centerline Jet trajectory

The depth–averaged velocity fields obtained for a channel momentum ratio, (U_R) equal to 0.92 and 1.80 are shown in Fig. 1. The velocity vectors representing the magnitude and direction of the stream flow reflects the general flow condition in the channel. Far upstream of the side discharge outlet, the velocity profile is in a quiescent condition and is unaffected by the side discharge. At a distance of least two recirculating lengths downstream of the discharge outlet, the velocity profile is less affected by the momentum of the side discharge.

The centerline jet trajectories for different momentum ratios are shown in Fig. 2a. There are two dominant flow regimes shown in this figures. These are the side channel momentum–dominated region represented by the lower half of the trajectory curve and the crossflow–dominated region represented by the upper half of the trajectory curve. The upper regime occurs at a distance in which the influence of the side channel mometum has diminished. This distance can be estimated with the use of the characteristic length scale. In each flow regime, respective functional relationships can be derived to describe the centerline jet trajectories.

An additional useful relationships is shown in Fig. 2b for the centerline jet trajectory of the side discharge. The distance along the jet trajectory, η is used to relate the trajectories instead of the longitudinal distance (x). In this figure, it can be seen that the jet trajectories lie approximately on the same line in the momentum– dominated region and diverge from one another in the crossflow–dominated region.

Concentration Profile Along The Jet Trajectory

A relationship between the dimensionless pollutant concentration and the dimensionless distance along the jet trajectory is shown in Fig. 3a. In this figure, three stages of the mixing processes can be noted. The first stage is shown in the plot with an exponential decrease of pollutant concentration located in the momentum–dominated region. The second stage occurs at/or near the channel interface where mixing is enhanced by turbulent mixing of channel flow momentum. This mixing process is characterized by the abrupt change in the slope of the curve shown in the same figure. The third stage represents the crossflow–dominated region beyond the second stage and extends farther downstream. A gradual exponential decrease of the pollutant concentration is observed which is analogous to the mixing process in a long open channel reach. Another useful relationship involving the channel discharges is shown in Fig. 3b; this relationship resultes in similar characteristic curves to those shown in Fig.3a.

CONCLUSION

An experimental investigation has been conducted on the flow hydrodynamics and the mixing of a neutrally buoyant pollutant constituent in the nearfield region of a compound open channel. The measured data were analysed using appropriate characteristic length scales and relationships were obtained for both the jet trajectory and the dilution of pollutant constituents along the centerline trajectory. Dilution in the nearfield region has three mixing stages. The first stage is the momentum–dominated region where dilution is governed by side discharge momentum influx. The second stage occurs at the main channel where mixing is enhanced by turbulent mixing of channel flow momentum. This mixing process is characterized by the sudden increase in dilution. The third stage represents the crossflow–dominated region where a gradual exponential decrease of the pollutant concentration is observed which is analogous to the mixing process in a long open channel reach.

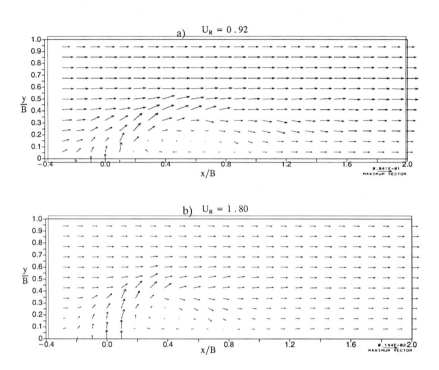

Fig.1 Measured velocity vectors in the nearfield region

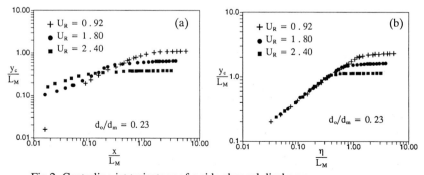

Fig.2 Centerline jet trajectory of a side channel discharge.

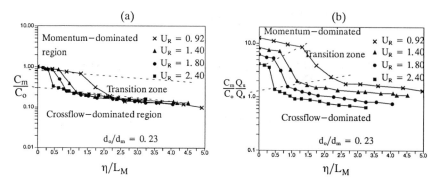

Fig.3 Depth–averaged pollutant concentration along the centerline of jet trajectory.

REFERENCES

Agg, A.R. and Wakeford A.C. (1972). Field studies of jet dilution of sewage at sea outfalls, Institution of Public Health Engineers Journal. 71, 126–149.

Anestis, I.D. and Chu, V.H. (1985). Entrapment characteristics in a recirculating eddy, 21st IAHR Congress, Melbourne, Australia. 2, 7–11.

Abdel–gawad, S.T. and McCorquodale, J.A. (1985). Initial mixing of cross–flowing jets in trapezoidal channels. 21st IAHR Congress, Melbourne, Australia. 2, 139–144.

Bennett, N.J. (1981). Initial Dilution: A practical study on the Hastings Long Sea outfall, Proc. of the Institution of Civil Engineers, Part 1, 70, 113–122.

Fischer, H.B, List, E.J., Koh, R.C.Y., Imberger, J. and Brooks, N.H. (1979). Mixing in Inland and Coastal Waters. Academic Press. New York.

Lee, J.H.W. and Neville–Jones, P. (1987). Initial dilution of horizontal jet in crossflow. J. Hyd. Div. ASCE, 113(5), 615–625.

McGuirk, J.J. and Rodi, W. (1978). A depth averaged mathematical model for the nearfield of side channel discharges into open channel flows, Journal of Fluid Mechanics. 86(4), 761–781.

Mikhail, R., Chu, V.H. and Savage, S.B. (1975). The reattachment of a two–dimensional turbulent jet in a confined crossflow, Proc. 16th Congress of IAHR. 3, 414–419.

Rouse, H. (1957). Diffusion of the lee of a two dimensional jet, 9th Congress International Applied Mechanics. 1, 307–312.

Wright, S.J. (1977). Mean behavior of buoyant jets in a cross–flow. J. Hyd. Div. ASCE. 103(5), 499–513.

1A7

MULTIPORT DIFFUSER PLUME TRAJECTORIES IN A CO-FLOWING CURRENT

María M. Méndez-Díaz and Gerhard H. Jirka
DeFrees Hydraulics Laboratory
School of Civil and Environmental Engineering
Cornell University, Ithaca, NY 14853 USA

INTRODUCTION

Waste water disposal in the ocean by means of submerged multiport diffusers is an effective and economically feasible water quality management solution for many coastal cities, in particular if it is combined with some form of primary or secondary treatment (National Research Council, 1993). In this study, we are concerned with the behavior of the merging jets/plumes originating from the central portion of a long diffuser that is laid perpendicularly to the shoreline and to the prevailing ambient currents. The diffuser is assumed to be located in "deep" unstratified water, giving rise to a distinct ascending plume of buoyant water without localized near-field instabilities (Jirka and Harleman, 1979).

The near-field dynamics of the buoyant jets originating from the ports of a unidirectional multiport diffuser are governed by several flow parameters and associated length scales that are summarized in Table 1. The ambient conditions (taken as unstratified in this study) are given by the average current velocity u_a and the water depth H. The discharge conditions are defined by the initial (port) discharge velocity U_o, the port diameter D, and the buoyant acceleration $g_o' = g(\rho_o - \rho_a)/\rho_a$ where ρ_o is the effluent density and ρ_a is the ambient density. Two kinds of flux variables and length scales are important, the three-dimensional ones relating to the individual jets, and the two-dimensional ones per unit diffuser length. The latter represent conditions after merging when a two-dimensional plume is formed. In that stage the flow dynamics may be represented by the concept of the *two-dimensional equivalent slot diffuser* (see Jirka, 1982). The concept neglects the details of the initial three-dimensional jets from the individual nozzles (ports) up to the distance of their merging. Instead, it is assumed that the merged two-dimensional plane jet comes initially from a slot of width $B = D^2 \pi/(4\ell)$ in which ℓ is the spacing between ports. This slot width definition preserves all dynamic characteristics of the diffuser and has been shown to be a reliable presentation for diffuser mixing provided attention lies outside the initial zone of merging (Jirka and Harleman, 1979). The significance of the different length scales given in Table 1 has been discussed by Jirka and Doneker (1991) and Jirka and Akar (1991). In addition, new discharge/buoyancy length scales, L_q^* and ℓ_q^*, respectively, are introduced herein as important measures of the trajectory characteristics of the lifting plume.

Trajectory relationships: For deep water discharges, the dynamic effect of the initial momentum flux is small as compared to the buoyancy flux, and the behavior of the flow is that of a plane plume (buoyancy driven flow). In the presence of an ambient current the plume rises with a trajectory governed by the magnitude of the ambient current, u_a, and by

Table 1: Flow Parameters and Length Scales of Multiport Diffuser		
	Round buoyant jet (D)	Plane buoyant jet (B)
Volume flux	$Q_o = U_o \pi D^2/4 \ [L^3/T]$	$q_o = U_o B \ [L^2/T]$
Momentum flux	$M_o = U_o^2 \pi D^2/4 \ [L^4/T^2]$	$m_o = U_o^2 B \ [L^3/T^2]$
Buoyancy flux	$J_o = U_o g_o' \pi D^2/4 \ [L^4/T^3]$	$j_o = U_o g_o' B \ [L^3/T^3]$
Jet/plume transition length scale	$L_M = M_o^{3/4}/J_o^{1/2}$	$\ell_M = m_o/j_o^{2/3}$
Jet/crossflow length scale	$L_m = M_o^{1/2}/u_a$	$\ell_m = m_o/u_a^2$
Plume/crossflow length scale	$L_b = J_o/u_a^3$	undefined
Discharge/buoyancy length scale	$L_q^* = Q_o^{3/5}/J_o^{1/5}$	$\ell_q^* = q_o/j_o^{1/3}$
Densimetric Froude number	$F_o = U_o/(g_o'D)^{1/2}$	$F_s = U_o/(g_o'B)^{1/2}$

the buoyancy flux per unit diffuser length, j_o. In general, the slope of the trajectory for a plume deflected by an ambient current is expected to follow the relation $dz/dx = u_c/u_a$ where z is the vertical coordinate and u_c is the centerline velocity of the plume. But, as illustrated in Fig.1, there are two expected flow configurations:

For the *weakly deflected* plume the centerline velocity is a constant, proportional to $j_o^{1/3}$, and therefore the slope can be expressed as $dz/dx = C_1(j_o^{1/3})/u_a$. Defining an ambient/discharge Froude number as $F_a = u_a/j_o^{1/3}$ and integrating, from a virtual origin, the equation for the trajectory of a weakly deflected plume is obtained as

$$\frac{z}{\ell_q^*} = \frac{C_2}{F_a^{1.5}} \frac{x}{\ell_q^*} \tag{1}$$

in which ℓ_q^* has been intoduced as a normalizing scale for both sides. The case of a *strongly deflected* plane plume is somewhat more complex. Akar and Jirka (1991) have proposed the plume is rising as an areally distributed source of buoyancy, j_a/u_a, generated when the ambient flow sweeps over the diffuser line. In that case, dimensional analysis shows the centerline velocity as proportional to $(j_a/u_a)^{1/2}$ and thus the strongly deflected trajectory relationship, upon integration, is

$$\frac{z}{\ell_q^*} = \frac{C_2}{F_a^{1.5}} \frac{x}{\ell_q^*} \tag{2}$$

Both Eq.1 and 2 correspond to straight line trajectories. The weakly deflected case is always associated with upstream intrusion as the mixed effluent flow, after surface impingement, experiences buoyant spreading against the weak ambient current (see Fig.1a).

Past laboratory experiments on diffuser plumes in deep co-flow have been limited in different ways. Cederwall (1971) and Roberts (1977) investigated plane buoyant jets issuing from a slot located in the bottom of the ambient water (thus, $h_o/H = 0$). Davidson (1989), on the other hand, studied multiport discharges with different spacings, but with very large port height values $h_o/H \approx 0.25$ to 0.5. A considerable discrepancy has been found in these past studies, as regards the transition between the regimes as well as the detailed trajectory behavior in each regime.

EXPERIMENTS

A series of experiments were performed in which a diffuser manifold model was towed at constant speed in 20m long x 0.77m wide x 0.88cm deep tank while a negatively buoyant salt solution was discharged horizontally through 24 ports, 4.72 mm in internal diameter. Three different port spacings were used and the submergence of the ports (equivalent to the height of the ports for the actual inverted situation) was kept constant at 16 cm. The towing speed varied between 1 and 7 cm/s. The trajectories were recorded using a video camera and then traced from the video monitor screen, superimposing images from repeated frames according to the towing (ambient current) velocity. Experimental parameters ranged between the following limits: Reynolds number, R_e, 1500 to 4700; port densimetric Froude number, F_o, 8 to 30; velocity ratio, $R = U_a/u_o$, 5 to 60; ambient/discharge Froude number, F_a, 0.17 to 2. In summary, the experimental program was designed so that (1) the discharge jets were fully turbulent with Reynolds numbers above 1200, (2) the port spacing was varied while the discharge per unit diffuser length was kept constant, i.e. keeping the overall diffuser length constant in an actual design, and (3) the port height value, $h_o/H \approx 0.18$, was typical for field cases.

Three distinct flow configurations were observed in the experiments:
a) For $F_a < 0.60$, a distinct plume that rises towards the surface of the water. After interaction with the surface, a hydraulic jump was observed followed by a buoyant layer spreading downstream and upstream of the source.
b) For $0.60 < F_a < 1$, an intermediate plume-like flow, together with blocking at the upstream side, was observed. A hydraulic jump and a spreading surface layer occurred downstream of the source. A thick intruding wedge, the length of which tends to decrease as the ambient current increases, was observed upstream of the source. During the experimental investigation, the thick upstream intruding wedge (1/3 to 1/2 the water depth) interfered with the trajectory of the plume causing it to be deflected towards the water surface (ambient bottom for actual situations). As a result, no trajectory measurements were made for this range of F_a.
c) For $F_a > 1$, the flow was swept downstream of the source and no upstream intruding wedge was observed. For the limiting case, $F_a = 1$, an incipient intruding wedge was observed and the plume seemed to rise towards the surface at a long distance downstream. Turbulent mixing was observed at both upper and lower boundaries of the plume, but the lower boundary showed a much more gradual growth due both to the constraining effect of the physical boundary and to buoyant damping.

Weakly Deflected Plane Plumes, $F_a < 0.6$: We use here the discharge/buoyancy length scale ℓ_q. is the dynamically important flow parameter that measures the interplay between the buoyancy flux, j_o, and the volume flux, q_o, of the line diffuser. This length scale is useful when comparing different diffuser discharges (high or low buoyancy, large or small flows), and when comparing small-scale laboratory discharges with large-scale field applications. In Fig.2, the normalized trajectories for weakly and strongly deflected cases can be seen to be grouped into two categories with the stronger deflections located lower in the double-logarithmic plot). Moreover, the data for individual experiments show some ordering within each category. For the weakly deflected cases this ordering is removed when replotting the ordinate as $(z/\ell_q.)F_a$ as has been done in Fig.3. The solid line in Fig.3 confirms the trajectory prediction, Eq.1. The trajectory constant, found from the experimental data is C_1 = 0.32 ± 0.007, thus somewhat smaller than the value C_1 = 0.36 calculated by Akar and Jirka (1991) from theoretical considerations as well as Davidson's (1989) data.

Strongly Deflected Plane Plumes, F_a > 0.60: When the trajectory data of strongly deflected plane plumes from Fig.2 are replotted against $(z/\ell_q)F_a^{1.5}$, the results, see Fig.4, are in good agreement with Eq.2. In contrast to C_1, however, there is no unique value of the trajectory coefficient C_2. Rather C_2 was found to depend on the discharge geometric characteristics, that is, with a parameter that combines port spacing, and port height, together with the buoyancy characteristics. Such a parameter is h_o/ℓ_q^*. The data shown in Fig.4 all correspond to runs with $h_o/\ell_q^* = 32$, giving $C_2 = 0.08$. An alternative series of experiments not shown here (see Méndez-Díaz, 1992) with $h_o/\ell_q^* = 67$ yielded a larger $C_2 = 0.17$. Davidson's (1989) experiments, on the other hand, had a much higher range $h_o/\ell_q^* = 79$ to 138 with corresponding C_2 values ranging from 0.13 to 0.24. On the basis of this, albeit limited, data base, we propose a variable trajectory relationship

$$C_2 = 0.0024 \, \frac{h_o}{\ell_q^*} \quad (for \quad \frac{h_o}{\ell_q^*} < 100) \qquad C_2 = 0.24 \quad (for \quad \frac{h_o}{\ell_q^*} > 100) \tag{3}$$

The parameter h_o/ℓ_q^* can be looked upon as a measure for the "leakiness" of the initially spaced and bottom separated diffuser jets. It is a measure for the passage of the ambient in between and under the diffuser jets in order to satisfy the entrainment demand at the lower boundary of the rising plane plume. The asymptotic value indicated by Eq.3, $C_2 \approx$ 0.24 as derived from Davidson's data, is adopted here as representative for unconfined flows, i.e. with large leakiness $h_o/\ell_q^* > 100$. On the other extreme, for $h_o/\ell_q^* = 0$, Eq.3 indicates a non-rising bottom-attached plume, $C_2 = 0$, consistent with Cederwall's (1971) and Roberts' (1977) observations on plumes from slots located in the bottom. Most actual diffuser designs would have smaller values leading to smaller C_2 as indicated by Eq.3. These results, though preliminary, show the great sensitivity of the two-dimensional diffuser dynamics to some local three-dimensional design details.

CONCLUSIONS
High degrees of dilution can be achieved with a properly designed multiport diffuser, ensuring that excessive pollutant concentrations are limited to small zones close to the discharge. However, highly complicated flow configurations can arise under different design or operating conditions. The dynamics of the plane plume that results from the merging of multiple buoyant jets within such a section depend largely on the ambient/discharge Froude number F_a. The three major flow configurations range from a weakly deflected plume regime with surface interaction and upstream spreading, to a complex intermediate regime with significant blocking and stagnant wedge formation, to a strongly deflected regime in which the plume gradually rises to the free surface.

The centerline plume trajectories in both extreme regimes are straight lines, but with distinctly different dependencies on F_a. In the strongly deflected regime, in particular, the plume behavior is crucially dependent on three-dimensional design details, such as port spacing and port height. It is these very details, measured by the "leakiness" parameter h_o/ℓ_q^*, that determine whether the plume attaches to the bottom or rises more or less quickly and finally mixes over the full water column. A survey of typical sewage diffuser installations by Méndez-Díaz (1992) gives a range of ℓ_q. from 0.05 to 0.15 m and of port height values from 0.5 to 3.0 m, thus a range of h_o/ℓ_q^* from about 5 to 60. The variability of the trajectory constant C_2 given by the tentative relationship, Eq.3, clearly shows the sensitivity of the ultimate plume rise to these design details. This detailed behavior, in turn, may have significant biological or ecological impacts, such as disturbances of benthic communities.

The present results have been incorporated into the Cornell Mixing Zone Expert System (CORMIX, Version 2.10 or higher, see Jirka and Hinton, 1992) that, among other discharge types, predicts the mixing of effluents from multiport diffusers under a wide range of ambient conditions and discharge design options.

Acknowledgments: The experimental research was partially supported by the U.S. Environmental Protection Agency (USEPA, Grant No. 813093) in the course of the CORMIX development. The first author gratefully acknowledges the support of the Venezuelan Fundación Gran Mariscal de Ayacucho and the Consorcio Inproman-VBL C.A. Many discussions with Prof. Ian R. Wood are gratefully acknowledged.

REFERENCES

Akar, P.J., and Jirka, G.H., 1991, "*CORMIX2: An expert system for hydrodynamic mixing zone analysis of conventional and toxic multiport diffuser discharges*", Tech. Rep. EPA/600/3-91/073, Environmental Research Lab., U.S. EPA, Athens, Georgia.

Cederwall, K., 1971, "*Buoyant Slot Jets into Stagnant or Flowing Environments*", Report No KH-R-25, W.M. Keck Lab. Hydraulics and Water Resources, California Institute of Technology, Pasadena.

Davidson, M.J., 1989, "*The Behavior of Single and Multiple, Horizontally Discharged, Buoyant Flows in a Non-Turbulent Coflowing Ambient Fluid*", Ph.D. Thesis, Report 89-3, Department of Civil Engineering, University of Canterbury, New Zealand.

Jirka, G.H., 1982, "Multiport Diffusers for Heat Disposal: A Summary". *J. Hydraulics Division*, ASCE, Vol. 108, HY12, 1425-1468.

Jirka, G.H., and P.J. Akar, 1991, "Hydrodynamic classification of submerged multiport diffusers discharges", *J. Hydraulic Engineering*, 117, No.9, 1113-1128.

Jirka, G.H., and R.L. Doneker, 1991, "Hydrodynamic classification of submerged single-port discharges", *J. Hydraulic Engineering*, 117, No.9, 1095-1112.

Jirka, G.H. and D.R.F. Harleman, 1979, "Stability and Mixing of Vertical Plane Buoyant Jet in Confined Depth", *J. of Fluid Mechanics*, Vol.94, 275-304.

Jirka, G.H., and S.W. Hinton, 1992, "*User's Guide to the Cornell Mixing Zone Expert System (CORMIX)*", Tech. Bulletin No.624, National Council for Air and Stream Improvement Medford, Mass.

Jirka G.H. and J.H.-W. Lee, 1994, "Waste Disposal in the Ocean", in "Water Quality and its Control," Vol.5 of *Hydraulic Structures Design Manual*, M. Hino, Ed., Balkema, Rotterdam.

Méndez-Díaz, M.M., 1992, "*Experimental Investigation on Unidirectional Multiport Diffuser Discharges in Coflowing Deep Water*", M.S. Thesis, Cornell University, Ithaca, NY.

National Research Council, 1993, *Managing Wastewater in Coastal Urban Areas*, National Academy Press, Washington, DC.

Roberts, P.J.W., 1979, "Line Plume and Ocean Outfall Dispersion", *J. Hydraulics Division*, ASCE, Vol.105, HY4, 313-331, (see also discussion by G.H. Jirka, HY12, 1573-1575).

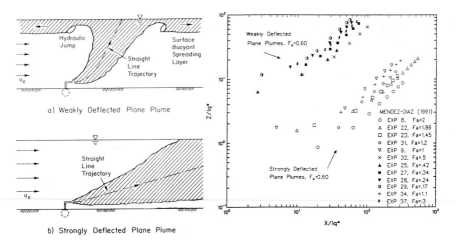

Fig.1: Trajectory behavior for buoyant multiport diffuser plumes in unstratified co-flow.

Fig.2: Plane plume centerline trajectories diffuser trajectories normalized by ℓ_q^*.

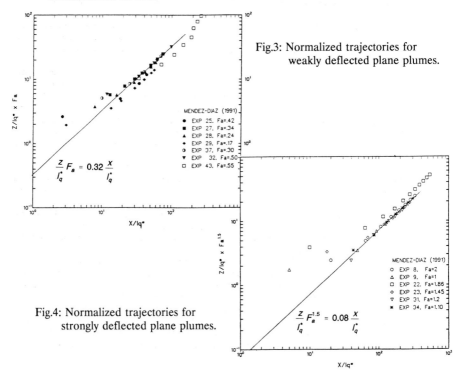

Fig.3: Normalized trajectories for weakly deflected plane plumes.

$$\frac{z}{l_q^*} F_a = 0.32 \frac{x}{l_q^*}$$

Fig.4: Normalized trajectories for strongly deflected plane plumes.

$$\frac{z}{l_q^*} F_a^{1.5} = 0.08 \frac{x}{l_q^*}$$

1A8

SURFACE INTERMEDIATE ZONE OF SUBMERGED TURBULENT BUOYANT JET IN CURRENT

Hai-Bo Chen
Binnie & Partners, Redhill, Surrey, U.K.
Torben Larsen
University of Aalborg, Aalborg, Denmark

ABSTRACT

This paper deals with the intermediate zone between the jet and plume stages of a submerged buoyant discharge from sea outfall in current. The stability criteria, plume width and height after the intermediate zone and the dilution within the intermediate region have been studied theoretically and experimentally. The study indicates that a stable upstream wedge may be existed when the densimetrical Froude number (F_{Δ}) is less than 0.45, an unstable wedge is likely to appear when F_{Δ} is between $0.45 \sim 1.0$, and no upstream wedge exists once F_{Δ} is greater than 1.0; the plume height and width could be readily estimated and finally the dilution in the intermediate zone is found to be insignificant.

1. INTRODUCTION

A turbulent buoyant jet discharged into a homogeneous flowing ambient is illustrated in Fig. 1. Basically, this process can be classified into three stages, namely jet, intermediate and plume stages.

In the jet stage, the discharged flow is mainly governed by the jet momentum and buoyancy, and the self-generated turbulence plays a dominating role in the path and dilution of the jet; In the plume stage, the plume is chiefly controlled by the buoyancy and the ambient turbulence.

Between the jet and plume stages, there exists a transition zone called intermediate stage/zone. In the intermediate stage, the effluent flow is governed by both momentum and buoyancy. It appears to be very difficult to describe the distribution of velocity and concentration in details in the intermediate stage because of the fact that both buoyancy and remained momentum will cause the jet to move horizontally with radial velocities as soon as the jet reaches the water free surface. Generally speaking, the buoyancy is intending to stabilize the upstream wedge and the cross flow momentum is to destructure the upstream intrusion.

In the past, the studies on turbulent buoyant jets and plumes in current have been mainly focused on the jet and plume zones, Larsen(1994) and Larsen, Petersen & Chen(1990), a little attention has been payed to the intermediate zone which links the jet and plume stages as a bridge.

Cederwall (1971) studied the stability of the upstream wedge using a two layer flow system which is essentially similar to saline intrusion at estuary. The flow was classified into three types, namely supercritical, subcritical and jet/plume like flows using the momentum flux ration and the source Froude number. A plant submerged buoyant jet with arbitrary angle into a stagnant, shallow fluid was studied theoretically and experimentally by Jirka & Harlemam (1973). The flow regions were divided into submerged, surface impingement, hydraulic jump and stratified counterflow regions. The stability and mixing characteristics were investigated. A round vertical buoyant jet in still water was studied by Lee & Jirka (1981). The same principles were used for the 2-D case with two differences, (a) a detailed treatment of the zone of flow establishment and (b) the transition from the surface impingement region to the outer flow is taken as a combination of a radial surface jet and a hydraulic jump.

In summary of the previous studies, it seems that the studies carried out were mainly concentrating on stagnant receiving waters except Cederwall's study, but Cederwall's study included only the stability analysis.

For a buoyant jet with arbitrary angle into a flowing ambient, in the author's view, in the stage of intermediate, three main aspects are of interests to be discussed. Firstly, the stability criterion for the upstream wedge induced by a turbulent submerged buoyant jet; secondly the initial plume height and width for the plume stage, and thirdly, the dilution in the intermediate zone.

The objective of this study is to investigate the intermediate stage covering the three aforementioned aspects, in order to shed some lights on this largely neglected area and later to develop a model which is able to describe the whole process in one code, including the jet, intermediate and plume stages.

2. THEORY

2.1. Stability Criteria for Upstream Wedge

An upstream wedge created by a submerged turbulent buoyant jet in a homogenous flowing ambient may be formed on the water surface and sketched in Fig. 1.

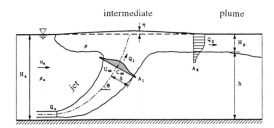

(a) Side View

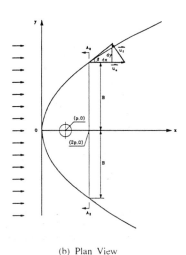

(b) Plan View

Fig. 1. Schematic Diagram of upstream wedge

It is assumed that the Bernoulli equation applies to a surface and a bottom streamline, Fig.1(a), the following equations are obtained immediately, Hansen & Jensen (1981).

$$\eta = \frac{u_a^2}{2g}$$

$$H_a + \frac{u_a^2}{2g} = h + \rho \frac{H_p}{\rho_a} + \frac{u_a^2}{2g}(\frac{H_a}{h})^2$$ (1)

in which, η : surface excess water lever, u_a : ambient current velocity, H_a : ambient water depth; h : water depth under the plume; ρ_a : ambient density; ρ : plume density; g : gravitational acceleration;

since $H_a + \eta = h + H_p$ and $\eta / H_a << 1$, then $h/H_a \approx 1-h_p/H_a$, thus, the Bernoulli equation finally yields:

$$2 (1-\frac{H_p}{H_a})^2 \frac{H_p}{H_a} = \frac{u_a^2}{\frac{\Delta\rho}{\rho_a}gH_a}$$ (2)

In fact the right hand side of Eq.(2) is a densimetrical Froude number,the maximum value of which is 0.296 for H_p/H_a in the range of 0.0 ~ 1.0,therefore, the criterion to form a stable upstream wedge depends on the value of the densimetrical Froude number, in other words, a stable upstream wedge may be established if

$$F_{\Delta s} = \frac{u_a^2}{\frac{\Delta\rho}{\rho_a}gH_a} \leq 0.296 \approx 0.30$$ (3)

The above analysis of stability condition for the upstream wedge is based on the two-layer flow system without taking the shape or the width of the surface plume into account.

Another approach to establish the stability criterion for the upstream wedge and to find the height of the plume is to consider the front velocity driven by buoyancy on the surface, Schrøder(1990), see Fig. 1.(b).

$$tg\beta = \frac{u_f}{\sqrt{u_a^2-u_f^2}}$$ (4)

in which, the front velocity u_f can be expressed as

$$u_f = (1-\frac{H_p}{H_a})\sqrt{\frac{\Delta\rho}{\rho_a}gH_p}$$ (5)

substituting Eq.(5) into Eq.(4), one obtains:

$$tg^2\beta [u_a^2-(1-\frac{H_p}{H_a})^2 \frac{\Delta\rho}{\rho_a}gH_p] = (1-\frac{H_p}{H_a})^2 \frac{\Delta\rho}{\rho_a}gH_p$$ (6)

where tg $\beta = \sqrt{2}/2$ by assuming the shape of the surface plume as a parabolic and written as $y^2 = 4px$, defining the location of the width of the plume at x = 2p, where p is the focal distance of the parabolic. Rearranging Eq.(6), it yields

$$3 (1-\frac{H_p}{H_a})^2 \frac{H_p}{H_a} = \frac{u_a^2}{\frac{\Delta\rho}{\rho_a}gH_a}$$ (7)

This leads to a similar equation to Eq.(3) with a different factor of 3 instead of 2. Hence, the criterion for a stable upstream intrusion created by a submerged buoyant jet in current states that the densimetrical Froude number has to be smaller than 0.45 instead of 0.30.

2.2. Plume Height and Width
In a stable condition, the plume height can be calculated from either Eq.(2) or Eq.(7).

The relation between the plume height and width can be derived by applying the continuity equation of mass in the intermediate zone. By substituting the Gaussian profile of velocity and integrating across the sections A_1 and A_2, the continuity equation of mass reads:

$$\iint_{A_1} [\, u_m \exp(-\frac{r^2}{b^2}) + u_a\cos\theta \,]\ dA = \iint_{A_2} u_a\ dA \tag{8}$$

After the integration, it becomes,

$$\pi b^2(\, u_m + 2u_a\cos\theta\,) = 2u_a B H_p \tag{9}$$

Finally, the half width of the plume is found to be

$$B = \frac{\pi b^2(u_m + 2u_a\cos\theta)}{2u_a H_p} = \frac{S_o Q_o}{2u_a H_p} \tag{10}$$

in which, S_o is the initial dilution defined at the end of the jet stage and Q_o is the initial jet discharge rate at the nozzle.

2.3. Dilution in Intermediate Zone
The conservation equation of density deficiency in the intermediate stage can be written as:

$$\frac{d}{ds}[\iint_A u\ \Delta\rho\ dA] = 0 \tag{11}$$

Substituting the Gaussian profiles of velocity and density, then integrating across the sections A_1 and A_2, Eq.(11) becomes:

$$\pi\lambda^2 b^2[\frac{u_m}{1+\lambda^2} + u_a\cos\theta]\Delta\rho_{m1} = \frac{\pi}{2}u_a\Delta\rho_{m2}BH_p \tag{12}$$

Finally, the dilution S_2 in the intermediate stage can be estimated as:

$$S_2 = \frac{\Delta\rho_{m1}}{\Delta\rho_{m2}} = \frac{u_a H_p B}{2\lambda^2 b^2(\frac{u_m}{1+\lambda^2} + u_a\cos\theta)} \tag{13}$$

It is believed that the excess jet velocity (u_m) is negligible at this stage in comparison to the ambient velocity and the angle θ is less than 45 degrees in the presence of the ambient current in most cases. Therefore it is perhaps reasonable to assume that $u_m \approx 0$ and $\cos\theta \approx 1.0$, then Eq.(13) can be approximated as

$$S_2 = \frac{\Delta\rho_{m1}}{\Delta\rho_{m2}} \approx \frac{u_a H_p B}{2b^2\lambda^2 u_a} \approx \frac{2}{5}\frac{H_p B}{b^2} \tag{14}$$

by taking the spreading coefficient $\lambda = 1.16$. The dilution could be estimated in the order of $2 \sim 4$ times based on the assumption that the plume height H_p has the same order of magnitude as the jet radius b and the plume width B is about $5 \sim 10$ times of b.

3. EXPERIMENT

Laboratory experiments were carried out in a flume with a dimension of 20 meter long, 1.5 meter wide and 0.8 m deep. Heated water discharged from a nozzle located at 10 cm above the flume bed and both flesh and salt receiving waters were used, The detailed experiment set-up and results are provided in Chen (1991) and Chen,Larsen & Petersen(1991).

3.1. Stability of Upstream Wedge
A series of experiments were performed in order to verify the theoretical stability criteria derived for the upstream wedge using both dye observation and photographic technique. It has been found in the laboratory studies that a stable upstream wedge is formed under the condition that the surface plume densimetrical Froude number ($F_{\Delta o}$) is less than 0.45 which coincides with the theory. The upstream wedge created by the submerged turbulent buoyant

jet will be completely expired if $F_{\Delta r} > 1.0$ and an unstable upstream wedge is likely to appear when $F_{\Delta r}$ is in the range of $0.45 \sim 1.0$. It seems that the flow regimes found in the experiment could be categorized into three types with stable, unstable and no upstream wedges.

3.2. Plume Height and Width

The experimental data on plume heights and widths are analyzed by calculating the variances of the measured cross-sections:

$$\sigma_y^2 = \frac{\sum_{i=1}^{m} \sum_{j=1}^{n} (y-\bar{y})^2 \Delta \rho_{ij} \Delta y_i \Delta z_j}{M_o} \qquad \sigma_z^2 = \frac{\sum_{i=1}^{m} \sum_{j=1}^{n} z^2 \Delta \rho_{ij} \Delta y_i \Delta z_j}{M_o} \tag{15}$$

in which,

$$M_o = \sum_{i=1}^{m} \sum_{j=1}^{n} \Delta \rho_{ij} \Delta y_i \Delta z_j \qquad \bar{y} = \sum_{i=1}^{m} \sum_{j=1}^{n} y_i \Delta \rho_{ij} \Delta y_i \Delta z_j \tag{16}$$

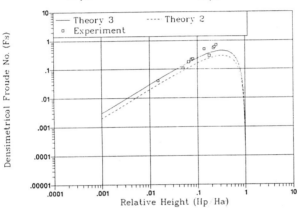

Fig. 3. Plume Height after Intermediate Zone

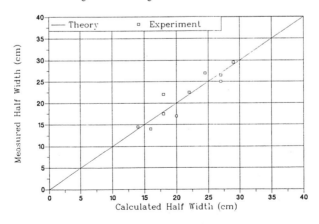

Fig. 4. Plume Width after Intermediate Zone

RESEARCH APPROACHES AND APPLICATIONS

In order to ensure the measurements correct the total mass budget is checked against the initial discharged mass by comparing the terms of $\iint_{A_o} \Delta \rho_o u_o dA$ and the zeroth moment M_o. Theoretically, the following equation should be valid:

$$\iint_{A_o} \Delta \rho_o u_o \, dA = \iint_A \Delta \rho u_a \, dy \, dz \qquad (17)$$

based on the assumption that the temperature T, and density difference obey a linear relation, i.e. $\Delta \rho_o = \beta_a \Delta T$, where, β_a is a constant.

The experimental data on the plume heights are plotted in Fig. 3 and compared with the theory. It shows that the experimental data agree better with the Theory 3 (factor of 3) derived by applying the front velocity driven by buoyancy force than the Theory 2 (factor of 2) based on the Bernoulli equations. It also indicates that the data in the stable region ($F_{4r} < 0.45$) fit better with the theories than that in the unstable region ($0.45 < F_{4r} < 1.0$).

The experimental data on plume width are demonstrated in Fig. 4 and compared with the theory. It seems that the measured widths are reasonably well correlated with the calculated widths. It should be pointed out that the dilution in the theory is the average dilution defined as $S_o = Q/Q_o$, but in the analysis of the experimental data on the plume width the minimum centerline dilution $S_{min} = \Delta \rho_o / \Delta \rho_m$ was used instead.

4. CONCLUSION

The intermediate zone of a turbulent buoyant jet in flowing ambient has been investigated theoretically and experimentally, the following conclusions may be draw from this study:

A stable upstream wedge could be formed if the densimetrical Froude number (F_{4r}) defined as Eq. (3) is less than 0.45; an unstable upstream wedge may be appeared when the densimetrical Froude number is in the range of $0.45 \sim 1.0$ and no upstream wedge exists when F_{4r} is greater than 1.0.

With a stable upstream wedge, the plume height and width at the end of the intermediate zone could be readily estimated using Eq.(7) and Eq.(10), respectively.

The dilution in the intermediate zone is found to be insignificant and estimated to be in an order of magnitude of $2 \sim 4$ times, provided that the upstream wedge is relatively stable.

5. REFERENCE

[1]. Cederwall, K. (1971). "Buoyant Slot Jets into Stagnant or Flowing Environments." W.M.Keck. Lab. of Hydr. & Water Resource. Calif. Inst. of Tech., Rep. No. KH-R-25, April.

[2]. Chen, H.B. (1991). "Turbulent Buoyant Jets and Plumes in Flowing Ambient Environments." PhD thesis, University of Aalborg, Denmark.

[3]. Chen, H.B. Larsen,T. and Petersen, O. (1991). "Turbulent Buoyant Jets in Flowing Ambients", Environmental Hydraulics, Lee & Cheung(eds).

[4]. Hansen, A.J. & Jensen, P. (1981). "Hydrographical and Hydraulic Marine Outfall Design." WHO Course, Copenhagen.

[5]. Jirka, G.H. & Harlemam, D.R.F (1973). "The Mechanics of Multiport Diffusers for Buoyant Discharges in Shallow Water." MIT, Ralph M. Parsons Lab. for Water Resource & Hydrodynamics, Tech. Rep. No. 169.

[6]. Larsen, T. (1994). "Numerical Modelling of Jets and Plumes, A Civil Engineering Perspective", Recent Research Advances in the Fluid Mechanics of Turbulent Jets and Plumes (eds.P.A.Davies and M.J.Valente),Kluwer Academic Publishers, Dordrecht.

[7]. Larsen,T. Petersen,O.and Chen, H.B.(1990). "Numerical Experiments on Turbulent Buoyant Jets in Flowing Ambients",Proc. International Conference on Computational Methods in Water Resource, Venice, June.

[8]. Lee, J.H.W. & Jirka, G.H. (1981). "Vertical Buoyant Jet in Shallow Water." J. of Hy. Div. ASCE 107, HY12.

[9]. Schrøder H. (1990) "Vandforurening" edited by Harremoes, P. (in danish).

1A9

Turbulent Flow Adjustment over a Stabilising Gravel Sediment Bed

S.J. TAIT [+] B.B. WILLETTS [*]

[+] Department of Civil & Structural Engineering University of Sheffield, Sheffield UK.

[*] Department of Engineering University of Aberdeen, Aberdeen UK.

INTRODUCTION

Change in the form and character of any river channel can be initiated by the movement of bed load. The problem of defining critical flow conditions associated with the initial entrainment and movement of sediment is of great importance. In the field, non-uniform grain size gravel sediment beds have been observed to exhibit rather erratic behaviour in that wide variations in the threshold conditions for bedload transport rate exist and do not appear to have a direct dependence on the prevailing flow conditions (Reid et al., 1985). In channels into which no sediment is introduced it has been found that as a mixed grain size bed degraded, coarse material accumulated on the bed surface and the bedload transport rate reduced significantly, thus indicating an enhanced degree of bed stability. This stability is due to the decreased probability of grain entrainment. The entrainment of sediment is thought to occur when the mobilising fluid forces exceed the forces that the grain can mobilise to resist motion. Traditionally the determination of incipient motion thresholds has been accomplished by the use of empirical relationships of mean, time averaged, variables to represent the mobilising and resisting forces. This approach was not able to explain the wide scatter of experimental data that has been produced. However when the grain entrainment process is studied at a grain scale it is the size and range of the turbulent fluid fluctuations that are important in determining the ability of the flow to entrain sediment (Grass, 1970). Observations of the mechanism of sediment entrainment have indicated that the flow structures that are significant are those which interact with the bed and are of high momentum. Grass (1970) postulated that the determination of bed movement was dependant on the overlap between two probability distributions representing respectively the critical resisting shear stress associated with the bed material and the range of the fluid shear stress distribution being applied to the bed.

This paper reports on flow velocity measurements made over a stabilising bed, in particular the differences in the near bed flow environment during periods of intense sediment activity and of virtually no sediment activity. Particular consideration was given to changes in the magnitude of the range of velocity fluctuations and the effect this had on the ability of the flow to entrain grains. It provides indications that grain scale turbulence features are an important factor in determining grain entrainment thresholds and thus transport and readjustment of river channel forms at a larger scale.

EXPERIMENTAL PROCEDURE

The objective was to examine the development of the turbulent flow over a stabilising mixed grain size sediment bed. The experiment was carried out in a recirculating tilting flume which was 0.3 metres wide and 12.5 metres long. In the experiment there was no upstream sediment supply and the flow rate was held constant. These conditions resulted in the bed surface coarsening over time so that the bedload transport rate reduced significantly. The sediment transport rate was measured using a bedload trap located in the base of the flume, 10.7 metres downstream from the flume inlet. All the velocity measurements were made at a position 1.0 metres upstream from the bedload trap on the centreline of the flume. During the course of the experiment continuous observations were made of the composition and transport rate of the bedload, of the composition and texture of the bed surface and of changes in the hydraulic environment.

FLOW ENVIRONMENT MEASUREMENTS

The near bed flow environment was measured using a Dantec 55X modular laser doppler anemometry system. It was capable of measuring two fluctuating orthogonal velocities at effectively a single point. The LDA system was set up so that measurements could be made close to the bed. The measurement system could be translated vertically so that velocity profiles could be measured at different times throughout the experiment as the armour layer developed. These were measured on the centreline of the flume at a point 1 meter upstream from the bedload trap. Each point was sampled at rate of 100Hz for a sampling period of approximately 100 seconds.

EXPERIMENTAL RESULTS
BEDLOAD TRANSPORT RATE

The bedload transport rate indicated a dramatic drop during the course of the experiment (fig 1). The rate of the transported material appeared to be split into three distinct time zones. The initial period which lasted for approximately 100 minutes was characterised by a high sediment transport rate, this was followed by a period in which the sediment transport rate dropped dramatically. The bed than exhibited a stable state in which the sediment transport rate had now acquired a low and stable value. Particular interest in this experiment was focused on the initial stage in which the sediment was highly mobile and the final stage in which the sediment was effectively static.

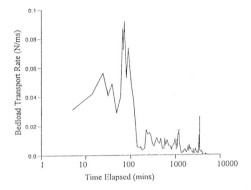

Figure 1 Development of the bedload transport rate with time.

MEAN VELOCITY PROFILES AND REYNOLDS STRESS DISTRIBUTION

Figure 2 indicates the measured velocity profiles after 70, 1000 and 3000 minutes had elapsed. This corresponds to periods in which the bedload activity can be described as active, declining and stable respectively. This shows a progressive deceleration of the flow during the experiment, the flow velocities having dropped in general throughout the flow depth. A semi-logarithmic plot of each of the velocity profiles appears to contain two segments, in the one closer to the bed the flow velocity is more uniform with depth than would be anticipated from the traditional log-law profile. This division of the velocity profile is reflected in the pattern of the vertical distribution of the Reynolds stress in which these two regions of flow are also observed (fig. 3)

Figure 3 show the distribution of the measured Reynolds Stress ($-\rho\overline{u'v'}$, ρ = density of the fluid, u' and v' are the streamwise and normal velocity components of the deviation from the mean velocity) measured at different times throughout the experiment. Both plots indicate that the Reynolds stress was suppressed in the flow region close to the bed consistent with the near bed flow regions observed in the velocity profiles. Nakagawa et al., (1989) obtained a similar pattern of measurements over fixed beds of glass beads and defined it as a "roughness sublayer". As the experiment progressed the depth of the layer in which the Reynolds stress is suppressed appears to change in that as the sediment bed becomes more stable the roughness sublayer thickened.

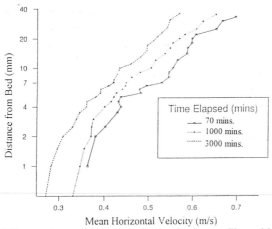

Figure 2 Comparison of the mean flow velocity profiles with time.

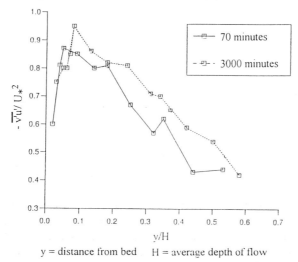

y = distance from bed   H = average depth of flow

Figure 3 Distribution of the Reynolds stress with distance from the bed.

TURBULENT FLOW PATTERN WITHIN THE "ROUGHNESS SUBLAYER"

The distribution of the velocity fluctuations (u' and v') were examined by distributing then into four quadrants. It was seen that at all times the flow was dominated by bursts of flow downwards and forwards and flow back and up in relation to the mean flow. Although the average size of the fluctuations was demonstrated to change close to the bed the percentage of the fluctuations in each direction remained approximately constant.

The magnitude of the velocity fluctuations in the fourth quadrant in which u' is in the streamwise direction and v' is towards the bed was examined more closely. This was done as it was believed that type of fluctuation was important in the entrainment of sediment (Gyr et al., 1989) in that it represented "high" speed fluid moving towards and therefore interacting with the bed. The average size of the velocity fluctuation and the proportion of the overall number of flow fluctuations that were in this quadrant were both observed to remain approximately constant throughout the experiment. However further examination of the distribution of magnitudes of the measured velocity fluctuations in this quadrant revealed a much more significant pattern (fig 4).

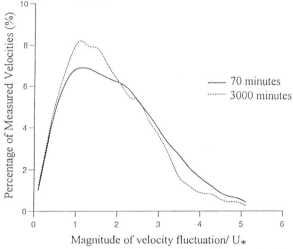

Figure 4 Distribution of the size of the flow fluctuations in Quadrant IV.

This figure indicates that the turbulent flow field was very different during the initial high transport stage of the experiment and after the bed had stabilised. Although the average size of the velocity fluctuations had remained similar there were great differences in the distribution of fluctuation size. At the end of the experiment there were fewer of the very large velocity fluctuations. Assuming that only the very large velocity fluctuations were likely to entrain grains it seemed reasonable to conclude that these changes in the turbulent flow field close to the bed reduced the likelihood of grain entrainment.

DISCUSSION AND CONCLUSIONS

These laboratory measurements are strong evidence that as the bed armour layer develops the turbulent structure of the flow changes. The thickness of the flow layer in which the Reynolds stresses are suppressed thickens. The sweep and bursts sequence of fluid behaviour has been linked to the presence of high values of Reynolds stress close to solid boundaries (Grass, 1971) in that the effects of a sweep coherent structure is mainly confined to a region close to the bed. Therefore the difference in the Reynolds stress distribution in the initial and final stages of the experiment reflects changes in the structure of the turbulent flow near to the bed (fig. 3). As the flow changed from being grain laden to essentially a grain free flow the suppression of the Reynolds stress was not expected to rise; the thickening of the "roughness sublayer" was therefore thought to be caused by the observed surface re-organisation of the sediment boundary. The most significant result is not the averaged changes in the "roughness sublayer" close to the bed but the changes in the instantaneous flow patterns within the layer (fig. 4). One of the major factors determining the initiation of sediment movement is the response of the bed to intense downward moving sweeps of fluid. Observations have indicated a link between the intense sweeps or groups of sweeps and the initiation of grain movement (Best, 1992; Gyr et al., 1989; Grass, 1970). The results have indicated that as the bed became more stable the turbulent velocity fluctuations change and that these changes involve the size and frequency of fluid "sweeps". The flow structures appeared to became more uniform in size in that the number of the more intense sweeps was reduced. If these are the flow structures most likely to entrain sediment then any reduction in their prevalence will lead to the ability of the flow to entrain sediment being reduced. An understanding of the structure of the turbulent flow close to a sediment bed will now allow the consistent definition of at least one of the probability distributions whose partial overlap determines the grain entrainment rate.

REFERENCES

Best J. (1992) On the entrainment of sediment and the initiation of bed defects: an insight from recent developments within turbulent boundary layer research; Sedimentology 39 797-811.

Grass A.J. (1970) Initial instability of fine bed sand; Proc. A.S.C.E Jour. of Hyd. Div. HY3 619-632.

Grass A.J. (1971) Structural features of turbulent flow over smooth and rough boundaries; Jour. of Fluid Mech. 50(2) 233-255.

Gyr A. Muller A. Schmid A. (1989) Observations of self stabilisation processes in sediment transport linked to coherent structures; Proc. XXIII Congress I.A.H.R. Ottawa A31-A38.

Nakagawa H. Tsujimoto T. Shimizu Y. (1989) Turbulent flow with small relative submergence; Int. Workshop on Fluvial Hydraulics of Mountain Regions Trent Italy I.A.H.R. A19-A30.

Reid I. Frostick L.E. Layman T.J. (1985) The incidence and nature of bedload transport during flood flows in coarse grained alluvial channels; Earth Surface Processes and Landforms 10 33-44.

Instantaneous shear stress on the bed in a turbulent open channel flow

A. KESHAVARZY and JAMES E. BALL

Water Research Laboratory, School of Civil Engineering
The University of New South Wales, Sydney, NSW, 2093, Australia

ABSTRACT

The entrainment of sediment from a bed of an open channel is a function of mean shear stress and instantaneous shear stress. The importance of the sweep event on sediment motion from the bed has been reported by many researchers. In this study, the shear stress of the sweep event was investigated in open channel flow with a rough bed. The instantaneous velocities of flow were measured in a laboratory flume using a small electromagnetic velocity meter and the magnitude of forces on the bed determined. The analysis of data showed that in sweep events, the mean shear stress is approximately more than two times the total time averaged Reynolds shear stress. The mean angle of the sweep events with the bed was determined also and found to be about 28° from the main direction of the flow toward the bed. The results presented form part of a major study into the turbulence and sediment transport in open channels.

INTRODUCTION

Bed load entrainment by river and stream flow is recognised as one of the most important problems in the field of sediment transport. The mechanics of sediment entrainment is one of the process where turbulence imposes a dominant influence. The turbulent motion strongly influences the rate of entrainment, deposition and transport of sediment particles. The frequencies at which sand particles are entrained or deposited are associated with instantaneous hydrodynamic shear forces on the sediment particles. In turbulent open channel flow, the hydrodynamic forces are not constant, but have a temporal variation. The entrainment and deposition of the particles not only depends on the mean shear stress at the bed but also on the instantaneous magnitude of the shear stress.

One aspect of recent research in turbulence has been the study of the bursting process which is a sequence of events including sweep, ejection outward and inward interactions. Kline *et al.* (1967) introduced the concept of bursting phenomena as a process by which momentum transfers between the turbulent and the laminar region near the bed. The bursting process consists of four events such as: sweep (downward front of flow velocity) $u' > 0$ and $v' < 0$, ejection $u' < 0$ and $v' > 0$, outward interaction $u' > 0$ and $v' > 0$ and inward interaction $u' < 0$ and $v' < 0$.

Grass (1971) used a hydrogen bubble technique to visualise the turbulent flow in open channel over smooth and rough bed. He concluded that the sweep and ejection events are responsible for most of the energy production and for the major contribution to Reynolds shear stress. He also found that beyond a certain distance from the boundary the turbulence intensity becomes independent of bed roughness. Thorne *et al.* (1989) and Nakagawa and Nezu (1978) pointed out that the sweep and ejection events occur more often than outward and inward interactions. William (1990) and Thorne et al. (1989) have shown that sediment entrainment occur from the bed mostly during the sweep event or flow with high velocity toward the bed.

Most of the existing critical shear stress models in sediment transport model are based on an average channel shear stress, which is defined in terms of depth, density of the flow and energy gradient. The particles on the bed however are subjected to instantaneous shear stresses much higher than the critical shear stress base a time averaged values. Raudkivi (1963) examined the sediment entrainment along the ripples on the bed and found that while the average shear stress is zero, entrainment occurs due to instantaneous shear stress. Novak and Nalluri (1975) studied the incipient motion of isolated sediment particles in circular and rectangular channel. They indicated that the effect of channel shape results higher turbulence intensity and critical shear stress as compared rectangular to circular cross section. They indicated that the critical shear stress obtained for entrainment of particles was appreciably lower than those generally adopted due to higher turbulence intensity in rectangular channel.

Previous investigations on bursting events have mostly concentrated on the study of the turbulent structure in air flow by means of wind tunnels. However, due to differences between air flow and water flow it is necessary to investigate the characteristics of the bursting process in water flow and particularly in open channels with natural bed roughness. In the study presented here, the characteristics of the bursting process have been investigated by measurements of instantaneous velocity within the flow. In this paper the magnitude of shear stresses and angles during the sweep event will be presented and discussed.

EXPERIMENTAL DETAIL

An experimental investigation was carried out in a non-recirculating tilting flume used for turbulence and sediment studies which consisted of a channel 0.61 m wide, 0.60 m high and 35 m in length. The bed was covered with 2 mm sand particles while the side walls were made of glass. The longitudinal and vertical velocity components were measured by means of a small electromagnetic velocity meter. These measurements were performed at a distance of 7 m from upstream end of the flume. The velocity components of the flow were recorded and analysed. More details of laboratory flume are presented by saiedi (1993).

RESULT AND DISCUSSION

The magnitude of shear stress in the sweep event is different from other events and also differs from mean shear stress of the flow in open channel. The event shear stress was determined after categorising the experimental data into the four regions. The shear stress of the sweep event was calculated and then normalised by the total shear stress at that point of the flow. This analysis is shown algebraically as equation 1. The variation of normalised shear stress (RSS) for the sweep event is shown in Figure 1. It can be seen that the shear stress in the sweep event increases with the depth from the bed to the water surface.

$$\tau = -\rho\overline{u'v'} \quad \text{and} \quad RSS = \frac{\left|\overline{u'v'}\right|(event)}{\left|\overline{u'v'}\right|(total)} \tag{1}$$

From the analysis of the experimental data, it was found that the magnitude of the shear stress in the sweep event was more than twice that the total shear stress near the bed. The normalised shear stress for the sweep event was determined as equal as 2.3-2.4 for d/H=0.07-0.15 as expressed in equation 2 . Consequently it would be expected that some particles which are not moved at the total shear stress are, in fact, able to move during the sweep event.

$$RSS_{(Sweep)} == 2.3 - 2.4 \quad \text{for} \quad d/H = 0.07\text{-}0.15 \tag{2}$$

where; u' , v' = velocity fluctuations in horizontal and vertical direction of flow.

Using regression analysis, the magnitude of the shear stress during a sweep event as a function of normalised depth is:

$$RSS = 2.4 + 1.72(d/H), \quad (r=0.57, SEE=0.41, n=57, P<0.001) \qquad (3)$$

Where; r=correlation coefficient.

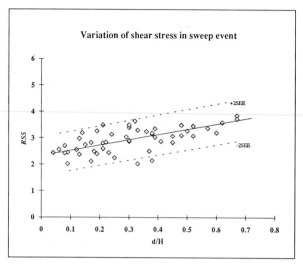

Fig. 1. Normalised shear stress in sweep event with depth

The applied force on the bed particles depends on the inclination angle of the sweep force to the bed. The instantaneous angle of the sweep and the ejection events was determined from the experimental data using equation 4. The mean angle of the events was calculated by taking the average at a defined depth of flow within the sample period of time.

$$\theta(sweep) = \tan^{-1}\left[\frac{v'(sweep)}{u'(sweep)}\right] \qquad (4)$$

Where; θ = angle of the events from horizontal direction of the flow and
         u' , v' = velocity fluctuations in horizontal and vertical direction.

Shown in Figure 2 are the average angles determined in this manner as a function of the normalised depth of flow for all experimental runs. It can be seen in Figure 2 that the angle of the sweep event in an open channel increases with depth from the bed to the free water surface. The angle of the sweep event is about 28° close to bed defined as d/H=0.07-0.15 and increases to 45° at the water surface.

Using regression analysis, the sweep angles as a function of normalised depth was determined as;

$$(Y=29+15.7X, r = 0.76) \qquad (5)$$

which is very significant. This equation is plotted in Figure 2.

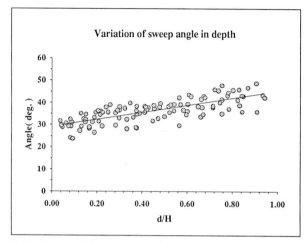

Fig. 2. Variation of the sweep angle with depth in bursting process

CONCLUSION

The magnitude of the shear stress in each event including; sweep, ejection, outward and inward interactions is very different from the mean shear stress at a certain point of the flow. In the sweep event the shear stress is approximately 2.3-2.4 times of the overall shear stress in the region near the bed. These temporal variations of the shear stress, have a distinct impact on the motion of sediment particles due to the high instantaneous force applied to sediment particles. The average angle for the sweep events from horizontal direction of the flow was determined and found to be approximately 28° at the bed and 45° at the water surface.

REFERENCES

Antonia, R. A. and Krogstad P. A., (1993), Scaling of the bursting periods in rough wall boundary layers, Experiments in Fluids, 15, pp. 82-84.

Bridge, J. S., (1981), Hydraulics interpretation of grain-size distributions using a physical model for bed load transport, J. of Sediment Petrology, Vol. 151, 4.

Bridge, J. S. and Bennet, S. J., (1992), A model for entrainment and transport of sediment grains of mixed sizes, shapes, and densities, Water Res. Res., 28, 2.

Grass, A. J., (1971), Structural features of turbulent flow over smooth and rough boundaries, J. of Fluid Mechanics, Vol. 50, Part 2.

Grass, A. J., (1982), The influence of boundary layer turbulence on the mechanics of sediment transport, Euromech 156, Mechanics of sediment transport, Istanbul.

Kline, S. J., Reynolds W. C., Schraub F. A. and Runstadler P.W., (1967), The structure of turbulent boundary layers, J. of Fluid Mechanics, Vol. 30, Part 4.

Nakagawa, H., Nezu, I., (1978), Bursting phenomenon near the wall in open channel flows and its simple mathematical model, Mem. Fac. Eng., Kyoto University, Japan, XL(4), Vol. 40, pp. 213-240.

Nezu, I. and Nakagawa, H., (1993), Turbulence in open-channel flows, IAHR-Monograph, Rotterdam: Balkema.

Novak, P. and Nalluri, N. (1975), Sediment transport in smooth fixed bed channels, Proc. ASCE, Vol. 101, No. HY9.

Raudkivi, A. J., (1990), Loose Boundary Hydraulics, 3rd Ed. Pergamon Press.

Raudkivi, A. J., (1963), Study of sediment ripple formation, ASCE, J. of Hydraulics Division, Vol. 89, HY6.

Saiedi, S., (1993), Experience in design of a laboratory flume for sediment studies, International Journal of Sediment Research, Vol. 8, No 3, Dec..

Thorne, P. D., Williams, J. J. and Heathershaw, A. D., (1989), In situ acoustic measurements of marine gravel threshold and transport, Sedimentology, 36.

Williams, J.J., (1990), Video observations of marine gravel transport, Geo. Mar. Lett., 10, pp. 157-164.

Yalin, M. S., (1992), River Mechanics, Pergamon Press, Ltd. Inc..

1A11

Velocity measurements in open channel flow over a rough bed

S. G. WALLIS and A. MOORES
Heriot-Watt University, Edinburgh, UK

INTRODUCTION

In this paper we report one part of a study concerned with solute dispersion in an open channel with a rough bed. The project is designed to investigate solute transport over a rigid bed containing holes or dead zones, and aims to allow a mechanistic interpretation of observed longitudinal dispersion in terms of the major advective and mixing mechanisms. The channel bed was designed to reproduce better the sort of dead zones found in rivers than had previously been possible with a gravel bed. The present paper focusses on measurements of the primary flow field and longitudinal turbulence intensities. The aims are to define the hydrodynamic characteristics of the flow in the channel and to compare them with established open channel flow hydraulics.

EXPERIMENTAL SET-UP

Experiments were performed in a re-circulating flume, of length 28 m and of uniform rectangular cross-section of width 0.75 m, in the Department of Civil & Offshore Engineering at Heriot-Watt University. The flume has a fixed bed slope, glass sides and for the work reported here, the bed was made of hardened sand impressed with a close packed formation of "hemispherical" depressions or dimples (average density of 450 per m^2; average diameter of 38 mm; average depth of 14 mm). The original sand bed had an average longitudinal bed slope of 1:1279. During the creation of the dimples, local deviations (of the order of $\pm$ 3 mm) of the surface level of the "islands" between the dead zones were formed. Thus as well as large scale roughness elements and textural surface roughness, there was some spatial non-uniformity of depth. The size of the dead zones was chosen to scale with natural dead zones found in typical stream beds. A tilting tail-gate at the exit of the flume was set to be approximately horizontal at the start of the work and was left in this position throughout the experiments. Fow rates in the channel were derived from a stage-disharge curve, established at the beginning of the work.

BULK FLOW HYDRAULICS

Table 1 summarises hydraulic data for the channel, where the hydraulic radius, R, depth of flow, d (measured to centre-line "island"), local water surface slope, S, and shear velocity, u_*, refer to a location 15.5 m from the flume entry. S was obtained from a second order regression line fitted to the backwater profile along the channel, and u_* was determined from $(gRS)^{0.5}$. Other parameters in the table are Reynold's number (Re), evaluated as $4UR/v$ where U is flow rate/flow area at 15.5 m and v is kinematic viscosity, and Darcy-Weisbach friction factor (f). Subcript b refers to side wall corrected values (French, 1986).

Q (l/s)	d (mm)	S (mm/m)	R (mm)	u_* (mm/s)	Re x10^4	f	R_b (mm)	f_b	u_{*b} (mm/s)
2.5	25.7	1.14	24.1	16.4	1.24	0.129	24.1	0.129	16.4
4.6	34.7	1.32	31.8	20.3	2.24	0.106	33.9	0.113	20.9
7.0	43.9	1.26	39.3	22.2	3.32	0.087	42.3	0.094	22.9
9.9	52.8	1.47	46.3	25.8	4.61	0.086	50.8	0.094	27.1
12.9	61.7	1.42	53.0	27.2	5.92	0.076	58.9	0.084	28.6
17.1	71.3	1.48	59.9	29.5	7.67	0.068	67.5	0.077	31.3
21.9	80.9	1.60	66.5	32.3	9.60	0.064	76.4	0.074	34.6
26.4	89.7	1.64	72.4	34.1	11.4	0.060	84.0	0.070	36.8

Table 1. Summary of hydraulic data.

Open channel flow over rough beds is described by a rough law resistance equation of the form shown below (ASCE, 1963), where k is a roughness

$$\frac{1}{\sqrt{f}} = a \log\left[\frac{bR}{k}\right] \tag{1}$$

parameter and a & b are constants. For two-dimensional flow, logarithmic velocity profiles and a von Karman's constant, κ, of 0.4, a=2.04. The value of b depends on the nature of the roughness and the physical interpretation of k, but is also influenced by the cross-sectional shape of the channel and the value of a.

Since neither b nor k are necessarily directly measurable quantities, and since it can be argued that for two-dimensional flow, equation (1) should use d or R_b instead of R, and f_b instead of f, experimenters have exercised a certain amount of freedom in interpreting their data. Two analyses were considered. Firstly, following customary practice, b was taken as 12 (ASCE, 1963; French, 1986).

and k was interpreted as the Nikuradse sand grain roughness, k_S, evaluated as 10.5 mm by regression of equation (1) to the data. Secondly, k was taken as the average dimple depth, R and f were replaced by d and f_b, in equation (1) and b was found via regression to be 11.5. This latter method is equivalent to that of Sayre & Albertson (1963) who replaced k/b with a single variable, χ, to represent the size, spacing and shape of the roughness elements (evaluated here as 1.22 mm). The data are shown in the form of the second method in Figure 1.

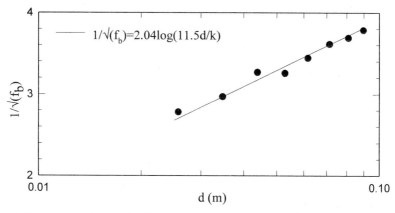

Figure 1. Resistance data in the form of equation (1), with k=14 mm.

VELOCITY PROFILES

Vertical profiles of longitudinal velocity were measured using an Airflow Developments pitot static tube (dynamic port diameter of 1mm) with an inclined manometer. Measurements were made at seven transverse positions at the 15.5 m location for flows of 7, 13.5 and 26 l/s. Measurements closer to the bed than 5 mm could not be made because of the minimum specification velocity of 0.2 m/s. Hence, we have only studied this data in terms of a velocity defect law, applicable to the outer region of the boundary layer, given by:

$$\frac{u_{max} - u}{U_*} = \frac{1}{\kappa} \ln\left(\frac{y - y_0}{d}\right) + \text{constant} \qquad (2)$$

where for each profile u_{max} is the maximum velocity in the profile, u is the velocity at a distance y above the bed, y_0 is the location of zero velocity and U_* is the local shear velocity. κ (=0.4) and d are as previously defined. A log-linear plot of each profile yielded U_* and y_0 following which the profiles were plotted in the form of equation (2). Figure 2 shows all 21 profiles plotted in this form.

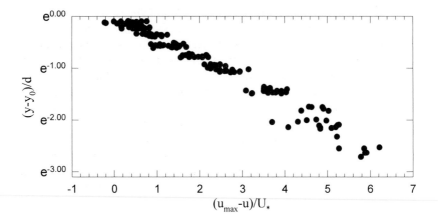

Figure 2. Velocity data in the form of equation (2).

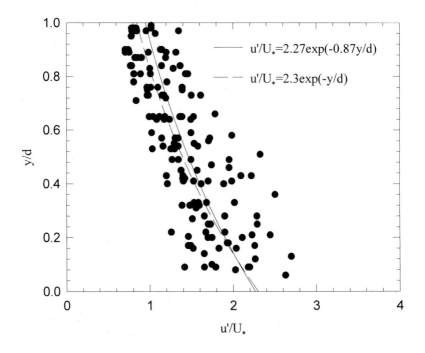

Figure 3. Turbulence intensity data in the form of equation (3).

TURBULENCE INTENSITIES

Longitudinal turbulence intensities were measured using a Dantec 55M hot film anemometer system with a conical probe. Measurements were made at several transverse positions at the 15.5 m location for the same flows used for the mean velocity measurements. The pitot static tube was used to calibrate the anemometer for turbulent mean velocity, and the instantaneous velocities were calculated using the usual linear calibration and stationarity requirements (Perry, 1982). The latter implied that data taken at less than $0.1y/d$ was unreliable, due presumably to wake interference from turbulent eddies trapped in, and shed by, the dimples. Figure 3 shows the reliable data in the form of equation (3), where the intensity is represented by the rms value of the velocity fluctuation, u', U_* is the local shear velocity and C & D are constants.

$$\frac{u'}{U_*} = C\exp(-Dy/d) \tag{3}$$

A regression line through the data gives C=2.27 and D=0.87. The equation due to Nezu & Nakagawa (1993) is also shown, for which C=2.3 and D=1.0.

DISCUSSION

The linear trends of the resistance data shown in Figure 1 confirm that fully rough turbulent flow exists in the channel. This is consistent with the relatively large values of f in Table 1 and the implied values of the Reynold's roughness number (u_*k_S/v being well in excess of 100). Based on the f-Re range and k_S value, the channel is rougher than those used by Kironoto & Graf (1994), but falls within the data discussed by Knight & MacDonald (1979). The results indicate that the bulk flow hydraulics of the channel containing roughness depressions is consistent with the more usual case of roughness projections. Since only one dimple arrangement has been examined, the effect of spacing is not yet known. It is likely that the behaviour will mirror that of isolated roughness elements and strip roughness (Sayre & Albertson, 1963; Knight & MacDonald, 1979).

Two checks on the velocity data were performed by comparing the flow rate obtained from integration of the velocity profiles with that derived from the stage, and comparing the width average of the local shear velocity derived from each profile (using $\kappa=0.4$) with the side wall corrected bulk shear velocity derived from the local water surface slope. The former agreed within ±10%, depending on the assumption made for the velocity gradient close to a boundary: the latter agreed within ±5%. This indicates that there were no gross errors in the pitot measurements. Figure 2 shows that the velocity data fits the velocity defect law: the data collapses on to a narrow band when non-dimensionalised with the local shear velocity, U_*, but shows more scatter if either u_* or u_{*b} is used.

The turbulence intensity data shown in Figure 3 show a general trend of decreasing intensity with increasing height above the bed. This, and the degree of scatter, is consistent with previous work, see e.g. Kironoto & Graf (1994). A comparsion of published values for C and D for flows over rough and smooth beds shows that they lie within the following ranges: $2.0 < C < 2.3$ and $0.8 < D < 1.0$. The authors' data lies at the upper range for C and the lower range for D.

CONCLUSIONS

Flow resistance, turbulent mean velocity and turbulence intensity data from an open channel with bed depressions has been presented. The flow in the channel is of the fully rough turbulent category, and the coefficients of the rough law are consistent with those found in other channels operating in a similar range of Reynold's number and friction factor, despite the different nature of, and the variety of scale of, the boundary roughness. Velocity profiles are logarithmic and collapse on to a velocity defect law when scaled with the local shear velocity and flow depth. Turbulence intensities also scale with the local shear velocity, are consistent with previous data and support the idea that the distribution of turbulent intensity in open channel flow does not vary with the nature of the boundary roughness.

REFERENCES

ASCE Friction factors in open channels, Task Force Report, Journal of the Hydraulics Division, ASCE, **89, HY2**, 1963, pp97-143.

French, R.H. Open-channel Hydraulics, McGraw-Hill, Singapore, 1986.

Kironoto B.A. & Graf W.H. Turbulence characteristics in rough unifrom open-channel flow, Proceedings of the Institution of Civil Engineers, Water Maritime & Energy, **104(4)**, 1994, pp 333-344.

Knight D.W. & MacDonald J.A. Hydraulic resistance of strip roughness, Journal of the Hydraulics Division, ASCE, **105, HY6**, 1979, pp 675-690.

Nezu I & Nakagawa H. Turbulence in Open Channels, Balkema, Rotterdam, 1993.

Perry A.E. Hot Wire Anemometry, Clarendon Press, Oxford, 1982.

Sayre W.W. & Albertson M. L. Roughness spacing in rigid open channels, Transactions, ASCE, **128**, 1963, pp 343-427.

1A12

AN EQUATION FOR
CLEAR-WATER BRIDGE-PIER SCOUR

CHONG-GUANG SU
Dept. of Hydraulic Engineering
Feng Chia University
Taichung, Taiwan, R.O.C.

INTRODUCTION

According to Hjorth (1975), Melville and Raudkivi (1977), Ettema (1980), and Breusers and Raudkivi (1991), the flow pattern around a pier is complicated. It may be segregated into five components: (1) Downflow in front of pier, (2) Horseshoe vortex, (3) Cast-off vortices and wake, (4) Bow wave, and (5) Trailing vortex (as shown in Figure 1). The complex flow structure will, as a rule, scour the sediment around the pier in a movable bed. This may result in failures that cost lives, money and the interruption of road networks. In the existent literature, there are many observations, equations and models for predicting the scour depth around a bridge pier. One major problem with this is those different equations and models often produce a wide range of scour depths under same circumstances. The fact evinces the need of continuous study on this topic.

In this paper a new equation for predicting the equilibrium scour depth is presented. The equation is derived from dimensional analysis and is checked by experimental data. It has been shown that the equation incorporated with the upstream flow depth of the bridge pier can better predict the scour depth.

DIMENSIONAL ANALYSIS

Generally speaking, the equilibrium scour depth, h_s, is related to the fluid property, the channel property, the flow property, and the bridge pier property. For this study, the following assumptions are adopted:

(I) The bed material for the test runs is uniform and noncohesive.

(II) The channel is straight and not constricted.

(III) The roughness of the channel bed is ignorable.

(IV) The unsteadiness of the flow is not considered.

(V) The bridge pier is cylindrical, without any protective device against bed scouring, and smooth on its surface.

From the previous statement concerning the scour depth one can write

$$h_s = f(\rho, \ g, \ \mu, \ \rho_s, \ d_{50}, \ U_c, \ h, \ U, \ S, \ D) \qquad (1)$$

where ρ is the water density, g is the gravity of acceleration, μ is the coefficient of viscosity of water, ρ_s is the density of the sediment, d_{50} is the median sediment size, U_c is the mean incipient velocity of the sediment, h is the upstream flow depth, U is the upstream mean velocity of river flow , S is the channel slope, and D is the diameter of the bridge pier. It is noted that Shen et al. (1963), and Bonasoundas (1973) use the Euler equation and the continuity equation to describe the downflow and arrive at an expression relating the downflow velocity, which is considered as the principal erosive flow mechanism, to the horizontal velocity of river flow and the flow depth measured from the bed upwards. The relationship is now interpreted in a slightly different fashion. The downflow velocity may be regarded as a function of the upstream mean velocity of river flow and the upstream flow depth. The interpretation is reflected on the right hand side of the equal sign in equation (1). As regards the term on the left hand side, it can be changed to $(h_s + h)$ to represent the total path for the downflow to develop. Thus,

$$h_s + h = f(\rho, \ g, \ \mu, \ \rho_s, \ d_{50}, \ U_c, \ h, \ U, \ S, \ D) \qquad (2)$$

By use of Buckingham Pi theorem and ρ, U, D as the three repeating variables, one can perform dimensional analysis, and have the results expressed in the form

$$\frac{h_s + h}{D} = f(\frac{h}{D}, \ \frac{D}{d_{50}}, \ \frac{U^2}{gD}, \ \frac{\rho U D}{\mu}, \ \frac{U}{U_c}, \ \frac{\rho_s}{\rho}, \ S) \qquad (3)$$

On the basis of Breusers' et al. (1977) finding, the effect of Froude number can be represented by the mean velocity and the sediment size. The Reynolds number is commonly considered a lesser parameter in open channel flow, and its effect can be neglected. In addition, the sediment density may be held as a constant equal to 2.65. Therefore, the functional relationship of equation (3) can be simplified as follows:

$$\frac{h_s + h}{D} = f(\frac{h}{D}, \ \frac{D}{d_{50}}, \ \frac{U}{U_c}, \ S) \qquad (4)$$

This is similar to those proposed by Breusers and others, except that the upstream flow depth is present in the term on the left hand side of the equation.

EXPERIMENTS

The experiment of this study is conducted in a recirculating type flume with a length of 22 m, a width of 1 m, and a depth of 0.75 m. The test section where graded sediment is filled up to a thickness of 15 cm, contained within two end sills, and perfectly leveled around the pier model measures 6.4 m. A ramp is provided at the entrance of the test section to keep the flow smooth, while a sluice gate at the end of the flume is used to control the flow depth. The condition of the experiments is rather extensive. It consists of

(I) diameter of pier, D: 8.8 cm, 7.5 cm, 6 cm, 4.8 cm, 4.2 cm, and 3.4 cm,

(II) median sediment size, d_{50}: 2.5 mm, and 3 mm,

(III) discharge, Q: 0.08 cms, and 0.15 cms,

(IV) upstream flow depth, h: from 17 cm to 43.5 cm,

(V) channel slope, S: 0, 0.004, 0.007, and 0.01.

The duration of the experimental runs depends on the approximation of the equilibrium condition in which the change of scour depth is no more detectable. It usually lasts two hours. Table 1 shows the experimental results.

ANALYSIS & DISCUSSION OF TEST RESULTS

The data set in Table 1 has been used to determine an expression for predicting the equilibrium scour depth according to the functional relationship shown as Equation (4). The nondimensional form which introduces flow depth, h on the left hand side to improve accuracy, is obtained by regression analysis and may be expressed in the following exponential form:

$$\frac{h_s + h}{D} = 1.378 \left(\frac{h}{D}\right)^{0.83} \left(\frac{D}{d_{50}}\right)^{0.045} \left(\frac{U}{U_c}\right)^{0.24} (1 - S)^{-1.47} \tag{5}$$

The correlation coefficient for this regression equation (r = 0.99) and its standard deviation (= 0.29) indicate an excellent fit.

Figure (2) shows the comparison between the calculated (scour depth + flow depth) from Equation (5) and the observed ones. It can be seen that most of the data points fall on the 45° line. It should be noted that the data set including 40 data points covers a wide spectrum of flow depth.

To test the universality of Equation (5) some published measurements of equilibrium scour depth, a total of 147 observations under a variety of flow conditions and bridge pier diameters, were assembled (Shen, Yanmaz, Yen, Tsai). Figure (3) compares the observed $(h_s + h)$ with the calculated ones from Equation (5). The standard deviation is 0.298. It is obvious that the agreement is also stringent.

It is worthy to mention at this point that the velocity profile of the downflow into the scour hole at the nose of the pier has its maximum velocity near or below the level of the channel bed (Ettema, 1980). In other words, the downflow commences near the flow surface (exclusive of the bow wave) and accelerates to its maximum velocity at a point near or below the channel bed. Once the maximum velocity is reached, the downflow begins its deceleration because of the retardative effect of the bottom. The total path of development, which consists of the decelerating and accelerating limbs, i.e., $(h_s + h)$, is vital to the bridge pier scour mechanism.

CONCLUSIONS

A new equation is presented in this paper that renders the prediction of the equilibrium scour depth at bridge piers more accurate. It is based upon experimental data with cylindrical piers under clear water flow. The new form of expression seems to be applicable to a wider spectrum of flow conditions. It is found that the incorporation of the flow depth in the equation has greatly improved the accuracy of predicting the equilibrium scour depth. This leads to the conclusion that the path for the downflow to develop is an important parameter when it comes to dealing with pier scour.

REFERENCES

Bonasoundas, Dr. Ing., 1973, "Flow Structure and Scour Problem at Circular Bridge Piers," Oskar V. Miller Institute Report No. 28, Munich Technical U.

Breusers, H. N. C. et al., 1977, "Local Scour around Cylindrical Piers," J. of Hydr. Research, Vol. 15, pp.211-252.

Breusers, H. N. C. and A. J. Raudkivi, 1991, "Scouring," A. A. Balkema, Rotterdam, the Netherlands.

Ettema, R., 1980, "Scour at Bridge Piers," School of Eng. Report No. 216, University of Auckland, New Zealand, pp.487-497.

Hjorth, P., 1975, "Studies on the Nature of Local Scour," Dept. of Water Res. Eng., Lund Institute of Technology, Bulletin Series A, No. 46.

Melville, B. W. and A. J. Raudkivi, 1977, "Flow Characteristics in Local Scour at Bridge Piers," J. of Hydr. Research, Vol. 15, pp.373-380.

Shen, H. W. et al., 1963, "Time Variation of Bed Deformation near Bridge Piers," Proc. of 10th Congress, I.A.H.R., Leningrad, Paper No. 3.14.

Shen, H. W. et al., 1969, "Local Scour around Bridge Piers," J. of the Hydr. Division, ASCE, Vol. 95, No. HY6, November, pp.1919-1940.

Tsai, W. C., 1989, "Study on Scour around Cylindrical Bridge Piers," M.S. thesis presented to the Graduate Institute of Hydraulic and Ocean Eng., Natl. Cheng Kung University, R.O.C., 93pp. (in Chinese).

Yanmaz, A. M. and H. D. Altmbilek, 1991, "Study of Time-Dependent Scour around Bridge Piers," J. of Hydr. Engineering, ASCE, Vol. 95, No. 10, Oct., pp.1247-1268.

Yen, R. F., 1986, "Scour around Bridge Piers on Gravel Bed," M.S. thesis presented to the Graduate Institute of Civil Eng., Natl. Taiwan University, R.O.C., 86pp. (in Chinese).

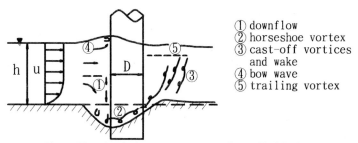

① downflow
② horseshoe vortex
③ cast-off vortices and wake
④ bow wave
⑤ trailing vortex

Fig 1. Sketch of Flow Pattern around a Cylindrical Pier

Table 1. Summary of Experimental Date

Run number	D cm	d_{50} mm	h cm	Q cms	U m/s	U_c m/s	U/U_c	S %	h_s cm
1	8.8	2.5	38.7	0.15	0.646	0.759	0.85	0	7.5
2	7.5	2.5	38.5	0.15	0.649	0.758	0.86	0	6.1
3	6.0	2.5	35.1	0.15	0.712	0.748	0.95	0	5.6
4	4.8	2.5	32.8	0.15	0.762	0.740	1.03	0	4.1
5	4.2	2.5	35.0	0.15	0.714	0.747	0.96	0	2.1
6	3.4	2.5	36.3	0.15	0.690	0.751	0.92	0	1.5
7	8.8	3.0	43.5	0.15	0.575	0.835	0.69	0	5.0
8	7.5	3.0	43.0	0.15	0.581	0.833	0.70	0	3.8
9	6.0	3.0	43.0	0.15	0.581	0.833	0.70	0	2.7
10	4.8	3.0	43.0	0.15	0.581	0.833	0.70	0	2.1
11	4.2	3.0	43.0	0.15	0.581	0.833	0.70	0	1.3
12	8.8	3.0	22.5	0.08	0.593	0.753	0.79	0	5.2
13	7.5	3.0	20.0	0.08	0.667	0.738	0.90	0	4.8
14	6.0	3.0	20.5	0.08	0.650	0.741	0.88	0	4.1
15	4.8	3.0	22.0	0.08	0.606	0.750	0.81	0	3.4
16	4.2	3.0	22.0	0.08	0.606	0.750	0.81	0	1.7
17	8.8	3.0	16.5	0.08	0.808	0.714	1.13	0	9.5
18	7.5	3.0	16.5	0.08	0.808	0.714	1.13	0	8.5
19	6.0	3.0	17.5	0.08	0.762	0.721	1.06	0	6.6
20	4.8	3.0	17.0	0.08	0.784	0.712	1.10	0	5.4
21	4.2	3.0	17.0	0.08	0.784	0.712	1.10	0	3.8
22	3.4	3.0	17.0	0.08	0.784	0.712	1.10	0	3.0
23	8.8	3.0	21.0	0.08	0.635	0.744	0.85	0.4	8.5
24	7.5	3.0	20.5	0.08	0.650	0.741	0.88	0.4	6.8
25	6.0	3.0	21.5	0.08	0.620	0.747	0.83	0.4	4.5
26	4.8	3.0	22.5	0.08	0.593	0.753	0.79	0.4	4.0
27	4.2	3.0	22.5	0.08	0.593	0.753	0.79	0.4	2.2
28	3.4	3.0	22.0	0.08	0.606	0.750	0.81	0.4	1.7
29	8.8	3.0	21.5	0.08	0.620	0.750	0.83	0.7	7.2
30	7.5	3.0	21.5	0.08	0.620	0.750	0.83	0.7	7.0
31	6.0	3.0	21.7	0.08	0.614	0.748	0.82	0.7	5.1
32	4.8	3.0	21.5	0.08	0.620	0.747	0.83	0.7	4.3
33	4.2	3.0	21.5	0.08	0.620	0.747	0.83	0.7	3.3
34	3.4	3.0	21.0	0.08	0.635	0.744	0.85	0.7	2.2
35	8.8	3.0	21.5	0.08	0.620	0.747	0.83	1	8.0
36	7.5	3.0	21.5	0.08	0.620	0.747	0.83	1	5.5
37	6.0	3.0	21.5	0.08	0.620	0.747	0.83	1	5.0
38	4.8	3.0	21.5	0.08	0.620	0.747	0.83	1	3.4
39	4.2	3.0	21.5	0.08	0.620	0.747	0.83	1	2.8
40	3.4	3.0	21.5	0.08	0.620	0.747	0.83	1	2.0

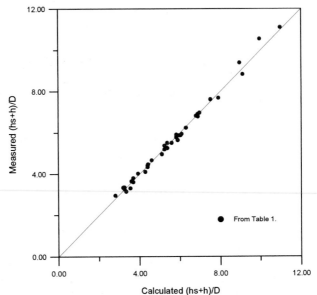

Fig. 2. Comparison of Measured and Calculated (hs+h)/D

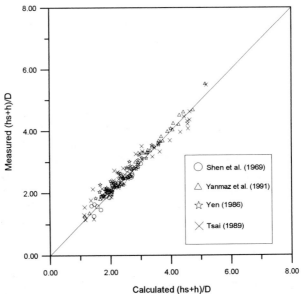

Fig. 3. Comparison of Measured and Calculated (hs+h)/D

1A13

The Surface Flow Structure of Large Scale Vortices at the rear of a Cylindrical Pier for the Flow of Transcritical Reynolds Number

Kazuo Ishino, Taisei Corporation, Yokohama, Japan
Kouji Kawaguchi, Honshu-Shikoku Bridge Authority, Kobe, Japan

1. Introduction

Ishino et al.[1] carried out the field measurement of the surface flow structure around a cylindrical pier (a main cylindrical pier 3P of the Akashi Kaikyo Ohashi bridge, which is now under construction) in the region of Reynolds number (hereinafter referred to as Re number)10^8, for the first time in the world, and examined the flow structure, though large scale vortices could not be observed at the rear of 3P. After this, one of the authors got a chance to photograph and videotape the occurrence of large scale vortices at the rear of 2P and 3P from the nearly 300m-high main tower constructed on 3P. In this study, the authors discussed the occurrence of large scale vortices at the rear of 2P and 3P, using pictures and images taken from top of the main tower.

2. Main pier foundation 2P, 3P and the main tower at the Akashi Kaikyo Ohashi bridge

As shown in Figs. 1 and 2, the Akashi Kaikyo Ohashi bridge is a 3910m-long suspension bridge between Tarumi-ku, Kobe City and Awajishima Island with the center span of 1990m. The main pier foundation 3P is a 62m-long steel caisson of diameter 78m which was installed on the sea bed of 57m deep as shown in Fig.1, 3.and 2P is a 70m-long steel caisson of diameter 80m which was installed on the sea bed of 60m deep as show in Fig 2. As shown in Fig. 2, the overall height of the main tower is 283m, whose top is TP + 297.2m above the sea level.

3. Field observation

The field observation was carried out at the spring tide on May 24, 1994, from 10:30 to 14:45. The tidal current was flowing east into Osaka Bay, and

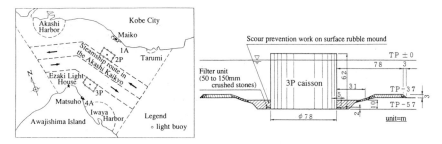

Fig.1 Location of the Akashi Kaikyo Ohashi bridge

Fig.3 3P caisson and scour prevention work

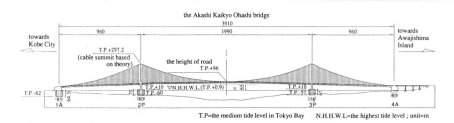

T.P=the medium tide level in Tokyo Bay   N.H.H.W.L=the highest tide level ; unit=m

Fig.2 Vertical cross sections of the Akashi Kaikyo Ohashi bridge

according to the tidal current estimation, the peak flow velocity of 3.2m / sec (6.2kt) was obtained around 12:32.

From the top of the main tower of 3P, the surface flow structure at the rear 3P was recorded by a camera and a 8mm video camera; while, the surface flow structure of the rear 2P was recorded only by a camera.

4. Discussion on the surface flow structure recorded by a camera and video camera

4.1. Surface flow structure at the rear of 2P

Pictures 1 (a), (b), (c), (d), and (e) show the surface flow structure at the rear of 2P caisson observed from the main tower of 3P. These pictures indicated the following:

- in Picture (a), large scale vortices with counterclockwise rotation, equivalent to Karman vortices, were distinct in the region between 1.75D to 3.5D downstream from the center of 2P.(D:the diameter of pier)
- in Picture (b), 40 seconds after (a), counterclockwise vortices were flowing down and clockwise vortices occurred.

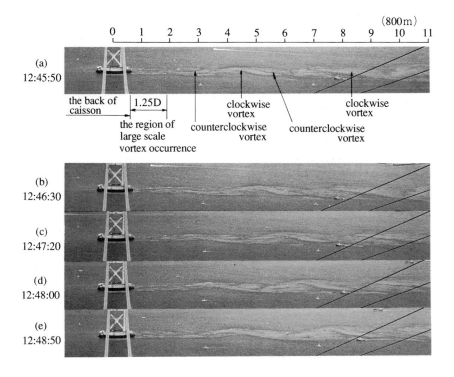

Picture.1   Surface flow structure at the rear of 2P observed from the main tower of 3P

- in Picture (c), 90 seconds after (a), clockwise vortices were distinct.
- in Picture (d), 130 seconds after (a), clockwise vortices were also flowing down and new counterclockwise vortices occurred.
- in Picture (e), 180 seconds after (a), counterclockwise vortices similar to those in Picture (a) occurred again.

From these pictures, it was concluded that the period of the large scale vortex occurrence was approximately 180 seconds.

As shown in Picture 1, counterclockwise vortices were more distinct than clockwise vortices, which can be explained by the coriolis effect pointed out by Nishimura et al.[2].

4.2. Surface flow structure at the rear of 3P

Pictures 2 (f), (g), (h), (i) and (j) show the surface flow structure at the rear of 3P at a distance of 7D from its center, observed from the main tower of 3P. Pictures 2 and the video images indicate the following:

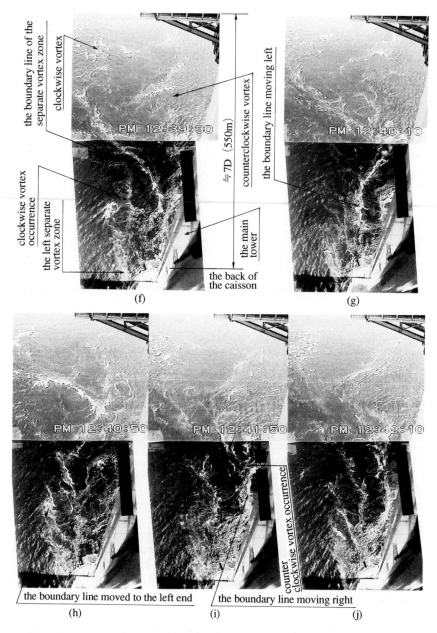

Picture.2   Surface flow structure at the rear of 3P observed from the main tower of 3P

- in Picture (f), a clockwise vortex occurred downstream and to the left of 3P, counterclockwise vortex was observed downstream, and another clockwise vortex was observed further downstream. In the center of a picture, a boundary line between the right and left separate vortex zones (hereinafter referred to as boundary line) at the rear of 3P was clearly visualized by white bubbles.
- in Picture (g), 40 seconds after (f), the boundary line at the rear of 3P started to move left. Meanwhile, a separate vortex from 3P was flowing down at 20-second intervals.
- in Picture (h), 80 seconds after (f), while the boundary line was moving left, a counterclockwise vortex occurred on the right of the boundary line. A clockwise vortex occurred where a counterclockwise vortex had been in Picture (f), In the left separate vortex zone, about four separate vortices were confined.
- in Picture (i), 120 seconds after (f), as the energy in the left separate vortex zone was confined into a smaller area compared to the right zone at the rear of 3P, the boundary line started to move right due to the repulsion of this energy.
- in Picture (j), 160 seconds after (f), flow structure similar to that in Picture (f) was formed again.

From the above observation, it can be summarized as follows:
- within the region 1.75D from the center of 3P, the energy of separate vortices occurred from the caisson at 20-second intervals became imbalanced between the right and left zones then, the separate vortices were converged at 160-second intervals in the region further than 1.75D downstream, forming large scale clockwise and counterclockwise vortices, which were as large as Karman vortices.

6. Discussion on the period of large scale vortex occurrence and the Strouhal number

Based on the data such as the water surface configuration around 3P, flow velocity spectrum and the period of the separate vortex occurrence form 3P (30 sec), which were obtained from the field observation in October 1989 when the incident flow velocity was 3.3m/sec in the westward current, Ishino et al.[1] indicated that the period of large scale vortex occurrence at the rear 3P was four times as large as that of separate vortex occurrence which was 30 seconds. From the result of this observation, the Strouhal number was obtained as $S_t=0.21$, which almost agreed to the measured value $S_t=0.19$ to 0.26 with $R_e=10^7$.

According to the results from the field observation in May 1994 when the incident flow velocity was 3.2m/sec in the eastward current, the period of separate vortex occurrence from the left side of 3P was 20 seconds, and the period of large scale vortex occurrence, equivalent to Karman vortex, was 160 seconds, which was eight times as large as that of separate vortex occurrence.

From the result of this field observation, the Strouhal number was obtained as $S_t=0.15$. The following are the differences between the above two field observations:

- The season differed; one was carried out in May, while the other was in October.
- The flow direction was opposite; one was westward, while the other was eastward.
- In October 1989, there was no structure around 3P which might cause large roughness, while in May 1994, scaffolding struts were constructed around 3P 5m below the water surface, which accelerated the roughness and turbulence on the water surface.

As Kimura [3] pointed out, when Re is large, R_{eT} can be estimated as in the following equation using coefficient of eddy viscosity ($\nu_T = 10^7 cm^2/sec$) instead of kinematic viscosity (ν), since the incident flow was turbulent.

Here, R_{eT} changes largely in accordance with the changes in ν_T. Also, the Strouhal number changes significantly when R_{eT} is in an order of 10^2. We assume that this is why S_t values obtained from the observational data in October 1989 and in May 1994 differed.

7. Conclusion

Through the field observation from the top of the main tower on 3P of the Akashi Kaikyo Ohashi bridge, large scale vortex occurrences equivalent to Karman vortex in the rear side of 2P and 3P were examined, where Pier Reynolds number using kinematic viscosity was $Re=2.5\times10^8$, and Pier Reynolds number for turbulent flow using eddy viscosity ($\nu_T = 10^7 cm^2/sec$) instead of kinematic viscosity (ν) was $R_{eT}=250$.

References

(1) Kazuo Ishino, Hideo Otani, Ryota Okada, Yoshitaka Nakagawa (1993): The flow structure around a cylindrical pier for the flow of transcritical Reynolds number, Proc. of XXV Congress of International Association for Hydraulic Research, Vol. V, pp. 417-424.

(3) Ryuji Kimura (1985): Flow Science (Revised), Published by Tokai Univ. pp. 2-25.

(2) Tsukasa Nishimura, Sotoaki Onishi, Sotaro Tanaka (1981): Application of Coherent Structure Model to Natural Flow and the Remote Sensing as Visualizing Method, Proc. of 36th Conference on Hydraulics, pp. 187-192.

1A14

Design of riprap for scour protection around bridge piers

TAE HOON YOON, SUNG BUM YOON & KWANG SEOK YOON
Department of Civil Engineering, Hanyang University, Seoul, Korea

ABSTRACT : For the determination of critical size of riprap particle used to protect the base of bridge pier, new design criteria are presented. In the new equations a number of flow characteristics such as the effects of flow depth, particle size, and placement depth on the stability of riprap are taken into consideration.

INTRODUCTION

For the protection of the river bed near bridge piers from scouring, various kinds of protection measures have been presented and studied. Posey(1974) studied the effect of inverted filter to protect bridge pier from leaching action. Chiew(1992) investigated the effects of slot, collar and the combinations placed at/around piers on reducing scour depths. Fotherby(1993) evaluated the performance of protective materials including grout mats, grout bags, footings and ripraps. The scour depths were reduced exponentially with the increase of protective area. Ruff & Nickelson(1993) examined the depth of local scour resulting from different percentages of area coverage using a single layer of riprap. The studies, described above, presented the effectiveness of protective measure on reducing scour depths and are not closely related to the stability of riprap.

As cited by Parola(1993), Breusers et al.(1977), Bonasoundas(1973) and Quazi & Peterson(1973) proposed the equations for the determination of riprap size for bed protection around circular and round-nosed piers. Parola(1993) conducted extensive experiments to investigate the stability of riprap and presented the design criteria. However, the use of his equation is limited to the cases of rectangular piers with the placement elevation slightly below the approaching stream bed. In this study, the new design criteria which can be applied for general practices are proposed.

DESIGN CRITERIA OF RIPRAP

The determination of critical condition, under which the riprap particle with given diameter becomes stable, is essential for the design of riprap to protect the bridge pier from scouring. According to Parola(1993) the design criteria for stable riprap are relatively few in comparison with the equations predicting scour depths.

Parola(1993) conducted a series of experiments for the determination of riprap size around the circular and rectangular piers. As shown in Fig.1 of Parola(1993), the stability number N_c defined by Eq.(1) has a weak correlation with D_o/y_o.

$$N_c = \frac{u_o^2}{(S_s-1)gD_p} \qquad (1)$$

where all the notations are defined in the end of this paper. Taking account of the effects of relative placement depth d/b, and relative rock size b/D_p (see Fig.1), he proposed the design criteria for the most severe cases when the riprap was placed slightly below the bed level around rectangular pier.

He left the detailed design criteria for the different cases of pier dimensions and relative placement depths (see Fig.3 of Parola;1993) as a further research.

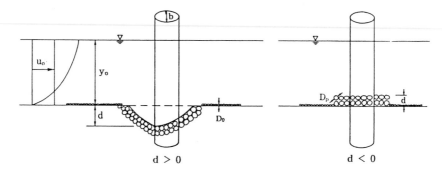

d > 0   d < 0

Figure 1. Schematic representation of riprap around pier.

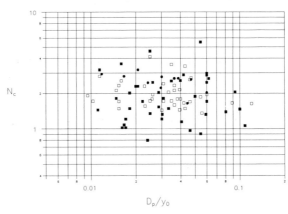

Figure 2. Relationship between particle stability number N_c and D_p/y_o:
□ ; rectangular pier(d<0),   ■ ; rectangular pier(d≥0),
● ; circular pier(d≥0)

Motivated by Quazi & Peterson(1973), Parola's experimental results are replotted in Fig.2 to show the relationship between N_c and D_p/y_o. It showes no improvement in correlation due to the neglection of important factors other than D_p/y_o. Similarly to Melville & Sutherland(1988), three correction parameters to the particle stability number of Eq.(1) are introduced to take account of other improtant factors as:

$$N_c^* = N_c K_y^2 K_D^2 K_d^{-2} \tag{2}$$

where N_c^* is the modified particle stability number and K_y, K_D and K_d are the correction parameters for flow depth, particle size and placement depth, respectively, and defined by

$$\text{for } d \geq 0, \qquad K_y = \begin{cases} 0.2(\dfrac{y_o}{b}) & \text{if } \dfrac{y_o}{b} \leq 5.0 \\ 1 & \text{otherwise} \end{cases} \tag{3}$$

$$\text{for } d \leq 0, \qquad K_y = \begin{cases} 0.2(\dfrac{y_o+d}{b}) & \text{if } \dfrac{y_o+d}{b} < 5.0 \\ 1 & \text{otherwise} \end{cases} \tag{4}$$

$$\text{and} \qquad K_D = (\dfrac{b}{25D_p})^{1.065} \tag{5}$$

$$K_d = \begin{cases} 1+\dfrac{d}{2.5b} & \text{if } d > 0 \\ 1 & \text{if } d \leq 0 \end{cases} \tag{6}$$

The correction due to flow depth K_y is given by Melville & Sutherland(1988), and is modified slightly to include the effects of relative placement depth d/b. The placement depth factor K_d accounts for the separation of approaching flow from the bottom due to horseshoe vortex.

The particle size factor K_D represents the combined effects of approaching flow displacement due to pier and downflow absoption by porous riprap layers. Using the correction parameters, Eqs.(3)-(6), the modified stability number N_c^* given by Eq.(2) is calculated for the experimental data of Parola(1993) and plotted against D_p/y_o. As shown in Fig. 3, strong correlations between N_c^* and D_p/y_o for each different case of experiments are observed. The empirical design criteria are then proposed as:

for rectangular pier with $d \leq 0$

$$N_c^* = 4.2 \times 10^{-5}(\dfrac{D_p}{y_o})^{-2.22} \tag{7}$$

for rectangular pier with $d \geq 0$

$$N_c^* = 9.5 \times 10^{-5}(\dfrac{D_p}{y_o})^{-1.92} \tag{8}$$

and for circular pier with $d \geq 0$

$$N_c^* = 5.3 \times 10^{-5} \left(\frac{D_p}{y_o}\right)^{-2.24} \tag{9}$$

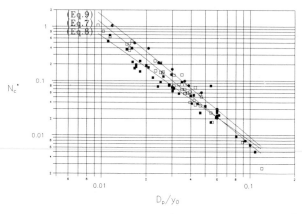

Figure 3. Relationship between modified particle stability number N_c^* and D_p/y_o:
□ ; rectangular pier(d<0), ■ ; rectangular pier(d≥0),
● ; circular pier(d≥0)

EXPERIMENTAL VERIFICATION

For the verification of design criteria Eqs.(7)-(9) proposed in this study, a series of experiments are conducted for the case of circular pier. The parameters set in experiments are

$$u_o = 0.29 \text{m/s}, \quad y_o = 0.166 \text{m}, \quad b = 0.05 \text{m}, \quad D_o = 0.4 \text{mm}, \quad S_s = 2.65 \tag{10}$$

A riprap is placed around the circular pier with its surface elevation equal to that of stream bed (i.e., d=0). The particle size of riprap used in the experiments is given in Table 1. Various combinations of different coverage and thickness of riprap layers are tested for each size of riprap particles. The particles between No.1 and No.4 (0.84mm < D_p < 2.38 mm) are unstable for all the combinations of coverage and thickness of riprap layers. The particles at the surface layer of NO.5 move, but remain without leaving the riprap area. This seems to be the critical conditions defined by Parola. The particles larger than 3.36 mm in diameter (i.e. No.6 and No.7) are unconditionally stable. Thus the critical size of riprap particle under the given flow condition is found in the range between 2.38mm and 3.36mm.

For the case of circular pier with d=0, Eq.(3) gives $K_d=1$ and Eq.(9) is simplified as:

$$D_p = \frac{0.795 u_o^{2.25} b^{0.15}}{(S_s - 1)^{1.12} g^{1.12} y_o^{0.27}} \tag{11}$$

Substituting all the experimental parameters of Eq.(10) into Eq.(11), $D_p=0.0023$m or

2.3mm is obtained, which underestimates the critical condition in comparison with the experimintal value.

If the lower envelope of Parola's experimintal data shown in Fig.3 is chosen as design curve, then Eq.(9) is modified as:

$$N_c^* = 4.6 \times 10^{-5} \left(\frac{D_p}{y_o}\right)^{-2.24} \tag{12}$$

and

$$D_p = \frac{0.906 u_o^{2.25} b^{0.15}}{(S_s-1)^{1.12} g^{1.12} y_o^{0.27}} \tag{13}$$

$$= 2.6 \text{ mm}$$

This shows a reasonable agreement with experimental critical condition. The discrepancy between experimental and predicted critical conditions could be explained by the following two facts. Firstly, there is no clearcut between stable and unstable conditions when these are based on eye-judgement. Secondly, the shape of riprap stones used in this study is not round, but it is round in Parola's experiments.

Table 1. Particle size of riprap used in experiments.

particle No.	Sieve No.		D_p(mm)			remarks
	pass	remain	min.	max.	average	
1	16	20	0.84	1.19	1.02	
2	12	16	1.19	1.68	1.44	
3	10	12	1.68	2.00	1.84	
4	8	10	2.00	2.38	2.19	S_s=2.65
5	6	8	2.38	3.36	2.87	
6	4	6	3.36	4.76	4.06	
7	-	4	4.76	9.52	7.14	

CONCLUSION

New design criteria are proposed for the determination of critical size of riprap particle placed around bridge pier. The new empirical formulas are obtained by modifying the equation presented by Quazi and Peterson(1973) and by applying those to the experimental data of Parola(1993). The effects of flow depth, particle size, and placement depth on the stability of riprap particles are included in the new equations. A series of experiments to test the new equations are performed and show a good agreement between predicted and observed values of riprap particle size. A further research is going on to improve the equations by including the roughness effect of approaching channel bed and coverage of riprap.

ACKNOWLEDGMENT
This study was financially supported, in part, by Samwoo Engineering, Seoul, Korea.

REFERENCES

(1) Bonasoundas, M. (1973) Flow structure and problems at circular bridge piers. Report No.28, Oskar V. Miller Inst., Munich Tech. Univ., Munich, West Germany.

(2) Breusers, H.N.C., Nicollet, G. & Shen, H.W. (1977) Local scour around cylindrical piers. J. Hydr. Res., Vol.15, No.3, pp.211-252.

(3) Chiew, Y.-M. (1992) Scour protection at bridge piers. J. Hydraulic Engineering, ASCE, Vol.118, No.9, pp.1260-1269.

(4) Fotherby, L.M. (1993) The influence of protective material on local scour dimensions. Proc. Hydraulic Engineering '93, ASCE, pp.1379-1384.

(5) Melville, B.W. & Sutherland, A.J. (1988) Design method for local scour at bridge piers. J. Hydraulic Engineering, ASCE, Vol.114, No.10, pp.1210-1225.

(6) Parola, A.C. (1993) Stability of riprap at bridge piers. J. Hydraulic Engineering, ASCE, Vol.119, No.10, pp.1080-1093.

(7) Posey, C.J. (1974) Test of scour protection for bridge piers. J. Hydraulic Engineering Div., ASCE, Vol.100, No.12, pp.1773-1783.

(8) Quazi, M.E. & Peterson, A.W. (1973) A method for bridge pier riprap design. Proc. First Canadian Hydraulics Conf., Univ. of Alberta, Edmonton, Canada, pp.96-106.

(9) Ruff, J.F. & Nickelson, J.R. (1993) Riprap coverage around bridge piers. Proc. Hydraulic Engineering '93, ASCE, pp.1540-1545.

NOTATION

b = pier diameter(m)

d = placement depth measured downward from stream bed(m)
  ($d>0$, below bed; $d<0$, above bed)

D_o = particle diameter of approaching stream bed(m)

D_p = minimum particle diameter of riprap for stability(m)

g = gravity acceleration(m/s^2)

K_D = correction factor for particle size

K_d = correction factor for placement depth

K_y = correction factor for flow depth

N_c = particle stability number

N_c^* = modified particle stability number

S_s = specific gravity(dimensionless) of riprap particle

u_o = depth averaged unobstructed flow velocity(m/s)

y_o = approach-flow depth(m)

1A15

Scaling of Hydraulic Roughness for Small and Large Scale River Models

C. NAISH & R. H. J. SELLIN
Department of Civil Engineering, University of Bristol, UK

INTRODUCTION

The construction of a major trunk road along the River Blackwater valley in Hampshire has resulted in the need to relocate a length of the River Blackwater. This gave the National Rivers Authority an opportunity to construct and study the hydraulic performance of an environmentally acceptable, doubly meandering, two-stage channel.

Investigations are being carried out in the laboratory at 1/5 and 1/25 scales, but in order to calibrate these and future hydraulic models accurately, and identify any detrimental influence that the effect of scaling up results from model to prototype may have, prototype river data is required. Because of the expense and difficulties involved in collecting field data, especially at times of flood, very few river studies have been undertaken. Myers (1990) investigated the modelling at a scale of 1/20 of a reconstructed length of the River Main formed as a straight compound channel. The only detailed field investigation into the hydraulic performance of meandering compound channels was carried out on the River Roding in Essex by Sellin, Giles and van Beesten (1990)

To the authors knowledge, this project will make available for the first time meandering compound channel data from prototype, 1/5 and 1/25 scale and will give the best opportunity to date to establish scale effects in hydraulic modelling.

RIVER BLACKWATER DATA COLLECTION

The test reach of the relocated River Blackwater channel has a meandering inner channel with medium sinuosity and a wider upper channel (the floodplain or berm) of low sinuosity. Figure 1 details the planform of the test reach, Figure 2 details a typical cross-section.

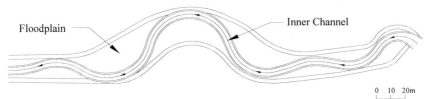

Figure 1 - Planform of the inner channel and floodplains (prototype scale)

Figure 2 - Typical channel cross-section (prototype scale)

FIELD DATA

A Sarasota electromagnetic coil gauging station has been installed immediately upstream of the test reach to measure the discharge to 5% accuracy at low flows and 1% accuracy at high flows. Five submersible pressure transducers with a 0-3m range and ±3mm accuracy have been deployed in manholes, connected to the river via 100mm dia. pipes, at strategic intervals along the test reach to measure the water stage. The gauging station and pressure transducers are linked to a TG1150 V35 data logger sited in a purpose built instrument hut adjacent to the gauging station. The data logger has a 90 day storage capacity and currently logs the river data at 15 minute intervals. An EPROM has been fitted which enables this time interval to be adjusted remotely. The logger is mains powered but is fitted with a battery backup and a lightning protection unit. The logger is being interrogated remotely at Bristol University via a modem and DTS Logmaster software.

MODEL DATA

An undistorted 1/5 scale model was constructed in the 56m long, 10m wide SERC Flood Channel Facility in the Laboratories of HR Wallingford. An undistorted 1/25 scale model was built in the Hydraulics Laboratory of Bristol University. Experimental arrangements, procedures and scaling laws are detailed in Naish & Sellin (1994). Currently the 1/25 model is being modified to incorporate a 2:1 vertical scale exaggeration.

DEPTH-DISCHARGE RELATIONSHIPS

Depth-Discharge data was collected from both the large and small scale models under a variety of surface and vertical roughening conditions. The 1/5 scale model was tested under various combinations of mortar finish, 8mm and 13mm dia. gravel surface finishes and 25mm dia. dowel vertical roughness, which were surface penetrating at all flow depths and supported by a trellis which was always above the water surface. The 1/25 scale model was tested under combinations of mortar finish, 0.4mm and 1.4mm dia. sand surface finishes and 5mm dia. dowel vertical roughness, with the dowel spacing scaled down from the similar frames used on the 1/5 scale model.

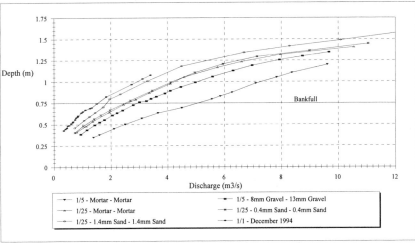

Figure 3 - Depth-Discharge performance at prototype scale

It can be seen from the Depth-Discharge curves in Figure 3 that the 0.4mm dia. sand used on the 1/25 scale model was producing a higher scaled roughness than the 8mm dia. gravel used on the 1/5 scale model (0.4mm dia. x 5 scale ratio = 2mm dia. at 1/5 scale). The prototype floodplain vegetation has been left to develop naturally, a floodplain maintenance policy will be decided upon after a full years field data collection. The inner channel has a surface finish complying to the following specification :-

% passing	Sieve size (mm)
100	75
50-85	37.5
30-55	20
0-5	5

The prototype Depth-Discharge data for December 1994 suggests that the inner channel roughness is higher than the scaled up roughness of the highest surface roughness modelled to date (1.4mm dia. x 25 scale ratio = 35mm dia. at prototype scale). This is probably due to a combination of scale effects and inner channel vegetation.

REYNOLDS NUMBERS

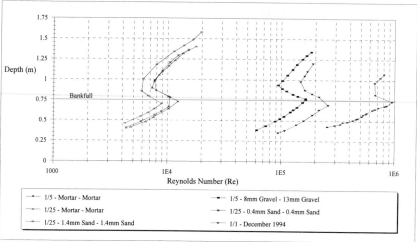

Figure 4 - Reynolds Numbers for prototype, 1/5 and 1/25 scale models

The Reynolds numbers detailed in Figure 4 are average values for the total cross-sectional area calculated with Reynolds number defined as :-

$$Re = \frac{\rho 4VR}{\mu} = \frac{4VR}{\nu}$$

where   V = mean flow velocity μ = absolute viscosity

           R = hydraulic radius ν = kinematic viscosity

It can be seen that the 1/5 scale model operates at Reynolds numbers in excess of 60000 therefore correctly modelling the prototype flow regime, but the 1/25 scale model operates at Reynolds numbers close to the turbulent regime limit.

MODIFIED MOODY DIAGRAM FOR OPEN CHANNEL FLOW

The Moody diagram is currently employed as the best method of predicting values of the Darcy-Weisbach friction factor 'f' for pipes knowing the Reynolds number of the flow and the height of surface roughness k_S. Using the data collected from the 1/5 and 1/25 scale models the values of the 'f' and Re have been calculated for the range of surface roughness used in the tests. Only inbank data has been used as overbank flow data includes the effects of main channel/floodplain interaction energy losses which cannot, at present, be accurately accounted for. The inbank data includes the effects of the main channel sinuosity of 1.13 but these can be eliminated by applying the Linearized SCS Method proposed by James (1994)

For sinuosities less than 1.7, $\quad \left(\dfrac{f'}{f}\right)^{1/2} = 0.43s + 0.57$

where
$\quad$ f $\;=\;$ Darcy-Weisbach friction factor
$\quad$ f' $\;=\;$ Darcy-Weisbach friction factor including bend losses
$\quad$ s $\;=\;$ channel sinuosity

therefore $\qquad\qquad f = \dfrac{f'}{1.115}$

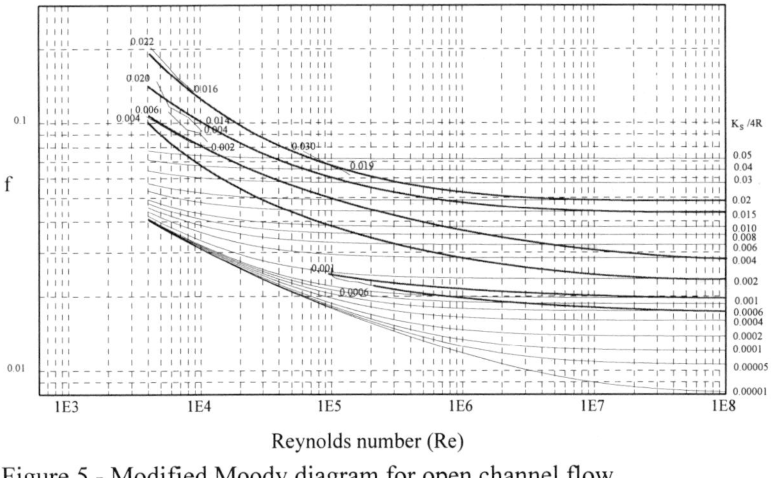

Figure 5 - Modified Moody diagram for open channel flow

Figure 5 details the Moody diagram for pipe flow with the 1/25 and 1/5 scale data, corrected for main channel sinuosity, superimposed. Points of equal $k_S/4R$ value have been highlighted and extrapolated to higher Reynolds numbers. Prototype data will be added to this diagram when a full hydraulic analysis of the test reach has been undertaken. It is anticipated that this diagram will prove valuable in the future when, knowing the prototype friction factor and surface roughness values, the desired small scale model surface roughness can be evaluated to accurately reproduce full scale data.

CONCLUSIONS

Depth-discharge results from a 1/5 and 1/25 scale model study of a compound channel with an environmentally sensitive design have been presented.

The depth-discharge curves for both model scales and prototype, with the difference in surface roughness taken into account, closely follow the same profile.

Care must be taken when scaling hydraulic roughness as the 0.4mm sand used on the 1/25 scale model produced a higher scaled roughness than the 8mm gravel used on the 1/5 model.

A modified Moody diagram for open channel flow has been proposed to assist in the selection of small scale model surface roughness values to accurately reproduce prototype scale data.

ACKNOWLEDGEMENTS

The authors with to acknowledge the financial support of SERC and the NRA in this research and the work carried out on the 1/5 scale model by Dr M Lambert of the University of Newcastle, Australia.

REFERENCES

Myers W R C (1990), Physical modelling of a compound river channel, Proceedings of international conference on River Flood Hydraulics, Wallingford, Ed. W R White, pp381-390, September 1990.

Sellin R H J, Giles A & van Beesten D P (1990), Post-implementation appraisal of a two-stage channel in the River Roding, Essex, Journal of the IWEM, No 4, pp119-130, April 1990.

C Naish & R.H.J. Sellin (1994), Scale Effects in the Hydraulic Modelling of Compound River Channels, Proceedings of the 2nd International Conference on Hydraulic Modelling, Stratford-upon-Avon, UK, pp361-377, 14-16 June 1994.

James C S (1994), Evaluation of methods for predicting bend loss in meandering channels, Journal of Hydraulic Engineering, ASCE, Vol 120, No 2, pp245-253, February 1994.

1A16

CALIBRATION OF MOVABLE-BED SCALE MODEL FOR THE NILE AT EL-KUREIMAT POWER PLANT

A.F. AHMED
Senior Researcher, HSRI, Delta Barrage, Egypt

ABSTRACT
In 1992 a decision was made to construct a new 4 x 600 MW thermal power plant on the east bank of the Nile River near El-Kureimat City (about 100 km south of Cairo). At the moment the construction works were started it was observed that an efficient cooling water intake system could not be designed. This because the available flow depth in front of the selected site will be too small to provide a sufficient discharge for the cooling system, (which ultimately will be equivalent to 80 m³/s). To cope with this problem a movable-bed scale model representing about 9 Km of the river including selected plant site was built and tested. Flume tests were carried out first on the bed material of the model to determine Chézy's C value as well as the appropriate magnitude of distortion for the movable-bed model. The purpose of this paper is to reveal the applied scaling of the processes involved and to show the results of the calibration runs.

INTRODUCTION

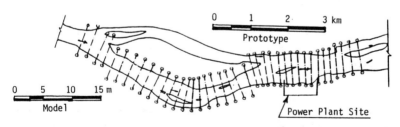

Fig. 1: General Layout of the Modelled Reach

Fig.(1) shows a general layout of the reach which elucidates that the selected site contains many islands. The largest one, which is called El-Kureimat Island, is about 4300-m long which divides the river into two branches. Further 500-m downstream of El-Kureimat Island, where the plant site is located, the river is widening and a small island is formed, which divides the river into two channels. The right one, where the intake structure of the power plant will be placed, has a width range between 250-m and 450-m and carries about 30 % of the total discharge. The collected data revealed that more than half of the width of the right channel at some locations along the plant site is above the water surface level during the winter closure period. Consequently, the required flow depth in front of the intake structure will not be sufficient to provide the required discharge for the cooling system. Therefore in order to solve this problem substantial river training works are needed.

PROTOTYPE CONDITIONS

In order to design the model, the hydraulic and morphological parameters that represent the problem area have to be determined by a survey. Most of the tests were done with the dominant flow rate which produces the yearly average sediment transport discharge. Since the duration of the high flow situation is about 100 days per a year, it was also decided to conduct some runs with high flow which represents the condition during field survey. The two mentioned prototype conditions are summarized in the following Table :

Parameter		Flow case	
		Dominant	High
Discharge	Q [m³/s]	1086.0	1464.0
Average top width	B [m]	531.0	531.0
Average water depth	h [m]	2.73	3.54
Water surface slope	i [cm/Km]	8.5	8.5
Average flow velocity	u [m/s]	0.75	0.78
Chezy coefficient	C [m¹ᐟ²/s]	49.0	45.0
Mean particle size	D [mm]	0.37	0.37

MODEL SCALING

For rivers with shallow friction-controlled flow and dominant bed load transport, as in the Nile River, Struiksma (1980) showed that the Froude number is of secondary importance for the reproduction of the flow pattern since the flow field is mainly governed by the bed topography and the roughness distortion ratio gL/C^2h. Here C is the Chezy coefficient, h is the average flow depth, g is the acceleration due to gravity and L is a characteristic length (for instance the meander length). For reproducing the flow field this ratio has to be reproduced at full scale which leads to the so-called roughness condition :

$$n_c^2 = n_L \: / \: n_h \qquad (1)$$

In which the scale factor n_x of any parameter X is defined as the ratio of the value of X in the prototype to that in the model. For alluvial streams the roughness condition leads, in most cases, to distorted models ($n_L > n_h$) because $n_c > 1$. In addition, a correct reproduction of the bed topography is expected if the sediment transport scale is constant in space. According to De Vries (1973), this is achieved when the ideal velocity scale is fulfilled. This scale follows from the assumption that there is a unique function between the transport parameter $\Psi = s/(D^3 \, \Delta g)^{1/2}$, the flow or Shields parameter $\theta = hi/\Delta D$ and the chézy coefficient. In which $\Delta = (\rho_s - \rho)/\rho$ is the relative density of the sediment, ρ_s is the density of the sediment, ρ is the density of water, i is the water surface slope, D is the grain size, and s is the sediment transport in volume (including pores) per unit width. Since similarity of the transport is aimed at, the ideal velocity scale can be obtained from the condition that $(n_\theta = 1)$ which together with Chezy's relation and using sand in the model $(n_\Delta = 1)$ yields :

$$n_u = n_c \, (n_D)^{1/2} \qquad (2)$$

Eq.(2) reveals that Froude number in the model will be, in most cases, larger than that in the prototype because $(n_u < n_h^{1/2})$. This means that the condition of the ideal velocity scale leads generally to a deviation from the Froude condition. The exaggerated model velocity is needed to provide sufficient bed material transport. Hence, the water surface slope is exaggerated in the model. In order to compensate such effect the model has to be constructed with the technique of a tilted datum plane. To calculate the correct water and bed levels the tilting angle i_t can be derived as follows:

$$i_t = i_m - i_p \, (n_L \, / \, n_h) \qquad (3)$$

In which i_m and i_p are the water surface slope in the model and prototype, respectively. The morphological time scale can be deduced by applying the scale rules to the continuity equation of the sediment transport and together with the transport formula derived by Engelund and Hansen (1967) the following condition is obtained:

$$n_t = (n_L \, n_h) \, / \, (n_D^{3/2} \, n_c^2) \qquad (4)$$

FLUME TESTS AND RESULTS

Tests were carried out in a 0.73-m wide and 22.0-m long recirculating flume to obtain the scaling ratios for depth, slope and flow velocity. The length scale $n_L = 150$ was first chosen so as to fit the available laboratory space. Since the scale of bed material ($n_D = 0.37/0.23 = 1.68$) is known, one can calculate all other scales from the preceding equations if the Chezy coefficient of the model bed C_m is known. To start with for the case of dominant discharge a first estimate of $C_m = 24$ $m^{1/2}/s$ was assumed which for $C_p = 49$ $m^{1/2}/s$ gives $n_C = 2.04$. Applying Eq.(1) for $n_L = 150$, the value of $n_h = 36$ was determined then the value of $h_m = 0.0755$ m was obtained. Apply Eq.(2) for $n_D = 1.68$ the value of $n_u = 2.64$ was determined which was utilized to obtain the mean flow velocity in the model as $u_m = 0.284$ m/s. The scale of the water surface slope can be obtained as $n_i = n_u^2/n_C^2 \, n_h$ which gives $n_i = 0.056$ and the corresponding model value was obtained as $i_m = 0.00155$. The discharge scale value was then determined as $n_Q = 14256$, which was utilized to calculate the model discharge as $Q_m = 76.18$ l/s. As the flume is 0.73 m wide, the corresponding discharge that required to run it was determined as $Q_F = 15.71$ l/s.

The parameters Q_F, i_m, and h_m were then utilized to operate the flume and the corresponding C_m was determined. As the measured C_m value was different from that assumed, the measured C_m was utilized to obtain new values for Q_F, i_m, and h_m which can be consequently utilized to run the flume and determining such new value for C_m. The above trial-and-error testing procedure was repeated till the calculated and measured values of the flow depth, velocity, slope and Chézy value were in agreement with the scaling conditions which resulted in $C_m = 22 m^{1/2}/s$. The same testing procedure was applied to the high flow case and the corresponding scaling ratios for both cases are as follows:

Parameter	Dominant discharge			High discharge		
	Prototype	Scale	Model	Prototype	Scale	Model
Width [m]	531.0	150	3.54	531	150	3.54
Depth [m]	2.73	30.2	0.09	3.54	35.9	0.099
Velocity [m/s]	0.75	2.89	0.26	0.78	2.65	0.29
Discharge [m³/s]	1086.0	13093	0.083	1464	14251	0.1027
Slope [-]	8.5×10^{-5}	0.055	0.00155	8.5×10^{-5}	0.047	0.0018
Chezy C [m$^{1/2}$/s]	49.0	2.23	22	45	2.05	22
Grain size [mm]	0.37	1.68	0.23	0.37	1.68	0.23

MODEL CONSTRUCTION

The boundary of the model was shaped according to the field survey. The bed consists of 0.45-m thick layer of medium sand with a mean particle size of 0.23 mm and a geometric standard deviation of 1.4. Separation of sand from the underground achieved by a plane concrete layer of 0.15 m thick which covered with an impervious material. To adjust the water surface level a flap gate of 5.0 m wide was installed at the lower end of the model. A total number of 39 pairs of supports, located at distances ranging between 1 m and 1.5 m apart, were firmly erected along both banks of the model as shown in Fig.(1). These supports were utilized to accommodate a portable bridge which was used to measure the bed profile of each cross-section along the model. A mixture of recirculating water and sediment were fed into the model through a manifold. Adjustable vertical salts and vanes were used to control the lateral variation of velocity and sediment load. More details are presented in our Laboratory Report (Gasser & Ahmed 1993).

MODEL CALIBRATION AND RESULTS

Tests were conducted, for each of the two flow conditions, in four phases. The first was carried out to adjust the mid-depth velocity distribution at the inflow in such a way as to simulate that measured in prototype. Using EMS current-meter many attempts were tried and the final results are shown in Fig.(2).

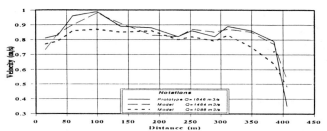

Fig. 2: Results of Calibrating the Inflow Condition

Attempts were then made, during the second phase, to adjust the discharge distribution through the two branches located around El-Kureimat Island. Mid-depth velocity profiles at a certain locations across the two branches were measured and the percentage of discharges were determined. As some deviation from the prototype data was observed, small adjustment was made at the upstream end of El-Kureimat Island until the correct discharge distribution was achieved. The third phase was conducted to calibrate the mean hydraulic parameters which were previously determined during the flume tests. The model was continuously running, day and night, during which the sounding bridge was applied twice a day. During this test, the flow discharge and slope were regularly monitored and the necessary adjustments were made. After a working time of about two weeks, during which the sounding bridge was applied for a total number of 29 and 30 for the high and dominant flow conditions respectively, the equilibrium was deemed to be established. This stage was assigned when the analysis of measured data show no significant variation in the moving averaged which was carried out on the base of 3 and 5 values as shown in Fig.(3).

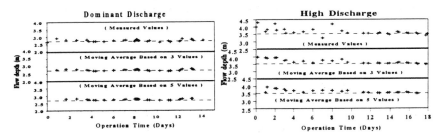

Fig. 3: Results of Calibration of the Averaged Parameters

It can be concluded, up to this stage, that the averaged flow depth and slope as well as the roughness condition are fulfilled. However, comparison of the bed topography in the model with that of the prototype, showed a local discrepancies in some cross sections. Therefore, some necessary modifications were locally introduced in the model during the fourth stage of the calibration then the bed profiles were measured. At this stage the model was considered to be calibrated for each of the two tested flow conditions. The final comparison for the model and prototype bed levels for the case of dominant discharge is shown in Fig.(4).

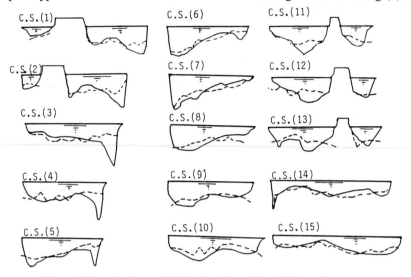

Fig. 4: Comparison of Bed Profiles after Calibration

CONCLUSIONS

The technique applied to derive the scale conditions which govern the hydraulic and morphological conditions and the use of the flume to establish the distortion and other relevant scales are very beneficial in designing the movable bed model for the river reach. The method of scale selection and calibration employed herein is more systematic than that of Hecker et al. (1989) but is still laborious.

REFERENCES

De vries, M. (1973)."Application of Physical and Mathematical Models for River Problems", Delft Hydraulics Laboratory, Publication No.112.

Engelund, F. and Hansen, E. (1967). " A Monograph on Sediment Transport in Alluvial Streams". Technisk Forlag, Copenhagen, Denmark

Hecker, G.E., Larsen, J., and White, D.K., (1989). "Calibration of a movable bed river model." Proc. of the Int. Symp. on Sediment Transport Modelling, ASCE, New Orleans, USA.

Gasser, M.M. and Ahmed, A.F. (1993)."El-Kureimat Power Plant, Set-up and Calibration Results ". Report No.3, HSRI, Delta Barrage, Egypt.

Struiksma, N. (1980)."Recent Developments in Design of River Scale Models with Mobile-Bed", IAHR Symposium on River Eng. and its Interaction with Hydrological and Hydraulic Research, Belgrad, May 26-28.

1A17

EROSION IN VEGETATIVE CHANNEL LININGS

JAMAL M. V. SAMANI[1] AND NICHOLAS KOUWEN[2]

[1])Graduate Student, Department of Civil Engineering, University of Waterloo, Waterloo, Ontario, N2L 3G1, Canada.
[2])Professor, Department of Civil Engineering, University of Waterloo, Waterloo, Ontario, N2L 3G1, Canada.

ABSTRACT

Vegetation is the most cost-effective form of erosion control. Since vegetation prevents erosion, the relationship between vegetation and erosion rate in vegetated open channels has been highlighted and analyzed. The deformation of vegetation caused by the flow's momentum has been taken into consideration. We obtain results that are in good agreement with the experimental erosion rate data for vegetated channels. The purpose of this paper is to show that the force equilibrium method is an acceptable method to estimate the tractive stress on the bed of vegetated channels and therefore to estimate the erosion rate, and that the amount of deformation of a vegetative channel lining due to flow is also a useful method to determine the erosion rate.

INTRODUCTION

Grass is the most effective type of plant for erosion control in many areas. A method to control the erosion or sediment load in water by channels lined with vegetation is difficult to deal with analytically and has not previously received a great deal of attention in the literature.

The purpose of this paper is to develop a method that is capable of determining the tractive stress at the soil boundary caused by the flow, since this stress is the main factor that affects the erosion rate, and to introduce a method that shows that the amount of deformation is an effective indexing variable in the erosion process. The approach used in this paper incorporates shear stress obtained from the force equilibrium approach to model the velocity profile in and outside vegetation in vegetated open channels. In this paper, the erosion rates from data collected at the University of Waterloo will be utilized.

BACKGROUND

The following review is a brief summery of developments in the area of estimating erosion rates.

Erosion Rate Estimation

Various elements of the soil erosion process were extensively discussed by Ellison (1947). A number of methods for assessing soil loss have been developed. The most common relationship is the Universal Soil Loss Equation (USLE) which is an empirical model developed by the US Dept. of Agriculture, USDA. Detailed processes are not reflected in the USLE and this limits its use to annual soil loss in shallow flows only.

Several sediment transport equations and soil loss relationships have been developed, both from experimental studies in laboratories with simulated rainfall and from statistical and regression analysis using field data which has been reviewed in Julien and Simons (1985). For the present analysis, detachment processes are considered to dominate and the erosion rate is assumed to be proportional to the effective stress in excess of some critical value. Meyer and Monke (1965) modified the Duboys sediment transport equation to estimate detachment of soil by flowing water as:

$$\frac{dEr}{dt} = kk(\tau_e - \tau_c)^a \tag{1}$$

in which kk is an erosion rate coefficient, a is an erosion rate exponent, τ_e is the effective shear stress, τ_c is the critical shear stress, and $\frac{dEr}{dt}$ is the erosion rate in volume per unit area per unit time. Both kk and τ_c are assumed to be soil properties.

The most important component in Equation (1) is τ_e which will be calculated using the mean of the force equilibrium method. The erosion rate is considered as the bedload transport rate.

Force Equilibrium Equation

The force equilibrium approach has been studied by a number of authors, the most recent being Murota et al. (1984) who modelled the velocity distribution inside the vegetated zone.

The force balance for a layer of fluid parallel to the bed can be written as:

$$F_G - F_D + d\tau / dy = 0 \tag{2}$$

where F_G is the force per unit area due to gravity, F_D is the drag force per unit area, τ is the shear stress. On the basis of the force balance equation, the drag force is given by:

$$F_D = NC_D A_s \rho \frac{U^2}{2} \tag{3}$$

where N = the number of grass elements per unit area, c_D = drag coefficient for a grass element, A_s = frontal projected area of a grass element, ρ = density of water. The gravity force for one unit width is given by:

$$F_D = \rho g \sin \alpha \tag{4}$$

where g is the gravitational constant, y is the depth, and α is the channel slope angle. The shear stress can be written in terms of the Boussinesq expression:

$$\tau = \rho \varepsilon \frac{dU}{dy} \tag{5}$$

where ε is the eddy viscosity. Using the Prandtl mixing - length hypothesis for ε, we get:

$$\varepsilon = \frac{\mu}{\rho} + \ell^2 \left(\frac{dU}{dy} \right) \tag{6}$$

where μ is the dynamic viscosity and ℓ is the mixing length. Tsujimoto et al. (1993) suggested a parameter to make it possible to predict the structure of flow above and inside the flexible vegetation due to the organized motion. When the flow is uniform, the force equilibrium equation becomes:

$$F_G - F_{Dx} + \frac{d\tau}{dy} - \phi = 0 \tag{7}$$

Using Equations 3 - 7 we will be able to estimate the effective shear stress. The boundary conditions for the system of N equations are $\tau = 0$ at the water surface and $U_{bed} = 0$ at the bottom. Assuming that there is a shear stress at the bottom, $\tau_1 = \rho C_{Df} A' \frac{U_1^2}{2}$, where A' is the portion of the bed area not occupied by the grass elements, Equation (7) can be solved numerically. To solve (7), the mixing-length, ℓ, was specified using the same expression used by Muorota et al. (1984), and the deflected height of the artificial grass was computed using the large deflection theory. This approach has been applied to the data base collected in this study at the University of Waterloo.

Experimentation

The basic component of the experimental facility is a 0.62 m wide, 0.40 m high, and 9.00 m long variable slope flume. Two sets of plastic roughness elements were cut, these being 100 mm and 150mm lengths, all elements were 5 mm wide with 5 mm lateral spacing and 28 mm longitudinal spacing. Lexan film of .25 mm thickness was embedded in 25 mm Kledgcell foam. The sediment was silica sand with a specific gravity of 2.65 . Two sizes were employed, $D_{50} = 0.435$ mm and $D_{50} = 0.244$.

To measure the erosion rate, a flow of clear water was allowed to erode the sand. The average height of the bed after a particular time and the associated velocity profile were measured. The experiments were repeated for increasing flows, from 4 to 16 ℓ / s. The whole procedure was done for channel slopes from 1 to 4%. Velocity profiles were measured using a 3 mm diameter pitot-static tube connected to a sensitive differential pressure transducer.

Results and Analysis

The velocity profiles and erosion rate measurements were both used to calibrate the force equilibrium and the erosion models. Effective shear stress calculated from the velocity profiles in this approach was used to estimate the erosion rate for each run. The results for the different grain sizes are as follows:

Coarse Sand; the dimensionless erosion rate is:

$$Er^* = 0.00075(\tau_e^* - \tau_c^*)^{1.35} \tag{8}$$

where Er^* is the dimensionless erosion rate and equal to Er / u_{*cr}, τ_e^* is the dimensionless effective shear stress calculated from velocity profiles program, τ_c^* is the dimensionless critical shear stress based on Shield's diagram is equal to 0.05, u_{*cr} is the critical shear velocity which is equal to τ_c / ρ.

Fine Sand: the dimensionless erosion rate is:

$$Er^* = 0.00072(\tau_e^* - \tau_c^*)^{1.58} \tag{9}$$

Figure 1 shows a typical comparison between the measured and computed values of the dimensionless erosion rate for the coarse sands. This figure shows that it is possible to calculate the effective shear stress on the sand inside a flexible roughness layer using a force equilibrium approach.

Figure 1 Measured & Calculated D. Erosion Rates for D50=.453 mm

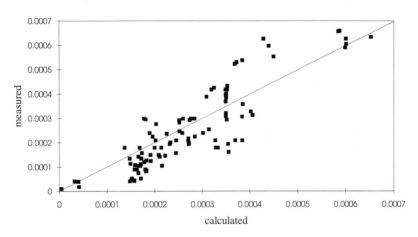

Other work has indicated that it is also possible to determine erosion rates using the compression of grass lining as an index (Kouwen, 1992; Samani and Kouwen, 1994). The method in this paper is based on a more fundamental fluid mechanics approach and yields similar results. Because it is based on physically based equations, it can be expected to be more versatile in its applicability to other problems.

CONCLUSIONS

A force equilibrium approach coupled to a modified Dubois sediment transport equation may yield adequate estimation of the soil erosion rate in vegetated channels. Earlier experiments have indicated that natural grass can be modeled using plastic strips. This paper reports that a similar approach can be used to model erosion rates in grassed channels.

APPENDIX I - REFERENCES

Ellison, WD. (1947). "Soil Erosion Studies". *Agriculture Engineering,* 28:145-146,197-201,245-248,297-300,349-351,402-405,442 -444.

Julien, D. Y., D. B. Simons (1985). "Sediment Transport Capacity of Overland Flow". *Transactions of ASAE,* 28(3): 725-762.

Kouwen, N., (1992), "Modern Approach to Design of Grassed Channels", *Journal of Irrigation and Drainage Engineering,* ASCE, 118(5): 733-743.

Murota, A., T. Fukuhara and S. Masaru (1984). "Turbulence Structure in Vegetated Open Channel Flows". *J. of Hydroscience and Hydraulic Engineering.* 2(1): 47-61.

Samani, J. M. V., N. Kouwen (1994). " Erosion in Vegetative Channel Linings". *Journal of Hydraulic Engineering,* ASCE. Submitted for Review.

Tsujimoto, T., T. Okada and K. Kontani, (1993). "Turbulent Structure of Open-Channel Flow over Flexible Vegetation", *KHL Progress Report, Hydraulics Laboratory, Kanazawa University, Japan. 1-15.*

APPENDIX II - NOTATION

a	an erosion rate coefficient,
C_D	drag coefficient for the grass;
Er	symbol indicating the erosion rate;
F_D	drag force;
F_G	gravity force;
g	acceleration due to gravity;
k	deflected roughness height;
kk	an erosion rate coefficient,
ℓ	mixing length;
U	mean velocity in the x direction;
y	distance from bed channel;
ε	eddy viscosity;
ρ	density of the water;
τ_o	average boundary shear stress;
τ_c	critical shear stress ; and
τ_e	the effective shear stress at the soil-water interface;
α	slope angle of the channel;
$\gamma_{\cdot 1}, \gamma_{\cdot 2}$	parameters to be determined by optimization, for the regions above the grass and below;
ϕ_M	implies an additional force due to the waving motion.

1A18

A STUDY OF THE DISPERSION OF MUD
IN OPEN CHANNEL FLOW

HIN-FATT CHEONG, ENG-SOON CHAN and SOE THEIN
Department of Civil Engineering
The National University of Singapore
10 Kent Ridge Crescent
Singapore 0511

ABSTRACT
Experiments were carried out on the dispersion of a cohesive sediment under steady flow conditions in a laboratory flume. The settling velocities of the fine sediments in flowing water were obtained by issuing the sediments into a steady uniform open channel flow and matching the measured sediment concentration profiles with the those obtained from a three dimensional mathematical approach incorporating· the standard two-equation k-ε turbulence model.

INTRODUCTION
In the light of increasing engineering activities, such as land reclamation in the coastal areas of Singapore, the prediction of siltation in tidal channels has assumed importance. This study is a direct result of the need to understand the siltation mechanism in tidal flows. Estuary muds, however, possess behaviours that are known to be very different from those of cohensionless sediments and they have constituents which are particles of sizes less than about 60 microns.

For particles in the clay fraction, the physico-chemical surface forces become significant by comparison with the gravitational and hydrodynamic forces acting on them and, under the influence of certain amounts of ionic constituents such as salt, the surface forces can be strongly attractive leading to the inherent tendency for the particles to cohere into mud flocs [Mehta and Parteniades (1975)].

The flocculation process of these cohesive particles in flowing water depends on two separate conditions [Odd and Owen (1972)]. The first condition arises from their collisions caused mainly by the turbulence of the flow and the differential settling of the flocs with the frequency of collisons being greater with higher suspended concentration. The second condition requires the cohesive bonds (which are dependent on the physical and mineralogical properties of the clay and the salinity of the water) between particles to be sufficiently large to prevent separation either under impact of the collision or under the shearing action of the flowing water. Attempts to determine the size and settling velocity of the mud flocs in still or flowing water have been based on empiricism.

One approach in ascertaining the in-situ settling velocities could be through a steady, two-dimensional sediment dispersion experiment where the settling velocities could be evaluated by matching vertical sediment profiles with those generated by a numerical model. Such experiments have been performed by Jobson and Sayre (1970) and Celik and Rodi (1988) for fine sand under turbulent flows. Mud from the Tanjong Pagar Container Terminal (TPCT) in Singapore was used for the experiments in this study. The full two dimensional continuity and momentum equations incorporating the two-equation k-ϵ turbulence model were used to derive the concentration profiles of the suspended cohesive sediments in a steady open channel flow.

MATHEMATICAL MODEL

The respective time-averaged equations for the streamwise x-momentum, the vertical z-momentum, continuity, and the k-ϵ turbulence model for steady two dimensional turbulent open channel flow are given as [Demuren and Rodi (1986)]:

$$U\frac{\partial U}{\partial x} + W\frac{\partial U}{\partial z} = -\frac{1}{\rho}\frac{\partial P}{\partial x} + 2\frac{\partial}{\partial x}(v_t\frac{\partial U}{\partial x}) + \frac{\partial}{\partial z}(v_t\frac{\partial U}{\partial z}) + \frac{\partial}{\partial z}(v_t\frac{\partial W}{\partial x}) \tag{1}$$

$$U\frac{\partial W}{\partial x} + W\frac{\partial W}{\partial z} = -\frac{1}{\rho}\frac{\partial P}{\partial z} + \frac{\partial}{\partial x}(v_t\frac{\partial W}{\partial x}) + 2\frac{\partial}{\partial z}(v_t\frac{\partial W}{\partial z}) + \frac{\partial}{\partial x}(v_t\frac{\partial U}{\partial z}) \tag{2}$$

$$\frac{\partial U}{\partial x} + \frac{\partial W}{\partial z} = 0 \tag{3}$$

$$U\frac{\partial k}{\partial x} + W\frac{\partial k}{\partial x} = \frac{\partial}{\partial x}(\frac{v_t}{\sigma_k}\frac{\partial k}{\partial x}) \frac{\partial}{\partial z}(\frac{v_t}{\sigma_k}\frac{\partial k}{\partial z}) + G - \varepsilon \tag{4}$$

$$U\frac{\partial \varepsilon}{\partial x} + W\frac{\partial \varepsilon}{\partial z} = \frac{\partial}{\partial x}(\frac{v_t}{\sigma_\varepsilon}\frac{\partial \varepsilon}{\partial x}) + \frac{\partial}{\partial z}(\frac{v_t}{\sigma_\varepsilon}\frac{\partial \varepsilon}{\partial z}) + C_1 G\frac{\varepsilon}{k} - C_2\frac{\varepsilon^2}{k} \tag{5}$$

where G is expressed by

$$G = v_t [2 (\frac{\partial U}{\partial x})^2 + 2 (\frac{\partial W}{\partial z})^2 + (\frac{\partial U}{\partial z} + \frac{\partial W}{\partial x})^2] \qquad (6)$$

U and W are time-averaged velocities in streamwise and vertical directions respectively; v_t is the eddy viscosity; ρ and P are the fluid density and pressure respectively. The empirical constants take on values according to $C_\mu = 0.09$, $C_1 = 1.44$, $C_2 = 1.92$, $\sigma_\epsilon = 1.3$ and $\sigma_k = 1.0$ as suggested by Rodi (1980).

The distribution of the time-averaged suspended sediment concentration C is

$$U\frac{\partial C}{\partial x} + W\frac{\partial C}{\partial z} = \frac{\partial}{\partial x} (\frac{v_t}{\sigma_s} \frac{\partial C}{\partial x}) + \frac{\partial}{\partial z} (\frac{v_t}{\sigma_s} \frac{\partial C}{\partial z} + W_s C) \qquad (7)$$

where W_s is the settling velocity and σ_s is the turbulent Schmidt number for suspended sediment flow and it is assumed to be constant with depth and whose value has to be estimated.

SOLUTION PROCEDURE
The initial profiles for U, k and ϵ were assumed to be uniform with depth and the calculations were carried out using the full form of the hydrodynamic equations, including the unsteady term, until fully developed conditions were reached. The fully developed hydrodynamics for each experimental run were used with the transport equation to give the sediment profiles. Grid independence was checked and a grid with 17 nodes in the vertical, 17 in the transverse and 80 in the longitudinal direction was finally adopted. The rigid lid approximation for the free surface and the appropriate boundary conditions followed closely those described by Demuren and Rodi (1988). The SIMPLE algorithm with a power law scheme described by Patankar (1980) was used to solve the governing and transport equations. The settling velocity W_s was prescribed in the general form $W_s = KC^n$ where K and n are constants and these were determined by trial and error through matching with the experimental results.

EXPERIMENTAL RESULTS & DISCUSSION
The experiments with mud from the TPTC were performed by Tan (1989) and the hydraulic conditions are given in Table 1. The dispersion tests were conducted using non-saline water in a rectangular glass-lined channel, 30m long, 0.9m wide with a water depth of 0.2m. Water supply is via a recirculating system consisting of a large sump and a constant head tower. Uniformity of flow across the width was achieved by trial and error with wire meshes. Velocity profiles were measured with a miniature current meter (Delft Hydraulics WVM) with resolution of 2mm/s for a mean flow of 0.2 m/s. The cohesive sediments were introduced in the form of a well mixed slurry from a

50 litre tank with a injection system via an array of 6mm copper tubes spaced 25mm apart with an average flow rate of 0.0148 l/s of the dispersants through each tube. Sampling was by 4 vertical arrays of 6mm copper tubes spaced at 4 cm and the 4 tubes at the same relative depth were channeled into one sampling bottle for analysis.

The measured concentration profile at $x = 15h$, where h is the water depth, was chosen as the upstream boundary condition. The computed concentration profiles for Runs 2, 3, 4 and 5 are shown with the measured values in Fig 1. The ordinate represents the depth normalised with h and the abscissa represents the concentration normalised with the mean concentration at each section. In all the experiments, no deposition of the mud was observed. Had there been any deposition on the bed, the mathematical treatment of the boundary condition for the sediment will invariably be very complex. Celik and Rodi (1988) reported from a similar study using the dye and fine/coarse sand experiments of Jobson and Sayre (1970) that a value of $\sigma_s = 0.5$ seemed to provide a better representation for the concentration profiles. However, the values of $\sigma_s = 0.75$ and $K = 0.00015$ (C expressed in g/l and W_s in m/s) are found to give the best visual match with the experimental results of Tang (1989). There is good agreement between the computed results and experiments. The largest deviations between theory and experiment are at the lowest measurement location where the velocity of flow is low and the sampling technique by fluid withdrawal is likely to be the cause in spite of the great care that have been taken. Thein and Cheong (1993) attempted to provide a relationship between σ_s and W_s/U_* where U_* is the shear velocity from the experiments of Jobson and Sayre and their results are shown in Fig 2 which also shows the virtual single point plots (effect of plotting scale) from the present mud dispersion study. The W_s used for the plotting positions were based on $W_s = 0.00015C^{0.6}$ with C taken as the mean concentration for the section. No clear trend can be discerned and further studies would be needed to establish the inter-relationship between σ_s and W_s.

CONCLUSIONS

The Schmidt Number and an empirical relationship of the form $W_s = KC^n$ for the insitu settling velocity of mud from the TPCT in Singapore were obtained through a matching of observed mud concentration profiles in a series of laboratory experiments with the results of the numerical approach using the full hydrodynamic equations and the two equation k-ε turbulence model. This relationship is expected to be valid for the concentration levels less than 20,000 mg/l.

REFERENCES

Celik I. and Rodi W.,"Modeling suspended sediment transport in nonequilibrium situations", Journal of Hydraulic Engineering, ASCE, Vol 114, No.HY10, pp. 1157-1190, October 1988.

Demuren A.O. and Rodi W.,"Calculation of flow and pollutant dispersion in meandering channels", Journal of Fluid Mechanics, Vol.172, pp-63-92, 1986.

Jobson H.E. and Sayre W.W.,"Vertical transfer in open channel flow", Journal of the Hydraulics Division, ASCE, Vol. 96, No HY3, pp. 703-724, March 1970.

Mehta A.J. and Partheniades E.,"An investigation of the depositional properties of flocculated fine sediments", Journal of Hydraulic Research, Vol. 12, No. 4, December, 1975.

Odd N.V.M. and Owen M.W.,"A two-layer model of mud transport in the Thames estuary", Proceedings of the ICE, Paper 7517S, Supp. (ix), London, U.K., 1972.

Patankar S.V.,"Numerical heat ransfer and fluid flow", McGraw Hill Book Company, New York, 1980.

Rodi W.,"Turbulence models and their applications in hydraulics", International Association for Hydraulic Research, Delft, The Netherlands, 1980.

Tan H.K., "Dispersion of fine sediments in open channel flow", Thesis submitted for the degree of Master of Engineering, National University of Singapore, 1989.

Thein S and Cheong HF, "Turbulent Schmidt Number is suspended sediment distribution in open channel flow", Proceedings of 9th Congress of the IAHR-APD, 24-26 August 1994, Vol. 2, pp. 509-518.

Table 1 Flow Conditions and Sediment Concentration Levels

Run No.	Depth (cm)	Mean Velocity (m/s)	U. (m/s)	Concentration In Supply Tank(g/l)
RUN 1	20	0.23	0.0098	Dye Experiment
RUN 2	20	0.23	0.0098	22
RUN 3	20	0.23	0.0098	20
RUN 4	20	0.23	0.0098	10
RUN 5	20	0.43	0.0175	27

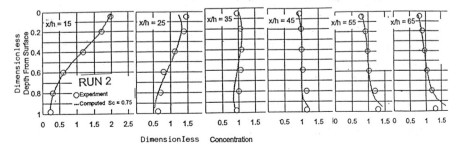

Fig. 1 Dimensionless Suspended Mud Concentration Profiles

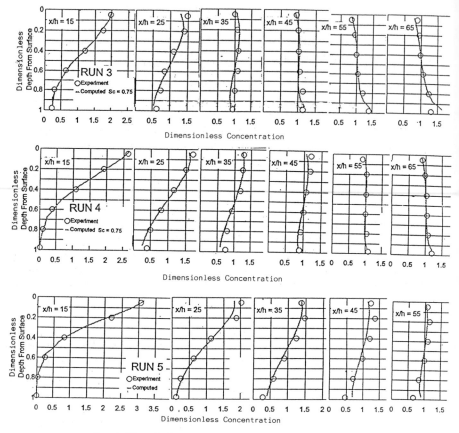

Fig. 1 Dimensionless Suspended Mud Concentration Profiles

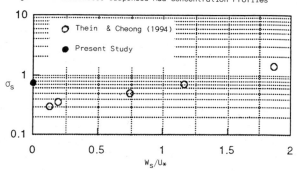

Fig. 2 Relationship between Turbulent Schmidt Number
for Suspended Sediment and W_s/U_*

1A19

NUMERICAL MODELLING OF SUSPENDED SEDIMENT FLUXES USING AN ACCURATE BOUNDED DIFFERENCE SCHEME

BINLIANG LIN and ROGER A. FALCONER
University of Bradford
Bradford, UK

ABSTRACT

Details are given of the application of a highly accurate bounded finite difference scheme, i.e. the ULTIMATE QUICKEST scheme, for the modelling of suspended sediment fluxes. The operator-splitting algorithm, and a conventional discretization (i.e. no splitting) algorithm, have been investigated for using this scheme. The scheme has been applied to a practical study to predict sediment fluxes in the Humber Estuary, England. Comparisons between numerical predictions and field measurements are presented herein.

INTRODUCTION

Numerical models based upon higher order upwind finite difference schemes have become increasingly popular in solving the advective-diffusion equation for sediment transport and water quality studies for estuarine and coastal waters, and these models are accurate for most simulation studies. However, for advection dominated flows, numerical oscillations, i.e. overshoot and undershoot, will still occur in regions where the gradients of the concentration are relatively large. Some physically unrealistic phenomena, such as negative sediment concentrations, can arise as the result of such undershoot or overshoot. These features also effect predictions of water quality constituents which may be related to sediment concentrations, e.g., heavy metals. The best way of eliminating numerical oscillations is to introduce a limiter function to bound the solution, such that undershoot and overshoot are first detected and then prevented (Cahyono, 1993). In this paper a 2-D depth-integrated estuarine sediment transport model is presented. A highly accurate finite difference scheme, based on the ULTIMATE QUICKEST scheme given by Leonard (1991) and Leonard and Niknafs (1990), was used to represent the advective terms in the suspended sediment transport equation. The operator-splitting algorithm, and a conventional discretization (i.e. no splitting) algorithm, have been used to solve the sediment transport equation. A boundary fitted curvilinear co-ordinate grid, originally developed by Chandler-Wilde and Lin (1992) but since refined, has been employed. This model has been applied to predicting the sediment fluxes in the Humber estuary, sited along the north-east coast of the UK. Comparisons between numerical predictions and corresponding field measurements are presented herein.

GOVERNING EQUATIONS

In the conformal boundary-fitted co-ordinate system the depth-integrated form of the advective-diffusion equation for suspended sediment transport was expressed in the following form:

$$
H\left(\frac{\partial C}{\partial t}+U_{\xi}\frac{\partial C}{h_1\partial\xi}+U_{\eta}\frac{\partial C}{h_2\partial\eta}\right)-\frac{\partial}{h_1\partial\xi}\left(HD_{\xi\xi}\frac{\partial C}{h_1\partial\xi}+HD_{\xi\eta}\frac{\partial C}{h_2\partial\eta}\right)
$$
$$
-\frac{\partial}{h_2\partial\eta}\left(HD_{\eta\xi}\frac{\partial C}{h_1\partial\xi}+HD_{\eta\eta}\frac{\partial C}{h_2\partial\eta}\right)=E \tag{1}
$$

where H = depth of flow, U_{ξ}, U_{η} = depth mean velocity components in ξ, η directions, h_1, h_2 = transformation scale factor in ξ, η directions, t = time, C = depth mean sediment concentration, $D_{\xi\xi}$, $D_{\xi\eta}$, $D_{\eta\xi}$ and $D_{\eta\eta}$ = depth mean dispersion coefficients in ξ and η directions, and E = net rate of erosion or deposition per unit area of bed and can be given as (Lin and Falconer, 1994):-

$$
E = W_s\frac{C_{ae}}{C_e}(C_e-C) \tag{2}
$$

where W_s = particle settling velocity, C_e = depth mean equilibrium concentration, C_a = sediment concentration at reference level 'a' near the bed, and C_{ae} = equilibrium sediment concentration at reference level 'a'. C_e and C_{ae} were calculated using the relationships given by van Rijn (1984).

NUMERICAL METHODS

In solving equation (1) two algorithms were investigated. Firstly, an operator-splitting scheme was considered, which divides the advective-diffusion equation according to its physical meaning and, in particular, also divides it along different directions. Secondly, the traditional algorithm which does not use the operator-splitting technique was also considered. More details of the numerical scheme are given in Lin and Falconer (1994).

SPLITTING SCHEME

For equation (1), the splitting scheme is given as follows :-

$$
C^{n+2} = L_{\xi}L_{\eta}L_{\xi\xi}L_{\eta\eta}L_sL_sL_{\eta\eta}L_{\xi\xi}L_{\eta}L_{\xi}C^n \tag{3}
$$

where L_ξ and L_η are one-dimensional finite difference operators that solve the one-dimensional pure advection equations in the ξ and η co-ordinate directions respectively, $L_{\xi\xi}$ and $L_{\eta\eta}$ are one-dimensional finite difference operators that solve the one-dimensional pure diffusion equation in the ξ and η co-ordinate directions respectively and L_s is a one-dimensional finite difference operator that solves the pure reaction equation.

In solving the advective component of equation (3) the QUICKEST scheme was first considered, with this being an explicit third-order upwind algorithm designed for highly advective unsteady flows. Since this scheme is third-order accurate, the leading truncation error is a dissipative (but not diffusive) spatial fourth order derivative. Numerical tests for pure advection in one-dimensional flows have shown that this scheme generally gives better results than second-order central (e.g. the Lax-Wendroff scheme) and second-order upwinding schemes. Nevertheless, unphysical overshoot and undershoot still occur near sharp changes of concentration in purely advective flows. Thus a bounded scheme designed by Leonard (1991) for the pure advective flow was also included, with this being a universal limiter and which can be applied to arbitrarily higher order transient interpolation models of the advective transport equations. The pure diffusion equation was solved using the explicit second order central difference scheme.

NON-SPLITTING SCHEME
In the non-splitting scheme, the finite difference formulation for equation (1) was written in the following form :-

$$C_{i,j}^{n+1} = C_{i,j}^n + \text{Advection} + \text{Dispersion} + \text{Sources (or Sinks)} + \text{Decay} \quad (4)$$

where i, j = grid square locations in ξ, η directions. For consistency, the advection terms in equation (4) were again formulated using the ULTIMATE QUICKEST scheme, with the dispersion terms being represented using the explicit second-order central scheme, and the source and decay terms being represented by the Euler method.

MODEL APPLICATIONS
The two schemes were first applied to two test cases for which analytical solutions exist, including: (i) simulating the pure advection of a circular column concentration distribution in a rotating flow, see Figure 1; and (ii) predicting the transport of an initial Gaussian concentration distribution in a rotating fluid flow. Results from these tests have shown that both the splitting and non-splitting algorithms produce stable and non-oscillatory solutions, that compare very favourably with the analytical solutions when the ULTIMATE limiter scheme was used to discretize the advection terms.

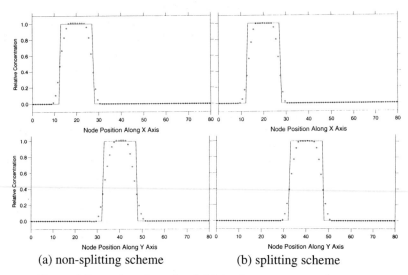

(a) non-splitting scheme   (b) splitting scheme

Figure 1 Pure advection of an initial column concentration

The non-splitting ULTIMATE QUICKEST scheme was then applied to predict the suspended sediment fluxes due to tidal motion in the lower reaches of the Humber Estuary. Data relating to both the hydrodynamic conditions and suspended sediment concentration distributions in the estuary were supplied by Associated British Ports (formerly British Transport Docks Board, 1980). Field measurements, including water elevations, tidal velocities and sediment concentrations, were taken at three survey sites, namely Halton Middle, Middle Shoal and Sunk Channel (see Figure 2). Comparisons of the sediment fluxes were made for the spring tidal and mid tidal ranges at Middle Shoal, Halton Middle and Sunk Channel. The magnitude of the predicted and measured concentrations were very close in terms of sediment transport modelling. Figure 3 illustrates the results at Halton Middle and Sunk Channel for the spring tide.

It is interesting to note that there appeared to be a phase lag between the predicted and measured concentrations for all simulations. Apart from the velocity lag caused by the lack of detailed bed roughness data, the assumption that the ratio of the depth averaged sediment concentration and the near-bed reference concentration was equal to the corresponding ratio of the equilibrium concentrations was thought to be another possible cause of this discrepancy. The results were anticipated to be more accurate if the vertical sediment transport profile could be adjusted with the tidal phase, with this dynamic process currently being considered in an extension of this research programme.

CONCLUSIONS

A highly accurate bounded finite difference scheme has been refined and extended to predict sediment transport fluxes. Both operator-splitting and conventional discretization algorithms have been investigated to solve the suspended sediment transport equation, with the schemes being tested for idealised test cases and the Humber Estuary. The introduction of a boundary fitted co-ordinate system has enabled more grid points to be included near the landward boundary in the Humber Estuary study, thereby leading to an increased accuracy relative to a rectangular grid with a comparable number of grid squares. The predicted velocities and water elevations for this study compared favourably with the measured data, thereby leading to the scope for accurate predictions of the sediment concentration distributions. Comparisons of the numerical predictions of sediment fluxes and field data were also made at three sites along the estuary where field data existed. Overall the quality of the comparisons was encouraging, particularly since no tuning was necessary for the simulations as for the previous study by Falconer and Owens (1990).

REFERENCES

British Transport Docks Board. (1980). Humber Estuary Sediment Flux, Part I: Field Measurements. Report No. 283, July, pp1-91.

Cahyono. (1993). Three Dimensional Numerical Modelling of Sediment Transport Processes in Non Stratified Estuarine and Coastal Waters. Ph.D. Thesis, University of Bradford, Bradford, UK, pp302.

Chandler-Wilde, S. N. and Lin, B. L. (1992). Finite Difference Method for the Shallow Water Equations with Conformal Boundary-Fitted Mesh Generation. In Hydraulic and Environmental Modelling : Coastal Waters (eds. Falconer, R.A. et al.), Ashgate publishing Co., Aldershot, pp539.

Falconer, R.A. and Owens, P.H. (1990). Numerical Modelling of Suspended Sediment Fluxes in Estuarine Waters. Estuarine, Coastal and Shelf Science, Vol. 31, pp745-762.

Leonard, B. P. (1991). The ULTIMATE Conservative Difference Scheme Applied to Unsteady One-Dimensional Advection. Computer Methods in Applied Mechanics and Engineering, Vol. 88, pp17-74.

Leonard, B. P. and Niknafs, H. S. (1990). Cost-Effective Accurate Coarse-Grid Method for Highly Convective Multidimensional Unsteady Flows. Proceedings of the CFD Symposium on Aeropropulsion, NASA Lewis Research Centre, Cleveland, OH, USA, pp227-240.

Lin, B. L. and Falconer, R. A. (1994). Modelling Sediment Fluxes in Estuarine Waters Using a Curvilinear Co-ordinate Grid System. Estuarine, Coastal, and Shelf Science (in Press).

van Rijn, L.C. (1984). Sediment Transport Part 2: Suspended Load Transport. Journal of Hydraulic Engineering, ASCE, Vol. 110, pp1613-1641.

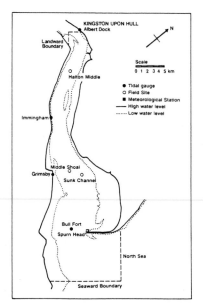

Figure 2 Plan of the Humber Estuary

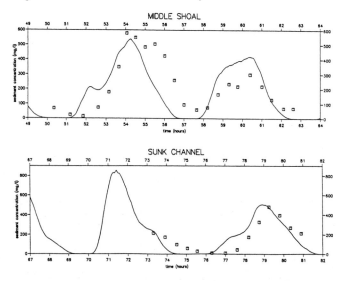

Figure 3 Comparison of Predicted and Measured Sediment Concentrations

1A20

PERIODIC BEHAVIOUR OF PARTLY VEGETATED OPEN CHANNEL

Dan Naot
Center For Technological Education
P.O.B. 305, Holon 58102, Israel

Iehisa Nezu and Hiroji Nakagawa
Division of Global Environmental Engineering
Department of Civil Engineering
University of Kyoto, Kyoto 606,Japan

ABSTRACT

The flow through the clear passage adjacent to the edge of a permeable vegetated domain in partly vegetated open channels is characterized by a free shear layer, which is exposed to a lateral flow due to the secondary currents. In certain configurations, these currents emerge out of the vegetated domain, push away the shear flow and alter its energy level. A feedback mechanism is formed as the longitudinal vorticity sources depend on the energy level. The instability which was observed numerically is discussed here on the background of recent experimental evidence for periodic behaviour.

INTRODUCTION

The flow in the partly vegetated open channel shown in Fig.1 was the subject of experimental (8) and theoretical (3,4) studies and is of particular interest when three dimensional flow effects, such as secondary currents, are sought.

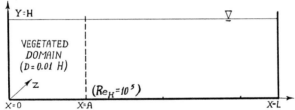

Fig.1 Partly Vegetated Rectangular Open Channel

Due to the vegetated domain permeability, the secondary currents produced in the free main channel manage to penetrate and also emerge the vegetated zone in spite of its attenuating character. The shear layer which develops along the vegetated zone edge, $x = A$, due to the attenuation of the streamwise velocity in the vegetated domain, is exposed to lateral motion. This shear layer is either pushed towards the vegetation, where the dissipation length is relatively small and the energy of turbulence is restricted, or pushed away from the vegetation and turned into a free shear layer with higher turbulence energy. As a result, the longitudinal vorticity sources, which produce the secondary currents, are affected and a feedback mechanism is formed. To cope with this situation the turbulence model includes grid invariant damping method in the $k - \varepsilon$ model equations, and time dependence in the algebraic stress model.

RESEARCH APPROACHES AND APPLICATIONS

Basically, the numerical simulations are based on the algebraic stress model (6,1,2) with the vegetation modelled as an internal resistance exerting drag force, produces energy of turbulence and interferes with its anisotropy and length scale as was partially reported in Refs.(3,4). A full length report was submitted for publication (5). Here, only the relevant parts of the model are given.

GRID INVARIANT DAMPING

Following the experience with grid dependent damping (7) and invariant damping (9), the energy equation was modified here as:

$$\frac{D}{Dt}k = \frac{\partial}{\partial x_i}\left(\frac{\upsilon_t}{\sigma_k}\frac{\partial k}{\partial x_i}\right) + \pi - \varepsilon - T \ , \tag{1}$$

with a correcting term T added note that the diffusion type model $\dfrac{\partial}{\partial x_i}\left(\dfrac{\upsilon_t}{\sigma_k}\dfrac{\partial k}{\partial x_i}\right)$ poorly describe $\dfrac{\partial}{\partial x_i}\left(\dfrac{1}{2}\overline{u_iu_\ell u_\ell} + \dfrac{1}{\rho}\overline{pu_i}\right)$ in regions where the production term π exceeds the energy dissipation rate ε. With $T = 0$ the standard form is maintained. Assuming that the distribution of T among the components of the Reynolds stress tensor take the form:

$$T_{ij} = \frac{T}{k}\left[\overline{u_iu_j}\Lambda_T + \frac{2}{3}k(1-\Lambda_T)\delta_{ij}\right] \ , \tag{2}$$

a modification to the algebraic stress model is obtained. However, in the absence of better knowledge, $\Lambda_T = C_1$ is assumed and the exact result for such a case is implemented:

$$\upsilon_t = \upsilon_t(T=0)\left(\frac{\varepsilon}{\varepsilon+T}\right) \ . \tag{3}$$

In addition, assuming that the corrected term associated with the triple correlation $\overline{u_iu_\ell u_\ell}$ is also related to the term that describes production of dissipation due to the stretching of the vortices: $-2\overline{\left(\dfrac{\partial u_i}{\partial x_m}\dfrac{\partial u_\ell}{\partial x_m}\dfrac{\partial u_\ell}{\partial x_i}\right)}$ a modification to the ε equation was also implemented:

$$\frac{D}{Dt}\varepsilon = \frac{\partial}{\partial x_i}\left(\frac{\upsilon_t}{\sigma_\varepsilon}\frac{\partial \varepsilon}{\partial x_i}\right) + \frac{\varepsilon}{k}\left[C_{\varepsilon1}\pi - C_{\varepsilon2}\varepsilon + C_{\varepsilon3}T\right] \ . \tag{4}$$

In the present study a threshold mode correcting term was used:

$$T = \max.\left[0 \ ; \ (\pi C_{\varepsilon4} - \varepsilon)\right] \ , \tag{5}$$

and $C_{\varepsilon3} = C_{\varepsilon1}$. $C_{\varepsilon4} = 0.90$ was selected high enough to "restrict" the energy peak at the free shear layer and low enough to avoid "over correction" close to the walls where the wall function technique was used.

142

TIME DEPENDENT TURBULENCE MODEL

Including also the substantial derivatives in the local equilibrium form of the stress transport model equations, a quasi local equilibrium time dependent model is obtained. Using the model

$$\varepsilon_{ij} = \frac{\varepsilon}{k}\left[\overline{u_i u_j}C_1 + \frac{2}{3}k(1-C_1)\delta_{ij}\right] , \qquad (6)$$

for the dissipative terms one may write:

$$\overline{u_i u_j} = \left[\pi_{ij} + P_{ij} - \frac{2}{3}\varepsilon(1-C_1)\delta_{ij} - \frac{D}{Dt}\overline{u_i u_j}\right]\frac{k}{\varepsilon C_1} . \qquad (7)$$

Note, however, that the solutions of the time independent equations

$$\overline{u_i u_j} = \left[\pi_{ij} + P_{ij} - \frac{2}{3}\varepsilon(1-C_1)\delta_{ij}\right]\frac{k}{\varepsilon C_1} , \qquad (8)$$

compose the standard (6) algebraic stress model we name it "Prompt A.S.M.", and rewrite equation (7) as:

$$\overline{u_i u_j} = \overline{u_i u_j}(\text{prompt})\theta + \overline{u_i u_j}(\text{old})(1-\theta) , \qquad (9)$$

where $\overline{u_i u_j}(\text{old})$ are the stresses at a former calculation step, and

$$\theta = 1\bigg/\left(1 + \frac{k}{\varepsilon C_1 \Delta t}\right) , \qquad (10)$$

is a "natural" relaxation factor. In time dependent fully developed state Δt is the time step and in steady state situation Δt is $\Delta z/w$, with Δz being a step in streamwise direction. Applying equation (9) only to the stresses which control the secondary currents, is referred to here as a "Delayed A.S.M.". If, in addition, equation (10) is replaced by a fixed typical value for θ, a "Vorticity Sources Relation" method is obtained.

PERIOD AVERAGED BEHAVIOUR

Calculated streamwise velocity contours obtained with the vorticity sources relaxation mode, assumed to represent period averaged results, are given in Fig.2 for a $2 \cdot 5H \times H$ rectangular clean main open channel compound with a $H \times H$ vegetated open channel. $(A = H; L = 3 \cdot 5H; D = 0.01H)$. The non dimensional vegetation density is given by: $N = 100 nHD$ where n is the number of rods per area unit, H is the channel depth and D is the rods diameter. Some trends become apparent. The vegetated subchannel becomes an "Ultra Rough" bank of the main channel and pushes the streamwise velocity maxima far away off the central line of the clean main channel. The secondary currents at the vegetated side of the main clean open channel are accelerated, pushing the streamwise velocity maximum below what is expected in a non vegetated case.

The identification of the free shear layer at the edge of the vegetated zone is not clear. At low vegetation density the secondary currents shift low level energy into the main channel and relatively high level energy into the vegetated domain, with the main free shear layer shifted deep into the main channel. At intermediate vegetation density, a second free shear layer is built up at the vegetated zone edge. With further increase in the vegetation density, this layer spans all over the vegetation zone edge, showing energy level similar to that of a rough wall.

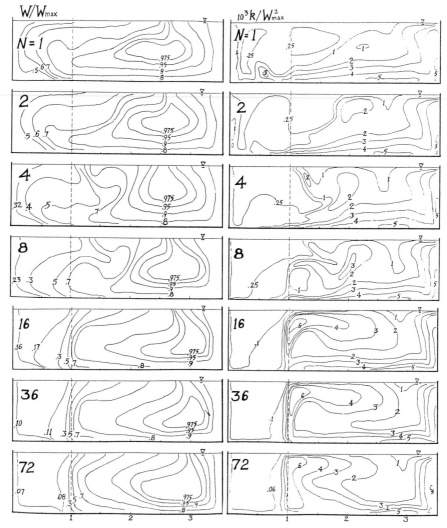

Fig.2 Partly Vegetated Rectangular Open Channel
(a) Streamwise Velocity(b) Energy of Turbulence

PERIODIC BEHAVIOUR

Control values for the streamwise veloicty and for the lateral stream function picked at fixed locations close to their maximae are given in Fig.3 for the three modes of operation: Prompt A.S.M., Delayed A.S.M. and Vorticity Sources Relaxation. The averaging effect of the last mode is clearly apparent. It smooths the high frequency oscillations and keeps the long term ones which are presumably associated with mass displacement. Both the prompt and the delayed algebraic stress models show periodic behaviour. However, both the period and the amplitude of the oscillations obtained with the delayed mode are substantially smaller. Note that a reduction in the forward step size of fifty percent changed the oscillatory results obtained with the delayed mode by less than five percent only, and that the errors in the continuity equations monitored as Mass Sources were reduced by a factor of ten, indicating that this is indeed an oscillatory solution which can be made accurate at will.

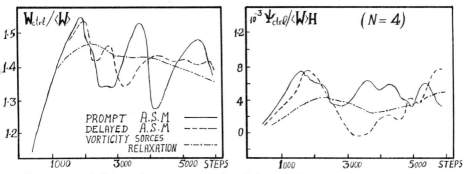

Fig.3 Control Values for (a) Streamwise Velocity (b) Lateral Stream Function

Typical periods, nondimensionalized with respect to H, are about 47 for $N = 4$ and are smaller for higher vegetation density. Indeed, Tsujimoto and Kitamura (8) report measured values ranging between 20 to 30 for N between 8.35 to 16.7, calling for further work preferably with full time dependent models.

To get better insight, a part of the shear layer formed in a stratified symmetric wide open channel is shown in Fig.4. The channel consists of a

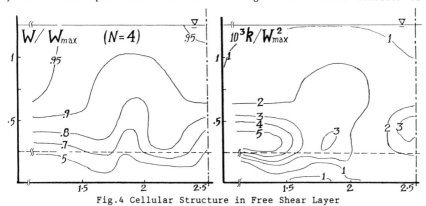

Fig.4 Cellular Structure in Free Shear Layer

RESEARCH APPROACHES AND APPLICATIONS

$5H \times H$ clear open layer positioned above a $5H \times 0.25H$ vegetated layer. The calculation, now performed with vorticity sources relaxation, showed a cellular behaviour at the center of the free shear layer $1.5 < x/H < 3.5$. The arrows shown in the figure indicate the direction of the secondary currents. With upward flow the shear layer is drifted into the clean channel interior and an energy peak is built up, and with downward flow the shear layer is drifted into the vegetated domain and the energy is attenuated substantially. Since the longitudinal vorticity sources depend on the normal turbulent stresses which, in turn depend on the energy, an unstable situation is formed.

CONCLUSIONS

Although the main objective of the study was the establishment of a data base for the hydrodynamic behaviour of compound open channels with partly vegetated flood plains, it also led to the observation that a shear layer that develops adjacent to a permeable domain edge, may show oscillatory or cellular structure composed of longitudinal vortices. The proportion between the length of the free shear layer and the perimeter of the solid wall dominated shear layers determine the intensity of the oscillations. While compound channel shows relatively weak oscillatory behaviour successfully smoothed with the vorticity sources relaxation, stratified wide open channel shows pronounced oscillations.

REFERENCES

1. Naot, D., Nezu, I., and Nakagawa, H., "Hydrodynamic Behaviour of Compound Rectangular Open Channels," Journal of Hydraulic Engrg., ASCE, Vol. 119, 1993, pp. 390-408.

2. Naot, D., Nezu, I., and Nakagawa, H., "Calculation of Compound Open Channel Flow," Jour. of Hydraulic Engrg., ASCE, Vol. 119, 1993, pp. 1418-1426.

3. Naot, D., Nezu, I., and Nakagawa, H., "Towards the Modelling of the Hydrodynamic Forces in Compound Open Channel with Vegetated Flood Plain," Proc. Hydraulic Engrg., JSCE., Vol. 38, 1994, pp. 437-442.

4. Nezu, I., Naot, D., and Nakagawa, H., "Simulation of Hydrodynamic Behaviour of Vegetated Rivers by Using Turbulence Model," 9th Congress of Asian Pacific Div., IAHR, Singapore, 24-26 August 1994.

5. Naot, D., Nezu, I., and Nakagawa, H., "Hydrodynamic Behaviour of Partly Vegetated Open Channel," submitted for publication, 1995.

6. Naot, D., and Rodi, W., "Calculation of Secondary Currents in Channel Flow," Journal of Hydraulic Engrg., ASCE, Vol. 108, 1982, pp. 948-968.

7. Patel, V.C., Rodi, W., and Scheurer, G., "Turbulence Models for Near-Wall and Low Reynolds Number Flows: A Review," AIAA Journal, Vol. 23, 1985, pp. 1308-1319.

8. Tsujimoto, T., and Kitamura, T., "Appearance of Organized Fluctuations in Open Channel Flow with Vegetated Zone," KHL Progressive Report, December 1992, pp. 37-45, Hydraulic Laboratory, Kanazawa University.

9. Yang, Z., and Shih, T.H., "A Galilean and Tensorial Invariant $k-\varepsilon$ Model for Near Wall Turbulence," NASA Technical Memorandum 106263, IAAA-93-3105, ICOMP-93-24, COMTT-93-10, 1993.

1A21

Turbulent Transport Measurement in Open Channel Flow Using Laser Doppler Velocimetry Combined with Laser Induced Fluorescence Technique

FENG,T. AND SHIONO, K.
Department of Civil & Building Engineering, Loughborough University of Technology Leicestershire LE11 3TU, UK

INTRODUCTION

In recent years the growing public interest in pollution levels related to environmental problems in river, estuarine and coastal waters has focused attention on the need to understand the mixing mechanisms which govern the transport rate of pollutants. There has therefore been an increasing interest among water engineers and scientists in improving water quality with the aid of three dimensional numerical models in conjunction with the turbulence models such as $k - \varepsilon$ model, the algebraic stress model, the Reynolds stress model and the large eddy simulation. There exists a number of numerical coefficients in the models, and in particular the diffusion coefficients are vitally important in the prediction of the solute transport rate. The value of the diffusion coefficients varies significantly depending on the turbulent levels, secondary current and shearing flows. However there are only limited data available to validate the computational results with respect to the diffusion coefficients due to the lack of the simultaneous measurements of the Reynolds stresses and fluxes in the real-time sequence.

The technique of laser induced fluorescence (LIF) known as fluorescence spectrometry in analytical chemistry [1] is able to do non intrusive measurement of concentration of fluorescent dye in water flow. Combining with laser Doppler velocimetry (LDV), the simultaneous velocity and concentration measurements have been undertaken by previous investigators in different flow circumstances [2],[3]. Although unlike laser Doppler velocimetry, this laser induced fluorescence technique is able to do only the relative measurement, but can be calibrated by using a small amount of dye sample with a known concentration. To justify the feasibility of concentration measurement, the property of the fluorescent dyes has been investigated and Rhodamine 6G is found to be a suitable fluorescent tracer for a laboratory test, It is also low cost, low toxicity and chemical stability.

EXPERIMENTAL APPARATUS

For the task of simultaneous velocity and concentration measurement, the existing TSI three component fibre optic laser Doppler velocimeter together with the data acquisition system (of Bradford University) has been adapted to a combined LDV and LIF system. The laser beams, e.g. green, blue and violet are transmitted through the fibre optics cables, are reflected in 90 degree using a 45 ° angle mirror and focused at the same point with the focal length of 80 mm.

The LIF device is based on the fact of photoluminescence of fluorescent solution. According to spectrometry theory, the wavelength of the fluorescence emission is always longer than that of the exciting light at room temperature condition (Stokes Law). The colour of the fluoresced light will be yellow-orange with the wavelength of 570 nm if Rhodamine 6G solution is shined by a green colour laser beam with wavelength of 514 nm. Therefore it is possible to separate out the orange light through an appropriate receiving optic filter.

In the present study, the green laser beam provided by the LDV system was used as a shinning beam at the point of the LDV measurement volume, simply the strongest. The fluoresced light together with the all scattered light was simply collected from the measurement volume by the probe head. A colour separator filtered the orange light. Two optical filters, a long wave pass filter (Corion LL550F7325) and a band pass filter (Corion S10570F7325), have been tested for their efficiencies, and the long wave pass filter was then chosen for the flow measurements.

A multi-channel Analogue to Digit convertor, DataLink, was also installed into the data acquisition system to sample the analogue signals of concentration with each LDV measurement. Together with three signal processors controlled by the FIND computer programme, this adapted system enabled us to record simultaneously velocity and concentration in the real-time sequence.

A flume which has dimensions of 13 m long, 0.31m wide and 0.5m deep was used for the experiments. A bed slope of 1:2,000 was set. The flume has been completely paved with toughened glass plates together with the glass side walls to ensure smooth uniform roughness all along the channel. An aerofoil shaped pier (NACA0024) for the dye injection has been designed and manufactured to minimise the effect of the disturbances caused by the wake behind the pier at the measurement point. The dye was pumped from the storage container into the steady pressure head tank then injected through the nozzle in the aerofoil shaped pier into the main stream with constant flow rate.

EXPERIMENTAL RESULTS

1. PRELIMINARY TEST RESULTS
A series of preliminary tests has been performed to calibrate the relationship between the voltage output of the receiving unit and the concentration of Rhodamine 6G. It can be seen from Fig. 1 that the linear relationship exists in the range of the concentration within 0.1 mg/l. It can also been seen that the slope of the calibration curves varies slightly due to the receiving optics adjustment. It therefore suggests that the routine calibration is necessary to adjust this variations.

To justify the effect of fluorescence absorption occurred along the receiving path, a series of tests have been carried out. A small beaker with Rhodamine 6G solution was put at the focal point of the laser beams in a large container with fresh water or the Rhodamine solution. The photo multiplier output voltage was recorded and analysed. The results suggest that the ambiguity caused by absorption could be eliminated in the fluorescence measurement if a test is carried out with low concentration of the dye bellow 0.1 mg /l all the time.

2. FLOW MEASUREMENT RESULTS
Measurements were carried out at a 10m downstream section with two water depths of 0.128m and 0.189m. Three dye injection points from the measurement section were chosen at 0.5m, 1.0m and 1.5m with injection levels of 0.039m and 0.155m. Typical instantaneous velocity fluctuations and concentration fluctuation for a 30 second duration are shown in Fig.2.

The energy spectrum is used for considering the nature of the energy transfer from the mean flow to the turbulence eddies and the nature of the energy transfer from the large energy-containing eddies to the much smaller dissipating eddies. Examples of the raw spectra of velocity and concentration are also shown in Fig. 2. The all spectra cover the most important ranges, for turbulence energy production, from the large containing eddies to smaller isotropic eddies(-5/3 power law). But the viscous subrange, which is proportional to the -7th power law, is not recognised in Fig. 2.

The vertical profiles of the mean velocity and concentration and the Reynolds stress and flux are shown in Fig. 3. The vertical profile of velocity shows the maximum velocity at just above z=0.1m. This feature is typically seen in a relative small aspect ratio (depth/width), in this experiment, a aspect ratio of 2~3 is. The profile of the concentration has a large gradient in the upper half of the depth. The Reynolds stress,$-\rho\overline{uw}$, decreases linearly from the bed, becomes zero at the place where the maximum velocity occurs, and becomes negative where the velocity gradient becomes negative. At the bed, the value of extrapolation of the $-\rho\overline{uw}$ profile agreed with the bed shear stress obtained by a Preston tube.

The Reynolds flux, $\overline{cw}$, is the mean rate of transport due to turbulence of quantity whose concentration is c across unit area normal to the component of velocity w. The distribution of $\overline{cw}$ in Fig. 3 shows that the value of the flux becomes negative below the injection point and positive above. This suggests that the vertical diffusion process takes place upwards and downwards from the injection of the dye. In order to verify the magnitude of the measured flux, $\overline{cw}$, the analogy to Fick's law can be used and, that is

$$\overline{cw} = -D_z \frac{\partial \overline{C}}{\partial z},\text{where } \frac{\partial \overline{C}}{\partial z} = \text{concentration gradient and } D_z = \text{eddy diffusivity. The}$$

results of the Reynolds flux, $\overline{cw}$, and the mean concentration in Fig. 3 give that the eddy diffusivity varies from -1.0^{-4} to $+1.0^{-4}$ depending on either side of the injection point. The magnitude is the same order of that of the depth average eddy viscosity $0.07u_*H = 2.0 \times 1.0^{-4}$. Fig. 4 shows $\overline{cw}$ and $\overline{cw}/\overline{uw}$ at various lengths of the injection point. Therefore the Reynolds stress, $-\rho\overline{uw}$ and flux, $\overline{cw}$ seems to be successfully measured.

CONCLUSIONS

Simultaneous velocity and concentration measurement is attempted by the existing three component laser Doppler velocimeter system with a laser induced fluorescence technique. Applying this system, the ambiguity caused by absorption could be eliminated in the fluorescence measurement when ensuring low concentration of the dye below 0.1 mg/l. The linear relationship between the voltage and the concentration was found to be within the dye concentration of 0.1 mg/l. The mixing process was found to be upwards and downwards from the dye injection level. The Reynolds stresses and fluxes were successfully obtained.

ACKNOWLEDGEMENT

The authors gratefully acknowledge the financial support by Engineering and Physical Sciences Research Council, the loan of a LDA system by Bradford University, the technical supports by Bristol Industrial and Research Associates Limited.

REFERENCES

1. C.Parker, 1968, Photoluminescence of Solutions, Elsevier Publishing Company, London.

2. P.N.Papanicolaou and E.J.List, 1988, Investigations of Round Vertical Turbulent Buoyant Jets, Journal of Fluid Mechanics, Vol.195, pp. 431-391.

3. Y.Enokida and A.Suzuki, 1989, Axial Turbulent Diffusion in Fluid between Rotating Concentric Cylinders. Annual Report of the Engineering Research Institute, Faculty of Engineering, University of Tokyo, Sept.

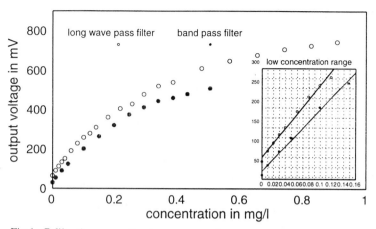

Fig.1 Calibration curve for dye concentration

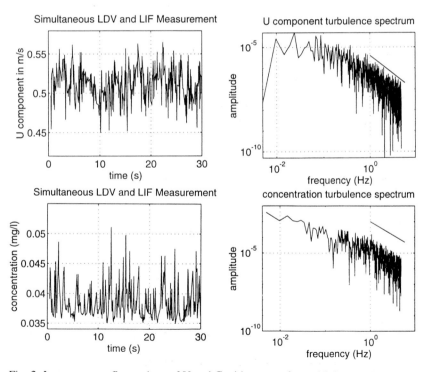

Fig. 2 Instantaneous fluctuations of U and C with spectra for y=15.6cm ,z=11.0cm and H=18.5cm

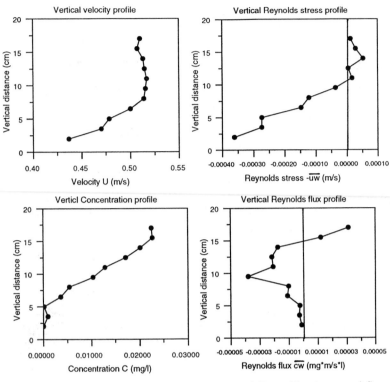

Fig. 3 Vertcal profiles of Mean parameters and Reynolds stress and flux at x=1.5m, y=0.156m with dye injection at z=0.155m (H=0.185m)

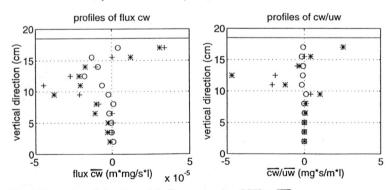

Fig. 4 Examples of the Reynolds flux and ratio of $\overline{cw}$ to $\overline{uw}$, o : at x=0.5m, +: x=1.0 m and * :1.5m.

SPATIALLY-DEVELOPING TURBULENT FLOW IN OPEN CHANNELS WITH PILE DIKES

SYUNSUKE IKEDA and FEI-YONG CHEN
Department of Civil Engineering,Tokyo Institute of Technology
Ookayama 2-12-1, Meguro-ku, Tokyo 152, Japan

INTRODUCTION

In fluvial rivers, pile dikes are effective as river training structures. They are used to retard the near-bank flow and protect the river banks, or be used in flood plains to get a balance of lateral sediment transport and maintain channel width not to be changed. Therefore, it is practically meaningful to study the mechanism of flow in open channels with pile dikes.

Some works have been documented on this subject (Akikusa et al., 1961; Raupach et al., 1982; Ikeda et al., 1991 and 1992; Nadaoka & Yagi, 1993). However, few studies have been performed on the development of transition region in open channels with pile dikes. The purpose of this paper is to apply a numerical technique to analyze the flow in the transition region and to reproduce the distributions of two-dimensional flow field, vorticity distribution, turbulence energy as well as the Reynolds stress.

BASIC EQUATIONS AND COMPUTATIONS

The two dimensional sub-depth scale (SDS & 2DH) turbulence model (Nadaoka & Yagi, 1993) is adopted to describe the flow quantities,

$$\frac{\partial h}{\partial t} + \frac{\partial uh}{\partial x} + \frac{\partial vh}{\partial y} = 0 \tag{1}$$

$$\frac{Du}{Dt} = -g\frac{\partial \eta}{\partial x} - f_x - \frac{c_f}{h}u\sqrt{u^2+v^2} + \frac{1}{h}\frac{\partial}{\partial x}(2\nu_t h\frac{\partial u}{\partial x} - \frac{2}{3}kh) + \frac{1}{h}\frac{\partial}{\partial y}[\nu_t h(\frac{\partial v}{\partial x} + \frac{\partial u}{\partial y})] \tag{2}$$

$$\frac{Dv}{Dt} = -g\frac{\partial \eta}{\partial y} - f_y - \frac{c_f}{h}v\sqrt{u^2+v^2} + \frac{1}{h}\frac{\partial}{\partial y}(2\nu_t h\frac{\partial v}{\partial y} - \frac{2}{3}kh) + \frac{1}{h}\frac{\partial}{\partial y}[\nu_t h(\frac{\partial v}{\partial x} + \frac{\partial u}{\partial y})] \tag{3}$$

The depth-averaged velocity components in directions, x and y, are u and v, respectively, the bed slope is s_0, the water depth is h, and the local free surface elevation is η. The symbol k is the depth-averaged turbulence energy due to sub-depth scale turbulene, ν_t is the depth-averaged eddy viscosity, c_f is the bottom friction coefficient, and f_x and f_y are the components of drag force induced by the pile dikes in the directions x and y, respectively. According to Raupach & Shaw(1982), they are expressed in the following manner:

$$f_x = \frac{ac_d}{2}u\sqrt{u^2 + v^2}, \quad f_y = \frac{ac_d}{2}v\sqrt{u^2 + v^2} \tag{4}$$

where c_d is the drag coefficient of a single pile dike, and $a = d/2l_xl_y$ is the density of pile dikes, in which d is the diameter of pile dikes, l_x and l_y are the intervals of pile dikes in x and y directions, respectively.

The distribution of sub-depth scale turbulence energy, k in the flow field is determined from the following semi-empirical transport equation:

$$\frac{Dk}{Dt} = \frac{1}{h}\frac{\partial}{\partial x}(h\frac{v_t}{\sigma_k}\frac{\partial k}{\partial x}) + \frac{1}{h}\frac{\partial}{\partial y}(h\frac{v_t}{\sigma_k}\frac{\partial k}{\partial y}) + p_{kh} + p_{kv} - \varepsilon \tag{5}$$

where $v_t = c_\mu k^2/\varepsilon$ is the depth-averaged eddy viscosity, $\varepsilon = c_\mu^{3/4}k^{15}/l$ is the depth-averaged turbulence energy dissipation, and $l = \alpha h$ is the characteristic turbulence length scale. The numerical constants are taken to be: $c_\mu = 0.09$, $\sigma_k = 1.0$, $\alpha = 0.1$. p_{kh} is the production of turbulence energy induced by the horizontal shear stress, and it can be written as,

$$p_{kh} = v_t[2(\frac{\partial u}{\partial x})^2 + 2(\frac{\partial v}{\partial y})^2 + (\frac{\partial u}{\partial y} + \frac{\partial v}{\partial x})^2] \tag{6}$$

The other term, p_{kv} is the production of turbulence energy due to the vertical shear, it is concerned with two factors: the bottom friction and the drag of pile dikes. Nadaoka & Yagi (1993b) proposed the following formula to express it:

$$p_{kv} = (\frac{c_f}{h} + \frac{ac_d}{2})(u^2 + v^2)^{1.5} \tag{7}$$

When we applied Eq. 7 to the fully developed region, it is found that it is not accurate enough as described subsequently.

Since the driving force gs_0 is equal to the resistance to flow at the undisturbed areas outside of the lateral shear layer, we have the following equation from the momemtum balance:

$$[\frac{c_f}{h}(u^2 + v^2)]_{main-flow-region} = [(\frac{c_f}{h} + \frac{ac_d}{2})(u^2 + v^2)]_{pile-dike-region} \tag{8}$$

Considering Eq. 5 at equilibrium state, the following relation is derived from an energy balance:

$$p_{kv} = \varepsilon \tag{9}$$

Then, we have

$$\frac{\varepsilon_m}{\varepsilon_p} = \frac{(\sqrt{u^2 + v^2})|_m}{(\sqrt{u^2 + v^2})|_p} \tag{10}$$

in which the variables denoted by the subscript "m" are for the main flow region, and "p" for the pile dikes region. As far as the fully developed region is concerned, the flow velocity in the main flow region is much larger than that in the pile dikes region. It is therefore seen from Eq.10 that the turbulence energy dissipation ε_m is much larger than ε_p. It leads to a conclusion that the turbulence energy in the main flow region is much larger than that in the pile dikes region, because k is proportional to $\varepsilon^{2/3}$. Hence, when Eq. 7 is employed, the location of the largest value of the turbulence energy at the fully developed region must be found to be in the undisturbed area of the main flow region far from the pile dikes. This is apparently incorrect, and Eq. 7 should be modified properly.

Considering the main contribution of the production of turbulence energy due to the vertical velocity gradient, we have,

$$p_{kv} = \frac{1}{h}\int_0^h \tilde{v}_t (\frac{\partial U}{\partial z})^2 dz \tag{11}$$

in which z is the vertical coordinate, U is the composition of the horizontal velocity components at height z and $\tilde{v}_t$ is the depth-averaged eddy viscosity. In terms of the shear stress hypothesis proposed by Prandtl, the above expression can be correlated to the local kinematic shear stress, $\tau_i = -u'w'$, where u', w' are the longitudinal and vertical components of the velocity fluctuations due to the subdepth scale turbulence, such that,

$$p_{kv}v_t = \frac{1}{h}\int_0^h (\tilde{v}_t \frac{\partial U}{\partial z})^2 dz = \frac{1}{h}\int_0^h \tau_i^2 dz \tag{12}$$

The depth-averaged shear stress can be described by (Raupach & Shaw,1982)

$$\tau = \tau_b + \tau_d = (c_f + \frac{ac_d h}{2})(u^2 + v^2) \tag{13}$$

where τ_d is the shear stress due to pile dikes, τ_b is the shear stress due to the bottom friction. Replacing the local shear stress, τ_i, in Eq. 12 by the depth-averaged value, τ, the following expression is obtained:

$$p_{kv}v_t = \frac{1}{h}\int_0^h \tau_i^2 dz = [(c_f + \frac{ac_d h}{2})(u^2 + v^2)]^2 \tag{14}$$

Considering Eq. 9 and the definition of $v_t = c_\mu k^2 / \varepsilon$, we derive

$$p_{kv}v_t = v_t \varepsilon = c_\mu k^2$$

From the above expression and Eq. 14, a formula for k is reduced to

$$k = (c_f + \frac{ac_d h}{2})(u^2 + v^2)/c_\mu^{1/2} \tag{15}$$

Then, a formula for the production of turbulence energy due to the vertical shear is obtained, such that,

$$p_{kv} = [(c_f + \frac{ac_d h}{2})(u^2 + v^2)]^{1.5} / l \tag{16}$$

Assuming the vertical velocity distribution takes a similar profile everywhere, Eq. 16 is applied to the transition area.

The boundary conditions should be specified to solve the governing equations. Since the ratio of the river width to the depth is assumed to be large, the effect of the river banks is neglected herein, and therefore slip conditions are adopted at the river banks. The periodic conditions are chosen for the inlet boundary, and the first order derivative of the variables in x-direction is taken to be zero at the outlet boundary. The initial conditions are calculated from a simplified model which assumes that the eddy viscosity is constant everywhere in Eqs. 2 and 3. Other flow parameters are determined according to the experiments performed by Ikeda

et al. (1992, Run 1) and Tsujimoto et al.(1994, Run B1), which are defined as
Case 1 and Case 2, respectively.

Table 1 Major Parameters for Calculations

	width of channel B(cm)	lenth of channel L(cm)	water depth h(cm)	bed slope S_0	bottom friction c_f	drag of pile-dikes c_d	density of pile-dikes a(cm^{-1})	pile-dikes area width B_s(cm)
Case 1	96	1200	6	0.001	0.007	1.8	0.01	30
Case 2	40	1500	4.28	0.00149	0.004	2.9	0.019	12

VELOCITY PROFILES

The depth-averaged flow fields are calculated by the numerical model described
in the above. The results indicate that the flow is retarded in the pile dikes region
(x: 500~1250 cm, y: -30~0 cm for Case 1, and x: 450~1500 cm, y: -12~0 cm for
Case 2) and transition regions with strong lateral diffusion are generated. It is
seen that some distances (Case 1, x=500~900 cm; Case 2, x=450~820 cm) are
required to reach a fully-developed region. For Case 1, the computed velocity
profiles of some cross-sections in the transition region are shown in Fig. 1 to
compare with the existing experimental data (Ikeda et al., 1992, Run 1). Since the
width of channel is not large enough, the lateral diffusion affects the whole main
flow region; i.e. the main flow is accelerated as the flow in the pile dikes region is
decelerated. The comparison of velocity profile between the calculated results
and the experimental data shows a good agreement.

VORTICITY

For Case 2, Figs. 2 and 3 illustrate the instantaneous vorticity distributions at 30,
50 seconds from the start of the calculation, respectively. In the inlet section (x=0
~450 cm), no vorticity source exists on the horizontal plane, because we adopt
the non-slip condition for the river banks, and a uniform inflow condition for the
inlet boundary. Hence, as shown in the figures, the vorticity does not exist in the
inlet section. On the contrary, the pile dikes induce retardation to the flow, and a
difference of flow velocity is generated at the interface between the pile-dike
region and the main flow region. This is the source of vorticity.

It is seen that the vorticity field begins to fluctuate at about x=820 cm, suggesting
to roll up to discrete vorticities. This is spatially developing instability in the
down-stream direction. As shown in Figs. 2 and 3, the longitudinal scale of the
large eddies in this region reaches to about 84 cm. The period of vorticity is about
1.3 seconds.

TURBULENCE ENERGY AND THE REYNOLDS STRESS

By substituting Eq. 16 into Eq. 5, the SDS turbulence energy equation is solved.
In the fully developed region the peak value of the turbulence energy in a cross-
section exists near the boundary between the main flow region and the pile-dikes
region (see Fig. 4). The turbulence energy (the vertical component is not

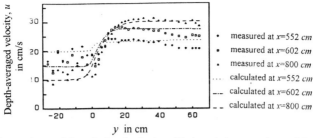

Fig. 1 Comparisons of the calculated velocity with the existing experimental data (Case 1)

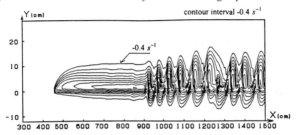

Fig   The instantaneous distribution of vorticity at t=30 seconds (Case 2)

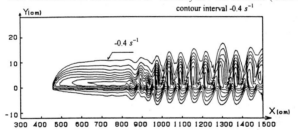

Fig. 3 The instantaneous distribution of vorticity at t=50 seconds (Case 2)

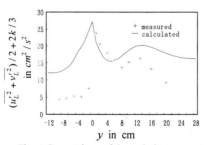

Fig. 4 Comparison of the turbulence energy

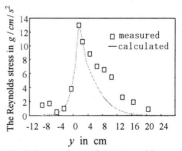

Fig. 5 Comparison of the Reynolds stress

included, because it was not measured by Tsujimoto) is compared with the measurement (Case 2). The predicted profile is similar. However, the absolute values of the prediction are slightly larger than the measurement. In Fig. 4, u'_L and v'_L are the components of velocity fluctuation due to the horizontal large eddies, and k is the sub-depth scale homogenous turbulence energy.

The computed distribution of the Reynolds stress compnent, $-\overline{u'v'}$, in fully developed region is compared with the measured data (Case 2) in Fig. 5. The skewed distribution is reasonably reproduced, and the peak value of the Reynolds stress agrees with the measurement.

CONCLUSIONS
This paper employed SDS & 2DH turbulence model to represent the turbulent flow in open channels with pile dikes. The velocity and vorticity distributions were reproduced reasonably well.

This paper modified the expression for the vertical production of turbulent energy due to the bottom friction and the drag of pile dikes. By using this formula, a reasonable distribution of turbulence energy was obtained.

The calculated velocity profiles, the lateral distributions of the Reynolds stress and the turbulence energy were tested by the existing experimental data. The results are found to be reasonable.

REFERENCES
Akikusa, I., Kikkawa, H., Sakagami, Y., Ashida, K., and Tsuchiya, A. (1961) Study on dikes, *Report of Publics Works Res. Inst.*, Ministry of Construction of Japan, Tokyo, Japan, Vol. 107, pp.61-153, (in Japanese).

Ikeda, S., Izumi, N. and Ito, R. (1991) Effect of pile dikes on flow retardation and sediment transport, *J. of Hydraulic Engineering*, ASCE, Vol. 117, No. 11, pp.1459-1478.

Ikeda, S., Ohta, K., and Hasegawa, H. (1992) Periodic vortices at the boundary of vegetated area along river bank, *Proc. of JSCE*, No.443, pp.47-54, (in Japanese).

Nadaoka, K. and Yagi, H. (1993) Horizontal large-eddy computation of river flow with transverse shear by SDS & 2DH model, *Proc. of JSCE*, No. 473, pp.35-44, (in Japanese).

Rastogi, A. K. and Rodi, W. (1978) Predications of heat and mass transfer in open channels, *J. of Hydraulic Division*, ASCE, Vol.104, pp.397-420.

Raupach, M. R. and Shaw, R. H. (1982) Averaging procedures for flow within vegetation canopies, *Boundary-Layer Meteorology*, Vol.22, pp.79-90.

Tsujimoto, T. and Kitamura, T. (1994) Experimental study of transverse mixing in open-channel flow with longitudinal zone of vegetation along side wall, *Proc. of JSCE*, No.491, pp.61-70, (in Japanese).

1A23

Some Turbulence Characteristics of Open-channel Flow over Rough Bed with Different Slopes

Dong Zengnan, Li Xinyu, Chen Changzhi, Mao Zeyu
(Department of Hydraulic Engineering, Tsinghua University
Beijing 100084,CHINA)

ABSTRACT

Experiments on the turbulence characteristics of flow over rough bed were conducted in a tilting flume. By using 1-D LDV, mean velocity distribution and longitudinal turbulence intensity were measured. Different series of experiments were performed in flows over two different sizes of glass marbles and under different steep bed slopes respectively to investigate the effects of slope on turbulence structures. The experimental results showed that there is no obvious influence of slope on flow velocity and turbulence intensity profiles.

1. INTRODUCTION

In the existing literature about study of turbulence characteristics of open channel flow over rough bed, many efforts have been made to study the mean velocity distribution (O'Loughlin & Annambholta [1],Dong et al. [3], Nakagawa et al. [2]). Dong et al. carried out lots of experiments in flows over gravels, and their results showed that the roughness sublayer existed and the relative roughness ratio H_t/K_s was recommended as the criterion to sort the flows into three different categories. Many other researchers also pointed out that the velocity distribution near the bed was uniform rather than the log-law profile(e.g. Nakagawa et al. [2]).

However, up to now most of the available data were collected from channels with small Froude number, and the slope was milder compared to that of mountain river, whose bed roughness is usually very large together with steeper slope, consequently, application of those conclusions was restrained. In this paper, the experiments on flows over rough bed channel with different steep slopes were conducted, and the relative roughness ratio were all less than 5, thus the flow was over large or medium scale roughness as pointed out by Dong [3]. The purpose of the present paper is to study the effects of channel slopes on

The project was supported by the National Science Foundation of China some turbulence characteristics such as velocity profile and turbulence intensity distribution.

2. EQUIPMENT AND EXPERIMENTS

All the experiments were carried out in a tilting flume of 7m long, 0.25m wide and 0.3m high. The roughness was made by tightly gluing glass marbles with diameters of 2.5cm and 1.5cm respectively so that the effect of irregularity of the shape of the roughness was avoided. Velocity was measured by using a one-dimensional LDV system, which was developed at the Hydraulic Laboratory of Tsinghua University. The advantages such as high resolution and high noise-signal ratio make it possible to measure the velocity very near the bed. The photoelectric signals from the LDV receiver were converted into digital signals by an A/D converter, which then were processed by a data processor JSP-1 to get the mean velocity and root mean square of velocity fluctuations and all the statistics.The measuring station was 4.5 meters far from the entrance. Since the water was very shallow, the fully developed flow occurred far upstream of the measuring station,which was already verified by Dong's careful experiments [3].

Table 1 Hydraulic conditions of different experiments (D=2.5cm)

Run	Water depth H(cm)	U_{max} (cm/s)	$Re = \frac{4U_{max}H}{v} \times 10^4$	$Fr = \frac{U_{max}}{\sqrt{gH}}$	H/D	$J(\%)$	Symbol in the figures
RS11	2.42	59.68	5.07	1.22	0.97	1.15	-○-
RS12	4.14	98.72	14.35	1.55	1.66	1.15	-◇-
RS13	3.28	71.71	8.25	1.26	1.31	1.15	-□-
RS21	1.22	59.81	2.56	1.72	0.49	2.10	-△-
RS22	3.12	81.36	8.91	1.47	1.25	2.10	-▽-
RS31	3.10	79.46	8.45	1.44	1.24	1.80	-○-
RS32	2.12	64.87	4.83	1.42	0.85	1.80	-◇-

Table 2 Hydraulic conditions of flows over glass marbles(D=1.5cm)

Run	Water depth H(cm)	U_{max} (cm/s)	$Re = \frac{4U_{max}H}{v} \times 10^4$	$Fr = \frac{U_{max}}{\sqrt{gH}}$	H/D	$J(\%)$	Symbol in the figures
S11	2.70	84.64	7.68	1.64	1.80	2.10	-○-
S12	2.61	83.78	7.67	1.65	1.74	2.10	-◇-
S13	2.99	94.67	9.94	1.75	1.99	1.93	-□-

At the measuring station,mean velocity and longitudinal turbulence intensity were measured under ten different hydraulic conditions including four different channel slopes.The parameters of each runs were listed in Table 1 and Table 2.

3. MEAN VELOCITY DISTRIBUTION

To study flow over rough bed, the first problem is how to determine the position of theoretical zero point (y_0) of the channel bed. Many researchers proposed a lot of different empirical formulae to calculate the position of theoretical zero point. Yet due to the complicate roughness forms, it is almost impossible to unify the existing formulae. In this paper, the theoretical zero point y_0 was determined from $y_0/D=0.2$ which was proposed by Einstein [4]. As shown in Fig. 1, both (Y_t) and (Y_t+y_0)were used in order to study the effect of the zero point on the velocity distribution. Typical mean velocity profile of flow over the glass marbles was plotted in Fig. 2. Referring to Fig. 2, the mean velocity U was non-dimensionalized by friction velocity U∗ which was deduced from the following equation

$$U_* = \sqrt{gJH_t} \tag{1}$$

where J is the hydraulic slope and H_t is water depth measured from the top of roughness. From this figure, it can be found that, by introducing the theoretical zero point y_0, the relationship between the non-dimensional velocity U/U∗ and Log((Y_t+y_0)/D) can be expressed as follows:

$$\frac{U}{U_*} = 5.75\log\frac{(Y_t+y_0)}{D} + A \tag{2}$$

while the pattern of U/U∗ vs. Log(Y_t/D) deviated from the standard log-law. From this comparison, it was highly recommended to use the distance (Y_t+y_0) to analyze the mean velocity distribution. Moreover, it can be found that the magnitude of U/U∗ is much smaller than that of flow over smooth bed as shown in Fig. 3. This result confirmed Nakagawa's conclusion [2].

In order to study the effects of channel slope on the velocity distribution, data from three runs with approximately equivalent relative roughness ratio were plotted in Fig. 4. The flows were also under three different bed slopes. From this figure, it can be seen that there is no obvious influence of slope on velocity distribution because all the data coincide rather well. Also in Fig. 5, three cases of experimental data of flows over a smaller size of glass marbles (D=1.5cm) were presented, and there is also no effects of channel slope on non-dimensional velocity distribution. It is very useful to demonstrate that there is no effect of slope on dimensionless velocity distribution, so, all the conclusions drawn in the experiment can be applied directly in practice without considering the channel slopes. This result was also found in flows over smooth bed(Li et al [5]).

The dimensionless velocity U/U_{max} of the seven runs were given in Figure 6, which also shows no obvious effect of channel slope on non-dimensional velocity distribution. This means that the maximum velocity U_{max} is also a proper non-dimensional velocity scale. Compared with Fig. 2, it seems that U_{max} is a better non-dimensional velocity scale than U*. When water flows over rough bed, it is usually difficult to measure accurately the water depth and hydraulic slope. Therefore, it is unlikely to guarantee the accuracy of U* if it was obtained from equ.(1). On the other hand, U_{max} was measured directly in the experiments, so it is more reliable.

4. TURBULENCE INTENSITY DISTRIBUTION

In these experiments, only the longitudinal turbulence intensities, u'/U*, were obtained by using 1-D LDV system. All the seven runs of data were presented in Fig. 7. Also in this figure, typical distribution of turbulence intensities of flow over smooth bed were illustrated in order to investigate the influence of roughness on turbulence characteristics. It can be noticed from this figure that all the turbulence intensities of the seven sets almost fall on the same curve, which indicates that the non-dimensional turbulence intensities are not influenced by the channel slopes either. The scale of the turbulent intensity is about 2.0 near the rough bed, which is nearly the same as that found by both Graf [7] and Nezu [6] and is slightly smaller than that of flow over smooth bed. The turbulent intensity increases slowly from 2.0 (near the rough bed) to about 2.2 at $(Y_t+y_0)/(H_t+y_0)=0.2$, then decreases to 1.5 near the water surface that are higher than those measured in flow over smooth bed, which means flow over large roughness was more turbulent than flow over smooth bed. Since the water is very shallow and the submergence is small, so the influence of surface wave on turbulence intensity may become significant near the water surface. Except the regions near the rough bed $((Y_t+y_0)/(H_t+y_0)<0.2)$ and near the water surface, the turbulence intensities of flow over rough bed are almost the same as those of flow over smooth bed. The turbulence intensity in the cross section was nearly not changed based on velocity scale U*, which implies that the velocity fluctuation was enlarged by the roughness.

The experimental data of three runs used in Fig. 4 were represented in Fig. 8 to investigate the effects of channel slope on turbulence intensities distribution. It can be seen that all the turbulence intensities of three sets agree well, which indicates that there is no influence of channel slope on turbulence intensities distribution.

5. CONCLUSIONS

From the above analyses, the following conclusions can be drawn:

(1). In flows over rough bed with $H_t/D<5$ (or small submergence), the magnitude of the non-dimensional velocity $U/U*$ is smaller than that in flows over smooth or small rough bed. The theoretical zero point is recommended as $y_0/D=0.2$ in these experiments.

(2). No effects of channel slopes on the velocity distribution and turbulence intensity are found in these experiments, so the results can be applied in flows over large roughness without considering the influence of channel slopes.

(3). The non-dimensional turbulence intensities in the region near the top of the large roughness are almost the same as those in flows over smooth or small rough bed, while near the water surface, turbulence intensities become larger. From the distribution of turbulence intensity, no evident effect of channel slope is found.

REFERENCES

1. O'Loughlin, E.M. and V.S.S. Annambhotla (1969): Flow phenomena near rough boundaries. Journal of Hydraulic Research, IAHR, Vol.7, No.2, PP.231-250.

2. Nakagawa, H.,T.Tsujimato and Y. Shimizu: Turbulent flow with small relative submergence, Proc. Itn'l Workshop on Fluvial Hydraulics of Mountain Regions, Trent, Italy, PP.A19-30,1989.

3. Dong,Z., Wang,J., Chen,C. and Xia, Z. Turbulence characteristics of open-channel flows over rough beds. Proceedings of XXIV IAHR Congress, 1991, Madrid, Spain

4. Einstein, H.A., and El-Samni, E.A., Hydrodynamic forces on a rough wall, Review of Modern Physics,Vol.21, No.3, 1949, PP.520-524.

5. Li,X., Dong,Z. and Chen,C., Some turbulence characteristics of supercritical flows, Proc. Int. Symposium on Hydraulic Measurement, Beijing, China, pp 107-111,1994

6. Nezu,I. and Nakagawa,H., Turbulence in open-channel flows, IAHR Monograph, A.A.Balkema, 1993.

7. Graf,W. Turbulence characteristics in rough uniform open-channel flows, Annual Report, 1991.

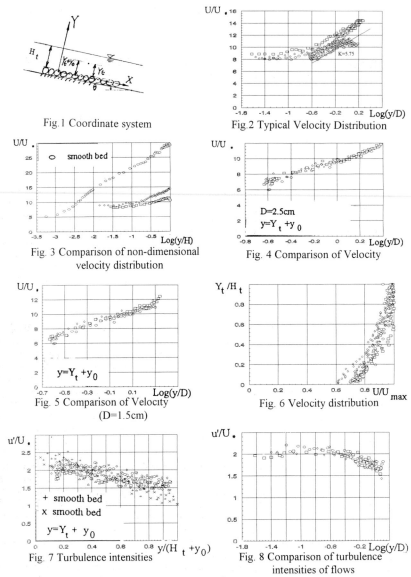

Fig.1 Coordinate system

Fig.2 Typical Velocity Distribution

Fig. 3 Comparison of non-dimensional velocity distribution

Fig. 4 Comparison of Velocity

Fig. 5 Comparison of Velocity (D=1.5cm)

Fig. 6 Velocity distribution

Fig. 7 Turbulence intensities

Fig. 8 Comparison of turbulence intensities of flows

(Legends in all above figures can be found from Tab.1 and 2)

1A24

NUMERICAL STUDY OF TURBULENT STRUCTURE OVER STRIP ROUGHNESS IN OPEN CHANNEL FLOW

JIAN LIU, AKIHIRO TOMINAGA and MASASHI NAGAO
Nagoya Institute of Technology, Nagoya, Japan

ABSTRACT

Turbulent structure over strip roughness in open channel flow with various roughness spacing is separately simulated by using the standard and modified low Reynolds number k-ε turbulence models. The turbulence characteristics in response to roughness spacing is predicted. The effect of the free surface is taken into account in the process of the calculation. Comparisons of the velocities calculated with two turbulence models with the experimental data are very good. For turbulent shear stresses and resistance coefficient, the standard k-ε model gives the excessive high predictions except for the case of L/k=∞, whereas low Reynolds number k-ε model presents the adequate results. The values of the turbulent kinetic energy increase apparently in the neighbor of the upstream corner of the roughness element. The mixing effects of the flow grow stronger and stronger with the decrease of the roughness spacing.

INTRODUCTION

Some studies on the flow over rough boundaries have been carried out experimentally and analytically using various kind of roughness elements for basic and practical purposes. On occasion when the size of roughness element is larger compared with the water depth, the flow far from the walls and the water surface are greatly changed, general expression of the law of such flow is more apparently difficult. Hence, the study of the flow in which roughness element is smaller compared with water depth has been made in the present paper. The resistance of the turbulent flow is generally determined by the size of the dissipation rate of turbulent kinetic energy, the dissipation rate stems from unsteadiness in the vicinity of the outer region of the viscous sublayer for the flow in a smooth open channel, however, its source is the wake flow due to the roughness for the flow over rough bed. Generally, the height of roughness element is considered to be a definite factor for the absolute scale, but, the spacing between roughness elements has important influence on resistance coefficient in the flow over strip roughness. To discuss the turbulent structure of the flow over two-dimensional strip roughness whose spacing is changing is one of the most basic approach to understand the generating mechanism of turbulent kinetic energy over rough boundary. Many experimental data have been obtained for the flow over strip roughness (See Perry, et al. (1), Futuya, et al.(2), Antonia, et al. (3) and Tominaga, et al. (4), (5)). The exprimental results show that the hydraulic resistance of strip roughness is characterized by the relative spacing L/k which is defined as the ratio of the spacing L and height k of strip roughness.

Although a number of the experimental studies of the flow over strip roughness have been made, few of them has been numerically accomplished. In order to fully find out the flow characteristics, the flow over strip roughness in open channel is evaluated in this study by employing the standard k-ε model (that is high Reynolds number (HRN) k-ε model) and low Reynolds number (LRN) k-ε model, respectively. In the part of the vertical boundary layer closest to the wall, gradients in the flow are large and require the use of many computational grid points. In the calculation using the standard k-ε model, the solution in this part of the boundary layer is approximated by the logarithmic wall functions (WF) to make the least possible use of computational grid points. Use of WF as boundary conditions when solving turbulent flows using the k-ε model has been relatively successful under local equilibrium conditions. But when the production term is not equal to the dissipation rate of turbulent energy, the use of WF presents a fundamental problem. Therefore, it is necessary to modify WF for predicting the flow as concise as possible. The modified WF of Nezu et al. (6) will be used in this study. Unlike the WF approach, a LRN closure attempts to model the turbulent transport across the entire near-wall region. This is especially important for predicting turbulent structure in the neighbor of strip roughness. In the present paper, the LRN turbulence model of Jones and Launder (7) is used to predict the turbulent flow over strip roughness in open channel.

MATHEMATICAL MODELS

MEAN TRANSPORT EQUATIONS AND TURBULENCE MODELS

The two-dimensional turbulent flow over strip roughness in open channel is described by Reynolds equations. The variables in the turbulent flow are modeled by a control volume method in the present study. The x and y coordinates are chosen in the streamwise direction and the direction normal to the wall, respectively. The turbulence models are the standard k-ε model and LRN k-ε model of Jones and Launder, with some constants modified by Launder and Sharma (8). The set of elliptic partial differential equations governing mass, momentum, and energy conservation for a steady incompressible flow can be written as follows:

$$\frac{\partial}{\partial x}(u\phi) + \frac{\partial}{\partial y}(v\phi) = \frac{\partial}{\partial x}\left(\Gamma_\phi \frac{\partial\phi}{\partial x}\right) + \frac{\partial}{\partial y}\left(\Gamma_\phi \frac{\partial\phi}{\partial y}\right) + S_\phi \tag{1}$$

where ϕ represents different dependent variables (u, v, k and ε), Γ_ϕ stands for generalized diffusion coefficients, S_ϕ is source terms.

The governing equations for the flow are represented in Table 1. The effective viscosity v_{eff} is the sum of the molecular and turbulent contributions, that is $v_{eff}=v+v_t$, v is the kinematic viscosity, the turbulent viscosity v_t is given by $v_t=f_\mu C_\mu k^2/\varepsilon$.

For the standard model, the functions f_μ, f_1 and f_2 tend to unity, the destruction term D and generation term E added to the transport equation of k and ε are equal to zero, and five empirical coefficients C_μ, $C_{\varepsilon 1}$, $C_{\varepsilon 2}$, σ_k, σ_ε for the model are employed by Rodi (9). For the LRN model, the f_μ, f_1, f_2, C_μ, $C_{\varepsilon 1}$, $C_{\varepsilon 2}$, σ_k, σ_ε and so on obtained by Launder and Sharma are adopted:

Table 1   Conservation equations

Equation	ϕ	Γ_ϕ	S_ϕ
Continuity	1	0	0
x- momentum	u	v_{eff}	$-\frac{\partial P}{\rho\partial x} + \frac{\partial}{\partial x}\left(v_{eff}\frac{\partial u}{\partial x}\right) + \frac{\partial}{\partial y}\left(v_{eff}\frac{\partial v}{\partial x}\right)$
y- momentum	v	v_{eff}	$-\frac{\partial P}{\rho\partial y} + \frac{\partial}{\partial x}\left(v_{eff}\frac{\partial u}{\partial y}\right) + \frac{\partial}{\partial y}\left(v_{eff}\frac{\partial v}{\partial y}\right)$
k	k	$v+v_t/\sigma_k$	$P_k-\varepsilon+D$
ε	ε	$v+v_t/\sigma_\varepsilon$	$C_{\varepsilon 1}\varepsilon P_k/k-C_{\varepsilon 2}\varepsilon^2/k+E$

$$P_k = v_t\left[2\left(\frac{\partial u}{\partial x}\right)^2 + 2\left(\frac{\partial v}{\partial y}\right)^2 + \left(\frac{\partial u}{\partial y} + \frac{\partial v}{\partial x}\right)^2\right],$$

$$D = -C_3 v\left[\left(\frac{\partial k^{1/2}}{\partial x}\right)^2 + \left(\frac{\partial k^{1/2}}{\partial y}\right)^2\right]$$

$$E = C_4 v v_t\left[\left(\frac{\partial^2 u}{\partial x^2}\right)^2 + \left(\frac{\partial^2 u}{\partial y^2}\right)^2 + \left(\frac{\partial^2 v}{\partial x^2}\right)^2 + \left(\frac{\partial^2 v}{\partial y^2}\right)^2\right]$$

$C_\mu=0.09$, $C_{\varepsilon 1}=1.44$, $C_{\varepsilon 2}=1.92$, $\sigma_k=1.0$, $\sigma_\varepsilon=1.3$, $f_1=1.0$, $f_2=1.0-0.3\exp(-Re_t^2)$

$$f_\mu = \exp\left[\frac{-3.4}{(1+Re_t/50)^2}\right] \tag{2}$$

where $Re_t=k^2/v\varepsilon$ is turbulent Reynolds number.

In Table 1, P_k represents the production of kinetic energy by interaction of mean velocity gradients and turbulent stresses. The empirical coefficients C_3 and C_4 in the expression of D and E are equal to 2 in accordance with Jones and Launder. Nezu, et al. modified the value of C_3 based on a number of experimental data, and advocated that $C_3=1.8$ was most suitable for the turbulent flow in open channel. The modified value of C_3 of Nezu, et al. will be adopted in the present study.

BOUNDARY CONDITIONS

Because Eq. (1) is elliptic, it is necessary to define boundary conditions for all variables on all boundaries of the flow domain: inlet, exit, solid wall and free water surface. At the inlet, the mean and fluctuating velocity can be taken from measurements thus that 1/7th power law profile or log-law profile being used, while zero gradients can set at the outlet. But, at the surface, attention must be paid to effects of water surface, thus, it is necessary to modify zero gradients at water surface. Nezu, et al. and Handler et al. (10) introduced dumping coefficient D_W in consideration of the damping effects of kinetic energy in water surface, and recommended that $D_W=0.8$ was suitable for the open-channel flows. For the dissipation rates near the free surface, although Handler et al. made detailed explication by direct numerical simulation, they yielded no clear conclusions, thus boundary condition of dissipation rate near free surface is still adopted zero gradient in the present study. Near the solid wall, low Reynolds number approach enables the extension of the k-ε turbulence model all the way to the wall. For the standard model, the WF (6) used to obtain boundary conditions for k and ε equations at the first inner grid point are employed:

$$\frac{k_p}{U_*^2} = \frac{1 - dU^+/dy^+}{\sqrt{\alpha C_\mu}}\Big|_{y_p^+} \tag{3}$$

$$\varepsilon_p = \frac{C_\mu k_p^2}{\nu}\left(\frac{dU^+}{dy^+}\right)\Big/\left(1 - \frac{dU^+}{dy^+}\right)\Big|_{y_p^+} \tag{4}$$

where U_* is the friction velocity, $U^+ = u/U_*$ is the dimensionless velocity, $y^+ = yU_*/\nu$ is the dimensionless wall distance, y_p^+ is the dimensionless distance from the first inner grid point to the wall. The empirical coefficient $\alpha = P_{kp}/\varepsilon_p$ is the function of y^+, P_{kp} is the generation term at the first inner grid point.

In order to avoid the entrance region problem, the periodic boundary condition has been used in the present study. A fixed number of inner iterations is performed for given inlet conditions, and calculated outlet values of velocities, kinetic energy, and dissipation rates are substituted as inlet conditions for the next outer iteration. A log-law profile is given for the streamwise velocity at the inlet for the first outer iteration. The computational domain shown in Fig.1 is chosen such that the inlet locates at the right end of the strip roughness. The calculated range in streamwise direction is 35-40 times of water depth, where the flow is fully developed.

DISCRETIZATION AND CALCULATED PROCEDURE

The spatial derivatives in Eq.(1) are discretized with the finite-volume method on the staggered grid system predicted by Patankar, et al. (11). For LRN model, the appearance of boundary layer requires a non-equidistant grid that gives a strong grid refinement along the walls. The sides of the finite volumes near the solid walls are positioned in y direction and in the front and back of roughness according to geometrical progression, and 67% of grids locate between the wall and the level of two times of roughness height in the y-direction. The pressure field is calculated by using the SIMPLE method of Patankar, et al. (11). The discretized

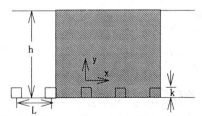

Fig.1 Definition of computation cell for flow over strip roughness in open channel

equations is solved by a tridiagonal matrix algorithm (TDMA). Some under-relaxation is required to prevent divergence of the iteration process. The region located in strip roughness is treated by means of the method proposed by Patankar, et al. (11). The criterion of convergence of numerical solution is on the basis of the absolute normalized residents of the equations that is summed for all control volumes in the computation. The solutions are regarded as convergence when these normalized residuals become less than 10^{-6} for all variables.

PRESENTATION AND DISCUSSION OF RESULTS

The various cases shown in Table 2 are numerically simulated by using the standard and modified LRN turbulence model. In Table 2, roughness elements are square with cross section of 0.5cm×0.5cm. Relative spacing L/k is changed from 2,4,8,16 up to infinity which means that only one element exists in the flow. Relative water depth h/k (h is the water depth measured from the bed) is set to be about 10 except for h/k=16 for the case of L/k=∞. The Reynolds number Re ($=U_m h/v$, U_m is the mean bulk velocity) and the Froude number Fr ($=U_m/\sqrt{gh}$, g is the gravitational acceleration) are separately set to be about Re=1.4×10^4 and Fr=0.35 for the cases of L/k≤16.

Without specific comment in this section, the predicted and measured values are non-dimensional, and the predicted ones are the calculated results when outer iterations equal to 20 for the case of L/k=16 and 35 for the others except for the case of L/k=∞, respectively. The calculated results are compared with the experimental data of Tominaga (5).

The streamwise velocity and Reynolds stress distribution based on the calculated values computed with the HRN and LRN k-ε turbulence models and experimental data of Ra15 and Ra04 are depicted in Fig.2 and Fig.3 (the origin of the x coordinate axis is at the center of the roughness element), respectively. No matter which one of the HRN k-ε turbulence model and LRN k-ε turbulence model is applied, the computed velocities are in good agreement with the measurements. The agreements of the calculated Reynolds stresses using the two models with the experimental data are generally good for the case of L/k=∞. For the case of L/k=16, the calculated Reynolds stresses obtained with LRN k-ε turbulence model agree with the experimental data, except for the under-prediction of about 15% in the vicinity of strip roughness; the HRN k-ε turbulence model gives high values, but the results are reasonably close to the experiment. The distributions of the Reynolds stresses $-\overline{uv}/U_m^2$ vary lineally far from the bed. The values of $-\overline{uv}/U_m^2$ near the bed change violently. This illustrates that stronger eddy disturbance occur in the vicinity of the roughness, especially, the variation of $-\overline{uv}/U_m^2$ for the case of L/k=∞ is remarkable.

Fig.4 shows the turbulent kinetic energy k/U_m^2 contours calculated with the LRN k-ε model. Because the computed range is different in one inner iteration, the starting values of the x-coordinates in Fig.4 are not same. It is found that the variation of k/U_m^2 in streamwise direction distributes over the region of y/h<0.2; the

Table 1 Hydraulic condition for calculations

Case	Relative spacing L/k	Relative depth h/k	Mean velocity (cm/s)	Reynolds number Re	Froude number Fr
Ra04	16	10	23.8	1.4×10^4	0.33
Ra07	8	10	23.2	1.3×10^4	0.32
Ra10	4	10	28.7	1.6×10^4	0.39
Ra13	2	10	28.1	1.5×10^4	0.39
Ra15	∞	16	28.7	2.3×10^4	0.32

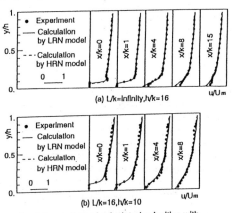

Fig. 2 Comparison of calculated velocities with experimental data for Ra15 and Ra04

values of k/U_m^2 is almost no change in streamwise direction for y/h>0.2. The maximum values of k/U_m^2 occur on nearby upstream corners of the roughness where flow impinges and a highly turbulent shear layer is generated except for the case of L/k=2, in which the normalized turbulent kinetic energy k/U_m^2 of the case of L/k=8 is the largest among the various cases. Moreover, with the decrease of the relative roughness spacing L/k for L/k ≤8, the values of the turbulent energy k/U_m^2 will swiftly decrease, and no apparent variation in the upstream and downstream sides of the roughness for the case of L/k=2 is presented; this phenomenon indicates that the dead water basins have been formed between the roughness elements.

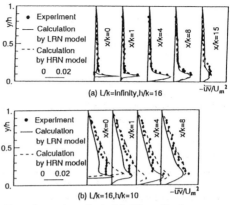

Fig. 3 Comparison of calculated Reynolds stresses with experimental data for Ra15 and Ra04

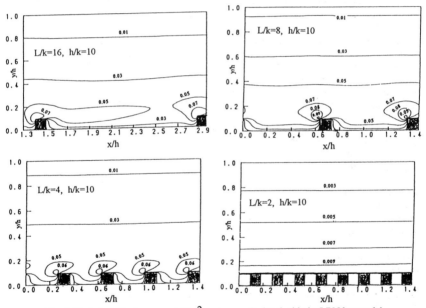

Fig. 4 Turbulent kinetic energy k/U_m^2 contours calculated with the LRN k-ε model

The distributions of the interval averaged Reynolds stresses $[-\overline{uv}]/U_m^2$ calculated with the LRN k-ε model for the cases of L/k≤16 are shown in Fig.5. The maximum and minimum values of $[-\overline{uv}]/U_m^2$ appear at L/k=8 and L/k=2, separately. This result is identical with the experimental data.

Fig.6 shows a comparison of the resistance coefficient f (=$8(u_*/U_m)^2$) computed with the two different k-ε models with the experimental data, together with previous data obtained by Tominaga et al.

(4) for the strip roughness with circular cross section of diameter ϕ=8mm for the relative water depth h/ϕ=10. The prediction of the f obtained with LRN k-ε model is in good agreement with the experiment, but the computed values are lager than the experimental data. The values of the f attain peak at the case of L/k=8. The varying tendency of the f for square roughness is in coincidence with the one for the circular roughness.

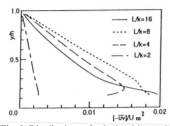

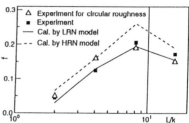

Fig. 5 Distributions of calculated interval averaged Reynolds stresses $[-\overline{uv}]/U_m^2$

Fig. 6 Comparison of calculated resistance coefficients with experimental data

CONCLUSIONS

Two different turbulence models have been successfully developed for the turbulent structure over strip roughness in open channel flow. Predictions for the velocities obtained with the two models are fairly good. The results of the Reynolds stresses and resistance coefficient with LRN k-ε model are better than the ones calculated with the standard k-ε model, which gives very large values. Further modification to turbulence models is needed to predict the Reynolds stress and resistance coefficient at the best.

From above-mentioned results, with the variation of roughness spacing, the parameters of the hydraulic characteristics in the vicinity of roughness vary distinctly. The turbulent structure over strip roughness in open channel flow turns gradually into that of smooth flow with the decrease of the relative spacing. The mixing effects of the flow grow stronger and stronger with the decrease of the roughness spacing.

REFERENCES

(1) Perry, A.E., Shofield, W.H. and Joubert, P. N.: Rough wall turbulent boundary layers, J.Fluid Mech., Vol.37-2, pp.383-413, 1969.

(2) Futuya, Y.: Miyata, M. and Fujita, H.: Turbulent Boundary.layer and Flow Resistance on plates rougened by wires, J. Fluid Eng. ASME, No,76-FE-6, pp.635-644, 1976.

(3) Antonia, R.A. and Luxton, R.E. : The response of a turbulent boundary layer to a step change in surface roughness, J.Fluid. Mech., Vol.48-4, pp.721-761, 1971.

(4) Tominaga, A. and Nezu, I.: Turbulent structure past strip roughness in open channel flows, 24th IAHR Congress Madrid, pp.c43-c50, 1991.

(5) Tominaga, A.: Effect of relative spacing of strip roughness on turbulent structure in open channel flows, Proc. Hydraul. JSCE Vol.36, pp.163-168, Feb. 1992 (in Japanese).

(6) Nezu, I. and Nakagawa, H.: Numerical calculation of turbulent open-channel flows by using a modified K-ε turbulence model, Proc. of JSCE, No.387/II-8, 1987 (in Japanese).

(7) Jones, W.P. and Launder, B.E.: The calculation of low-Reynolds number phenomena with a two-equation model of turbulence, Int. J. Heat Mass Transfer, Vol.16, pp.1119-1130, 1972.

(8) Launder, B.E. and Sharma, B.I.: Application of the energy dissipation model of turbulence to the calculation of flow near a spinning disc, Letters In Heat Mass Transfer, Vol.1, pp.131-138, 1974.

(9) Rodi, W. : Turbulence Models and Their Application in Hydraulics, IAHR, monogragh, Delft, the Netherlands, 1980.

(10) Handler, R. A., Swean, Jr., T. F., Leighton, R.I. and Swearingen, J.D.: Length seales and the energy balance for turbulence near a free surface, AIAA J., Vol.31, No.11, pp.1998-2007,Nov., 1993.

(11) Patankar, S.V.: Numerical Heat Transfer and Fluid Flow, Hemisphere, Washington, D.C., 1980.

1A25

3D TURBULENT SIDE DISCHARGE INTO OPEN CHANNEL FLOW

W. Czernuszenko[1], Yafei Jia, Sam S.Y. Wang
The University of Mississippi,
Center for Computational Hydroscience and Engineering
University MS 38677, USA

ABSTRACT

The three dimensional computational model (CCHE3D) has been applied to simulate the velocity field in the open channel flow near a side discharge. The interaction of the primary flow and the side discharge generates a large recirculation flow. The flow is described by set of equations: Navier-Stocks, continuity and k-ε equations. The equations are solved by using the SMART numerical procedure.

The numerical solution describes the three dimensional recirculation velocity field in the vicinity of the side discharge faithfully. The sizes of the recirculation zone (height and length) are correctly predicted in comparison with laboratory measurements. The numerical simulations have shown that the relationship between the recirculation zone shape factor and the momentum flux ratio is not universal as it was suggested by other investigators.

MATHEMATICAL MODEL

For steady flow situation, to which our attention is restricted, the equations governing the mass and momentum in tensor notation are as follows:

$$\frac{\partial U_j}{\partial x_j} = 0 \qquad (1)$$

---

[1]on leave The Institute of Geophysics, PAS, Warsaw, Poland

$$\frac{\partial U_j U_i}{\partial x_j} = \frac{\partial}{\partial x_j}[\nu_*(\frac{\partial U_j}{\partial x_i}+\frac{\partial U_i}{\partial x_j})] - \frac{1}{\rho}\frac{\partial p_*}{\partial x_i} \tag{2}$$

where: subscripts i and j stand for horizontal, vertical and lateral coordinates, also notation x, y and z is used, U_i - i-component of time averaged velocity, also for velocity components notation U, V and W is used, ν_* is the sum of laminar and turbulent viscosity and $p_* = p + \gamma y$.

The first equation presents the continuity equation for fluid with constant density but the second one - the momentum equation with assumption that the turbulent stresses are proportional to the mean velocity gradient (Boussinesq's eddy-viscosity concept).

TURBULENCE CLOSURE
The description of the turbulence viscosity is described by the ratio:

$$\nu_t = c_\mu \frac{k^2}{\epsilon} \tag{3}$$

Where the kinetic turbulence energy, k and the dissipation rate of turbulence energy, ϵ are calculated from the full, elliptic form of 3D transport equations [1].

BOUNDARY CONDITIONS
The boundary conditions must be specified at all boundaries because of the elliptic nature of the equations. All variables must be known at the first and last cross sections of the open channel as well as the bottom and side walls including the outlet of the side discharge channel.

The inflow plane

The semi-empirical relationships for turbulence intensities developed by Nezu and Rodi [6] were adopted for estimating the kinetic energy at the inlet cross-section. Turbulence eddy viscosity distribution is assumed to be the parabolic one and the turbulence energy dissipation rate ϵ were obtained from equation (3). This approach has been proven to be appropriate to produce accurate results [3].

The inlet section was located 14h (h-water depth) upstream of the discharge outlet. A transversal distribution of flow discharge per unit width was specified as a power function of distance from the wall with a power 1/7, and the vertical distribution of the longitudinal velocity component at the inlet section was assumed to be logarithmic.

The outlet cross-section and water surface

At the outlet cross-section longitudinal gradients of the variables u, v, w, p_*, k, ϵ, were set to be zero.

On the water surface $z=\eta$, the gradients in vertical direction for u, v, p_*, k, ϵ, were zero. The vertical velocity component can be obtained using the kinematic condition and the relationship $p_*=\rho g \eta$.

The rigid walls

In regions very close to the channel bed as well as the vertical walls, the law of the wall approach was applied to provide approximations of shear stress, and turbulence properties k and ϵ. The wall boundary values of k and ϵ were estimated assuming the local equilibrium of turbulence energy

$$k = \frac{u_*^2}{C_\mu^{1/2}}, \qquad \epsilon = \frac{u_*^3}{\kappa y} \qquad (4)$$

At the discharge channel, boundary values for k and ϵ were specified in the same way as Demuren and Rodi [1].

$$k_d = 0.004\,V_o^2, \qquad \epsilon_d = C_\mu^{3/4}\,\frac{k_d^{1.5}}{0.09b} \qquad (5)$$

NUMERICAL APPROACH

The model equations were solved using the control volume method and SIMPLE procedure [8]. Staggered grid and ADI method were adopted in computation. Special tests were performed for choosing a numerical scheme to solve the advective part of the equations. Numerical schemes: HYBRIB, QUICK and SMART were considered. HYBRID scheme shows unacceptable numerical diffusion. SMART scheme avoided the oscillating nature of QUICK when concentration gradient is large and preserved high order of accuracy for moderate variation of concentration.

Tests for transport of a concentration block and a Gaussian cone showed that the SMART scheme is good in handling shock wave fronts and it has no problem of oscillation and predicting negative concentration [4]. These comparisons confirmed the superiority of SMART and it is therefore adopted to the CCHE3D code to simulate non-linear three dimensional problems of side discharge to open channels.

COMPUTATION RESULTS AND DISCUSSION

The mathematical model described in previous sections had been verified with experimental measurements conducted by Rodi and Weiss [8] and Mihail et al [5].

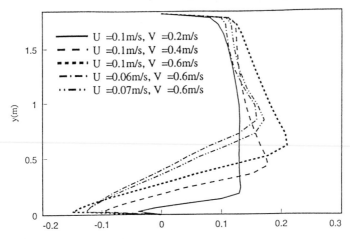

Figure 1. Calculated lateral distribution of velocity u in m/s at water surface across center of recirculation zone

Fig.1 shows the distribution of the longitudinal component of velocity (u) on the water surface along the sections cross the center of the recirculation. It is easy to note the influence of the side discharge on the lateral velocity distributions. When the discharge velocity is larger, the location of the maximum velocity is farther from the discharge wall. Velocity near the opposite wall also increases in this case.

In general, all the numerical simulations are in good agreement with the data, except for the rough bed channel at low side discharge which did not show any recirculation whereas the laboratory experiments [8] had a very small one. The sizes of the recirculation eddy predicted by the model were the same as those measured. All the simulated eddies have only a single cell, however, those in the physical model are usually more complicated and have more than one cell. Handling these small eddy structures may need extra fine grid and more complicated turbulence model.

All measured data and computational results for smooth bed are plotted in Fig.2. It can be seen that the characteristics of the recirculation eddy observed in physical modeling are faithfully being reproduced by the numerical model. There is one

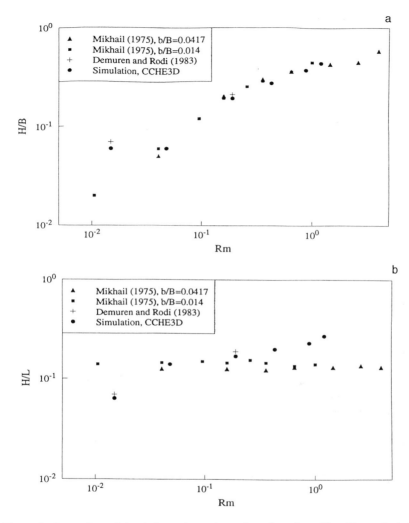

Figure 2. Comparison of simulation and experimental results, where: H and L are the height and length of the recirculation eddy, B is the width of main channel and R_m is the momentum flux ratio

pair of data (measurement and calculation) representing the case with a very wide discharge channel. This pair of data does not follow the main trend which is

characterized by narrow (point) discharge channel. Ratio H/B grows with the momentum flux ratio (discharge flow to channel flow) R_m almost linearly and the growth slows down gradually with higher value of R_m.
This is perhaps caused by the influence of the opposite wall: the larger the recirculation height, the stronger the constrain from the opposite wall would be. It can be seen that the wall constrain effect starts at about $R_m = 0.2$.
Mikhail et al [5] suggested that the eddy shape factor, H/L, would remain a constant 0.13 for all the values of R_m (Fig.2b). However the numerical simulation results show that this relationship is not universal: H/L increases with R_m. The shape factor might be related also to other hydraulic conditions in addition to R_m.
In order to fully verify the applicability of the correlations in Fig.2 to general practices, further research work are needed including both physical experiments and numerical simulation.

REFERENCES

[1]. Demuren, A.O., and Rodi, W., "Side Discharge into Open Channels: Mathematical Model", ASCE. Journal of Hydraulic Engineering, Vol. 109, No. 12, 1983, pp 1707-1722.

[2]. Gaskell, P.H., and Lau, A.K.C., "Curvature-Compensated Convective Transport: SMART, A New Boundedness-preserving Transport Algorithm", International Journal for Numerical Methods in Fluids, Vol. 8, 1988, pp 617-641.

[3]. Jia, Y. and Wang, S.Y., "3-D numerical simulation of flow near a spur dike", Proceedings, Advances in Hydroscience and Engineering, Vol. 1, part 2, 1993, pp 2150-2156.

[4]. Jia, Y., Czernuszenko W. and Wang S.Y.,"Simulation of 3D Side Discharge into Open Channel", CCHE Report, June 1994, University of Mississippi.

[5]. Mikhail, R., Chu, V.H., and Savage, S.B.,"The reattachment of the two-dimensional turbulent jet in a confined cross flow", Proceedings, 16th International Association for Hydraulic Research Congress, Sao Paulo, Brazil, Vol. 3, 1975.

[6]. Nezu, I., and Rodi, W., "Open-Channel Flow Measurements with a Laser Doppler Anemometer", Journal of the Hydraulic Engineering, ASCE, 1986, Vol. 112, No. 5, pp. 335-355.

[7]. Patankar, S.V.,"Numerical heat transfer and fluid flow, Hemisphere Publishing Corporation.

[8]. Rodi, W., and Weiss, K.,'Laboruntersuchunngen des nahfelds von Warmwassereinleitungen in rechteckgerinne", Report Nr. 574, Institute for Hydromechanics, University of Karlsruhe, W. Germany, 1980.

1A26

Statistical model of turbulent nature flow

E.N. DOLGOPOLOVA
Water Problems Institute, Russian Academy of Sciences, Moscow, Russia

INTRODUCTION

The characteristic property of turbulence is the chaotic fluctuations of velocity in given point. The fluctuations of all the three components of velocity can not be denoted random processes in the strikt sense. But we can state with assurance that turbulent movement is an undeterministic process and it ought to be considered by application of statistic approach [6]. For wide class of flat rivers the ratio of depth h to width B much less than 1. It allows one to assume the flow in river to be two-dimensional when the processes near banks are of no concern. The proposed model was formed for plane flow.

MEAN VELOCITY PROFILE

Let us consider an open nature flow to be related to cartesian coordinate system: x- streamwise direction, y - vertical direction (y=0 - bottom), z - horizontal lateral direction. The time averaged components of velocity corresponding these directions are u, v, w, and the fluctuations of these components are u', v', w'. The analysis of experimental results which were obtained in flumes, channels and rivers [1, 3, 8] shows that the mean velocity profile can be described with the help of power law [9]:

$$ u = (1 + \alpha)\overline{u}\,\eta^{\alpha} \qquad (1) $$

where $\overline{u}$ - everage vertical line velocity; $\eta = y \ / \ h$, h - depth of the flow; α - velocity profile exponent.

The origin of coordinates y=0 is chosen at the level of sand dunes crests on the bottom. Zero values of mean velocity and maximum of velocity fluctuations are consistent with this horizon in the flow. In such reference some information about change of mean velocity profile along a bottom dune can be lost, but this approach enables us to develope general description of the open flow.

The exponent α in (1) is related to Darcy-Weisbach coefficient f as follows [4]:

$$ \alpha = \sqrt{f} \ / \ k\sqrt{m} \qquad (2) $$

where k-Karman's constant, m - coefficient of cross-sectional shape.For plane flow (h/B<<1) m=2.

STANDARD DEVIATIONS AND CORRELATON FUNCTIONS

The distribution of standard deviations of fluctuating velocity components along the depth can be described by [9]:

$$\sigma_i \, / u_* = a_i + b_i \sqrt{\eta} \qquad (3)$$

where $\sigma_1 = \sqrt{\overline{u'^2}}$, a_i, b_i - empirical coefficients; u_* -shear velocity; $u_* = k \; \alpha u$

Coefficients a_i and b_i were obtained in [9] with the help of a great body of experimental results as follows: $a_1 = 2.1$, $b_1 = -1.2$, $a_2 = 1.3$, $b_2 = -0.6$, $a_3 = 1.7$, $b_3 = -1.0$.

Different authors suggest various relationships for change of turbulent intensity along the depth. For example, in [10] there were considered logarithmic and exponential expressions for turbulent intensity. These dependences are in good correspondence with experimental data as (3) is.The advantages of the expression (3) are: 1. The (3) holds for three components of the flow velocity, the empirical coefficients being found for the whole of the components; 2.The correlation coefficient for streamwise and vertical velocity components r_{12} calculated with the help of (3) is in good agreement with experimental data. Using the expression for shear stress [9]:

$$\overline{u'w'} \, / u_*^2 = 1 - \eta \qquad (4)$$

we obtain for r_{12} :

$$r_{12} = (1 - \eta) \Big/ \big(a_1 + b_1 \sqrt{\eta}\big)\big(a_2 + b_2 \sqrt{\eta}\big) \qquad (5)$$

The autocorrelation function of fluctuations of longitudinal velocity component can be described by the accepted expression:

$$r(\tau) = \exp(-\tau/\theta) \qquad (6)$$

where τ- time interval; θ- Eulerian scale of turbulence.

SCALES OF VELOCITY FLUCTUATIONS AND TURBULENT MIXING COEFFICIENTS

The Eulerian time scale can be easily defined from experiment in flumes and in rivers. The length scale L can be estimated with the help of Taylor's hypothesis.

The range of length scales measured in rivers is - 0.5h ÷ 3h. As the length scale is connected with the boundary layer which can be less then the depth of the nature flow we use for θ the following expression:

$$\theta = 0.5h/\overline{u} \qquad (7)$$

The (7) was confirmed by our results of measurements in rivers [9].

To calculate the turbulent mixing coefficients ε_i we use Taylor's relationship:

$$\varepsilon_i = \sigma_i^2 T \qquad (8)$$

here T - Lagrangian time scale of velocity fluctuations. Since it is impossible to measure Lagrangian characteristics of the flow in nature we have to derive them from Eulerian characteristics. T may be found with the help of following relationships:

$$r_E(\tau) = r_L(\beta\tau) \ [2], \ \ \overline{u}\theta = d\overline{\sigma_1}T, \ \ \beta\theta = T \ \ , \ \ d=2.5 \quad (9)$$

Using the relationships for u_* and σ_i and averaging u and σ_1 over the depth we derive:

$$\overline{\sigma_1} = 1.4 k\alpha\overline{u} \qquad (10)$$

The expressions (9) and (10) give us the relationship between Eulerian and Lagrangian scales in plane flow:

$$T = \theta/1.4\alpha \qquad (11)$$

To estimate Lagrangian scale we can use the range of change of α in rivers, that is 0.1 ÷ 0.2. In addition, for longitudinal component of fluctuating velocity it was found in flume by visualization [5]: $T = 4\theta$ .It confirms the possibility of estimation of T by (11). Thus using (11) we derive depth-averaged mixing coefficients:

$$\overline{\varepsilon_i} = c_i u_* h \qquad (12)$$

where c_i - empirical coefficients. The calculated value of c_3 is 0.18, which is in sufficiently good agreement with experimental result c_3=0.21. The suggested relationships were used in order to calculate the propagation of impurity over the surface of the river. The fundamental characteristics of the river are: B = 17m, h = 0.5m, $\overline{u}$ = 0.5m, Re = 2.5 $\cdot 10^4$, Fr = 5 $\cdot 10^{-2}$. The floats with neutral buoyancy were used as an impurity. The density function of floats was calculated with the help of suggested expressions for plane flow. The minor discrepancy between experimental and probability densities can be explained by the nonuniformity of the real river flow.

DIFFUSION OF IMPURITY IN INHOMOGENEOUS FIELD OF VELOCITY

The experiment was carried out at the River Volga. The length of the section was 1 km and the mean depth was 4 m. At the place of measurements the channel line goes from the left bank to the right one causing the inhomogeneity of the flow. The results of the measurement showed that the impurity blot stretched in x-direction and was approximately unchanged in lateral direction. The mechanism of the impurity propagation can be explained as follows: all water particles have mean velocity component directed to the right bank; the particles deviating from the center line to the left move faster, and to the right - slower. The turbulent diffusion in this case was modelled by Monte-Carlo method. In the test sites there were measured and calculated the values of $\sigma_x, \sigma_z, \bar{z}$. The comparison of results of measurement and the results of simulation of trajectories of particles by method of random walks shows that these results are in good correspondence. This fact demonstrates that the suggested statistical model of turbulent flow does work and the supposed mechanism of propagation of impurity blot in inhomogeneous field of velocity is correct.

DISTRIBUTION OF VELOCITY FLUCTUATIONS

The distribution of velocity fluctuations of the longitudinal component of the velocity can be described (as a first approximation) with the help of first four statistical moments: mean velocity u, variance σ^2, skewness s and kurtosis k. Analysing the results of measurements of u' in flumes and rivers [3] we can suggest the following distribution of s and k along the depth of the flow:at $\eta \in [0, 0.2)$ - $s = 0.28$, $k \in [2.3, 2.7]$; at $\eta \in [0.2, 0.6)$ - $s = 0$, $k = 2.7$, at $\eta \in [0.6, 1.0]$ - $s = 0$, $k = 3$. Using the Pearson's equation for probability density function of velocity fluctuations $f(u')$ [7]:

$$\frac{df}{du'} = \frac{(u' - \delta_1) f}{\delta_0 + \delta_1 u' + \delta_2 u'^2},\qquad(13)$$

one can calculate $f(u')$, the coefficients $\delta_0, \delta_1, \delta_2$ in (13) being derived from the statistical moments of the flow velocity fluctuations. As a result of analysis of $f(u')$ for three layers of the flow there was obtained [3] that for bottom layer the distribution of the velocity fluctuations considerably differs from the Gaussian one, and probability density function $f(u')$ is the function of beta distribution. At $\eta \in [0.2, 1.0]$ $f(u')$ is practically normal. This research enables one to model the initial movement of particles elevating above the bottom by method of random walks.

CONCLUSIONS

The model of turbulent plane flow in nature enables one to calculate the most of fundamental characteristics of the river including the coefficients of turbulent diffusion. The detailed research of distribution of fluctuations of velocity longitudinal component confirms the opportunity for using the Gaussian law in order to calculate the propagation of impurity over the surface of the flow. The investigation of probability density function f(u') inside the bottom layer showed that it is the function of beta distribution. Thus turbulent fluctuations of velocity longitudinal component in this region make the elevation of solid particles from the bottom easier. This fact should be taken into account when one considers the beginning of movement of sediment particles.

REFERENCES

1.Batchelor G.K. An Introduction to Fluid Dynamics. Cambridge, University Press 1970, p.757.

2.Czernuszenko W., Lebiecki P. Turbulence and diffusion in open channel flow. Archiwum hydrotechniki, vol. 32, N 3/4, 1985, pp.377-391.

3.Dolgopolova E.N., Orlov A.S. Estimation of the distribution of the longitudinal component of the velocity fluctuations of the channel flow. Water resources N 2, 1989, pp.85-90.

4.Dolgopolova E.N. The dependence of Darcy-Weisbach resistance coefficient on the shape of river bed. Proceedings of the Int. Symp. "Runoff and Sediment Yield Modelling", Warsaw, Poland, Sept.14-16, 1993, pp.285-290.

5.Fidman B.A., Lyatcher V.M. Investigation of turbulence by photo- and cine-filming. Collected vol.: Dynamics and Thermics of River Flow, "Science", Moscow pp. 109-125 (in Russian).

6.Frost W., Moulden T.H. Handbook of Turbulence. vol.1 Fundamentals and Applications, PLENUM PRESS, New York and London, 1977, p.535.

7.Kendall M.G., Stuart A. The Advanced Theory of Statistics, vol.1 - Distribution Theory, London, 1962.

8.McQuivey R.S. Summary of turbulence data from rivers, conveyance channels and laboratory flumes. Geol.Surv.Prof.Paper N 802 B, 1973.

9.Orlov A.S., E.N. Dolgopolova, Debolsky V.K. Some empirical relationships of river turbulence. Water Resources, N 6, 1985, pp. 85-90

10.Wang J., Dong Z., Chen C., Xia Z. The effects of bed roughness on the distribution of turbulent intensities in open-channel flow. J. of Hydraulic Research, vol. 31, N 1, 1993, pp.89-96.

1A27

Coherent Structure near The Side-wall in Open Channel Flow

TAISUKE ISHIGAKI and HIROTAKE IMAMOTO
Associate Professor   Professor
Ujigawa Hydraulics Laboratory, D.P.R.I.
Kyoto University, Kyoto, Japan

KOJI SHIONO
Associate Professor
Department of Civil Engineering
Loughborough University of Technology
Loughborough, UK

ABSTRACT

Velocity, boundary shear stress and flow structure near the corner in rectangular open channels are investigated. Boundary shear stress was estimated from velocity distributions obtained by LDA measurements and by using Preston tube, and also calculated by Naot–Rodi model. Flow structure near the side–wall was visualized by the hydrogen bubble method. Results shows the secondary flow cells closely related to the distribution of boundary shear stress and the coherent structure in the wall region of the side–wall.

INTRODUCTION

Flow structure near the corner in an open channel flow is three dimensional (Imamoto etc. 1993) and modifies the velocity and boundary shear stress distributions. The modifications are connected with secondary flow cells or vortex motions in the region by many researchers. For instance, Nezu and Nakagawa (1985) explained the relation between the secondary flows and the wavy distribution of boundary shear stress on the channel bed near the corner by using their precise data obtained by LDA measurements. And they also showed the parabolic distribution on the side–wall of channel and pointed out that the mean value was decreasing in accordance with the increase of the aspect ratio of channel section. However, the flow structure near the side wall has not yet been fully explained, it is discussed in this paper by using the data of velocity and boundary shear stress measurements and flow visualization.

Table 1 Hydraulic conditions

Case	B (cm)	H (cm)	Q (l/s)	I	Re	Fr
B/H= 5 (LDA)	20.0	3.9	3.97	1/1480	9300	0.42
B/H=10 (PT)	40.0	4.0	3.12	1/680	9300	0.69
B/H= 1 (FV)	4.0	4.0	0.40	1/600	10300	0.40
B/H= 2 (FV)	8.0	4.0	0.79	1/840	10200	0.39
B/H= 3 (FV)	12.0	4.0	1.20	1/890	10300	0.40
B/H= 5 (FV)	20.0	4.0	2.02	1/1140	10500	0.40
B/H= 7 (FV)	28.0	4.0	2.80	1/1230	10300	0.40
B/H=10 (FV)	39.0	4.0	3.90	1/1300	10300	0.40

LDA:Laser doppler anemometer, PT:Preston tube, FV:Flow visualization

EXPERIMENTAL METHODS

The experiments were carried out in an open channel with glass walls. The channel was 8 m long, 40 cm wide and 23 cm deep. The measurements were performed at 2.5 m upstream from the channel end on the hydraulically smooth bed. The channel width was changed by setting a glass wall in the channel. Longitudinal and vertical components of velocity were measured by a two color 2 W Argon Laser system, and lateral component was measured by a 15 mW He–Ne Laser system. The boundary shear stress was figured out from the velocity distribution with the logarithmic velocity distribution in which Karman constant κ was 0.41. The boundary shear stress was also measured with a Pitot tube (outer diameter 3 mm, inner diameter 1.5 mm) as a Preston tube, and also calculated by the algebraic stress model which was developed by Naot and Rodi(1982). The calibration curve for the tube was in good agreement with Patel's curve (1965), therefore, we used his curve as the calibrating curve (Imamoto & Ishigaki 1989).

To visualize the flow in the wall region of side wall, we used the hydrogen bubble method. Platinum wire with the diameter of 0.05 mm was set along the side wall of channel at the height of 0.5 mm which is in the lower part of the buffer layer. Time lines of hydrogen bubbles were produced at each 0.05 second, and recorded by a 35 mm still camera or a video camera. By using a computer-aided picture processing system (Imamoto & Ishigaki 1990) the velocity distribution at each 0.1 second was obtained. The velocity was figured on the distance of the adjacent time lines. In experiments, using a straight open channel of rectangular cross-section, 13 m long, 39 cm wide and 19.5 cm depth, we set a partition in the channel and changed the aspect ratio of channel section between 1 to 10. All experiments were carried out in fully developed turbulent flow over smooth bed. Hydraulic conditions as shown in Table 1.

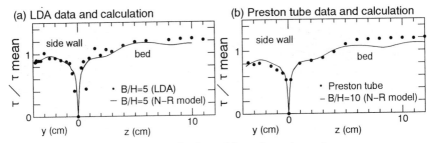

Figure 1 Distribution of boundary shear stress

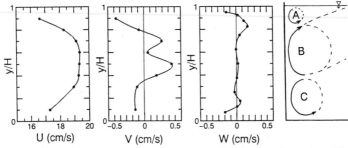

Figure 2 Distributions of three components of velocity along the side–wall and secondary flow cells.

BOUNDARY SHEAR STRESS AND VELOCITY DISTRIBUTIONS

Distributions of the boundary shear stress, τ, normalized by the mean value, $\tau_{mean} = \rho\,gRI$, are shown in Figure 1 in which both measured and calculated distributions are illustrated. Distribution on the bed is wavy, the results being similar to those reported by Nezu and Nakagawa (1985). Distribution along the side–wall is parabolic, showing peak around the half of the water depth. These distributions are considered to be related to the three dimensional structure of the flow. The point of down–ward secondary flow to the bed or the side–wall is closely related to the point where the boundary shear stress shows maximum value, and the point of up–ward flow from the wall coincides with the minimum point of the shear stress. Measured and calculated data shows good agreement with each other.

Figure 2 shows the relation between the distributions of three velocity components along the side–wall and secondary flow cells. U is the longitudinal component of velocity, and V the vertical, W the lateral. Positive direction of V

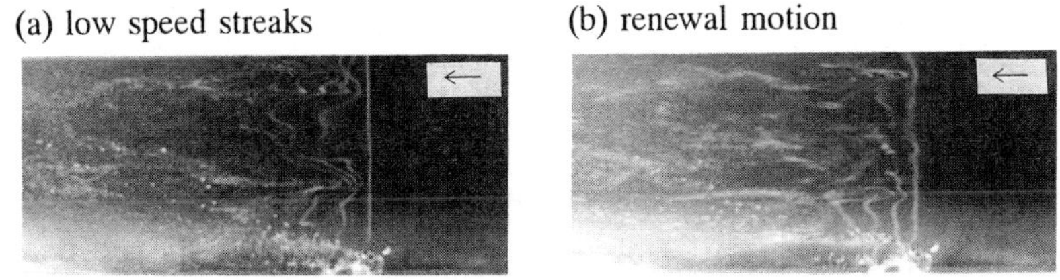

(a) low speed streaks

(b) renewal motion

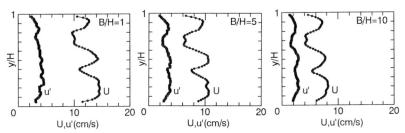

Photo. 1 Example photographs of time lines near the side–wall
(B/H=5, z=0.5mm, z⁺=7.4)

Figure 3 Mean velocity U and turbulence intensity u' along the side–wall
obtained by the hydrogen bubble method.

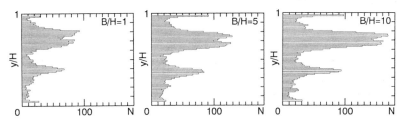

Figure 4 Distribution of frequency N of the position where the velocity is
lower than 60 percents of mean value along the side–wall.

is up–ward, and the right–ward direction is positive for W. To consider the three
distributions, three secondary flow cells could be illustrated as shown in the
figure. These cells affect the distribution of boundary shear stress.

COHERENT STRUCTURE NEAR THE SIDE–WALL
Time lines obtained by the hydrogen bubble method extremely close to the side–
wall indicate a wavy distribution of velocity. Two low speed streaks can be
detected easily as shown in Photo. 1 (a), and they are flushed out periodically
as shown in Photo. 1 (b). This is the bursting phenomena as same as those

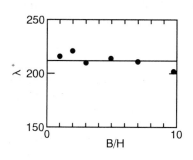

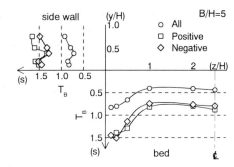

Figure 5 Normalized spacing of low–speed streaks λ^+ and aspect ratio of channel section.

Figure 6 Distribution of bursting period along the perimeter by VITA method (k=0.4, T^+=45)

observed in the wall region by many researchers, however, these low speed streaks present almost the same positions and the renewal period observed in the record of video tapes is longer.

Distribution of mean velocity, U, and turbulence intensity, u', for three kinds of channel are shown in Figure 3. Two low speed parts are clearly distinguished in any cases. The position where the velocity is lower than 60 percents of mean value along the side–wall are counted and illustrated in Figure 4. It shows that there are two peaks at y/H=0.37 and y/H=0.73 of relative depth. These positions are nearly the same of up–ward and down–ward positions of secondary flow in Figure 2. This implies that the characteristics of secondary flow cells relate these phenomena. Normalized spacing of the two low speed parts, $\lambda^+=u.L_y/\nu$, is almost constant, λ^+=210, as shown in Figure 5, and it is two times of the value, λ^+=100, in previous works for the wall turbulence. Where u. is the friction velocity, L_y the vertical spacing and ν the kinematic viscosity.

To evaluate the bursting period, VITA (variable–interval time–averaging technique) method (Blackwelder & Kaplan 1976) was used. The period obtained by the method is depend on the constant k and the normalized averaging time T^+, where k is constant for defining the threshold for VITA variance, u'^2, as ku'^2, and T^+ is defined as $T^+=T_m u.^2/\nu$, T_m is the averaging time. According to the work by Komori etc. (1989), we used the value of 0.4 for k and 45 for T^+. Figure 6 shows the distribution of the bursting period, T_B, obtained by VITA method with 102,400 data for 512 seconds at each point of which distance from the wall is 5 mm. The period of all events, and of the positive gradient du/dt>0, and of the negative gradient du/dt<0 are shown in the figure. The period on the side–wall is about two times longer than that of the channel center. This implies that the longitudinal scale of coherent structure near the side–wall is different from the scale near the center of channel.

CONCLUSIONS

The relation between the distribution of boundary shear stress and secondary flow and the bursting motion near the side–wall of channel were investigated in several rectangular open channels of different aspect ratio of cross–section. The following results were obtained:

1) There are three secondary flow cells near the corner and they affect the distribution of boundary shear stress. From the result it is considered that the characteristics of the cells relate to the bursting phenomena on the side–wall.

2) Two low speed streaks are observed in the buffer layer on the side–wall, and they present almost the same position. Normalized spacing is larger and the renewal period is also longer than those in previous works for the wall turbulence. This implies that the longitudinal scale of coherent structure near the side–wall is larger than the scale near the center of channel.

REFERENCES

Blackwelder, R.F. & Kaplan, R.E. 1976 On the wall structure of the turbulent boundary layer, J. Fluid Mech. 76, 89–112.

Imamoto, H. & Ishigaki, T. 1989 Turbulence, secondary flow and boundary shear stress in a trapezoidal open channel, Proc. of 23rd IAHR, A, 23–30.

Imamoto, H. & Ishigaki, T. 1990 Visualization of secondary flow in a compound open channel, Proc. of 7th APD–IAHR, Vol. III, 485–490.

Imamoto, H., Ishigaki,T. & Shiono, K. 1993 Secondary flow in a straight open channel, Proc. of 25th IAHR, Vol. I , 73–80.

Komori, S., Murakami, Y. & Ueda, H. 1989 The relationship between surface–renewal and bursting motions in an open–channel flow, J. Fluid Mech., 203, 103–123.

Naot, D. & Rodi, W. 1982 Calculation of secondary currents in channel flow, J. ASCE, HD, Vol.108, HY8, 948–968.

Nezu, I. & Nakagawa, H. 1985 Three–dimensional turbulent structure (Longitudinal vortex) in open channel flow and effect of free surface on its structure, Annuals of D.P.R.I., Kyoto Univ., No.28B–2, 499–522.

Patel, V.C. 1965 Calibration of the Preston tube and limitations on its use in pressure gradient, J. Fluid Mech., Vol.23, 185–208.

1A28

A COMPUTER CODE FOR TURBULENCE CALCULATIONS

Giancarlo ALFONSI, Giuseppe PASSONI, Lea PANCALDO
and Domenico ZAMPAGLIONE
Dipartimento di Ingegneria Idraulica, Ambientale e del Rilevamento
Politecnico di Milano, Piazza L. da Vinci 32, 20133 Milano, ITALY

ABSTRACT
A computer code for the numerical integration of the Navier-Stokes equations in three dimensions, is presented. The system of the governing equations in non-dimensional form of the time-dependent flow of a viscous incompressible fluid in a channel, is handled - following the Direct Simulation approach - by means of mixed technique, a spectral-finite difference numerical scheme; the elliptic pressure problem, cast in the Helmholtz form, is solved with the use of a cyclic reduction procedure. Results are presented mainly to demonstrate the ability of the model in performing rather accurate calculations. The availabilty of computer codes with such characteristics is of primary importance for the analysis of any modified version of the system of non-linear partial differential equations of viscous fluid flow for turbulence calculations, including Large Eddy Simulation models.

1. INTRODUCTION
The contemporary development of advanced numerical techniques and the simultaneous availability of fast digital computers, have given the possibility of integrating the system of non-linear partial differential equations of viscous fluid flow, with increasing accuracy and decreasing CPU times. These technical progresses have offered new instruments to the investigation of one of the most demanding problems in fluid sciences, the calculation of turbulent flows. In a preceding work ([1]), the methods that have been followed in the past to model turbulent flows, have been critically reviewed; starting from the so-called Reynolds averaging procedure and the problem of the closure of the Navier-Stokes equations, the different ways reported in the literature to appropriately model the turbulent components of the velocity vector, have been analyzed. Recent times have been characterized by the advent of new approaches in

turbulence research; in particular, Full Turbulence Simulation and Large Eddy Simulation ([1]) represent advanced computational techniques, both of them following the so-called Direct Turbulence Simulation approach, which basically consists in abandoning any time averaging procedure, in order to avoid all the related closure problems.

In this work, a new computer code for the numerical integration of the Navier-Stokes equations, is presented; the results that have been reported are mainly directed in demonstrating the reliability of the code in performing accurate calculations. The aim of the whole research is twofold: i) to provide a computer code for the fast and accurate numerical integration of the three-dimensional, time-dependent system of the non-linear partial differential equations of viscous fluid flow, ii) to utilize an appropriate computing tool for the execution of extended numerical simulations of high Reynolds numbers flow cases, by following the Large Eddy Simulation approach. In § 2, the numerical techniques that have been used in building the code, are outlined; the program uses a mixed spectral-finite difference algorithm in its time advancement portion, while a cyclic reduction procedure is used for the pressure problem; § 3 contains the results of the first runs executed by the code, while in § 4 the concluding remarks are drawn.

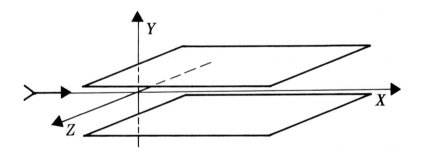

Figure 1. Calculation domain.

The problem that has been considered is the time-dependent flow of a viscous incompressible fluid in a three-dimensional channel, characterized by two solid walls at $y = \pm 1$ respectively (fig. 1); this problem ([2],[3], among others) is a reference case in turbulence research and offers the possibilty of comparing different results of both numerical and experimental nature.

2. NUMERICAL TECHNIQUES

The system of non-linear partial differential equations — in non-dimensional form and index notation — governing the flow of a viscous incompressible fluid, is considered:

$$\partial_i V_i = 0 \tag{1a}$$

$$\partial_o V_k + \partial_i (V_k V_i) = -\partial_k p + \frac{1}{Re} \partial_i \partial_i V_k \tag{1b}$$

$(i, k = 1, 2, 3)$, where the subscript "o" denotes the partial derivative with respect to time and the coordinates x_1, x_2, x_3 become x, y and z, with velocity components u, v and w respectively (Re is the Reynolds number). For the numerical integration of system (1), the following procedure is implemented ([4] among others). By admitting the flow fields to be periodic in the x and z directions, the equations are expanded in Fourier series, obtaining:

$$ik_x \hat{u} + \frac{\partial \hat{v}}{\partial y} + ik_z \hat{w} = 0 \tag{2}$$

$$\frac{\partial \hat{u}}{\partial t} + ik_x(\widehat{u^2}) + \frac{\partial(\widehat{uv})}{\partial y} + ik_z(\widehat{uw}) + ik_x p = -\frac{1}{Re} k^2 \hat{u} + \frac{1}{Re} \frac{\partial^2 \hat{u}}{\partial y^2} \tag{3a}$$

$$\frac{\partial \hat{v}}{\partial t} + ik_x(\widehat{vu}) + \frac{\partial(\widehat{v^2})}{\partial y} + ik_z(\widehat{vw}) + \frac{\partial p}{\partial y} = -\frac{1}{Re} k^2 \hat{v} + \frac{1}{Re} \frac{\partial^2 \hat{v}}{\partial y^2} \tag{3b}$$

$$\frac{\partial \hat{w}}{\partial t} + ik_x(\widehat{wu}) + \frac{\partial(\widehat{wv})}{\partial y} + ik_z(\widehat{w^2}) + ik_z p = -\frac{1}{Re} k^2 \hat{w} + \frac{1}{Re} \frac{\partial^2 \hat{w}}{\partial y^2} \tag{3c}$$

where the superscript "$\wedge$" denotes the Fourier transformed variables and $k^2 = k_x^2 + k_z^2$. By incorporating the convective terms and the diffusive terms along x and z in the definition of the following further expressions:

$$C_u = ik_x(\widehat{u^2}) + \frac{\partial(\widehat{uv})}{\partial y} + ik_z(\widehat{uw}) + \frac{1}{Re} k^2 \hat{u} \tag{4a}$$

$$C_v = ik_x(\widehat{vu}) + \frac{\partial(\widehat{v^2})}{\partial y} + ik_z(\widehat{vw}) + \frac{1}{Re} k^2 \hat{v} \tag{4b}$$

$$C_w = ik_x(\widehat{wu}) + \frac{\partial(\widehat{wv})}{\partial y} + ik_z(\widehat{w}^2) + \frac{1}{Re}k^2\widehat{w} , \qquad (4c)$$

a mixed algorithm for the time advancement is devised: a fourth-order Runge-Kutta scheme is used for the convective terms and for the diffusive terms along x and z, while a second-order implicit Crank-Nicolson procedure is implemented to handle the diffusive terms along y. The following equations hold for each Fourier mode:

$$\widehat{u}\,(t+\Delta t) = \widehat{u}\,(t) - \tfrac{1}{6}\,\Delta t\left(C_{u_1} + 2C_{u_2} + 2C_{u_3} + C_{u_4}\right) +$$

$$\frac{1}{Re}\,\Delta t\,\tfrac{1}{2}\left(\frac{\partial^2\widehat{u}(t+\Delta t)}{\partial y^2} + \frac{\partial^2\widehat{u}(t)}{\partial y^2}\right) - ik_x\phi \qquad (5a)$$

$$\widehat{v}\,(t+\Delta t) = \widehat{v}\,(t) - \tfrac{1}{6}\,\Delta t\left(C_{v_1} + 2C_{v_2} + 2C_{v_3} + C_{v_4}\right) +$$

$$\frac{1}{Re}\,\Delta t\,\tfrac{1}{2}\left(\frac{\partial^2\widehat{v}(t+\Delta t)}{\partial y^2} + \frac{\partial^2\widehat{v}(t)}{\partial y^2}\right) - \frac{\partial\phi}{\partial y} \qquad (5b)$$

$$\widehat{w}\,(t+\Delta t) = \widehat{w}\,(t) - \tfrac{1}{6}\,\Delta t\left(C_{w_1} + 2C_{w_2} + 2C_{w_3} + C_{w_4}\right) +$$

$$\frac{1}{Re}\,\Delta t\,\tfrac{1}{2}\left(\frac{\partial^2\widehat{w}(t+\Delta t)}{\partial y^2} + \frac{\partial^2\widehat{w}(t)}{\partial y^2}\right) - ik_z\phi \qquad (5c)$$

which result in a system of linear algebric equations with tridiagonal matrix of coefficients; $C_{u_{1,2,3,4}}$, $C_{v_{1,2,3,4}}$ e $C_{w_{1,2,3,4}}$, are the expressions of C_u, C_v e C_w, for what the implementation of the fourth-order Runge-Kutta scheme is concened, while ϕ is the integral of the - Fourier transformed - pressure at each time step. In particular, the evaluation of ϕ is performed in three steps (index l), according to the expression ([4]):

$$\phi_l = \psi_l - \tfrac{1}{2}\,\frac{1}{Re}\,\Delta t_l\,\frac{\partial^2\psi_l}{\partial y^2} \qquad (6)$$

and invoking mass conservation (ψ is an additional intermediate variable and the superscript "*" denotes intermediate values of the velocities):

$$\frac{\partial^2 \psi_l}{\partial y^2} - k^2 \psi_l = ik_x u_l^* + \frac{\partial v_l^*}{\partial y} + ik_z w_l^* . \tag{7}$$

No-slip boundary conditions have been imposed at the solid walls, while along the other directions periodic boundary conditions have been implemented.

3. RESULTS

In this section the first calculations - still at low values of the Reynolds number - performed by the code are presented; in order to reduce computational errors ([4]), the velocity has been considered as the sum of a mean component and a perturbation, where the mean component is defined as to obey the low-Reynolds-number approximation of the momentum equations.

In figure 2 calculations at Reynolds number 1000 are shown: an initial condition in the velocity of parabolic type is given to the program and this profile remains stable through several time steps (non-dimensional times $t = 5$ and 10). Figure 3 reports calculations also performed at $Re = 1000$. An initial condition of the type $u = 0.75(1 - y^8)$ is given to the program and the computations are advanced in time; the resulting velocity profile clearly reveals its evolution towards the theoretical profile $u = (1 - y^2)$.

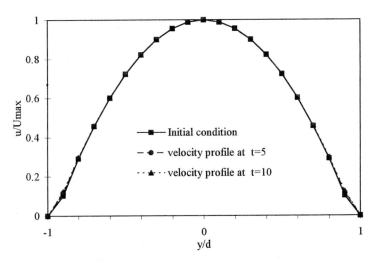

Figure 2. Steady state solution.

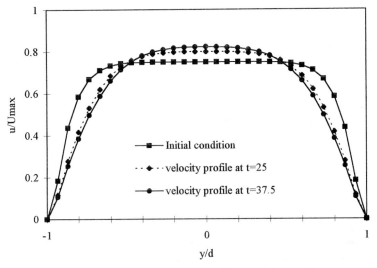

Figure 3. Evolution of the initial velocity profile.

4. CONCLUSIONS

The first results of the calculations executed by a new three-dimensional Navier-Stokes solver, are presented. The aim of the work is to provide a reliable computer code, in order to perform numerical simulations of high Reynolds number viscous fluid flows, by following the Large Eddy Simulation approach for turbulence calculation.

REFERENCES

[1] Alfonsi, G., Passoni, G., Zampaglione, D., "Integrazione numerica delle equazioni di Navier-Stokes in tre dimensioni: primo test di calcolo sull'equazione di Poisson per la pressione", Proceedings National Conference "Turbolenza e Vorticità", Capri, 1993.

[2] Moin, P., Kim, J., "On the numerical solution of time-dependent viscous incompressible fluid flows involving solid boundaries", Journal of Computational Physics, 35, 1980.

[3] Moin, P., Kim, J., "Numerical investigation of turbulent channel flow", Journal of Fluid Mechanics, 118, 1982.

[4] Wray, A., Hussaini, M.,Y., "Numerical experiments in boundary-layer stability", Proceedings Royal Society of London, A, 392, 1984.

1A29

Three-layer model of turbulent currents in covered channels

V.K. DEBOLSKY, E.N. DOLGOPOLOVA, R.V. NEYMARK
Water Problems Institute, Russian Academy of Sciences, Moscow, Russia

In this paper the three-layer model is generalized for the covered channels.
Let us consider a flow above rough bottom and with developed roughness of the top cover. That is we consider a steady flow without laminar viscous layer near the both boundaries, and we don't take into account change of ice thickness. Both boundary flows are considered as being formed independently, their interaction is manifested in formation the flow core of a layer with the shear stress which is equal to zero at the outer boundaries of the layer. Thus the thickness of this layer and velocity distribution in it defined by bottom and top surfaces as well as by water discharge. We examine the values of discharge which do not destroy the top surface but can change the roughness of both boundary surfaces. The velocity gradients of near bottom and near cover are proposed to be equal zero on the boundaries of the flow core layer. The definition of two trivial conditions at the intermediate layer boundaries does not imply that turbulent viscosity coefficient equal to zero inside of this layer [2-4]. In flows under consideration the transfer can be realized in both directions - to the cover and to the bottom. The results of experiments in [1] confirm the possibility of such description of underice flow.
Three-layer flow model implies the existence of the layer near bottom, the layer near top surface and the core of the flow between them on which external boundaries

$$\frac{\partial u}{\partial y}\bigg|_{y=h_1} = \frac{\partial u}{\partial y}\bigg|_{y=h-h_2} = 0$$

The conventional flow velocity power profile

$$u = u_{01} + (u_1 - u_{01})\left(\frac{y}{h_1}\right)^{\alpha_1} \tag{1}$$

near the bottom or

$$u = u_{02} + (u_2 - u_{02})\left(\frac{h-y}{h-h_2}\right)^{\alpha_2} \tag{2}$$

near the top surface don't ensure conversion to zero on the external layer's boundary partial derivative $\dfrac{\partial u}{\partial y}$, therefore let us modify velocity profile as:

$$u = u_{01} + (u_1 - u_{01})\left(\frac{y}{h_1}\right)^{\alpha_1} + \alpha_1\left(1 - \frac{y}{h_1}\right)\frac{y}{h_1} \tag{3}$$

near the bottom and

$$u = u_{02} + (u_2 - u_{02})\left(\frac{h - y}{h - h_2}\right)^{\alpha_2} + \alpha_2\left(1 - \frac{h - y}{h - h_2}\right)\frac{h - y}{h - h_2} \tag{4}$$

near the top surface.
Such modification of the flow velocity power profile reduce to equalities

$$\frac{\partial u}{\partial y}\bigg|_{y=h_1} = 0 \text{ and } \frac{\partial u}{\partial y}\bigg|_{y=h-h_2} = 0$$

Maximum derivation of the velocity values calculated by (1),(2) from calculated by (3),(4) is equal $\dfrac{\alpha_1}{4}$ if $\dfrac{y}{h_1} = 0.5$ and $\dfrac{\alpha_2}{4}$ if $\dfrac{h - y}{h - h_2} = 0.5$, i.e. doesn't exceed 5% for $\alpha_1 = \alpha_2 = 0.2$

Relative to flow velocity in the flow's core let's assume that it smoothly changes from the one boundary to the other and has maximum inside of the core. The polynomial of minimum fourth power satisfies this condition.
Hence we look for velocity inside flow's core as:

$$u = a(y - h_1)^4 + b(y - h_1)^3 + c(y - h_1)^2 + d(y - h_1) + u_1 \tag{5}$$

If $y = h_1$ then $\dfrac{\partial u}{\partial y} = 0$, consequently, $d = 0$

If $y = h_2$ then $u = u_2$ and

$$u_2 - u_1 = a(h - h_1 - h_2)^4 + b(h - h_1 - h_2)^3 + c(h - h_1 - h_2)^2 \tag{6}$$

also $\dfrac{\partial u}{\partial y} = 0$ and

$$0 = 4a(h - h_1 - h_2)^3 + 3b(h - h_1 - h_2)^2 + 2c(h - h_1 - h_2) \tag{7}$$

Therefore

$$b = -2\left[\frac{u_2 - u_1}{(h - h_1 - h_2)^3} + a(h - h_1 - h_2)\right] \tag{8}$$

$$c = -\frac{3}{2}b(h - h_1 - h_2) - 2a(h - h_1 - h_2)^2 \tag{9}$$

Let us suppose $a = k\dfrac{u_2 - u_1}{(h - h_1 - h_2)^4}$ $\qquad$ (10)

$$b = -2\frac{u_2 - u_1}{(h - h_1 - h_2)^3}(1 + k) \tag{11}$$

$$c = \frac{u_2 - u_1}{(h - h_1 - h_2)^2}(3 + k) \tag{12}$$

Extreme point inside the core has ordinate

$$y = h_1 + \frac{3 + k}{2k}(h - h_1 - h_2) \tag{13}$$

It is necessary to fulfill the condition $\dfrac{\partial^2 u}{\partial y^2} < 0$ the maximum of the flow velocity to be inside the flow's core, i.e. it is necessary to carry out

$$\frac{(k - 3)(k + 3)}{k}(u_2 - u_1) > 0 \tag{14}$$

Let's consider to cases:

$u_2 - u_1 > 0$ in this case must be $k > 3$

$u_2 - u_1 < 0$ in this case must be $k < -3$

Therefore we may write

$$u_{max} = u_1 + (u_2 - u_1)(3 + k)^3\frac{(k - 1)}{16k^3} \tag{15}$$

If $u_1 = u_2$ then sought for polynomial degenerates into expression $u = u_1 = u_2 = const$.

Coefficient by $u_2 - u_1$ in expression for u_{max} when varying k from 3 to 10 changes from 1 to 1.236. It may expect the value k will be changed few along the stream when smooth varying of the channel depth and width.

Let us assume that u_{01} and u_{02} are constants.

$u_{01} = 0.005U_0 \div 0.05U_0$, $u_{02} = 0.005U_0 \div 0.05U_0$

As a result the problem of flow velocity calculation involves 6 unknowns $h_1, \alpha_1, u_1, h_2, \alpha_2, u_2$. For them definition let's write impulse and energy equations in the layer near bottom and in the layer near cover [5].

$$\int_0^{h_1} \frac{\partial}{\partial x}[u(u_1 - u)]dy + \frac{du_1}{dx}\int_0^{h_1}(u_1 - u)dy = \frac{\tau_b}{\rho} \tag{16}$$

$$\frac{\rho}{2}\frac{d}{dx}\int_0^{h_1} u(u_1^2 - u^2)dy = \int_0^{h_1} \mu\left(\frac{\partial u}{\partial y}\right)^2 dy \tag{17}$$

$$\int_{h-h_2}^{h} \frac{\partial}{\partial x}[u(u_2 - u)]dy + \frac{du_2}{dx}\int_{h-h_2}^{h}(u_2 - u)dy = \frac{\tau_t}{\rho} \tag{18}$$

$$\frac{\rho}{2}\frac{d}{dx}\int_{h-h_2}^{h} u(u_2^2 - u^2)dy = \int_{h-h_2}^{h} \mu\left(\frac{\partial u}{\partial y}\right)^2 dy \tag{19}$$

In the flow's core to allow for hydrostatics equation we'll obtain

$$u_1\frac{du_1}{dx} + g\frac{dh_1}{dx} = u_2\frac{du_2}{dx} + g\frac{d(h-h_2)}{dx} \tag{20}$$

Let us take as the sixth equation the equation received to differentiate with respect to longitudinal coordinate the equation of discharge Q through the river's cross-section.

$$\frac{d}{dx}\int_0^{h} u\,dy = \frac{dQ}{dx} \tag{21}$$

Let us assume $\dfrac{dQ}{dx} = 0$ and make dimensionless eq.(16)-(21).

Assume that $\dfrac{\tau_b}{\rho} = \dfrac{c_{fb}|u_1|u_1}{\text{Re}}$ and $\dfrac{\tau_t}{\rho} = \dfrac{c_{ft}|u_2|u_2}{\text{Re}}$, where $c_{fb} = K\alpha_1^2$ and $c_{ft} = K\alpha_2^2$ are coefficients of bottom and cover friction, accordingly ($K\sim0.8$). When integrating of the right parts of eq.(17) and eq.(19) to define μ is used von Karman's model [5]

$$\mu = \rho\kappa^2\left(\frac{\partial u}{\partial y}\right)^3 / \left(\frac{\partial^2 u}{\partial y^2}\right)^2 \tag{22}$$

After making dimensionless and substituting (22) both integrals assume

$$\frac{\rho\kappa^2}{Re}\int_0^{h_1}\frac{\left(\dfrac{\partial u}{\partial y}\right)^5}{\left(\dfrac{\partial^2 u}{\partial y^2}\right)^2}dy \quad \text{and} \quad \frac{\rho\kappa^2}{Re}\int_{1-h_2}^{1}\frac{\left(\dfrac{\partial u}{\partial y}\right)^5}{\left(\dfrac{\partial^2 u}{\partial y^2}\right)^2}dy \tag{23}$$

As a result of the substitution velocity's profiles in eq. (16)-(19) and (21) let's receive the system of linear equations for the derivatives

$\dfrac{dh_1}{dx}, \dfrac{du_1}{dx}, \dfrac{d\alpha_1}{dx}, \dfrac{dh_2}{dx}, \dfrac{du_2}{dx}, \dfrac{d\alpha_2}{dx}$. Each of them is the function of $h_1, u_1, \alpha_1, h_2, u_2, \alpha_2$.

The obtained system of equations was solved numerically by Gauss's method.
It is necessary to give values $h_1, u_1, \alpha_1, h_2, u_2, \alpha_2$ when $x = 0$ to integrate this system. Because we have no apropos information about these values it is proposed to find them from:
1) condition that computed discharge is equal to initial discharge;

2) let us assume that the derivatives $\dfrac{dh_1}{dx}, \dfrac{d\alpha_1}{dx}, \dfrac{dh_2}{dx}, \dfrac{du_2}{dx}, \dfrac{d\alpha_2}{dx}$ are equal to zero.

As a result the problem of the initial conditions definition to solve the system of differential equations is reduced to the calculation of the non-linear equations system. Let's solve this system by Newton's method . The main difficulty is the selection of the first approximation. The initial values are considered to be found if the maximum difference of consecutive approximations for every value is occurred less then 10^{-4}. After definition the initial values of $h_1, u_1, \alpha_1, h_2, u_2, \alpha_2$ are substituted into the system for solving it relative to the derivatives. The varification of the model was made with the help of measurements of the flow velocity profiles at the Moskva river in winter. The mean characteristics of the river were $B = 70$ m, $h = 2$ m, $Q = 56 \text{m}^3 / \text{s}$, Re $\sim 10^6$. Bottom and ice surfaces had developed roughnesses. Bottom structures consisted of sand dunes of large scale with small ripples of ~ 4 sm height. Ice roughness can be described as galls of $2 \div 3$ sm depth and about 5 sm in diameter were situated in chess order. The comparison of the experimental data and the calculation showed that they are in sufficiently good agreement. There is extremely good correspondence between the three-layer model results and experiment for the core of thr stream, while the two-layer models give much worse description of this part of the flow [1].
Thus the three-layer model has an advantage over the two-layer model in description and calculation of covered flows.

Notations

u -longitudinal flow velocity

u_{01} -bottom wall velocity

u_1 -flow velocity on external boundary of the near bottom layer

h_1 -thickness of the near bottom layer

u_{02} -top wall velocity

u_2 -flow velocity on external boundary of the near ice layer

h_2 -thickness of the near ice layer

h -river depth

y -vertical coordinate

x -longitudinal coordinate

α_1 -power in the velocity profile near the bottom

α_2 -power in the velocity profile near the ice

L -length of the river section

U_0 -average velocity in river's initial cross-section

ν -laminar flow viscosity

μ -turbulent flow viscosity

τ_b -flow friction stress on the bottom surface

τ_i -flow friction stress on the ice surface

ρ -fluid density

Q -flow discharge

g -gravity acceleration

References

1.Debolsky V.K., Dolgopolova E.N., Orlov A.S. Statistic characteristics of dynamics of river flows. In: Hydrophysics processes in rivers, reservoirs and marginal seas", Moscow , Nauka Publ. 1989.

2.Debolskaya E.I. Investigation of turbulence structure of river flows under ice. In : Dynamics of streams and lithodynamics processes in rivers, reservoirs and marginal seas, Moscow , Nauka Publ. 1991.

3.Debolsky V.K., Zyryanov V.N., Mordasov M.A. On turbulent exchange in a tidal river-mouth under ice-cover. In: Dynamics and thermic of rivers and water reservoirs, Moscow: Nauka Publ. 1984.

4. Landau L.D., Lifshits E.M. Hydrodinamics, Moscow , Nauka Publ., 1986.

5.Schlichting H. Boundary-layer theory, Moscow, Mir Publ., 1971

1B1

A MULTI-LAB INVESTIGATION OF ICE-COVERED DUNE-BED FLOW

R. Ettema, V.C. Patel, B.T. Smith, J.Y. Yoon and F.M. Holly
Iowa Institute of Hydraulic Research, Iowa City, USA

I. Mayer, P. Bakonyi and O. Starosolszky
VITUKI, Budapest, Hungary

ABSTRACT

This paper gives an overview of a collaborative research program concerning flow in ice-covered channels, and presents an insight into one aspect of research findings produced to date. That aspect is the effect of cover presence on flow field over a dune, and relatedly dune response to altered flow field.

INTRODUCTION

Iowa and Hungary share common interests in the behavior of dune-bed rivers that become ice-covered during frigid winters. Those interests have prompted IIHR and VITUKI to pursue development of computational procedures for modeling water flow, sediment and contaminant-transport processes, as well as ice-cover stability, in ice-covered alluvial channels. To date, the collaboration has entailed a comprehensive set of "multi-lab" studies with the overall objective of elucidating how ice-cover presence modifies flow distribution and, thereby, affects flow resistance and bedload transport. The studies undertaken by IIHR and VITUKI may be described as multi-lab in two senses. They involve two research institutes in different parts of the world, and three forms of laboratory: a lab flume, a numerical (cfd) flume, and the river.

The occurrence of an ice cover on a wide open-channel flow approximately doubles the wetted perimeter of flow, consequently substantially increasing flow resistance. The flow responds with an increase in depth and a concomitant decrease in average flow velocity. By and large, the channel's alluvial bed holds its slope over long distances when the cover is imposed. The numerical modeler of such a flow, though, now faces a double dilemma common to composite channels generally. One dilemma is to evaluate the overall resistance encountered by flow. The other arises in determining flow properties, such as eddy diffusivity, if the intent of modeling is to trace the transport of sediment, or of a contaminant conveyed in the body of the flow. The presence of an ice cover substantially modifies the distribution of temporal mean flow velocities over the flow depth and, relatedly, alters the turbulence characteristics of the flow. Turbulence is generated and diffused from an additional boundary, the ice cover. Complicating those dilemmas is the likelihood that the form of the alluvial bed may change in response to the changed flow field.

COMPONENT STUDIES
IIHR and VITUKI prepared a research plan comprising the following studies:

1. Laboratory flume experiments to illuminate how cover presence modifies the flow field, bed morphology, and bedload rate, of a dune-bed channel;
2. Numerical flume experiments to further illuminate how cover presence modifies flow field;
3. Characterization and modeling of ice and alluvial bed conditions in rivers.

The insights gained from these studies would form the basis of a procedure for use in computational hydraulics models to route water, sediment, or contaminant through ice-covered rivers. IIHR has undertaken studies 1 and 2. IIHR and VITUKI are jointly pursuing study 3 and interpreting the findings from the three studies to facilitate computational hydraulics modeling.

LAB-FLUME STUDY
The following two sets of experiments were conducted with a 30m-long, 0.91m-wide and 0.45m-deep flume:

1. Loose-bed experiments to investigate how a cover affects flow depth, bedform geometry and bedload transport rates for constant discharge.
2. Fixed-bed experiments to investigate how cover presence modifies the flow field over a dune bed. Flow parameters mapped, using LDV, over two dunes include: distribution of temporal mean velocities in the streamwise and vertical directions, Reynolds stress, and turbulence kinetic energy.

The loose-bed experiments used a 1.3mm diameter uniform sand, with plywood sheets simulating level ice covers. Three cover roughnesses were simulated. One was smooth. The rougher covers had equivalent grain roughness heights estimated as 3 and 25mm. To map the flow field over a dune in the requisite detail required that dunes in the flume be fixed. Thus, dunes formed in openwater flow over the loose bed were fixed with a dusting of cement. Full details of the flume experiments are given by Smith and Ettema (1995).

As anticipated, and shown in Fig. 1, cover presence and increasing roughness increased flow depth and decreased mean velocity. The abcissa reflects cover presence and roughness as a ratio of average shear stress on the cover and the bed, τ_{cover}/τ_{bed}. For the conditions examined, cover presence and increasing roughness flattened the dunes, which increased their wave length proportionately with flow depth (flow depth/wave length remained constant), but reduced in height. Fig. 2 shows representative values of dune length and height for the four conditions. The most dramatic consequence of cover presence was on bedload rate. With the roughest cover present bedload dropped to about 30% of its openwater value, as evident from Fig. 1.

The mappings plotted in Figs 3a,b show the measured streamwise component of velocity, u, and Reynolds stress, $-\overline{u'v'}$, for flow over a dune in openwater and covered flow conditions. Similar mappings for vertical component of velocity, v, and turbulence kinetic energy are given by Smith and Ettema. The major changes of increased flow depth and altered u profiles are readily apparent. Noteworthy, though, are the essentially constant values of u near the bed; if anything values of u for the rougher cover are a whisker greater than for the smoother cover and openwater conditions. This finding suggests that the covers did not substantially reduce skin friction over the dune. Fig. 3b, though, shows significant reductions in peak Reynolds stress all along the dune crest, suggesting that reduced turbulence energy is an important mechanism through which the cover may alter bedform geometry and reduce bedload

NUMERICAL-FLUME STUDY

A numerical flume was developed to examine flow in a 2-D channel with prescribed dune geometry and roughness, and a floating cover of known roughness. The model involves solution of the Reynolds-averaged Navier-Stokes equations with the k-ω turbulence model which accounts for wall roughness. Work only now is underway to validate the model against the flume data described above, but early results are available from a numerical experiment with the dune geometry used in an openwater flume experiment published by Meirlo and de Ruiter (1988). Openwater results from the numerical flume were compared with their data. Subsequently, a smooth ice cover was imposed on the openwater flow in the numerical flume, and the cover's effect on the flow calculated. The process of enquiry is the same as for the fixed bed results described above. Full details of the numerical flume and its results are presented by Yoon and Patel (1994).

The results generally corroborate those from the lab flume, but slight differences presently are under discussion. Imposition of the cover increased water depth by 16%. Similar reductions in turbulent kinetic energy and Reynolds stress were found for the numerical flume as for the lab flume, but, as shown in Fig. 4, the numerical flume suggests more definitely that cover presence reduces skin friction along the dune crest. In Fig. 4, u_τ = shear velocity and λ = dune length.

RIVER STUDIES

The studies focus on extending and implementing results, such as those presented, from the lab and numerical flumes to enable accurate computational modeling of flow, sediment and contaminant transport in extensive lengths of ice-covered alluvial rivers. A subject of particular interest at present is development of a procedure for estimating the bedform-drag component of flow resistance. Tasks include: acquiring river data on flow, bedform geometry, and bedload under ice; and, adapting, to ice-covered conditions, existing computational codes for calculating flow resistance and bedload transport.

ACKNOWLEDGMENTS
The set of studies described here were supported by the US National Science Foundation under Grant Numbers CTS-90-0267 and INT-9006140, and by the Hungarian Academy of Sciences.

REFERENCES
Mierlo, M.C.L.M, and de Ruiter, J.C.C. (1988), "Turbulence Measurements Above Artificial Dunes," Report Q789, Delft Hydraulics Laboratory, Delft, The Netherlands.
Smith, B. T., and Ettema, R., (1995), "Flow in Ice-Covered Dune-Bed Channels," IIHR Report in press, IIHR, Iowa City, IA.
Yoon, J.Y. , and Patel, V.C., (1993), "A Numerical Model of Flow in Channels With Sand Dune Beds and Ice Covers," IIHR Report No. 362, IIHR, Iowa City IA.

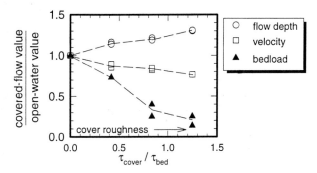

Figure 1. Flow and Bedload Data

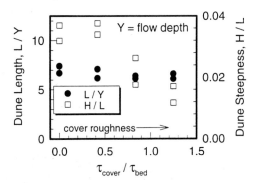

Figure 2. Dune Length and Steepness Data

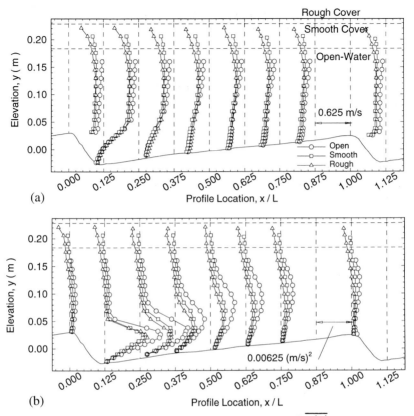

(a)

(b)

Figure 3. (a) Profiles of u; (b) Profiles of $-\overline{u'v'}$.

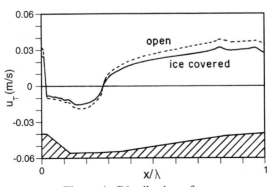

Figure 4. Distribution of u_τ.

1B2

NUMERICAL SIMULATIONS OF DUNES USING PHYSICAL MODEL CONCEPTS

KHIN NI NI THEIN

International Institute for Infrastructural, Hydraulic and Environmental Engineering, Delft, The Netherlands.

ABSTRACT

This paper describes the stream-flow velocity relationships that are essential for simulations of dunes in rivers. Simulations of dunes, both experimentally and computationally, must take into account the two different regimes of plane bed and dune-covered bed. Although both have to be considered in the development of a generic river plan-form movement simulation model, the phenomena of both regimes under the condition of a prescribed set of discharge, slope and grain size can only be properly addressed in a numerical model (Thein, 1994). The necessary relationships and corresponding computations are presented.

CO-ORDINATE SYSTEM

The differential equations describing the conservation of mass and momentum are formulated using a co-ordinate system (s,n,z), where s and n are orthogonal curvilinear coordinates in the horizontal plane and where the z-axis is vertical and positive in the upward direction. The local curvature of the s-line is taken as positive when the positive n-lines diverge, and inversely.

GOVERNING DIFFERENTIAL EQUATIONS

When relating the horizontal coordinates to the channel geometry, the equation of continuity for steady incompressible flow reads:

$$\frac{\partial U_s}{\partial s} + \frac{U_s}{R_n} + \frac{\partial U_n}{\partial n} + \frac{U_n}{R_s} + \frac{\partial U_z}{\partial z} = 0 \tag{1}$$

The momentum equations can be written as

$$\frac{\partial U_s^2}{\partial s} + \frac{\partial U_n U_s}{\partial n} + \frac{\partial U_z U_s}{\partial z} + 2.\frac{U_n U_s}{R_s} + \frac{U_s^2 - U_n^2}{R_n} + \frac{1}{\rho}\frac{\partial P}{\partial s} + F_s = 0 \tag{2}$$

$$\frac{\partial U_s}{\partial s}\frac{U_n}{} + \frac{\partial U_n^2}{\partial n} + \frac{\partial U_z}{\partial z}\frac{U_n}{} + 2.\frac{U_s U_n}{R_n} + \frac{U_n^2 - U_s^2}{R_s} + \frac{1}{\rho}\frac{\partial P}{\partial n} + F_n = 0 \tag{3}$$

$$\frac{\partial U_s}{\partial s}\frac{U_z}{} + \frac{\partial U_n U_z}{\partial n} + \frac{\partial U_z^2}{\partial z} + \frac{1}{\rho}\frac{\partial P}{\partial z} + F_z = -g \tag{4}$$

in which P = pressure, ρ = mass density, g = acceleration due to gravity, U_s, U_n, U_z = velocity components in s-, n- and z-directions respectively and, F_s, F_n, F_z = friction terms in s-, n- and z- directions respectively. The friction terms can be expressed in terms of velocities in their respective directions by introducing the Reynold's stress concept and the Prandtl mixing length hypothesis. Therefore bed shear stresses in both s- and n-directions are calculated during the simulations. Further, dune dimensions are calculated and local bed roughnesses are also calculated accordingly.

PROBLEM DESCRIPTION

The alluvial roughnesses in s- and n- directions are determined by calculating the shapes and dimensions of dunes in rivers. Fredsøe's (1982) method provides the geometry of the bed patterns from a knowledge of the flow parameters, the suspended sediment load and the bed load. The accent is on the mass flux (q = U.h m^2/s) as this determines the sediment transport. The total load is split up into a bed load and a suspended load. The calculated bed load, suspended load and their gradients with respect to the dimensionless shear stress are introduced into the dune dimension formulae. The shear stress due to the form friction is then calculated. Finally, the total shear stress is calculated and converted to an actual water depth corresponding to the prevailing situation. There is only one depth corresponding to a given situation and another cannot be assumed. As mentioned above, obtaining a unique water depth makes it possible to predict a channel roughness. This method determines automatically the regime of the flow according to the knowledge of the dune's dimension (eg. for a plane bed situation the dune height-to-length ratio tends to zero).

A basic problem arises even after further elaborations and developments have been carried out. The bed topography model becomes unconditionally unstable after coupling the roughness predictor model to the horizontally two-dimensional bed-topography model which simulates the flow and bed-topography in a meandering river. This problem arises from out of the physics of the river bed. The equations applied in the calculation of the dunes' dimension were abstracted from results obtained from physical models. In physical models, simulations of dunes were realised by running experiments with plane bed conditions and then increasing the discharge in order to form the dune-covered bed and, while maintaining this condition, the velocity was kept the same. In this way following equations were established:

$$\frac{U_d}{U_{fp}} = 6. + 2.5 \ln \left(\frac{h_p}{K_d}\right) \tag{5}$$

$$U_{fp} = \sqrt{gh_p I} \tag{6}$$

in which

U_d	=	mean velocity corresponding to the dune-covered bed
U_{fp}	=	friction velocity due to the skin friction (grain roughness)
h_p	=	water depth corresponding to the plane bed situation
K_d	=	total roughness (dune-covered bed)
g	=	acceleration due to gravity
I	=	water surface slope

These equations suggest that the total roughness of a dune-covered bed can be calculated by adding form roughness and grain roughness. Form roughness is calculated from the dimensions of the dunes, which in turn presumes the existence of a dune-covered bed, while the grain roughness is calculated under plane bed conditions. These two regimes are linked by a common factor, which is a constant flow velocity which in turn necessitates a varying discharge. Now let us look upon our numerical simulation using a given set of discharge, slope and grain size: it is obvious that whenever total roughnesses are calculated and applied for the next time step in the numerical model, the corresponding calculated discharges must increase according to the water depth over the dunes, which is mostly about 1.5 to 2 times higher than in the plane bed condition. It is this which causes the unconditionally unstable solution. Hence, further investigations were carried out, as described in the following section.

SOLUTION METHOD

For the first step of the calculation, the river bed is assumed to be plane, i.e. no sand wave is present at the bottom. In this case the bed roughness K_p is approximately 2.5 times d_{50} in which d_{50} is the mean grain diameter. The flow equation, the resistance equation (Colebrook-White) and the bed shear stress equation can be solved simultaneously by using three specified variables i.e. discharge, slope and roughness K_p. Thus water depth, shear parameter and bed shear stress are calculated for the plane bed situation. Additionally, the sediment transport is split up into two parts and calculated as bed load and suspended load. Engelund and Fredsøe's equation is used for the bed load transport. For suspended sediment transport an integration procedure is applied, which combines a vertical velocity profile and an eddy viscosity distribution over the vertical plane. Since the two-dimensional model in the curvilinear coordinate system has been developed in such a way that the width of the river is split into many strips (sub-channels), forming a compound channel composed of many sub-channels, the eddy viscosity distribution over the vertical plane models can be applied at each strip across the river (Thein, 1994). The suspended sediment load is calculated in this way.

The length and height of a dune are functions of the bed load, the suspended load and the gradients of the loads, and from these the dune dimensions are obtained. The dimensionless shear stress due to the expansion loss can be calculated through the ratio between the loss and the dimensionless shear stress of the plane bed. This ratio is a function of the dune height, the dune length, and the ratio of the mean velocity to the friction velocity due to the skin friction. It should be mentioned that the ratio of mean velocity to the friction velocity due to the skin friction is taken from the plane-bed condition. To be more precise, U_d/U_{fp} should be determined by Equation (5). However, because U_d and h_d are unknown this requires a complicated iteration procedure and the introduction of another assumption that the ratio of plane bed U_p/U_{fp} can be used instead of the ratio U_d/U_{fp}. This assumption is valid because, in a physical model, the mean flow velocity is kept the same for both regimes, ie. $U_d = U_p$, assuming that small variations do not alter the flow velocity significantly. However, it is impossible to use this assumption in a numerical model because this is a major cause of numerical instability in the case of a constant discharge accompanied with a given slope and grain size diameter. This phenomena is explicated in Figs. 1 and 2.

The important parameters such as the mean velocity, the friction velocity, the water surface slope (or strictly speaking the energy slope), and the friction factor are each split into two parts, namely, (1) the grain-influenced parameter and (2) the form-influenced parameter. The integrated roughness factor is then computed for both regimes. The results are presented as a family of curves which are a family of paraboloids the focal lengths of which vary with the discharges. One family of curves serves as a graphical solution for a given discharge, slope and the properties of the bed material for the maximum range of water depths, as exemplified by Fig. 3. For several slopes and different grain sizes, a large number of families of curves are produced. After determining numerous curves like those in Fig. 3, the difference between these two velocity ratios was determined using numerous simulations, as presented in Fig. 4. This figure shows that using the velocity ratio (U_p/U_{fp}) for the plane bed instead of U_d/U_{fp} is not justified for the case of a constant discharge accompanied by a given slope and sediment grain diameter - which is exactly the situation that occurs in the bed-topography model of meandering rivers. The other

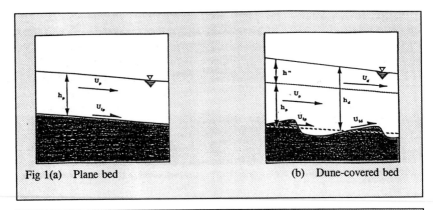

Fig 1(a)   Plane bed   (b)   Dune-covered bed

Fig. 2 The comparison of flow conditions between Figs. 1(a) and 1(b).

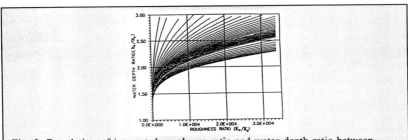

Fig. 3 Correlation of integrated roughness ratio and water depth ratio between plane bed and dune-covered bed.

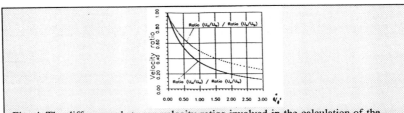

Fig. 4 The differences between velocity ratios involved in the calculation of the roughness factor in an alluvial-river bed

parameters, such as the total dimensionless shear stress, the dimensionless shear stress due to the skin friction, the friction factor and the friction velocity are similarly simulated. Therefore the relationships between each parameter mentioned above and the shear parameter are found for both conditions. To find the dimensionless ratios of the parameters for two different conditions, the corresponding value of each parameter in the dune-covered bed case is divided by the corresponding value of each parameter as determined for the plane-bed case. These dimensionless ratios of the parameters are used to draw up the general relationship shown in Eqn. 7.

$$(\frac{U_d}{U_{fp}}) = (\frac{U_p}{U_{fp}})^{1/3} * (\frac{U_d}{U_{fd}})^{2/3} \tag{7}$$

The range of validity of the velocity relationship is limited by three variables: (1) water discharge Q from 130 m^3/s to 88000.0 m^3/s ; (2) energy slope I from 10^{-6} to $51 * 10^{-6}$; (3) mean grain diameter d_{50} from 0.47 mm to 1.25 mm.

CONCLUSIONS

Eqn. 7 gives the U_d/U_{fp} ratio in terms of the U_d/U_{fd} and U_p/U_{fp} ratios. By using this relationship, the roughness coefficient and the dune dimensions are satisfactorily predicted. The roughness predictor model becomes compatible with the bed-topography simulation model's specifications. After coupling these two models, a verification for the consistency of the discharge is made. The deviation of water discharge before and after the roughness has been updated proved to be only 0.35% . Therefore, the discharge which is inherent in the current simulation time step is hardly affected by the updating procedure of the roughness and is still representative of the prevailing condition. Thus the calculation can proceed reciprocally for both regimes.

REFERENCES

1. Fredsøe, J. 1982, "Shape and dimensions of stationary dunes in rivers", J. of Hyd. Vol. 108, No. HY8, Aug. 1982.

2. Thein, K.N.N., 1994, "River plan-form movement in an alluvial plain", Ph.D thesis, IHE Delft, The Netherlands.

1B3

Genetic Model Induction Based on Experimental Data

VLADAN BABOVIĆ

International Institute for Infrastructural, Hydraulic and Environmental
Engineering, IHE-Delft, The Netherlands

ABSTRACT

Genetic programming (GP) has been used to perform an analysis of sediment transport data and to induce the relationship between the bed concentration of suspended sediment c_b and the hydraulic conditions. The results obtained through the application of this novel technology are shown to be of similar accuracy to the ones obtained by more traditional techniques.

INTRODUCTION

Darwinian theory is today a widely accepted doctrine that provides an explanation of evolutionary processes as the adaptation of lineages of information to their changing environment. The fundamental idea in evolutionary algorithm technology is that of plagiarising the process of Darwinian evolution and capturing its essence in an artificial media, such as a computer, ergo providing the possibility of using this creative power to construct entities that are useful in our own engineering applications. This paper reports on the success of the power of simulated evolution applied to the domain of sediment transport.

EVOLUTIONARY ALGORITHMS

Almost forty years of research and application have led to the independent development of four main streams of *evolutionary algorithms* (EA): *Evolution Strategies* (Schwefel, 1981), *Evolutionary Programming* (Fogel, 1966), *Genetic Algorithms* (Holland, 1975) and *Genetic Programming* (Koza, 1992).

The EAs must provide the processes that will emulate evolutionary adaptation. Firstly, it is necessary to define a way of transferring the genetically encoded information from parents to offspring so that the offspring resemble their ancestors. Secondly, some errors should occur in this copying process, so that some variability is introduced in the material that is being inherited. In nature these errors are referred to as mutations. And, thirdly, the better adapted entities should have better chances to reproduce more and better-surviving offspring. Obviously, these processes together correspond to the actions of natural selection.

As even their common name applies, evolutionary algorithms are based on the simulation of natural evolutionary processes. Although a crude simplification of organic evolution, EAs have proven to be robust and efficient search algorithms. All these algorithms are based on arbitrarily initialised populations of individuals (chromosomes, search points, potential solutions) in a multi-dimensional search space. These individuals undergo a sequence of unary (e.g. *mutation*-type) and higher-order (e.g. *crossover*-type or sexual-reproduction-like) transformations. Promoting the more fit through appropriate *selection* criterion, the next generation of individuals is designated. To summarise, EAs maintain a population of structures that undergo a series of modifications that collectively loosely correspond to the natural evolution.

Natural evolution can be understood as a process of adaptation. Adaptation, at the same time can easily be conceived as a growth of knowledge. Clearly, as an entity adapts to its environment, its knowledge about this environment constantly grows. Thus, in the circumstances of artificial evolution, evolving entities learn about the problem domain in which they develop, and after long enough simulation evolve the solution to the problem. In some previous studies, the creative power of evolution has been used to automatically calibrate hydro-dynamic models (Babović *et al.*, 1994), induce salt-intrusion models (Babović and Minns, 1994) or even generate a new class of rainfall-runoff models (Babović, 1995).

EVOLVING ENTITIES

In so-called canonic genetic algorithms (GA), the evolving entities are simply modelled as binary strings. The extension of this, most elementary model, depicts evolving individuals as real-valued utilities (see, for example, Babović, 1993). In Genetic Programming (GP), however, the entities undergoing evolution are represented as algebraic expressions in so-called reverse polish notation (RPN). The goal of the evolutionary process within a GP framework is to establish an algebraic expression that will characterise the experimental data sets. The formulae that produce less deviation when compared to experimental data are considered to be more fit, and consequently produce more offspring. Accordingly, the 'information' that is coded in these 'successful' formulae is propelled into subsequent generations, driving the entire process towards more appropriate formulations that better model the problem domain. For a more detailed description of genetic programming see Babović (1995).

SYSTEMS IDENTIFICATION

The problem of creating an artificial environment, to which evolving entities adapt, can be equated to a system identification problem, since modelling in the sense of 'systems identification' can be defined as the process of generating closed-form mathematical expressions from observed data. System identification techniques are applied in many fields in order to predict the behaviour of unknown (or not-so-well-known) systems when given sets of input-output data (Babović, 1995).

Following Iba *et al*, (1993) the problem may be defined in the following way. Imagine that a system produces an output value, y, and that this y is dependent on m input values, thus:

$$y = f(x_1, x_2, x_3, ..., x_m) \tag{1}$$

Given a set of N observations of input-output tuples, as shown in Table 1,

INPUT				OUTPUT
x_{11}	x_{12}	...	x_{1m}	y_1
x_{21}	x_{22}	...	x_{2m}	y_2
...	...	...	...	...
x_{N1}	x_{N2}	...	x_{Nm}	y_N

Table 1. An input-output example set

the system identification task is to approximate the 'true' function f with $\bar{f}$. Once this approximate function $\bar{f}$ has been estimated, a predicted output y can be found for any input vector $(x_1, x_2, ..., x_m)$, i.e.

$$\bar{y} = \bar{f}(x_1, x_2, x_3, ..., x_m) \tag{2}$$

The $\bar{f}$ is called the 'complete form' of f. This is to say that we are *searching*, using evolutionary algorithms, for a complete form of f. So far, we have described only a form of artificial environment to which self-reproducing entities will try to adapt themselves. The self-reproducing, evolving entities, will be described in more detail in the sequel.

EXPERIMENTAL DATA

The present study makes use of experimental flume data utilised by Zyserman and Fredsøe (1994). The experimental data consisted of total, steady state sediment load for a range of discharges, bed slopes and water depths. Zyserman and Fredsøe used the Engelund-Fredsøe and Einstein (for more details, see Zyserman and Fredsøe, 1994) formulation to calculate the bed concentration of suspended sediment c_b and used these values in conjunction with hydraulic parameters to perform system identification and formulate the expression for bed concentration of suspended sediment c_b. The hydraulic conditions were represented by *Shields parameter*, θ, defined as:

$$\theta = \frac{u_f}{(s-1)gd} \qquad \text{and} \qquad \theta' = \frac{u_f'}{(s-1)gd} \qquad (3) + (4)$$

where:
u_f -shear velocity $=(gDI)^{0.5}$
s -relative density of sediment
d_{50} -median grain diameter
D -average water depth
I -water surface slope
u_f' -shear velocity related to skin friction $=(gD'I)^{0.5}$
D' -boundary layer thickness defined through:

$$\frac{v}{u_f'} = 6 + 2.5\ln\left(\frac{D'}{k_N}\right)$$

$$(5)$$

v -mean flow velocity
k_N -bed roughness $=2.5d$
The study makes use of the so-called Rouse number z, defined as:

$$z = \frac{w_s}{\kappa u_f} \qquad (6)$$

where:
w_s -settling velocity of suspended sediment
κ -von Karman's constant (≈ 0.4)
After dimensional analysis, Zyserman and Fredsøe (1994) formulate the following expression:

$$c_b = \frac{0.331(\theta' - 0.045)^{1.75}}{1 + \frac{0.331}{0.46}(\theta' - 0.045)^{1.75}} \qquad (7)$$

It is quite obvious from equation (7) that Zyserman and Fredsøe make use of only θ'. Nevertheless, comparative analysis with some other, more complex expressions indicate that formula (7) is of comparable, if not better accuracy.

MODELS INDUCED BY THE MEANS OF GENETIC PROGRAMMING

A genetic programming environment was set-up with all measured data and corresponding parameters based on the measurements: θ, θ', u_f, u_f', d_{50} and w_s. The evolutionary process resulted in a number of expressions, but here, due to the space limitations, only the best-performing ones will be presented. The most accurate expression induced during the evolutionary process:

$$c_b = 0.71 U_f \sqrt{U_f'} \left(\theta' - \sqrt{e^{d_{50}}} \right) \tag{8}$$

gives an absolute average error of 0.0269, as compared to 0.02665 generated by the Zyserman and Fredsøe (1994) formula.

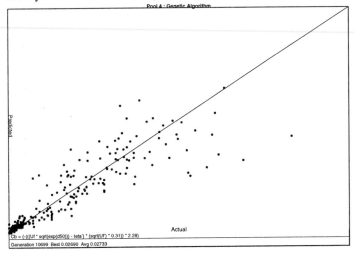

Figure 1 *Scatter plot of bed concentrations for GP induced expression (8) and Zyserman and Fredsøe formula (7)*

A somewhat simpler expression:

$$c_b = 0.74(\theta' - 1.56 U_f) \sqrt{U_f'} \tag{9}$$

gives a negligibly larger error of 0.02691. In addition to these, a slightly less accurate formula producing an error of 0.02898 is presented as well:

$$c_b = (0.02 - 0.16\theta')^{0.89} \tag{10}$$

In addition to considering the accuracy, sensitivity analysis should also be performed before choosing the definitive formula. In many cases, it is required that the formula should not be sensitive to noise and inaccuracies in input parameters, and accuracy is of only secondary interest.

CONCLUSIONS
It has been demonstrated that Genetic Programming is a novel technique which facilitates the process of experimental data analysis and systems identification. GP-based induction leads to various models from which the final choice for an application depends not only on accuracy, but on some other factors, like sensitivity.

REFERENCES
Babović, V., 1993, Evolutionary algorithms as a theme in water resources, *Scientific presentations AIO meeting '93*, Delft University of Technology
Babović, V., 1995, Genetic programming and its application to rainfall-runoff modelling, *Journal of Hydrology*, in press.
Babović, V., and Minns, A.W., 1994, Use of computational adaptive methodologies in hydroinformatics, in Verwey, A., Minns, A.W., Babović, V., and Maksimović, Č., *Proceedings of the First International Conference on Hydroinformatics*, Vol. **1**, Balkema, Rotterdam, pp.201-210
Babović V., Larsen, L.C. and Wu, Z., Y., 1994, Calibrating hydrodynamic models in by means of simulated evolution, in Verwey, A., Minns, A.W., Babović, V., and Maksimović, Č., *Proceedings of the First International Conference on Hydroinformatics*, Vol. **1**, Balkema, Rotterdam, pp.193-200
Fogel, L.J., Owens, A.J. and Walsh, M.J., 1966., *Artificial Intelligence through Simulated Evolution*, Ginn, Needham Heights
Holland, J., 1975, *Adaptation in Natural and Artificial Systems*, The University of Michigan Press, Ann Arbor
Iba, H., de Garis, H., and Sato, T., 1993, Solving identification problems by structured genetic algorithms, *Technical Report ETL-TR-93-17*, Electrotechnical Laboratory, Tsukuba
Koza, J.R., 1992, *Genetic Programming: On the Programming of Computers by Means of Natural selection*, The MIT Press, Cambridge
Schwefel, H.-P., 1981., *Numerical Optimisation of Computer Models*, Wiley, Chicester
Zyserman, J.A., and Fredsøe, J., 1994, Data analysis of bed concentrations of suspended sediment, *Journal of Hydraulic Engineering*, ASCE, Vol. **120**, No. **9**, pp.1021-1042

1B4

Analysis of Experimental Data Using Artificial Neural Networks

A.W. MINNS
International Institute for Infrastructural, Hydraulic and Environmental
Engineering, Delft, The Netherlands

ABSTRACT
An artificial neural network (ANN) was used to analyse the relationships between experimental data with a minimum of preconception. The ANN that makes use of all of the available measured data and does not require any preliminary analysis to select or disregard any hydraulic parameter. The results of this new methodology are shown to be much more accurate than results from the more traditional method of dimensional analysis.

INTRODUCTION

A common approach to the analysis of experimental data is through dimensional analysis and curve fitting. Although the empirical formulae thus derived often fit the experimental data to a high degree of accuracy, these formulae often present the aspect of extremely complex, non-linear combinations of parameters and constants that do not really give much insight into the physical system being described. Also, the form and accuracy of the formulae are often very sensitive to the choice of dimensionless parameters. In many cases, for the sake of simplicity, several dimensionless parameters, and hence measured data, may be disregarded entirely without significantly affecting the final accuracy of the resulting formula. The fact that the *exact* form of the empirical relation is not as important as the ability of the formula to map the experimental data accurately implies that this kind of analysis may be very efficiently carried out using artificial neural networks (ANNs).

Generally, the objective of such an analysis is simply to relate quantities that are very difficult to measure outside of a laboratory to parameters that can be easily measured in the field. The study of sediment transportation provides a good example of this problem. In particular, the measurement of the concentration of suspended sediment is extremely difficult near the bed due to the large vertical gradients in the concentrations that occur very close to the bed. Another problem is that of trying to separate the bed load from the suspended load.

ARTIFICIAL NEURAL NETWORKS

ANNs are computational tools inspired by the natural sciences of neurology and psychology. A typical ANN consists of a set of predefined *nodes* (representing biological neurons, see Hopfield, 1994), quite often arranged in layers, connected to each other by weighted *links* (representing biological synapses and dendrites). The nodes act as summation devices that produce an output signal (the *response*) by passing the sum of the weighted inputs (the *stimuli*) through some threshold function. The distribution of weights over the links occurs during a so-called *supervised learning* phase or 'mode' in which the ANN is *trained* to reproduce a given set of output vectors for a corresponding set of input vectors.

Many publications describe in much greater detail the architecture of various types of ANNs (for example Beale and Jackson, 1990; Hertz *et al.*, 1991). However, it is rapidly becoming apparent that a certain class of ANN, known as multi-layer, perceptron-like networks (Rumelhart and McClelland, 1986), are capable of representing quite satisfactorily almost any function encountered in applications so far in real-world applications. Indeed, standard multi-layer feed-forward networks, with only one hidden layer have been found capable of approximating any measurable function to any desired degree of accuracy (Hornik *et al.*, 1989). Networks of this type are trained using a method referred to as error back-propagation in which, for a given input pattern, the desired output pattern is compared to the output given by the network and the difference between these two patterns (the error) is minimised by adjusting the weights in the network by an amount proportional to the weight itself and to the total measure of the error. Although this method does not guarantee convergence to an optimal solution, since local minima may exist, it appears in practice that the back-propagation method leads to solutions in almost every case (Rumelhart *et al.*, 1994). Errors in representation would appear to arise only from having insufficient hidden units or the relationships themselves being insufficiently deterministic. Given the encouraging results obtained to-date in the fields of hydrology and salt-intrusion modelling (see Hall & Minns, 1993; Babovic & Minns, 1994; Minns and Hall, 1995) it was decided to test the applicability of ANNs to sediment transportation modelling.

EXPERIMENTAL DATA

The data sets used in this study were taken from the work of Zyserman and Fredsøe (1994), which were in turn derived from Guy *et al.* (1966). This work involved the determination of the bed concentration of suspended sediment c_b from flume experiments. The experimental data provided the total, steady-state sediment load for a range of hydraulic conditions including varying discharge, bed-slope and water depths. Zyserman and Fredsøe calculated the suspended sediment load q_s by subtracting the bed-load, calculated from the Engelund-Fredsøe (1976) formulation, from the total load. c_b could then be determined from q_s by applying Einstein's (1950) formulation in which the suspended sediment concentration profile is integrated over the depth down to the lower limit of the suspended sediment layer located at a distance of twice the grain diameter from the bed.

The hydraulic conditions of each experiment could be represented by the derived *Shields parameter*, θ, given by:

$$\theta = \frac{u_f}{(s-1)gd} \qquad \text{and} \qquad \theta' = \frac{u'_f}{(s-1)gd} \qquad (1)+(2)$$

where
u_f = shear velocity = $\sqrt{gDI}$
s = relative density of sediment
d = d_{50} = median grain diameter
D = average water depth
I = water surface slope
u'_f = shear velocity related to skin friction = $\sqrt{gD'I}$
D' = boundary layer thickness defined by

$$\frac{v}{u'_f} = 6 + 2.5\ln\left(\frac{D'}{k_N}\right) \qquad (3)$$

v = mean flow velocity
k_N = bed roughness = $2.5d$

Various combinations of the Shields parameter were considered together with various forms of the Rouse number, z, given by

$$z = \frac{w_s}{\kappa u_f} \qquad (4)$$

where
w_s = settling velocity of suspended sediment
κ = von Karman's constant (≈ 0.4)

The resulting relationship derived by Zyserman and Ferdsøe (1994) was then

$$c_b = \frac{0.331(\theta' - 0.045)^{1.75}}{1 + \frac{0.331}{0.46}(\theta' - 0.045)^{1.75}} \tag{5}$$

THE NEURAL NETWORK APPROACH

Equation (5) implies that the only significant parameter for determining the bed concentration is θ'. The results given by Zyserman and Fredsøe indicate that Equation (5) gives comparable, if not better, accuracy than several other, more complex empirical formulations. The question remains, however, whether this result could be improved by including *all* of the measured data. The parameters that incorporate all of the measured data are θ, θ', u_f u_f', d_{50} and w_s.

An ANN was set up with the six parameters given above as input and with the bed concentration c_b as the single output. The results are shown in Fig. 1 as a scatter plot of the measured bed concentrations (actual) compared to the bed concentrations as given by the ANN and by Equation (5) (predicted).

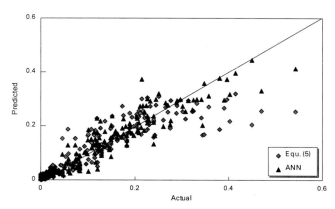

Figure 1   Scatter plot of bed concentrations for an ANN with six input parameters compared to results from Equation (5)

It can be seen from Fig. 1 that the ANN has a smaller scatter from the diagonal compared to the results from Equation (5), especially for the higher values of bed concentration. To compare the accuracy of the ANN to that of Equation (5), the root-mean-square (RMS) error was calculated for each method. The RMS error for the ANN was 0.034 and the RMS error for Equation (5) was 0.049. This indicates a significant improvement in the predicting capability of the ANN compared to the more traditional approach. Furthermore, the ANN makes use of all of the available data and hence provides a solution that will be sensitive to variations in any of the hydraulic parameters and not just θ' as in Equation (5).

If Equation (5) is accepted as having no deeper physical meaning (i.e. it has no *semantic content*), so that it is only a computational tool to calculate c_b, then the only major difference between this tool and the ANN is that Equation (5) can be written down *exactly* on paper while the ANN is stored, usually electronically, as a series of weights and connections between nodes. The evaluation of Equation (5), however, also requires the use of some sort of modern computational device and so the restriction of ANNs to be used only on computers is not considered a major limiting factor here.

Interestingly enough, if a simplified ANN is used to calculate c_b from only one input parameter, namely θ', then results of very similar accuracy to Equation (5) are obtained as shown in Fig. 2. In this case, the ANN results have an RMS error of 0.048, which is comparable to the results from Equation (5).

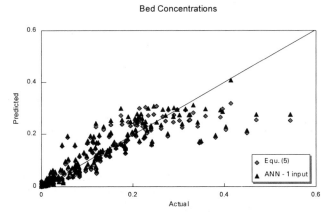

Figure 2   *Scatter plot of bed concentrations for an ANN with one input parameter*

CONCLUSIONS

An ANN is capable of learning the relationship between easy-to-measure hydraulic flow parameters and the bed concentration of suspended sediment. The ANN gives a more accurate result than that achieved by dimensional analysis. Furthermore, whereas dimensional analysis may reject some hydraulic parameters in order to simplify the resulting expressions, the ANN makes use of all of the available measured data. An ANN is therefore a very powerful tool for analysing experimental data and deriving very accurate representations of the measured parameters.

REFERENCES

Babovic, V. and Minns, A.W. (1994), Use of computational adaptive methodologies in hydroinformatics, *Proc. 1st International Conference on Hydroinformatics,* Verwey, A., Minns, A.W., Babovic, V. and Maksimovic, C (eds.), Volume **1**, Balkema, Rotterdam, pp. 201-210.

Beale, R. and Jackson, T. (1990), *Neural Computing: An Introduction*, Institute of Physics, Bristol.

Einstein, H.A. (1950), The bed-load function for sediment transport in open channel flow, *Technical Bulletin No. 1026*, U.S. Dept. of Agriculture, Washington, D.C.

Engelund, F. and Fredsøe, J. (1976), A sediment transport model for straight alluvial channels, *Nordic Hydrology*, **7**, pp. 293-306.

Guy, H.P., Simons, D.B. and Richardson, E.V. (1966), Summary of alluvial channel data from flume experiments: 1956-61, *Professional Paper 462-I*, U.S. Geological Survey, Washington, D.C.

Hall, M.J. and Minns, A.W. (1993), Rainfall-runoff modelling as a problem in artificial intelligence: experience with a neural network, *Proc. 4th National Hydrology Symposium*, Cardiff, British Hydrological Society, London, pp. 5.51-5.57.

Hertz, J., Krogh, A. and Palmer, R.G. (1991), *Introduction to the Theory of Neural Computation*, Addison-Wesley, Redwood City, Ca.

Hopfield, J.J. (1994), Neurons, dynamics and computation, *Physics Today*, Vol. **47**, No. **2**, pp. 40-46.

Hornik, K., Stinchcombe, M. and White, H. (1989), Multilayer feedforward networks are universal approximators, *Neural Networks*, **2**, pp. 359-366.

Minns, A.W. and Hall, M.J. (1995), Artificial neural networks as rainfall-runoff models, *Journal of Hydrology*, in press.

Rumelhart, D.E., McClelland, J.L. *et al.* (1986), *Parallel Distributed Processing. Explorations in the Microstructure of Cognition, Volume 1, Foundations*, MIT Press, Cambridge, Ma.

Rumelhart, D.E., Widrow, B. and Lehr, M.A. (1994), The basic ideas in neural networks, *Communications of the ACM*, Vol. **37**, No. **3**, pp. 87-92.

Zyserman, J.A. and Fredsøe, J. (1994), Data analysis of bed concentration of suspended sediment, *Journal of Hydraulic Engineering*, ASCE, Vol. **120**, No. **9**, pp. 1021-1042.

1B5

The Use of Global Random Search Methods for Models Calibration

D. P. SOLOMATINE

International Institute for Infrastructural, Hydraulic
and Environmental Engineering, Delft, The Netherlands

THE PROBLEM OF MODEL CALIBRATION

The fact, that hydraulic and hydrologic models can be used effectively only if properly calibrated, needs no special justification. The present paper deals with universal methods for such calibration, and provides some results on the use of these methods on an example of a hydrologic conceptual model.

The objective of calibrating any model of a physical system is to identify the values of some parameters in a model which are not known a priori. This is achieved by feeding this model with input data, and comparing the computed output variables to the variable values measured in the physical system. Their difference should be minimized by solving an optimization problem, with independent variables being the unknown model parameters. Among the widely used evaluation criteria for goodness of fit (with the vector OBS_i of observed output variables values, and the corresponding vector MOD_i of modelled values), the following ones may be mentioned: the simple least squares (SLS)

$$F^2 = \sum_{1}^{n} (OBS_i - MOD_i)^2$$

and dimensionless coefficient of determination (varying from $-\infty$ to $+1$):

$$R^2 = \frac{\frac{1}{n}\sum_{1}^{n}(OBS_i - \overline{OBS})^2 - \frac{1}{n}\sum_{1}^{n}(OBS_i - MOD_i)^2}{\frac{1}{n}\sum_{1}^{n}(OBS_i - \overline{OBS})^2}$$

where

$$\overline{OBS} = \frac{1}{n} \sum_1^n OBS_i$$

CHARACTERISTICS OF HYDRAULIC MODELS

Since complex models of water-related assets, like hydrodynamic models of open channels, unsteady processes in water distribution networks, or groundwater models incorporate numerical schemes, normally it is impossible to represent their response function in analytical form, and even to assess its shape in the parameter space. Normally, the model is represented in the form of an *algorithm*, and the response can be achieved only through its execution.

Among hydrologic models, the lumped conceptual rainfall-runoff (CRR) models, for example, are based upon the simpler algebraic and step-wise relations between various components of rainfall, precipitation-evaporation and the resulting runoff from the catchment; most model parameters cannot be assessed from field data but have to be obtained through calibration. Often, such models represent the catchment as a system of inter-connected tanks with additional outlets; e.g., Sacramento (soil moisture accounting part of the National Weather Service River Forecast System, USA) with 17 parameters (see, e.g., *Dung er al. 1992*), NAM (part of MIKE-11 modelling package) with 11 parameters, and Sugawara model (*Sugawara 1978*) with 8 parameters. Inspite of seeming simplicity of equations used in CRR models, they are, like hydrodynamic models, highly non-linear, and contain discontinuities, resulting in discontinuous and non-differentiable functions used for calibration evaluation criteria, which are to be minimized (*Dung et al. 1992*).

It should be mentioned, that apart from the minimization of the formal criteria for goodness of fit presented above, there are some recommended non-formal conditions in determining goodness of fit, e.g., for hydrologic models, these are agreements between average simulated and observed flows, and for the peak flows. Often, the 'overall agreement' for response function is taken into account. These rules can be easily applied together with the automatic calibration routines in the iterative search for the proper parameters values.

SINGLE-EXTREMUM FUNCTION MINIMIZATION

The problem of single-extremum minimization(or *local search problem*) has been extensively studied for the last 50 years (references and numerical codes can be found in *Press et al. 1991*). If the function is non-linear and is expressed analytically, and derivatives can be computed, the gradient methods are used: steepest descent method, conjugate gradient methods, and methods using second derivatives values or assessments.

If no analytical expression is provided, then the *direct search* methods like downhill simplex of Nelder and Mead, rotating directions by Rosenbrok, or direction set method by Powell. In the case of non-smooth and discontinuous functions the methods of global search are used.

MULTI-EXTREMUM FUNCTION MINIMIZATION (GLOBAL SEARCH)

Methods of global search are not based on the assumption of single-extremum, and some of these are referred below. For an extensive coverage of the topic see, e.g., *Zhigljavski 1991*.

Methods based on the use of local search techniques include such methods as multi-start (when multiple local searches are launched), candidate points with cluster analysis, tunnelling, and 'heavy ball' methods. The main representative of the group of deterministic 'covering methods' is the grid search. There are also methods based on statistical and Bayesian or information-statistical models.

The group of global *random* search methods includes random sampling (idea used by other methods), random covering, evolutionary (genetic) algorithms, simulated annealing (see *Press et al. 1991*), and the randomized version of multi-start.

GLOBAL OPTIMIZATION TOOL *GLOBE*

In order to test the applicability of global optimization to problems of hydraulic models calibration, the PC-based system GLOBE has been built. It is fully user-configurable, has six-window graphical interface displaying the progress of minimization in different projections, and can be used with an external program in the form of executable module. Currently, it includes three algorithms:

- *adaptive cluster covering (ACCO)*, proposed here by the author, and based on the ideas of automatic classification of prospective points into clusters with the subsequent random covering with the 'floating' shrinking area of search adapting to the gradually revealed topology of the minimized function; ideologically this algorithm is between the clustering algorithm, and the algorithms of random sampling and random covering;

- *Multis*, a version of a typical algorithm of random multi-start with the local search, developed by the author. For local search we use the Powell non-derivative minimization, line minimization by bracketing, and Brent quadratic interpolation are used (see *Press et al., 1991*); it has been modified for the use in constrained optimization problems by the introduction of a penalty function;

- *genetic algorithm (GA)* (see, e.g., *Wang 1991*) with some modifications.

Because of the limited size of the paper, here we present the essence of the *ACCO* algorithm only.

1. Sample *m* points in feasible domain $\Omega \subset \mathbb{R}$, compute the function value f_i at each point and choose *p* best points (with lowest f_i). Identify k_m clusters, such that the points inside cluster are close to each other, and clusters are far from

each other. For each cluster, identify the smallest n-dimensional interval K for prospective search, containing all points from the cluster.

2. Sample r_k points inside each K and choose t_k best points in each. Rearrange the clusters so that each would include the corresponding best points only. If stopping criterion (e.g., if $u\%$ of best points do not differ more than $w\%$ from the overall best) is achieved, stop. Remove clusters in which last e iterations the best point have been worse than the worst point in any other cluster.

3. Identify the 'center of gravity' of each cluster. Adapt clusters by composing around these centers the new search intervals, size of which is $v_k\%$ less than in the corresponding previous one. Go to 2.

The presented version of the algorithm serves as the basis for more sophisticated versions currently tested, with periodical global resampling, non-

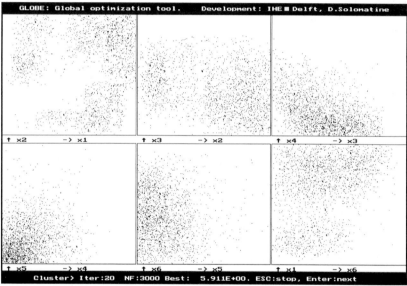

Figure 1. *GLOBE* screen with projections of 3000 evaluated points on six coordinate planes (on screen, clusters and best points are shown in color)

uniform sampling distributions, various versions of clusters generation and stopping criteria. The feasible domain Ω is usually an n-dimensional interval; the constraints are accounted by introducing the penalty function outside of the feasible domain.

The three algorithms have been tested on the following functions: random ellipsoids on top of each other (upto 8 dimensional space); several single- and multi-extremum functions used as traditional benchmarks in optimization (2- and

3-dimensional); CRR two-tank Sugawara, 1978 model (8-dimensional). The output screen after ACCO algorithm running the calibration problem of the Sugawara model in one of the studies (Awach Kaboun river basin in Kenya) is shown in Fig.1.

CONCLUSIONS

1. The traditional single-extremum optimization methods are often inadequate to the complexity of calibration problems.

2. Test runs of the global optimization tool *GLOBE* with three algorithms, and the results of other studies (e.g., *Dung et al., 1992*) shown the high potential of global optimization methods in calibration. The choice between various methods may depend on the type of the problem, and more research is needed in order to compare various methods used in various problem areas. Tentative comparisons results for three compared methods follow:

- *adaptive cluster covering* algorithm is potentially efficient, effective and robust. More experiments are needed for fine-tuning its adaptive features;

- *Multis* algorithm is efficient, effective and very accurate, especially in smooth functions, but it may be less reliable on discontinuous functions;

- *genetic algorithm* is less efficient, but potentially robust.

The following table shows the relative efficiency (how small the number of function evaluations is) and effectiveness (how small the found function value is) of the three algorithms, obtained on the calibration of the Sugawara, 1978 CRR model in the study of Awach Kaboun river mentioned above. Initial random sample contained 1000 uniformly distributed points.

Algorithm	Number of function evaluations (model runs)	Model run, at which function value got into 10% proximity of the minimum	SLS function value found
Adaptive cluster covering	3 000	1 600	58.09
Multis algorithm	6 312	1 300	57.04
Genetic algorithm	11 408	2 700	57.51

The efficiency problem can be very important since there are models to calibrate with dozens of parameters, and the volume of the search space depends exponentially on the number of parameters.

3. In our opinion, the lower efficiency of *GA* may be attributed to the highly redundant strategy of search with the high percentage of extra function evaluations. This feature is inherent to this class of algorithms, since they follow the ideas of natural evolution, appealing but highly redundant. Apart from that, the type of 'crossover' used in genetic algorithms (exchange of some of the parents' coordinates' values) leads often to the appearance of 'offsprings' in the search space quite far from their highly fit parents, and hence normally with the lower fit; so, the fit gained by parents is not inherited by offsprings. This seems to be inconsistent with the natural evolution theory, which GAs are said to follow. However, evolutionary algorithms are gaining popularity in hydraulic research (*Wang 1991*; *Babovic et al. 1994*).

High efficiency of *ACCO* can be explained by such features as the randomized start, using clusters of prospective points, and the adaptive strategy of subsequent randomized search. In our opinion, this class of algorithms is very promising for solving multi-extremum optimization problems.

4. Due to the high capacity of modern computers, the computationally intensive global search methods can be widely and effectively used in model calibration and other optimization problems, and the results of this and other studies provide justification for that.

REFERENCES

Babovic, V., Wu, Z. and Larsen L.C. (1994). Calibrating Hydrodynamic Models by Means of Simulated Evolution. Proc., First Intern. Conference on Hydroinformatics, Delft, The Netherlands. Balkema, Rotterdam, 193-200.

Duan, Q., Sorooshian, S., Gupta, V. (1992). Effective and Efficient Global Optimization for Conceptual Rainfall-Runoff Models. Water Resources Research, 28, No.4, 1015-1031.

Press, W.H., Flannery, B.P., Teukolsky, S.A., Vetterling, W.T. (1989). *Numerical Recipes. The Art of Scientific Computing*. Cambridge University Press, Cambridge, 702p.

Sugawara, M. (1978). Automatic Calibration of the Tank Model. Intern. Symposium on Logistics and Benefits of Using Mathematical Models of Hydrologic and Water Resource Systems. Preprints. Italy, October 24-26. IIASA, Laxenburg, Austria.

Wang, Q.J. (1991). The Genetic Algorithm and Its Application to Calibrating Conceptual Rainfall-Runoff Models. Water Resources Research, 27, No.9, 2467-2471.

Zhigljavsky, A. (1991). *Theory of Global Random Search*. Kluwer Academic Publishers, Dordrecht, 341p.

1B6

COMPOSITE MODELLING -
AN OLD TOOL IN A NEW CONTEXT

J.W. KAMPHUIS
Queen's University, Kingston, Ontario, Canada, K7L 3N6

PHYSICAL (HYDRAULIC) MODELS

Hydraulic problems have traditionally been studied with physical (hydraulic) models. Model scales are based on the assumptions that the equations and dimensionless functional relationships apply equally to a prototype and model. Since it is not possible to satisfy all the scaling relationships most physical models contain **Scale Effects** and because physical models are expected to simulate complicated prototype conditions with simple modelling tools, they also contain **Laboratory Effects.** Physical models are used because they simulate the prototype better than either the equations or the dimensionless functions on which they are based. Dimensional analysis describes a process in general form - physical models give detail. Physical models also automatically include many non-linear and complex processes and boundary conditions that are not clearly understood and therefore cannot yet be adequately expressed by equations.

Physical models of coastal phenomena may be classified as **Design Models** and **Process Models** (Kamphuis, 1991 and 1995). Design models may be further subdivided into **Long Term** and **Short Term**. Design models simulate actual complex prototype situations in order to provide specific information for design or retrospective study of failures. Sediment transport studies near harbour entrances and tidal inlets are long term design studies; many years of wave climate and sediment transport are simulated. Breakwater stability models are short term design studies; they model the response of a structure to a few (usually extreme) storm conditions. The process model studies a physical process in detail, such as how waves and currents interact, how bedform vortices move sediment up into the water column or how wind waves influence diffusion.

The **long term design model** suffers most from scale and laboratory effects, since a particular complex prototype must be correctly simulated in a model basin of limited size and capability. Scaling theory becomes mainly academic and is only relevant in the broadest sense for the planning of such models.

Technically, such models should only give **qualitative** results but the value of such results cannot be underestimated. They describe and often re-define the problem and give many indications toward possible solutions. On the other hand, any **quantitative** results derived from such models must normally be directly attributed to the understanding, experience and ingenuity of the modeller.

Historically, the long term design model has been the most common type and because of its requirements for large laboratory space, it gave birth to the great laboratory facilities that were built earlier this century. Often long term design models are still a good shortcut to results when answers are needed and the intermediate steps (equations and interactions of various parts of the model) are not clearly understood. But because of their limitations, the use of such models has waned and as a result the great modelling laboratories have experienced difficult times since the 1970's.

For the **short term design model** the input prototype conditions are better defined and simpler which means that laboratory effects are usually smaller. The sections of the prototype tested are often also smaller and thus larger models (smaller model scales) become economically possible. This has given rise to the popularity of large wave flumes. Proper scaling methods become much more relevant for these models and the model tests are usually shorter which makes these tests economically more attractive.

Because the **process model** is really an abstraction of prototype conditions, it offers the modeller most alternatives. It can be set up specifically to minimize laboratory and scale effects and thus the process model will yield the best quantitative results. A thorough understanding of the scaling methods and scale effects obviously becomes very relevant for these models.

IMPLICATIONS FOR PHYSICAL MODELLING

For many researchers, particularly those mainly involved in numerical modelling and field studies, the shortcomings of the hydraulic model are sufficient reason to reject this truly incredible tool. Field experiments are looked to for "real" data. Numerical models are not fraught with scale effects and produce appealing graphics which will convince the client that the new methods are better. The researcher familiar with physical modelling, on the other hand, first makes use of the very real and important **qualitative** impressions provided by a physical model. Although these images may not correspond directly to reality, at least they are based on a degree of physical similarity so that many of the complex processes, their interactions and the complicated boundary conditions are all reasonably simulated. These qualitative results represent the real difference between a physical and numerical model. The numerical model simply **echoes** the concepts and ideas of the modeller; it does not go beyond the input equations and coefficients. A physical model, on the other hand, **adds** to the modeller's input. The actual physical similarity of the model and the prototype fills in many details that the modeller did not or could not consider in the original model design.

A physical model study cannot produce **quantitative** results successfully unless the modeller understands the model scaling principles and knows which parameters are important and what are the scale and laboratory effects. However, good quantitative physical modelling results surpass both numerical modelling and field experimentation results. Kamphuis (1991) argues that the hydraulic model is the pivotal study tool that can link field experimentation and numerical modelling symbiotically. Clearly hydraulic modelling is not simply a matter of applying some **recipes** based on equations and dimensional analysis. Dimensional analysis can best be used to organize and understand the problem and to set up a scaling framework[1]. Any available equations can then be applied to incorporate scaling for what is known theoretically or empirically. The modeller must then carefully sift through the resulting scale effects and build a model that best simulates those aspects of the prototype which are of greatest interest. This may mean that several models are necessary to study different aspects of one single prototype.

To understand scale effects better, a **scale series** (models of the same prototype, but built at different scales) may be used. Scale series testing is also very useful if inadequate prototype information is available. The usually sparse prototype results may then be filled in from the physical model results.

The **trend** for better quantitative answers to more precise questions requires that the model results can be trusted without putting undue faith in the modeller. Because of the scaling and laboratory limitations, physical modelling has therefore recently tended to move away from the long term design model toward the process model. This trend will continue. It means proper understanding of the details of model scaling as demonstrated above will become ever more important and relevant. One indication of this is the recent publication of four large volumes which deal primarily with scaling of physical models - Martins (1989), Shen (1990), Hughes (1993) and Chakrabarti (1994).

Along with this move toward more precise process modelling comes the tendency to reduce laboratory effects as much as possible by improving

---

[1] In the same manner, dimensional analysis should also be used to plan, organize and understand field experiments and exercises in numerical modelling.

laboratory equipment and methodology. For a long term design model, the many simplifying assumptions meant that the actual model input conditions could be quite crude. Since for a process model the input conditions can be quite accurately specified, more care is taken to simulate the input conditions correctly. This has resulted in rapid improvements to modelling equipment. For example, in only a few years, wave generation has gone from paddle-generated regular waves, through long crested irregular waves to directional seas. It is now also possible to suppress unwanted long wave activity and Dalrymple (1989) talks about "designer" waves in which the computer program driving the directional wave generator takes into account the reflections off the side walls of the basin to produce a prescribed wave exactly at the structure to be tested.

COMPOSITE MODELLING

Process models alone cannot provide direct solutions to practical problems. They are building blocks used as steps toward a solution. The mortar that holds these building blocks together is computer calculations. The complete modelling task produces what could be called a **Composite Model** consisting of three distinct phases - physical modelling, the analysis of the physical model results and computation. The computation can be a complete numerical model in which case the physical process models simply provide appropriate coefficients and transfer functions. It is more likely, however, particularly for engineering studies (to provide useful answers within a limited time and budget for situations with complicated boundary conditions) that the computation will be a relatively simple statistical summation of a number of carefully determined process modelling results or of relatively simple empirical relationships derived from such model results. Non-linear interactions between the various building block units need to be carefully investigated from the process model results.

As an **example**, consider long term scour near tidal inlet structures. This process is impossible to study effectively with a long term design model. Short term design models could perhaps provide practical answers but only with respect to limiting scour conditions. It is possible, however, to conduct a large number of relatively similar, easy to set up process studies, each simulating scour at a single location for individual combinations of one current direction and velocity, one wave height, one wave period and one incident wave angle. Non-linear interaction can be derived from the time evolution of the scour holes for each of these imposed conditions. Ideally empirical relationships can be derived from the individual model studies to be combined in computer calculations which reflect wave and tidal current statistics to produce short term and long term erosion rates and volumes, limiting states, etc. To understand scale effects better, scale series tests can also be done.

Another example of composite modelling is the 50 hydraulic model studies carried out at Queen's University, at various scales and for various input conditions, simulating erosion of sacrificial beach islands used as drilling platforms in the Canadian Beaufort Sea. This set of rather complicated process models was successfully linked with a relatively simple computation to reconstruct the sequence of erosion events leading up to the eventual destruction of a prototype island and its equipment.

ADVANTAGES OF COMPOSITE MODELLING

Composite modelling has a number of distinct advantages over either pure physical modelling or pure numerical modelling. Because scale and laboratory effects are limited in process models, the main drawback of physical modelling has been reduced. Because physical modelling results are included in any numerical calculations, the output goes far beyond simply echoing input equations. Because both modelling concepts are combined, the method is immediately useful for problems that cannot be solved by either, by drawing on the strengths of both methods.

Several aspects of composite modelling make it economically attractive. First, the physical models are **simple** (with respect to the scaling relationships), **inexpensive** and **easy to understand**. They are repeatable and because the tests are very similar to each other, the experience gained with the first studies is immediately used in the later, similar studies resulting in a high efficiency.

Second, the physical model results used in composite modelling are **generic**; they are not very site specific and may be used to solve many similar problems for totally different layouts and locations. One could visualize, in time, complete libraries of such physical process modelling results which can be combined computationally to solve many different problems, greatly reducing the number of new model tests actually required to solve a new problem.

A third interesting aspect is that model **calibration** takes place within the computation phase. Thus calibration, verification and all the "what-if" scenarios are carried out at **low cost**.

Finally, the physical modelling and the computation phases of a composite model study need not be carried out by the same organization. For example, an informed **client** could also do the scenario computations once the laboratory has provided the building blocks.

CONCLUSIONS

Physical models are a unique tool. Because of the close physical similarity to the prototype, a physical model immediately provides valuable qualitative results and impressions. However, physical models are plagued by scale and laboratory effects and therefore common sense, intuition and experience of the modeller as well as the modeller's knowledge of scale and laboratory effects are needed to convert qualitative results into quantitative answers.

The requirements for more scientific results in the future will continue a trend in physical modelling away from the overall long term design models of immediately practical situations toward more abstract process models. Process models are accurate analogues of relatively simple prototype conditions. Accuracy is achieved by minimizing scale and laboratory effects and introducing relatively simple boundary conditions more correctly. The use of process models will require a more thorough understanding of model scaling and scale effects than was needed in the past.

To obtain practical answers from the rather abstract process models, it is necessary to link many individual process model results by computer calculations (composite modelling). This effective modelling technique combines the strengths of both physical and numerical modelling. The models are simple and readily interpreted. They are also not site specific and thus can be used for several projects. Composite modelling is economical since the calibration and "what-if" scenario work is done in the relatively inexpensive computation phase.

ACKNOWLEDGMENTS

The author thanks the Natural Sciences and Engineering Research Council of Canada for sponsoring his research on physical modelling research since 1970.

REFERENCES

Chakrabarti, S.K. (1994), "*Offshore Structure Modeling*", Advanced Series on Ocean Engineering, Vol 9, World Scientific Publishing, Singapore, 470 pp.

Dalrymple, R.A. (1989), "Physical Modelling of Littoral Processes", *Recent Advances in Physical Modelling*, R Martins (Ed), Kluwer Academic Publishers, pp 567-588

Hughes, S. (1993), "*Physical Models and Laboratory Techniques in Coastal Engineering*", Advanced Series on Ocean Engineering, Vol 7, World Scientific Publishing, Singapore, 567 pp.

Kamphuis, J.W. (1991), "Physical Modelling", *Handbook of Coastal and Ocean Engineering*, J. Herbich (Ed), Gulf Publishing, Vol 2, Ch 21, pp 1049-1066.

Kamphuis, J.W. (1995), "Physical Modelling of Coastal Processes", *Advances in Coastal Engineering, Vol 2*, World Scientific Publishing, Singapore, 567 pp.

Martins, R.(1989), "*Recent Advances in Physical Modelling*", R. Martins (Ed), Kluwer Academic Publishers, 620 pp.

Shen, H.W. (1990), "*Movable Bed Physical Models*", H.W. Shen (Ed), Kluwer Academic Publishers, pp 1-12.

1B7

Numerical and physical modelling of dynamic impact on structures from a flood wave

ASLAK LOVOLL and DAGFINN K. LYSNE
Department of Hydraulic and Environmental Engineering, The Norwegian
Institute of Technology, 7034 Trondheim, Norway

NILS R. B. OLSEN
SINTEF NHL, 7034 Trondheim, Norway

INTRODUCTION

Dam break and flash floods cause severe damage to people and infrastructure each year. Due to the rapid increase in discharge, bores and shock waves can form. The dynamic forces from the impact and the accelerating flow have to be known to assess the damage from such floods. This study describes a physical model study and a numerical model that have been used to examine the dynamic forces from a flood wave with steep front.

PHYSICAL MODEL STUDY

To create a steep flood wave, a gate was installed in a horizontal flume with dimensions 22x0.6x0.8 meter in the streamwise, cross- streamwise and vertical direction, respectively. Water was fed continuously at a rate of 0.055 m³/s behind the gate. When the water level reached 0.5 meter above the bed the gate was removed and the water created a flood wave propagating down the flume. The velocity and water depths were measured at two locations in the flume. The discharge approached steady state of 0.055 m³/s after about 40 seconds, as the reservoir was emptied. Figure 1 shows the laboratory flume and the set-up.

The structure was modelled by a rectangular cylinder with dimensions 0.1x0.07x0.3 meter, placed in the centre of the flume. The forces on the cylinder in the streamwise and cross-streamwise directions were measured by a force transducer based on shear strain measurements. Samples were taken at different frequencies to be sure to measure the shock or impact effects. The sampling frequencies were varied from 133 to 4000 Hz. Each experiment was repeated at least 5 times. The measured values varied between 5 and 10 %.

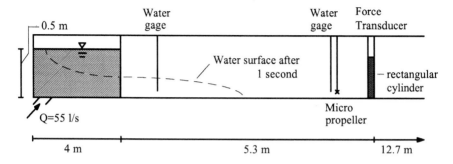

Figure 1. Laboratory flume.

The released water forms a steep wave (bore) about 0.1 m deep which propagate downstream with a speed of about 2.3 m/s. The flow is supercritical until the wave reaches the outlet. Then the flow becomes subcritical and gradually approaches steady flow condition, 0.13 m deep with a velocity of 0.7 m/s.

NUMERICAL MODELING

The flow around a cylinder is complex with secondary currents and formation and shedding of vortexes. It is not the intention to make accurate simulations of these effects in this study, but rather to simulate the mechanisms that are important for the force on the structure in steep flood waves. For detailed studies on the simulation on flow around a circular cylinder, see Olsen, N. R. B. and Melaaen, M. C. (1993) and (1995).

A 2-D or 3-D unsteady model is needed to simulate the complex flow of the present case. Several models exist, see e.g. Onishi, Y. (1994) and Wang, S. S. -Y. (1991). This study uses a 3-D unsteady model called SSIIM, developed by N. R. B. Olsen (1991) for water and sediment flow in channels and rivers. The numerical model is based on the Reynolds averaged Navier-Stokes Equations. The turbulent stresses are modelled with the standard k-ε turbulence model. A further description of the k-ε model is given by Rodi, (1980). A control volume method is used for discretization, using an upstream implicit method. Constant density is used, with non-compressible fluid. The SIMPLE method is used for the pressure coupling (Patankar, 1980). This method calculates the pressure based on the continuity defect in a cell. In the present study the SIMPLE method is not used for the cells closest to the surface. Instead, the continuity defect for these cells is used to calculate the change in the water surface. The grid is updated after each time step based on the changes in the water surface.

RESULTS

The laboratory flume is modelled by structured grids of different sizes. The size of the grid cells is of crucial importance for simulation of large gradients. The gradients are particularly large in the wave front and around the cylinder. The effect of the grid and the time step on the force is seen in Figure 2. The different simulations are explained in Table 1.

Simulation	19	21	23	24
Δx (m)	0.11-0.25	0.10-0.50	0.025-0.50	0.025-0.25
Δt (s)	0.005	0.02	0.02	0.005

Table 1. Characteristics of different simulations.

To get an accurate simulation a grid was used with 57x13x7 cells in the streamwise, cross- streamwise and vertical direction, respectively (simulation 24).

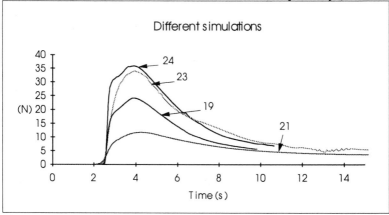

Figure 2. Dependency upon selected values of Δx and Δt.

A small time step (Δt) is necessary to keep the wave front steep and to accurately simulate the impact of the wave front hitting the cylinder. This explains the difference between simulation 23 and simulation 24 in Figure 2, between 2 and 3 seconds. Simulation 19 and 24 were stopped after about 10 seconds because of the long computational time required.

A comparison between simulation 24 and the measured force on the cylinder is given in Figure 3. The force in Figure 3 is an average of 5 identical runs. The force can be divided in three parts: drag, added mass (fluid inertia) and dynamic impact. The dynamic impact is caused by the first part of the wave front hitting the cylinder. The force increases from 0 to 26 N in 0.085 seconds. The water level still rises and after one second the force reaches its maximum of 35 N. The force is now dominated by a combination of drag and added mass. From about 10 to 40 seconds the velocity and the water level are approaching steady condition and drag is starting to dominate. From about 40 seconds and on, the flow is steady giving a force of about 4.5 N.

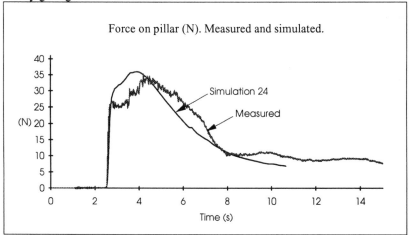

Figure 3. Measured and simulated force on structure due to a flood wave.

DISCUSSION

The total wave force on a structure is given by:

$$F = \frac{\rho}{2} \cdot C_D \cdot A \cdot u|u| + \rho \cdot C_M \cdot \forall \cdot \frac{Du}{Dt} + \frac{\rho}{2} \cdot C_S \cdot B \cdot u^2 \tag{1}$$

where:

F=Total force per meter of structure ρ=density of fluid
C_D=drag coefficient   A=projected area in the flow direction
u=parallel undisturbed velocity C_M=inertia mass coefficient
$\forall$=volume of the structure   t=time
C_s=impact (slamming) coefficient   B=with of structure

The first two terms are given by the Morison equation, developed for wave forces on vertical cylinders, Morison et. al. (1953). The last term is the impact force as defined by Miller (1977). Equation (1) is difficult to apply in this case. As mentioned above and illustrated in Figure 3, there is an effect from each term, but it varies with time. The wave front impact is of course present only in the beginning. The drag and inertia are present all the time, but the coefficient varies with time as discussed by Sarpkaya (1981). In addition the coefficients depend upon the shape of the structure, Reynolds number, relative roughness and dynamic response from the structure. Therefore the complexity of the situation is better handled by a numerical model than Equation (1).

Table 1 and Figure 2 show different simulations. In addition to the numerical algorithms, the resolution in time and space is essential for the result. Small Δt is necessary to simulate the impact. The peak force is depending on the spatial resolution of the grid around the rectangular cylinder. Comparing simulation 21 and 23 in Figure 2 shows the effect of reducing Δx from 0.1 m to 0.025 m close to the cylinder. To obtain a good result, Δx should also be relatively small between the reservoir and the cylinder. This avoids diffusion of the wave front.

In Figure 3 there is some deviation between simulated and measured force. This may partly be due to air entrainment in the front of the wave. This effect was observed in the physical model, but it was not included in the numerical model. The flow is also very complex going from supercritical to subcritical. In the physical model shock waves form along the wall. Due to this and reduced possibility for finer grid resolution because of limited time, some deviations are expected. Despite this the result is considered to be satisfactory.

CONCLUSION

The 3-D numerical model is capable of modelling a complex unsteady open channel flow. Even though the long computational time limits the possibility for accurate simulations of every detail, the model is able to predict the dynamic forces on a structure from a flood wave with steep front.

The results indicate that dynamic impact and the effects from added mass is important in addition to drag. The numerical model gives the possibility to study the effects of dynamic impact, added mass and drag in unsteady open channel flow in more detail.

REFERENCES

Miller, B. L. (1977): "*Wave Slamming Loads on Horizontal circular Elements of Offshore Structures*", Jour. Roy. Inst. Naval. Arch., RINA, Paper No. 5.

Morison, J. R., Johnson, J. W. and O'Brien, M. (1953): "*Experimental studies on wave forces on piles*", Proceeding of Fourth Conference on Coastal Engineering.

Olsen, N. R. B. (1991): "*A Three-Dimensional Numerical Model for Simulation of Sediment Movements in Water Intakes*", Ph.D. Dissertation, Department of Hydraulic and Environmental Engineering, The Norwegian Institute of Technology, Norway.

Olsen, N. R. B. and Melaaen M. C. (1993): "*Three-Dimensional Calculation of Scour Around Cylinders*", Jour. of Hydraulic Eng., Vol. 119, No. 9.

Olsen, N. R. B. and Melaaen M. C. (1995): "*Three-dimensional numerical modeling of transient turbulent flow around a circular cylinder*", Submitted to XXVI'th IAHR Congress, HYDRA 2000.

Onishi, Y. (1993): "*Sediment Transport Models and their Testing*", Computer Modelling of Free-Surface and Pressurised Flows. Edited by Chaudhry, M. H. and Mays, L. W. Proceedings of the NATO Advanced Study Institute. Pullman, WA, U.S.A., 1993.

Patankar, S.V. (1980): "*Numerical Heat Transfer and Fluid Flow*", Hemisphere Publishing Corporation, New York.

Rodi, W. (1980): "*Turbulence models and their application in hydraulics*", IAHR State-of -the-art paper.

Sarpkaya, T. and Isaacson, M. (1981): "*Mechanics of Wave Forces on Offshore Structures*", Van Nostrand Reinhold Company, New York, U.S.A.

Wang, S. S. -Y. (1991): "*Environmental Free-Surface Flow Modeling and Verification*", Computational Mechanics, Proc. Asian Pac. Conf., Publ. by: A.A. Balkema, Rotterdam, Netherlands.

1B8

NUMERICAL SIMULATION IN THE MANIFOLDS OF HYDRO POWER SCHEMES

R. KLASINC, H. KNOBLAUCH
Department of Hydraulics Structures and Water Resources Management
Graz University of Technology, AUSTRIA

ABSTRACT

In the design of powerhouse projects for hydro power schemes, increasing importance is attached to the hydraulic design of penstock manifold systems. The purpose of hydraulic optimation is in this case to minimise the system's energy losses. Especialy the design of manifolds for pumped storage systems - flow in both direction - is of main interesst. Determination of such losses has practically mostly been based on scale model tests. A physical model - a pipe distribution for two turbines - for measuring of head losses was constructed. The numerical model based on the method of finite volume elements was applied on the same model. As comparison of the results from the numerical model with the values measured on the physical model showed good agreement, the use of a numerical model of this kind can be regarded as an important and valuable supplement to the studies on a physical model, as this allows easy variation of model geometry for the study of alternative projects.

INTRODUCTION

To minimise the total loss in water conduits, it is necessary to know the hydraulic loss caused by the manifold. Increasing awareness of this fact has led designers to pay more attention to the determination of the hydraulic losses in the pipe junction. Determination of such losses has practically been mostly based on scale model tests, although in this case the expense of time and money involved in the variation of the geometry of such models tends to be high. If it were possible, however, to reproduce by mathematical means the extremely complex processes taking place in a pipe junction, variation of the geometrical boundary conditions would be easier. Although this could not replace the scale model tests altogether because the numerical model need's calibration, the computation is an ideal preparation for the model test. At our department, a comparative calculation using a numerical model was performed in parallel with a scale model test.

Test Set-up

A physical scale model was constructed to simulate the flow patterns of the prototype and measure pressure levels and velocity heads, using water as a flow medium (see Fig. 1).
Pressures were measured in six instrument cross-sections upstream of the dividing junction and in four cross-sections in each branch downstream of the junction. An instrument cross- section consisted of four drillings diametrical distributed over the circumference of the pipe and connected by a ring line. A transducer measured the pressure differential between an instrument cross-section above the junction and a cross-section below the junction.

Velocity heads were computed from the flow rates measured by magneto-inductive flow meters at the end of the branches. The losses due to the pipe junction were calculated by comparing the energy upstream and downstream of the junction (Bernoulli).

$$h_v = h_g - h_f$$

h_v -> shape loss

h_g -> total loss

h_f -> friction loss

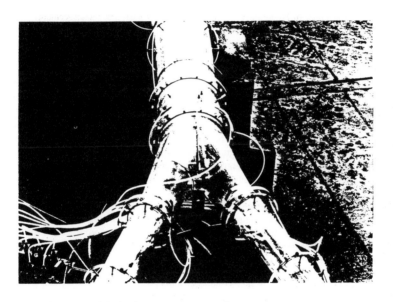

Fig.1: Physical model with instrument cross sections

Theoretical Calculation of Shape Losses
In flows affected by viscosity forces, the Reynold's number, Re, apart from the geometrical characteristic, must be kept equal for model and prototype. This ensures not only geometrical characteristic similarity, but also the observance of Reynold's law (Kobus, 1978):

$$Re_r = \frac{Re_n}{Re_m} = \frac{\rho_r v_r L_r}{\eta_r} = 1$$

$$v_r = \frac{v_n}{v_m} = \frac{\eta_r}{\rho_r L_r}$$

Indices: n = prototype

m = model

L_r = model scale

For laboratory tests using the same fluid, ($\rho_r = 1$) and ($\eta_r = 1$), and Reynold's law may be reduced to the requirement that the velocity scale should be inversely proportional to the length scale:

Therefore, on a reduced scale model, velocities should be adjusted so as to be larger than those on the prototype. The model studied was constructed to a scale of 1:13. That means that the velocities

$$v_r = \frac{1}{L_r} = \frac{v_n}{v_m} = \frac{L_m}{L_n}$$

used must be 13 times larger than those on the prototype to satisfy the requirements of Reynold's law of similarity. However, velocities of this magnitude cannot be accomplished in a model test. Therefore, it is necessary to use the model results obtained from certain Reynold's numbers for inferring much higher Reynold's numbers and to use these for computing the corresponding losses for the prototype.

For this purpose, measured values of pressure losses, (a series of points on the flow - head loss diagram) were approximated by means of a polynomial. The measured velocity heads were treated in the same way. This gave two functions to be used for extrapolation to larger Reynold's numbers. The function of differential velocity head versus flow can be determined accurately. Head loss function was approximated by writing the polynomial $\Delta p = c_0 + c_1.Q_a + c_2.Q_a^2$. This function was determined following extensive studies using different polynomials (Klasinc, 1992).

This method (simulation of losses for the prototype) was also simulated with numerical model.

NUMERICAL MODEL
The Bases for the Computer Program

The program used, solves the conservation equations of mass, impulse (three Cartesian components), thermal energy, turbulence and passive transport equations in general nonorthogonal moving systems of coordinates. Spatial discretising was accomplished by use of the finite volume element method (Hinze, 1987).

Mathematical treatment of the turbulences was based on the k - ε turbulence model. This is a two-equation model using the principle of dynamic viscosity for turbulent flow and a transport equation for the rate of dissipation, ε, associated with a certain length. This is the widely-used computer program.

The k - ε turbulence model employs additional partial differential equations for the turbulent kinetic energy k defined by

$$k = \frac{u_i^2}{2} \qquad \text{and its dissipation} \qquad \varepsilon = c_\mu^{\frac{3}{4}} \frac{k^{\frac{3}{2}}}{L}$$

In the above equation L stands for a turbulent length scale and c_μ is an empirical coefficient. There are two important features that distinguish near-wall regions from other portions of the flow field: First there are steep gradients of most of the flow properties, and secondly, the turbulent Reynold's number is low so that the effects of molecular viscosity can influence flow energy.

Actually, these problems could also be solved by means of the above method provided a very dense mesh of elements is established near the boundary of the model. But it is easier to approximate the boundary layer by semi-empirical relationships. Thus, the near-wall velocity (u_c - point C) variation is described by the logarithmic relationship:

$$\frac{u_C}{u_\tau} = \frac{1}{\kappa} \ln\left(-\frac{yu_\tau\rho}{v}\right) + C$$

where κ and C are experimentally determined constants. The velocity, u_τ, can be written as:

$$u_\tau = \left(\frac{\tau_W}{\rho}\right)^{\frac{1}{2}} \qquad \tau_W = \text{wall shear stress}$$

The computer program (AVL, 1993) is based on the assumption of a hydraulically smooth pipe. With a roughness k_S of 0,0015 mm (plexiglass) and a diameter of 128 mm and 292 mm, respectively, with maximum Reynold numbers of 5 x 10^5, the model closely approaches hydraulically smooth conditions, so that the numerical model is adapted to this conditions.

The Model Used
Before undertaking the computation, we had first to determine the geometry of the bifurcation and then generate a mesh of finite volume elements (with about 30 000 elements, Fig. 2).
The boundary of the model is decided by the geometry of the pipe. Inlet and outlet openings must be defined and boundary conditions have to be found out for the inlet opening. These are head, flow and temperature. In the case under discussion, a temperature of 20° Celsius was adopted. Adiabatic processes do not play a decisive role in case where water is used as a fluid.

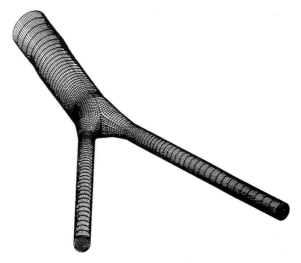

Fig. 2: Geometry and generated volume elements for the comparative calculation

In the distribution of elements over the cross sectional area, there are certain critical elements for which calculation presents problems. The ideal condition for the calculation is that all the angles of an element are approximately 90 degrees. Then more the angles differs from 90 degrees, the larger are the resulting inaccuracies in the calculation. The problem can be mitigated by establishing a closer mesh in these areas.

COMPARISON
Turbine Mode - Symetrical Dividing Flow
For comparing the results of the numerical and physical models we chose the pressure loss pattern in the dividing junction as being best suited for this purpose.
As can be seen from Fig. 3, the pressure loss patterns of the two models are almost the same. The deviations from the straight line can in part be explained by the problems involved in data collection on the physical model. The measured differential pressures as well as the velocities are large values in absolute terms, but the difference to be found between the two curves is very small, so that even slight inaccuracies tend to have large effects on the determination of the loss.

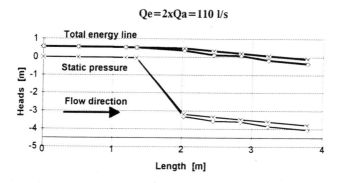

Fig. 3: Comparison of pressure losses between physical model (◊) and numerical model (x)

As shown from figure 4, the velocity distribution returns to uniformity already at a short distance from the dividing junction. This agrees with the results of the physical model, where the pressure distribution was also fairly uniform in the instrument cross sections downstream of the junction.

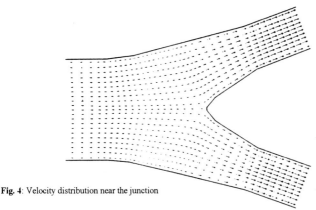

Fig. 4: Velocity distribution near the junction

Pumping mode - Symetrical Combining Flow

The calculated head losses are much bigger as in turbine mode. See fig. 5.

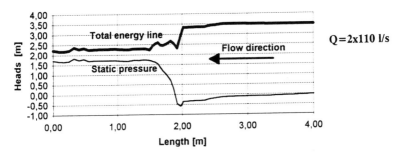

Fig. 5: Head losses in numerical model

CONCLUSION

Summarising, we may say that the estimating of head losses in numerical resp. physical models showed relatively good agreement. Further studies are required with regard to test the numerical models on different manifolds. The application of a numerical model of this type can be regarded as an important supplement and improvement for studies on physical models in ensuring supplementary representation of measured values as well as facilitating the study of alternatives by means of easily variable model geometry.

REFERENCES

Kobus. H.   Wasserbauliches Versuchswesen
                   Deutscher Verband für Wasserwirtschaft
                   (DVWW), 1978

Klasinc, R   Power Losses in Distribution Pipes
Knoblauch, H.   HYDROSOFT 92', Valencia, Spain
Dum, T.

Hinze, O. J.   Turbulence
                   McGraw-Hill Classic Texbook Reiussue, 1987

AVL   Flow In Reciprocating Engines (FIRE), Program Manual, Ver. 5.0, Graz 1993

1B9

HYDRAULIC MODELLING OF COMPLICATED FLOW AT AN INTAKE
- A CASE STUDY OF FLOW PHENOMENA -

CHEN HUIQUAN
Prof., China Institute of Water Resources and Hydropower Research (IWHR)

ABSTRACT

In the final phase of development of the Lamma Power Station in Hong Kong, an ash lagoon will be constructed which has to be located in close proximity to the future No. 3 Cooling Water Intake due to the geographical and environmental constraints. This unusual lay-out will lead to complicacy of flow and ill effects on operation.

A series of physical models have been built for simulation of the future seawater flow pattern with the lagoon in place to ensure healthy operation of the CW Intake[1]. The paper presents the model methodology used for realizing flow and sedimentation simulation and yielding main test results. Emphasis is laid on the line of thought to tackle such flow problem with some interesting hydraulic phenomena obtained in the experiments given.

I. INTRODUCTION

The Lamma Power Station is located on the north coast of the Ha Mei Wen Bay in the west of the Lamma Island, Hong Kong with a total generated capacity of 2,500MW developed in 3 stages. The first and second stages providing 1,800MW were completed. Fig. 1 and Fig. 2 indicate the general location of the power station.

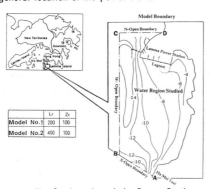

	Lr	Zr
Model No.1	200	100
Model No.2	400	100

Fig. 1 Location of the Power Station

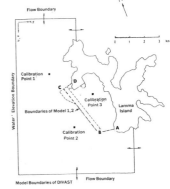

Fig. 2 Location of Model Boundaries

In the final stage of the plant's development, an ash lagoon will be built for the current disposal arrangement of barging the ash damped. Taking into account of the geographical and environmental constraints, the lagoon has to be situated on the southern seawater in front of the station in close proximity to the future intake No. 3 (see Fig. 3). The presence of the lagoon will greatly complicate the ambient flow and deteriorate the thermal-hydraulic behaviors of the intake. Thus appear the questions as: could the water intaking capacity under such unfavorable boundary conditions be the same as that originally designed with no ash lagoon and the siltation blockage of the intake culvert be unlikely to happen.

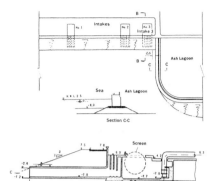

Fig. 3   Layout of Intake 3

The previous studies on the effect of the ash lagoon were merely conducted by mathematical simulation, the 2 − D model DIVAST [2], which indicated that the flow pattern in the north of the bay will be altered greatly in the immediate vicinity of the lagoon. Truly, the complexity and the high 3 − D behaviors of the actual flow near the intake does necessitate the physical modelling to afford the correct localized flow pictures and the detailed information about the dynamic effect of the lagoon on the intake.

The main objectives of the physical modelling of this involved flow region is to furnish a sound base for evaluating the hydraulic influence of the lagoon upon the No. 3 intake capability and the potential of sediment clogging or air-entering at the entrance of the intake.

The paper focuses on the methodology to approach the problem for anticipating the flow characteristic of the water region. Some newly found hydraulic phenomena are also briefly illustrated.

II. METHOD OF APPROACH

1. CONCEPTIONAL ANALYSIS

The goal of the experiments could be finalized as the testing on the variation of water intaking capability $\triangle Q = \triangle (V \times A)$ of intake 3 after the construction of the lagoon where V and A are the average velocity and the control sectional area of the intake entrance or culvert. As presented in Sketch 1, many physical parameters come into play, such as tide and so on, in determining the intaking capability. To reproduce faithfully the $\triangle Q$ in the model, it is theoretically necessary to have all the five individual parameters simulated. But each physical motion has its own similarity requirement and there exist such an amount of scale contradictions that it seems not feasible to simulate all the phenomena meanwhile.

It follows, in order to make the practice of hydraulic modelling in reality and to have the physical pheomena interested correctly reflected and ascertained in modelling, that the following line of thought has to be adopted:

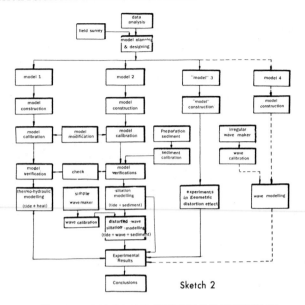

Sketch 2

III. MODEL DESIGN AND EXPERIMENTAL EQUIPMENTS

1. MODEL DESIGN

The three open boundaries AB , BC , CD of the Model 1 , 2 and the water region studied are indicated in Fig.1. Fig.4,5 illustrate the general layout of the Model 1 and Model 4 respectively.

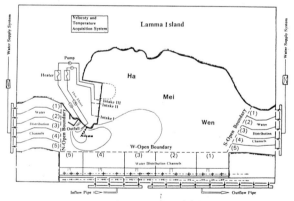

Fig. 4   General Layout of Model No. 1

2. MEASURING APPARATUS

(1) Water temperature measurement: Temperature scanning system was used, With 150 semi-conductor probes mounted on frames controlled by water-following point gauges. Water surface temperature involved thermo-vido system with 216 color levels.

(1) The experiments are carried on parallelly in several models. Each model is designed and fabricated in strict accordance with the reproduction of the very physical features to be studied;
(2) Investigations are focussed on the quantitative comparisons of the experimental data obtained in the case of with or without the lagoon.
(3) The final conclusion is deduced from synthetic analyses of the informations obtained from the different models.

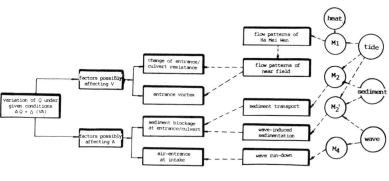

Sketch 1

2. PLANNING OF EXPERIMENTS

A total of 4 models was used with the functions and features of each model in the listed as following table. The coordination and the experimental procedure of the models could also be seen in Sketch 1, 2

Table

Model	M_1	M_2	M_3	M_4
Functions (main points studied)	velocity field temperature field vortex	flow pattern siltation comparison of deposition vortex	flow pattern temperature field scale effect of geometric distortion	wave reflection wave refraction wave run-down
Features	tidal flow simulation heat flow simulation small geometric distortion good for low velocity measurement (tide + heat) experiment	tide simulation siltation simulation easy for modification (tide + sediment) experiment (tide + wave + sediment) experiment	steady flow systematic flume tests	wave simulation geometric undistortion wave experiment
Length Ratio L_r	200	400		100
Depth Ratio Z_r	100	100		100
Geometric Distortion $\varepsilon = L_r / Z_r$	2	4	1, 2, 4	1

251

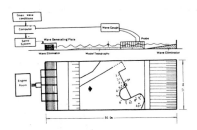

Fig. 5   Model No. 4 and the Wave Basin

(2) Velocity measurement: Photoing with digital image processing and other tracing method used.

3. SIMULATION SYSTEM OF TIDAL CURRENT

Both open-and closed-loop tidal open boundary control systems were used and well coupled in the same model in presence. Due to the complexity of the spatial and time-dependent variations along the three open boundaries, each boundary was subdivided into smaller sections and 14 channels in total were provided with the discharges given by the relevant mathematical model DIVAST. A combination of solenoid valves with electric butterfly valves was used so as to have the closed-loop control system supplemented by a cheaper open-loop control system. The system has proved its higher adaptability, shorter adjusting period and good performance.

IV. MODEL CALIBRATIONS AND VERIFICATIONS

Either in Model 1 or in Model 2, good agreement is achieved between the given and simulated data both in spring and neap tides. Fig. 6, 7 demonstrates a part of comparison curves.

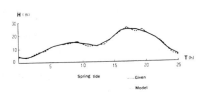

Fig. 6   Water Elevation Verifications

V. EXPERIMENTAL RESULTS

1. FLOW PATTERNS

The flow patterns for the flood and ebb periods are in overall similar to those of the DIVAST model realizations. But at the high and low water turns, the flow presents its complexity with the plane eddies taking place in great extent. With the presence of the constructed lagoon, there exists an eddy flow in the corner restricted by the southern seawall and the lagoon's projected west wall. The low variation in the early stage of the ebb period should deserve special attention. The coastal flow in approaching the seawall instead of hitting the latter headlong gradually turns southward; its upper layer having mixed with the effluent from the outfalls does not dip underneath, threatening of its entrance into the

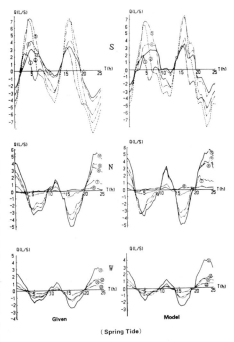

Fig. 7   Comparisons of Hydrographs

intake asconventionally thought unavoidable and guarded against. The induced flow again leads to an interesting phenomena, that the intake water temperature increment at this period is even smaller than the case without lagoon.

2. VORTEX

Both in Model 1 and Model 2, no penetrating swirling vortex in proximity of the intake 3 was found. At the very corner of the water region surrounded by the seawall and the lagoon sea wall, there is frequently a vortex but in small strength of circulation.

The correct simulation of the vortex should follow the requirement of geometric original. In order to have a quantitative understanding of the scale effect of the geometric distortion on the vortes and other phenomena observed in the Model I and Model 2, a systemic study in Model 3 was conducted. The results manifest, for $\varepsilon \leqslant 4$, flow and temperature patterns including the location of the vortex certer is similar in general. The experiments verify the correctness and accuracy of the conclusions drawn from Model 1 and Model 2.

3. SILTATION

White bakelite powder was chosen as the model sediment. Comparison of the degree of deposition after running 60 tidal cycles with and without the ash lagoon was made. The test results disclosed the low siltation in front of the intakes, which could be attributed to the influence of the high intaking velocity there. Only in the dead corner between the seawall and lagoonwall, there is some degree of silltation owing to the presence of the local vortex.

4. WAVE

In order to get some pictures about the effect of the wave action on sedimentation, wave was simulatedly generated and mounted to the Model 2. The test reveals the fact that the surface fluctuations in the corner region of the intake 3 is particularly intricate on account to the wave reflection and refraction both taking place in this region. In the case of low water level and high wave height, the entrance of the intake could be exposed to the air during the period of wave run-down.

The undistorted Model 4, strictly designed for wave simulation, was constructed and tested. It was discovered that even under extreme lowest water level whence the wave trough elevation was 1m below the top plate of the entrance, no air admission was observed, with consequently no air passing through the culvert. This comes from the fact that when waves run down the slope of the seawall at the entrance of the intake, the water mass of the previous wave accumulated also rushes down the slope and well forms a water curtain which screens the entrance from air admission and keeps the culvert always in a state of full pipe flow as shown in Fig.8.

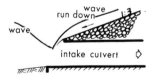

Fig.8 Water Curtain

VI. CONCLUSIONS

The construction of the ash lagoon leads to a very complicated flow. The several constitutents include irregular boundaries, tidal current, vortex, siltation, wave and heated effluent circulation. The flow characteristics wlth lagoon do not effect the water-intaking capability of the intake and induce no sediment clogging, the chief care of the designer being eased. Experience with the present study shows that well-considered and well-hybridized physical models could be successfully used for studing such complex flow regime.

REFRENCES

[1] Chen Huiquan, Physical Modelling Study for Intake 3 of Lamma Power Station, Report of IWHR, 1994.

[2] Binnie and Partners, Hydrodynamic Model, Lamma Island Power Station Ash lagoon, Sept, 1990.

1B10

AERATOR LOCATION IN DAM SPILLWAYS:
PROPOSAL FOR A METHOD

Ramón GUTIERREZ SERRET
Civil Eng. Ph.D., Head Hydraulic Structures Dept. Centre
for Research and Experimentation (CEDEX), Ministry
of Public Works, Transport and Environment - SPAIN

A method is presented which, based on numerical solution of gradually aerated differential flow equations, makes it possible to determine whether aerators are needed, the number required and the approximate location in chutes, with a view to preventing cavitation damage. A computer program has been prepared for the application of this method.

The development of the method can be structured into three stages: a) analysis of self-aeration throughout the chute, b) siting and design of the first aerator, and c) analysis of the flow downstream from the first aerator and, where necessary, the positioning of the second and succesive devices. On completion of these stages, the process would, when required, continue with the design of further aerators. The modus operandi of each stage is:

a)   Analysis of self-aeration throughout the chute.

The "cavitation index (σ)", and the "air concentration on the channel bottom (C_o)", must be calculated throughout the chute, for the entire range of overflows. The design flows to be considered, can be set at intervals of 20% from the maximum flow $(0'2\ Q_{Max})$, with a view to determining the minimum value of σ, because this does not coincide with Q_{Max}.

The cavitation number (σ) is expressed by the well-known formula:

$$\sigma = \frac{P - P_v}{\dfrac{\rho V^2}{2}} \tag{1}$$

where P and V are the pressure at the channel bottom[1] and the flow velocity in each section, P_v is the water vapour pressure and ρ is the water density.

The concentration on the bottom (C_o) can be obtained using the expressions proposed by several authors, Rao & Gangadhariah (1971), Hager (1991), or from the transversal air distribution [C(y)] proposed by Wood (1991), making y = 0 (Gutiérrez Serret,

---

[1] P = h_w; h_w = "equivalent water depth" at which the water will theoretically flow, free of air, but flowing at the average real velocity of the air-water mixture ("equivalent flow").

1994). Bearing in mind the simplicity of the formulae proposed by Hager, it is thought that the following equations are the most suitable for the case of self-aeration:

$$C_o = 1.25 \left(\frac{\pi}{18D}\alpha\right)^3; \qquad 0^\circ \leq \alpha \leq = 40^\circ \qquad (2)$$

$$C_o = 0.65 \cdot \sin \alpha; \qquad 40^\circ \leq \alpha \leq 80^\circ \qquad (3)$$

where α is the angle between the chute bottom and horizontal.

However, the need to obtain P and V, in order to calculate σ and to determine C_o downstream from the aerators (Stage c), makes it necessary to use the two differential equations which govern the gradually aerated flows: "continuity for air" and "energy for aerated flow", These equations are expressed as follows:

- Continuity for air equation.

$$\frac{d\overline{C}}{dx} = (1-\overline{C}) \left[\frac{(V_T)_u \cos\alpha}{q_w} (k_e\overline{C}_u - K_T\overline{C}) (1-\overline{C}) + \frac{\overline{C}}{b}\frac{db}{dx} \right] \qquad (4)$$

- Energy for aerated flow equation.

$$\frac{dh_w}{dx} = \frac{\sin\alpha \cdot (1 + h_w\frac{d\alpha}{dx}) + E\frac{q_w^2}{gh_w^2}\frac{1}{b}\frac{db}{dx} - I}{\cos\alpha - \frac{E}{g}\frac{q_w^2}{h_w^3}} \qquad (5)$$

where the following are:

x : Chute distance from the "Inception Point".
b : Chute width, generally variable; $b = b(x)$
σ : Angle between chute and horizontal, generally variable; $\alpha = \alpha(x)$
q_w: Specific water flow $(m^3/s.m)$.
q_a: Specific air flow $(m^3/s.m.)$.
h_w: Equivalent water depth.
$\overline{C}$: Average air concentration in each section; $\overline{C} = q_a/q_a + q_w$.
V_e: Entrainment velocity.
V_T: Terminal bubble velocity in turbulent flows. Usually considered as 0.17 m/s (Wood, 1991) < 0.40 m/s (Chanson, 1992).
$K_e = V_e/(V_e)_u$; $K_T = (V_T)/(V_T)_u$
subindex u: Refers to the values of $\overline{C}$, V_e and V_T in the "uniform aeration zone"
E : Kinetic energy correction coefficient

I : Friction slope: $I \sim q_w^2 \cdot f_w / 4g \, k_w^3$ (Darcy-Weisbach)
where:
f_w: Roughness coefficient of the "equivalent flow";

$$\frac{f_w}{f} = \frac{1}{1+10\overline{C}^4} \tag{6}$$

f: Roughness coefficient of the non-aerated flow;
g : Apparent gravity.

Numerical methods (finite differences) are generally used to solve this somewhat complex equation system, and the average concentration and the equivalent depth throughout the chute: $\overline{C} = \overline{C} \,(x)$ and $h_w = h_w \,(x)$, can be determined. It is then possible to obtain the remaining parameters that characterize the aerated flow (emulsified depths, velocities, etc.) and the longitudinal distribution of the cavitation number $[\sigma = \sigma \,(x)]$. The following simplifications are usually taken: $E \sim 1$, $V_e \sim (V_e)_u$ and $V_T \sim (V_T)_u$. Thus $K_e = K_T \sim 1$.

It often occurs that the slope and width of the chute are constant. In such cases, the differential equation system becomes two independent equations. The longitudinal distribution of the average concentration $\overline{C} = \overline{C} \,(x)$, is obtained by integrating the continuity equation (4). The energy equation (5), can be solved on the basis of the flow conditions at the Inception Point (depth $h_{Crit.}$ and velocity $V_{Crit.}$) through the finite differences method, thereby determining the development of the equivalent depth throughout the chute $h_w = h_w(x)$ (Chanson, 1991 and Gutiérrez Serret, 1994). Along the same lines, an approximate method for calculating the emulsified depth of the water-air mixture throughout the length of the chute, was proposed by Sinniger and Hager in ICOLD Bulletin Nº 28.

b) Siting and design of the first aerator

Once the longitudinal distribution of the cavitation number and the concentration at the bottom are known $[\sigma = \sigma(x); C_o = C_o \,(x)]$, the first aerator must be placed as far as possible upstream where the following conditions simultaneously hold: $\sigma < \sigma_{Crit.}$ and $C_o < 7-8\%$. As values of the critical cavitation number (σ_{Crit}) 0.2-0.25 can be adopted (Falvey, 1990).

When the position of the first aerator has been determined, it must be designed. It is advisable to predimension the aerator, beginning by selection of the typology and initial dimensions. Experience gathered from aerators already in operation is extremely useful at this phase (Pinto, 1991).

A mathematical analysis of the aerator operation must then be undertaken, determining the jet path, entrained air discharges, the pressure in the cavity and the aerator operation point. Such an analysis must be conducted for different discharges - low, medium and high - so that all operational hypotheses are covered. Checks must also be made to ensure that no flooding takes place with low discharges and that aeration is not excessive.

c)   **Analysis of the flow downstream from the first aerator. Positioning of the second and successive devices.**

In accordance with the similarity that can be established between the flow downstream from an aerator and self-aeration (Chanson, 1992), the use of the continuity equations for air (4) and for the energy of the aerated flow (5), taking the depth h_i and the average concentration $\bar{C}_i$ at the outlet jet impact zone as initial conditions, makes it possible not only to characterize the flow downstream from the aerator, but also to obtain the concentration at the bottom and the cavitation number throughout the chute [C_o = C_o (x); σ = σ (x)], then, if once again σ < 0.2 - 0.25 and C_o < 7-8%, the siting of a second aerator can take place.

The aforementioned procedure will then be repeated from the second aerator, and so forth, as many times as necessary, until the entire chute is suitably aerated.

The initial flow conditions immediately after the jet impact - depth h_i and average concentration $\bar{C}_i$ - can be determined using the formulae proposed by Chanson (1992)[2]. These expressions are:

$$\frac{h_i}{h_{impact}} = 1.92 - 1.35\theta \qquad (7)$$

$$\bar{C}_i = \frac{Q_a - Q_a^{DEAER}}{Q_w + Q_a - Q_a^{DEAER}} \qquad (8)$$

$$Q_a = Q_a^{DUCT} + Q_a + Q_a^o \qquad (9)$$

$$Q_a^{DEAER} = 0.0762 \cdot \theta \cdot Q_a \qquad (10)$$

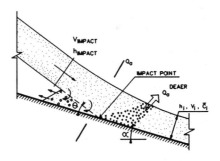

Fig.1.- Impact zone

---

[2] (Falvey, 1990) also proposes an expression to calculate h_i

where the following are:

Q_w: Water chute discharge; Q_a^{DUCT}: Air ducts discharge
Q_a: Air flow at end of impact zone before deaeration starts
Q_a^{DEAER}: Air discharge lost on impact
Q_a: Air discharge at the aerator begining
θ : Angle between jet impact and bottom. Determined by
calculating the jet path beforehand (Stage b).

Calculation of the concentration at the bottom (C_o) downstream
from the aerator after jet impact, is made by applying
differential equations (4) and (5) of the gradually aerated flow,
to the above-mentioned concentration C_o, under the hypothesis
that the longitudinal distribution of the concentration at the
bottom [$C_o = C_o$ (x)] and the average concentration [$C = C$ (x)]
are roughly governed by analogous laws.

As the three stages described must be performed with the aid of
software, a computer program has been developed. The application
of this program to the Canales Dam spillway, in the south of
Spain (Granada), is shown in a summarized way in Fig. 2.

When the aeration devices to be placed have been defined, these
should generally be checked by means of scale model tests and,
where necessary, with prototype tests, in spite of the problems
entailed in aeration tests.

Finally, it should be mentioned that the method presented is of
a preliminary nature for guidance purposes, and it must also be
emphasized that when planning the final project for aerators,
reference should be made to practical experience gained from
projects already in operation.

BIBLIOGRAPHY

CHANSON, H. (1992). "Air Entrainment in Chutes and Spillways". Research Report
Nº CE 133, Dept. of Civil Eng. University of Queensland, Australia.

FALVEY, H.T. (1990). "Cavitation in Chutes and Spillways". Engineering
Monograph Nº 42, Bureau of Reclamation, Denver, U.S.A.

GUTIERREZ SERRET, R. (1994). "Aireación en las Estructuras Hidraúlicas de las
Presas: Aplicación a los Aliviaderos". (Aeration on Hydraulic Structures of
Dams: Application to Spillways). Doctoral Thesis. Polytechnic University of
Madrid, Spain.

HAGER, W.H. (1991). "Uniform Aerated Chute Flow". ASCE, Journal of Hydraulic
Division, Vol. 117, HY4, April, pp. 528-553.

ICOLD (1992). "Spillways. Shockwaves and Air Entrainment". Bulletin 81.

PINTO, N.L.S. (1991). "Air Entrainment on Free Surface Flows". Design Manual
Nº 4. IAHR, A.A. Balkema Published, Rotterdam, Netherlands, Chap. 5.
"Prototype Aerator Measurements". pp. 115-130.

RAO, N.S.L. and GANGADHARAIAH, T. (1971). "Self-Aerated Flow Characteristics
in Wall Region". ASCE. Journal of Hydraulic Division, Vol. 97, HY9, Sept., pp.
1285-1303.

WOOD, I.R., (Ed.) (1991). "Air Entrainment in Free-Surface Flows". Design
Manual Nº 4, IAHR, A.A. Balkema Published, Rotterdam, Netherlands, Chap. 3.-
"Free Surface Air Entrainment on Spillways", pp. 55-84.

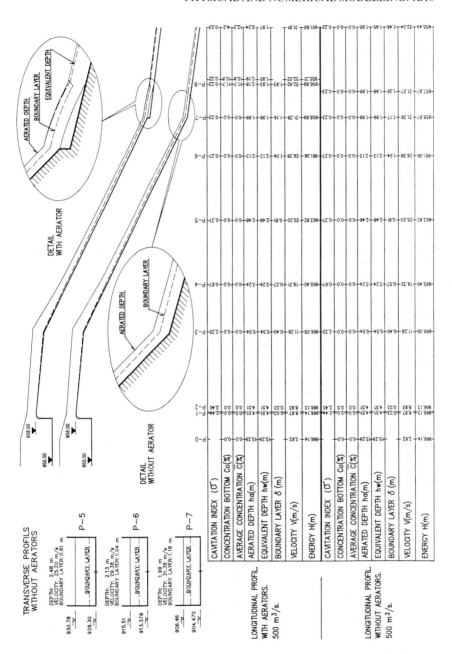

Fig 2.—Aerator Location. Design Process. Application to the Canales Dam Spillway (Spain).

1B11

Stepped spillways. Design for the transition between the spillway crest and the steps

Cristobal Mateos Iguacel, Director, and Victor Elviro Garcia, Head of Division, Hydraulics Laboratory, (CEDEX), SPAIN

The Hydraulic Works Administration has commissioned the Hydraulics Laboratory of the CEDEX to carry out research into a methodology for designing stepped spillways, in which emphasis is placed on the main factors that influence operation: flow rate, step dimensions, air distribution in the flow, energy dissipation, rapidly varying pressures, etc. Until now, a specific scale model has been used for each case, when studying the transition between the threshold and the uniform chute of a stepped spillway. The aim of this paper is to determine a suitable design of the transition on the basis of the surface flow height with which the strict profile of the crest is defined.

1. INTRODUCTION.

The system whereby large dams are constructed using the rolled concrete technique, was developed in the 80's, and there are now over 150 such dams in operation or under construction throughout the world. An additional advantage of this system, is that it allows for the spillway to form part of the dam, so it is possible not only to retain the step created by the frame during construction, but also to achieve considerable energy dissipation throughout the chute, thus reducing the volume of the stilling basin.

Until now, not enough research work has been carried out into stepped spillways, and there is an element of doubt concerning their long-term operation, the maximum flow rate to be discharged and the design methodology. With a view to obtaining a greater working knowledge regarding such doubts, the Hydraulic Works Administration commissioned the Hydraulic Laboratory of the Hydrographic Studies Centre (CEDEX) to conduct general research into this type of spillway. An analysis is made of the main variables that affect operation, with a view to providing a correct hydraulic definition of such devices.

This paper presents a summary of the work carried out on one of the points contained in the above-mentioned study, which concerns an analysis of the transition between the spillway threshold and the stepped chute.

2. CONSIDERATIONS PRIOR TO DESIGNING A STEPPED SPILLWAY.

A typical stepped spillway is divided into two distinct zones: the more or less conventional crest profile and the stepped chute. As will be shown, the two zones must be joined by a transition zone. The crest profile is dimensioned for a surface flow close to or at maximum and, for these high flows continuity problems do not generally arise even if no special attention is given to the transition. Problems do occur with small discharges, because the flow falls like a jet onto the first step, the full impact hitting the tread of the step, and thus leaves horizontally, thereby missing the intermediate steps and so causing greater problems with successive impacts. If the first step is relatively large by comparison with the surface flow and the latter is swift, a springboard effect takes place and the surface flow tends to jump.

As has already been mentioned, when the discharge increases the jump disappears, but depressions begin to appear, together with fluctuations and rapidly variable waves, both in the high and low pressure zones, and these are impossible to eliminate. Cavitation is a particularly serious problem, and can only be prevented by allowing air to enter the surface flow. The air enters naturally from the surface, but it may not reach all the zones in which it is needed, or might do so in insufficient quantities. It is known that the flow rate determines where the aereation starts, and when such rates are high, swift and unaerated zones appear, together with considerable disturbance at the bottom, all of which means that there are practical limits to the maximum flow rate that it is advisable to discharge.

The tests performed in the Hydraulic Laboratory with different chutes, plus data obtained from tests conducted in other laboratories, as order of magnitude and for steps between 80 and 100 cm. high, yield the following values when aereation begins:

Flow rate	Elevation difference between crest and inception point
4 m³/sec. m.	7.0 m.
6 m³/sec. m.	10.0 m.
8 m³/sec. m.	13.0 m.
10 m³/sec. m.	16.0 m.
16 m³/sec. m.	24.0 m.

This data shows that serious cavitation problems can arise if 10 m³/sec. m. is exceeded. In this case, it is necessary to carry out scale model tests which measure rapidly varying pressures on the steps, basically on their outer edges and nearby areas.

The aim of the design criterion used in these works is to obtain a uniform transition that regulates the flow for the entire range of discharges, preventing the water depth from separating for all of them, especially for the smaller discharges.

The criterion for dimensioning the transition is such that aeration in the chute takes place quickly for large discharges, even though the flow is uneven for smaller discharges. Therefore, if the transition project is carried out when there is a need to aereate the water depth, the values indicated in the above table must be taken into account, so that total aereation occurs in the sections closest to the crest, thereby reducing cavitation risk.

3. TRANSITION STUDIES.

The main aim of the tests conducted, is to find a design which reduces to a minimum both the magnitude of the jump and the range of discharges for which it takes place.

In order to systematize the tests, work is carried out with a chute having a gradient of 0.75:1, to which the tangent is a Bradley profile given by the following formula:

$$\frac{Y}{H} = 0.5 \left(\frac{X}{H}\right)^{1.85}$$

where the initial point of the coordinates is the spillway threshold, "X" being the horizontal axis and "Y" the vertical one in relation to the spillway crest, and "H" the maximum static height of the surface in the reservoir in relation to the crest.

The tests are conducted in the laboratory, with H = 0.20 m., 0.40 m. and 0.80 m. Any scale may be applied to the results obtained, Froude number being considered invariable, but taking into account the restrictions that the scale effects impose on hydraulic similarity. Reynolds' number must also be sufficiently large, not only for the nominal design discharge, but also for the discharges that cause the jump, which are much smaller. The surface tension, as represented by the Weber number, also affects this type of test.

Nine different designs were tested, all of which worked satisfactorily for the high discharges, but the performance was different for small discharges. Only the tests whose solutions were considered to be the most suitable, are referred to in this paper.

4. MAIN DESIGNS TESTED.

4.1. Design 1.

After some trials, the design shown in Fig. 1 was finally tested. In this design, the steps start at a horizontal distance H/2 from the crest, at 3H/20 horizontal intervals, so it was necessary to increase this magnitude after the eighth one, until the steps were of the same size as those of the chute.

Both the point where the steps begin and the dimensions of the first steps are deduced from the test, and are the two basic variables which determine the performance in transition. Hardly any jump occurs with this solution, except at H/20 water surface elevation, in which case the surface flow flips slightly from the first step to the third one. The jump disappears by increasing the discharge.

Such a smooth transition is the only drawback to this design, due to the great length required. This increases constructional problems.

4.2. Design 2.

Fig. 2 shows the design which aims to overcome the main shortcomings of the previous one by progressively increasing the transition dimensions. The steps begin at 2H/3 from the crest, the step dimensions being defined by the horizontal length with the following progression, H/10; H/8; 3H/20; H/5; H/4 .... until uniform steps are achieved.

The operation of this solution is generally agitated, and in the case of small water surface elevations, a jump takes place, not from the 1st step, but from the 2nd. In the same way that the first solution might be indicate a smooth transition elevation, this second solution would indicate a sudden transition elevation.

Tests performed on other designs show the advantages of starting the steps at a point closer to the spillway crest in order to make the transition smoother. Models whose theoretical profile is defined by 20 and 40 cm. water surface elevation, are used in these tests. In both models, the smaller surface elevations perform in a different way for the same definition, thus showing the importance of both the viscosity and the surface tension, so it was decided to use the following models for surface flows of 40 cm. and 80 cm. in the designs, as these make it possible to consider tests with proportionally reduced discharges as being reliable.

5. PROPOSED DESIGN.

The test for this design is conducted using a theoretical profile defined for 80 cm., whose transition is defined in Fig. 3. The steps start at H/3 from the crest, and they are horizontally defined by values of H/8, H/7, H/6.5, H/6, H/5.5 ... At the onset, the water runs from step to step up to levels of 3.35 cm. (H/24), over the spillway crest, then there is a jump from the 1st step to the 3rd one, which skims the edge of the 2nd step. With levels of 4.2 cm. (H/19), a further disturbed jump takes place from the 3rd step to the 6th, and the flow ceases to be continuous. The range for this process is very limited, because the jump disappears for 5.0 cm. (H/16) flows owing to the flooding of the 2nd step, a further jump taking place to the 4th step; this phenomenon occurs up to flows of 6.4 cm. (H/12.5), at which point the jumps disappear altogether. The small size of the jumps makes it difficult to observe the process from the front of the spillway, from where the operation appears to be regular for the entire range of flows, and for practical purposes this is the case.

The above definition was adapted to the Alcollarín dam, which has a flatter theoretical profile. Values were applied as a function of water nappe H, and the desired results were obtained, thus further endorsing the validity of the definition.

Without any basic modifications being made to the steps, slight adjustments were made to them, with a view to completely eliminating the small jumps. Allusion will only be made to the best result, in which the edge of the first steps was skimmed, and cylinders with a radius of 1.6 cm. (H/50) were formed; this design succeeded in completely eliminating the jumps, but their real dimensions and the range of flows for which this occurs, mean that the corresponding improvement may not make up for the constructional difficulties entailed.

As has already been mentioned, the design has considerable advantages for low discharges, but there is no appreciable difference for the high ones, so the precautions currently taken must be applied. They are not to be used with nappes exceeding 3 m., that is to say, for flow rates greater than 11 m³/sec. m., unless extra measures are taken.

6. CONCLUSIONS.

The following conclusions may be drawn from the scale model tests carried out on the transition between the crest and the stepped spillway:

1- The works are carried out with Bradley-type guiding profiles. It has been possible to show that the results can be applied to another type of guiding profile, and also to tangents with gradients of 0.75:1, normal for this type of dam.

2- For small surface flows, well below those of the project, the start of the steps has a springboard effect, causing the surface flow to jump. The transition must be studied in detail in order to prevent this detrimental effect.

3- The scale effect is extremely important in this type of test, because the study must concentrate on small surface flows. An analysis must be made not only of how the Reynolds and Weber numbers are affected for the design surface flows, but also for the small discharges "in situ". The Hydraulics Laboratory has finally decided to use 1/4 or 1/3 scale models for the final tests.

4- The design defined in Fig. 3 is proposed for the transition. The steps start at height H/3 of the spillway crest, and the steps progressively increase in size until they reach the uniform steps of the chute.

5- With the proposed design, the jump takes place for a small range of slight discharges, which can only be observed on large-scale models. The transition operates in an entirely satisfactory way for the rest of the flows. All types of jump are prevented by rounding off the edges of the first five steps.

BIBLIOGRAPHY

1- M.Bindo.The Stepped Spillway of M'Bali Dam.Water Power & Dam Construction.January.1993.

2- H. Chanson. Comparison of Energy Dissipation between Nappe and Skimming Flow Regimes on Stepped Chutes. Journal of Hydraulic Research. Vol. 30. 1994. Nº 2.

3- G.C. Christodoulou. Energy Dissipation on Stepped Spillways. Journal of Hydraulic Engineering. Vol. 119. May 1993.

4- Essery and Horner. The Hydraulic Design of Stepped Spillways. CIRIA Report 33. London.

5- C. Mateos Iguácel and V. Elviro García. The Use of Stepped Spillways in Energy Dissipation. 60th Executive Meeting ICOLD. Sept. 1992. Granada.

6- C. Mateos Iguácel and V. Elviro García. Regularidad del Flujo en Aliviaderos Escalonados. XVI Congreso Latinoamericano de Hidraúlica (Chile). October 1994.

7- N. Rajaratnam. Skimming Flow in Stepped Spillways. Journal of Hydraulic Engineering. Vol. 116. Nº 4. April 1990.

8- D. Stephenson. Energy Dissipation Down Stepped Spillways. Water Power & Dam Construction. Sep. 1991.

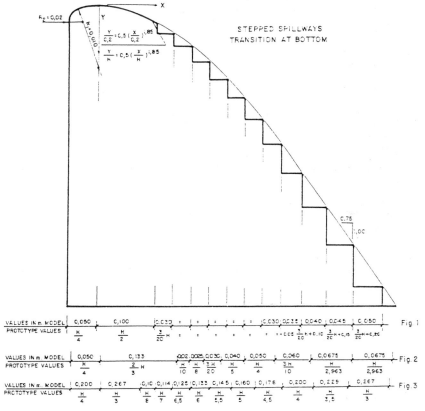

FIGURE - 1 - GENTE TRANSITION DESIGN.

FIGURE - 2 - RAPID TRANSITION DESIGN.

FIGURE - 3 - PROPOSED DESIGN

1B12

DISCHARGE COEFFICIENTS OF BROAD AND SHARP CRESTED SIDE WEIRS ON CHANNEL BENDS

Dr. J.G. HERBERTSON
Department of Civil Engineering
University of Glasgow
Glasgow, U.K.

Dr. H.K. JASEM
Water Industry Department
Cardonald College
Glasgow, U.K.

ABSTRACT

Discharge coefficients were obtained for both broad and sharp crested side weirs on a wide gentle channel bend under subcritical flow conditions. In each case equations were derived for the relationship between the discharge coefficients and the main channel Froude number. The equations were compared with those of other authors.

INTRODUCTION

The work reported originated from the modelling of the partial neck cut-off of the meander loop shown in Fig. 1. The cut-off was simulated by a broad crested side weir placed on the outside of a wide, gentle channel bend. The bed topography changes which occurred in the prototype have been reported by Herbertson and Fares (1991) and the results of the physical model study by Fares and Herbertson (1993). On completion of the cut-off study it was decided to investigate further the behaviour of side weirs, both broad and sharp crested, on channel bends. Many parameters have been studied but the present paper concentrates on discharge coefficients. The study has relevance to the behaviour of irrigation intakes and in the case of the broad crested weir to the modelling of flows overtopping flood banks.

APPARATUS

The layout of the recirculating laboratory flume is shown in Fig. 2. The flume consisted of a 60° bend working section with centreline radius of 1.5m lying between two straight channel reaches. Both straight and curved sections were of rectangular cross-section, 0.5m wide and 0.12m deep.   A gap was left in the outer wall of the bend between angles 25° and 35° to accommodate the side weirs.   Weir heights and lengths were varied.   The total flow supplied to the flume was measured by means of an orifice plate and the discharge passing over the side weir was measured by a 90° V-notch weir in a tank placed between the end of the weir channel and the sump tank.

DISCHARGE COEFFICIENTS

Although a side weir is placed parallel to the channel flow in contrast to a conventional weir which is placed normal to the flow, the discharge over the weir is usually obtained by a conventional weir equation in the form proposed by De Marchi (1941)

$$q_s = C_d (2g)^{0.5} (Y_c - C)^{1.5} \tag{1}$$

in which q_s is the discharge per unit length of weir
             C_d is the discharge coefficient
             Y_c is the water depth in the channel at the upstream end of the weir
             C is the height of the weir crest.

The coefficient of discharge may be expressed in functional form as -

$$C_d = f\left(F_c, \frac{C}{Y_c}, \frac{L}{Y_c}, \frac{L}{B} \right) \tag{2}$$

in which F_c is the Froude number of the main channel flow
             Y_c is the water depth in the main channel
             L is the length of the weir crest
             B is the main channel width.

Herbertson and Jasem (1994) and Singh and Satyanarayana (1994) amongst others have shown that satisfactory relationships for C_d can be obtained using only F_c and C/Y_c.   The following relationships have been proposed by Herbertson and Jasem (1994)

$$C_d = 0.49 \sqrt{\frac{2 - 2P}{F_c^2 + 2 - 2P}} \qquad \text{[sharp crested weir]} \tag{3}$$

$$C_d = 0.49 \sqrt{\frac{1 - P}{F_c^2 + 1 - P}} \qquad \text{[broad crested weir]} \tag{4}$$

in which F_c is the Froude number in the main channel and $P = \dfrac{C}{Y_c}$.

Using experimental values of F_c, C and Y_c, the discharge coefficients for sharp and broad crested weirs were calculated by equations 3 and 4. The C_d values were then plotted against F_c as shown in Figs. 3 and 4. Best fit equations were then obtained by regression techniques. These are

$$C_d = 0.52 - 0.175\,F_c \qquad \text{[sharp crested weir]} \qquad (5)$$

and

$$C_d = 0.53 - 0.25\,F_c \qquad \text{[broad crested weir]} \qquad (6)$$

A comparison of the results obtained using Eq. 5 with those obtained from the equations proposed by Hager (1987) and Cheong (1991) is shown on Fig. 5.

Similarly for broad crested weirs and results obtained using Eq. 6 are compared with those obtained from equations proposed by Swamee and Pathak (1994) and a modified version of the Subramanya and Awasthy (1972) equation (Herbertson and Jasem (1994)).

CONCLUSIONS

The object of the experimental study reported herein was to investigate the performance of sharp and broad crested side weirs on the outer side of a wide, gentle channel bend. From the study of the discharge coefficients it has been possible to draw the following conclusions -

1. Sharp crested weirs are more efficient than broad crested weirs, hence for the same crest height and length the discharge coefficient of a sharp crested weir is greater than that of a broad crested weir.

2. Discharge coefficients decrease with increasing Froude number in the main channel, the relationships being -

$$C_d = 0.52 - 0.175\,F_c \qquad \text{[sharp crested weir]}$$

and

$$C_d = 0.53 - 0.25\,F_c \qquad \text{[broad crested weir]}$$

REFERENCES
Cheong, H.F. (1991), "Discharge Coefficient of Lateral Diversion from Trapezoidal Channel". Journal of the Irrigation and Drainage Division, ASCE, Vol. 117, No. 4, pp. 461-475.

De Marchi, G. (1941), "Channels with Increasing Discharge". L'Energia Elettrica, Milano, Italy, Vol. 18, No. 6, pp. 351-360 (in Italian).

Fares, Y.R. and Herbertson, J.G. (1993), "Behaviour of Flow in a Channel Bend with a Side Overflow (Flood Relief) Channel". Journal of Hydraulic Research, IAHR, Vol. 31, No. 3, pp. 383-402.

Hagar, W.H. (1987), "Lateral Outflow over Side Weirs". Journal of Hydraulic Engineering, ASCE, Vol. 113, No. 4, pp. 491-504.

Herbertson, J.G. and Fares, Y.R. (1991), "Bed Topography Changes produced by Partial Cut-off of a Meander Loop". Advances in Water Resources Technology, G. Tsakiris (ed.)., Balkema, Rotterdam, pp. 113-120.

Herbertson, J.G. and Jasem, H.K. (1994), "Performance of Broad and Sharp Crested Side Weirs on Channel Bends". IAHR Congress, APD-9, Singapore, Vol. 2, pp. 117-124.

Subramanya, K. and Awasthy, S.C. (1972), "Spatially Varied Flow over Side Weirs". Journal of the Hydraulics Division, ASCE, Vol. 98, No. HY1, pp. 1-10.

Swamee, P. and Pathak, S.K. (1994), "Subcritical Flow over Rectangular Side Weirs". Journal of the Irrigation and Drainage Division, ASCE, Vol. 120, No. 1, pp. 212-217.

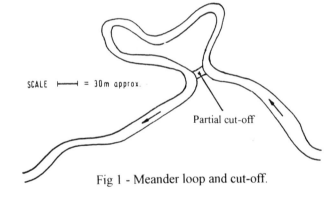

SCALE ⊢——⊣ = 30 m approx.

Partial cut-off

Fig 1 - Meander loop and cut-off.

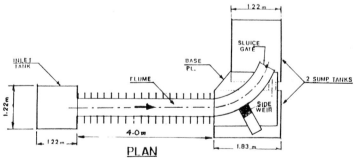

Fig 2 - Flume layout.

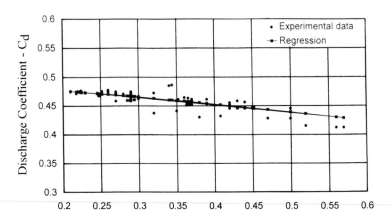

Fig 3 - C_d versus F_c relationship
Sharp crested weirs.

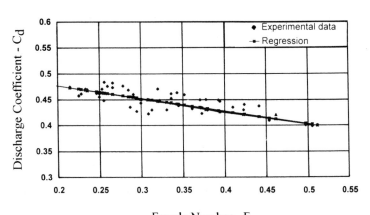

Fig 4 - C_d versus F_c relationship
Broad crested weirs.

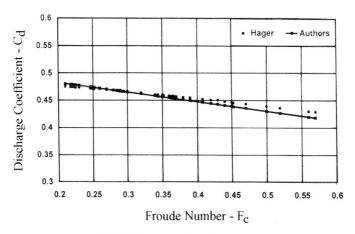

Fig 5 - C_d versus F_c sharp crested weirs
Comparison of data.

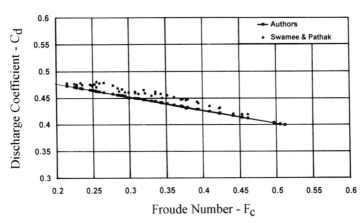

Fig 6 - C_d versus F_c broad crested weirs
Comparison of data.

1B13

Dam-break in a valley with tributary

A. PAQUIER
Cemagref, Lyon, France

One of the problems dealing with public safety is dam-break. When a narrow valley lies downstream, the wave created by the water coming trough the broken dam advances rapidly (velocity as high as 50 km/h for an instantaneous failure). The development of 1-D numerical models solving shallow water equations leads nowadays to reasonable results at a cost much lower than physical models. Yet, 1-D numerical models cannot be easily applied in any case. Particularly, we shall be interested in the following case : when the front reaches a junction, at the same time, it continues downstream but also can go upstream in the tributary. This problem could be modelled by 2-D shallow water equations but, in order to simplify computations, 1-D modelling is often kept. Then, the problem stays in sharing the discharge between the 2 valleys. Usually, flow in the tributary is not computed and it is supposed that, in the downstream valley, water depth is the same as without tributary and that, in the tributary, maximum water elevation is the same as at the junction.

This kind of computation gives water depths larger than real ones except when the tributary is in the same direction as the downstream valley. In that case, in the tributary, the flow can reach higher level than the maximum water elevation at junction and, after reflection, in the main valley, it is even possible to reach higher water elevation than without tributary.

Our objective is to obtain a method which leads to a 1-D computation precise enough in that case. Previously, for little angles of junction, we have found by comparing 1-D methods with 2-D models that it was relatively easy to find good results in terms of water depth in the tributary. Same comparison could be used for larger angles. However, we wished to compare computational results with real dam-break cases. As no one was available, we used results of laboratory experiments.

DESCRIPTION OF EXPERIMENTS
A lot of experiments have been carried out in order to simulate situations very close to dam-break problems. We looked for experiments in which we could

find precisely the consequence of junction. Such experiments were carried out in the National Hydraulic School of Toulouse (France) in the sixties. They described briefly their experimental installation and the various cases they treated [Escande et al., 1961, 1963]. We examine particularly the experiments which deals with a junction with an angle close to 180° and the corresponding cases without tributary.

The experimental canal was built in order to obtain a view at scale 1/300 of the French Truyère river downstream Sarrans dam. The particular features of this valley are its narrowness and its steep slope (about 2.8 %). The dam failure was replaced by the rapid removal of a gate. The reservoir of the dam was constituted by some part of the valley at scale 1/300 but, upstream, the valley was replaced by a tank with horizontal bottom. During the experiments, the tank was kept at constant level which implied that, following the unsteady period immediately after the removal of the gate, the flow converged slowly to a steady state.

With the tributary at 180° angle, the experiments showed that in the main valley, the front arrival was delayed and the maximum water elevation reached immediately later was lowered ; yet, after some time, the reflection inside the tributary led to secondary waves with oscillation of water that reached higher elevation than without tributary.

DESCRIPTION OF NUMERICAL SCHEMES

The code RUBAR 20 that we used solved 2-D shallow water equations written in conservative form :

$$\frac{\partial h}{\partial t} + \frac{\partial Q_x}{\partial x} + \frac{\partial Q_y}{\partial y} = 0 \tag{1}$$

$$\frac{\partial Q_x}{\partial t} + \frac{\partial\left(\frac{Q_x^2}{h} + g\frac{h^2}{2}\right)}{\partial x} + \frac{\partial\left(\frac{Q_x Q_y}{h}\right)}{\partial y} = -gh\frac{\partial Z}{\partial x} - g\frac{Q_x\sqrt{Q_x^2 + Q_y^2}}{K^2 h^{7/3}} \tag{2}$$

$$\frac{\partial Q_y}{\partial t} + \frac{\partial\left(\frac{Q_x Q_y}{h}\right)}{\partial x} + \frac{\partial\left(\frac{Q_y^2}{h} + g\frac{h^2}{2}\right)}{\partial y} = -gh\frac{\partial Z}{\partial y} - g\frac{Q_y\sqrt{Q_x^2 + Q_y^2}}{K^2 h^{7/3}} \tag{3}$$

where h is the water depth, Z the bed elevation, Q_x and Q_y the "discharges" (water depth multiplied by velocity) along respectively the x-axis and the y-axis, t the time, g the gravitational acceleration constant and K the Strickler's friction coefficient.

In order to deal with front on dry bed, RUBAR 20 use an explicit finite volume scheme of Godunov type. In order to reach second-order accuracy, the variables (z water elevation, Q_x and Q_y) are supposed varying linearly (Van Leer method) in space inside 1 cell (independently on x-axis and y-axis) and, moreover, for computation of fluxes through the edges, an approximation at intermediate time step $(t + 0.5\ \Delta t)$ is used [Paquier].

For 1-D computation, we disposed of the code RUBAR 3 which was developed for dam-break computation. The explicit finite difference scheme solving 1-D shallow water equations [Vila] was based on the same methods as 2-D numerical scheme. The equations are written in conservative form:

$$\frac{\partial A}{\partial t} + \frac{\partial Q}{\partial x} = q \tag{4}$$

$$\frac{\partial Q}{\partial t} + \frac{\partial \left(\dfrac{Q^2}{A} + p \right)}{\partial x} = -gA\left(\frac{\partial Z}{\partial x} + S_f \right) + B \tag{5}$$

in which Q is the discharge, A the wetted area, q the lateral inflow, p the pressure, Z the bottom elevation, x the distance along the waterway, t the time, g the gravitational acceleration constant, S_f the friction slope computed from the Strickler's relationship and B the lateral pressure.

For treatment of junction, the 1-D equations should be completed by special equations in that region. Basically, the integration of 2-D shallow water equations on the cross-direction for each one of the 3 branches will give equations similar to 1-D shallow water equations if we project the equation for the conservation of the quantity of movement to an x-axis corresponding to the direction of the average velocity. In our case, we decided to simplify the computation by considering that this main axis was still the one of the dam valley. In that junction cell, the discretization for 1-D scheme is then adapted as follows :

$$\frac{\Delta A}{\Delta t} + \frac{Q_3 - Q_1 - Q_2}{\Delta x} = 0 \tag{6}$$

$$\frac{\Delta Q}{\Delta t} + \left(\frac{Q_3^2 \cos a_3}{A_3} - \frac{Q_2^2 \cos a_2}{A_2} - \frac{Q_1^2 \cos a_1}{A_1} \right) \bigg/ \Delta x + \frac{P_3 - P_2 - P_1}{\Delta x} = gA(S_0 - S_f) + B \tag{7}$$

with A the flow area, Q the discharge, p the pressure, S_0 the bottom slope, S_f the friction slope, B the lateral pressure, g the gravity acceleration, a_i the angle of the branch i at junction, Δt and Δx the time and space steps.

1-D AND 2-D COMPUTATIONS

The numerical schemes were used on a simpler case than experiments. In order to obtain 1-D and 2-D computational results as close as possible, we decided to select rectangular valleys with constant width.

For main valley, width was 100 meters and slope 2.8% (except 200 m upstream horizontal). For tributary, width was 20.4 meters and slope 3%. The angle between tributary and downstream main valley was 10° resulting to an angle of junction of 170°.

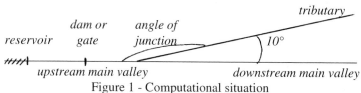

Figure 1 - Computational situation

The 1-D equations were slightly changed in order that friction relationship was the same in 1-D and 2-D ; so, hydraulic radius was replaced by water depth which means that the friction along the banks were neglected. In order to balance this point, Strickler's coefficient was taken as 20 in the tributary instead of 40 in the main valley.

As in the experiments selected, inflow in the tributary was zero, inflow in the main valley was determined by the water level kept constant and initially, we had dry bottom downstream the dam. As along main valley, flow remains supercritical, no condition was imposed downstream.

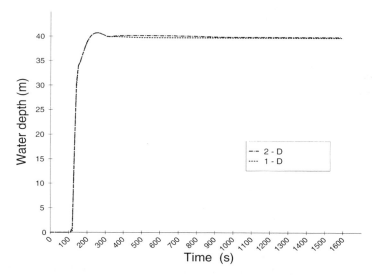

Figure 2 - Comparison of computed water depths without tributary

In figure 2, we show the results at a cross-section 6 kilometres downstream the dam. As expected, without tributary, the 1-D and 2-D results are very close. We can notice a breaking in the rising level. This corresponds to the limit point between the 2 phases :

      - first, the discharge is obtained by emptying of the upstream valley

      - secondly, the discharge is coming through the upstream boundary of our model i.e. the discharge is determined in order to keep water elevation constant in the upstream reservoir.

This breaking point was also noticed in the experiments. However, after this point, water level was rising regularly and did not show such a maximum as computations show. This maximum comes from an oscillation of discharge at the upstream boundary which revealed that the numerical treatment was not perfect ; similarly, the slight difference in 1-D and 2-D steady elevation is due to a slight difference in upstream discharge coming from different treatment of upstream boundary.

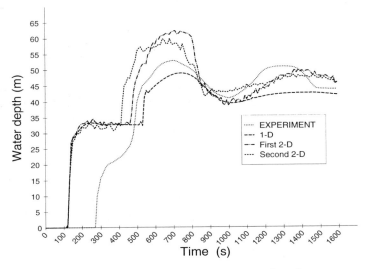

Figure 3 - Water depths downstream junction

Figure 3 gives the results at the same cross-section as figure 2 ; we show 2 different 2-D computational results with different geometry at junction. Also, we show approximate experimental results at a cross-section where water depths are also about 40 meters. Computations give oscillations such as in experiments but damping is higher in numerical model. The time of arrival of main wave (even without tributary) is also different between experiments and computations. These 2 last points are certainly due to the difference in cross-sections of the valley between experiments and computations ; another possible cause, the inadequacy of the treatment of front by shallow water equations, has been rejected as comparison of numerical results with other experiments (those of [Chervet and Dalleves]) in simple canal do not reveal such differences even with reverse slope. However, the more important point in figure 3 is the large differences in the amplitude of oscillations (the large ones, the smaller ones in 2-D computations are numerical ones amplified by reflections on walls) and even between the 2 computations with 2-D code.

CONCLUSION

For a junction where the tributary is nearly in the direction opposite to the upstream dam valley, we show that 1-D computation cannot (up to now) give results similar to 2-D computations because the angle of junction seems insufficient to define the junction completely. Moreover, as the use of shallow water equations in the front area should also be discussed, additional detailed laboratory experiments will be necessary to precise if results of 2-D computations are acceptable for any topography of junction.

In such problems as dam-break flow, physical models cost a lot and field measurements are very seldom. So the use of numerical models based on shallow water equations is the only generalised method for practical applications. 2-D numerical models may be used to validate 1-D models in a lot of cases. However, laboratory experiments are obligatory for validation of both 1-D and 2-D models in particular conditions ; in such cases, it is also clear that improvement of numerical methods is still necessary in order to obtain more precise results.

REFERENCES
A. CHERVET, P. DALLEVES
Calcul de l'onde de submersion consécutive à la rupture d'un barrage.
Schweizerische Bauzeitung 88(19), (Mai 1970), pp. 420-432.

L. ESCANDE, J. NOUGARO, L.CASTEX, H. BERTHET
Influence de quelques paramètres sur une onde de crue subite à l'aval d'un barrage.
AIRH 9ème Assemblée générale (Dubrovnik 1961), pp. 1198-1206.

L. ESCANDE, J. NOUGARO, L.CASTEX, S. BACQUIE
Propagation d'une onde de crue subite à la suite de l'effacement d'un barrage.
AIRH 10ème Assemblée générale (1963).

A. PAQUIER
New methods for modelling dam-break wave.
Proceedings of the Speciality Conference on "Modelling of Flood Propagation Over Initially Dry Areas" (1994) pp. 229-240.

J. P. VILA
Modélisation mathématique et simulation numérique d'écoulements à surface libre.
La Houille Blanche 6/7 (1984) pp. 485-489.

1B14

DRIFTWOOD BEHAVIOR IN HORIZONTAL TWO-DIMENSIONAL BASINS

HAJIME NAKAGAWA and KAZUYA INOUE
Disaster Prevention Research Institute,
Kyoto University.
Uji, Kyoto, 611, JAPAN

INTRODUCTION

When a clump of driftwood debouches with river water into a protected low-lying area due to levee breaking, collision usually accelerates the destruction of wooden houses. A large amount of timber may directly attack and destroy houses when storm surges break a timber holding pond and the wood flows out. Moreover, many pieces of wood are debouched into bays from rivers and are spread widely by flooding, hindering the navigation of ships and damaging fishery industries. These phenomena indicate that flood and storm surge disasters are brought about not by flood water alone but by suspended objects, such as driftwood and ships, as well.

To clarify the part driftwood has in causing disasters, we need to know its flow behavior characteristics. A numerical simulation model for computing the diffusion of pieces of driftwood debouching into a horizontal two-dimensional flow field is presented that is based on the hydraulic characteristics of the driftwood. Calculations are made with an interacting combination of Eulerian fluid and Lagragian driftwood equations. The results are compared with those of hydraulic model experiments.

BASIC EQUATIONS OF FLUID AND DRIFTWOOD MOTION
BASIC EQUATIONS OF FLUID MOTION

The basic equations used to calculate the behavior of an overland flood flow are the horizontal 2-D momentum and continuity equations:

$$\frac{\partial M}{\partial t} + \frac{\partial(uM)}{\partial x} + \frac{\partial(vM)}{\partial y} = -gh\frac{\partial H}{\partial x} - \frac{1}{2}fuv\sqrt{u^2 + v^2} + \nu\Big(\frac{\partial}{\partial x}h\frac{\partial u}{\partial x} +$$

$$\frac{\partial}{\partial y}h\frac{\partial u}{\partial y}\Big) + \frac{\partial}{\partial x}\Big\{h\Big(2A_h\frac{\partial u}{\partial x} - \frac{2}{3}k\Big)\Big\} + \frac{\partial}{\partial y}\Big\{hA_h\Big(\frac{\partial u}{\partial y} + \frac{\partial v}{\partial x}\Big)\Big\} + \frac{\tau_{sx}}{\rho} \quad (1)$$

$$\frac{\partial N}{\partial t} + \frac{\partial(uN)}{\partial x} + \frac{\partial(vN)}{\partial y} = -gh\frac{\partial H}{\partial y} - \frac{1}{2}fv\sqrt{u^2+v^2} + \nu\Big(\frac{\partial}{\partial x}h\frac{\partial v}{\partial x}+$$
$$\frac{\partial}{\partial y}h\frac{\partial v}{\partial y}\Big) + \frac{\partial}{\partial y}\Big\{h\Big(2A_h\frac{\partial v}{\partial y}-\frac{2}{3}k\Big)\Big\} + \frac{\partial}{\partial x}\Big\{hA_h\Big(\frac{\partial v}{\partial x}+\frac{\partial u}{\partial y}\Big)\Big\} + \frac{\tau_{sy}}{\rho} \quad (2)$$

$$\frac{\partial h}{\partial t} + \frac{\partial M}{\partial x} + \frac{\partial N}{\partial y} = 0 \tag{3}$$

where M, N = the water discharge per unit width for x and y, i.e., $M = uh$ and $N = vh$; u and v = the respective depth-averaged velocity components in the x and y directions; h = the water depth; H = the water level, i.e., $H = h + z_b$, z_b = the elevation of the bed; ρ = the density of the fluid; g = the acceleration due to gravity; f = the resistance coefficient; ν = the kinematic viscosity; A_h = the kinematic eddy viscosity horizontally; k = the turbulent energy; τ_{bx} and τ_{by} = the respective shear stresses on the bed in the x and y directions; τ_{sx} and τ_{sy} = the shear stresses at the water surface in the x and y directions; t = time; and x and y = the coordinate axes horizontally. According to Hosoda and Kimura (1) the horizontal kinematic eddy viscosity, A_h, and turbulent energy, k, are

$$A_h = ahu_* \quad ; \quad k = 2.07u_*^2 \tag{4}$$

where u_* = the friction velocity; and a = a numerical constant with the value 0.3. For the resistance coefficient, f, alternatively the laminar or turbulent flow resistance formula based on the Reynolds number is used;

$$f = 6/Re \ (Re < 400) \quad ; \quad \sqrt{2/f} = 3.0 + 5.75\log Re\sqrt{f/2} \ (Re \geq 400) \tag{5}$$

where Re = Reynolds number defined by $Re = \sqrt{u^2+v^2}h/\nu$.

We assume that the shear stresses, τ_{sx} and τ_{sy}, at the water surface are generated by the reaction of the drag force acting on the driftwood;

$$\tau_{sx} = \frac{1}{A}\sum_{k=1}^{N_t}\Big\{\frac{1}{2}\rho C_{Dx}W_k(u_k - U_k)A_{kx}\Big\} \ \Bigg\}$$
$$\tau_{sy} = \frac{1}{A}\sum_{k=1}^{N_t}\Big\{\frac{1}{2}\rho C_{Dy}W_k(v_k - V_k)A_{ky}\Big\} \ \Bigg\} \tag{6}$$

where u_k and v_k = the respective driftwood velocity components in the x and y directions; U_k and V_k = the respective local velocity components of the fluid in the x and y directions at the position of the centroid of the driftwood; $W_k = \sqrt{(u_k - U_k)^2 + (v_k - V_k)^2}$; A_{kx} and A_{ky} = the respective projected areas of the submerged part of the driftwood in the x and y

directions; and C_{Dx} and C_{Dy} = the respective drag coefficients in the x and y directions; A = the area of the water surface, written $A = \Delta x \Delta y$ (Δx and Δy = the grid sizes of the finite difference equation); and N_t = the number of pieces of driftwood in area A.

BASIC EQUATIONS OF DRIFTWOOD MOTION

It is assumed that the pieces of driftwood are sufficiently dispersed so that collisions between them are infrequent. The drag forces produced by the velocity difference between the fluid and driftwood and the body force component produced by the gradient of the water surface are considered. On the basis of these assumptions, each piece of driftwood, individually labeled by subscript k, is assumed to obey the equations

$$\frac{dX_k}{dt} = u_k \quad ; \quad \frac{dY_k}{dt} = v_k \tag{7}$$

$$(m_k + mC_M)\frac{du_k}{dt} = m(1 + C_M)\frac{U_k}{dt} - m_k g \frac{\partial H_k}{\partial x} - \frac{1}{2}\rho C_{Dx} W_k(u_k - U_k)A_{kx} \tag{8}$$

$$(m_k + mC_M)\frac{dv_k}{dt} = m(1 + C_M)\frac{V_k}{dt} - m_k g \frac{\partial H_k}{\partial y} - \frac{1}{2}\rho C_{Dy} W_k(v_k - V_k)A_{ky} \tag{9}$$

where X_k and Y_k = the position of the centroid of the driftwood; m_k = the mass of the driftwood; m = the mass of the fluid occupied by the volume of a piece of driftwood; C_M = the virtual mass coefficient; and H_k = the water level at the position of the centroid of the driftwood. We neglect the horizontal component of the buoyancy force term in the horizontal two-dimensional coordinate systems because it acts vertical to the water surface.

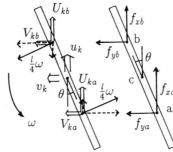

Fig.1 Definition sketch of the rotational angle of a piece of driftwood

On the supposition that the driftwood can be divided into two pieces at the centroid, "c", and that the drag force acts on both centroids, "a" and "b", of these pieces (Fig.1), the rotational motion of the driftwood is written

$$I d^2\theta_k/dt^2 = \Sigma N_0 = (\ell/4)\{(f_{xa} - f_{xb})\sin\theta_k - (f_{ya} - f_{yb})\cos\theta_k\} \tag{10}$$

where

$$f_{xa} = \tfrac{1}{2}\rho C_{Dx}\sqrt{(U_{ka} - u_k - u_{rka})^2 + (V_{ka} - v_k - v_{rka})^2}(U_{ka} - u_k - u_{rka})\tfrac{A_{kx}}{2}$$

$$f_{ya} = \tfrac{1}{2}\rho C_{Dy}\sqrt{(U_{ka} - u_k - u_{rka})^2 + (V_{ka} - v_k - v_{rka})^2}(V_{ka} - v_k - v_{rka})\tfrac{A_{ky}}{2}$$

$$f_{xb} = \tfrac{1}{2}\rho C_{Dx}\sqrt{(U_{kb} - u_k - u_{rkb})^2 + (V_{kb} - v_k - v_{rkb})^2}(U_{kb} - u_k - u_{rkb})\tfrac{A_{kx}}{2}$$

$$f_{yb} = \tfrac{1}{2}\rho C_{Dy}\sqrt{(U_{kb} - u_k - u_{rkb})^2 + (V_{kb} - v_k - v_{rkb})^2}(V_{kb} - v_k - v_{rkb})\tfrac{A_{ky}}{2}$$

$$u_{rka} = (\ell/4)(d\theta_k/dt)\sin\theta_k \quad , \quad v_{rka} = -(\ell/4)(d\theta_k/dt)\cos\theta_k$$

$$u_{rkb} = -(\ell/4)(d\theta_k/dt)\sin\theta_k \quad , \quad v_{rkb} = (\ell/4)(d\theta_k/dt)\cos\theta_k$$

θ_k = the rotational angle of the piece of driftwood; I = its moment of inertia; and ℓ = its length. The rotational angle of the piece of driftwood can be evaluated deterministically from these equations, but a statistical analysis of the experiments are also considered (described later).

As the motion of the piece of driftwood is restricted near the water surface, the depth averaged velocity components of the fluid, U_k and V_k, must be transformed to the surface velocity components. For example, when $Re > 400$, the following relations are used

$$U_k \rightarrow \frac{5.5 + 5.75\log(u_*h/\nu)}{3.0 + 5.75\log(u_*h/\nu)}U_k \quad ; \quad V_k \rightarrow \frac{5.5 + 5.75\log(u_*h/\nu)}{3.0 + 5.75\log(u_*h/\nu)}V_k \quad (11)$$

As it is very difficult to solve these equations analytically, numerical calculations are made by approximating the equations to finite differencing equations. We here have introduced the Adams–Bashforth scheme in the time integration, and the QUICK scheme in the integration of the convection term in the momentum equations. For the driftwood motion, time–forward and space–centered schemes are adopted, the velocity components of the driftwood being obtained explicitly.

FLUCTUATION OF DRIFTWOOD POSITION

Driftwood position can be evaluated by integrating the equations $dX_k/dt = u_k$ and $dY_k/dt = v_k$, deterministically under suitable initial conditions, but they fluctuate due to disturbances at the water surface. The fluctuation components ΔX_k and ΔY_k are considered to be evaluated by referring to Dukowicz (2)

$$\Delta X_k = \sqrt{4K_x\Delta t}\,\mathrm{erf}^{-1}(\alpha) \quad ; \quad \Delta Y_k = \sqrt{4K_y\Delta t}\,\mathrm{erf}^{-1}(\beta) \quad (12)$$

where α, β = random variables independently and uniformly distributed in the range [0,1], and erf^{-1} = the inverse function of the error function, erf, given by

$$\mathrm{erf}(s) = \{1 - \Phi(\sqrt{2}s)\} = \frac{1}{\sqrt{\pi}}\int_s^\infty \exp(-\eta^2)d\eta \quad (13)$$

$$\Phi(s) = \frac{1}{\sqrt{2\pi}} \int_{-\infty}^{s} \exp(-\eta^2/2)d\eta \tag{14}$$

The relations $K_x/u_*h = 0.629$ and $K_y/u_*h = 0.208$, obtained experimentally for the diffusion coefficient of the driftwood (3), are introduced. Consequently, the driftwood position is estimated by adding the turbulent fluctuation value to the value obtained deterministically from the equations of motion;

$$X_k^{n+1} = X_k^n + u_k^n\Delta t + \sqrt{4K_x\Delta t}\,\mathrm{erf}^{-1}(\alpha) \tag{15}$$

$$Y_k^{n+1} = Y_k^n + v_k^n\Delta t + \sqrt{4K_y\Delta t}\,\mathrm{erf}^{-1}(\beta) \tag{16}$$

ROTATIONAL ANGLE OF DRIFTWOOD

The rotational angle of a piece of driftwood can be evaluated deterministically by solving Eq.(10), but it must be effected by fluctuation in the fluid flow. Actually, in a uniform flow field, driftwood has been shown to run down with rotational motion having the mean angular velocity $\overline{\omega} \simeq 0$ and the standard deviation $\sigma_\omega = 81.6Fr$ (Fr = Froude number defined by $Fr = \sqrt{(u^2 + v^2)/gh}$) (3). This suggests that the characteristics of the fluctuation of the rotational angle of the driftwood are expressed by the characteristics of the angular velocity. Accordingly, we consider two cases; one in which the rotational angle, θ_k, is decided only by the angular velocity, ω_d, obtained deterministically, the other in which it is decided by both ω_d and ω_p. These two cases are shown as

$$d\theta_k/dt = \omega_d \quad ; \quad d\theta_k/dt = \omega_d + \omega_p \tag{17}$$

where ω_d = the angular velocity of the piece of driftwood obtained from Eq.(10) deterministically; and ω_p = the fluctuation of the angular velocity of the driftwood evaluated stochastically.

Assuming the rotational angular velocity of the fluctuating component of a piece of driftwood follows a normal distribution (3), its distribution function, Φ, is given by

$$\Phi\left(\frac{\omega_p - \overline{\omega}}{\sigma_\omega}\right) = \frac{1}{\sqrt{2\pi}} \int_{-\infty}^{\frac{\omega_p - \overline{\omega}}{\sigma_\omega}} \exp\left(-\frac{\eta^2}{2}\right)d\eta \tag{18}$$

and $\gamma(= (\omega_p - \overline{\omega})/\sigma_\omega)$ is obtained from the inverse function, Φ^{-1}, for uniformly distributed random numbers within [0,1]. After γ is obtained, ω_p is estimated from $\omega_p = \gamma\sigma_\omega + \overline{\omega}$.

COMPARISON OF THE RESULTS

Fig.2 shows comparisons of the experimental and calculated positions and the rotational angles of the pieces of driftwood. The left column gives the calculated results in the case of $d\theta_k/dt = \omega_d$, the right column those in the case of $d\theta_k/dt = \omega_d + \omega_p$, and the center column the experimental results. The calculated results for $d\theta_k/dt = \omega_d + \omega_p$ show very good agreement with the experimental ones, especially at $t = 20$ sec.

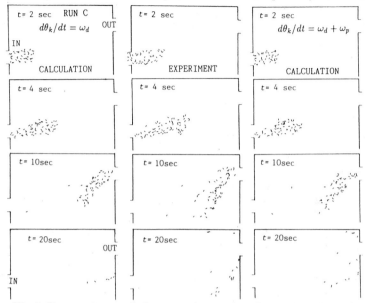

Fig.2 Comparisons of the experimental and calculated results

CONCLUSIONS

The equation of the rotational motion of driftwood was evaluated dynamically as was the equation of translational motion in the Lagrangian form. An interacting combination of Eulerian fluid and Lagrangian driftwood equations were used, in which the turbulent diffusivity and the fluctuation component of the rotational angular velocity of the driftwood were given stochastically. The positions and rotational angles for pieces of driftwood were well explained by the calculations.

REFERENCES

1. Hosoda. T. and I. Kimura : Vortex Formation with Free Surface Variation in Shear Layer of Open Channel Flows Near Abrupt Expansion, Proc. of Hydraulic Engg., JSCE, Vol.37, 1993, pp.463–468 (in Japanese).
2. Dukowicz, J.K. : A Particle-Fluid Numerical Model for Liquid Sprays, Jour. of Comp. Physics, Vol.35, 1980, pp.229–253.
3. Nakagawa, H., T. Takahashi and M. Ikeguchi : Driftwood Behavior by Overland Flood Flows, Jour. of Hydroscience and Hydraulic Engg., JSCE, Vol.12, No.2, 1994, pp.31–39.

1B15

Numerical and Image Analyses of Turbulent Flow in Open Channel Trench

ICHIRO FUJITA[*], TOHRU KANDA[**], TAKAMITSU MORITA[***], and MASAO KADOWAKI[****]
* Dept. of Civil Eng. Gifu Univ. Yanagido, Gifu 501–11, JAPAN
** Dept. of Civil Eng. Kobe Univ. Rokkodai, Nada, Kobe 657, JAPAN
*** Graduate student of Gifu Univ., **** Graduate student of Kobe Univ.

ABSTRACT
Turbulent flow structures in a trench section of an open channel flow are examined by a numerical analysis based on the three dimensional large eddy simulation and by a two dimensional image analysis. Mean and turbulence structures and time dependent characteristics of turbulent shear flows in a longitudinal vertical section are compared. Both methods revealed the average and space–time structures of the flow satisfactorily.

1. INTRODUCTION
Open channel flows in a trench, a shallow concave section, exhibit complicated flow features such as flow separation, reattaching flow, reverse flow, boiling flow, etc., which are schematically illustrated in Fig.1[1]. The flow can be considered as the combination of the flow after a backward facing step and the flow before a forward facing step. The flow field is dominated by separated vortices that varies considerably in time and space; therefore, the time–averaged Reynolds equations are not appropriate for the numerical analysis of those unsteady flow features. Experimentally, conventional measurement techniques using point probes are almost impossible to reveal the spatial evolution of separated vortices. In this research, the large eddy simulation is used for the numerical approach and the particle imaging velocimetry(PIV) is used for the experimental approach in order to examine the time varying phenomena at the trench section.

2. IMAGE ANALYSIS
2.1 EXPERIMENTAL SETUP
Figure 2 shows the experimental setup[2]. A trench section, 2cm in depth H and 20 cm in length L, is set in the middle of a straight channel made of transparent material. The water depth changes abruptly from 2.0 to 4.0 cm at the trench

section keeping almost horizontal water surface level. The Froude number and the Reynolds number at the inlet section are 0.49 and 4000, respectively. No supercritical flow appears in the entire region. The flow field seeded by tracer particles is visualized by a laser–light sheet and the images are recorded using a high–speed video camera. The

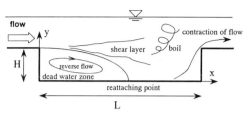

Fig.1 Flow patterns in a trench

nylon 12 particle with a mean diameter of about 0.2mm and a specific gravity of 1.02 is used as a tracer.

2.2 IMAGE PROCESSING SYSTEM

The image processing system is composed of a personal computer installed an image memory board, a video tape recorder, a monitor, and a magneto optical disk(MO) for storing image data. The analogue data are converted into digital quantities of 512x512 pixels with eight–bit resolution. The cross correlation method, one of the pattern matching techniques, is used as a PIV. The template for pattern matching is 15 pixels square and the interval between images is 1/125 seconds. Since the size

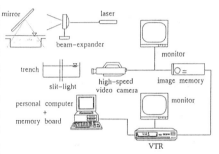

Fig.2 Experimental setup

of one pixel is about 0.055cm, the resolution of velocity becomes a few centimeters per second. The mean and turbulent characteristics of the flow are analyzed by using two thousand images.

3. NUMERICAL ANALYSIS
3.1 FUNDAMENTAL EQUATIONS
The equations used for the numerical analysis are the continuity equation

$$\frac{\partial u_i}{\partial x_i} = 0 \qquad (1)$$

and the momentum equations

$$\frac{\partial u_i}{\partial t} + \frac{\partial u_i u_j}{\partial x_j} = -\frac{\partial p}{\rho \partial x_i} + \frac{\partial}{\partial x_j}[(\nu + \nu_t)(\frac{\partial u_i}{\partial x_j} + \frac{\partial u_j}{\partial x_i})] + F_i \qquad (2)$$

285

where u_i is the grid–scale velocity component in the x_i direction, p the pressure, ν the kinetic viscosity, and F_i the external forces. Smagorinsky's subgrid–scale model is used as a turbulence model, which can be expressed as

$$\nu_t = (Cf\Delta)^2[(\frac{\partial u_i}{\partial x_j}+\frac{\partial u_j}{\partial x_i})\frac{\partial u_i}{\partial x_j}]^{1/2} \qquad (3)$$

where

$$\Delta = (\Delta x_1 \Delta x_2 \Delta x_3)^{1/3} \qquad (4) \qquad F_i = g(\sin\theta, -\cos\theta, 0) \qquad (5)$$

C is the model constant equal to 0.1, f is the Van Driest's damping function effective only near wall boundaries, θ is the slope of the bed, and Δx_1, Δx_2, and Δx_3 are the grid sizes in the longitudinal, vertical, and spanwise directions, respectively. The above equations are discretized on the staggered grid of non uniform mesh system. The convection terms are expressed in terms of the QUICKEST scheme, third–order accuracy in space and second–order accuracy in time. The velocity and pressure fields are coupled by the HSMAC method[3].

3.2 INITIAL AND BOUNDARY CONDITIONS

Initially, the velocity field in the concave region is assumed to be stagnant and the velocity profile of the power–law distribution is given over the entire region above the inlet base level. The water surface is treated as a rigid–lid boundary and the Spalding's wall–function law is introduced for the wall boundary condition; the free–outflow condition is applied at the outlet boundary. The region for the development of the inflow boundary layer, 6H in length, is attached in front of the trench section, where the periodic boundary condition for each velocity component is applied. The non–dimensional time step is 0.001 and the turbulence statistics are accumulated using the data from one hundred to two hundred non–dimensional time.

4. DISCUSSION

4.1 MEAN VELOCITY

The mean velocity distributions at the central vertical section are compared in Fig.3. The flow separated at the entrance of the trench reattaches the bottom wall accompanying a large recirculating region behind it. The agreement between the two methods is quite satisfactory except that the calculated results show a little

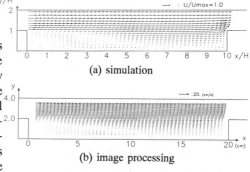

(a) simulation

(b) image processing

Fig.3 Mean velocity distributions

larger separation bubble.

4.2 TURBULENCE INTENSITIES

The turbulence intensities obtained from the longitudinal velocity fluctuations are compared in Fig.4. The local increase of the intensity along the shear layer is simulated fairly well, although the simulated result shows smaller values particularly near the entrance of the trench. The insufficiency of the inflow turbulence created by the periodic boundary condition and the rigid–lid assumption applied at the water surface are considered as the cause of the difference. The similar tendency can be recognized for the turbulence intensity of the vertical velocity component as shown in Fig.5.

4.3 REYNOLDS SHEAR STRESS

The Reynolds shear stress distributions shown in Fig.6 demonstrate a fair good agreement where the separated vortices predominates, although the simulated distribution near the free–surface in the entrance region indicates smaller values due to the above–mentioned reasons.

4.4 INSTANTANEOUS DISTRIBUTION

Figure 7 shows an example of an instantaneous velocity distribution. Flow patterns of considerable similarity are obtained; the shear layer is subjected to a wavelike deformation due to the instability of the flow that accompanies separated

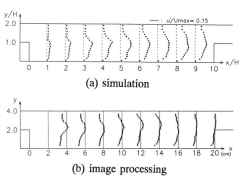

(a) simulation

(b) image processing

Fig.4 Turbulence intensity in the longitudinal direction

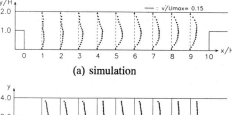
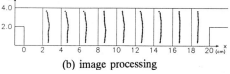

(a) simulation

(b) image processing

Fig.5 Turbulence intensity in the vertical direction

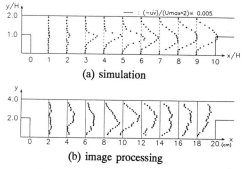

(a) simulation

(b) image processing

Fig.6 Reynolds stress

vortices of a large scale. In addition, some of the fluid particles are convected toward the water surface after the reattaching zone. It may suggest the creation of a boil vortex.

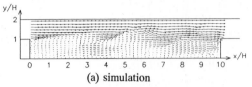

(a) simulation

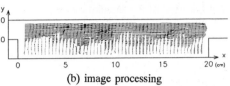

(b) image processing

Fig.7 Instantaneous velocity distribution

4.5 TIME DEPENDENT STRUCTURE

The space–time structure of the flow in a trench can be expressed as the evolution of a quantity on a check–line. Figure 8(a) demonstrates the structure of the downstream momentum on a horizontal check–line at y/H=1.2 which is taken just above the shear layer. The figure shows several inclined white-streaks generated just after the entrance of the trench; they indicate several high speed masses of fluid are convected downstream at nearly a constant speed. In addition, the figure visually suggests that the assumption of frozen–turbulence is locally satisfied. Furthermore, the intermittent production of the thick white streaks implies that a large–scale motion at a low frequency is created along with the development of the shear layer with a frequency of about two hertz. It should be noted that the motion accompanies the flow structure of a smaller scale with higher frequency, which is visualized as a bundle of thin white streaks.

Figure 8(b) indicates the result

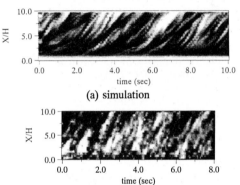

(a) simulation

(b) image processing

Fig.8 Evolution of downstream momentum on check–line at y/H=1.2

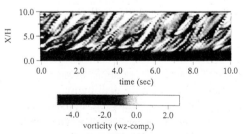

Fig.9 Evolution of vorticity distribution at y/H=1.2

for the image analysis. Although the spatial resolution is rather low in the image analysis, the convection of the momentum at a low frequency can be clearly recognized. Its frequency is nearly the same as the simulation, whereas the convection velocity is a little different from the simulation. It should be noted that the white streaks appear already from the entrance which suggests the abundance of the inflow turbulence in the real

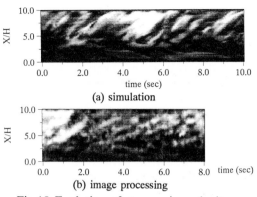

(a) simulation

(b) image processing

Fig.10 Evolution of streamwise velocity distribution at y/H=0.5

flow field. The same expression for the simulated vorticity component is indicated in Fig.9. Clearly, vorticity is intermittently convected downstream at an almost constant speed except near the entrance region where the influence of the low inflow turbulence is appreciable. Another comparison of the streamwise velocity variation is shown in Fig.10, in which the check line is taken horizontally at y/H=0.5. It shows quite an analogous space–time structure. The flow boundary between the positive and negative regions of convection is located almost at x/H=2.5 for the both cases.

5. CONCLUSION

The numerical analysis based on the large eddy simulation showed a considerable agreement with the image analysis as long as the mean velocity field and the large–scale motion of fluid are concerned. The reasons for the discrepancy in the turbulence characteristics are: (1) the insufficient treatment of the inflow and water surface boundary conditions in the numerical analysis and (2) the rather low spatial resolution in the image analysis. Nevertheless, the general agreement is satisfactory and these methods can be strong tools for the understanding of the time dependent structure of complex turbulent flows.

REFERENCES

[1] Fujita,I.,Kanda,T.,Komura,S.,et.al.(1993), Image Processing of Flows in a Trench Using a High–Speed Video Camera and Numerical Analysis by LES, Proc. of Hydraulic Engineering,JSCE,37,481–486.(in Japanese)
[2] Fujita,I.,Komura,S., and Kanda,T.(1993), Measurements of Turbulent Flow in a Trench Using an Image Processing Technique, Proc. 5th Int. Symp. on Refined Flow Modelling and Turbulence Measurements,309–316.
[3] Fujita,I.,Kanda,T.,and Morita,T.(1994), Numerical Analysis of Cavity Flow with Free–Surface by LES, Proc. 26th Symp. on Turbulence,334–337.(in Japanese)

Verification of a Three-dimensional River Flow Model with Experimental Data

G. Pender
University of Glasgow
UK

D. Keogh
Napier University
UK

J. R. Manson
Bucknell University
USA

P. Addison
Napier University
UK

INTRODUCTION

The paper reports on the use of high quality LDA measurements of shear layer driven circulations in verifying a three-dimensional numerical model for the simulation of open channel flows. In recent years, much effort has gone into verifying numerical models of flow in straight channels, where the flood plains are parallel to the main channel. In this type of flow a vertical shear layer generates secondary circulations in the channel and flood plains. The main problem facing researchers has been the selection of a turbulence model to reproduce these circulations, see for example Prinos[1]. The difficulty with verifying numerical models for fully three-dimensional flow in river channels is that as soon as the main channel is skewed to the flood plain the complexity of the flow increases considerably, Sellin et al[2]. In regions where the flood plain flow passes over the main channel there are large vertical velocities and horizontal shear layer generated circulations occurring within the main channel. In an attempt to isolate some of these features this paper presents comparisons of physical model results of horizontal of shear layer generated circulations in slots with three-dimensional numerical model results. The geometry used here differs from previous slot flow simulations, as the slot walls are vertical with 90° corners.

SHEAR LAYER DRIVEN CIRCULATIONS

The basic mechanisms causing secondary circulations when flow takes place over a slot are that the main flow passing over the slot generates a shear layer which drives the secondary circulation cell. Two types of slot circulations are possible, a circulation that completely fills the slot or a circulation where flow reattachment occurs with in the slot.

PREVIOUS EXPERIMENTAL STUDIES OF SLOT FLOWS

Ideally data for comparison with numerical model results should be obtained by LDA measurements. This is particularly true when the data is to be used for turbulence model evaluation. There appears to be little of this type of data available for shear layer driven circulations in slots. Jasem[3] collected water surface elevation data and velocity profiles using a mini-propeller meter. Aspect ratios of 20, 10, and 2 were employed. Jasem recognised that the patterns of recirculation differed depending on the B/H ratio, although the accuracy of the velocity measurements were insufficiently detailed to provide comprehensive conclusions and too sparse to draw firm conclusions on the position of reattachment points. Fujita, Michiue and Hinokidani[4], Fujita, Komura and Kanda[5] and Fujita, Kanda, Yano and Morita[6] have studied open channel flow over a slot using a flow visualisation technique combined with a correlation method. Their results further indicate a dependency of the circulation pattern on the slot aspect ratio. Once again however, the data published in insufficiently detailed to draw firm conclusions or be used for numerical model verification.

PHYSICAL MODEL

Slots with aspect ratios of 5, 10 and 20 were set up in a glass sided laboratory flume 4000 mm long by 100 mm wide and 300 mm deep. A one component Dantec LDV system was used to obtain mean and fluctuating velocity components in the centre of the flow for both the x and z planes. Measurements were taken on a mesh with $\Delta x = 10mm$ and $\Delta z = 2mm$. U and V velocities were measured by rotation of the laser probe through 90° at each point on the experimental mesh. Movement of the laser probe was achieved by using a purpose made traverse system accurate to 1/10th of a millimetre in both planes.

PHYSICAL LAWS

The flows of interest in this study are described by the law of conservation of momentum applied in the vertical and streamwise direction and conservation of mass, see Prinos[1]. The turbulence model applied in the following is the standard (or linear) k -ε model as described by Rodi[7]. Where the turbulence quantities are k, the kinetic energy of turbulent fluctuations per unit mass, and ε its rate of dissipation. Turbulent stresses are represented by Boussinesq's eddy viscosity assumption. The model the constants are $c_\mu= 0.09$, $\sigma_k =1.0$ $\sigma_t =1.3$, $c_1 = 1.44$ and $c_2 = 1.92$.

NUMERICAL SOLUTION

The equations are solved by a fractional step projection method on a Cartesian grid using a method similar to that proposed by Viollet et al[8]. A full description of the numerical technique is provided in Manson[9]. The velocities and all scalar quantities are defined at the computational cell corners while the pressure is defined at the cell centre. The turbulence quantities are advanced in a similar manner but require an additional computation for the turbulence source terms. Boundary conditions must be prescribed at the inlet, outlet, free surface, bed and side walls. At the bed and side walls a wall law is applied as described by Rodi[8].

SIMULATION OF FLOW OVER A SLOT

The inlet conditions employed are

$$u_{in} = \frac{u^*}{\kappa} \ln\left(\frac{z}{z_o}\right)$$

$$w_{in} = 0$$

$$k_{in} = \frac{u^*}{\sqrt{C_\mu}}\left(1 - \frac{z}{h}\right)$$

$$\varepsilon_{in} = \frac{|u^*|}{\kappa z}\left(1 - \frac{z}{h}\right)$$

where u^* was estimated from $\sqrt{gS_o h}$ and z_0 is chosen to reproduce the experimental inlet velocity profile as closely as possible. The depth of flow is 1.6 cm with S_0 equal to 0.0014 giving an average friction velocity of 1.48 cm/s. A grid of 101 x 29 was used with $\Delta x = 0.5$ cm and $\Delta z = 0.2$ cm.

RESULTS AND DISCUSSION

Figure 1(a) to 1(f) show comparisons of the computed and measured velocities at selected sections through a slot with an aspect ratio 5. Overall the comparisons are encouraging. In this case, however, the recirculation fills the slot and no reattachment takes place, see Figure 2. An early analysis of the results from the slots with aspect ratios of 10 and 20 indicate that in these cases flow reattachment does take place. One interesting feature that the numerical results suggest is that the presence of the downstream step stretches the length of the recirculation zone beyond the X/S = 7 typical for backward facing step problems. A more detailed analysis of the physical model results is required to confirm this. Figure 3(a) and (b) shows the distribution of turbulent kinetic energy in the slot.

As expected the computed turbulence field possess greater uniformity than the measured. Overall, however, the general pattern of the computed and measured turbulence fields is in agreement.

CONCLUSIONS

Detailed comparisons of physical model and computer model results for a shear layer generated recirculation in a slot with an aspect ratio of 5 are in good agreement. The work is continuing with larger aspect ratios which will provide greater insights in to the predictive capacity of three-dimensional models of river flow.

REFERENCES

1. PRINOS P. *Three-dimensional flow in compound open channels*, 5th Int. Symp. on Refined Flow Modelling and Turbulence Measurements, Paris, France, 1993.
2. SELLIN R.H.J., ERVINE D.A., WILLETTS B.B. *Behaviour of meandering two-stage channels*, Proc. Inst. Civil Engrs Water Maritime and Energy, Vol 101, June, 1993.
3. JASEM H.K. *Flow in two-stage channels with the main channel skewed to the flood plain direction*. PhD Thesis, Department of Civil Engineering, University of Glasgow, 1990.
4. FUJITA M., MICHIUE M. and HINOKIDANI O. *Mathematical Simulation of Flow and Suspended Load in Open Channels with a Trench*, Proc. of Hydraulic Engineering, JSCE, Vol. 35.
5. FUJITA M., KOMURA S. and KANDA T. *Measurements of Turbulent Flow in a Trench Using Image Processing Technique*, Refined Flow Modelling and Turbulence Measurements Proc. of the 5th International Symposium, Paris, France, Sept, 1993.
6. FUJITA I., KANDA S., KOMURA S., YANO Y. and MORITA T. *Image Processing of Flows in a Trench Using a High Speed Video Camera and Numerical Analysis by LES.* Proc of Hydraulic Engineering, JSCE, Vol 37.
7. RODI W. *Turbulence models and their application in hydraulics.* IAHR state of the art report, 1980.
8. VIOLLET P.L., BENQUE J.P. and GOUSSEBAILE J. *Two-dimensional numerical modelling of non isothermal flows for unsteady thermal-hydraulic analysis.* Nuclear science and engineering,1983, **20**, No 3.
9. MANSON J.R. *The Development Predictive Procedure for Localised Three Dimensional River Flows.* PhD Thesis, University of Glasgow, Glasgow, UK, 1994.

Figure 1
Comparison of Velocity Profiles Aspect Ratio = 5

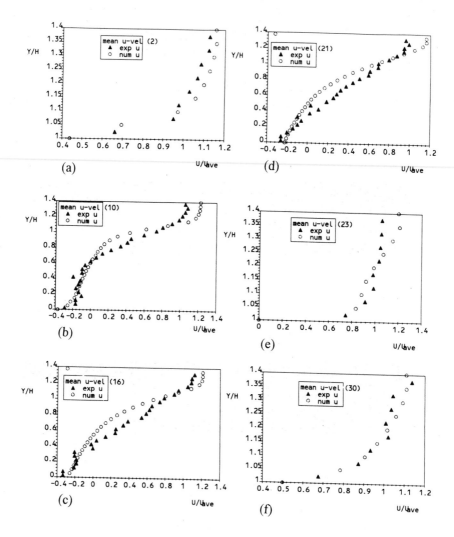

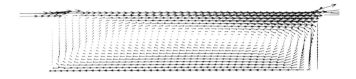

Figure 2
Velocity Circulation in Slot Aspect Ratio = 5

(a)

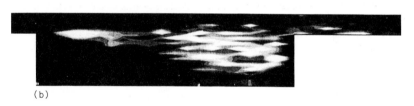

(b)

Figure 3
Comparison of Turbulence Fields Aspect Ratio = 5

1B17

HORIZONTAL FLOW STRUCTURE IN CHANNELS WITH DENSE VEGETATION CLUSTERS AND THE NUMERICAL ANALYSIS THEREOF

SHOJI FUKUOKA
Professor, Hiroshima University, Higashi-Hiroshima, 724, Japan
AKIHIDE WATANABE
Senior Research Engineer, Public Works Research Institute,
Ministry of Construction, Tukuba, Ibaragi, 305, Japan

ABSTRACT
This paper, focusing on horizontal large-scale eddies in channels with dense vegetation clusters and on the regular surface fluctuation that occurs downstream as a result, explains the structure of horizontal flow conditions using analysis based on laboratory experiments and a shallow-water equation devised with a model of horizontal mixing. The results of numerical analysis of horizontal velocity fields, water surface fluctuation fields and predominant wavelengths -- determined through shallow water equation -- accurately correspond to experimental results and enable predictions of sufficient accuracy.

1. INTRODUCTION
Velocity in rivers of dense vegetation clusters is extremely low, resulting in large differences in velocity with the main current. This, in turn, results in vigorous momentum exchange between the flow in vegetation clusters and the main current and in horizontal mixing with a large transverse water-level gradient around the vegetation. Mixing with this type of horizontal large-scale eddies becomes a resistance factor of the flow. [1)2)3)]

In this research, we performed a hydraulic model experiment on a flow field with typical patterns of vegetation growth in order to determine the characteristics of the flow field. Next, we developed a simple, two-dimensional (horizontal) model of horizontal mixing that is caused by vegetation, then used this model to determine the predominant wavelength as a nonlinear stability problem. Finally, we investigated the structure of horizontal mixing and horizontal flow conditions by solving this model and comparing these results with those of our experiments.

2. EXPERIMENT ON HORIZONTAL MIXING CAUSED BY VEGETATION CLUSTERS
(1) EXPERIMENTAL PROCEDURE
For our cases of rivers with conspicuous horizontal mixing caused by vegetation clusters, we performed channel experiments with uniform, continuous vegetation growing in the center of the channel and uniform, continuous vegetation growing along the banks of the main channel in a course with a compound section.

The channel used in these experiments was a straight channel, made of steel, with a uniform rectangular section, 15m in length, 1.2m in width, and a bed slope of 1/1000. The Manning roughness coefficient of the channel bottom is 0.011. The experiment in which vegetation was placed in the center of the channel is named "Case A". In "Case B", a flood channel was created by affixing pieces of plywood 40cm wide and 2cm high along the length of the channel to perform an experiment on a compound section channel with a ratio of main channel width to flood channel height of 20 and a ratio of main channel width to total width of 1/3. The flood channel was covered with geotextile sheet in order to give it a higher roughness coefficient than the main channel. A porous vegetation model made of plastic (91% porosity, permeability coefficient K = 0.38m/s) was used.

(2) RESULTS OF EXPERIMENT ON CASE A

Velocity and water level fluctuation were measured under the following conditions: vegetation cluster width b' = 10cm; vegetation height h = 4cm; discharge Q = 11 l/s (flow depth h = 4.5cm).

Figure 1 shows velocity fluctuations u' and v' and water level fluctuation η' measured near the vegetation cluster. Velocity and depth were measured respectively with a two-component electromagnetic current meter and a capacitance-type wave gage, which were attached opposite from each other next to the vegetation cluster. Both velocity and water level fluctuated nearly periodically, with the phases of u' and v' diverging by one-half period. Consequently, Reynolds stress also fluctuated periodically, showing a large peak value. The fact that the phases of η' and v' nearly coincide near the vegetation and that the phases of η' on either side of the vegetation cluster are reversed[2)3)] shows that a flow from higher to lower areas in the water surface is being produced on either side of the vegetation cluster.

Figure 2 shows power spectrum density S(ω) at various frequencies ω for the waveforms of velocity fluctuation u' as measured at a position 15 cm from the vegetation boundary towards the main current. (Line spectrums are also shown.) In Figure 2 the waveforms are represented with measured values (thin line) and the dominant fluctuation component of line spectrum (thick line). S(ω) has a large low-frequency range and a clear peak, while the frequency indicated by the line spectrum diagram generally complies with the measured waveform. Figure 3 shows a contour of surface level; high and low parts in the surface regularly appear in the longitudinal direction, and that the two sides separated by the vegetation have crests and troughs in the surface that have laterally opposite phases. We see that horizontal eddies occur in depressed parts of the surface; more specifically, that horizontal large-scale eddies appear alternately on the left and right sides of the vegetation cluster and maintain this configuration as they move downstream.

(3) RESULTS OF EXPERIMENT ON CASE B

As in Case A, in Case B water surface profile, velocity fluctuation and other factors were measured and displayed visibly under conditions at which mixing is the most prominent. With a constant vegetation width b' of 3.5cm, measurements were performed with a continuous configuration (height, 6.5cm). Figure 4, a contour of surface level measured with a discharge of 9 l/s (depth h = 6.8cm) shows that crests and troughs in the surface appear alternately in the downstream direction in the main channel. In Photo 1 aluminum powder was used to make the horizontal flow

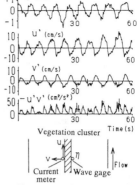

Fig. 1 Changes over time in water level and velocity near the vegetation cluster

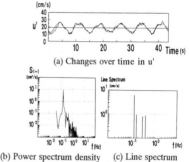

(a) Changes over time in u'

(b) Power spectrum density   (c) Line spectrum

Fig. 2 Spectrums of velocity fluctuation u'

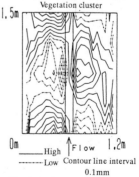

Contour line interval 0.1mm

Fig. 3 Surface level contour

conditions visible: aluminum powder meanders through the main channel, while lines of aluminum powder can also been seen in the flood channel. Figure 5, a graphic representation based on Figure 4 and Photo 1, shows that in areas of low water level in the main channel there are horizontal eddies around which the aluminum powder is concentrated. It is likely that the aluminum powder meanders because of these eddies, which proceed, alternately left then right, down the main channel. With the horizontal eddies that form in the flood channel revolving in the opposite direction of those in the main channel, these eddies, alternately revolving in opposite directions, cause the characteristic whiskerlike lines to appear at Location A, the position in the flood channel where the flow concentrates. In this case, the phases of these whiskerlike lines on the left and right sides are out of synchronization by π.

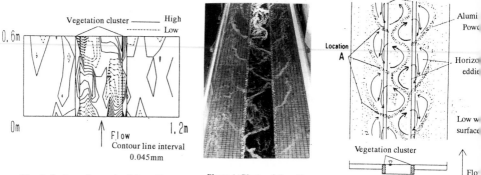

Fig. 4 Contour of water level (case B)   Photo 1: Photo of Case B

Fig. 5 Illustration of Case B

3. NUMERICAL ANALYSIS OF HORIZONTAL FLOW FIELD
(1) METHOD OF ANALYSIS

The basic equations for flows where vegetation is present is represented with (1), (2), and (3).

$$\frac{\partial u}{\partial t}+u\frac{\partial u}{\partial x}+v\frac{\partial u}{\partial y}=gI-g\frac{\partial \eta}{\partial x}-\frac{c_f u\sqrt{u^2+v^2}}{h_o}-g\frac{u\sqrt{u^2+v^2}}{k^2}+v_t\left\{\frac{\partial}{\partial x}\left(2\frac{\partial u}{\partial x}\right)+\frac{\partial}{\partial y}\left(\frac{\partial u}{\partial y}+\frac{\partial v}{\partial x}\right)\right\} \tag{1}$$

$$\frac{\partial v}{\partial t}+u\frac{\partial v}{\partial x}+v\frac{\partial v}{\partial y}=-g\frac{\partial \eta}{\partial y}-\frac{c_f v\sqrt{u^2+v^2}}{h_o}-g\frac{v\sqrt{u^2+v^2}}{k^2}+v_t\left\{\frac{\partial}{\partial x}\left(\frac{\partial v}{\partial x}+\frac{\partial u}{\partial y}\right)+\frac{\partial}{\partial y}\left(2\frac{\partial v}{\partial y}\right)\right\} \tag{2}$$

$$\frac{\partial \eta}{\partial t}+\frac{\partial(\eta u)}{\partial x}+\frac{\partial(\eta v)}{\partial y}=0 \tag{3}$$

The following equation was used for the eddy viscosity coefficient (v_t).

$$v_t=\frac{1}{6}\kappa u_* h_o \tag{4}$$

where u is the depth-averaged velocity in the downstream direction; v, the depth-averaged velocity in the transverse direction; η, depth; g, gravitational acceleration; c_f, friction resistance coefficient; κ, von Karman's constant (0.4); u_*, friction velocity; h_o, average depth. K, permeability coefficient, represents the permeation of the vegetation. U_w, the apparent velocity inside the vegetation cluster, is represented with permeability coefficient K and energy gradient I_e as shown in equation (5)[2)3)].

$$U_w=KI_e^{1/2} \tag{5}$$

Next, we shall discuss the development of a model for horizontal mixing. Velocities u and v and water depth η are divided into their average values and the amount of fluctuation therefrom.

298

Fluctuation quantities are represented by superimposing the first-order mode of wavelength L and the second-order mode of wavelength $L/2$.

$$u(x,y,t,)=u_0(y,t)+\sum_{m=1}^{2} u_m(y,t)\cos\frac{2m\pi}{L}(x-ct+\alpha_m(y)) \tag{6}$$

$$v(x,y,t,)=v_0(y,t)+\sum_{m=1}^{2} v_m(y,t)\cos\frac{2m\pi}{L}(x-ct+\beta_m(y)) \tag{7}$$

$$\eta(x,y,t,)=h_0(y,t)+\sum_{m=1}^{2} \eta_m(y,t)\cos\frac{2m\pi}{L}(x-ct+\gamma_m(y)) \tag{8}$$

The reason for incorporating the secondary mode into fluctuations is because of the mixing phenomenon is determined by mixing of a scale corresponding to that of large-scale horizontal mixing.

Substituting u, v, and η in equations (6), (7), and (8) into basic equations (1), (2), and (3) produces 15 simultaneous partial differential equations for unknown variables$_1$, u_0 v_0, and h_0 ; fluctuation amplitudes u_1, u_2, v_1, v_2, η_1, and η_2; and phase differences $\gamma_1 - \alpha_1$, $\gamma_1 - \alpha_2$, $\gamma_1 - \beta_2$, $\gamma_1 - \gamma_2$ and wave speed c. An arbitrary value is given to γ_1. The solution is obtained by assigning the small disturbance of 1/100 of measured depth to an initial water level and then time-integrating these unsteady terms explicitly. The restrictive condition assigned is a discharge-constant condition. In this model, the predominant fluctuation components of u, v, and η are extracted and a wave-motion shape is assigned for the downstream direction in order to make this a one-dimensional problem for the transverse direction.

(2) CALCULATING THE PREDOMINANT WAVE NUMBER AS A NONLINEAR STABILITY PROBLEM

We will first determine, through numerical analysis, the predominant wave number of a periodic eddy. Ikeda et al[5]. have performed a linear stability analysis on periodic eddies that form near vegetation boundaries, and have attempted to determined the predominant wave number of small disturbance in a flow with an assigned velocity distribution. However, because wave number and velocity distribution are interrelated, they should be determined simultaneously as a nonlinear relationship.

The analytic method described in section 3.(1) was used to determine the predominant wave number through nonlinear stability analysis in which velocity distribution is calculated with different wave numbers. The occurrence of predominant wave number is assumed when the value of Reynolds stress due to large-scale eddy near the vegetation boundary is maximum with respect to wave number.

To determine this, we calculated the maximum value of Reynolds stress at various wave numbers using the experimental values of Fukuoka and Fujita[2][3] and the authors' experimental values. The conditions used in calculations by Fukuoka et al. are in Table 1 and the results of these calculations are in Figure 6. Reynolds stress for the velocity field of the first-order mode has a peak at wave number k, indicating the existence of a wave number region where horizontal mixing is most vigorous. However, it appears that this peak wave number has a certain width. The time required for Reynolds stress to achieve a stabilized value after the beginning of progression is expressed in the diagram as Tg. It can be seen that the higher the frequency is, the shorter the time required for stabilization is.

In other words, in the group of wave numbers at which Reynolds stress peaks, those with higher frequencies progress faster. It is possible that a stable field is formed at the stage at which this eddy fully progresses, after which eddies with smaller wave numbers can no long progress. In actual experiments performed by Fukuoka and Fujita with a 50-m-long channel, a wavelength of 3.5m was observed, and was situated in the region with the largest wave number in this gentle peak.

Similarly, predominant wave number of calculation for Case B agree well with the result of experiment as shown in Figure 7.

Table 1 : Experimental conditions used in calculations[2][3]

Discharge (l/s)	Bed slope	Uniform depth(cm)	channel width (cm)	Vegetation width (cm)	Bed roughness	Permeability coefficient(m/s)
37	1/1000	4.5	300	30	0.011	0.38

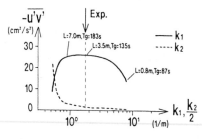

Fig. 6 Relationship between Reynolds stress and wave number, (Fukuoka and Fujita)

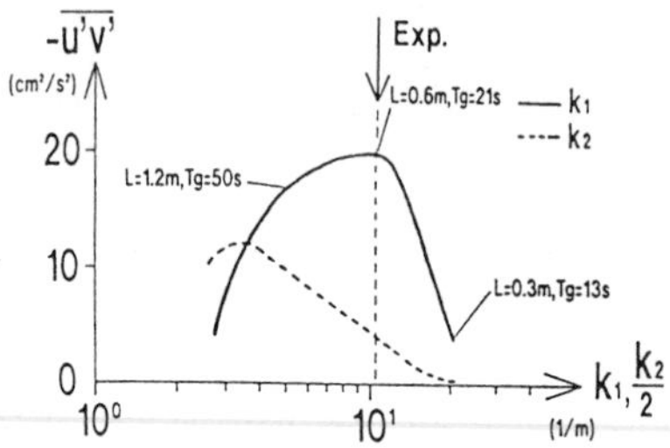

Fig. 7 Relationship between Reynolds stress and wave number (Case B)

(3) RESULTS OF CALCULATIONS FOR CASE B

Here, we shall present the results for the stage in which the flow has developed sufficiently. Figures 8 and 9 respectively show the velocity vector in a stationary coordinate system, and the contour of the surface. They show that laterally alternate large-scale horizontal eddies have formed in the main channel, and that the surface is lower in these areas. The flow meanders along the periphery of these horizontal eddies in the main channel, accurately corresponding with the experimental results shown in Fig. 4 and Photo 1. The periphery of these eddies passes through the vegetation into the flood channel. Figures 10 and 11 respectively show the depth-averaged velocity distribution and Reynolds stress distribution. The calculated and measured values for depth-average velocity distribution generally correspond. The fact that the velocity in the main channel is roughly the same as in the flood channel in spite of the lower roughness coefficient and shallower depth in the main channel indicates that in the main channel, velocity is being reduced by horizontal mixing. The distribution of Reynolds stress has a larger peak value on the main channel side. This indicates that in a compound channel, horizontal mixing in the main channel predominates.

Next, we visibly represented the calculated flow in order to compare it with the distribution in Photo 1. Here, we examined the movement of markers released downstream into the obtained velocity field. These results are shown in Figure 12.

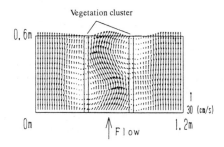

Fig. 8 Calculated velocity vector (Case B)

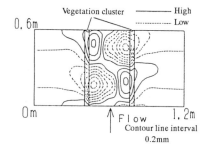

Fig. 9 Calculated water level contour (Case B)

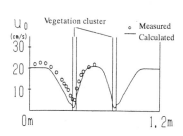

Fig. 10 Depth-averaged velocity distribution (Case B)

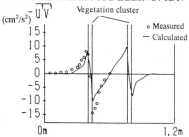

Fig. 11 Reynolds stress distribution (Case B)

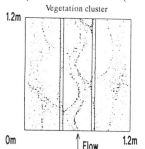

Fig. 12 Marker distribution by calculation (Case B)

As in Photo 1, the markers show a meandering path in the main channel and take on the characteristic whiskerlike lines near the shoulders of the flood channel. As in the experiment, these lines indicate a laterally alternating configuration of horizontal eddies in the main channel, with a resultant π shift in the right and left phases.

4. CONCLUSION

The primary conclusions of this research are as follows.

1. Large-scale horizontal eddies occur near longitudinally continuous vegetation, and their downstream progression causes periodic continuous fluctuation in the surface. Because this eddy formation alternates in the downstream direction between the two sides separated by vegetation, a large surface gradient forms in the transverse direction inside the vegetation cluster. These horizontal eddies and transverse surface gradient cause fluid mixing in the transverse direction.

2. We developed a model of horizontal mixing caused by the presence of vegetation and incorporated it into the basic equations in order to perform nonlinear stability analysis on the flow field. These results showed that it is possible to estimate the predominant wavelength of horizontal eddies.

3. Upon analyzing the flow conditions, horizontal flow conditions and shearing force at the vegetation boundary agreed with the values obtained in experiments to a significant degree, thus demonstrating that such horizontal mixing is determined by predominant wavelengths, such as large-scale horizontal eddies, and that two-dimensional horizontal flow analysis is capable of arriving sufficient solutions.

REFERENCE

1) Fukuoka, S. and Fujita, K. : Water level prediction in river courses with vegetation, Proc. 25th IAHR Congress, Vol. 1, pp. 177-184, 1993.

2) Fukuoka, S. and Fujita, K. : Hydraulic effects of luxuriant vegetation on flood flow, Report of Public Works Research Institute, Ministry of construction, Vol. 180, pp.1-64, 1990.(in Japanese).

3) Fukuoka, S. and Fujita, K. : Flow resistance due to lateral momentum transport across vegetation in the river course, Int. Conf. on Physical Modeling of Transport and Dispersion, 12B, Boston, pp. 25-30, 1990.

4) Fukuoka,S. Watanabe, A. and Tsumori, T.: Structure of plane shear flow in river with vegetations, J. of Hydraulic, Coastal and Environmental Engineering, JSCE, No.491, pp.41-50, 1994. (in Japanese).

5) Ikeda, S. Ohta, K. and Hasegawa, H. : Periodic vortices at the boundary of vegetated area along river bank, Proc.of Hydraulics and Sanitary Engineering, JSCE, No.443, pp.47-54, 1992. (in Japanese).

1B18

A Comparative Study of the Use of the Lateral Distribution Method in Modelling Flood Hydrographs in a Compound Channel

J F Lyness	University of Ulster, Newtownabbey, UK
W R C Myers	University of Ulster, Newtownabbey, UK
J B Wark	Babtie, Shaw and Morton, Glasgow, UK

INTRODUCTION

River channels can often be described as two stage or compound channels with flow in the full channel cross section only occurring during the passage of flood waves. A reach of the River Main, Northern Ireland, has been reconstructed as a compact compound channel. The reach, shown in Fig 1, is 805 m long and has been surveyed at 9 cross sections. Flow gauging and steady flow studies by Martin and Myers [1] have provided stage discharge curves, unit width discharge profiles and Manning's n for bankfull flow, nbf, and floodplain flow, nfp. Diaphragm pressure transducers installed at section 14, upstream, and section 6, downstream, have provided synchronized water level hydrographs during the passage of flood waves.

The Lateral Distribution Method has been used to compute unit width discharges, stage discharge curves and conveyance functions permitting comparison between computed and measured values. The Lateral Distribution Method has been described by Wark, Samuels and Ervine [2] and is based on calculating the distribution of flow within the channel. The flow is assumed to be steady and uniform in the longitudinal direction and the water surface is assumed to be horizontal. The governing partial differential equation in terms of the unit flow, q, is given below

$$gDS_{xf} - \frac{Bf|q|q}{8D^2} + \frac{\partial}{\partial y}\left[\nu_t \frac{\partial q}{\partial y} \right] = 0 \qquad (1)$$

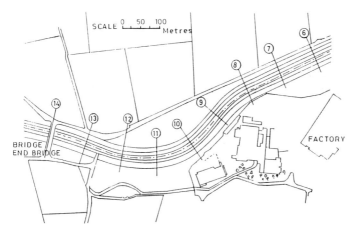

FIG 1: PLAN VIEW OF EXPERIMENTAL REACH OF RIVER MAIN

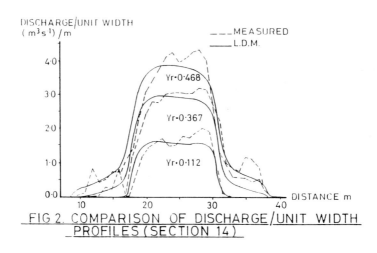

FIG 2. COMPARISON OF DISCHARGE/UNIT WIDTH PROFILES (SECTION 14)

$B = (1 + S_x^2 + S_y^2)^{\frac{1}{2}}$ x = longitudinal co-ordinate direction

D = flow depth y = lateral co-ordinate direction

f = Darcy friction factor = $8gn^2/D^{\frac{1}{3}}$ q = longitudinal unit flow

g = gravitational acceleration υ_t= lateral eddy viscosity

S_x = longitudinal slope of channel bed n = Manning's n

S_y = lateral slope of channel bed

The lateral eddy viscosity υ_t relating to bed roughness generated turbulence is given by

$$v_t = \lambda \, U_* \, D \tag{2}$$

U_* = shear velocity = $(\tau_b/\rho)^{\frac{1}{2}}$ ρ = fluid density

λ = nondimensional eddy viscosity τ_b = bed shear stress

A discussion of the derivation of equation (1) from the general 2-D shallow water equations is given by Wark [3]. The governing equation has been solved iteratively using the finite difference method to provide profiles of discharge/unit width for prescribed section flow depths. The nondimensional eddy viscosity, λ, was taken as 0.16. Following the steady flow studies of the compound channel reach the Manning's n for the floodplain was nfp = 0.0450 and for the main channel zone was taken as that for bankfull flow, nbf = 0.0386.

STEADY FLOW COMPARISONS

Measured unit flow profiles were compared with those computed using the Lateral Distribution Method and are shown in Fig 2. The computed profiles are smoother than those obtained from flow gauging and do not show the same peak on the right hand side of the section. This is probably due to this section (14) having a slight bend upstream with the measured peaks occurring on the outside of the bend. The solution using the Lateral Distribution Method assumes a straight river reach.

Integration of unit flow profiles provided the discharges in the main channel and floodplains. Fig 3 shows the LDM discharges compared with measured discharges for a range of overbank flows. Overbank flow depths are given in terms of the relative depth Yr (depth of flow on floodplain/total depth of flow).

It can be seen that the computed and measured discharges show good agreement except in the region 0.17 < Yr < 0.32, the maximum error in total discharge in this region being approximately 10%.

Using the LDM conveyance functions were calculated for five sections along the reach. The Single Channel Method and Divided Channel Method described by Ackers [4] were also used to calculate conveyance functions for comparison

FIG.3 COMPARISON OF DISCHARGES
UPSTREAM Section 14

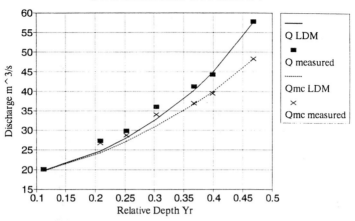

FIG.4 CONVEYANCE COMPARISON
UPSTREAM Section 14

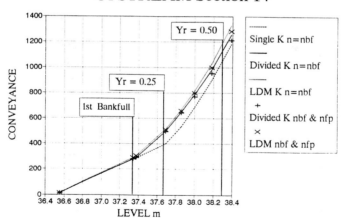

FIG.5 COMPUTED WATER DEPTHS vs TIME
UPSTREAM Section-30hr flood-max Yr=0.45

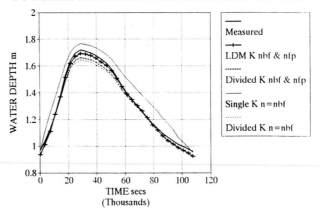

FIG 6 ERROR IN COMPUTED
FLOODPLAIN DEPTH vs Yr
((COMP−OBS)/OBS)x100%

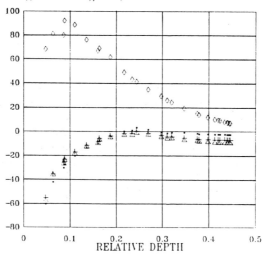

purposes. Fig 4 shows that the LDM conveyances with main channel and floodplain roughness zones are similar to the Divided Channel Method with main channel and floodplain roughness zones. At small floodplain depths, Yr < 0.3, differences are not significant and at large floodplain depths, Yr > 0.6, the LDM becomes closer to the Single Channel Method.

UNSTEADY FLOW COMPARISONS

Several overbank floods were recorded in the reach by synchronized diaphragm pressure transducers installed at the upstream section (14) and downstream section (6) of the reach. A 30 hour overbank flood was selected for one dimensional unsteady flow modelling comparisons. The unsteady flow model was based on the Preissmann finite difference "box" scheme solution of the St Venant equations and is described by Amein and Fang [5]. The model used 5 cross sections.

Fig 5 shows the depth hydrograph obtained at the upstream boundary of the reach (section 14) compared with the measured depth hydrograph. The Single Channel Method overestimates depth and the Divided Channel Method underestimates depth. For this flood, with a peak at relative depth Yr = 0.45, the conveyance functions from the LDM provide the closest underestimation of the depth hydrograph. Fig 6 shows the errors in computed floodplain flow depth for the overbank flood. The LDM errors follow the same distribution as the Divided Channel Method and show the LDM to have greater accuracy than the Divided Channel Method in the region Yr > 0.3.

REFERENCES
[1] Martin L.A. and Myers W.R.C., Measurement of overbank flow in a compound river channel. Proc Inst Civ Engrs, Part 2, 1991, 91, Dec, 645-657.
[2] Wark J.B., Samuels P.G. and Ervine D.A., A practical method of estimating velocity and discharge in compound channels. Proc Int Conf on River Flood Hydraulics, ed W.R. White, John Wiley and Sons, 1990, 163-172.
[3] Wark J.B., The equations of river and floodplain flow, Research Report, Dept of Civil Engineering, University of Glasgow, 1988.
[4] Ackers, P. Hydraulic design of two-stage channels. Proc Inst Civ Engrs, Wat Marit and Energy, 1992, 96, Dec, 247-257.
[5] Amein M. and Fang C.J., Implict flood routing in natural channels. J Hydr Div, ASCE, 96, Hy12, Paper 7773, Dec, 1970, 2481-2500.

Un cas d'intégration modélation physique - modélation numérique au support informatique

M.Popescu
Institut de Recherches
pour l'Ingéniérie de
l'Environnement - Bucarest
et Université de Constantza
Roumanie

F. Ionescu
Université de Génie
Civil
Bucarest , Roumanie

P.Vlase
PEME , France
51 , Rue de la Libération
62920 , Gonnehem

1. Introduction

La modélation physique et la modélation numérique représentent des méthodes courantes d'étude dans la recherches hydraulique pour une vaste classe de problèmes .

Traditionnellement les problèmes de l'ingéniérie hydraulique étaient abordés par des méthodes analytiques et expérimentations sur modèles physiques .

La modélation numérique , considérée au commencement comme une nouvelle modalité de dépassement des méthodes analytiques qui ont de possibilités limitées de résolution , s'est imposé rapidement dans le traitement de presque tous les problèmes de l'ingéniérie hydraulique ; en arrivant même à poser le problème si les méthodes numériques ne peuvent pas résoudre complétement les problèmes en question , en renonçant aux modéles physiques .

La pratique et les développements ultérieurs ont montré que l'intégration de ces deux modalités d'aborder ces problèmes représente une voie d'amélioration continue des méthodes de résolution des problèmes de l'ingéniérie hydraulique .

2. Modélation physique et modélation numérique

Afin d'établir le cadre de présentation de certains aspects relatifs à l'intéraction modélation physique - modélation numérique , on présente tout d'abord quelques éléments fondamentaux de la modélation physique et de la modélation numérique pour le cas d'un problème , étudié par les auteurs , du domaine des mouvements nonpermanents des systèmes hydrauliques sous pression .

2.1 Modèle physique

L'installation de laboratoire , de dimensions relativement grandes , pour l'étude des mouvements nonpermanents des systèmes hydrauliques sous pression , réalisée à l'Institut de Recherches Hydrotechniques de Bucarest , a été tellement conçue pour permettre la simulation et l'étude dans des conditions de laboratoire , pratiquement de tous les types de mouvements nonpermanents des systèmes hydrauliques sous pression rencontrés dans le domaine hydrotechnique , [4] , [5] , [6] .

L'installation peut fonctionner autant en régime gravitionnel (Schéma hydraulique I) qu'en régime de pompage (Schéma hydraulique II) situations de régime rencontrées en pratique aux usines hydroélectriques et respectivement stations de pompage .

On mentionne quelques données technico-fonctioneles concernant les pricipales parties de l'installation de laboratoire (fig.1) , [4] : A - réservoir d'eau , en forme cylindrique , ayant H = 7,75 m et D = 1 m ; B - Conduite de pression φ 125 mm et

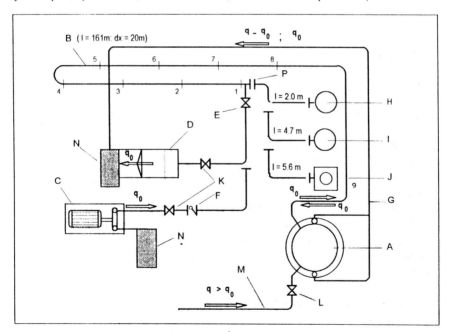

Fig.1 Schéma général , en plan de l'installation expérimentale de laboratoire

L - 161 m ; C -deux pompes de type centrifuge liées en parallèle , ayant les charactéristiques de régime suivantes : $H_0 = 20$ m , $Q_0 = 50$ l/s ; D - Bassin final à déversoir de mesure ; E - vanne à fermeture rapide ; H , I , J - chambre d'équilibre , réservoir d'air , dispositif de protection antichoc ; K , L - vannes de réglage ; P - section de raccord à l'installation pour les moyens de protection ci dessus mentionnées .

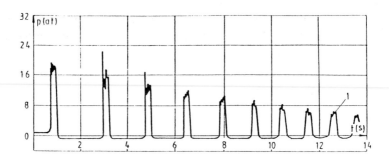

Fig.2 Pression mesurée dans la section 1 , à la fermeture brusque de vane
($q_0 = 16,5$ l/s)

Parmi les mesures effectuées sur cette installation de laboratoire , dans la figure 2 , on présente une mesure de pression dans la section 1 de la conduite , sur le schéma hydraulique I (réservoir conduite-vanne) , à la fermeture brusque de vanne , [4], [5].

2.2 Modèle numérique
En utilisant les performances actuelles des ordinateurs électroniques, l'intégration numérique des équations du choc hydraulique est devenue possible presque dans tous les cas rencontrés en pratique , c'est à dire pour des systèmes hydrauliques sous pression parmi les plus complexes et aux conditions à limite très diverses , [1] , [5] . La méthode des caractéristiques est employée fréquemment pour le calcul pratique du choc hydraulique par simulation numérique .
C'est ainsi que le système d'equations du mouvement :

$$\frac{\partial H}{\partial x} + \frac{1}{g}\frac{\partial V}{\partial t} + \frac{\lambda}{2gD}V|V| = 0$$

$$\frac{\partial H}{\partial t} + \frac{c^2}{g}\frac{\partial V}{\partial x} = 0$$

(1)

qui forme un système aux dérivés partielles quasiliniaire se transforme en equations aux dérivés totales :

$$\frac{dx}{dt} = \pm c$$

$$\pm \frac{g}{c}\frac{dH}{dt} + \frac{dV}{dt} + \frac{\lambda}{2D}V|V| = 0$$

(2)

où on a employé les notations habituelles .

En utilisant la forme d'écriture en différences finies sur un réseau (x,t) entre les points P et A et respectivement P et B , les équations (2) ont les formes suivantes :

$$V_p - V_A + \frac{g}{c}(H_p - H_A) + \frac{\lambda}{2D}V_A|V_A|\Delta t = 0$$

$$x_p - x_A = c(t_p - t_A)$$

(3)

pour la courbe caractéristique directe c+ et

$$V_p - V_B - \frac{g}{c}(H_p - H_B) + \frac{\lambda}{2D}V_B|V_B|\Delta t = 0$$

$$x_p - x_B = -c(t_p - t_B)$$

(4)

pour la courbe caractéristique indirecte c- .

En partant des conditions initielles commes pour t = t0 et en utilisant les relations pour les conditions à limite spécifiques au schéma hydraulique analysé , le modèle numérique permet le calcul des H et V dans tous les noeuds du réseau (x,t) - c'est à dire le calcul du coup de bélier dans toutes les sections caractéristiques du système hydraulique analysé , [5] .

Afin d'exemplifier dans la figure no. 3 on présente les résultats du calcul numérique pour la variante présentée au point 2.1.

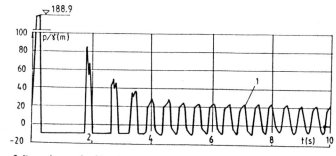

Fig. 3 Pression calculée dans la section 1 , à la fermeture brusque de vanne (q0=16,5 l/s)

3. Possibilités d'intéraction modélation physique - modélation numérique au support informatique

Du point de vue de l'intérêt pratique dans l'ingéniérie hydraulique , on peut distinguer les étapes suivantes dans l'évolution de l'intéraction modélation physique - modélation numérique :

I. Le modéle physique de laboratoire contrôlé par le calculateur , qui assure le contrôle des dispositifs de réglage et de mesure , le monitoring et le traitement des données mesurées et la rédaction des résultats (par tableaux et graphiques) , [6] ;

II. Les expérimentations sur modèle physique , effectuées dans les conditions de l'étape I , sont dirigées en parallèle aux simulations des mêmes variantes d'étude , sur des modèles numériques .

Les résultats tellement obtenus permettent d'une part le contrôle , et d'autre part le perfectionnement du modèle numérique .

III. De l'utilisation en parallèle du modèle physique et du modèle numérique (étape II) on passe à leur utilisation intégrée , dans le sens que le modèle numérique est utilisé afin de compléter la série d'expérimentations nécessaires à une étude la plus complete possible du problème analysé , dans des conditions optima de temps et de coût .

4. Etude de cas

Dans ce rapport [6] on a présenté un example d'intéraction modélation physique - modélation numérique pour l'installation de laboratoire ci-dessus mentionnée(fig.1).

L'installation a été munie d'un système électronique automatique de commande , mesure , centralisation et traitement des données expérimentales , qui comprend de l'équipement électrique et électronique adéquat , un calculateur de processus et un microcalculateur .

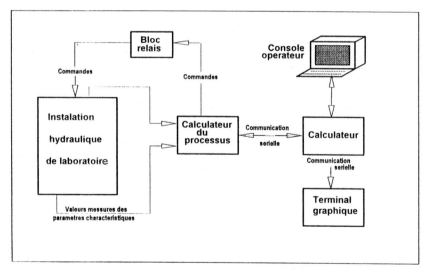

Fig.4 Schéma général de l'intéraction modèle physique - modèle numérique

Le schéma générale de fonctinnement de l'installation expérimentale de laboratoire contrôlé par le microcalculateur est présenté dans la figures 4 , [6] .

5. Conclusions

La modélation physique et la modélation numérique sont arrivées à présent à des performances qui peuvent être améliorées difficilement d'une manière radicale .

Une modalité de réaliser une étape importante afin de résoudre les problèmes de l'ingéniérie hydraulique peut être ·obtenue par l'interaction modélation physique-modélation numérique .

Au support informatique accessible à présent , cette intéraction peut être réalisée par plusieures modalités .

Dans cette communication on propose une modalité pratique d'intéraction modélation physique - modélation numérique qui permet trois niveaux d'intéraction: I - modèle physique contrôlé par l'ordinateur , II - utilisation en parallèle de ces deux modèles et III - utilisation intégrée , dans le sens que le modèle numérique reprend et complète la série d'expérimentations nécessaires afin de résoudre le problème d'une manière la plus complète .

Bibliographie
1. Chaudhry , H.
Applied Hydraulic Transients ,(Second Edition) , Van Nostrand Reinhold Company , New York, 1987.
2. Kalkwijk , J.
Investigation into cavitation in long horizontal pipelines caused by waterhammer , Delft hydraulics laboratory , Publication No 115 , 1974 .
3. Popescu M.
Computation of hydraulic transients in complex hydro-schemes , Water Power and Dam Construction , England , No. 9 , 1986 .
4. Popescu , M.
Résults expérimentaux dans l'étude des mouvements non-permanents des systèmes hydrauliques sous pression avec des applications aux usines hydroélectriques et aux stations de pompage , XX Convegno di Idraulica e Construzioni Idrauliche , Padova , Italia , 8-10 Settembre , 1986 .
5. Popescu M. , Arsenie D.
Méthodes de calcul hydraulique pour des usines hydroélectriques et des stations de pompage , Edition Technique , Bucarest , 1987 (en roumain) .
6. Popescu , M. , Jelev, I., Vlase , P., Dragomir , C.
L'étude des mouvements non-permanents des systèmes hydrauliques sous pression sur une installation de laboratoire contrôlée par microcalculateur , XXII Convegno di Idraulica e Construzioni Idrauliche , Cosenza , Italia , 4-7 octobre , 1990 .

1B20

THE USE OF INFORMATICS FOR HYDRAULIC APPLICATIONS

Prof. Dr. Rawya Monir Kansoh

Hydraulics and Irrigation Enginering Dept,
Faculty of Engineering, Alexandria University, EGYPT

ABSTRACT

Due to the pressing demands on information and decision, the revolution on informatics has taken place. A model has been developed to help in making decisions in the area of water resources.

The main aim used in developing the proposed model was for unsteady flow conditions using method of characteristics and the finite difference method.

INTRODUCTION

The aim of this paper is to describe the importance of developing decision support system technique in water resources management with applicationn in studying decision support system technique and mathematical modelling.

The importance of such study is influenced by the severe changes that take place on weather conditions, and the continuous decrease in rain fall. Consequently, the decrease of the share of EGYPT from River Nile and the drastic droughts that occurred in Africa can be evaluated.

Information technology becomes apparent, which helps in providing Decision Support System (DSS) in water resources.

INFORMATION TECHNOLOGY

Information technology is an all-embracing term which is applied collectively to the modern techniques associated with advanced computing systems and data communications.

In Great Britain, for example, the proportion of the labor force employed in information related sectors of the economy has increased from about 30% to nearly 45% between 1961 and 1987. (Figure 1) shows the underlying trends. In United States the trends are similar, as indicated by (Figure 2).

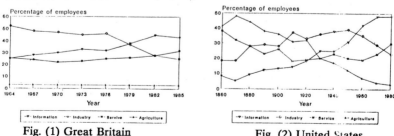

Fig. (1) Great Britain **Fig. (2) United States**

WATER RESOURCES MANAGEMENT DECISION SUPPORT SYSTEM

New methods for addressing water resources problems are being developed and implemented which incorporate advanced computer based capabilities for data management, analysis, and direct information communication to water system managers and operators for decision support.

Advanced data collection systems measure the full range of environmental and other system parameters including water flow, pressure, and quality.

HYDRALEX

The propagation of flood waves in river and open channel beds is a typical problem of gradually varied unsteady flow. To study the propagation of flood waves in open channels, the one dimensional differential equations of gradually varied unsteady flow, established by Barre de Saint Venant is being used.

Essentially, two different approaches can be followed.

The first, the well known method of characteristics, consists in converting the original system into an equivalent system of ordinary differential equations and then replacing the total derivatives by the corresponding incremental ratios.

The second in replacing directly the partial derivatives by quotients of finite differences.

The solution of such systems consis in the computation of the quantities Y and Q at a given time, from the known values of the same quantities at the previous time and from the given boundary conditions.

Hydralex is newly developed computer model (refering to hydraulic-Alexandria). The model is designed to simulate flood wave propagation along watercourses. The system is based on the solution of the St Venant equations in their complete form using an explicit finite difference scheme and method of characteristics for some boundary conditions. Knowledge of the geometrical characteristics and of the roughness of the river bed is required. The geometrical description of the river or channel bed is given by means of cross-section, to express the passive resistance using the Manning formula. The system is built up in an interactive mode. HYDRALEX's structure is very flexible to accept new boundary conditions or specifying new parts of the watercourse for example pipe, syphons, ... etc.

Numerical Methods Used

1. Finite difference method with explicit scheme for the interior nodes.
2. Characteristic method for boundary conditions.
3. Improved Euler method for the ordinary differential equations to determine the back water curve for steady flow using the formula.

$$\frac{dy}{dx} = \frac{S_o - S_f}{1 - \dfrac{BQ^2}{gA^3}}$$

4. Bisection method for nonlinear algebraic equations to

calculate the critical water depth and normal water depth.

General Description

Starting from the known values of water levels and discharges (initial conditions) and from the boundary conditios, HYDALEX is designed to determine the values of the same quantities, levels and discharges in any cross-section at any time, provided that geometry and roughness of the river or channel is given.

For boundary condition, water levels or discharges at the upstream section, and water levels at the downstream section can be chosen.

The St Venant equations are being adapted in the following form:

$$\frac{\partial Q}{\partial X} + B\frac{\partial Y}{\partial t} = 0$$

$$\frac{\partial Q}{\partial t} + \frac{\partial}{\partial X}(\frac{Q^2}{A}(+ gA\frac{\partial Y}{\partial X} - gA(So - Sf) = 0$$

in which:
x = current abscissa (positive in the direction of flow)
t = time   Q = discharge   B = width of water surface
Y = water depth   g = acceleration due to gravity
A = cross-sectional area   Sf= friction head losses
So = slope of channel bed in lingitudinal direction

Wetted area, wetted perimeter and superficial width are determined, for each cross-section, as function of depth.
To express the passive resistance Manning formula is being used
 In order to solve the St Venant equations the finite difference explicit scheme explained earlier has been adapted.

Input-Output Files

HYDRALEX leaves total freedom in naming input files. Input file must be in ASCII format and free of any special characters or tabs. Most word processors can create text files for use by HYDRALEX.

The HYDALEX Input File Description

First line contains the number of river's or channel's reaches and then for every reach three lines required.
*   The first line contains ((reach number, its beginning starting (m), its end station (m), number of grid points, bed width, side slope "S horizontal: 1 vertical", longitudinal slope" positive when sloping downward in the direction of flow", Manning's roughness coefficient, initial discharge and initial water depth)).

*   The second line contains the upstream boundary condition and its data.
*   The third line contains the downstream boundary condition and its data.
The last line contains the time interval Δt which is suggested in the calculation.

Boundary Conditions

For upstream end one of the following lines can be used:

0   for non control structure, used in the interior reaches where the change only happen in the cross-section or slope or discharge and it is not followed by any thing.
1.   for constant water depth at the upstream and is followed by upstream water depth and the calculation time.
2.   for constant discharge at the upstream and is followed by upstream discharge and the calculation time.
3.   for inflow hydrograph and is followed by the number of changes in discharge, its time interval "equally" and its values m/s.

For the downstream end one can choose one of the following conditions:

0   for non control structure at its end, used in the interior reaches where the change only happens in the cross-section or slope or discharge and it is not followed by any thing.
1.   for known and constant downstream water depth and ist followed by the downstream water depth.
2.   for known downstream discharge and is followed by the downstream discharge.
3.   for free end and is not followed by any thing.
*** and it can be expanded such that the program will adopt more cases.

MODEL RESULTS

In order to introduce the results of our model the next different cases are selected to show the model capabilities in dealing with the different boundary conditions.

Case 1

In this case, the model will be used to predict the water wave propagation and the maximum water depth along the channel. Unsteady flow is produced by instantaneously closing a downstream gate at t=0 . Working with a channel that has the next properties. Bed width = 20.00 m . Side slop = 2 . Longitudinal slope = 0.0001 m/m Manning roughness coefficient = 0.013. Channel length = 1000 m Initial discharge = 110 m³/s
The boundary condition at upstream is defined as constant water depth = 3.069 m and at the downstream defined as close end (Q=0) the initial condition is considered as constant water depth along the channel equal to the normal depth = 3.069 m, figures (3 & 4).

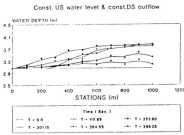

Fig. (3) Water Depth Versus Station

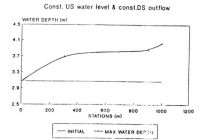

Fig. (4) Initial Water Depth Vs Maximum Water Depth

<u>Case 2</u>

The model to predict the water propagation and the maximum water depth along a channel considering the boundary condition at upstream is defined as constant inflow discharge and at the downstream defined as constant water depth. Working with a channel with the following properties. Bed width = 20.00 m. Side slope = 2. Longitudinal slope = 0.0001 m/m. Manning roughness coefficient = 0.013. Channel length = 1000 m . Initial discharge = 110 m^3/s. The boundary condition at upstream is defined as constant inflow discharge = 110 m^3/s and at the downstream defined as constant water depth = 4.1 m. The initial condition is considered as constant water depth along the channel equal to the normal depth = 3.07 m.

From Figure (5) , one can see that the initial condition is far from the real situation, but the model begins changing it until it reaches a reasonable condition. In the field, the disturbance must be considered due to its existence. In other word we can say that if the downstream water depth is required to reach a certain value = 4.1 m, we must keep in mind that the corresponding upstream water depth will be nearly = 4.75 m. It is very important because if the land level around the channel is less than 4.75 m, meaning that this area will be inundated. Figure (6) shows the maximum water depth which permissible along the channel.

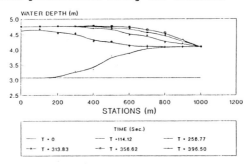

Fig. (5) Water Depth Vs. Station

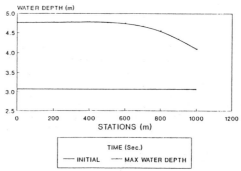

Fig. (6) Initial Water Depth Vs.
Maximum Water Depth

CONCLUSIONS

From the model results it can be seen that a wide spectrum of usable and reasonably accurate mathematical model for flood measuring is available. This variability can be attributed to the fact that the phenomenon is very important from a physical point of view and rigorous mathematical schematization can be readily simplified in many different ways.

The model developed which was based on the complete St Venant equations is, as other similarly developed models, undoubtedly the most descriptive.

The field of computer technology has proven to be of valuable support to hydraulics problems especially practical problems. It helps in decision making which sometimes can be of crushal values. Some purposes and objectives of using the model can be summarize in functions as:

1) Calibration of the model for flood routing on a specific reach of watercourse,
2) Simulating the flood wave propagation under different conditions,
3) predicting its state and its effect in the area around it.
4) predicting the water profile in a new watercourse under design.
5) predict the worst case such as a sudden failure to a control structure which controls the discharge into watercourse.

REFERENCES

1. Chadwick, A.J. and Morfett, J.C. Hydraulic in. Civil Engineering 2nd ed. Allen & Unwin, London 1989.
2. Chao, J.L. and R.R. Trussell. Design of Flow Distribution Channels. Journal of the Environmental Engineering Division, ASCE. Vol. 106, o EE2, Proc. Paper 15317, April, 1980.
3. Chapra, S.C., and R.P. Canale. Numerical Methods for Engineers with Personal Computer Applications. McGraw-Hill, New York, 1985.
4. Chow, V.T. Open Channel Hydraulics. McGraw-Hill, New York, 1986.
5. Ciriani, Tito, A., Maione Ugo, Wallis, James R. Mathematical Models for Surface Water Hydrology John Willey & Sons, New York, USA, 1977.
6. Cuge, A., Holly, F.M., JR. and Verway, A. Practical aspects of computational River Hydraulics. Pitman, London, 1980.
7. HEC, User's Manual, US Army Corps of Engineers, U.S. Army, Corps of Engineering, 1976.
8. HYDRO, User's Manual, Ch2m Hill, Inc, Oreon, USA, 1991.
9. Johnson, L.E. Water Resource Management Decision Support Systems. Jour. of Water Resources Planning and Management, Vol. 112, No. 3, July, 1986.
10. Mahmood, K. and Yevjecich, V. Unsteady Flow in Open Channels. Water Resources Publications, Fort Collins, Colorado, 1975.
11. Simonovic, S.P., Intelligent Decision Support and Reservoir Management and Operations. Jour. of Computing in Civil Engineering, Vol. 3, No.4, October, 1989.
12. West, C., and Lording, R. Information Technology Application, Heinemann Newnes, Oxford UK, 1989.

1B21

FRICTION-DOMINATED FLOW COMPUTATION IN LAKE SAINT-PIERRE, QUEBEC

ANDRÉ GAGNON, Eng., M.Eng.
SNC■LAVALIN, Montreal, Quebec, Canada

ABSTRACT

The velocity field of lake Saint-Pierre, a partial lake and river system, is calculated using an elementary two-dimensional method for the first approximation of the stream function based on the numerical computation of friction-dominated flow. Precision of the results is obtained by putting emphasis on the determination of the significant variables, i.e. water depth and friction using extended computer resources. A very high resolution grid is generated in order to counterbalance the degree of approximation introduced as well as to provide an appropriate graphical frame for the determination of the streamlines. The comparison of results with measurements and with another, more refined, type of modelling confirms the validity of the method.

INTRODUCTION

The lake Saint-Pierre is a shallow enlargement in the Saint-Laurence river, Eastern Canada's largest river which discharges into the Atlantic ocean. Because of the presence of the Saint-Laurence seaway's shipping canal which runs in the central region of the lake, it can be considered as a large, straight two-stage channel. In addition, the flow distribution is also determined to a great extent by the lake's abundant and various vegetation. Such peculiar friction conditions of the river bed, along with lateral inflow from minor tributaries makes the prediction of the flow distribution much difficult. In channels where the width is large as compared to the depth, the flow field calculation can be approximated without compromising too much accuracy by assuming it is friction-dominated. It will be shown that the results of this approximation method compare very well with the measured velocities, along with Leclerc's [INRS-Eau, 1992] calculation using a two-dimensional finite element model. However care must be provided in designing a computation grid in which the finite difference is reduced to a minimum, as demonstrated below.

GOVERNING EQUATIONS

Figure 1 shows the main bathometric characteristics over the extent of lake Saint-Pierre. The flow field is governed by the usual pair of continuity and momentum equations. In shallow depths the Navier-Stokes equation is replaced by the simpler two-dimensional depth averaged Saint-Venant equation:

$$\frac{\partial u}{\partial t} + u\frac{\partial u}{\partial x} + v\frac{\partial u}{\partial y} + g\frac{\partial H}{\partial x} = F_x$$

$$\frac{\partial v}{\partial t} + u\frac{\partial v}{\partial x} + v\frac{\partial v}{\partial y} + g\frac{\partial H}{\partial y} = F_y$$

(1) a,b

The continuity equation is

$$\frac{\partial H}{\partial t} + \frac{\partial uh}{\partial x} + \frac{\partial vh}{\partial y} = 0$$

(2)

where F_x, F_y are the friction terms, h is the water depth, H the water elevation with respect to a fixed datum, u and v the depth average velocity components in the x and y directions respectively, g the acceleration due to gravity and t the time. As mentioned above, since the horizontal length scale of the lake (10 km) is significantly greater than the vertical length scale (2 to 5 metres), the friction terms within F_x and F_y on the right hand side are dominating the convective acceleration terms on the left hand side of equation (1) and the flow is called *friction-dominated*.

Significant simplification to the flow computation is possible for the friction-dominated steady flow. Based on Lagrange's concept of velocity potential, the flow is determined by a set of stream (ξ) and potential (η) functions in which η is equal to the piezometric head, that is the water level. Neglecting the convective acceleration terms on the left hand side and the Coriolis, wind and diffusion terms included in F_x, F_y on the right hand side, Saint-Venant's equation (1) becomes

$$\frac{\partial \eta}{\partial x} = S_x , \qquad \frac{\partial \eta}{\partial y} = S_y$$

(3) a,b

and the continuity equation (2) becomes

$$\frac{\partial(uh)}{\partial x} + \frac{\partial(vh)}{\partial y} = 0$$

(4)

where $(S_x, S_y) = \nabla\eta$ is the gradient of the piezometric head and is a function of the water depth h and the bed friction coefficient c_f as follows:

$$S_x = \frac{c_f u \sqrt{u^2+v^2}}{2gh} \quad , \qquad S_y = \frac{c_f v \sqrt{u^2+v^2}}{2gh} \quad . \qquad \text{(5) a,b}$$

GENERATING THE COMPUTATION GRID

Further simplification to the calculation procedure is possible if the channel is relatively straight so that a rectangular coordinate system may be selected to align with the general direction of the flow. By orienting its x axis along this direction (Figure 2), the potential lines are assumed to be straight lines and the v component of the velocity is negligible. The u component thus becomes the principal velocity and is determined by the Manning or the Chézy equation. For example, let Q_A be the total discharge and S_A the friction slope across a section of the channel of area "A", and the friction be determined by the Manning formula ($n = h^{1/6}(c_f/2)^{1/2}$) where n is the Manning coefficient. Therefore the solution of the flow field can be found through the numerical computation of the finite difference formulation of

$$Q_A = S_A^{1/2} \int_{y_{Left}}^{y_{Right}} \frac{h^{5/3}}{n} \, dy \qquad \text{(6)}$$

across a uniform computational grid, which density is prescribed by the average scale of the velocity gradients with respect to that of the entire domain. A relatively small grid spacing is designated (Δx=169 and Δy=30 metres: this is 200 transversal by 495 longitudinal sections to cover the lake's flood plain, or 99 696 nodes) in order to provide a sufficient number of nodes in the shipping canal and a high resolution in the computation field. High resolution will facilitate the determination of the streamlines and the stream function since the extrapolation between grid nodes is made easier.

Since no absolute reference level is implied in the formulation of equation (6) such a H in equation (2), some stage-discharge relationship must be established so that the raw bathometric input data become explicit. Using an appropriate averaging computer algorithm, the water depth h is carefully estimated around every grid node with respect to a referenced water datum. The condition of the vegetation (which varies from one season to another) at the time represented by the simulation must also be determined and combined with the effect of substratum friction such as to generate $n = f(x,y)$ across the domain. These two variables h and n, along with the coordinates of the nodes in both the grid (i, j) and the earth (x longitude, y latitude) coordinates are stored in a large database. This database becomes the final computation grid.

COMPARISON OF RESULTS AND DISCUSSION

The computer simulation of the friction-dominated model (FDM) and Leclerc's finite-element model (FEM) were aligned by using the same stage-discharge relationship and friction map for a given flow event (original data are a courtesy of INRS, see References). A general picture of the results is shown in Figure 3, outlining the stream function ξ for example in November of 1990. The lake Saint-Pierre's main hydrodynamic features are clearly observed, particularly the shipping canal. In Figure 4, the image represents the flow field computed by FDM (where each *pixel* is a grid node) in the middle of July 1990. It illustrates some secondary channels where vegetation, although at its growth peak, is scarce and where the friction is therefore reduced. In Figures 5, an example of the continuous velocity profile of FDM is compared with the measured depth averaged velocity along with the magnitude of the principal velocity by FEM at specific numbered stations in the lake which are aligned with the cross-section. The comparison is during the same event as in Figure 4.

In general, the friction-dominated flow calculation results tend to demonstrate that the velocity profile is a function of the bed and the friction profile along any given transverse section. This relation highly corresponds to the flow behaviour in lake Saint-Pierre and is well reflected by the results of the method, except very near the point of discharge of tributaries where much of the flow is not in the principal direction (not shown here).

CONCLUSION

The velocity field of a relatively straight channel with a high width to depth ratio such as lake Saint-Pierre was successfully calculated using an elementary method based on friction-dominated flow. The relative accuracy of the results was obtained by emphasizing the determination of the significant variables in the theoretical equations, that is to say the water depth and the friction coefficient, and by decreasing the finite difference down to a relatively small increment with respect to the dimensions of the physical domain. The numerical flow computation method presented in this paper is simple and can certainly be recommended as a tool to obtain the first estimate of the flow field in a river provided considerable digitized information relative to its bed. In return, the computation grid will provide a useful frame for the geographical reference of the numerical results if such data were initially attached during its generation. The computer resources required are important, however those required for a more refined flow calculation are very significantly greater than for the approximated flow solution and it is not always practical and necessary to make the additional step. Hence, for many practical application in rivers where the depth is shallow as compared with the width, the simulation based on the friction-dominated flow equations described here is quite an acceptable method.

REFERENCES

Gagnon, A. (1993), "Computation of friction-dominated flow in lake Saint-Pierre", *Thesis submitted in partial fulfilment of the requirements for the degree of Master in Engineering*, Faculty of Graduate Studies and Research, McGill University, Montreal, Canada, 56 p. + Appendices.

INRS-Eau (1992), "Modélisation hydrodynamique des écoulements en eau libre du tronçon Tracy - lac St-Pierre", *Modélisation intégrée du suivi de la qualité de l'eau du tronçon Tracy - Lac Saint Pierre*, rapport No 1, Vol. 1, plan d'action Saint-Laurent, centre Saint-Laurent, Montréal, Canada, 176 p.

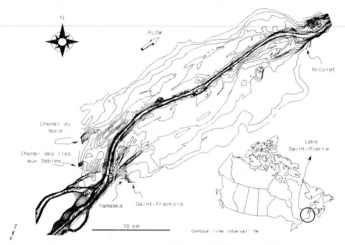

Figure 1 Bathometric chart of Lake St-Pierre (St-Laurence river).

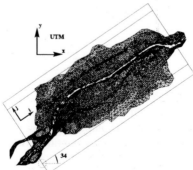

Figure 2 Alignment of the computation grid. A relation is set between the geographical (Universal Transverse Mercator map projection - x,y) and the grid (i,j) coordinates, which have different orientations.

Event November 1990
Computation by FDM (ξ in m3/s)

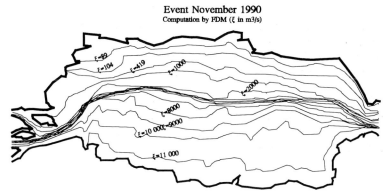

Figure 3 Streamlines generated using the results of the computation grid.

Event July 1990

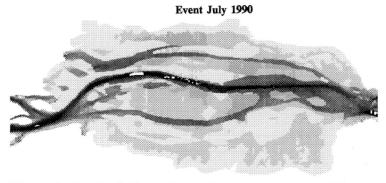

Figure 4 Velocity field: each grey shade correspond to a different range of velocities.

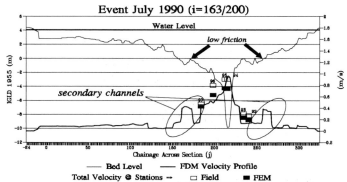

Figure 5 Example of a comparison of results at approximately 4/5 downstream of the computation grid.

1B22

APPLICATIONS OF REFINED NUMERICAL SCHEME FOR ADVECTION TO DIFFUSION SIMULATIONS IN A NATURAL BAY

T. KOMATSU, K. OHGUSHI* and S. YANO
Department of Civil Engineering, Kyushu University, Fukuoka 812, JAPAN
*Department of Civil Engineering, Saga University, Saga 840, JAPAN

ABSTRACT
We proposed a highly accurate numerical scheme for calculating advection transports of mass, heat and so on. This new scheme is based on the concept of solving a second order wave equation instead of a first order wave equation. The superior accuracy and stability of the scheme have already been proved by the previous comparisons with several other ones in many model calculations. In this study, the new scheme was applied to depth-averaged (2-dimensional) unsteady diffusion simulations in a natural bay. As a result of this research, it was found that this scheme also retains its high accuracy and simplicity even in applications to a natural bay which has complex flow fields that vary with time and space.

INTRODUCTION
Changes of water quality in rivers, oceans or enclosed water areas such as lakes and bays can be predicted by calculating both the flow field and mass transport numerically by using an appropriate mathematical modelling. The spread of mass is described by a partial differential equation that involves two kinds of transports, that is, advection and diffusion. The split operator approach allows to compute the advection and the diffusion independently for each time step. The advantage is that it allows the use of an different appropriate scheme for each process. An accurate calculation of the diffusion will be easily executed, while it is more difficult to attain a high degree of accuracy in calculating the advection.

Over the past few decades, several accurate numerical schemes for pure advection were proposed, such as the Six-point scheme [Komatsu et al. (1985)] and so on. Those schemes have high accuracy and stability, but, at the same time, they have also some difficulties with treatment in the vicinity of the boundary or with applications to multidimensional problems. Recently, we proposed a new refined numerical scheme for the pure advection, which is free from such problems

[Komatsu et al. (1995)].

In this study, the new scheme was applied to 2-dimensional unsteady diffusion simulations in a natural bay, which had unsteady and localized flow fields.

THE SOWMAC SCHEME

To calculate the 1-dimensional pure advection, which is expressed by

$$\frac{\partial C}{\partial t} + U \frac{\partial C}{\partial x} = 0 \tag{1},$$

where, C : concentration of diffusing solute and U : velocity, a number of finite difference schemes were developed and used, such as the Upwind difference scheme, the Lax-Wendroff scheme, the Leapfrog scheme, the QUICK scheme etc. However, as for their accuracies, most of them yield serious errors. On the other hand, a few schemes such as the Holly-Preissmann scheme [Holly and Preissmann (1977)], the Six-point scheme and the Improved six-point scheme [Komatsu et al. (1989)] are well known as accurate ones.

Recently, we developed a new scheme for calculating the pure advection based on a new concept that the second order wave equation should be calculated instead of the first order advection equation because of its higher accuracy and stability. We leave a detail about the derivation of this scheme to Komatsu et al. (1995) and give only its final expression for 1-dimensional pure advection equation as follows:

$$p_1 C_{i-1}^{n+1} + p_2 C_i^{n+1} + p_3 C_{i+1}^{n+1} = p_4 C_{i-1}^n + p_5 C_i^n + p_6 C_{i+1}^n \tag{2}$$

$$p_1 = 0.3776\ \alpha_{0+} + 0.3152\ \alpha_{0-} - 0.5467\ \alpha_{1+} + 0.4843\ \alpha_{1-} + 0.1691\ \alpha^2$$

$$p_2 = 1.3072 + 0.0624\ |\alpha| - 0.3382\ \alpha^2$$

$$p_3 = 0.3152\ \alpha_{0+} + 0.3776\ \alpha_{0-} + 0.4843\ \alpha_{1+} - 0.5467\ \alpha_{1-} + 0.1691\ \alpha^2$$

$$p_4 = 0.3776\ \alpha_{0+} + 0.3152\ \alpha_{0-} + 0.5157\ \alpha_{1+} - 0.4533\ \alpha_{1-} + 0.1381\ \alpha^2$$

$$p_5 = 1.3072 - 0.0624\ |\alpha| - 0.2762\ \alpha^2$$

$$p_6 = 0.3152\ \alpha_{0+} + 0.3776\ \alpha_{0-} - 0.4533\ \alpha_{1+} + 0.5157\ \alpha_{1-} + 0.1381\ \alpha^2$$

$$\alpha_{0+} = AINT\left\{ \frac{1+\alpha}{1+|\alpha|} \right\} \qquad \alpha_{1+} = \frac{\alpha+|\alpha|}{2} \qquad \alpha = \frac{U\ \Delta t}{\Delta x}$$

$$\alpha_{0-} = AINT\left\{ \frac{1-\alpha}{1+|\alpha|} \right\} \qquad \alpha_{1-} = \left| \frac{\alpha-|\alpha|}{2} \right|$$

where, α : the Courant number, Δt : time increment, Δx : computational grid interval, $AINT$: the function, which truncates decimals, in the FORTRAN, n : time step index and i : x-directional computational point index.

We named this scheme SOWMAC (Second Order Wave equation Method for Advective Calculation).

COMPARISON OF THE SOWMAC WITH OTHER SCHEMES

In order to show the high validity of the SOWMAC scheme, 1-dimensional problem of pure advection was adopted as a simple case. An initial condition for concentration distribution is given as the superposition of two kinds of Gaussian distributions. One is centered at $x = 1,400$m with maximum concentration $C = 10.0$ and another at $x = 2,400$m with peak value $C = 6.5$. Both have the same standard deviation of 264m. The concentration distribution is transported downstream for 9,600sec by constant velocity $U = 0.5$m/sec. The computational intervals are $\Delta x = 200$m and $\Delta t = 100$sec. FIG. 1 shows the results calculated by the SOWMAC scheme and other three schemes under the same conditions (see TABLE-1). The thick solid line in the figure indicates an exact solution. Several numerical results show large numerical diffusions or serious phase errors. On the other hand, the SOWMAC scheme has almost the same accuracy as the Six-point scheme without phase errors although only three computational grid points are used.

EXTENSION TO MULTIDIMENSIONAL PROBLEMS

It is straightforward to apply the SOWMAC scheme originally formulated for 1-dimensional base to 2- or 3-dimensional problem. 2-dimensional advection-diffusion equation is expressed as follows:

$$\frac{\partial C}{\partial t} + U \frac{\partial C}{\partial x} + V \frac{\partial C}{\partial y} = \frac{\partial}{\partial x}\left(D_x \frac{\partial C}{\partial x} \right) + \frac{\partial}{\partial y}\left(D_y \frac{\partial C}{\partial y} \right) \tag{3},$$

where V : velocity component in y-coordinate direction and D_x, D_y : turbulent diffusion coefficients for each direction. By applying the split operator approach to Eq. (3), this equation is divided into the pure advection equation,

$$\frac{\partial C}{\partial t} + U \frac{\partial C}{\partial x} + V \frac{\partial C}{\partial y} = 0 \tag{4},$$

and the diffusion equation,

$$\frac{\partial C}{\partial t} = \frac{\partial}{\partial x}\left(D_x \frac{\partial C}{\partial x} \right) + \frac{\partial}{\partial y}\left(D_y \frac{\partial C}{\partial y} \right) \tag{5}.$$

We leave a detailed discussion of the physical and mathematical meanings of the split operator approach for Komatsu *et al.* (1995).

By dividing the 2-dimensional pure advection process (Eq. (4)) into two successive steps, that is, the advective transports in the x-direction and the y-direction, Eq.(4) can be approximately replaced by a series of 1-dimensional pure advection equations (see FIG.2), which are described below.

$$\frac{\partial C}{\partial t} + U \frac{\partial C}{\partial x} = 0 \qquad (6), \qquad\qquad \frac{\partial C}{\partial t} + V \frac{\partial C}{\partial y} = 0 \qquad (7).$$

Eqs. (6) and (7) are easily solved by the SOWMAC scheme. It should be noted that these equations must be calculated in the same time step sequentially. Like the ADI method, this procedure drastically reduces the cost of 2-dimensional computation in spite of the scheme's implicity. For example, in the case of $N_x \times N_y$ computational grid system, implicit schemes usually need to solve simultaneous equations of $N_x \times N_y$ unknowns. On the other hand, the SOWMAC scheme requires only solving simultaneous equations of N_y unknowns N_x times for y-directional advection after solving simultaneous equations of N_x unknowns N_y times for x-directional one. These simultaneous equations will be solved very easily because the SOWMAC scheme constructed by three spatial grid points yields only tridiagonal matrices on computation. 3-dimensional problem is also solved in the same manner as the 2-dimensional one by adding z-directional advection as the third process.

APPLICATIONS OF THE SOWMAC SCHEME TO DIFFUSION CALCULATIONS IN A NATURAL BAY

The SOWMAC scheme was originally formulated by assuming constant velocity field. However, most of natural flow fields, such as bays and lakes, usually vary complicatedly with time and space. In order to show the validity of the SOWMAC scheme even in such a complex flow field, an attempt to apply three schemes, that is, the SOWMAC, the Upwind and the Six-point scheme, to advection calculations of diffusion simulations in a natural bay was made.

The numerical simulations of tidal current and diffusion of chlorinity, which is conservative solute, were carried out in Hakata Bay in Japan by using a depth-averaged (2-dimensional) model. Details of the simulations are omitted here [see Komatsu et al. (1994)]. In this study, locality of dispersion coefficient D was ignored, namely, D = const. First, we attempted a case of D = 80m²/sec. This is the most suitable value for Hakata Bay. FIG. 3 shows the distribution of chlorinity calculated by the SOWMAC scheme. FIG. 4 shows the comparisons of the calculated distributions of chlorinity by three schemes. Locations for comparison are shown as black dots in FIG. 3. Apparently, numerical diffusion of the Upwind scheme is larger than others and in spite of using only three computational grid points the SOWMAC scheme has almost the same accuracy as the Six-point scheme which uses six ones, as is analogized from the name. Next, we tried a case when smaller physical diffusivity is assumed, that is, D = 10m²/sec (see FIG. 5). It is clear from FIG. 5 that, as well as the case that D = 80m²/sec, the accuracy of the SOWMAC scheme is equivalent to that of the Six-point scheme. The numerical diffusion of the Upwind scheme appears more remarkably because of the small physical diffusivity.

CONCLUSIONS

In this study, we introduced a new accurate numerical scheme for advection. This scheme is called the SOWMAC scheme. Comparison of this scheme with various other schemes in model calculations proves its superior accuracy. It will be easily applied to multidimensional practical problems by decomposing the velocity vector into each directional component. It was confirmed from 2-D diffusion simulation in a natural bay that the SOWMAC scheme is still accurate enough for the advection calculation even in a complex flow field. The difference between the calculated result by the Upwind scheme and those by two other schemes tells us the importance of selection of appreciate scheme for advection.

REFERENCES

Holly, F.M., Jr. and Preissmann, A. (1977). "Accurate calculation of transport in two dimensions", J. Hydr.Div., ASCE, 103(11), 1259–1277.

Komatsu, T., Holly, F.M., Jr., Nakashiki, N. and Ohgushi, K. (1985). "Numerical calculation of pollutant transport in one and two dimensions", J. Hydroscience and Hydr. Engrg., JSCE, 3(2), 15–30.

Komatsu, T., Ohgushi, K. and Asai, K. (1995). "Refined Numerical Scheme for Advective Transport in Diffusion Simulation", J. Hydr. Div., ASCE (in submission).

Komatsu, T., Ohgushi, K., Asai, K. and Holly, F.M., Jr. (1989). "Accurate numerical simulation of scalar advective transport", J. Hydroscience and Hydr. Engrg., JSCE, 7(1), 63–73.

Komatsu, T., Yano, S. and Asai, K. (1994). "On the Estimation of Dispersion Coefficient for 2-Dimensional Simulations in a Bay", Proc. of 9th Congress of Asian and Pacific Division of IAHR, Singapore, Vol. 3, 345-352.

TABLE–1 Several schemes for calculations of advection and their accuracies

Case	Scheme for spacial derivative (1)	Scheme for temporal derivative (2)	Degree of accuracy of (1)	Degree of accuracy of (2)
1	Upwind scheme	Euler scheme	1	1
2	QUICKEST scheme		3	3
3	Six-point scheme		1	1
4	SOWMAC scheme		2	2

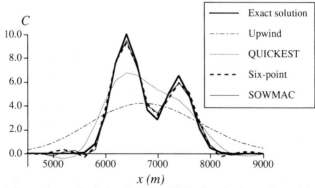

FIG. 1 Comparisons of calculated results of 1-D advection by various schemes

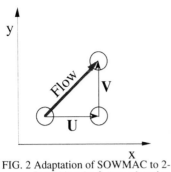

FIG. 2 Adaptation of SOWMAC to 2-D problem by 2-step advection calculations

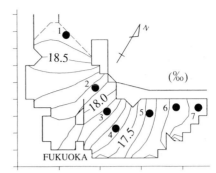

FIG. 3 Calculated chlorinity distribution in Hakata Bay by the SOWMAC scheme (in the case of $D = 80$ m^2/sec)

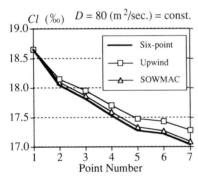

FIG. 4 Comparisons of calculated results of 2-D chlorinity diffusion simulations in Hakata Bay by three schemes (in the case of $D = 80$ m^2/sec)

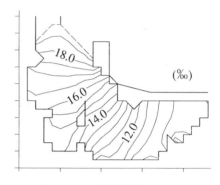

(a) using the SOWMAC scheme

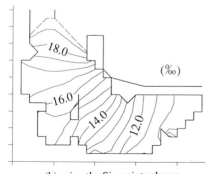

(b) using the Six-point scheme   (c) using the Upwind scheme
FIG. 5 Calculated chlorinity distribution in Hakata Bay (in the case of $D = 10$ m^2/sec)

1B23

ADCP AND S4 CURRENT METERS COMPARED IN WEAKLY TURBULENT STRATIFIED WATERS

A. EDWARDS[*], C. GRIFFITHS[*] AND L. FERNAND[#]

*NERC Dunstaffnage Marine Laboratory (DML), Oban, Argyll, PA34 4AD

Directorate of Fisheries Research (DFR), Lowestoft.

ABSTRACT

Sea-water in the Sound of Sleat is stratified in Summer. We measured currents in the 20 metres of the water closest to the surface with an RDI Acoustic Doppler Current Profiler (ADCP: 1200 kHz) and a vertical moored string of InterOcean electromagnetic S4 current meters in light winds for about 100 hours. The frequency of averaged measurement was every minute. Comparison of the records reveals a good correlation between the types of instrument that improves as averages are taken over longer periods of a few minutes. At higher wind speed the correlation deteriorates. Spectral speed analysis shows the noisier nature of ADCP records and the greater discrimination of the S4 in detecting spectral peaks. Differences may be related to difficulties of determining a zero calibration for the S4s and to lesser sensitivity of the ADCP to the effects of small scale turbulence in the natural flows.

INTRODUCTION

The ADCP is now a common instrument widely used in academic and commercial research. InterOcean S4s have been available since about 1985 and there is a large user community. Both instruments are more difficult to calibrate than rotor current meters because of the problems of finding a large enough tow tank.

Moorings are often designed with an S4 meter in the wave zone, exploiting the high sampling rate (2Hz) to average the orbital wave velocities, and a deeper ADCP to measure the bulk current. For confidence in such joined profiles it is useful to compare the instruments. In 1992, an experiment in the stratified waters of the Sound of Sleat produced several time series of data from both types of instrument: a string of S4s and an ADCP were in use at the same site. In this paper we compare the instrument characteristics using these time series.

The S4 is a vector averaging electromagnetic instrument. An alternating magnetic field in moving seawater generates a voltage of the same frequency. Orthogonal electrodes measure the voltage and the instrument calculates the two corresponding horizontal components of current. The instrument samples at 2 Hz and may be programmed to average over any given time interval, in this work 10 seconds (20 cycles). All instruments were corrected for tilt from the vertical, although the mooring ensured that the maximum tilt remained well within (usually <2°) the 5° limit suggested by the manufacturers for high latitudes.

A narrow band ADCP emits pings of sound and uses the Doppler shift of the backscattered returns to calculate the velocity of the scatterers in the water. A large standard deviation is associated [2] and several pings must be averaged if random errors are to be sufficiently reduced. The sampling rate and depth bin size are user definable, both affecting the accuracy of the resultant measurement. The instrument here was built by Research Development Instruments (RDI) and operated at 1200 kHz, giving a maximum range of 25 m. A more comprehensive explanation of the principles of these instruments is given by RDI [2].

There are two modes of operation: one is ship mounted - rigidly attached to the ship's hull and operational under way; the other is self-contained, used on static ships or moorings. A self contained zero speed calibration is difficult because, in the large tank needed even for high frequency ADCPs of limited range it is difficult to ensure that the water is static. This paper deals with self contained operation of the ADCP with one metre depth bins. There are be random and biased errors. The random error r is given by $r = 13.33 \ N^{-0.5}$ cm.s^{-1} where N is the number of pings contributing to an average. Over 5 seconds, r =5.8 cm.s^{-1} and, over 60 seconds, r = 1.6 cm.s^{-1}. The quoted bias of the ADCP is of the order of 0.5 to 1.0 cm.s^{-1}.

There have been many intercomparisons between Aanderra and S4 current meters and between Anderaas and ADCPs. Because of the low Aanderraa sampling rate, these comparisons could be performed only at low frequency. There have been very few instances where an ADCP has been compared at many depths in a high shear environment [4]. We know of only one comparison between an operational RDI instrument and an operational S4. Griffiths et al [5], in the Iceland - Faeroes region, deployed an upward looking 150 kHz SC ADCP with an S4 directly below and a conventional current meter some 25 m below that. For long term averages the instruments compared within the manufacturer's specifications. Previous comparisons were made on deep sea moorings where the usual mode of operation of self contained ADCPs is to record every 10 minutes or once a hour. New high frequency ADCPs are now used to study estuarine processes whose definition requires much higher sampling rates. Oceanic features such as internal waves also need to be sampled at relatively high frequency if aliasing of high frequency phenomena is to be avoided. There is therefore some interest in comparing instrument performance at these high sampling frequencies.

EXPERIMENTAL SET-UP

In July 1992, the vessel "Calanus" was moored in the Sound of Sleat, on the west coast of Scotland, in water 100 metres deep to a substantial four point mooring pointing southwest. The watch circle of the ship is believed to be better than 5 metres radius. To measure internal waves, current meters and CTD chains were deployed. Figure 2 shows the experimental arrangements. The 1200 kHz

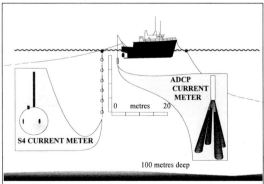

Figure 1: experimental setup in the Sound of Sleat

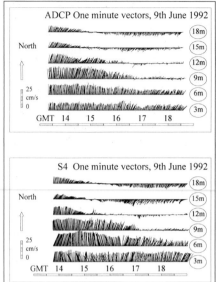

Fig. 2 : Time Series of one minute vectors from ADCP and S4 at 3,6,9,12,15 and 18 metres depth, Sound of Sleat, 9th June 1992. Winds were light, about 2 to 4 m.s^{-1}.

ADCP was hung from the after "A" frame with a weight to keep it vertical. In calm conditions the instrument was placed so that the first depth bin would correspond to about 2 metres depth. In rough conditions the instrument was lowered to around 5 m. 6 S4 meters were strung at 3 m intervals from a buoy between the after mooring lines. A small deep weight was used to keep the S4s vertical. This mooring ensured that the meter depths were constant so that they could be related to the ADCP depth bins.

Previous attempts to calibrate DML S4 meters in tow tanks were unsuccessful at low speeds (~ 10 cm.s^{-1}) because of small tank size. Zero offsets set in the factory in still conditions may drift through time and by knocks. Normal practice at DML checks offsets before deployment, using a large volume (27 m^3) settling tank. This is ideal for calibration because of its large size and, being underground, fairly constant temperature. It is therefore free from air and thermally induced currents. Each instrument is deployed for at least one hour. The data are checked for good noise threshold results and any apparent current. Instrument offsets are then adjusted. The test is repeated and the procedure is believed to be accurate to about 1 cm.s^{-1}.

The S4 meters at 3,6,9,12,15 and 18 metres depth were run at 2 Hz averaged over 10 seconds. The ADCP was run with 5 second averaging with one metre depth bins. The ADCP data were converted using the RDI Collect software with usual 25% good criteria quality control. Data were pre-filtered to remove frequencies above 1 cycle per minute (Filter response = 0 for 1 minute and 1 for 2 minutes). To unify the data sets from the ADCP and the S4, on-the-minute values were interpolated from the prefiltered time series.

TIME SERIES

Experiments lasted a few hours each day during June 1992. Figure 2 shows typical results on June 9th. Winds were light, about 2 to 4 m.s^{-1}. There is good general

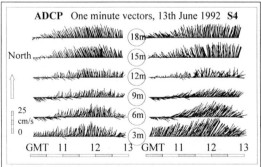

Fig. 3 : Time Series from ADCP and S4 at 3,6,9,12,15 and 18 metres depth, Sound of Sleat, 13th June 1992.

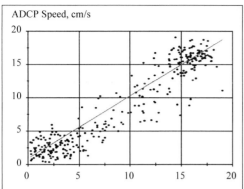

Fig. 4: Regression of ADCP and S4 speeds, June 9th 1992, 9m deep. The slope (m<1) reveals greater S4 sensitivity to small scale eddies smoothed by the ADCP's greater measurement volume. The offset reveals errors in zero calibrations.

agreement between the ADCP and the suite of S4s, although the direction of the S4 currents near the surface (3m) is more variable than the ADCP. There is important difference between the instruments: surface waves corrupted the ADCP results at all depths whereas they most affected the near surface S4s. This was more marked on 13th June (fig.3), when winds blew steadily at 8 to 10 m.s^{-1}.

REGRESSION

Excluding the windy June 13th, with a high rms difference of about 8 cm.s^{-1} between the S4 and ADCP at 3 metres, the ADCP speeds have been regressed on the S4 speeds on 7 other days and at all depths. For example, fig. 4 compares the speeds at 9 metres on 9th June. The correlation is one of the best obtained: $\text{Speed}_{ADCP} = 0.95*\text{Speed}_{S4} + 0.1$ cm.s^{-1} with a root mean square difference of 2.1 cm.s^{-1}. Table 1 summarises the regressions at each depth. Regressions with a computed offset or forced offset of zero are both shown.

TABLE 1: REGRESSIONS OF ADCP SPEED ON S4 SPEED. Dates: 8-12,19,20 June 1992

Computed offset: $\text{Speed}_{ADCP} = m.\text{Speed}_{S4} + c$					**Offset = 0:** $\text{Speed}_{ADCP} = m_0.\text{Speed}_{S4}$
Depth metre	regression coefficient	m	c, cm.s^{-1}	rms,cm.s^{-1} speed difference	m_0
3	0.57	0.44	2.1	4.3	0.64
6	0.78	0.51	1.7	3.9	0.66
9	0.82	0.75	0.9	2.5	0.85
12	0.73	0.77	0.9	1.9	0.91
15	0.55	0.48	1.7	2.4	0.74
18	0.62	0.56	1.6	2.4	0.78

The range of comparison is limited to the low speeds encountered. Fig. 5 shows the histogram of measured speeds. The comparison is clearly restricted to speeds well below 20 cm.s^{-1}.

Throughout, the ADCP is less sensitive than the S4s in the slope (m<1) of the regression. The ADCP estimates speed at each depth as a combination of four narrow 1m long bins. Thus the measurement averaging volume is significantly larger than that of the S4. As the beams diverge from the instrument the high frequency correlation between each beam "spot light" will decrease. In turbulent flow therefore, speeds measured by the ADCP will be less than spot S4 measurements that detect eddies small relative to the ADCP measurement volume.

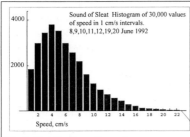

Fig 5: Distribution of speeds

The offset (c) of the ADCP speeds is positive, about 1 to 2 cm.s⁻¹. This is larger than the quoted ADCP bias so may owe to a zero error in the ADCP or to zero errors on all S4s. We cannot distinguish these possibilities, but note that the second is as likely as the first so long as there were weak convective motions of ~1 cm.s⁻¹ in the calibration tanks. This is possible. The rms differences are greatest near the surface, presumably reflecting the different susceptibility of the instruments to near surface motion.

SPECTRAL RESPONSE

Comparison of the spectral responses of the two instruments is interesting. Fig. 6 shows examples of speed spectra (at 0.02 min⁻¹ intervals). On the 10th June, between periods of 2 and 10 minutes, at 18 metres there is more energy throughout the ADCP spectrum than the S4 spectrum. At 6 metres, there is more energy in many parts of the ADCP spectrum, but the S4 shows a clear peak at 0.44 min⁻¹ (period 2.3 minutes) that is flattened and dispersed in the ADCP spectrum. This peak occurs to less at neighbouring S4s and is believed to reflect oceanographic processes in the neighbourhood of the pycnocline. The energy distributed throughout the ADCP spectrum may be associated with the random error of 1.5 cm.s⁻¹ on the one minute values. This whitening of the spectrum is a drawback not shared by the S4.

In view of the different spectral responses, it is illuminating to investigate the effect of longer

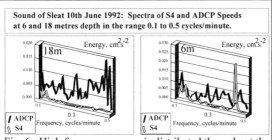

Fig. 6 : High frequency energy is distributed throughout the ADCP spectrum, whereas the S4 spectrum is generally lower and reveals distinct peaks that are obscure in the ADCP record.

averaging periods on the correlation. Fig. 7 shows the correlation between the S4 and ADCP speed time series on the 10th June when running means of 1,2 ...10 minutes duration are formed before correlation. The correlation increases markedly as the data are averaged over longer periods up to about 5 minutes. Beyond this, the increase slows or even reverses.

In summary, high frequency phenomena (period a few minutes) are not well correlated between the instruments and there is a deal of high frequency noise in the ADCP spectrum when compared to the S4. Records are most similar when averaged over a few minutes.

DISCUSSION

This was not a controlled test tank type of comparison. The measurement bins of the S4 are a few decimetres in each dimension: bins of the ADCP are vertically thicker (1 metre) and spread over a horizontal scale increasing with depth. The instruments were separated by about 10 metres horizontally. Among the natural phenomena were internal waves, eddies on various

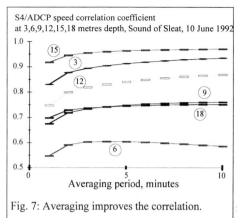

S4/ADCP speed correlation coefficient at 3,6,9,12,15,18 metres depth, Sound of Sleat, 10 June 1992

Averaging period, minutes

Fig. 7: Averaging improves the correlation.

scales, strongly sheared flows and surface waves. These all occur on scales that compare with the instrument measurement volumes and the separation of the instruments. We expect correlation to be correspondingly downgraded, as we have demonstrated.

Nevertheless, our low speed comparison has revealed important differences that bear on several problems. Regression showed the ADCP to be less sensitive (m<1) than the S4, perhaps because of the different measurement volumes of the instruments in turbulent flow. The rms difference between the two was a significant fraction, 2 to 4 cm.s^{-1}, of the working range, about 0 to 20 cm.s^{-1}. The zero offset (c) of the ADCP relative to the S4s may reveal the difficulty of obtaining zero flow in the calibration tanks. The spectral differences between the instruments are non-trivial: random noise in an ADCP time series whitens the spectrum; the S4 seems better able to discriminate spectral peaks.

The widespread practice of combining an S4 meter at the surface with a deeper ADCP may produces composite current profiles of disjoint parts. Because of the spectral differences, some improvement of the correlation of the instruments is possible by averaging speeds over a few minutes rather than one minute. The statistical estimation of extremes of current depends on the spectrum of the current and it therefore matters what type of meter is used to measure for extreme current estimation. These practical applications exemplify the need for sceptical care that has to be taken in the use of such disparate instruments.

ACKNOWLEDGEMENTS

We thank the officers and crew of R.V. "Calanus". Experiments were paid for by a contract managed by B.C.Barber of the Space Department, Defence Research Agency, Farnborough.

REFERENCES

[1] J. H. Trageser (1986). Calibrating "Smarter" current meters. Sea Technology, Feb. 1986.

[2] RDI (1989) ADCP Principles of Operation: A practical Primer. RDI Inst. San Diego, CA.

[3] Pollard R. and J. A. Read (1989) Method for calibrating ship mounted acoustic Doppler profilers and the limitations of gyro compasses. Journal of Atmospheric and Oceanic Technology Vol 6 No.6 pp 859-865.

[4] Belliveau, D.J. and J. W. Loder (1990) Velocity errors associated with using short pulse lengths in a 150kHz acoustic Doppler current profiler. Proceedings of the IEEE Fourth Working Conference on Current Measurement. April 1990: pp231-243.

[5] Griffiths, G., N.A. Crisp & L.A. Povey. (1991) Moored current measurements made in the Iceland Faeroes region during the July-August 1990. Institute of Oceanographic Sciences Deacon Laboratory, Wormley, Godalming, Surrey. Report No. 286.

1B24

Three-dimensional numerical modeling of water flow through a gate plug

NILS R. B. OLSEN and OLE OLDERVIK
SINTEF NHL, 7034 Trondheim, Norway

INTRODUCTION

Today there exist several numerical models that are able to calculate the water flow field in a general three-dimensional geometry. The present study investigates the performance of some of these models applied for a gate plug. The gate plug case is chosen as an example from hydraulic engineering. Physical model tests for this case exist [1], which is used for validation of the numerical models.

THE GATE PLUG

The gate plug is used in hydropower tunnels as a mean of supporting a gate. The gate plug decreases the cross-sectional area of the tunnel, and consists of a contraction and an expansion region. The gate is located between these regions. Fig. 1 a) and b) shows the grid of the gate plug. Two plugs are investigated. Plug A has a relatively long expansion region, similar to a diffusor. For Plug B this expansion region is cut after 1/3 of the length. In the simulated cases the plug is located in a rectangular tunnel, 7.4 m high and 5.76 m wide. Only one quarter of the geometry is simulated, and symmetry conditions are used on two planes. The length of the geometry is 37 meters. The inlet velocity is 1.5 m/s.

THE NUMERICAL MODELS

The following numerical models were used: Fidap [2], Fluent and SSIIM. All models solve the Navier-Stokes equations with the k-ε turbulence model. Fidap is based on the finite element method, while the other models are based on the finite volume method [3]. Fidap uses upwinding and artificial diffusion for stability. The finite volume methods uses upwinding, implicit solution methods and the Rhie and Chow interpolation method [4] for stability.

RESULTS FROM THE MODELS

The velocity flow field looks very similar for all the models. Velocity vector results from Fidap is shown in Fig. 2 for the two cases. A longitudinal profile is shown which cuts the corner and the centerline of the tunnel. Note the recirculation zone for case B. SSIIM gave a small recirculation zone also for case A. The energy loss is given in the table below.

	Physical model [1]	Fluent ver. 2.00	SSIIM ver. 1.4	Fidap ver. 7.04
Case A	1800	2200	1955	1470
Case B	1800	3700	3112	2830

Table 1. Energy loss throug gate plug in Pascal

DISCUSSION

The results seems reasonable for case A for all the models. For case B there is more deviation between the calculated and observed energy loss. This can be due to a relatively coarse grid at the cut of the expansion. The question of number of grid cells is for many cases dependent on the speed of the numerical model and the computers available. To check this, a new grid with 8 times as many cells were calculated for case B using SSIIM. This gave a head loss of 2496 Pascal, a reduction in error from 73 % to 39 %.

CONCLUSIONS

For a geometry where the grid can be made fairly smooth, the numerical models give a result with an accuracy of 10-20 %. For a more complex geometry with large recirculation zones and a coarse grid the deviation between calculated and measured head loss can be over 100 %.

REFERENCES

[1]   Noreng, K., "Energy loss in the waterways of hydropower plants", SINTEF NHL report 608544, 1984, (In Norwegian).

[2]   Fluid Dynamics International, "Fidap 7.0 User's manual", 1993.

[3]   Patankar, S. V. "Numerical Heat Transfer and Fluid Flow", McGraw-Hill Book Company, New York, 1980.

[4]   Rhie, C.-M, and Chow, W. L. "Numerical study of the turbulent flow past an airfoil with trailing edge separation", AIAA Journal, Vol. 21, No. 11, 1983

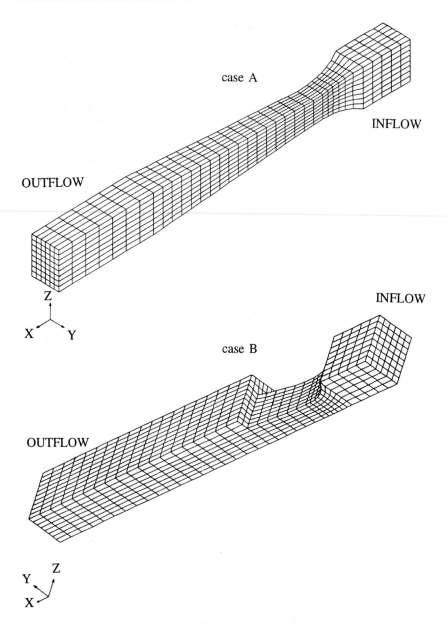

Figure 1   Three-dimensional view of the grids

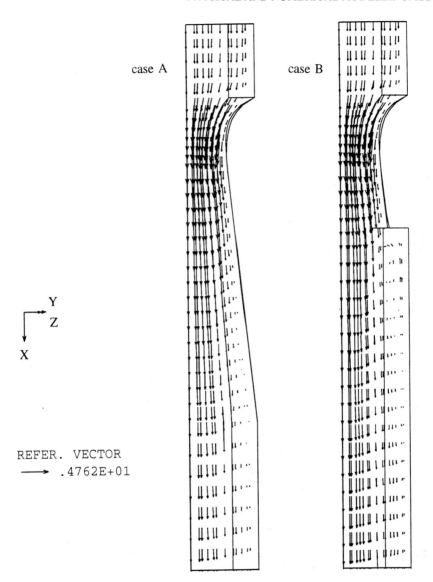

Figure 2   Velocity vector plot in a longitudinal section through the corner and the centerline of the gate plug

1B25

Field Examination of a Distribution System Water Quality Model

Sami G. Elmaalouf,[1] S.M. ASCE
Young C. Kim,[2] M. ASCE

Abstract

In the aftermath of the Safe Drinking Water Act (SDWA) of 1974, numerous models have been established to assess the quality of water and simulate its deterioration in distribution systems. Many of these models, however, are associated with a great deal of complexity especially when a practicing engineer is attempting to utilize them in order to structure a computer program that would help predicting water quality variations in a conveying system. Furthermore, many of these models have not been verified by field studies.

This paper presents and describes the development of a field study that examines a steady state distribution system water quality model. The field study requires a profound understanding of the hydraulic behavior in the system as the variation in delivered water quality is dependent upon the hydraulic mixing effects in this system. The study is conducted at the California State University, Los Angeles to verify a comprehensive model that was developed to predict the constituents and/or contaminants propagation, decay, and the overall quality variations in water distribution networks.

Introduction

A simple and reliable water quality model is essential for predicting and monitoring quality deterioration and constituents or contaminants behavior in water distribution systems as the 1974 Safe Drinking Water Act (SDWA) put various restrictions on the quality of delivered water to consumers. Examining this model experimentally is of equal importance. This paper presents and describes the development of a field study that examines a steady state distribution system water quality model. The model is based on algorithms that were reported by Elmaalouf et al. (1). The reported model is efficient and solves for quality parameters by direct substitution, yielding results that would aid engineers and planners to predict water quality variations in a conveying hydraulic network. Initial understanding of the hydraulic behavior in the distribution network is required as the variations in the delivered quality are dependent upon mass and energy conservation, and the overall hydraulic mixing effects in the distribution system.

The first part of this paper demonstrates the steady state water quality model, while in the second part the experimental program and a comparison of results will be presented and discussed.

----
[1]Engineer, Public Works Department, City of Los Angeles, 600 S. Spring Street, 15th Floor, Los Angeles, CA 90014
[2]Professor, Civil Engineering Department, California State University, Los Angeles, 5151 State University Drive, Los Angeles, CA 90032

Water Quality Analysis in Pipe Networks

Traditionally, the main interest of water supply managers and engineers was to hydraulically satisfy all the water demands at various different locations in their municipality. Quality assessment as well as contaminants/constituents propagation within a municipal water conveying system were not a leading interest, since the drinking water quality was basically measured immediately on its way leaving a water treatment plant.

We know (as of 1993) that this is not the case anymore. In the aftermath of the Safe Drinking Water Act (SDWA) of 1974, many concerns have increased over quality assessment, contamination and waterborne substances propagation and the overall blending effect, of different sources with different quality, in a distribution system.

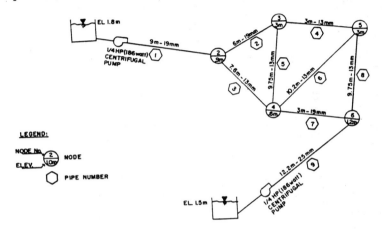

Figure 1. Experimental Distribution System

Before we start solving for quality parameters in a distribution system, it is very important to successfully identify the flow contribution, delivered to junction nodes, from supply sources. The contribution of supply sources to loads of discharge at various nodes in water distribution systems, can be determined when a nodal mass continuity takes place. The source contribution problem can be formulated as

$$[Q_{sc}] \cdot f_c - [Q_{fo}] \qquad (1)$$

where, $[Q_{sc}]$ is the source flow contribution matrix, f_c is the supply source contribution factor, and $[Q_{fo}]$ is the source outflow forcing matrix. Therefore

$$[Q_{sc}]^{-1} \cdot [Q_{fo}] - f_c \qquad (2)$$

343

where $[Q_{sc}]^{-1}$ is the inverse of the source flow contribution matrix. If the total inflow from a source node is equal to the demand at a particular node, then Q_{sc} is set to unity. Otherwise

$$Q_{sc} - \Sigma \, Q_i + q \qquad (3)$$

After a successful determination of the supply source contribution factor and the correct assembly of the source flow contribution matrix, $[Q_{sc}]$, we have to analytically formulate the contaminant/constituent concentration problem.

Given the source flow contribution matrix and knowing the concentration of a certain chemical at the supply source (simply by actual field quality measurement at the supply source), the contaminant/constituent concentration problem formulation is

$$[Q_{sc}] \, . \, C_c = [C_s] \qquad (4)$$

where C_c is the contaminant/constituent concentration at junction nodes and C_s is the contaminant/constituent concentration injected from a supply source(s). Or

$$[Q_{sc}]^{-1} \, . \, [C_s] = C_c \qquad (5)$$

Most contaminants/constituents decay as they travel from the supply source to a junction node or from one junction node to another. The rate of concentration decreases as water travels farther from a supply source, given a steady state condition.

From the contaminant/constituent concentration problem formulation that was discussed earlier, we can determine the concentration of chemicals at all junction nodes in question, knowing the contaminant/constituent concentration rate at a source supply node(s). After observing all the predicted concentrations at different locations in the network, we can formulate the decay problem from mass balance relationships as follows

$$\frac{\Delta C}{t_{i,j}} = \xi \, (C_{i,j}) \qquad (6)$$

or with a quadratic formulation

$$[C_i C_j] \, . \, (CD)_{i,j} = - \, \frac{1}{t_{i,j}} \, [\Delta C] \qquad (7)$$

hence

$$(CD)_{i,j} = - \, \frac{1}{t_{i,j}} \, [\Delta C] \, . \, [C_i C_j]^{-1} \qquad (8)$$

where C_i and C_j are the contaminant/constituent concentrations at junction nodes i and j, respectively; t_{ij} is the flow time from junction node i to junction node j, ξ is decay reaction rate which is dependent upon the time variation between one junction node and another, $(CD)_{ij}$ is the concentration decay rate from junction node i to junction node j, and ΔC is the change in the contaminant/constituent concentration between junction node i and junction node j.

Water ages when it is in the supply source or when it is traveling via pipes from one junction node to another. This aging process contributes to the quality deterioration of delivered water at different junction nodes in a distribution system. Therefore, studying this problem is of great importance in the field of water quality investigation. This problem can be formulated as

$$[T_{ave}] \cdot t_{ave} = [\omega] \tag{9}$$

where $[T_{ave}]$ is the average water age matrix, t_{ave} is the average water age at a junction node, and $[\omega]$ is the water age-flow contribution matrix.

If the total inflow from a source node is equal to the flow rate in an pipe, then T_{ave} is set to unity. That is, T_{ave} is equal to one when

$$Q_s(k) = -\Sigma f_{c_i} \cdot Q_{i,j} \tag{10}$$

where f_{c_i} is the source contribution factor at junction node i, and Q_{ij} is the flow rate, in a pipe, from junction node i to junction node j. Otherwise

$$T_{ave} = Q_s(k) + \Sigma f_{c_i} \cdot Q_{i,j} \tag{11}$$

The average age-flow contribution matrix is assembled as follows

$$[\omega] = \begin{bmatrix} \Sigma f_{c_i} \cdot Q_{i,j} \cdot t_{i,j} \\ \cdot \\ \cdot \\ \cdot \end{bmatrix} \tag{12}$$

where t_{ij} is the flow time through a pipe connecting junction node i to junction node j and is computed by the equation

$$t_{i,j} = \frac{l_{i,j}}{V_{i,j}} \tag{13}$$

where l_{ij} is the length of the pipe, and V_{ij} is the velocity of water traveling from junction node i to junction node j.

Experimental Program

Four different scenarios were analyzed for the distribution system shown in figure 1. The experimental system comprises nine pipes, five connecting junction nodes, two constant-head tanks (flow sources), five ball valves at each node to control the outflow at node points, and two 1/4 hp (186 Watt) centrifugal pumps in lines 1 and 9.

All four scenarios were verified experimentally. In the first scenario the ball valves at nodes 2 and 5 were calibrated to discharge 0.28 l/s and 0.57 l/s, respectively. All other valves were fully closed. A steady (constant) head was maintained at both feeding tanks by adding distilled water continuously. The system was operating for seven minutes to maintain a steady state flow in the network. The flow rates were checked at discharging nodes by means of a magnetic flow meter.

The distribution system was stopped at this point. A chemical (Benzene-C_6H_6) was introduced to the tanks at dissimilar proportions, making the constituent concentration 10 mg/l in supply tank No.1 and 3 mg/l in supply tank No.2. The system was again started and water samples were collected at discharging nodes 2 and 5. Chemical analyses followed.

In the second scenario the ball valves at nodes 3 and 4 were calibrated to discharge 0.57 l/s and 0.28 l/s, respectively. All other valves were fully closed. Exact steps were followed and similar data was gathered as in the first scenario. Water samples were collected at discharging nodes 3 and 4. Chemical analyses followed.

In the third scenario the ball valves at nodes 4 and 5 were both calibrated to discharge 0.28 l/s. All other valves were fully closed. Exact steps were followed and similar data was gathered as in the preceding scenarios. Water samples were collected at discharging nodes 4 and 5. Chemical analyses followed.

In the fourth scenario the ball valves at nodes 3, 4, and 5 were all calibrated to discharge 0.28 l/s. All other valves were fully closed. Exact steps were followed and similar data was gathered as in the preceding scenarios. Water samples were collected at discharging nodes 3, 4, and 5. Chemical analyses followed.

Table 1. Analytical and Experimental Hydraulic Analyses
(First Scenario)

Calculated				Measured			
Pipe	Flow (l/s)	Node	Pres. (KN/m^2)	Pipe	Flow (l/s)	Node	Pres. (KN/m^2)
1	0.28	2	437	1	0.30	2	—
2	0.28	3	441	2	0.30	3	—
3	0.00	4	441	3	0.01	4	—
4	0.28	5	430	4	0.30	5	—
5	0.00	6	437	5	0.01	6	—
6	0.28			6	0.30		
7	0.28			7	0.30		
8	0.28			8	0.30		
9	0.57			9	0.50		

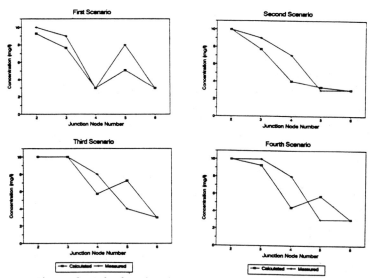

Figure 2. Calculated and Measured Chemical Concentrations

Conclusion

This paper presents a comparison of various experimental results to analytical quality results. The comparisons range from poor to good and do not provide a conclusive evidence that the comprehensive steady state quality model--based on explicit graph-theoretic algorithms for mixing problems in multiple-source networks--works. The agreement between the computer model and experimental results decreases somewhat when the traveling water velocities increase in pipes. This is quite expected most probably because of the short travel time between one node and another.

From a hydraulic point of view, the comparisons range from good to excellent and provide evidence that the distribution network used in this study was modeled adequately. The results of the computer model and experimental data slightly vary most probably because minor losses were not defined accurately in the computer model.

Water quality modeling in distribution systems is a complex phenomenon. Although all mass balance relationships are satisfied in the presented algorithm, yielding correct mathematical solutions for quality operating parameters, confidence in the reliability of the model and the utilization of reasonable data are essential. More models and experimental studies with more complex networks than the one presented here are recommended in order to further verify and validate the presented method's actual applicability.

Appendix-References

1.   Elmaalouf, S. G., "A Comprehensive Steady State Quality Model for the Assessment of Contaminants Behavior in Water Distribution Systems," Master of Science Thesis, Civil Engineering Department, California State University, Los Angeles, 1992.

1B26

Physical and Mathematical Modelling
of Alluvial Channels in Regime

HECTOR DANIEL FARIAS
Adjunct Professor, Departamento de Recursos Hídricos
Universidad Nacional de Santiago del Estero, Argentina

INTRODUCTION

The hydraulic geometry of alluvial channels is a controversial and complex topic. In the literature are distinguished three approaches for its treatment: empirical, semi-empirical and analytical. From the view point of hydraulic modelling, the first two can be enclosed in the physical modelling category, and the third can be classified as a mathematical model approach. Nevertheless, no method is substantially different from other, since the same interact mutually, sharing advantages and limitations. This is evidenced by the fact that the functions employed to calculate the frictional characteristics and the sediment transport rates possess a strong empirical component. In effect, most of that equations are the result of calibrations carried out using observations in laboratory and some few in man-made canals and natural channels.

The purpose of this paper is to show the capabilities of the physical and mathematical model approaches to estimate the geometry of regime alluvial channels, in the context of a generalized regime concept, and taking into account the physical relations governing alluvial channel dynamics.

GEOMETRY OF SANDY CHANNELS IN REGIME

If an unlined channel of given initial geometry (for example, trapezoidal) is excavated in alluvial soil, in first instance the solid fraction of the cross sectional boundary of the channel will be constituted by local soil. After an extended period of operation of the system (generally of the order of magnitude of a year), is generated a process of "maturing" of the channel, and the material of the perimeter starts to present comparable characteristics to that of the sediment transported by the flow, that at the same time is of nature akin to that of material transported by the river from which is taken the water for the channel in question. This maturing phenomenon is a consequence of the morphodynamic activity developed by the flow, through the erosion and sedimentation processes (Farias, 1994). If the channel is projected (and is built) in such a way that its three-dimensional geometry will be able of conveying adequately the practical ranges of liquid discharges and to transport efficiently the solid material concentrations associated with those liquid flows, the maturing process will be developed without producing large variations in the general morphology of the channel. As far as the channel is concerned, if the initial design does not result appropriate, it is highly probable the appearance of erosions and/or depositions until the flow reach to sculpt a geometry (transverse and longitudinal) adapted to lead the sediment and water discharges of the most efficient way possible (Farias, 1992).

Considering the above-mentioned comments, it can be said that an alluvial channel is in a stability condition, dynamic equilibrium, or regime, when the capacity of transporting sediments by the flow is balanced with the rate of solid material supplied to the channel reach. Introduced the regime concept, in the project phase the problem consists in sizing the channel in such a way that its configuration will be as approximate as possible to the stable geometry, what will minimize the erosion and sedimentation processes, and consequently the operation and maintenance costs.

From a macro-morphological point of view, it can be said that an alluvial channel is a physical system with three degrees of freedom, because it is capable to adjust its width, depth and slope in response to the imposed variables, such as, water and sediment discharges and boundary material features. However, only two well-established phenomenological relationships are available to solve the problem: the alluvial friction and sediment transport laws. The third relation, describing the lateral stability (width adjustment) is not yet available, so the problem of channel geometry appear as indeterminate.

PHYSICAL MODELLING OF REGIME ALLUVIAL CHANNELS
Straight micro-channels, self-formed in their own sediment, which is transported by the flow in a dynamic equilibrium condition, are over-simplified representations of field-size stable alluvial channels. However, this artificial laboratory facility is usually employed to study the dynamics and morphological features of regime alluvial channels. The main reasons to use these setups are quite obvious: better work environment than in field conditions, easy measurements and control of variables, accurate setting of boundary conditions, etc.. Nevertheless, a question arises immediately from the preliminary analysis of laboratory studies: Is a micro-channel actually a scale model of any larger channel which exists in nature?, and if the answer is 'yes', which are the scale ratios to transfer results from the model to the natural

system (prototype)?. This question, initially drawn by Lacey (1966), was analyzed in a previous research (Farias and Borsellino, 1992), considering that the laboratory studies were oriented to obtain behavioral information on stable alluvial channels, suitable for correlating with the regime concept. The investigations analyzed were those by Wolman and Brush (1961), Ackers (1964), Ranga Raju et al. (1977), Ikeda (1981), and Diplas (1990), and the conclusion was that "design equations derived from observations in laboratory micro-channels should be applied with care for sizing prototype alluvial channels, and would be preferable to apply formulas based in theoretical considerations." Another points detected were that the micro-channels have aspect ratios (Γ = T/H: width/depth) greater than field sandy canals and the slopes are in a reasonable agreement. A problem in experimental studies is that the materials used to form the channel are too uniform and completely cohesionless. In nature, the finer grains are transported as suspended load and deposited near the banks in a berming process that limits the channel widening. The analysis also indicated that the micro-channels tend to work better as models of gravel streams instead of sandy channels. Also, the only way to consider a micro-channel as a model of a sandy canal is by means of a distortion in the Froude condition.

In spite of the important quantity of laboratory studies, only Ackers attempted the derivation of regime type equations for the hydraulic geometry of those channels. The best fit regime equations obtained by Ackers show certain agreement with the Lacey relations. The geometry data of micro-channels are included in a plot, which will be presented and discussed below in this paper.

ANALYTICAL APPROACH (MATHEMATICAL MODELLING)

During the last two decades, several approaches have been used to provide the necessary relationship to solve the problem of alluvial channel stability, some of them relying on a type of variational argument in which the maximum or minimum of some quantity is sought. Definitions and discussions of such hypotheses have been presented in detail by Bettess and White (1987) and Hey (1988). Farias (1990) has shown that, in the case of straight sandy channels with uniform boundary material and constant dominant discharge, the most of the variational principles can be enclosed into a general one: maximum efficiency in sediment transport (MEST). The MEST principle admits that an alluvial channel has reached its regime condition when the total bed material transport is a maximum. In a series of studies (Farias, 1993a, 1994), it was obtained a menu of rational regime equations for the design of stable alluvial channels. The relationships, based on the employment of the MEST principle, have been presented in the following dimensionless format:

$$P_* = \Phi_{P*}(Q_*, Q_{S*}, d_*) \ , \ R_* = \Phi_{R*}(Q_*, Q_{S*}, d_*) \ , \ S_* = \Phi_{S*}(Q_*, Q_{S*}, d_*) \qquad (1)$$

with: $Q_* = Q/[g \ \Delta \ d^5]^{1/2}$, $Q_{S*} = Q_S/[g \ \Delta \ d^5]^{1/2}$, $d_* = [g \ \Delta/v^2]^{1/3}d$, $P_* = P/d$, $R_* = R/d$, $S_* = S/\Delta$, $\Delta = (\rho_s-\rho)/\rho$. Symbols are: Q: dominant water discharge, Q_S: sediment transport rate, d: is the median size of the channel boundary sediment, P: wetted perimeter, R: hydraulic radius, S: slope, g: gravitational acceleration, v: kinematic viscosity, ρ: water density, ρ_s: sediment density.

Farias (1993a) showed that the application of MEST hypothesis, by means of a computer algorithm, and a plot of the solution points in a Q_*-Γ (where $\Gamma = T/H$) diagram (logarithmic coordinates) exhibits a rising pattern, which suggests a relationship of the type : $\Gamma = k_1 \ Q_*^{k_2}$ (where k_1 and k_2 are nearly constants) as a good fitting curve. Taking into account this observation, a set of morphological, regime type equations, was worked out by performing a multiple regression analysis on a theoretical data base generated by means of a computer algorithm based upon the solution for extremal hypotheses, coupled with the application of suitable formulas for alluvial friction and sediment transport. The power-type equations are presented in the following format: $X_* = a_0 \ d_*^{a_1} \ C_{S*}^{a_2} \ Q_*^{a_3}$, where X_* may be P_*, R_* or S_*, and a_i (i = 0,1,2,3) are constants, and $C_{S*} = C_S = Q_S/Q$ is the total sediment concentration. A set of these equations is presented in the last column of Table 1, jointly with the sediment transport (T) and alluvial friction (F) relations, which were selected taking into account their accepted usefulness as reasonably accurate predictors.

Code	Theoretical Regime Equations
E-H F&T: Engelund & Hansen	$P_* = 0.950 \ d_*^{0.068} \ C_{S*}^{-0.009} \ Q_*^{0.497}$ $R_* = 0.954 \ d_*^{-0.216} \ C_{S*}^{-0.022} \ Q_*^{0.399}$ $S_* = 9.766 \ d_*^{-0.040} \ C_{S*}^{0.666} \ Q_*^{-0.200}$
A-W F: White et al T: Ackers & White	$P_* = 0.076 \ d_*^{0.666} \ C_{S*}^{-0.023} \ Q_*^{0.555}$ $R_* = 1.213 \ d_*^{-0.432} \ C_{S*}^{-0.120} \ Q_*^{0.357}$ $S_* = 0.918 \ d_*^{0.419} \ C_{S*}^{0.521} \ Q_*^{-0.193}$
B-B F&T: Brownlie	$P_* = 0.202 \ d_*^{0.253} \ C_{S*}^{-0.067} \ Q_*^{0.524}$ $R_* = 0.624 \ d_*^{-0.141} \ C_{S*}^{-0.110} \ Q_*^{0.368}$ $S_* = 11.097 \ d_*^{-0.098} \ C_{S*}^{0.605} \ Q_*^{-0.225}$
V-R F&T: Van Rijn	$P_* = 0.584 \ d_*^{0.210} \ C_{S*}^{-0.050} \ Q_*^{0.492}$ $R_* = 0.203 \ d_*^{-0.085} \ C_{S*}^{-0.139} \ Q_*^{0.398}$ $S_* = 4.680 \ d_*^{-0.210} \ C_{S*}^{0.504} \ Q_*^{-0.233}$
P-P F&T: Peterson & Peterson	$P_* = 0.201 \ d_*^{0.199} \ C_{S*}^{-0.104} \ Q_*^{0.520}$ $R_* = 0.709 \ d_*^{-0.131} \ C_{S*}^{-0.077} \ Q_*^{0.375}$ $S_* = 23.342 \ d_*^{0.130} \ C_{S*}^{0.779} \ Q_*^{-0.213}$
K-K F&T: Karim & Kennedy	$P_* = 0.488 \ d_*^{0.213} \ C_{S*}^{-0.043} \ Q_*^{0.500}$ $R_* = 0.239 \ d_*^{-0.134} \ C_{S*}^{-0.172} \ Q_*^{0.381}$ $S_* = 8.467 \ d_*^{0.010} \ C_{S*}^{0.642} \ Q_*^{-0.221}$

Table 1.- Rational Regime Equations based on MEST principle.

EVALUATION OF PHYSICAL AND MATHEMATICAL CRITERIA

An extensive comparison of theoretical results with field alluvial channel data was carried out. The data compilations used (355 data sets) cover the ranges: $Q = 0.07 - 407.76$ m^3/s; $d = 0.029 - 0.805$ mm; $P = 1.7 - 111.8$ m; $R = 0.180 - 3.718$ m; $S = 0.000058 - 0.000473$; $T = 1.5 - 112.8$ m; $H = 0.244 - 4.084$ m; $C_s = 16 - 908$ ppm.

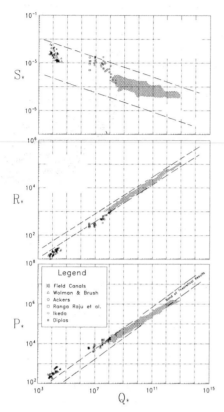

Figure 1.- Dimensionless Hydraulic Geometry Plots for Regime Sandy Channels.

Fig.1 show the dimensionless hydraulic geometry plots. The dashed lines are the envelopes resulting of all the computer generated rational regime formulas. Data from laboratory studies (physical models) are also included. In general, micro channels have a tendency to be wider and shallower than regime sand-silt channels, with exception of some of Ackers' class "ii" channels, which fall within the theoretical belts. This fact indicates that almost all the experiments may be affected by scale effects with regard to the hydraulic geometry. Nevertheless, it is opportune to remember that all the friction and transport functions employed to obtain the rational regime formulas by means of a computer aided solution of MEST principle also were strongly calibrated with laboratory data. Summarizing the concept, it result apparent that the use of analytical approaches operates as a filter of scale effects observed when the results obtained from small laboratory channels are transferred directly to prototype (field canals). This is a quite interesting result, because it can be used to explain the usefulness of extremal hypotheses when they are applied to predict channel geometry.

CONCLUSIONS

The geometry of regime alluvial channels is a good example of the integration of experimental and analytical research in fluvial hydraulics. The state of the art in this topic appears to be the result of the interaction of physical and mathematical modelling of alluvial channel processes. In fact, the rational approach to the problem uses predictors for alluvial friction and sediment transport whose accuracy is steadily increased by introducing calibration techniques based on laboratory observations. The application of variational principles as tools for predicting the stable channel width, coupled with the friction and transport functions, configures a set of relations suitable to solve with the aid of informatic tools (optimization algorithms). This paper has show that the final result can be of compact format (regime equations), and its application to field conditions is reasonably acceptable.

Acknowledgements. Funds for this study were provided by CICYT-UNSE and CONICET, Argentina.

REFERENCES

Ackers,P. (1964) "Experiments on Small Streams in Alluvium", Journal of the Hydraulics Division, ASCE, Vol. 90, No. HY4, July, pp. 1-37.

Diplas,P. (1990) "Characteristics of Self-Formed Straight Channels", Journal of Hydraulic Engineering, Vol. 116, No. 5, May, pp. 707-728.

Farias,H.D. (1992) "Maximum Efficiency Sandy Channels and their Morphology", Proc. V Intl. Symp. on River Sedimentation, Karlsruhe, FRG.

Farias,H.D. & Borsellino,M. (1992) "Analysis of Criteria for Modelling Regime Alluvial Channels", Proc. Intl. Symposium on Hydraulic Research in Nature and Laboratory, Wuhan, China, Vol.2, pp. 96-101.

Farias,H.D. (1993a) "Hydraulic Geometry of Sand-Silt Channels in Regime". Water: The Lifeblood of Africa, Intl. Symposium, Victoria Falls, Zimbabwe.

Farias,H.D. (1993b) "Morphology of Regime Alluvial Channels: A Review", Adv. in Hydro-science and Engrg., S. Wang (Ed.), Washington DC, USA., pp. 1423-1428.

Farias,H.D. (1994) "Geometry of Alluvial Channels in Dynamic Equilibrium", Proc. International Symposium on State of the Art in River Engineering Methods and Design Philosophies, St. Petersburg, Russia, Vol. 3 .

Hey,R.D. 1988. "Mathematical Models of Channel Morphology", Chapter 5 in M.G.Anderson (Editor): Modelling Geomorphological Systems, J.Wiley & Sons Ltd., Chichester, Sussex, U.K., pp. 99- 125.

Ikeda,S. (1981) "Self-formed Straight Channels in Sandy Beds", Journal of the Hydraulics Division, ASCE, Vol. 107, No. HY4, pp. 389-406.

Lacey,G. 1966. "Models Scales and the Regime Concept", Modern Trends in Hydraulic Engineering Research, Proc. of the Golden Jubilee Symposia, Pune, India, pp. 10-14.

Ranga Raju,K.G., Dhandapani,K. & Kondap,D.M. (1977) "Effect of Sediment Load on Stable Sand Canal Dimensions", J. Waterways Div., ASCE, Vol. 103, No. WW2, pp. 241-249.

Wolman,M.G. & Brush,L.M. (1961) "Factors Controlling the Size and Shape of Stream Channels in Coarse Non-Cohesive Sands", U.S. Geological Survey Prof. Paper No. 282-G.

1B27

A GEOMETRIC MODEL FOR SELF-FORMED
CHANNELS IN UNIFORM SAND

DONALD W KNIGHT

School of Civil Engineering
The University of Birmingham
Birmingham, England, UK

GUOLIANG YU

State Key Hydraulics Laboratory
Sichuan Union University
Chengdu, Sichuan, China

ABSTRACT

A geometric model of a straight self-formed channel is presented which is based on the depth-averaged flow equation by Shiono and Knight, the critical shear stress at the junction point by Shields, the side shape by Parker and the sediment discharge rate by Ackers&White. It models a channel with a flat bottom and two side banks at the threshold of motion.

INTRODUCTION

A regime channel geometry is formed by the interaction of water and sediment in alluvial rivers. Extremal hypotheses applied to river regime are summarized by Bettess and White (1987). The hydraulic geometry of threshold channels has been studied by many researchers, among them Diplas and Vigilar (1992). In rivers, the sediments on the banks are generally smaller than those on the river bed. Fine materials in graded bank materials combine with coarse materials under pressure to resist bank erosion. Recent experiments and research show that sediment concentration plays an important role in the channel shape. When the sediment concentration is large enough, the ratio of width to depth can be greater than 100. The shape and dimensions of straight self-formed channels in uniform sand bed are still the subject of further study.

GEOMETRIC MODEL FOR A REGIME CHANNEL

A typical self-formed straight channel shape, without sand waves, is illustrated in Fig. 1. It is symmetric about the channel centre line, and sediments move only on that part of the channel perimeter in the central region. There is a critical point where sand is at a critical state of incipient motion. Therefore the cross sectional shape of a straight channel can be made up two components that are associated with different states of sediment behaviour. The channel shapes on both sides of this critical point are crucially different. On the bank side there is no bed load motion, and its shape is mainly determined by the angle of repose and the shear force. In the central belt width, however, where there is sand in motion, its shape is determined predominately by shear stress, turbulence and the bed load. Thus when analysing channel shape, the two parts have to be considered separately.

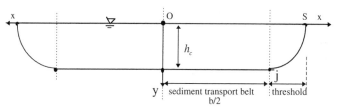

Fig. 1 Cross sectional shape of a self-formed channel

DEPTH-AVERAGED VELOCITY AND BED SHEAR STRESS

The depth-integrated form of the Navier-Stokes equation for a self-formed straight channel, neglecting secondary flow, may be written as

$$\rho g h S_0 - \rho \frac{f}{8} U_d^2 \sqrt{1 + \frac{1}{s^2}} + \frac{\partial}{\partial x} \left\{ \rho \lambda h^2 \left(\frac{f}{8} \right)^{1/2} U_d \frac{\partial U_d}{\partial x} \right\} = 0 \tag{1}$$

An analytic solution to equation (1) is given by Shiono and Knight (1988). For the case of h=constant, the depth-averaged velocity, is U_d given by:

$$U_d = \left[A_1 e^{\gamma x} + e^{-\gamma x} + \frac{8 g S_0 h}{f} \right]^{1/2} \tag{2}$$

and for linearly varying depth, U_d is given by:

$$U_d = \left[A_3 Y^{\alpha_1} + A_4 Y^{-\alpha_2} + \omega Y \right]^{1/2} \tag{3}$$

where s is the side slope(1:s, vertical:horizontal), S_0 is the longitudinal gradient, f is the local friction factor, g is gravitational acceleration, h is the water depth, λ is eddy viscosity, ρ is the water density, and the various constants are given as:

$$\gamma = \left(\frac{2}{\lambda}\right)^{1/2}\left(\frac{f}{8}\right)^{1/4}\frac{1}{h}$$

$$\alpha_1 = -\frac{1}{2}+\frac{1}{2}\left\{1+\frac{s\sqrt{1+s^2}}{\lambda}\sqrt{8f}\right\}^{1/2}$$

$$\alpha_2 = \frac{1}{2}+\frac{1}{2}\left\{1+\frac{s\sqrt{1+s^2}}{\lambda}\sqrt{8f}\right\}^{1/2}$$

$$\omega = \frac{gS_0}{\dfrac{\sqrt{1+s^2}}{s}\dfrac{f}{8}-\dfrac{\lambda}{s^2}\sqrt{\dfrac{f}{8}}}$$

and Y=depth function on the side slope, e.g. Y=h-((x-b/2)/s) for x>b/2 where there is a side slope domain. The local boundary shear stress is given by the ancillary equation:

$$\tau_b = \rho\frac{f}{8}U_d^2 \tag{4}$$

SHAPE OF CENTRAL SECTION

In rivers and experimental channels, bed load is transported in certain portions of the beds that are termed bed load transport belts. Obviously the bed load transport belts are flat and this conclusion can be proved as follows:

If a regime channel bed has a lateral slope, sand particles will always move away from any upper position to a lower position on the side slope under the action of gravity. With time and sediment transport, the transverse sloping bed will become flat. Furthermore, when a regime channel bed is flat, there are no forces to accumulate any moving sand to form a lateral slope. Therefore a regime channel has a flat bed, that is a bed load transport belt, on which sand particles move along the channel. Thus the perimeter originating from the channel centre to the critical point, j , can be considered as a flat bed, i.e.

$$y = h_c \qquad \text{when } x \le x_j \tag{5}$$

SHAPE OF BANK SIDE

On the bank side, the entire perimeter is assumed to be at threshold. Let μ be the submerged static coefficient of friction and β be the ratio of lift force to drag force for a sand, and $\kappa=\mu\beta$. In order for the entire bank side perimeter to be at the threshold of motion, the top width of bank side of the regime channel is given by Parker (1979) as

$$t = \frac{h_c\cos^{-1}\kappa}{\mu}\sqrt{\frac{1+\kappa}{1-\kappa}} \tag{6}$$

Then the half top width of a regime channel, T, is

$$T = x_j + \frac{h_c\cos^{-1}\kappa}{\mu}\sqrt{\frac{1+\kappa}{1-\kappa}} \tag{7}$$

and the boundary shape of the bank side is given by

$$y = \frac{h_c}{1-\kappa}\left[\cos\left(\mu\sqrt{\frac{1-\kappa}{1+\kappa}}\frac{x-x_j}{h_c}\right)-\kappa\right] \qquad \text{when } x \geq x_j \qquad (8)$$

Hence the shape of the entire channel can be represented by

$$y = \begin{cases} h_c & ,x \leq x_j \\ \dfrac{h_c}{1-\kappa}\left[\cos\left(\mu\sqrt{\dfrac{1-\kappa}{1+\kappa}}\dfrac{x-x_j}{h_c}\right)-\kappa\right] & ,x \geq x_j \end{cases} \qquad (9)$$

When there is no sediment motion on the channel bed, then $x_j = 0$. The whole perimeter is then at threshold of motion, and there is no flat central region. At the junction point, j, the bed shear must satisfy Shields's relationship

$$\tau_{bj} = \tau_*(\gamma_s - \gamma)d \qquad (10)$$

where τ_* is the Shields parameter at small incipient probability.

There are two variables, S and x_j, which need to be determined in order to obtain the shape and dimensions of a regime channel. The closure equations are those of fluid continuity and bed load transport :

$$Q = 2\int_0^T U_d y(dx) \qquad (11)$$

$$Q_b = 2\int_0^{x_j} q_b(dx) \qquad (12)$$

where q_b is the bed load discharge per unit width. The Ackers&White sediment transport formula (1993) is

$$\gamma\frac{Xh}{\gamma_s d}(\frac{V_{d*}}{U_d})^n = C(\frac{F_{gr}}{A}-1)^m \qquad (13)$$

in which X is the depth-averaged concentration, and the ancillary equations for channel resistance are:

$$\frac{F_{gr}-A}{F_{fg}-A} = 1.0 - 0.76[1 - \frac{1}{\exp(\log D_{gr})^{1.7}}]$$

$$F_{fg} = \frac{V_{d*}}{\sqrt{gd(s-1)}} \qquad V_{d*} = \sqrt{\frac{\tau_b}{\rho}} \qquad D_{gr} = d[\frac{g(s-1)}{v^2}]^{1/3}$$

The various coefficients in equation (13) are given by:

(I) for transitional size, $1 \leq D_{gr} \leq 60$

$$n = 1.0 - 0.56\log(D_{gr}) \qquad\qquad m = \frac{6.83}{D_{gr}} + 1.67$$

$$\log C = 2.79 \log(D_{gr}) - 0.98 \log^2(D_{gr}) - 3.46 \qquad A = \frac{0.23}{\sqrt{D_{gr}}} + 0.14$$

for coarse sediment, $D_{gr} > 60$: n=0.0, m=1.78, C=0.025, A=0.17.

CALCULATION PROCEDURE

The calculation procedure may be summarised as follows. For a given longitudinal bed gradient, S_0, an initial estimate is made of the channel dimensions, bottom width, b and centerline depth h_c. For the given S_0 and b, values of friction factor and dimensionless eddy viscosity are assumed and then the equations (2) to (6) are solved to give the lateral variation of depth-averaged velocity, U_d, and the boundary shear stress, τ_b, and hence channel discharge via equation (11). The bed shear stress at the junction point also has to satisfy equation (10). Equations (9) to (13) are then solved, using the boundary shear stress values, to give the sediment concentration across the channel and hence the total bed load rate via equation (12). The initial estimate of depth may be need to be adjusted to give the correct total discharge rate and the bed width likewise may need to be adjusted to give the correct sediment discharge rate. The bed and the bank will need to be divided into a number of elements, each element either having a constant depth domain or on the bank a linearly varying depth domain.

It is obvious that the choice of friction factor and eddy viscosity are important. They may vary along the wetted perimeter as the Reynolds number varies with the local depth and velocity. The issue of sediment resistance also needs to be considered carefully, as shown by Lovera (1969) and Vanoni & Nomicos (1960), amongst others.

VALIDATION OF MODEL

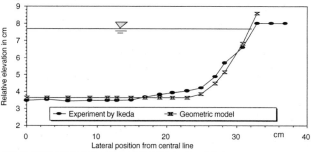

Fig. 2 Comparison between model and experimental shapes

The geometric model by the authors was tested against the experimental data of Ikeda (1981), using results from Run 16 in which $S_0 = 0.00209$ (1/479). A comparison between the model and experimental results is shown in Fig. 2, where the calculated results are seen to be in reasonable agreement with the data. In order to generate this solution λ was taken as 0.10 and f as 0.048. The calculated results gave bed width b=45 cm, centerline depth $h_c = 4.15$ cm, top width T=63.7 cm, cross-sectional area A=239 cm^2, mean velocity U=0.34 m/s, discharge Q=8.05 l/s, and $q_b = 0.46$ g/min/cm. The bed load rate compares with a measured value of 0.51 g/min/cm.

CONCLUSIONS

1. A self-formed channel has a flat bottom which is a sand transport belt. Sands move less actively close to the junction margin where sands are at threshold of motion. With a decrease in bed load, the flat bottom width reduces. The channel will be at threshold when there is no bed load.

2. The geometric model consists of the depth-averaged flow equation by Shiono and Knight, a junction point condition by Shields formula, side shape by Parker, sediment discharge by Ackers-White. The model is consistent with experimental data of Ikeda, but needs further rigorous testing.

REFERENCES

Ackers, P. , 1993, "Sediment transport in open channels: Ackers and White update", *Proc. Instn Civ. Engrs Wat. Marit. & Energy, 101, Dec., pp247-249.*
Bettess, R. and White W.R., 1987, "Extremal Hypotheses Applied to River Regime", *Sediment Transport in Gravel-bed Rivers, Edited by C.R. Thorne, J.C. Bathurst and R.D. Hey, John Wiley & Sons Ltd.*
Diplas P. and Vigilar, G., 1992, "Hydraulic Geometry of threshold channels", *ASCE, Vol. 118, HY4, pp597-614.*
Ikeda S., 1981, "Self-formed straight channels in sandy beds", *ASCE, Vol. 107, Hy4, pp. 389-406.*
Lovera, F. and Kennedy, J.F., 1969, "Friction factors for flat-bed flows in sand channels", *J. Hyd., Div. ASCE, Vol.95, HY4, pp1227-1234.*
Parker, G., 1979, "Hydraulic Geometry of Active Gravel Rivers", *ASCE, Vol. 105, HY9, pp.1185-1201.*
Shiono K. and Knight D.W., 1988, "Two dimensional analytical solution for a compound channel", *Proc. of 3rd Int. Symp. On Refined Flow Modeling and Turbulence Measurements, Tokyo, Japan, July, 1988, pp. 503-510.*
Vanoni, V.A. and Nomicos G.N., 1960, "Resistance properties of sediment-laden streams", *Trans, Amer. Soc. Civil Engrs., Vol.125, pp. 417-430.*
White, W.R., Bettess, R. and Paris, E., 1982, "Analytical approach to river regime", *HY10, Vol. 108, ASCE.*

1B28

Experimental Assessment of a Rational Regime Theory

BABAEYAN-KOOPAEI K.[1] *and VALENTINE E.M.*[2]
Department of Civil Engineering, University of Newcastle upon Tyne, Newcastle upon Tyne, NE1 7RU, UK.

INTRODUCTION

An important problem in the design and operation of alluvial channels is the determination of their geometrical parameters in a dynamic equilibrium condition.

To achieve a dynamically stable condition a channel adjusts its slope, depth and width to carry the required amount of sediment and water. There are two approaches for the prediction of stable channel geometry: (a) regime methods and (b) rational or physical methods.

Regime relationships are those that have been established between various geometric elements of rivers or laboratory channels and their sediment size and discharge when they are stable. Among pioneers in developing this kind of formula were Kennedy (1895), Lindley (1919), and Lacey (1929).

In rational approaches, research has focused on finding a theoretical basis for regime equations. Some of the most recent rational approaches to the design of stable channels involve the combination of a sediment transport equation, a flow resistance equation and some extremal hypothesis, such as the minimisation of energy dissipation rate [Yang and Song (1979)], stream power minimisation [Chang (1980)], and sediment transport maximisation [White, Paris, and Bettess (1981)].

In addition, Bettess and White (1983) proposed a framework for the classification of meandering and braiding channels based on the White et al (1981) [WPB] design theory for stable straight alluvial channels. Valentine and Shakir (1992) assessed this using existing field and flume data and suggested that a revised framework and further laboratory data were required.

---

[1] Ph.D. Student
[2] Lecturer in Hydraulic Engineering

The Bettess and White framework for the classification of channels is dependent on the prediction of the equilibrium slopes of the WPB theory for the design of stable straight channels.

The aim of this paper is to assess experimentally the applicability of the WPB theory for prediction of the physical parameters of stable straight channels. A set of experiments for the development of stable straight channels was performed in a 2.5 m wide and 22 m long flume constructed in the hydraulics laboratory of the Civil Engineering Department of the University of Newcastle upon Tyne. The sediment used in the experiments was uniform sand with $d_{50}=1mm$.

In the following sections, the procedure and the results are discussed.

LABORATORY CHANNEL DEVELOPMENT

For each test, a trapezoidal channel was carved in the sand bed of the flume using a screeding board fixed to a carriage. The initial dimensions of the channels were predicted by the WPB theory. A computer program was used to develop tables for the discharge range of the experiments. After carving the channel in the sand bed, water was introduced to the channel, and the flow rate gradually increased to the desired discharge. The water surface gradient was checked at the beginning of each experiment and was adjusted by using the downstream gate if necessary. The channel was then allowed to develop freely while all hydraulic parameters were kept constant. The experiment was stopped when the channel achieved an equilibrium plan form.

Typically, in the early stages of the test the channel width increased. In order to consider widening, the rate of width change of the channels was recorded. The measurements showed that most of the channels obtained 90% of their final stable width in the first two hours of the experiment run time. Afterwards this rate of change decreased and normally after eight hours, the rate of change of width reduced to less than 1% per hour. This criterion was accepted as the stable form.

GEOMETRIC CHARACTERISTICS OF THE CHANNEL

In the following sections the cross-sectional geometry of the developed straight channels is compared with the WPB theory.

WATER SURFACE WIDTH

Comparison shows that the WPB theory underpredicts the water surface width as shown in Fig. 1. The observed widths are approximately 46% greater than those calculated using the WPB theory.

MEAN DEPTH

The observed value of mean depth was estimated by dividing the observed cross-sectional area by the observed width. The WPB theory overpredicts the mean depths consistently at this scale, as shown in Fig. 2. The mean value of the plotted discrepancy ratios is 1.7 with the standard deviation of 0.3.

CROSS-SECTIONAL AREA

The cross-sectional area of the developed channels is also compared with those of WPB predictions. The results of the comparison is shown in Fig. 3. The observed cross-sectional areas are approximately 12% greater than those calculated using the WPB theory.

WATER SURFACE SLOPE

The measured water surface slope of the developed straight channels is compared with the calculated slopes using the WPB theory. The calculated slopes are found using the mean values of sediment concentrations measured in the experiments. This comparison reveals that the WPB theory overpredicts the slopes. See Fig. 4. The mean value of the plotted discrepancy ratios is 1.4 with the standard deviation of 0.3.

It is frequently indicated in the literature e.g. Ranga Raju et. al (1977) that the water surface slope of a stable straight channel is strongly dependent on the sediment concentration of the channel. The WPB theory is used to show the effect of sediment concentration changes on the water surface slope. The effect is shown for the discharge range of 1.0-5.0 l/s and the sediment size of $d_{50} = 1$ mm in Fig. 5. The slopes were steeper as the sediment concentration increased which is consistent with observation.

The sensitivity of water slope to sediment concentration indicates that precise prediction of sediment concentration is important. Thus the sediment transport formula plays an important role. For the design of stable alluvial channels, Babaeyan-Koopaei and Valentine (1995) used the sediment transport maximisation hypothesis of White et al. (1981) with different combinations of sediment transport and flow resistance equations. They concluded that the success of the method is quite dependent on the equations adopted.

The WPB theory is based on the Ackers-White (1973) sediment transport formula which is under review. Bettess (1991) discussed the impact of the modifications to the equation. He stated that for fine sediments near the threshold of motion around 0.06 mm in size, the transport rate is increased by a factor of 10, while for larger sediment mobilities the transport rate is reduced by a factor of up to 10. For coarse sediments around and above 3.6 mm, the changes in the transport rate are less than 10%.

It is thus clear that the modification of the formula will have a substantial effect on the predictions of the WPB theory.

SEDIMENT CONCENTRATION

The calculated sediment concentrations were found by using observed water surface slopes. The observed sediment concentrations are 60% higher than the calculated values of sediment concentrations for the observed slopes. The results are shown in Fig. 6. The difference between the observed sediment concentrations and the calculated ones is likely to be because of the sediment transport predictions. It is believed that the difference will be reduced after updating the Ackers-White (1973) sediment transport formula.

SUMMARY AND CONCLUSIONS

In this paper, the hydraulic and geometric characteristics of developed stable straight channels are compared with the WPB theory predictions. The results show that this theory underpredicts water surface width, cross-sectional area, and sediment concentration of these channels for the range of discharges. It is also shown that this theory overpredicts mean depth and water surface slopes. This difference is attributed to the predictions of the Ackers-White (1973) sediment transport formula which is now under review.

REFERENCES

1- Ackers, P. and White, W. R. (1973) " Sediment Transport : New Approach and Analysis ", Proc., Journal of Hyd. Division, ASCE, Vol. 99, No. 11, pp. 2041-2060.

2- Babaeyan-Koopaei, K. and Valentine, E. M. (1995) " A rational regime theory : different combinations of formulae ", The Second Int. Conf. on Hydro-Science and Eng., Beijing, China.

3- Bettess, R. (1991) "Updated sediment transport theory" Research Focus, No. 7 (October Volume).

4- Bettess, R. and White, W. R. (1983) "Meandering and braiding of alluvial channels" Proc. ICE, Part 2, 75, pp. 525-538.

5- Chang, H. H., (1980) "Geometry of Gravel Streams", Journal of Hydraulic Div., ASCE, Vol. 106, No. 9, pp. 1443-1456.

6- Kennedy, R. G. (1895) "The prevention of silting in irrigation canals" Paper No. 2826, Proc. ICE, Vol. 119, London.

7- Lacey, G. (1929-1930) "Stable channels in alluvium" Minutes of Proc. ICE, 229, pp 259-384.

8- Lindley, E. S. (1919) "Regime Channels" Proc. Punjab Engineering Congress, Vol.7.

9- Ranga Raju, K. G., Dhandapani, K. R. and Kandap, D. M. (1977) "Effect of sediment load on stable canal dimensions" Proc., ASCE, Journal of Waterways, Port, Coastal and Ocean Div., Vol. 103, WW2, pp. 241-249.

10- Valentine, E. M. (1986) "Application to braided rivers of the Wallingford regime theory" Proc. of the gravel bed rivers workshop, Hydrology Centre, Public. No. 9, Christchurch.

11- Valentine, E. M., and Shakir, A. S. (1992) "River channel planform : An appraisal of a rational approach" Eighth Congress of the Asia and Pacific Division of IAHR, Poona, India.

12- White, W. R., Paris, E., and Bettess, R., (1981) " River Regime based on Sediment Transport Concepts ", Report No. IT 201, Hyd. Res. Station, Wallingford, UK.

13- Yang, C. T. and Song, C. C. S. (1979) " Theory of Minimum Rate of Energy Dissipation ", Journal of Hyd. Div., ASCE, Vol. 105, No. 7, pp. 769-784.

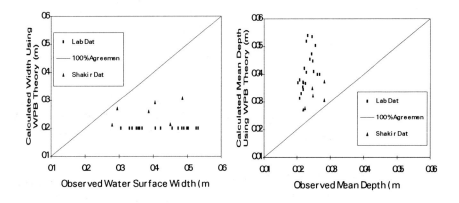

Figure 1. Comparison of the observed and calculated water surface widths.

Figure 2. Comparison of the observed and calculated mean depths.

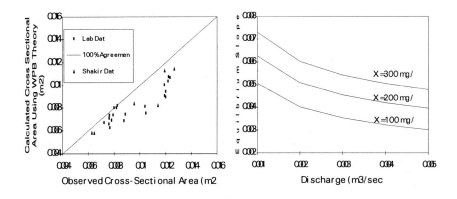

Figure 3. Comparison of the observed and calculated cross sectional area.

Figure 5. Effect of sediment concentration on the slope of stable straight channel.

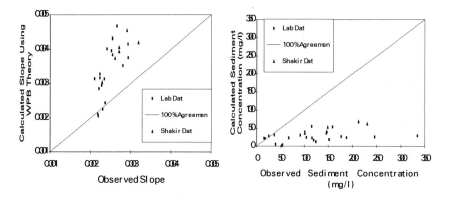

Figure 4. Comparison of the observed and calculated water surface slopes.

Figure 6. Comparison of the observed and calculated sediment concentrations.

THE ROLE OF AERATION IN REDUCING DAMAGE CAUSED BY CAVITATION

Ramón GUTIERREZ SERRET

Civil Eng. Ph.D., Head Hydraulic Structures Dept. Ministry of Public Works, Transport and Environment - Center for Research and Experimentation (CEDEX). SPAIN.

The reasons why flow aeration is probably the main way of rectifying the damage caused by cavitation can, in order of importance, be summed up as follows:

- The collapse pressure for cavities with air is lower than for those which lack this element.
- The velocity of the shockwaves generated during collapse is considerably reduced. Air concentrations [$C = Q_a/(Q_a + Q_w)$] of 8% cause at least a tenfold reduction in celerity. Furthermore, this circumstance attenuate the "domino" effect on the adjacent cavities, which serves to lessen the damage still further.
- Cavities with air, which are larger than those without, cause the two types to merge, thereby increasing cavity size and, either their collapse does not give rise to damage or this is less serious (sizes over 0,1 mm. cause virtually no damage).
- Pressure far below atmospheric pressure is not exerted, so the possibility of cavitation is reduced.

With regard to the air concentration that must exist close to the facings if cavitation damage is to be prevented, different authors have stated that small amounts of air adjacent to the concrete surface considerably reduce the damage caused. Figure 1 shows some of the results of two significant tests.

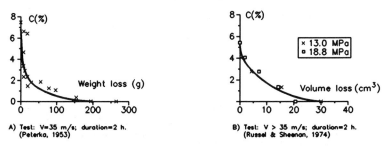

A) Test: V=35 m/s; duration=2 h.
(Peterka, 1953)

B) Test: V > 35 m/s; duration=2 h.
(Russel & Sheenan, 1974)

Fig. 1.- Relationship between air concentration in the flow and damage to concrete by cavitation.

It can be deduced from these and other works, that with an air concentration between 5% and 10% on the concrete surfaces, practically no cavitation damage is caused. In the author's opinion, 8% can be presented as an acceptable percentage for ensuring that with the concrete strength and finishes that are now "normal" in the hydraulic structures of dams no damage take place, as long as the flow velocities are not too great (v < 30 - 35 m/s. or even above).

This 8% threshold as regards the flow velocity and the degree of finishing of the surfaces, has meat that presents trends are to reconcile velocity and aeration of the flow with surfaces finishes not very smouth, whose construction and maintenance are not particularly complicated from a practical viewpoint.

With regard to this, Falvey (1990), on referring to surface unevenness, relates the finishes habitually used by the U.S. Bureau of Reclamation with the cavitation number (σ) and with flow aeration (Table C.1). No similar information is available where general roughness of the surfaces are concerned.

SURFACE TOLERANCES			SURFACE TOLERANCE - CAVITATION Nº - AERATION		
TOLERANCE TYPE (T)	OFFSET Max. Height (mm)	SLOPE (V:H)	CAVITATION Nº (σ)	TOLERANCE WITHOUT AERATION	TOLERANCE WITH AERATION
T_1 T_2 T_3	25 16 12	1: 4 1: 8 1: 16	> 0.6 0.4 - 0.6 0.2 - 0.4 0.1 - 0.2 < 0.1	T_1 T_2 T_3 Modify design Modify design	T_1 T_1 T_1 T_2 Modify design

C.1.- Surface tolerance. Requirements according to values of cavitation index and the presence or lack of aeration (Falvey, 1990).

The other aspect related to the aeration-damage binomial, concerns the required strength of the concrete if erosion due to cavitation is to be prevented. In this sense, the author deduces from tests carried out by several researchers (1) that, as an order of magnitude, concrete with a compression strength of 20 MPa will tolerate velocities in the region of 30-35 m/s., if there is an 8% air concentration in the vicinity.

Finally, and with reference to current practice, it is clear that when aeration at the bottom does not naturally reach the aforementioned levels (C $\approx$ 8%), and there is a risk of damage from cavitation (e.g. σ < 0.2 - 0.25), it now normal practice provide aerators for the artificial aeration of the.

Bibliography.-

(1)   Chanson, H. (1992). "Air Entrainment in Chutes and Spillways". Research Report Nº CE 133, Department of Civil Engineering, University of Queensland, Australia. (P. 6, Test Summaries).

(2)   Falvey, H.T. (1990). Cavitation in Chutes and Spillways". Engineering Monograph Nº 42. Bureau of Reclamation, Denver, U.S.A.

(3)   Gutiérrez Serret, R. (1994). "Aireación en las Estructuras Hidráulicas de las Presas: Aplicación a los Aliviaderos". (Aeration in Hydraulic Structures of Dams: Application to Spillways). Doctoral Thesis. Universidad Politécnica, Madrid, Spain.

(4)   PGOH (1989). "Pliego de Prescripciones Técnicas Generales para la Ejecución de Obras Hidráulicas". General Technical Specifications for Hydraulic Works). Arts. 32.49 and 41.35. "Facing Finishes". Ministry of Public Works, Transport and Environement, Spain.

1B30

Three Dimensional Flow Analysis in Open Channel

YOSHIHIRO MIYAMOTO[*1], KAZUO ISHINO[*2]
[*1]CFD group of SEA Corporation, Tokyo, Japan
[*2]Technical Research Institute of Taisei Corporation,
Yokohama, Japan

1. INTRODUCTION

Classic open channel flow analysis is based on one dimensional equation. However two or three dimensional analysis is required to predict the flow affected by geometry change. To investigate this subject mathmatically FLOW-3D[(1)] was used. FLOW-3D is a general three dimensional fluid flow program developed by Flow Science Inc. .

2. WALL FRICTION

Authors applied conventional quadratic wall shear($\tau_w = -0.5 f A_w \rho |u| u$) as a replacement of current wall function for three dimensional open channel flow analysis. A_w is shear area divided by control volume, f is friction-coefficient, ρ is fluid density, u is velocity. Wall friction coefficient of open channel problem described in next section is evaluated as 5×10^{-2} using equation $f = 0.25/Re^{0.125}$, (Re is local Reynolds number) which corresponds to wall function of power law used in turbulent model.

3. OPEN CHANNEL FLOW ANALYSIS

8.5m width of trapezoidal cross section 220m straight open channel (Fig. 1) is current problem with free discharge end. 0.25m thickness, 1.8m height of center board obstacle is setup. This open channel is designed to flow 60m^3/s without-center board obstacle.

Half region of open channel is divided into three dimensional mesh ($10 \times 90 \times 26$ cells in each direction). Fluid height distribution of FLOW-3D agrees with the converted data from small scale (1/20) experimental data (Fig. 2). Critical flow phenomena appear at free discharge exit (Fig. 3).

80m short length channel including the downstream end of center board is analysed in detail using fine mesh cells ($15 \times 150 \times 37$ cells in each direction). Boundary condition is given using the result of whole length analysis. Standing wave is formed in region of downstream of center board (Fig.4). Pressure loss occurs at downstream end of center board due to the flow separation (Fig. 5). Because of it, secondary flow appears in that region (Fig. 6).

4. CONCLUSION

Friction coefficient 5×10^{-2} used in this problem is the one of smooth wall. Further discussion may be a requirement for the influence of wall roughness. However computational simulation would be a useful prediction tool for this type of problem.

REFERENCE

1. "FLOW-3D computational modeling power for Scientists and Engineers", Flow Science Inc., Technical Manual

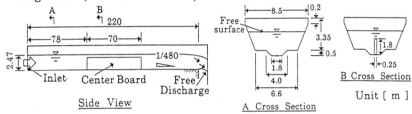

Fig.1 Schematic of Open channel flow with Free Discharge

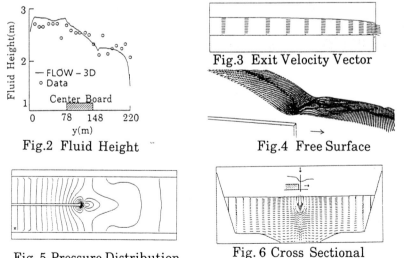

Fig.2 Fluid Height

Fig.3 Exit Velocity Vector

Fig.4 Free Surface

Fig. 5 Pressure Distribution

Fig. 6 Cross Sectional Velocity Vector

1B31

ENERGY DISSIPATION IN SKIMMING FLOW OVER STEPPED SPILLWAYS. A COMPARATIVE ANALYSIS

JORGE MATOS and ANTÓNIO QUINTELA
Department of Civil Engineering, Technical University of Lisbon, Lisbon, Portugal

INTRODUCTION

A reanalysis of published experimental data in skimming flows over stepped spillways is presented in this paper. The results obtained from the application of intrusive versus non-intrusive methods indicate clearly that great care should be taken in order to analyse possible significant underestimation of the residual energy at the spillway toe.

A suggestion to evaluate the total head loss on the spillway (or residual energy at the toe) for preliminary design is presented.

ENERGY DISSIPATION IN SKIMMING FLOWS. A COMPARISON

If uniform flow conditions are reached at the toe of the spillway, the energy loss ratio ($\Delta H/H_0$) can be written as a function of H_d/d_c as indicated by Stephenson (1991) and Chanson (1994). For an un-gated spillway $\Delta H/H_0$ equals (as Chanson (1994)):

$$\frac{\Delta H}{H_0} = 1 - \frac{\left(\dfrac{f}{8\sin\alpha}\right)^{1/3}\cos\alpha + \dfrac{E}{2}\left(\dfrac{f}{8\sin\alpha}\right)^{-2/3}}{\dfrac{H_d}{d_c} + \dfrac{3}{2}} \qquad (1)$$

where H_0 is the maximum head available, H_d the dam crest height above the spillway toe, α the spillway slope, f the friction factor of the aerated flow, d_c the critical depth, E the kinetic energy correction coefficient and ΔH the total head loss on the spillway.

Fig. 1 ilustrates values of $\Delta H/H_0$ obtained from skimming flow experiments in which different experimental measuring techniques were used. The indirect or non-intrusive method was applied herein using the momentum equation to obtain an estimate of the equivalent water depth upstream of the hydraulic jump, at the toe. Eq. (1) is also plotted in Fig. 1, for f=0.06, f=0.18 and f=1.00 (α=53°, E=1.00).

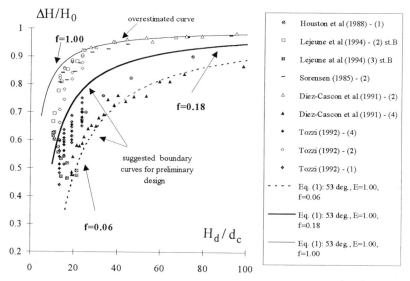

Experimental data: step height (scaled to prototype), $0.12 \leq h_d$ (m) ≤ 1.50; $51 \leq \alpha$ (deg.) ≤ 59.
(1) - velocity (Pitot); (2) - flow depth near the toe (point gauge/scale), *intrusive;*
(3) - equivalent water depth (obtained from flow depth and drag force on a little sphere);
(4) - conjugate flow depth of the hydraulic jump (point gauge/scale), *non-intrusive.*

Fig. 1. Energy dissipation in skimming flow over stepped spillways.

From Fig. 1 it can be concluded that: (i) the values of $\Delta H/H_0$ obtained from the non-intrusive method as well as those obtained from velocity data are considerably lower than the values obtained from flow depth measurements at the toe; (ii) the $\Delta H/H_0$ values obtained using the equivalent water depth estimated from velocity and air concentration experimental data by Lejeune et al (1994) are closer to non-intrusive calculations.

These observations are to be expected since ΔH can be considerably overestimated if it is calculated on the basis of the aerated flow depth instead of the equivalent water depth. Lejeune's data also confirms that a significant overestimation of $\Delta H/H_0$ can occur when using the aerated flow depth to compute the residual energy.

Considering the above analysis it seems adequate to apply Eq. (1) using a value for f between 0.06 and 0.18 as a first estimate of the energy loss in skimming flow over steep stepped spillways.

Although the results obtained with the non-intrusive method seem to be relatively accurate, a more thorough research taking into account air entrainment is considered of particular importance to obtain a more precise estimate of the residual energy at the toe of steep stepped spillways, common to RCC dams.

REFERENCES

CHANSON, H. (1994). Comparison of energy dissipation between nappe and skimming flow regimes on stepped chutes. *J. of Hydr. Res.*, IAHR, Vol. 32, N° 2, pp. 213-218.

DIEZ-CASCON, J. et al (1991), Studies on the behaviour of stepped spillways, *Water Power & Dam Construction*, Sept., pp. 22-26.

HOUSTON, K.L. and RICHARDSON, A.T. (1988). Energy dissipation characteristics of a stepped spillway for an RCC dam. *Proc. The Int. Symp. on Hydraulics for High Dams*, IAHR, Beijing, China, pp. 91-98.

LEJEUNE, A., LEJEUNE, M. and LACROIX, F. (1994). Study of skimming flow over stepped spillways. *Proc. Conf. on Modelling, Testing and Monitoring for Hydro Powerplants*, Int. J. on HP&D, Budapest, Hungary, July, 11 pp.

MATOS, J. and QUINTELA, A. (1995). Comparison of energy dissipation between nappe and skimming flow regimes on stepped chutes - Discussion. *J. of Hydr. Res.*, IAHR, Vol. 33, No. 1, pp. 135-139.

SORENSEN, M. (1985). Stepped spillway hydraulic model investigation. *J. of Hydr. Engrg.*, ASCE, Vol. 111, No. 12, Dec., pp. 1461-1472.

STEPHENSON, D. (1991). Energy dissipation down stepped spillways. *Water Power & Dam Construction*, Sept., pp. 27-30.

TOZZI, M.J. (1992), Caracterização/ Comportamento de escoamentos em vertedouros com paramento em degraus (Hydraulics of stepped spillways). *PhD Thesis*, University of São Paulo, Brazil (in Portuguese).

1C1

HYDRAULIC CHARACTERISTICS OF UNSTEADY FLOW IN OPEN CHANNELS WITH FLOOD PLAINS

A. TOMINAGA*, J. LIU*, M. NAGAO* and I. NEZU**

* Dept. of Civil Engineering, Nagoya Institute of Technology, Showa, Nagoya 466, Japan.
** Dept. of Civil & Global Engineering, Kyoto University, Sakyo, Kyoto 606, Japan.

Abstract

Hydraulic characteristics of unsteady flows were investigated experimentally in a rectangular open channel with symmetrical flood plains using a computer-controlled water supply system. Some noticeable features of unsteady flow structure in compound channels were revealed which were significantly different from those in single cross-sectional rectangular channels. When the unsteadiness is very large, the velocity and bed shear stress in the main channel becomes much larger in the rising stage than that expected from the steady flow. When the unsteadiness is mild, the main-channel velocity decreases by the effects of developed mixing between the main-channel and flood-plain flows.

Introduction

In recent years, flood plains become very useful as ecological and recreational spaces in urban rivers. When one consider such utilities of flood plains for multiple purposes, it is very important to understand hydrodynamic characteristics in compound open channel flows in flood. A number of researches have been conducted on resistance law and flow structures in compound channel flows. Three-dimensional turbulent structures of compound open channel flows associated with secondary currents have been recently revealed by the use of velocity measurements and flow visualizations (e.g. Tominaga & Nezu (1991) and Imamoto & Ishigaki (1991)). However, most of them treated the flood flow as a steady uniform flow with the peak discharge. It is known that the flow in a rising stage of flood is rather different from the flow in a falling stage as shown by Nezu et al. (1993a,b) and Tu & Graf (1992). They found that the bed shear stress becomes considerably larger in the rising stage than in the falling stage. It is very interesting to investigate the unsteady flood flow over compound open channels, since it encounters a sudden change of cross section. Tominaga et al (1994) conducted experiments of unsteady flow in a compound channel with an one-side flood plain and trapezoidal main-channel. They pointed out the unsteady flow characteristics of compound channels different from that of single cross-sectional rectangular channels. In this study, time-dependent three-dimensional flow structures are newly measured in a rectangular compound channel with symmetrical flood plains and hydraulic characteristics of unsteady flow in compound open channels are investigated.

Experimental Apparatus and Method

The experiments were conducted in a 13m long and 0.6m wide 0.4m deep tilting flume. Rectangular flood plains were set on both sides of the flume. The width was 0.20m and the height of was 0.059m. The channel slope S was set as S=0.001. Fig.1 shows a open-channel experiment system. The discharge is controlled by changing the rotation cycle of a water-pump motor by means of a transistor inverter. The rotating cycle is controlled by a personal computer using feedback from the signal of an electromagnetic flow meter through an A/D and D/A converter board. Arbitrary discharge hydrographs can be obtained from the computer. The water is supplied into an settling tank with mesh screens at the channel entrance. The test section was set at 7.5m downstream from the entrance. The base discharge Q_b was set to 0.003m³/s. In this case, the flow depth h was 3.9cm and the flow was limited in the main channel. The peak discharge Q_p was set to 0.02m³/s. The discharge is increased from Q_b to Q_p linearly in a rising time T_p. After keeping the peak discharge during $(1/6)T_p$, the discharge is decreased from Q_p to Q_b in a falling time same as T_p. In the present experiments, T_p was taken as 60s and 120s. An example of given discharge hydrograph when T_p=60s is shown in Fig.2. The identity of these repeated discharge hydrograph was reasonably good. In the present experiments, unsteadiness is rather large and the flood wave includes of dynamic-wave characteristics. In actual floods, the effects of unsteadiness are often observed in many Japanese rivers (e.g. Fukuoka et al. (1990)).

Velocity was measured by a micro propeller velocimeter 3mm in diameter. Simultaneously, water depth was measured by a condenser-type water-wave gage 0.1m down stream from the velocimeter. Since the sensing point of velocimeter submerges or unsubmerges with time, the propeller velocimeter is adequate for the measurements of unsteady flood flow. The output signal of the propeller velocimeter tends to be reduced and linear relation between the velocity and the output voltage is no longer sustained near the free surface. A calibration equation was made from the preliminary experiments as a function of the submerged depth of the propeller. Then, the velocity was corrected by this equation using the flow depth obtained by the water-wave gage. Both the output signals of the velocimeter and the wave gage were recorded on floppy disks and processed by computers. The sampling frequency was 50Hz. The sampling time was 180s and 320s, respectively for T_p=60s and 120s. In order to obtain the time-dependent three-dimensional flow structures in the whole cross section, the velocity was measured at each point repetitively during the flood event with the identical hydrograph. The velocity signals were arranged in time so as to be t=0 at the rising point of the flow depth. Then,

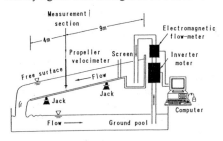

Fig. 1   Schematic description of
experiment system

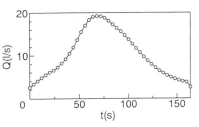

Fig. 2   Discharge hydrograph
(T_p=60s)

the time was divided into consecutive 4 seconds periods and the velocity was averaged in each period. These 4s interval-averaged values of velocity are used in the following discussions.

Fig.3 shows the examples of normalized time variations of flow depth in compound and single cross sections. The abscissa is normalized time divided by the arrival time of the peak flow depth T_{hp}. The ordinate is the dimensionless variation of flow depth Δh ($\equiv (h-h_b)/(h_p-h_b)$), in which h_b and h_p is the base and peak flow depth. In the rising stage, the flow depth of the compound channel becomes slightly large from the time when the over-bank flow has just started to the peak time. In the falling stage, the flow depth maintains relatively larger stage in the compound channel than the rectangular channel.

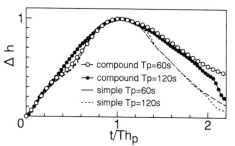

Fig. 3   Time variations of flow depth

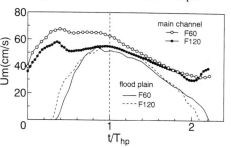

Fig. 4   Time variations of depth-averaged velocity

Time Variation of Depth-Averaged Velocity

Fig.4 shows the normalized time variation of depth-averaged primary velocity U_m at the representative sections in the main channel and the flood plain. The velocity U_m decreases with an increase of T_p in the main channel, whereas it is not so changed in the flood plain. The velocity in the main channel attains a peak much earlier than the peak time of the flow depth. This peak time was also relatively earlier than that in the single rectangular channel. After the

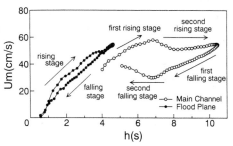

Fig. 5   Depth-velocity curves in compound channel

slight decrease from this peak, the high velocity is maintained till the peak time of the flood plain velocity. The flood-plain velocity attains a peak faster than T_{hp} when $T_p=60s$, whereas the peak time of the flood-plain velocity approaches to the peak time of flow depth when $T_p=120s$. These features of velocity variation implies that the turbulent mixing of the main-channel and flood-plain flows once decrease the main-channel velocity and this effects becomes to be sufficiently developed as increasing the flood period.

Fig.5 shows the depth-velocity curves at the main channel and the flood plain for $T_p=120s$. In the main channel, the velocity varies drawing a clear loop and the difference of the velocity is extremely large between the rising and falling stages. Total process of depth-velocity variation can be divided into four stages: first and second rising stage and first and second falling stage, as shown in Fig.5. In the flood plain, this loop property is less significant than in the main channel.

Time-Dependent Three-Dimensional Flow Structures

Fig.6 shows the isovels at each time period of the flood from the rising stage to the falling stage of T_p=60s with those of the steady flow. When t=28s in the rising stage, the velocity in the main channel becomes maximum and the interaction with the flood-plain flow has not been observed yet. When t=48s, the interaction between the main-channel and flood-plain flows becomes to be observed and the velocity near the free surface in the main channel is decelerated. The flow structure is almost compatible with that of the Q=0.01m³/s steady flow case, but the velocity is much larger in the unsteady case, especially in the main channel. At the stage peak of t=72s, the structure in the main channel is very similar to the steady one of Q=0.018m³/s. The effects of the inclined upflow from the flood-plain edge are clearly recognized. At t=120s in the falling stage, the velocity is much smaller than that at t=48s which has almost the same flow depth and it is also smaller than that of steady flow case. Fig.7 shows isovels of T_p=120s. When t=48s and t=76s in the rising stage, the flow structure is almost similar but the velocity is smaller, compared with the case of T_p=60s. The difference between the main-channel and flood-plain velocities becomes smaller than the case of T_p=60s. Of particular significance is that the flood-plain velocity becomes larger than the main-channel velocity around the peak stage in the case of T_p=120s. In comparison with the steady flow, the main-channel velocity decreases whereas the flood-plain velocity is not so changed. This may caused by the secondary currents which are considered to be more developed in the unsteady flow, as estimated from the isovels. When t=224s in the falling stage, the isovels are almost same as those in the case of T_p=60s.

Time Variation of Friction Velocity

It is indicated by Nezu et al. (1993a) that the log-law is applicable even in such an unsteady flow. Though the log law on the flood plain is not sufficiently established with respect to the integral constant in the present study, the friction velocity U∗ was roughly estimated using the log law distributions of the primary mean velocity. Fig.8 shows time variation of friction velocity of T_p=60s along the wetted perimeter. The distribution of steady flow with maximum discharge of Q=0.018m³/s and 0.010m³/s are also shown in this figure. At t=28s when the main-channel velocity becomes maximum, the value in the main channel considerably exceed the steady-flow value of Q=0.018m³/s. When t=48s, U∗ is much larger in the whole perimeter than the steady-flow value of Q=0.010m³/s with similar flow depth. On the flood plain, it becomes maximum at t=60s, the velocity slightly exceeds the steady value. In the falling stage, t=120s, U∗ becomes smaller in comparison with the steady flow. Fig.9 shows the time variation of friction velocity of T_p=120s. The peak value at t=44s in the main channel is not exceed the steady value. In the rising stage, t=72s, the distribution is almost the same as that of the steady flow of Q=0.010m³/s. At the stage peak, t=136s, flood-plain value of U∗ is similar to the steady value, whereas the main-channel value becomes relatively very small. In the falling stage, the feature is very similar to the case of T_p=60s.

It is known that the bed shear stress becomes larger in the rising stage in rectangular channels. This feature becomes more significant in the main channel with flood plain and the

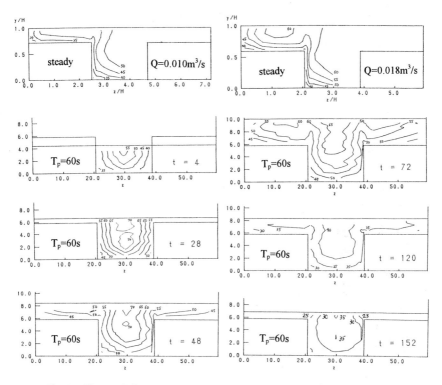

Fig. 6   Time variations of isovels in compound channel (steady and T_p=60s)

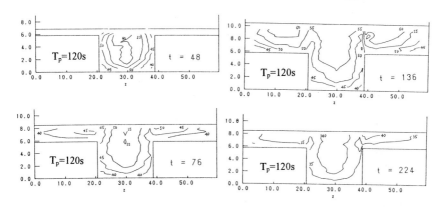

Fig. 7   Time variations of isovels in compound channel (T_p=120s)

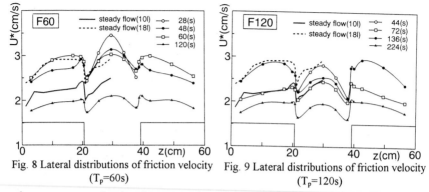

Fig. 8 Lateral distributions of friction velocity (T_p=60s)

Fig. 9 Lateral distributions of friction velocity (T_p=120s)

peak appears very earlier than the stage peak, when the hydrograph is very sharp, as shown by Tominaga et al (1994). On the other hand, when the hydrograph is mild, it is considered that the interaction between the main-channel and flood-plain flows could be developed near the stage peak and the main-channel velocity decelerated by this effect.

Conclusions

Unsteady flow structures in open channels with flood plain were investigated from the repetitive traversing measurements on the same discharge hydrograph using a computer-controlled water supply system. Unsteady flow structure in compound channels were significantly different from those in the single cross-sectional rectangular channels. When the unsteadiness is very large, the peak value of the velocity appears very earlier in the main channel than in single cross-sectional rectangular channels. In the flood plain, unsteadiness is smaller than that in the main channel. When the unsteadiness is mild, the main-channel velocity decreases in the second rising stage by the effects of the mixing developed between the main-channel and flood-plain flows.

References

Fukuoka et al. (1990): Study on flood flow and river-bed variation in the Hinuma River, Report of PWRI, vol.180-2, pp.1-94.

Hayashi,T., and Ooshima, M. (1988): Effects of the unsteadiness of flood waves on their turbulence structure, Proc. of 32nd Japanese Conf. on Hydraulics, pp.607-612.

Imamoto,H. and Ishigaki,T. (1991): Experimental study on the turbulent mixing in a compound open channel, 24th IAHR Congress, Madrid, vol.C, pp. 609-616.

Nezu,I. and Nakagawa, H. (1993a): Basic structure of turbulence in unsteady open-channel flows, 9th Symp. on Turbulent Shear Flows, Kyoto, vol.1, 7.1.1-7.1.6.

Nezu,I., Nakagawa,H., Ishida,Y. and Fujimoto,H. (1993b): Effects of unsteadiness on velocity profiles over rough beds in flood surface flows, 25th IAHR Congress, Tokyo, vol.A1, pp.153-160.

Tominaga,A. and Nezu,I. (1991): Turbulent structure in compound open-channel flows, J. Hydraulic Engrg., ASCE, vol.117(1), pp.21-41.

Tominaga,A., Nagao,M. and Nezu,I.(1994): Experimental study on unsteady flow in open channels with flood plains, Proc. of Symp. on Fundamentals and Advancements in Hydraulic Measurements and Experimentation, ASCE, pp.396-405.

Tu,H. and Graf,W.H. (1992): Velocity distribution in unsteady open-channel over gravel beds, J. Hydroscience and Hydraulic Engineering, vol.10, No.1, pp.11-25.

1C2

Unsteady Flow Characteristics in Compound Channels with Vegetated Flood Plains

JAYARATNE, B. L., H. TU .,N. TAMAI, *Dept. Civil Engrg., Univ. of Tokyo*
and K. KAN, *Dept. Civil Engrg., Shibaura Instit. of Tech., Japan*

INTRODUCTION

Flow structures in compound channels have complicated three dimensional behaviors, and strong interaction between the main-channel and flood-plain flows is the most important feature of compound channel flows. Roughness of the flood-plain, and flow depth ratio between flood-plain and main-channel, may have significant effects on this interaction. In natural rivers, roughness effect from the vegetation on the flood plain reduces the flow velocity over the flood plain and further enhances the velocity difference between main-channel and flood plain flows. Due to this different flow velocities on both sides of the interface, strong eddies will be produced, which hamper the main flow.

Knight & Hamed (1994), Nalluri & Judy (1985) and some other researchers have examined the effect of roughness difference between the main-channel and flood-plain on the flow resistance. Fukuoka & Fujita (1989), using experimental results, showed that the intensity of momentum transfer changed remarkably depending on the position, width and permeability of the porous media (vegetated zone). However, few experimental investigations have been done on unsteady compound channel flows with vegetated flood plain. In this study, detailed velocity and depth measurements were taken across a compound channel made in a well equipped large tilting flume, with vegetation on the flood plain.

FLOW MEASUREMENTS

The experiments were conducted in a 25m long, 1m wide, tilting flume. More details are given in Tu et al. (1994(a)). The flood plain has been made of aluminum plates, part of which (180 cm long, where the measuring section was located) was replaced by wooden boards to avoid possible disturbance on the electromagnetic probes from the aluminium. The width of the flood-plain is 60 cm and that of the main-channel is 40 cm (Fig.1). It simulates half of a symmetric compound channel. Rigid wooden cylinders were used as model plants, in order to make a rigid vegetation layer. The diameter and height are 0.2cm and 5cm, respectively. The model plants were set in a square pattern of

2.5cmx2.5cm on an acrylic resin plate, which was itself fixed on the flood-plain. The thickness of the resin plate is 1.45cm, such that the flood-plain thickness is 6.45cm. The measuring section was selected at a distance of 12m from the flume entrance while the vegetated zone was 8m long.

Hydrographs of desired shapes can be created by controlling the opening of a valve in the pipe system using relevant software. In the present paper, only the results of one hydrograph will be discussed due to space limitation. Water depth was measured using limnimeters, and the three velocity components were measured with electromagnetic probes, both having a diameter of 5mm. An I-type velocity meter (SFT-200-05) was used to measure the longitudinal and transverse velocity components (u and w). The other is of L-type (SFT-200-05L), for probing only the vertical velocity component (v). The flume's bottom slope was kept 1/2000 throughout this study.

The water depths were measured at 16 positions across the channel section at z=5cm, 10cm, 15cm, 25cm, 35cm, 45cm, 50cm, 55cm, 60cm, 65cm, 70cm, 75cm, 80cm, 85cm, 90cm and 95cm (see Fig.1). In each of these 16 positions, longitudinal and transverse velocity components were taken at every 5mm in the vertical up to y=8.5cm, the first point being 1cm from the channel bottom. The vertical velocity components were taken at z=50cm, 55cm, 60cm, 65cm, 70cm and 75cm, also every 5mm in the vertical, from y=3.5cm to y=8.5cm. The sampling rate was 5Hz and the five readings per second were averaged to render mean velocities. Since there were many measuring points and the flume is an outdoor facility, the experiments should be finished as early as possible before the changing weather conditions (specially wind) may have undesirable effects on the measurement. Therefore velocities at one point were measured by passing the hydrograph only once. The signals from the electromagnetic probes or the limnimeters were recorded with a NEC computer for later treatment.

WATER-DEPTH & WATER-SURFACE VARIATIONS

In Fig.2.(a) are presented the water depths measured at z=25cm, z=55cm (on the flood plain) and z=65cm, z=80cm (in the main-channel). The depths measured on the flood-plain are referenced to the top of the flood-plain. It can be recognized from our raw data (which is not discernable from Fig.2(a)) that the peak of the depth on the flood-plain occurs a little later than that in the main channel, indicating a slight time lag among the peaks of the depths in the present investigation.

Figure 2(b) shows that in the interactive contact zone, water surface elevation is significantly lower than those of other area. With shedding of vortices due to different flow velocities between main-channel and flood-plain, strong eddies are produced, which hamper the main flow, resulting in higher energy dissipation in this interactive contact zone compared to that in the main-channel leads to lower depths. On the flood-plain, it shows rather high water surface elevation, indicating that the flood-plain roughness (due to vegetation) increases the flow resistance and rises the water level.

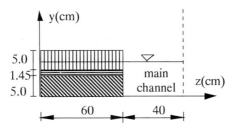

Fig.1 A sketch of the measuring section

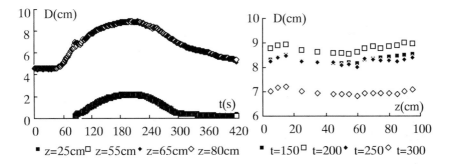

Fig.2(a) Depths' variations with time   Fig.2(b) Water surface profiles

EVOLUTION OF THE VERTICAL VELOCITY PROFILE

The profiles of longitudinal and transverse velocity components (u and w) at z=65cm, 80cm and 95cm are presented in Fig.3. Time instants are selected for every 50 seconds so that it covers the whole process of the hydrograph.

It can be clearly seen that u- components at z=65cm, close to the interface is significantly smaller than that at z=80cm and z=95cm. This is mainly due to lateral momentum exchange at the interface. Moreover, Fig.3 shows that w-components at z=65cm, below the flood-plain is positive implying the flow towards the flood-plain. It is decreased at z=80cm and is almost zero at z=95cm. This behavior is quite clear for the the rising stage. At the same time above the flood-plain height (specially at z=80cm), negative w- shows the flow towards main-channel, indicating a clockwise secondary flow. But it becomes weaker in the receding stage. This is similar to the secondary flow pattern described by TOMINAGA, et al. (1993).

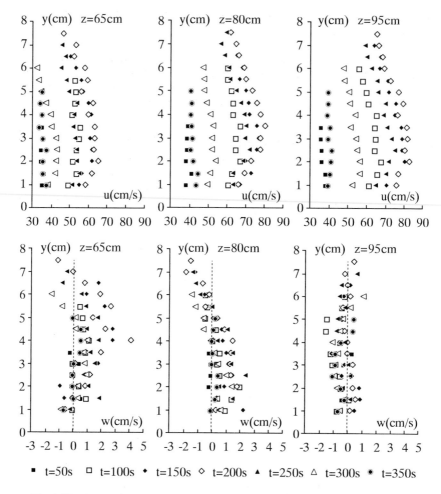

Fig.3 Vertical profiles of u- and w-, at selected time instants; (w is positive towards the flood-plain)

TRANSVERSE VELOCITY VARIATIONS

The variation of u- and w- components across the channel section corresponding to D=8.3cm and D=8.5cm in both the rising and receding stages are shown in Fig.4. These data are taken at y=7cm, i.e., 0.55cm above the flood-plain level. Figure 4(a) also shows that the longitudinal velocity close to the interface (i.e. in the interactive zone) is quite smaller compared with those outside the zone, because the enlargement of energy dissipation decelerates the flow over the flood-plain, thus enhancing the difference between the main-channel and flood-plain velocities.

Longitudinal velocity components in the main-channel are very much higher than those on the flood-plain. Furthermore, it shows that, for an equal water depth, the velocity difference between rising and receding stages is relatively smaller on the flood-plain as compared with that in the main-channel.

According to the velocity variations in unsteady compound open channel flows presented by Tu H. et al. (1994(b)), there is a slight variation of longitudinal velocity components across the main channel section in rising stage but in receding stage, it vanishes. The present investigation, with vegetated flood-plain shows that it is highly non-uniform for both rising and receding stages. Moreover, this velocity gradient across channel section, close to the interface is quite larger than that in unsteady compound open-channel flows.

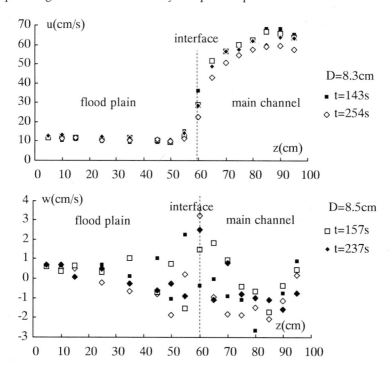

Fig.4. Variations of u- and w- across the channel section at y=7cm

From the cross sectional distribution of w-component, basically channel section can be divided into three parts as z=0 to 30cm on the flood-plain, z=30cm to 75cm (interactive contact zone) and z=75cm to 100cm in the main-channel. In contact zone, positive w in the rising stage, indicates that the flow is towards the flood-plain direction, whereas in the receding stage it is towards the opposite direction. It seems that flow pattern reverses from rising to receding stages.

CONCLUSIONS

From the results presented in this paper, it may be concluded that the roughness of the flood-plain due to vegetation increases the flow resistance and rises the water level on the flood-plain (excluding the interactive zone). Close to the interface, strong eddies may have been produced which hamper the main channel flow. It leads to lower surface elevation.

Longitudinal velocity close to the interface is significantly smaller compared with that of other regions in the main channel. Roughness elements decrease the flow over the flood plain and it enhances the difference between main channel and flood-plain velocities.

Unlike in the smooth compound channel flows, velocity distribution in the transverse direction is highly non-uniform in both rising and receding stages and velocity gradient close to the interface is quite high.

It indicates that in the contact zone, flow pattern reverses from rising to receding stage, and outside of the contact zone, there is a clockwise secondary flow in the main-channel.

REFERENCES

NUDING A. (1994), "Hydraulic resistance of river banks covered with trees and brushwood", 2nd International conference on River Flood Hydraulics, York, England pp427-438

TOMINAGA A., I.NEZU and M.NAGAO (1993), "Hydraulic characteristics of flow in compound channels with rough flood-plain", Proceedings of XXV Cong. of IAHR, pp89-96

FUKUOKA,S. and FUJITA,K. (1989), "Added flow resistance of flood flow due to lateral velocity discontinuity", Proc. of JSCE, vol.411, 63-72

KNIGHT, D.W. and Hamed, M.E (1984), "Boundary shear in symmetrical compound channels", J.Hydr.Eng., ASCE, vol.110, 1412-1430

NALLURI,C. and JUDY,N.D.(1985); "Interaction between main channel and flood plain flow", Proc. of XXI IAHR Cong., 378-382

KAWAHARA Y. and N.TAMAI (1989), "Mechanism of lateral momentum transfer in compound open channel flows", Proc.of the XXIII IAHR Cong., Ottawa, Canada, Vol.B, pp.B463-B470

TU H., N.TAMAI and K.KAN(1994(a)), "Unsteady flow velocity variations in near an embayment", Proc.of Hydraulic Eng., JSCE, Vol.38, pp.703-708

TU H., N. TAMAI and K. KAN (1994(b)), "Velocity variations of unsteady compound open channel flows", Proc.of 9th APD-IAHR Congr, Vol.2, pp.417-424.

1C3

Velocity Measurements in Unsteady Compound Open-Channel Flows

TU H., B. L. JAYARATNE, N. TAMAI, *Dept. Civil Engrg., Univ. of Tokyo*
and K. KAN, *Dept. Civil Engrg., Shibaura Instit. of Tech., Japan*

INTRODUCTION

Compound channels are widely seen in the middle- and lower-reaches of many natural rivers. Since flood plains are dry areas during low discharge seasons, they are explored either as entertainment parks, or even exclaimed by peasants as fertile farming fields in the case of some large rivers. When flood comes, however, amusement centers or crops are sometimes inundated. It is due to flood-relieving needs or environmental concerns that flows in a channel with flood plains (compound-channel flows) have drawn increasing attention in recent years.

In a laboratory flume, Rajaratnam and Ahmadi (1981) measured and investigated in great detail velocity variations across a compound channel section. Kawahara and Tamai (1989), using an algebraic stress model along with the k-ε model, calculated the 3-D flow fields and studied the mechanism of the momentum transfer between the main channel and the flood plain. Knight et al. (1994) reviewed past investigations on shear stress distributions in compound channel flows. One common conclusion from these studies is that, due to momentum exchange between the flood plain flow and the main channel flow, the velocities on the flood plain are increased compared to the otherwise undisturbed ones (if without the compound section), while the velocities in the main channel are reduced.

These studies, together with others, have provided necessary solutions for many practical problems that may be encountered in natural rivers. However, there are floods in natural rivers. When a flood occurs, there is a not only a big change of water depth at the transition of the flood plain and the main channel, but also the fact that the flow now becomes unsteady.

Unsteady flow experiments, however, are very difficult to undertake. There is, first of all, the lack of proper facility for unsteady flow generation in a conventional hydraulics laboratory, in addition to the fact that it is very much time consuming to conduct unsteady-flow experiments. In the present investigation, detailed velocity measurements across a compound channel, which was made in a well-equiped large tilting flume.

EXPERIMENTAL PROCEDURES

The same tilting flume as described in our previous studies (Tu et al. 1994a) was used. The flood plain here, however, was made of aluminium plates, except for the area (180cm long, which was covered by wooden boards to avoid possible disturbance on the electromagnetic probes from the aluminium plates) where the measuring section was located. The flood plain is 5cm high and 60cm wide, while the main channel is 40cm wide (Fig.1). Together, these simulate half of a symmetric compound channel. Depth and velocity measurements were conducted at a section 12m from the flume entrance, which guarantees a fully developed flow for the experiments. With a computer and relevant software, hydrographs of desired shapes can be easily created by controlling the opening of a valve in the pipe system. In the present study, due to time limit only one hydrograph (Fig.2) was investigated in the flume.

Water depth, D, was measured using limnimeters. The three velocity components were measured with two electromagnetic probes (by TKC), both of a diameter of 5mm. One is of I-type (SFT-200-05), for measuring the longitudinal and the transverse velocity components (u and w); the other is of L-type (SFT-200-05L), for probing only the vertical velocity component (v). The flume's bottom slope was constant (1/2000) for this study.

The water depths were measured at 16 positions across the channel section, namely, at (see Fig.1): z=5cm, 10cm, 15cm, 25cm, 35cm, 45cm, 50cm, 55cm, 60cm, 65cm, 70cm, 75cm, 80cm, 85cm, 90cm and 95cm. In each of these 16 positions, two of the three velocity components (longitudinal and transverse) were taken every 5mm in the vertical (up to a maximum of 15 points), the first point being 1cm from the channel bottom (below which the sensors can not probe the velocity fields correctly). While the vertical velocity component, considered not negligible near the interface, was taken at z=50cm, 55cm, 60cm, 65cm, 70cm and 75cm, also every 5mm in the vertical, from y=3.5cm to y=7cm.

During the passages of the hydrograph, velocity measurements were conducted at two different positions (at a certain water height) by using in parallel the two probes, which were then exchanged and the same procedures repeated. The sampling rate was 5Hz, and the five readings per second were averaged to render the time-mean velocities, though no information on the turbulent structure.

Since there were many measuring points and the flume is an outdoor facility, the experiments should be finished as soon as possible. So the velocities at one point were measured by passing the hydrogrpah only once. The water depth measurements were carried out separately from the velocity measurements. All the data were recorded with a NEC computer for later treatment.

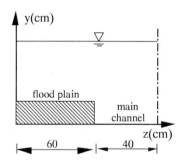

Fig.1 A Sketch of the Measuring Section

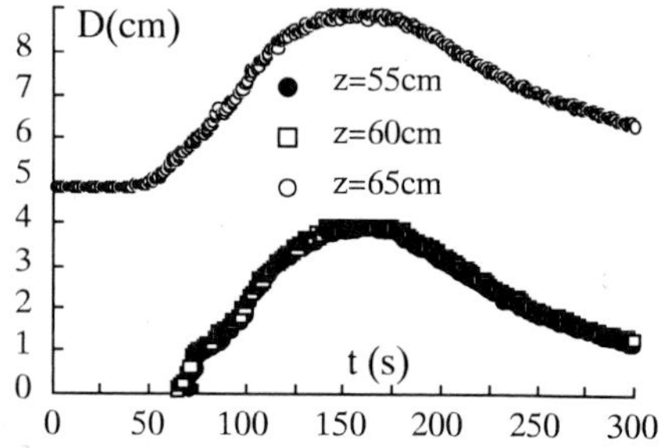

Fig.2 Depths Measured on the Flood Plain, at the Interface and in the Main Channel

EXPERIMENTAL RESULTS

WATER-DEPTH VARIATIONS

Shown in Fig.2 are the depths measured at z=55cm (on the flood plain), z=60cm (at the interface) and z=65cm (in the main channel). The one at the interface (referenced to the top of the flood plain) is a litter larger than that on the flood plain (about 5mm's difference, unrecognizable in Fig.2). However, there is no visible time lag among the peaks of the depths in the present investigation.

EVOLUTION OF THE VERTICAL VELOCITY PROFILES

Since the vertical velocity component, v, is very small even near the interface, and also due to space limit, in Fig.3 are presented only the profiles of the longitudinal and transverse velocity components, u and w, at selected time instants (covering thus the whole process of the hydrograph). For clarity, only those measured at three positions ($z=65$cm, close to the interface; $z=80$cm, in the main channel's center; and $z=95$cm) are shown. Not given here are the profiles on the flood plain, from which there are only limited data points.

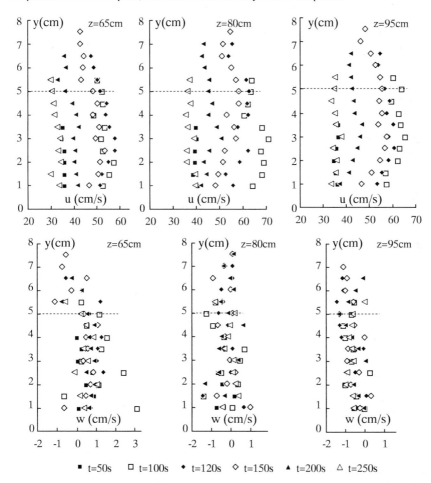

Fig.3 Vertical Profiles of u and w; at Selected Time Instants; Dotted Line Indicates the Flood Plain's Height (Note that positive w-values indicate flows to the flood plain direction)

We shall first examine the profiles of the u-component. One sees from Fig.3 that the velocities close to the interface (z=65cm) are smaller than those at z=80cm or at z=95cm. And for the profiles at each position (z=65cm, 80cm and 95cm here), it is observed that velocities are also reduced close to, and above, the flood plain height. This is considered as due to lateral momentum exchange.

As for the w-profiles, note that positive w-values designate flows toward the flood plain, while negative ones to the main channel direction. At the main channel's center (z=80cm), the average of all w-values is almost zero. Close to the interface (z=65cm) and below the flood plain height, w-values are positive; above the flood plain height, however, negative w-values become dominant, implying a clock-wise secondary flow. At z=95cm, w-values are negative virtually from the bottom to the surface, indicating flows toward the flume's right hand bank.

COMPARISON OF VELOCITY PROFILES IN THE RISING AND RECEDING STAGES

To examine the differences between the rising and receding stages, the vertical profiles for both the u- and w-components, all taken at D=7cm, are given in Fig.4. First for the u-profiles, even with the same water depth, velocities in the rising stage are about twice of those in the receding stage. It can be seen here too that the velocities close to the interface (z=65cm) are smaller compared with those at z=80cm or at z=95cm. Though velocity distribution below the flood plain height seems to follow log law, above the flood plain height it does not (this is also true at z=80cm and z=95cm), due to probably momentum exchange after the flood plain is inundated.

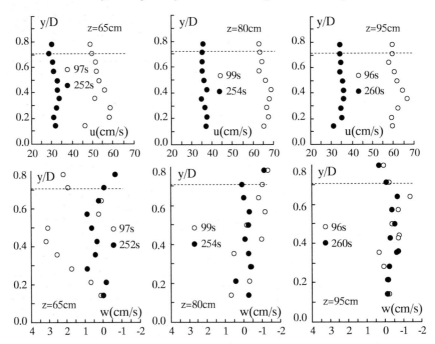

Fig.4 Vertical Profiles of u and w, in the Rising and Receding Stages; Dotted Line Indicates the Flood Plain's Relative Height; Depth D=7cm for All the Profiles

As for the w-profiles, at z=80cm and z=95cm, there seems little difference between rising and receding stages, both having about the same negative w-values. Close to the interface (at z=65cm), however, w-values are relatively large and positive in the rising stage, implying a somewhat strong secondary flow toward the flood plain. While in the receding stage, w becomes much smaller even close to the interface.

TEMPORAL VELOCITY VARIATIONS
Shown in Fig.5 are the temporal variations of the u-and w- velocity components, at z=65cm, 80cm and 95cm, all taken at y=5cm (the same height as the flood plain's). It is clearly seen that for the u-component, it is the largest at z=80cm, at the main channel's center; smallest at z=65cm, close to the interface. The velocity difference is the most evident near the hydrograph's peak. This further confirms our previous conclusions (Tu et al. 1994b). For the w-component, as has been discussed above, it is on the average zero at z=80cm; positive at z=65cm, implying flows toward the flood plain direction; and negative at z=95cm, indicating lateral flows toward the flume's right hand bank (see Fig.1).

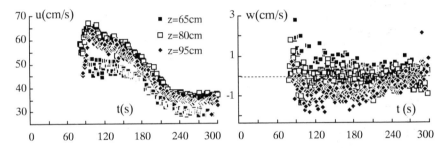

Fig.5 Temporal Variations of u and w, for z=65cm, 85cm and 95cm; All Data Taken at y=5cm

TRANSVERSE VELOCITY VARIATIONS
In order to see how the u- and w-components vary across the channel section, four sets of data, corresponding to D=7.5cm and D=8cm, in both the rising and receding stages, are shown in Fig.6. Note that all the data are taken at y=6cm, i.e., at the same height (and 1cm above the flood plain) across the channel section.
First, as would be expected, the u-values in the main channel are about twice of those on the flood plain. For an equal water depth (take D=8cm for example), the velocity difference on the flood plain between the rising and receding stages is relatively small as compared with that in the main channel. On the other hand, across the channel section, velocity distribution is highly non-uniform in the rising stage, while becoming more uniform in the receding stage.
From the cross-sectional distribution of the w-component, all the data shown in Fig.6, in the rising or in the receding stages, might be roughly divided into 4 zones across the channel section, although there exists certain discrepancy. In the first zone, 0< z <20cm, the average of w-component is almost zero, hence secondary flow influence from the interface seems rather limited here. In the second zone, 20cm< z <45cm, with w-values decreasing and in the third, 45cm< z <60cm, with w-values increasing, the transverse velocity component is negative. In the fourth zone (the main channel), w-values become again zero.
Note, however, that Fig.6 only shows the data from y=6cm, thus this division into 4 zones may not be considered as valid for the data measured at other heights. The simplest way would

be dividing the channel section into two zones, flood plain zone and main channel zone; and in the vertical direction, also into two zones, the one below the flood plain height (when there is only the main channel, which corresponds to the low stage flows seen in natural rivers), and the one above the flood plain height (including both the flood plain and the main channel).

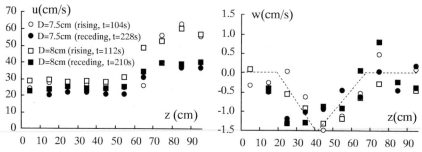

Fig.6 Variations of u and w across the Channel Section, at y=6cm; (Note that positive w-values indicate flows to the flood plain direction, and vice versa)

CONCLUDING REMARKS

Water depths and velocities across a compound channel section, under unsteady flow conditions, have been measured in detail. From the results presented in this paper, it may be concluded that: 1) velocities (mainly the longitudinal component) are reduced close to the interface between the flood plain and the main channel, this being particularly evident at about the flood plain's height; 2) results of the transverse velocity component (w) indicate that, in the main channel and close to the interface, there is a secondary flow toward the flood plain, which is much stronger in the rising stage; while at the center of the main channel the w-component is almost zero; 3) as for the vertical velocity distribution, log law seems to be valid only below the flood plain height; 4) for an equal depth, the longitudinal velocity component's distribution across the compound channel section is highly non-uniform in the rising stage, while it becomes much more uniform in the receding stage. Note, however, that the conclusions drawn here may or may not be valid for other hydrographs. It remains to be further investigated by studying more hydrographs of different shapes, including natural ones.

REFERENCES

KAWAHARA Y. and N. TAMAI (1989), "Mechanism of lateral momentum transfer in compound channel flows", Proc. of the 23th IAHR Congress, Ottawa, Canada, Vol.B, pp.B463-B470.

KNIGHT, D. W. and K. W. H. YUEN and A. A. I. AL-HAMID (1994), "Boundary shear stress distribution in open channel flow", in Mixing and Transport in the Environment, Ed. by K. J. BEVEN, P.C. CHATWIN AND J. H. MILLBARK, John Wiley & Sons LtD., pp.51-87.

RAJARATNAM, N. and R.M. AHMADI (1981), "Hydraulics of channels with flood plains", J. of Hydr. Res., Vol.19, No.1, pp.43-60.

TU H., N. TAMAI and K. KAN (1994a), "Unsteady flow velocity variations in and near an embayment", Proc.of Hydraulic Eng., JSCE, Vol.38, pp.703-708.

TU H., N. TAMAI and K. KAN (1994b), "Velocity variations in unsteady compound open-channel flows", Proc.of 9th APD-IAHR Congr, Vol.2, pp.417-424.

1C4

EXPERIMENTAL STUDY OF FLOWS IN EMBAYMENTS

Y. KAWAHARA*, K. NAKAGAWA*, K. KAN**
* Dept. of Civil Engineering, University of Tokyo
Hongo, Bunkyo-ku, Tokyo 113, Japan
** Dept. of Civil Engineering, Shibaura Institute of Technology
Shibaura, Minato-ku, Tokyo 113, Japan

INTRODUCTION

River improvement works employing conventional engineering methods may destroy the natural healthy environment in riparian areas and may cause adverse effects on the ecosystem. To prevent monotonous artificial environment and to preserve high level of natural condition in riparian areas, construction of embayments along the shore of rivers to provide ecosystem with shelters, foods and nutrients is highly needed.

Embayments along a river bank are basically dead-water regions. Such regions not only occur frequently in small dimension in natural rivers with irregular side and bed configuration, but also in large dimension in streams passing groynes, bays and harbors. In all the cases, mass exchange between the dead water zone and the main stream is of prime importance because it controls the water quality in both regions. Hence, prior to actual construction of man-made embayments, quantitative evaluation of water quality in embayments is essential. This requires a sufficient knowledge of the flow characteristics in an embayment and at its entrance.

The flow in an embayment is complex because of separation, recirculation, complex eddy structures and so on. Owing to the complexity of the flow, a limited number of studies have reported mean velocity field in similar geometric configuration (Booij,1991; Tingsanchali-Chinanont,1991; Langendoen,1992; Jalil et al., 1993,1994). Evidently, the flow complexity increases during flood events since the flow in a submerged embayment shows strong three-dimensionality and the vegetation surrounding it makes flow and sediment transport more complex. The understanding of flood flows is crucial to take measures for maintaining the functions of embayment at low water depth.

It has been shown that the depth-averaged flow model coupled with the k-ε model can reproduce the mean flow field in a square embayment with different entrance shape under low water conditions (Jalil et al.,1994). More experimental data are, however, necessary to verify the performance of a depth-averaged flow model because the flow in a rectangular embayment shows different recirculation pattern from that in a square case (Kawahara et al.,1994).

Thus this study has the objective to present the detailed measurement of mean velocity distribution in embayments with different shape, expecting to offer the reference data for numerical models. Due to the space limitation, this paper focuses on the experimental results under shallow water conditions.

EXPERIMENTAL SETUP AND CONDITIONS

Experiments were carried out in a 100 cm wide tilting flume having a compound cross section (Fig. 1). The width of the main channel was 40.0 cm and the height of the floodplain was 5.0 cm. The embayment built up in the floodplain had the width 20 cm and the length being changed. The bed slope of the flume was set at 1/2,000 for all the experiments. Table 1 summarizes the experimental conditions. The impermeable plate of 1.0 cm thickness was introduced along the channel-embayment interface to partially block the opening of the embayment. Case-6 had two lines of model tree whose diameter was 2 mm, the height 3.4 cm and the interval being 2.5 cm. Table 1 shows the total discharge (Q), the average velocity in the main channel (U_m), the water depth at the main channel center (H_m) and the water depth at the embayment center (H_e).

Water depth was measured by point gauge. The distributions of longitudinal and

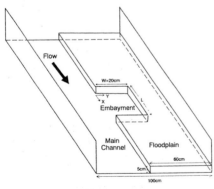

Fig. 1 Perspective view of the flume.

Table 1 Experimental conditions.

CASE	Geometry of Embayment (cm)	Q (ℓ/s)	Um(cm/s)	Hm(cm)	He(cm)
1	W=20, L=20	5.93	32.28	4.69	4.63
2	20, 60	3.88	26.80	3.63	3.62
3	20	3.99	27.48	3.67	3.63
4	40	3.90	27.23	3.58	3.54
5	40	3.78	26.32	3.61	3.59
6	40 vegetation	2.95	23.12	3.19	3.19

transverse velocity components were measured using I-type and L-type of two component electromagnetic currentmeter. The probe of the currentmeter was 5 mm in diameter and 1.8 cm in height. The sampling rate was 10 Hz and the sampling time was 1 minute. The measurements were started after the establishment of a steady state flow in the channel. The measuring stations were mainly concentrated in the embayment and its entrance.

RESULTS AND DISCUSSIONS
MEAN VELOCITY FIELD
Case-1 has a completely open interface and a square embayment. Fig. 2 compares the time-averaged velocity vectors at three different levels which located at 30 %, 50 % and 80 % of the water depth from the bottom. The large scale of recirculation in each plane is almost the same over a large portion of the embayment, although the instantaneous velocity shows three-dimensional features in particular near the downstream edge of the embayment. This result offers the grounds for depth-averaged flow models. The recirculation center is located at the downstream of the embayment center and nearer to the channel. Secondary recirculation is absent in Fig. 2 because the currentmeter probe is most likely too large to detect it.

In Case-2 the width to length ratio was 1/3 and the interface was entirely opened. The flow pattern is shown in Fig. 3. Two recirculation zones are formed. The larger gyre locates in the downstream zone and the smaller one at the upstream corner. A row of vortices, generated due to flow instability, develops along the interface.

In Case-3, one third of the interface was blocked from the upstream to model the effect of sediment deposition at the entrance. The measured results are depicted in Fig. 4. The impermeable plate restricts the size of the larger gyre whereas the other one expands behind the plate.

Fig. 5 shows the flow pattern in Case-4. In this case two thirds of the interface was blocked from the upstream. The main recirculation zone becomes smaller than that in Case-3 and its center shifts downstream. A calm region spreads over a half of the embayment.

Drastic change in the flow pattern occurs when the central part of the interface is blocked, which is Case-5. Fig. 6 shows that flow comes into the embayment from the upstream opening and goes out through the downstream interface. A strong and anticlockwise gyre is formed at the upstream portion of the embayment while a large recirculation develops behind the blocking wall. The third recirculation zone is present at the downstream corner although its size and the velocity are small. The flow pattern indicates active mass and momentum transfer in the

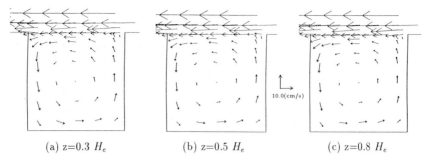

(a) z=0.3 H_e (b) z=0.5 H_e 10.0(cm/s)   (c) z=0.8 H_e

Fig. 2   Velocity distribution in Case–1.

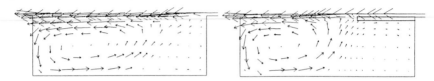

Fig. 3   Velocity distribution in Case–2.   Fig. 4   Velocity distribution in Case–3.

Fig. 5   Velocity distribution in Case–4.   Fig. 6   Velocity distribution in Case–5.

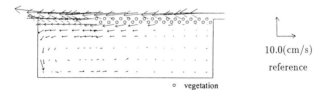

10.0(cm/s)

reference

o   vegetation

Fig. 7   Velocity distribution in Case–6.

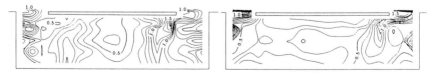

Fig. 8 Distribution of $\sqrt{\overline{u^2}}/U_*$ in Case–5.   Fig. 9 Distribution of $\sqrt{\overline{v^2}}/U_*$ in Case–5.

embayment.

In Case-6, shown in Fig.7, the impermeable wall was replaced with two lines of model tree. The difference in flow pattern among Case-2, Case-4 and Case-6 is perceptible. In this case the development of vortices along the interface is suppressed by the model vegetation. And the mass and momentum transfer occurs through the vegetated zone. Hence the size of the recirculation region lies between Case-2 and Case-4 while the magnitude of velocity is smaller than those in Case-2 and Case-4. The serene water zone is formed at the upstream region of the embayment.

TURBULENT FLOW FIELD

The currentmeter used in this study can not capture the full feature of turbulence but may offer its qualitative information. Fig. 8 gives the distribution of the 'turbulent intensities' u' and v' in Case-5 as examples. High intensity regions spread near the interface and in high velocity zone of the upstream gyre.

MOMENTUM TRANSFER

The momentum transfer between the main channel and the embayment is examined. Fig. 9 compares the convective and diffusive transfer rates measured along

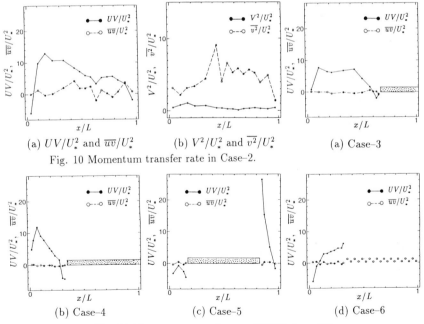

(a) UV/U_*^2 and $\overline{uv}/U_*^2$ (b) V^2/U_*^2 and $\overline{v^2}/U_*^2$ (a) Case-3

Fig. 10 Momentum transfer rate in Case-2.

(b) Case-4   (c) Case-5   (d) Case-6

Fig. 11 Momentum transfer rate UV/U_*^2 & $\overline{uv}/U_*^2$.

the line 2.5mm inside the embayment and parallel to the interface in Case-2. The positive values of UV and $\overline{uv}$ mean the momentum transfer from main channel to embayment. The friction velocity U_* in the figures is defined as $\sqrt{gH_mI}$. It is apparent that the values of $\overline{v^2}$ is much larger than that of V^2 and the quantities UV and uv have the same order of magnitude at the upstream part where the streamwise velocity shows the maximum gradient, although careful attention is needed to the results because of the resolution of currentmeter. Fig. 11 presents the results of UV and $\overline{uv}$ in Case-3 through Case-6. The diffusion term varies little and takes a small value in all the cases. In Case-5 a large amount of momentum is transported into the embayment through the upstream interface. The behaviors of convective transport in Case-4 and Case-6 show clear contrast. These results lead to the importance of a highly accurate convection scheme in the numerical simulation with a depth-averaged flow model.

CONCLUSIONS

The following conclusions can be drawn from the present study.
1. The mean flow is nearly two-dimensional over a large portion of the embayment when the water depth is small and a depth-averaged calculation is justifiable.
2. The geometry of an embayment and the entrance shape have considerable influence on the flow characteristics, such as the number, size and position of gyre in the embayment.
3. The convective transport play an important role in the momentum transfer through the interface between the main channel and the embayment.

ACKNOWLEDGMENTS

The authors gratefully acknowledge the financial support offered by the Maeda Memorial Engineering Foundation and the Foundation for River Environment Management.

REFERENCES

1. Booij,R.: Eddy viscosity in a harbour, Proc. XXIV IAHR Congress, Madrid, Vol.C, 81-90 (1991).
2. Jalil,M.A., Kawahara,Y., Tamai,N. and Kan,K.: Experimental investigation of flow in embayment, J. Hydraulic Eng., JSCE, 37, 503-510 (1993).
3. Jalil,M.A., Kawahara,Y. and Tamai,N.: Experimental and numerical investigations of flow in an embayments, Proc. 9th Asian & Pacific Div. IAHR, 223-230(1994).
4. Kawahara,Y., Jalil,M.A. and Tamai,N.: Momentum transfer in man-made embayments at low water depth, Proc. China-Japan Bilateral Symp. on Fluid Mechanics and Management Tools for Environment, Beijing, 106-113 (1994).
5. Langendoen,E.J.: Flow patterns and transport of dissolved matter in tidal harbors, Report No. ISSN 0169-6548, Delft University of Technology (1992).
6. Tingsanchali,T. and Chirananont,B.: Investigation of flow circulation in a channel side pool, Proc. Int. Conf. on Environ. Hydraulics, Vol.1, 467-472 (1991).

1C5

TURBULENT STRUCTURES OF FREE SHEAR LAYER OVER CAVITY USING FIBER-OPTIC LDA

Iehisa NEZU, Akihiro KADOTA and Hiroji NAKAGAWA

Department of Civil and Global Environment Engineering,
Kyoto University, Kyoto 606, Japan.

INTRODUCTION

An open-channel flow over discontinuous boundaries is often observed in the downstream of sand waves on river bed and man-made hydraulic structures such as water gates, weirs and trenches. Especially, the trench is installed in river in order to control sediment transport. It is very important in hydraulic engineering to investigate the structures over the discontinuous boundaries, because the turbulence production and dissipation become larger due to the rapid changes of shear stress.

Some investigations on coherent vortices produced at separation point, i.e., the upstream edge of trench, have been conducted in the recent years. Knisely and Rockwell (1982) have measured the cavity shear layer by using LDA and they carried out the visualization of coherent vortices by means of the hydrogen-bubble technique. They suggested that the total turbulence intensity was reduced by as much as 50%-80% comparison with step flow and they found the peak of low-frequency component which was affected by mean velocity and length of cavity layer in addition to the fundamental frequency f. Yagi (1984) has conducted the visualization by using a dye injection technique and examined instability of coherent vortices. It was also suggested that the peak of low-frequency was functions of mean velocity, length of cavity layer and thickness of separated shear layer. Recently, Fujita et al. (1994) have measured the open-channel flows with a trench by particle image velocimetry (PIV) which uses the correlation method and they examined the momentum and vorticity distribution.

These studies, however, have paid attention only to the structure of coherent vortices but did not examined statistical quantities and shear stress. In the present study, the turbulence measurements in and over the cavity were conducted by making use of a two component fiber-optic laser

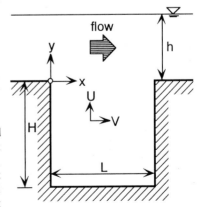

Fig. 1 Coordinate system

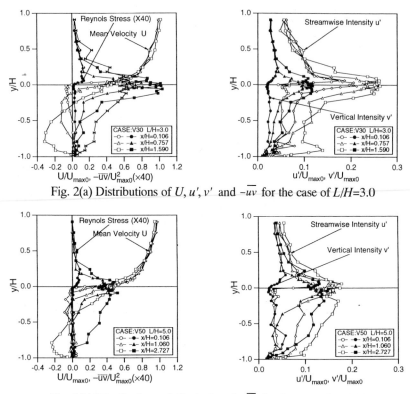

Fig. 2(a) Distributions of U, u', v' and $-\overline{uv}$ for the case of $L/H=3.0$

Fig. 2(b) Distributions of U, u', v' and $-\overline{uv}$ for the case of $L/H=5.0$

Doppler anemometer (FLDA). Statistical structures, shear stress, exchange process of momentum and pressure distributions for comparing with backward-facing step flows were investigated in detail.

EXPERIMENTAL EQUIPMENT AND PROCEDURES

The experiments were conducted in a $10m$ long, $40cm$ wide and $50cm$ deep tilting flume. A trench configuration which was made of iron plate was settled and adjusted smoothly on the bed at $7m$ distance from the upstream. The depth of cavity H and the water depth h were $6.6cm$ from the bed and $6.6cm$ from the trench, respectively (Fig.1). The channel bed was a flat and smooth wall. In order to examine the basic structure of cavity flows, 6 cases of length of cavity L ($L/H=0.5,1,2,3,5,7$) were set and the discharge Q was constant at $Q=2.5(\ell/s)$.

Two components of instantaneous velocities ($\tilde{u}$, $\tilde{v}$) were measured with four-beam and Argon-Ion fiber optic LDA system (DANTEC-made). This system is connected to a personal computer in order to control automatically the frequency shift, bias and so on. For measurements near the bed or free surface, the angle of beam was adjusted parallel to the boundary so that the instantaneous velocity very near the wall could be measured

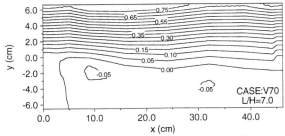

Fig.3 Streamlines of $\Psi /(U_{max}H)$ in cavity flows

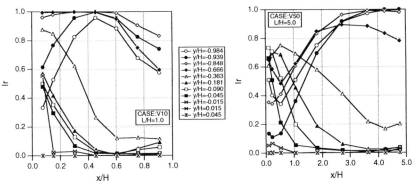

Fig.4 Time rate of inversed velocity I_r

accurately. Output signals from FLDA were recorded in a digital form with average sampling frequency 100 Hz and sampling time 60 sec. Statistical analyses were conducted then by using a workstation.

RESULTS AND DISCUSSION
Mean Velocity
and Turbulence Characteristics

Fig. 2(a) and (b) show mean velocity profiles U, turbulence intensities u', v' and Reynolds stress

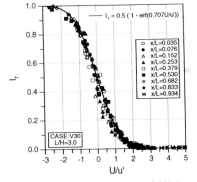

Fig.5 I_r as a function of U/u'

$-\overline{uv}$ for two different lengths of cavity, respectively. The origin, $x=0$, is defined at the upstream corner of cavity. These values are normalized by the maximum mean velocity U_{max0} at $x=0$. The streamwise mean velocity U immediately in the cavity become negative and reversed near the upstream edge. Then, away from the upstream edge, the region of reversed velocity ($U<0$) become wide and the location of inflection point approaches to the bed. Outside of the cavity ($y>0$), the Reynolds stress $-\overline{uv}$ corresponds

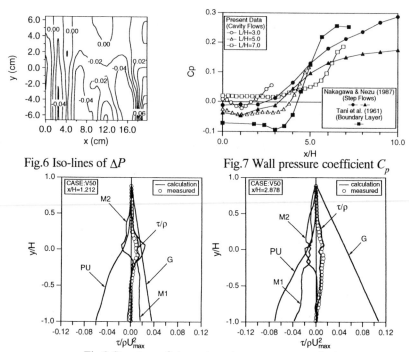

Fig.6 Iso-lines of ΔP

Fig.7 Wall pressure coefficient C_p

Fig.8 Structure of shear stress in momentum analysis

to the slope of U and seems to be triangular distribution which is satisfied in uniform flows, irrespective of x. Of particular significance is that the Reynolds stress becomes larger in the shorter length of cavity. There may be a criterion of L for the characteristics of turbulence. Yagi (1984) also suggested that the fundamental frequency f of coherent vortices increased rapidly and spatial growth rates of turbulence become larger in the some region of L. The same feature as Reynolds stress distribution is found in the turbulence intensity u',v'. Furthermore, the u' and v' show the same order in the longer length of cavity and the isotropic tendency can be seen.

The streamlines Ψ normalized by U_{max} and H for the longest cavity length are shown in Fig. 3. The negative region which indicate the circulation is presented in the downstream side of cavity. The streamline of $\Psi>0$ is almost parallel to the bed. The Ψ in cavity is different from that in the backward-facing step flow investigated by Nezu and Nakagawa (1987). They found the reattachment point ($\Psi=0$ and $y=-H$) and suggested that the reattachment point was about $x/H=5$. In contrast, there is no reattachment in the cavity flow. Fig. 4 shows the time rate I_r of reversed velocity in the total sampling time for streamwise direction. It is seen that the value of I_r is larger in the downstream of cavity and it becomes almost unity in the longer L. Fig. 5 shows the distribution of I_r against U/u'. The theoretical curve in the case of Gaussian distribution is also shown in the figure. It slightly deviates from the curve around $-1.0<U/u'<1.0$ and indicates that the $u(t)$ also deviates from Gaussian distribution.

Dynamic Pressure Distribution

The mean pressure P is calculated from integration of the Reynolds equation, as follows:

$$\frac{P}{\rho} = g(h - y) \cdot \cos\theta + \frac{\Delta P}{\rho} \quad (1) \quad , \quad \frac{\Delta P}{\rho} = \int_{y}^{h} \frac{\partial(UV + \overline{uv})}{\partial x} dy + \left[V^2 + \overline{v}^2\right]_{y}^{h} \quad (2)$$

in which, θ is the channel slope. The term $g(h - y) \cdot \cos\theta$ in Eq.(1) is the hydrostatic pressure, while ΔP is the dynamic pressure which is caused by the velocity variation. Fig. 6 shows iso-lines of the dynamic pressure which was calculated from Eq.(2) by using the measured values. The ΔP is normalized by $\rho U_{max}^2 / 2$. The high positive pressure is presented in the downstream of cavity where the reversed flow is more significant. In this figure, the different feature from the step flow is recognized. It is seen that the pressure which causes the circulation region is formed. The wall pressure coefficient $C_p \equiv 2(P - P_0)/\rho U_{max}^2$, where P_0 is the hydrostatic wall pressure at $x=0$, is shown in Fig. 7. The step flow data obtained by Nakagawa and Nezu (1987) is also included. The same variation of C_p as the step flow is indicated in the upstream region of cavity, while it jumps immediately before the opposite side of cavity. It suggests that the pressure distribution is not relaxed by the free surface due to the complex configuration of cavity.

Shear Stress Distribution

The shear stress $\tau/\rho = -\overline{uv} + \nu \cdot \partial U / \partial y$ can be obtained by integrating the Reynolds equation, as follows:

$$\frac{\tau}{\rho} = gI_e(h - y) - \int_{y}^{h} \frac{\partial U^2}{\partial x} dy - \left[UV\right]_{y}^{h} - \int_{y}^{h} \frac{\partial}{\partial x}\left(\frac{\Delta P}{\rho} + \overline{u^2}\right) dy \quad (3)$$

$$= \quad G \quad +M_1 \quad +M_2 \quad +PU$$

where, M_1 and M_2 are momentum terms. PU is the contribution of dynamic pressure and turbulence. Fig. 8 is variations of each term in Eq. (3). The measured values of shear stress are also included. In comparison with the other terms, the gravity term G is dominant over the cavity. The shear stress at $y>0$ shows almost linear distribution as mentioned before. On the other hand, the terms PU and M_1 cannot be neglected any longer inside of the cavity. However, the G cancels the PU and M_1. Consequently, the total shear stress becomes almost zero.

Exchange Process of Momentum and Mean-Flow Energy

Integration of Eq.(3) using the Green-Gauss theorem for the control volume leads to (Nakagawa and Nezu 1987):

$$[M]_1^2 = Volume \cdot g \cdot \sin\theta - \int_{1}^{2} \frac{\tau_0}{\rho} dx, \qquad M = \int_{-H}^{h}\left(U^2 + \overline{u^2}\right) dy + \frac{g}{2}h^2 \cdot \cos\theta \quad (4)$$

$$[H]_1^2 + \frac{1}{gU_m h}\left\{\iint Gen \cdot dx\,dy + \left[\int_{-H}^{h}\left(\left(\overline{u^2} - \overline{v^2}\right)U + \overline{uv}V\right) dy\right]_1^2\right\}, \quad H = \frac{\alpha U_m^2}{2g} + h \cdot \cos\theta \quad (5)$$

401

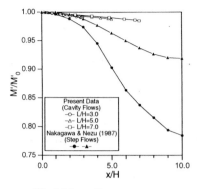

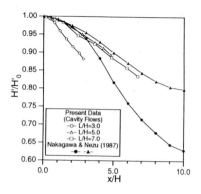

Fig.9 Mean-flow momentum   Fig.10 Mean energy

where, α is the energy correlation coefficient. Figs. 9 and 10 show the exchange process of momentum and mean-flow energy, respectively. The integration of the dynamic pressure was abbreviated in Eq. (4) because it is negligibly small. In order to describe the variation of flow depth clearly, the hydrostatic pressure term at $x=0$ was subtracted, as follows: $M' \equiv M - (gh_1 \cdot \cos\theta)/2$ and $H' \equiv H - h \cdot \cos\theta$ in Figs. 9 and 10, respectively. In comparison with step flow data, the momentum H decreases more slowly and uniformly. There is no difference by the length of cavity. However, the mean-energy decreases more rapidly in the cavity with the shorter length L. It is considered that the decay of mean-flow energy is almost same magnitude as the increase of generation, as seen in the step flows.

CONCLUSION

In this study, the turbulence measurements were conducted by making use of FLDA in the cavity shear flows and statistical structure such as mean velocity and Reynolds stress, dynamic pressure and shear stress were examined. The significant difference between the cavity shear flow and the step flow were seen clearly.

REFERENCES

1) Knisely, C. and Rockwell, D. (1982), J. Fluid Mech., vol.116, pp.157-186.

2) Yagi, S. (1984), Doctoral Dissertation, Kyoto University. (in Japanese)

3) Fujita, I., Kanda, T., Kadowaki, M. and Kaizu, T. (1994), 26th Symp. on Turbulence, pp.96-99. (in Japanese)

4) Nakagawa, H. and Nezu, I. (1987), J. Hydr. Res., IAHR, vol.25, no.1, pp.67-88.

1C6

FORM DRAG OF A SINGLE RECESS IN A FLOW BOUNDARY

BY: J M JORDAAN, JNR.   and   S P SCHUTTE
     Chief Engineer, Design Services, Final Year Student,
     Department of Water Affairs and   Civil Engineering
     Forestry, Pretoria, RSA   and   Department, University of
     Professor (part time, Civil Engineering   Pretoria, Pretoria, RSA.
     Department, University of Pretoria.

SUMMARY
A full scale flume-mounted model of a typical formed recess in a segmentally lined tunnel was instrumented and flush-mounted in the false bed of a hydraulic test channel. For prototype values of flow velocities, dynamometric measurements were made of the form drag induced by the flow on such an individual recess in order to integrate its contribution to the hydraulic flow resistance of an entire segmentally lined tunnel with (unfilled) recesses. These results obviated possible scale effects implied in the parallel scale-model tests (separately reported on) or the need for costly full-scale tunnel tests. Investigations were also made regarding the mutual influence of two and three such recesses in line, in order to establish whether any interference, and hence any additional cumulative effects due to the regularity of the repeating pattern of recesses, would take place.

It was found that the flow disturbance by a single recess was largely contained within itself and close to it. Control tests were also carried out in which the recess was filled in and its *surface* drag was measured (to be compared with the *form* drag of the unfilled recess). A significant *difference* was obtained, but due to the small aggregate surface area occupied by the recesses (about 1,5%) it was found that the gross effect over the tunnel lining was small. This agreed with the scale model tests separately reported on, and leads to the conclusion that *leaving the recesses unfilled would in fact save on capital cost and will not imply significant additional energy losses*. The quantitative results will be presented at the Congress.

THEORETICAL CONSIDERATIONS
The shapes of the recesses are either wedge-like or trapezoidal in form. Any individual recess allows the flow to expand momentarily within it and then again to recombine with the main flow (Fig. 2). A separation eddy would occur at A, the flow-dividing streamline would impact at B and a secondary eddy might form at C after which the main flow would be re-established and resumed at D. A pressure field would be set up over a small area at B adding to the drag forces, causing added hydraulic resistance and hence energy losses.

A succession of recesses, in line and parallel to the flow direction, could set up a wavy separation streamline near the surface (Fig. 3) that may or may not cause interference effects. Were the recesses sufficiently close together, the absolute hydraulic roughness, K_r in the resistance equation due to Colebrook and White (Rouse, 1950), could be readily associated with the depth of the recess, while were they far apart, the isolated increase in K_s or an effective value $K_{r\ (eff)}$ would only occasionally be superimposed on the much lower K_o for the smooth tunnel lining. A recess, of course, has a much smaller effect on adding to hydraulic friction than a projection of equal dimensions has, due to the reduced velocity gradient in the boundary layer which is less disturbed by a recess than by a projection.

A valid consideration would be to directly determine the drag force on each recess due to the flow field's expansion, reattachment and any successive separation including possible mutual interference. It is likely that the initial disturbance of each recess would make the largest contribution to the drag force and that wake-separation and mutual interference with the next recess would have less effect. At each recess a three-dimensional flow field would be set up (contrary to what the previous 70:30 criterion in the first paper presupposes). For laminar boundary flow it becomes self-healed immediately downstream, but for turbulent boundary flow, it could set up a lengthy turbulent wake. Measuring and integrating the pressure field on the interior surface of the recess, or directly determining the overall form-drag force on the recess, enables the net hydraulic resistance to be determined (form-drag force minus reduction in skin-friction drag, due to the absence of the portion of the cylindrical tunnel surface over the recess).

As a control, and as a direct measure of the *skin friction* drag to be subtracted from the *form* drag, the recess was then filled-in on the model, and the drag force then also directly determined. From theoretical considerations, involving boundary layer and "point of impact of free streamline" assumptions, a force of 1-2 N is estimated per recess under prototype flow velocities of 2m/s. Integrated over the tunnel interior this would increase the hydraulic head loss by 5 to 10%.

From the above it may be deduced a *priori* that the head loss increase could likely be between 0 and 10% for the unfilled recesses.

DESCRIPTION OF EXPERIMENTAL PROCEDURE

During the *full scale* tests B models of individual longitudinal and transverse recesses were suspended in depressions in the false floor of the channel and roller-mounted so as to be unrestrained in the flow direction, (Fig 5) except for restraining nylon lines attached to it and to a weight on a weighbridge mounted above the flume, which allows the drag force induced by the flow across the modelled recess to be directly measured. The models were lightly joined by flexible membranes to the false channel floor on all sides so that only longitudinal drag forces had any influence on the weighing system. Forces were determined for various flow velocities (made equal in magnitude to prototype velocities) under free-surface flow conditions. Three aspect attitudes were be examined: *longitudinal* (with taper along and against the flow) and *transverse* (with taper at right angles to the flow). The forces thereafter were also determined for the said attitudes with the model recesses infilled by a filler of the same density as water (to ensure that all forces other than the drag forces remain unaltered): the drag forces thus determined on the "unfilled" and "filled" recess models were then compared directly. As a result the net force was thus determined, and by summing that over the number of recesses per kilometre of tunnel, the Darcy-Wasbach friction coefficient f obtained for both a "smooth surface" tunnel and a "recessed surface" tunnel. From this determination of f the "effective roughness" k-value was obtained; and also, by formula substitution, the Manning "n" value was obtained by standard textbook methods. (Webber, 1971).

DESCRIPTION OF APPARATUS
The tests were conducted in a flow circuit and flume at the Civil Engineering Laboratory of the University of Pretoria (Fig.4 and Photo's 1 and 2) as was described in the previous paper.

The flow circuit consists of A, a pipeline and B, a flume (in which full scale models made of each of two types of recesses are installed i.e. the *tapered* recess tested in both directions (longitudinal to the flow) and the *trapezoidal* recess, tested tranverse to the flow).

The pipeline and the flume form connecting parts of the flow circuit leading from a sump through two portable submersible pumps via flexible tubing, flow control valves, a manifold, a magnetic flow meter. It leads then into the tunnel leg component, from there into a channel head box, then through a tiltable canal section to a Vee-notch weir and tailrace box, and back by flexible pipe connection to the sump.

SIGNIFICANCE OF FINDINGS
The results of tests A and B (suitably scaled) are now directly compareble. If they are in reasonable agreement the hypothesis: **whether hydraulic losses of recesses left unfilled will significantly differ from that of recesses filled-in,** will be answered. If there is a significant experimental or error-induced difference between the results obtained by means of tests A and B it will be necessary to determine the causes thereof and to allow for it, so as to obtain a significant common and conclusive result regarding the main hypothesis i.e. whether infilling can be profitably dispensed with.

ACKNOWLEDGEMENTS
The facilities of the University of Pretoria, Civil Engineering Department's hydraulic laboratory, was made available for student research projects. The support received from the faculty, staff and the students is acknowledged. The permission of the Director-General, Department of Water Affairs and Forestry to present the paper is acknowledged gratefully.

REFERENCES
Ackers, P. and Pitt, J.D., 1982. Hydraulic roughness of segmentally lined tunnels. Report No 96, C.I.R.I.A., London.

Colebrook C.F., 1958. The flow of water in unlined, lined and partly lined rock tunnels, Proc. Inst. CE, London, paper 6281.

Jordaan, J.M., Jnr, 1992. Hydraulic roughness increase and decline of the capacity of large diameter water tunnels, Proc. Tuncon 92, SANCOT, Maseru, Lesotho.

Metcalf, J.R., 1986. "Tunnel Roughness Report", Memo, H2 - 01, TCTA, HDTC, Internal communication. Randburg, RSA.

Experimental Equipment for the Full scale Recess Model tests, showing left to right: the approaching flow conduit, reducer and piping to head box with sluice gate (inside), glass walled channel for housing false floor and modelled recesses. Slope adjustment crank handle visible left of centre.

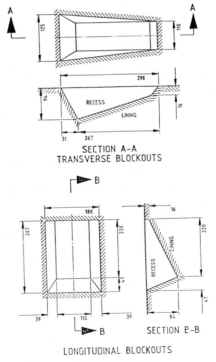

SECTION A-A
TRANSVERSE BLOCKOUTS

SECTION B-B

LONGITUDINAL BLOCKOUTS

FIGURE 1   : SHAPE OF RECESSES

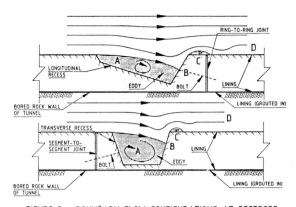

FIGURE 2 - BOUNDARY FLOW CONFIGURATIONS AT RECESSES

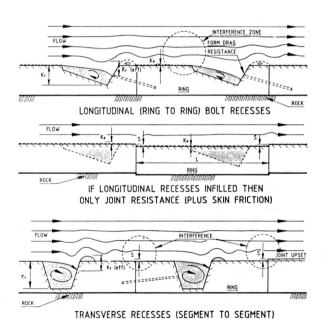

LONGITUDINAL (RING TO RING) BOLT RECESSES

IF LONGITUDINAL RECESSES INFILLED THEN
ONLY JOINT RESISTANCE (PLUS SKIN FRICTION)

TRANSVERSE RECESSES (SEGMENT TO SEGMENT)
(POSSIBLE FLOW INTERFERENCE)
FIGURE 3 - FLOW CONFIGURATIONS AT RECESSES WITH
REFERENCE TO JOINT UPSETS AND INTERFERENCE

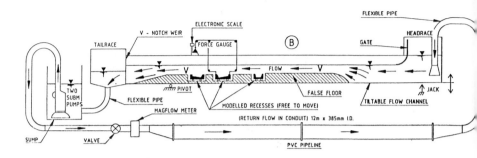

FIGURE 4   : EXPERIMENTAL SETUP FOR : B - THE FULL-SCALE RECESS TESTS
(SCHEMATIC)

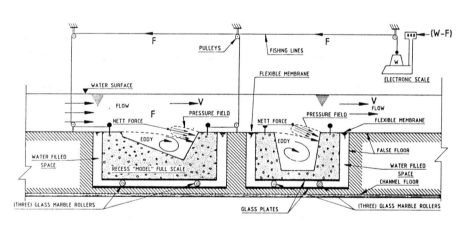

FIGURE 5 - DETAIL OF EXPERIMENTAL SETUP FOR DRAG
FORCE F MEASUREMENT IN OPEN CHANNEL
(ON LEFT) FULL SCALE RECESS MODEL, AND SIMILAR
FOR (RIGHT) RECESS MODEL, AND FOR INFILLED CASES)

1C7

Experimental Study on Internal Structures of Ultra-Rapid Flow (On Roll-Wave Characteristics)

AKIRA MUROTA*, MASAHIRO MIYAJIMA*, KOHJI MURAOKA**
* Osaka Sangyo University, Daito, Japan
** Osaka University, Osaka, Japan

1. INTRODUCTION

It is well known that thin sheet flow in steep slope channels has the phenomenon of self excited, periodical Roll-Wave trains if Froude number exceeds approximately 2 (Ultra-Rapid flow). In the engineering and environmental field, it is important to clarify the characteristics of the Roll-Wave so as to gain information for estimating channel design and to better understand the natural process of the movement of sediment.

Many papers exist concerning Roll-Wave generation: V.Vedernikov 1), Iwasa 2), Tamada 3) and so on. However, few papers have addressed the Roll-Wave's hydraulic characteristics that proceeds the generation phase, with the exception of Dressler 4), Isihara 5), Iwagaki and Iwasa 6), Vide 7) and Murota and Miyajima 8). It has been difficult to measure the pertinent data since water depth is shallow, flow velocity is high, and wave change is tremendous and rapid in time series and spatially. Furthermore, only our experiments have measured velocity, real time data of water changes and velocity, and clarified visualization of the Roll-Wave flow.

This paper attempts to clarify Roll-Wave characteristics: the behaviors and structures of the mean and internal flow. To pursue these experiments non-intrusive equipment was used: an ultra sonic levelmeter, a Laser-Doppler anemometer and an ultra frame speed camera.

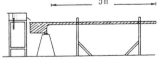

Figure 1 Experimental channel

2. EXPERIMENTAL CHANNEL AND METHOD

The experiments were carried out by using an open channel made of clear acrylic, 5m long by 20cm wide, with a variable slope device as shown in Figure. The measuring instrument was installed 3.8m from the upper stream and the velocity and water level were measured. Velocity was measured 0.5mm ~ 1.0mm from the channel bed to the wave crest. The photographs of

Roll-Wave fronts were taken under the condition of Run 15 of Table-1.

3. EXPERIMENTAL CONDITIONS

Table 1 shows the experimental conditions. Channel slope was controlled from 1/6 to 1/30 and the discharge rate was approximately 0.5 l/s to 1.3 l/s. This experiment was carried out in the condition of 3<Fr<7, 2000<Re<6000.
In this experiment: ho=(hmax+hmin)/2, Fr: Froude number; Re: Reynolds number.

4. ROLL-WAVE ASPECTS AND SCALE

A time series profile of a Roll-Wave is shown in Figure 2. It is observed that the water level spikes at the Roll-Wave front and then decreases smoothly until the next spike. Photo 1 indicates a Roll-Wave front sample taken by an ultra frame speed camera. The Roll-Wave front shown in this photo appears like a surging breaker and the jump

Table-1 Experimental conditions

Run No.	Channel Slope S	Discharge rate Q(l/s)	Mean flow depth ho(mm)	Froude number Fr	Reynolds number Re
1	1/5.97	0.504	2.5	6.5	2100
2	1/5.97	0.951	3.5	7.2	4100
3	1/8.25	0.648	3.1	6.0	3000
4	1/8.25	1.11	4.2	6.5	5100
5	1/8.25	1.34	4.7	6.7	6000
6	1/12.7	0.936	4.7	4.7	4000
7	1/12.7	1.18	5.1	5.2	5000
8	1/21.4	0.722	4.5	3.8	3200
9	1/21.4	0.921	5.5	3.7	3900
10	1/21.4	1.10	5.9	3.9	4800
11	1/31.4	0.858	5.3	3.6	3400
12	1/31.4	1.01	6.7	3.0	4100
13	1/10.2	1.01	4.7	5.1	5200
14	1/21.1	1.01	5.3	4.2	4200
15	1/10.0	0.840	4.2	4.9	4500

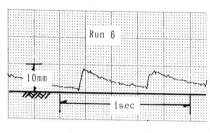

Figure 2 Roll-Wave's time series profile

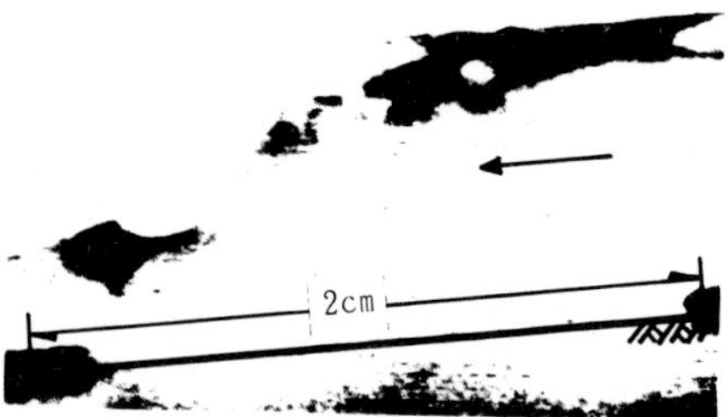

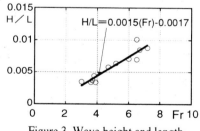

Photo 1 Roll-Wave front

point in Figure 2 corresponds to the Roll-Wave front shown in Photo 1.

The ratio of hmax/hmin within the experimental range increases to approximately twice to 3 times in accordance with the Fr's increase.

Figure 3 shows the relationship between the average wave steepness (H/L) and the Fr and it indicates that wave steepness increases in rela-

H/L=0.0015(Fr)-0.0017

Figure 3 Wave height and length scale

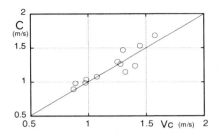

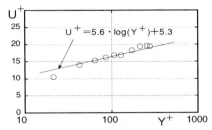

Figure 4 Observed wave celerity and
water velocity

Figure 5 Mean velocity distribution
(Run 7)

tion to the increase of the Fr.

5 .MEAN FLOW DISTRIBUTION

5.1 WAVE CELERITY AND MEAN WATER VELOCITY OF WAVE CREST

In Figure 4, the vertical axis is observed wave celerity C and the horizontal axis is LDA's mean velocity data Vc at the wave crest. From the viewpoint of coincidence of velocity and celerity at the wave crest, since C and Vc closely correspond, it is considered that at the Roll-wave front some breaking condition such as a surging breaker is satisfied.

5.2 MEAN VELOCITY DISTRIBUTION

Figure 5 shows mean velocity distribution by U^+ and Y^+. Here $U^+= \bar{u}/U_*$; $U_*= \sqrt{ghoS}$, $Y^+=U_* z/\nu$, g: gravitational acceleration; S: channel slope, ν ; coefficient of kinematic viscosity.

According to Figure 5, within this experimental range, Roll-Wave velocity distribution in a very shallow flow is quite similar to the log-law distribution of turbulence. This suggests that Roll-Wave flow also forms a mean velocity field where log-law distribution occurs and it is able to be treated as an open channel Uniform flow.

6. INTERNAL STRUCTURES OF ROLL-WAVE

6.1 VISUALIZATION OF FLOW FIELD

By using an ultra frame speed camera, Roll-Wave flow fields are observed under the specifications of 1125 frames/sec. As a tracer, polystyrene particles of 1.04 specific

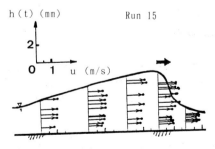

Figure 6 Velocity profiles of Roll-Wave
flow field (Run 15)

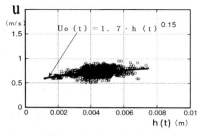

Figure 7 Water depth and velocity (Run 14)

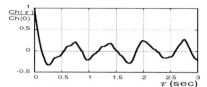

Figure 8-1 Auto correlation coefficient of h(t)

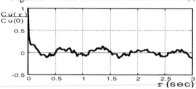

Figure 8-2 Auto correlation coefficient of u

gravity approximately 200 μ m diameter were used, and photos were taken 1.0cm from the side wall. It is observed that velocity profiles at the Roll-Wave front are considerably disturbed and in the whole field, the body of the Roll-Wave flow appears to be strongly influenced by the channel bed.

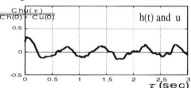

h(t) and u

Figure 8-3 Cross correlation coefficient

6.2 WATER DEPTH CHANGES AND MAIN FLOW VELOCITY (2mm from the bed)

Figure 7 shows the raw data of water depth changes in relation to velocity at 2.0mm from the bed. As the water depth increases, the velocity also increases in a clear pattern. Mean velocity Uo(t) is given as a power function of water depth h(t). Figure 7 suggests that the variance of turbulence fluctuations has different parameters depending on water depth.

6.3 CORRELATION COEFFICIENTS OF WATER DEPTH CHANGES AND VELOCITY

The data from the oscillation of water level and the fluctuation of velocity measured at 1.0mm from the bed is discussed here. The number of collected samples are 1024 and sampling time is 0.01sec. The results are shown in Figure 8-1~8-3 (Run 13). The vertical axis is auto and cross correlation coefficients, and the horizontal axis is lag time. The oscillation of water level has a remarkable periodical pattern and the oscillation pattern is very stable (shown in Figure 8-1). The auto correlation coefficient of main velocity also indicates a remarkable periodical pattern and its tendency coincides with the pattern of oscillation of water level (Figure 8-2). Figure 8-3 shows the cross correlation of water depth h(t) and velocity u. This result indicates that oscillation of water depth almost coincides with the time series pattern of the fluctuation of velocity and clarifies the strong relationship between water depth and the fluctuation of velocity.

6.4 TURBULENCE DISTRIBUTION

Figure 9 shows the relationship between water depth and velocity fluctuations. In figure 9, the vertical axis is the absolute of deviation u' (u'=u-Uo(t) shown in Figure 7) made dimensionless by Uo(t), the horizontal axis is water depth fluctuation made dimensionless by measuring location z. The scatter plot appears very characteristic in shape, triangular in type. By dividing the plot into 3 areas in accordance to the water depth difference: maximum water depth hmax, mean water depth ho and minimum water depth hmin, each turbulence scatter trend shows considerable difference according to area. Relative turbulence intensity is slightly less than 8% at the minimum water depth, it is 10% at the mean water depth and 9% at the maximum water depth. It is considered that the effect of free surface at a very shallow water depth strongly influences turbulence behavior.

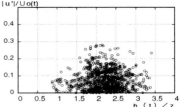

Figure 9 Water depth and velocity fluctuations (Run 14)

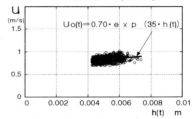

Figure 10 Water depth and velocity (at ho from the bed)

In this experiment, when we use Uo(t) as a typical velocity of flow, Reynolds number (Uo(t) h(t)/ ν) is approximately 2200, 3600 and 4800 when corresponding to a water depth of 3mm, 4.5mm and 5.6mm, respectively. Therefore, we think that at maximum water depth, Roll-Wave fronts generate relatively large turbulence with rapid growth. At mean water depth, active turbulence develops and in shallow water, turbulence is influenced by the shallow water depth and/or free surface effect.

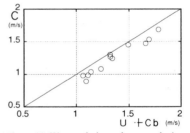

Figure 11 Water celerity and mean velocity

6.5 WATER DEPTH CHANGES AND MAIN FLOW VELOCITY (at ho from the bed)
Velocity behaviors that occurred at the mean water depth point are shown in Figure 10. The trend in figure 10 is quite different from the trend shown in figure 7. The fit-curve in figure 10 shows a concave curve compared to figure's 7 convex curve. This phenomenon suggests that there is no water bed friction near this area, and therefore, it is considered that Roll-wave must exist on mean flow.

7. MEAN STRUCTURES OF ROLL-WAVE
By analyzing the relationship between oscillation of water depth and main velocity from Photo 1, Figure 7 and Figure 10, it is assumed that Roll-Wave flow is a surge which progresses on mean velocity flow U. U; Q/(B ho), B; Channel width. Formula (1) and (2) follow :

$$C=U+C(bore) \text{ -----------------------------(1)}$$
$$C(bore)= \sqrt{g\ ho\ hmax/hmin} \text{ ---------(2)}$$

The results are shown in Figure 11. Figure 11 is a relationship between observed wave celerity and U+C(bore). This figure indicates the coincidence of C and U+C(bore); the Roll-Wave is considered as a surge which progresses on the mean flow.

8. CONCLUSION

The two main characteristics of the Roll-Wave flow structures in a steep slope channel are first, the body of the whole flow is strongly influenced by bed friction. Second, Roll-Wave front is formed by surging. We consider that the flow is formed by a balance between a body's bed friction and the surging at the Roll-wave front. We now plan to carry out more detailed experiments so as to research the dynamic characteristics of Roll-wave flow and fluctuation in a desire to contribute to the engineering and environmental fields.

Acknowledgments
The experiments discussed in this paper were performed by an ultra-frame high speed camera that Prof. T. Etoh's group of Kinki University had developed. The Laser-Doppler anemometer was obtained from a Ministry of Education's grant and various instruments were obtained from an Osaka Sangyo University's subsidy.
The authors are very grateful for the above support.

References
1) Vedernikov,V: "Condition at the Front a translation wave disturbing a steady motion of a real fluid", Comptes rendus(Doklady)de'l'Academie des Sciences de l'U.R.S.S., No4, Vol 48, pp 239-242, 1945
2) Yosiaki Iwasa: "The criterion for instability of steady uniform flow in open channels", Memoirs of Faculty of Engineering, Kyoto University, Vol.16, No.6, pp 264-275, 1954
3) Ko Tamada,Hirofumi Tougou: "Stability of Roll-Wave on Thin Laminar Flow down an Inclined Plane Wall", Journal of the Physical Society of Japan, Vol.47, No.6, December, pp 1992-1998, 1979
4) Dressler.R.F.: "Mathematical solution of the problem of roll-waves in inclined open channels", Communication on Pure and Applied Mathematics,Vol.2, No. 2/3, pp 149-194, 1949
5) Tojiro Ishihara and so on: "Theory of the Roll-Wave Trains in Laminar Water Flow on a Steep Slope Surface" , Trans., of JSCE, Vol.19, pp 46-57, 1954 (in Japanese)
6) Yuichi Iwagaki,Yoshiaki Iwasa: "On the Hydraulic Characteristics of the Roll-Wave Trains", Proceeding of JSCE, Vol.40, No.1, pp 5-12 (in Japanese)
7) Vide J.P.M.: " Open channel surges and roll waves from momentum principle", Journal of Hydraulic Research, Vol.30, pp.183-196, 1992, No.2
8) Akira Murota,Masahiro Miyajima: "Experimental Study on Internal Structures of Ultra-Rapid Flow (Mainly on Mean Flow and Wave Characteristics)", Proceedings of Hydraulic Engineering, JSCE, Vol37, pp 563-568, 1993 (in Japanese)

1C8

SHEAR CELL FOR DIRECT MEASUREMENT OF FLUCTUATING BED SHEAR STRESS VECTOR IN COMBINED WAVE/CURRENT FLOW

A J GRASS, R R SIMONS, R D MACIVER, M MANSOUR-TEHRANI
and A KALOPEDIS
University College London, London, UK.

INTRODUCTION

The superposition of tidal and wave induced currents is a typical feature of the coastal zone, producing a hydrodynamic environment of great complexity. These currents interact in a highly non-linear manner to produce turbulent boundary layer velocity and instantaneous bed friction force vectors that fluctuate both in magnitude and direction. Given this complexity, the development of effective analytical and numerical models for coastal process prediction must continue to depend heavily on data and observations from laboratory and field experiments for calibration and validation purposes.

Reliable prediction of seabed shear stress is a particularly important priority for coastal engineers wishing to estimate, for example, wave height attenuation and energy loss, current strength, sediment transport and patterns of sediment erosion and deposition, all matters of great practical concern. Time averaged bed shear stresses have in the past been indirectly inferred from near bed velocity profiles, wave height attenuation and water surface slopes. Whilst such data has proved valuable, development of more sophisticated models, particularly concerned with sediment entrainment and transport, requires information on instantaneous time variations in the local bed shear vector.

Measurements of fluctuating bed shear stress are very restricted to date primarily due to the inherent difficulties in designing suitable instrumentation. Reidel and Kamphuis (1973) used a shear plate device to measure wave induced stresses on rough beds in a two dimensional flume. More recently Arnskov *et al.* (1983) have reported measurements of the bed shear stress vector in a model wave basin using a surface mounted hot film element. The latter technique is however only suitable for hydrodynamically smooth bed conditions. The present paper addresses these measurement problems. It describes the successful development of a novel shear cell device, capable of measuring instantaneous bed shear stress vectors fluctuating both in magnitude and direction under the combined action of waves and currents on both smooth and rough beds.

THE UCL SHEAR CELL

The main design criteria for the shear cell were that it should be capable of measuring a rapidly and directionally varying horizontal force vector with sufficient sensitivity to resolve the small shear forces applied to smooth beds (typically 0.5 grammes force maximum on the shear plate, which highlights the design challenge) whilst having the capacity to cope with the much larger forces induced on extremely rough beds.

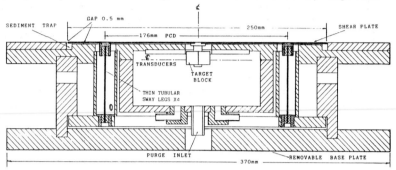

Figure 1: Cross-section of the UCL shear cell.

A sectional drawing of the UCL shear cell is shown in figure 1. The cell body is housed in a specially prepared bed cavity, such that its upper surface is flush with the local surrounding bed surface. A positive displacement approach is used whereby the applied shear forces cause the active shear plate to be displaced in the direction of the force vector. The shear plate consists of a circular disc of anodised aluminium, 250mm in diameter and 0.9mm thick, mounted flush with the top surface of the shear cell body and supported by four thin tubular legs. Bushes permanently fixed to the ends of the flexible legs are rigidly clamped to the shear plate at the top and the shear cell body at the base. Any horizontal force applied to the shear plate causes the legs to flex in sway mode and the shear plate to be displaced in the bed plane a distance directly proportional to the force. A 0.5mm clearance between the perimeter of the shear plate and the cell body was chosen to minimise the discontinuity effects on the bed flow, the chances of debris jamming the moving parts of the instrument and the magnitude of the displacement velocities. The cavities can be purged to remove contaminants. By careful choice of the stiffness of the support legs, the maximum displacement of the active element can be adjusted to less than 0.5mm under the action of the anticipated forces. The natural vibration frequency of the shear plate system, determined by its effective mass and the stiffness of the support legs, was designed to be at least an order of magnitude greater than the typical wave frequencies. The phase lag associated with the inertial response had also to be minimised. Keeping the shear plate mass to a minimum to achieve these system properties, whilst ensuring a very high degree of plate flatness proved to be the greatest challenge of the design process.

The shear plate is an interchangeable feature of the shear cell. Separate plates can be used, each with different roughness characteristics. In this way the instrument can be deployed in both rough and smooth boundaries.

The horizontal displacement vector of the shear plate is measured by two orthogonally mounted proximity transducers (Sensonics Eddy Current Probes) which produce an output voltage linearly related to the displacement. The transducers monitor the movement of a small steel block centrally mounted on the underside of the shear plate (see figure 1). The transducers are remarkably sensitive devices, having a linear response over an operating range of 1mm with a resolution of 0.1µm. Accurate calibration of the force/displacement relationship is achieved by inclining the shear cell at known angles and correlating the gravity force components acting on the known effective self-mass of the shear plate system (simulating the hydrodynamic forces) to the transducer output voltage.

COMPONENT FORCES

The displacement of the shear plate is not solely due to the shear stress acting on the top surface. So called Froude-Krylov forces, generated by wave induced pressure gradients, also act on the plate. These pressure forces are proportional to the effective plate volume. The plate thickness must therefore be kept to a minimum compatible with adequate structural stiffness. The plate area is governed by the need to maximise the total shear force vector. In addition, the pressure gradients also drive a small laminar Couette flow through the narrow parallel wall cavity, 0.5mm wide, between the shear plate and the shear cell body, generating a small shear stress on the underside of the shear plate. Account must be made of these forces when recovering the shear stress force from the measured total force vector.

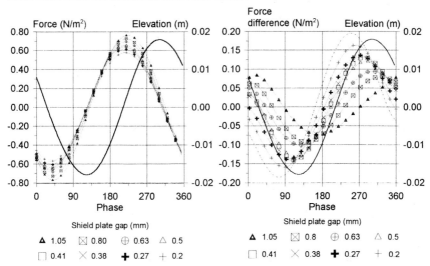

Figure 2: Wave induced total measured force as a function of the gap between the shield and the shear plate.

Figure 3: Difference between total force on unshielded and shielded shear plate as the gap is progressively reduced.

Preliminary calibration tests were carried out in a two-dimensional, smooth bed wave flume to investigate the interaction of the three contributions to the total measured force. The flume width was 450mm with a water depth of 300mm and a 1.18s monochromatic wave of height 36mm was used. A flat, sharp edged shield plate, 248mm in diameter and 0.9mm thick, was rigidly supported a small distance directly above the shear plate, progressively reducing the contribution of the shear stress on the top surface of the shear plate as the gap separating the two plates was reduced. As the gap tends to zero, the shear cell responds solely to the forces associated with the wave induced pressure gradients. The total force per unit area measured by the shear cell as the separation was reduced is shown in figure 2. The magnitude of the total force is seen to progressively decrease and most importantly, and as expected, the phase angle of the maximum force shifts towards that of a pure pressure force, 90° in advance of the wave crest (maximum velocity), as the gap decreases. For gaps of less than 0.5mm, the total force measured by the shear cell is closely in phase with the pressure force, indicating that for such small gaps the shear force on the upper and lower surfaces of the shear plate are closely in phase with the pressure gradient force. Hence, the shear force on the underside of shear plate can be interpreted as an effective increase in the thickness of the plate.

The trends in phase alignment of the forces is more clearly demonstrated by subtracting the measured total force obtained with the shield plate in place from a reference measured total force cycle obtained with the shield plate removed. The resulting force per unit area tends to the required shear stress on the top surface of the shear plate as the gap is reduced, with a peak magnitude 45° in advance of the wave crest (maximum velocity) in accordance with theory, as the separation tends to zero (figure 3).

In addition to the effective increase in the shear plate thickness, arising from the underflow, some reduction in the pressure force might also be expected due to possible pressure attenuation in the vertical 0.5mm gap surrounding the shear plate. Combining this attenuation effect and the effective increase in plate thickness into a single multiplication factor, to be applied to the Froude-Krylov pressure force, greatly simplifies extraction of the required shear stress force from the measured total force. The instantaneous total force vector measured by the shear cell can thus be represented by

$$F_{total}(t) = \tau_0(t) * \pi r^2 + \alpha * P(t) \qquad [1]$$

where $\tau_0(t)$ is the shear stress vector, $P(t)$ is the undistorted Froude-Krylov pressure gradient force, r the radius of the shear plate and α is the pressure correction factor. Rearranging gives

$$\tau_0(t) = \frac{F_{total}(t) - \alpha * P(t)}{\pi r^2} \qquad [2]$$

Thus to recover the bed shear stress from the measured total force it is necessary to calibrate the pressure correction factor and determine the undistorted instantaneous pressure force applied to the shear plate.

PRESSURE CORRECTION FACTOR CALIBRATION

For waves propagating over a smooth bed, linear wave theory provides a reliable solution for the amplitude and phase of the bed shear stress in terms of the velocity immediately outside the wave induced boundary layer

$$\tau_0(t) = \rho\sqrt{\eta\omega}\, u_0 \sin(\omega t + \pi/4) \qquad [3]$$

where ω is the wave frequency, u_0 is the velocity amplitude outside the boundary layer and ωt is the wave phase above the centre of the shear cell. The theoretical pressure force acting on the active element is given by

$$P(t) = 2rb\frac{\rho g H}{2\cosh kd} \int_{-\pi/2}^{\pi/2} \cos\theta \sin(kr\cos\theta)\, d\theta \cos\omega t \qquad [4]$$

where b is the thickness of the shear plate, k $(=2\pi/\lambda)$ is the wavenumber and H is the wave height. The pressure correction factor is obtained by optimising the agreement between the measured shear stress, recovered from equation [2] with equation [4], and the theoretical smooth bed shear stress, equation [3]. Values in the range 0.9 and 1.2 have been obtained for α depending on the wavelength used in the calibration.

By extracting the measured force and force difference values at particular phase angles from the traces presented in figures 2 & 3 respectively, the resulting data can be plotted as a function of the shield plate gap for fixed phase angles. Regression lines can then be plotted through the data points and extrapolated to obtain directly the pressure forces and shear forces as the gap tends to zero. This procedure has been applied to the data in figure 2 & 3 yielding a maximum shear stress of 0.188N/m^2 in excellent agreement with the theoretical value of 0.186N/m^2 established using the measured velocity just outside the wave boundary layer.

With a known pressure correction factor and velocity measurements immediately outside the wave induced boundary layer the shear stress can be recovered from the total force using equation [2].

SAMPLE EXPERIMENTAL RESULTS

For waves propagating over a smooth boundary, the estimates of the bed shear stress agreed with theoretical predictions to within 1-10%. This is felt to be satisfactory considering that, for a smooth boundary, the pressure forces are approximately a factor of two larger than the shear stress for the waves studied. At a rough boundary, the induced shear stresses are up to an order of magnitude larger than those at a smooth boundary and the error is therefore substantially reduced.

Measurements of the bed shear stress under combined wave and current on smooth and rough boundaries have been made with the shear cell in a complex three dimensional wave basin. Both monochromatic and random wave sequences propagating orthogonally over currents have been studied. A section of the time history of the recovered bed shear stress together with the near bed velocity of a random wave sequence on a rough bed is shown in figure 4.

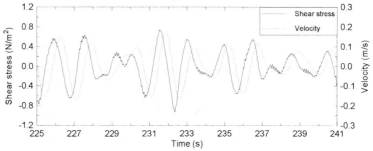

Figure 4: Wave induced velocity and cell measured shear stress for random waves.

This indicates excellent correlation between the velocity and force response registered by the shear cell. It is interesting to note that the shear force leads the velocity by approximately 75°. This suggests a dominant contribution from the Froude-Krylov forces acting on the bed roughness elements for this particular set of conditions.

CONCLUSIONS

A novel shear cell device has been developed and successfully tested and deployed. This instrument is capable of direct measurement of the time varying (in both magnitude and direction) bed shear stress vector induced by combined, non-colinear waves and currents. The shear cell can be used with both smooth and rough boundaries in an industrial laboratory environment.

Two instruments are currently in use at the newly opened UK Coastal Research Facility at Hydraulics Research Ltd., Wallingford. The research program forms part of a long term project supported by the UK Engineering and Physical Science Research Council (EPSRC). This includes measurements of the bed shear stresses under combined waves and currents at arbitrary angles, over a level bed and a sloping beach, for both smooth and rough boundaries. Initial results are reported by MacIver et al. (1995).

REFERENCES

. Arnskov MM, Sumer M and Fredsoe J (1993). Bed shear stress measurements over a smooth bed in three dimensional wave-current motion. Coastal Engineering, 20:277-316.

MacIver RD, Simons RR and Grass AJ (1995). Shear stresses and hydrodynamics of combined wave and currents. Submitted to Coastal Dynamics '95 Conference, Gdansk, Poland.

Reidel PH and Kamphuis JW (1973). A shear plate for use in oscillatory flow. J. Hydraulic Research, 11:137-156.

1C9

Flow-Induced Multiple-Mode Vibrations of a Prototype Weir Gate

PETER BILLETER
Laboratory of Hydraulics, Hydrology and Glaciology (VAW)
Swiss Federal Institute of Technology, Zurich, Switzerland

1. SITUATION

The Eglisau power plant is located on the river Rhine. It consists of a power house and a regulating weir with six openings, both lying on the same axis perpendicular to the flow direction of the river. The weir openings of width $L=15.5$ m are locked with double-leaf gates with a total height $H_0=12.74$ m (Fig. 1). The main bearing elements are horizontally latticed girders of parabolic shapes with a spanwidth of 16.94 m each. The gate leafs are lifted by means of suspension chains consisting of 10 chain-lamellae per link. The mass of the gate is 120 t for the upper leaf and 30 t for the lower one. The gate lip is furnished with an oak tree beam with width $d=0.34$ m. The gate is designed for either underflow or overflow or then for a combination of both operation states. The outflow under the gate is only submerged during flood periods, i.e. when the tailwater level is high and the opening s_u of the lower gate is small.

2. MEASUREMENT SET-UP AND PRELIMINARY INVESTIGATIONS

In order to assess the three-dimensional vibration modes, the gate leafs were equipped with 12 inductive accelerometers. The dynamic forces F_c on the suspension chains were measured by strain gauges which were attached to the rim of the chain lamellae. The optimum position of the strain gauges was evaluated on a physical model of one chain lamella and checked by Finite-Element-computation. It was found that the measured stress at the rim of the lamella is nearly proportional to the acting force (Fig. 1). Furthermore, the spring rigidity of the chain was estimated and an equivalent prismatic tensional rod of a rectangular cross sectional area A_S was determined. Interestingly enough, the ratio A_S/A_0 (A_0 = median cross section area of a chain lamella) is within the range of 0.21 to 0.22 for most chains used as gate suspensions.

The structural analysis of the gate showed two dominant groups of vibration modes: 1.) Rigid-mass oscillation of both the upper and the lower gate leaf with the spring force provided by the suspension chains. 2.) Flexural vibrations of the gate bodies which are two- or three-dimensional, continuous distributed-mass systems. For practical application, a restriction to the dynamic analysis of the lowest horizontal main girder was appropriate. The natural frequencies of both vibration modes were computed and compared to data collected at the prototype gate in medium air (Tab. 1).

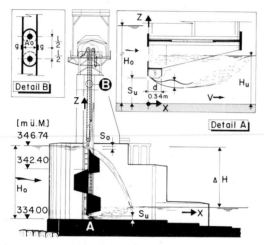

	z(t)	x(t)
$f_{n,c}$ [Hz]	5.32	28
$f_{n,m}$ [Hz]	5.3	30.5
β in air [%]	≈ 8.0	≈ 3.5
f_z, f_x [Hz]	4.6 - 5.5	22.2 - 21.1
V_{rz}, V_{rx}	7.5 - 9.0	1.96 - 1.86
Amplitude z_0, x_0 [mm]	0.5 - 1.7	0.1 - 0.16
Amplitude $\ddot{z}_0, \ddot{x}_0$ [m/s^2]	0.6 - 1.9	1.9 - 3.1
Amplitude Fc_0 [kN]	42 - 197	-

Fig. 1: Cross section of the Eglisau weir gate. Detail A: Area of the gate lip and of the lowest main girder. Detail B: Suspension chain of the gate bodies (g=gauges).

Tab. 1: Summary of the vibration data measured at the Eglisau weir gate. (notation see text and Fig. 1)

For this purpose, the lower gate leaf was excited by a kind of eccentric device which produced a periodical force. The device was attached either to the gate lip acting in the vertical direction, or to the lowest main girder acting in the horizontal direction. It was thus possible to determine the natural frequencies and the magnification factors as functions of the excitation frequency. The estimated frequencies $f_{n,c}$ agreed reasonably with the prototype data $f_{n,m}$. The viscous damping ratios β in air were deduced from the magnification factors of the extraneous excitation. Since these values were rather high, a considerable contribution of Coulomb-damping must be suspected.

3. INVESTIGATION UNDER OPERATION

The investigation took place during a flood period. The head difference ΔH varied between 9.5 m and 10.12 m, and the gate lip was submerged for small opening ratios of the lower gate ($s_u/d < 2$-3, depending on the opening of the upper gate s_0). Measurements were made for a wide range of possible gate leaf positions. In Fig. 2, the mean lift coefficient C_L of the lower gate is plotted versus s_u/d. With the upper leaf closed, C_L reaches a maximum at $s_u/d \approx 0.65$ and then decreases strongly. For the opening s_0 of the upper leaf between

2.0 m and 2.5 m, C_L shows an even stronger decrease. C_L is considered to be a stability indicator. A positive inclination of the C_L versus s_u-curve indicates a possibly unstable operation range (Thang, 1990). In fact, gate vibrations of the lower gate leaf were detected for underflow with $s_u = 0.36$ m ($s_u/d \approx 1.1$) and simultaneous overflow with s_o in the range of 0.9 m to 2.0 m. Further investigations were carried out with $s_u/d \approx 1.1$. ΔH was kept constant at 9.98 m and the outflow velocity v_0 was 14 m/s.

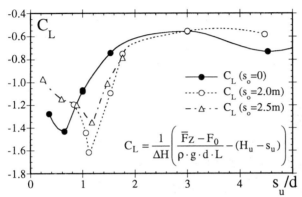

Fig. 2: *Mean lift coefficient C_L of the lower gate leaf plotted versus the opening ratio s_u/d for various openings s_o of the upper leaf.*

4. DESCRIPTION OF THE VIBRATION MECHANISM

In order to study the excitation mechanism, the dynamic system of the lower gate leaf was reduced to a single-mass oscillator with 2 degrees-of-freedom (z- & x-direction, see Fig. 1). For the hydrodynamic excitation force F_w, a linear approach was chosen with $F_{wz}, F_{wx} = f(x, \dot{x}, z, \dot{z})$. Eq. 1 is a possible solution for the equation of motion of the oscillator. F_w can then be written as Eq. 2:

$$z = z_z \cdot \sin(\omega_z t) + z_x \cdot \sin(\omega_x t) \; ; \; x = x_z \cdot \sin(\omega_z t - \varphi_{xz}) + x_x \cdot \sin(\omega_x t - \varphi_{xx}) \qquad (1)$$

$$F_{wz} = (v_0^2/2) \cdot \rho \cdot B \cdot d \cdot \{ C_{Fzz} \cdot \sin(\omega_z t + \phi_{zz}) + C_{Fzx} \cdot (\cos\omega_x t + \phi_{zx}) \}, \qquad (2a)$$

$$F_{wx} = (v_0^2/2) \cdot \rho \cdot B \cdot d \cdot \{ C_{Fxz} \cdot \sin(\omega_z t + \phi_{xz}) + C_{Fxx} \cdot (\cos\omega_x t + \phi_{xx}) \} \qquad (2b)$$

ϕ_{ii} is the phase shift between deflection and excitation force. All non-linear components of the self-excited system are included in the force coefficients C_{Fii}. Energy is transferred from the flow to the body if $0 < \phi_{ii} < 180°$.

The characteristic data of the vibrations is displayed in Tab. 1. A sample of time series and power spectra of both accelerations and forces is shown in Fig. 3. The vibration of the whole structure is dominated by a vertical rigid-mass oscillation of the lower gate leaf with the frequency f_z. f_z depends strongly on the opening of the upper gate (Fig. 4b). It is influenced by added mass effects which are caused by the varying submergence of the whole gate. The vertical rigid-mass oscillation of the lower leaf is being superimposed by streamwise flexural vibrations of the lowest horizontal main girder. Its basic frequency f_{x1} lies between the 4th and the 5th harmonic of f_z.

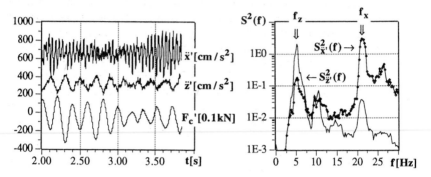

Fig. 3: *Typical time series and power spectra of the vibration mechanism.*
$\ddot{z}$ = vertical motion with f_z, $\ddot{x}$ = horizontal motion with frequency f_x,
$F_c{}'$ = suspension force; $s_u/d = 1.1$, $s_o = 1.5$ m , $(s_o+s_u)/H_o = 0.144$.

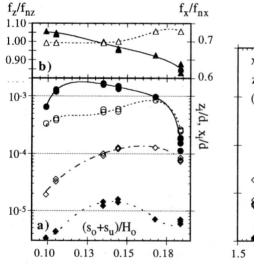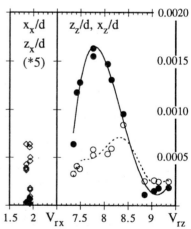

Fig. 4: a) *Relative amplitudes versus gate opening, notation see Fig. 5.* **b)** *Frequency ratio ($\blacktriangle$)=f_z/f_{nz}, ($\triangle$)=f_x/f_{nx}*

Fig. 5: *Relative vibration amplitudes versus reduced velocity V_{ri}. ($\bullet$)= z_z, ($\circ$)= x_z, ($\diamond$)= x_x, ($\blacklozenge$)= x_z.*

The frequency ration $f_x/f_{nx1} \approx 0.7$ agrees well with measurements and computations by Ishii & Knisley (1992). In Fig. 4a the relative vibration amplitudes z_i/d, x_i/d are plotted versus the relative opening $(s_o+s_u)/H_o$ of the gate. The phases shift φ_{xz} is always positive and φ_{xx} is always negative. This indicates that the gate has a press-open characteristic for f_z and a press-shut characteristic for f_x (see Eq. 1). Thus, the vibration x_x, z_x with f_x would still be sustained if no vibration with f_z occurred (Ishii & Knisley, 1992).

Fig. 5 shows the relative vibration amplitudes versus the reduced velocity $V_{ri} = v_0/(d \cdot f_i)$. For the z_z-oscillation, V_{rz} is in the range where the transition from instability-induced excitation (impinging vortices) to movement-induced excitation (galloping) occurs. An interaction between both excitation mechanism is thus expected (Naudascher & Rockwell, 1994). The horizontal oscillation is instability-induced due to impinging leading-edge vortices (ILEV) with n = 2 (Thang, 1990; Jongeling, 1988).

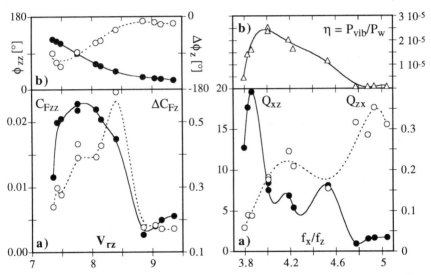

Fig. 6: *a) Force coefficients C_{Fzz} ($\bullet$) and ΔC_{Fz} ($\circ$) versus V_{rz}, b) phase angles ϕ_{zz} ($\bullet$) and $\Delta\phi_z$ ($\circ$).*

Fig. 7: *a) Magnification factors Q_{xz} ($\bullet$) & Q_{zx} ($\circ$) versus the frequency ratio f_x/f_z, b) effectiveness η ($\triangle$)*

The force coefficient C_{Fzz} for the vertical rigid mass oscillation with frequency f_z has a maximum for $V_{rz}=7.8$ and a phase shift $\phi_{zz}= +90°$ (Fig. 6). ϕ_{zz} runs from $+180°$ to $0°$. The vibration is therefore sustained over the whole range of V_{rz}. The coupling force between the vertical and horizontal motion for the frequency f_z is written dimensionless as $\{\Delta C_{Fz}\}=\{C_{Fxz}\}-\{C_{Fzz}\}$ with the phase angle $\Delta\phi_z$ ($\{X\}$ denotes a vector). A negative phase indicates that the horizontal motion is driven by the vertical one. The discontinuity at $Vr \approx 8.5$ is associated with a change in the flow pattern. It is assumed that the discontinuity marks the end of the range where the vertical galloping is influenced by impinging vortices from the horizontal motion. The increase of C_{Fzz} at $V_{rz}>9$ could be interpreted as the beginning of galloping without vortex influence involved.

Of further interest is the coupling between the dominant vibration modes with different frequencies f_z, f_x. In Fig. 7a, the magnification factors Q_{xz} ($\propto z_z/C_{Fzx}$) and Q_{zx} ($\propto x_x/C_{Fxz}$) are plotted versus the frequency ratio f_x/f_z. It seems that the horizontal vibration with f_x is amplified in a range close to the 5th harmonic of f_z, while the vertical vibration with f_x is sustained by the 1/4-subharmonic of f_x. Since at 10 - 11 Hz a peak - which could arise from period doubling of f_x - was found in all the power spectra, the above supposition is quite reasonable. The coupling with the 1/4th subharmonic is found to be much more efficient. The effectiveness η (ratio of the mean power consumed by the vibration and the water power supplied by the flow) reaches a distinct maximum at $f_x/f_z = 4$ and then decreases towards $f_x/f_z = 5$ (Fig. 7b).

It can be concluded that the following excitation mechanism is likely to occur:

I) The horizontal vibration x_x (f_x) due to an ILEV-mechanism is consistent and drives a vertical motion z_x with f_x. x_x and z_x are feeding turbulent energy to the free shear layer beneath the gate lip. The interaction of the shear layer with the gate lip is thus being fostered.

II) The vertical vibration z_z itself is caused by galloping-excitation. Between $V_{rz}= 7- 8.5$, the galloping is triggered and amplified by the impinging-vortex mechanism of the horizontal vibration (I). The coupling is highst for the 1/4-subharmonic of f_x.

5. FINAL REMARKS

The results of an extended prototype investigation carried out on a double-leaf gate were presented. The data provided a closer insight into the complex structural and fluiddynamic behaviour of the gate. Flow-induced vibrations with multiple degrees-of-freedom were detected for the lower gate leaf. For the analysis of the data collected, a simplified model of a linear single-mass oscillator was being used. A vibration mechanism was found, involving galloping in the crossflow direction and ILEV-excitation in the streamwise direction. Both vibration modes were strongly coupled. The vibration was removed by altering the gate lip geometry.

REFERENCES

ISHII, N., KNISELY, C.W. (1992), *J. Fluids & Structures*, **6**/6 p. 681.
JONGELING, T.HG. (1988), *J. Fluids & Structures*, **2**/6, p. 541.
NAUDASCHER, E., ROCKWELL, D. (1994), *Flow-induced Vibrations*, Balkema.
THANG, N.D. (1990), *J. Hydr. Eng.* ASCE,**116**/HY3, p. 342.

1C10

Wall Pressure Measurement in the Bottom Outlet of the Panix Gravity Dam

MICHAEL T. BENESCH
Laboratory of Hydraulics, Hydrology and Glaciology (VAW)
Swiss Federal Institute of Technology Zurich (ETH)

1. INTRODUCTION

In summer 1992, during the first impounding of the reservoir of the 53 m high concrete dam in the Panix valley, Switzerland, the Hydraulic Division of the Laboratory of Hydraulics, Hydrology and Glaciology at the Swiss Federal Institute of Technology Zurich was able to perform prototype measurements in the bottom outlet region. Fig. 1 shows a cross section of the Panix dam.

As already reported by Volkart and Speerli [1], water pressure near the service gate, air entrainment, air pressure and air velocity in the bottom outlet tunnel and the air concentration of the bottom outlet flow were measured at five different pressure heads and several different gate openings.

This paper is restricted to the water pressure fluctuations measured upstream and downstream the service gate.

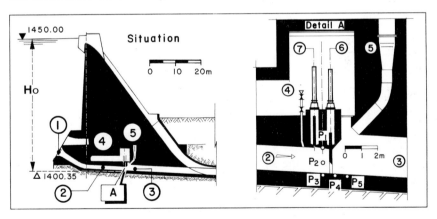

Fig. 1: *Cross section of the Panix dam: 1): Bottom outlet intake, 2): Bottom outlet pressure tunnel, 3): Bottom outlet free surface tunnel, 4): Gate chamber, 5): Air supply pipe, (Detail A): 6): Service gate, 7): Emergency gate, P1 - P5: Pressure cylinder locations.*

Gate openings could be varied between 0 and 1450 mm. Only at the lowest pressure head of 34 m could the gate be opened fully. To avoid flood damage in the valley downstream, gate openings were restricted at pressure heads of 40 and 45 m to 1350 mm, at 47.5 m to 1050 mm and at 50 m to 675 mm.

2. PRESSURE SENSORS AND INSTRUMENTATION

During dam construction, 5 special steel cylinders were welded into the armour plating of the bottom outlet near the service gate, in which the pressure sensors were installed. Three cylinders were situated 0.3 m upstream the service gate, one in the roof (P1), side wall (P2) and floor (P3) respectively. Two cylinders were also situated in the floor 0.4 m (P4) and 1.4 m (P5) downstream the service gate, as shown in the detail of Fig. 1. In total 22 pressure sensors of different size were mounted in the tops of the five cylinders, 18 of them flush-mounted. Four Kulite XTM sensors were mounted in 'pinhole' adapters to decrease the pressure sensitive diameter of the sensors. The diameter of the hole was 0.5 mm.

The use of pinhole adapters was unsuccessful, as air penetrated into the cavity during the measurements, although the cavities inside the adapters were filled carefully with water. The air reduced the resonance frequency of the Helmholtz resonator drastically from around 20 kHz to nearly 1 kHz.

Some technical data of the sensors are shown in Tab. 1.

Sensor type	No.	Diameter	Pressure range *)	Frequency range	Sensivity
Jensen PTE	3	19 mm	5 bar sg	0-0.68 kHz	0.83 V/bar
Jensen PE	2	19 mm	5 bar a	0-2 kHz	1.0 V/bar
Kulite ETM375	8	8.2 mm	7 bar a	0-8 kHz	0.71 V/bar
Kulite XTM190	9	3.7 mm	7 bar a	**)	1.1mV/V/bar

*): sg: sealed gage, a: absolute; **) depending on used amplifier.

Tab. 1: Technical data of the pressure sensors used.

The Jensen and the Kulite ETM sensors had build-in amplifiers, giving a full scale output of 5 Volts. Two ELAN carrier amplifiers were used for signal conditioning of the Kulite XTM sensors, without build-in amplifier. The carrier amplifiers limited the signal frequency range to 2 kHz. During the last measurements at a pressure head of 50 m, two of the flush mounted XTM sensors (P3 and P4) were connected to a DC-amplifier, based on Analog Devices' 1B31 signal conditioner chip, providing a frequency range up to 20 kHz.

The signals were filtered by a Precision Filter system 32 (6 channels) and a system 6000 (12 channels) low-pass-filter with variable cut-off frequency. The second filter had the capability of adding a DC-voltage at the input to enable offset compensation. Afterwards the signal could be amplified to increase the dynamic resolution, without overdriving the filter with a DC voltage. Analog-digital conversions were performed by two National Instruments' NB-MIO16 boards plugged into a Macintosh IIfx computer. Each board could sample 8 channel with a maximum sampling frequency of 7.3 kHz with 16 bit resolution. The sample time clocks of both boards were synchronised.

At each gate opening several measurements were made with different sample frequencies. The sampling frequency typically varied between 500 Hz and 15 kHz. At a pressure head of 50 m, measurements with a sample frequency of 50 Hz and 30 kHz were also carried out. Each time series consisted of 50,000 samples per channel. Before a time series was sampled, a test measurement was performed. A special computer program estimated the mean value and the standard deviation measured with each sensor connected to the system 6000 filter and automatically set the offset and gain value of the corresponding filter channel. All programs for data sampling and filter control were written in the graphical programming language LabView 2.

As only 16 channels could be sampled simultaneously with the computer, the remaining channels were recorded with a 8 channel PCM DAT-recorder (Heim DATaRec-E8), which has a frequency resolution up to 5 kHz.

3. WALL PRESSURE SPECTRA UPSTREAM THE GATE

The wall pressure measured upstream the service gate is strongly affected by the acoustical properties of the bottom outlet pressure tunnel (Fig. 2), which acts as an acoustic resonator, especially at low and moderate gate openings.

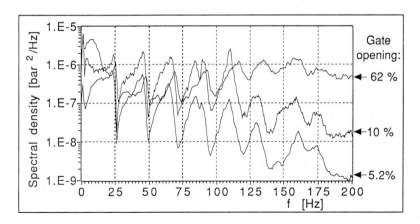

Fig. 2: *Power spectra of the wall pressure fluctuations measured at P3 at different gate openings (pressure head Ho = 34 m).*

Due to 'vortex stretching' underneath the gate, sound waves are generated, according to the theory of 'aerodynamically generated sound' [2], [3]. These travel upstream and are reflected by the change of the acoustic impedance in the region of the pressure tunnel intake. This is why destructive interferences can be seen at certain frequencies in the power spectra, as shown in Fig. 2. The more the gate is opened, the more the notches are masked by wall pressure fluctuations produced by turbulent velocity fluctuations.

The sound waves are reflected a second time differentially at the gate and in the region of contraction under the gate. As both reflected waves are not in phase, no clearly defined notches can be seen in the power spectra due to those reflections. A sound wave once generated at the gate, travels a few times up and down the pressure tunnel, being reflected at both ends, until its energy is dissipated. This can be seen from the auto correlation function $R_{xx}(\tau)$ shown in Fig 3, where three 'echoes' of the sound wave can be clearly identified.

Because of the complicated reflection mechanism in the gate region, no equally spaced sharp resonance peaks can be observed in the spectra, which would be expected, if the pressure tunnel acts as a simple acoustic resonator. As there seemed to be no feedback mechanism, which would compensate for the energy dissipation of the wave, the multiple resonance peaks are weak.

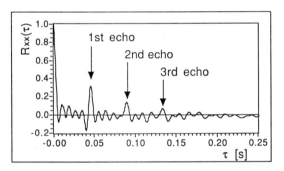

Fig. 3: *Auto correlation function $R_{xx}(\tau)$ of the wall pressure fluctuations measured at P3 at a gate opening of 7% (pressure head $H_o= 34$ m).*

4. FREE SURFACE FLOW DOWNSTREAM OF THE GATE

The free surface flow downstream the service gate is highly turbulent. During the measuring campaign 8 pressure sensors were destroyed, 7 of them mounted in P4 and P5, although they had a pressure range of 7 bars and a maximum pressure range of 14 bars.

At all gate openings, the pressure signals consist of a broad-band noise up to frequencies of several kilohertz, as shown in Fig. 4. Most of the pressure fluctuations have a convective character, i.e. the pressure fluctuations generated by local turbulent velocity fluctuations, move downstream with a convection velocity u_c in the order of the mean velocity.

The convection velocity can be calculated from the signals of two pressure sensors, based on the sensor distance d and the signal delay time $\Delta\tau$ in the cross correlation function Rxy(τ), by: $u_c = \Delta\tau/d$. An example is shown in Fig. 5a. As the sensor separation was d = 929 mm, u_c had a value of 24.1 m/s.

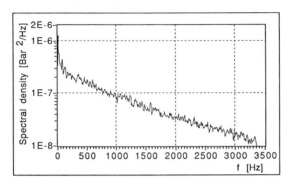

Fig. 4: *Power spectra of the turbulent broad band wall pressure fluctuations measured at P5 (gate opening of 21%, pressure head Ho = 34 m).*

Fig. 5b shows the convection velocities measured at a pressure head of 34 m for different gate openings. A smaller convection velocity was measured with a sensor distance of 70 mm, whereas a higher convection velocity was measured with a longer sensor distance of 929 mm. This characteristic is known from experiments in fully developed flow and boundary layer flow [4], and is explained in the following way: High frequency fluctuations have a shorter lifespan and are mainly produced in the near wall region, which has a lower local mean flow velocity. Low frequency fluctuations have a longer lifespan, are generated further away from the wall and move with a convection velocity which is nearly the mean velocity.

Therefore, it should be possible, to estimate the mean flow velocity in model and prototype investigations by the cross correlation of wall pressure signals, if the sensor separation is sufficiently long . As we could not measure the flow velocity directly during this field campaign, further experiments are necessary to test this.

5. FINAL REMARKS
The field measuring campaign showed the strong influence of the acoustic properties of the pressure tunnel on the pressure fluctuations inside the tunnel.

As no feedback mechanism could be observed, the acoustics of the pressure tunnel will, however, have no influence on the performance of the bottom outlet.

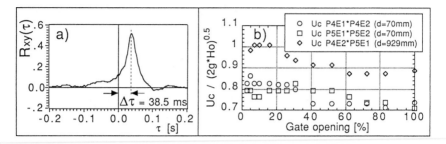

Fig. 5: *a): Cross correlation function $RP4E2*P5E1(\tau)$ of the wall pressure fluctuations (gate opening of 52%, pressure head Ho= 34 m), b): Convection velocities u_c for different gate openings at a pressure head of 34 m, measured with different sensors mounted in P4 and P5.*

Hydraulic model investigations of bottom outlets are often carried out according to Froude's law. Therefore frequencies will scale as: $f_{Prototype}/f_{Model}= \lambda_L^{-0.5}$ (λ_L = length scale). But frequencies f_A due to the acoustic response of the tunnel will scale as: $f_{AProtototype}/f_{AModel}= \lambda_L^{-1}$, if the velocities of sound are the same in model and prototype. This means, that power spectra measured in a hydraulic model of a pressure tunnel cannot be transferred to the prototype by simply applying Froude's law. In the prototype, the shape of the spectrum will change, because the frequencies of the notches and resonance peaks will scale to lower values, than would be expected from Froude's law.

ACKNOWLEDGEMENT
We would like to thank the "Nordostschweizerische Kraftwerke AG" (NOK) for the permission to carry out this field campaign and the "Kraftwerke Ilanz AG" (KWI) for their support during the measurements.

REFERENCES
[1] Volkart, P. U., Speerli, J. Prototype investigation of the high velocity flow in the high head tunnel outlet of the Panix gravity dam. *Trans. 15th congr. ICOLD Durban* **4** , (Q. 71 R. 6) 55-78 (1994)

[2] Lighthill, M. J. On sound generated aerodynamically. *Proc. R. Soc. London Ser. A* **211**, 564-587 (1954)

[3] Powell, A. Theory of vortex sound. *J Acoust. Soc. Am.* **36,**177-195 (1964)

[4] Blake,W. K. Mechanics of flow-induced sound and vibrations, *Ser. App. mathematics and mechanics* **17,** Academic Press Inc. Orlando (1986)

1C11

IMPROVEMENT OF SUPERCRITICAL PIER FLOW

Roger REINAUER

Versuchsanstalt für Wasserbau, Hydrologie und Glaziologie
ETH-Zentrum, CH-8092 ZURICH, Switzerland

INTRODUCTION

Chute piers as used on spillways or in bottom outlet junctions induce rooster tails, i.e. shockwaves originating from either pier end corner and transversing the tailwater channel. The disturbance by shock is not a hydraulic problem except for the asymmetry of the approach flow to the energy dissipator, but a considerable increase of freaboard may yield an uneconomic chute design (Golzé 1977).

In existing structures, rooster tails have been countered by reducing the lateral pier extent at the pier end, that is narrowing the pier width very gradually to yield an arrow-shaped pier termination. Also, the pier height was sometimes gradually reduced from the maximum flow depth to zero height, in order to guide the flows on either pier side into the tailwater. Due to negative pressures, cavitation occured sometimes at pier ends, particularly with junctions in bottom outlets. Also, the arrow-shaped pier is uncommon in design, must often be lined and can be quite costly. This paper aims at introducing the so-called chute *pier extension*, by which the pier wave can practically be removed with a simple device. The pier extension can also be added to existing chute piers.

HYDRAULICS OF CHUTE PIER WITHOUT EXTENSION

Fig.1 shows pier waves downstream of an abruptly ending pier for $F_o=2$. Right behind the pier, a standing wave of considerable height is developed, which is referred to as *wave 1*. At the tailwater end, shockwaves originate which eventually expand and impinge on the lateral sidewalls, where *wave 2* occurs. Compared to the configuration without pier, the side walls must be increased, depending on the pier geometry and the Froude number.

Fig.1   Rooster tail in the wake of rectangular pier, with impingement on chute side walls.

The flow pattern for a horizontal chute is defined in Fig.2. The piers have a width b_P and the axial interdistance between two adjacent piers is b_A, thus $b_o=b_A-b_P$ is the approach width. The distance of the pier axis from the side wall is b_W. In the wake of the pier, wave 1 of height h_{1m} at location x_{1m} and width b_{1m} develops, with x as longitudinal coordinate measured from the pier end. Wave 2 originates from the interaction of the pier next to the wall and the chute wall, with height h_{2m} at location x_{2m}, and of width b_{2m}. The approach flow depth is h_o, the approach velocity V_o and $F_o=V_o/(gh_o)^{1/2}$ is the approach Froude number with g as gravitational acceleration.

Experiments were conducted to find expressions for the main flow characteristics. Such information will hardly be obtained from a computational approach, given the highly spatial flow pattern. It was found that for $F_o>4$ to

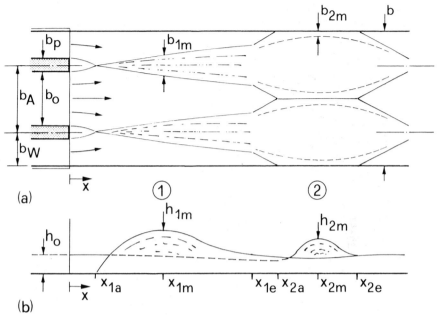

Fig.2   Definition of flow behind pier wave.

5, the height of pier wave did not further increase with F_o, and this domain of hypercritical flow was further studied. Subscript «M» refers to those plateau values.

For $0 < h_o/b_P < 6$ the characteristics for *wave 1* are (Reinauer and Hager 1994)

$$h_{1M}/h_o \quad = 2.5 \ (h_o/b_P)^{1/2} \ , \tag{1}$$

$$x_{1M}/(b_P F_o) = 2.3 \ (h_o/b_P)^{2/3} \ , \tag{2}$$

$$b_{1M}/b_P \quad = 1.8 \ (h_o/b_P)^{2/3} \ . \tag{3}$$

For *wave 2*, the additional parameter b_A/b_P is involved, and

$$h_{2M}/h_o \quad = 1+4 \ (h_o/b_P)^{1/2}(b_A/b_P)^{-1} \ , \tag{4}$$

$$x_{2M}/(b_P F_o) = 1.9 \ (h_o/b_P)^{2/5}(b_A/b_P)^{2/3} \ , \tag{5}$$

$$b_{2M}/h_o \quad = 0.12 \ (h_o/b_P)^{-1/2}(b_A/b_P) \ . \tag{6}$$

Accordingly, the waves can easily have a height several times larger than the approach flow.

EFFECT OF CHUTE PIER EXTENSION

The pier extension is a device by which the origin of shock wave development is significantly reduced. Fig.3 shows a definition sketch of the pier extension which is applied on one side of the pier end. It corresponds to a narrow wall usually made of stainless steel by which the symmetric jet impact of waves from either pier side is avoided and *wave interference* is applied instead. The wave originating from the side without extension is then 'covered' with the wave from the other pier side, and practically no tailwater disturbance may be seen. Of design interest is thus the length x_E of the pier extension.

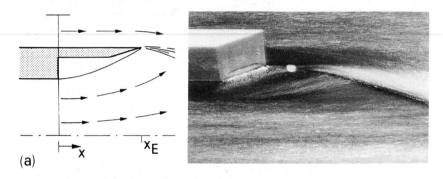

Fig.3   Pier extension a) definition of flow, b) corresponding photograph.

The effect of pier extension depends on the relative position of the waves developed, and are influenced by the pier geometry and the approach Froude number F_0. From experiments, it was found that

$$x_E/(b_P F_0) = 0.65(h_0/b_P)^{2/3} \qquad (7)$$

yields maximum heights h_{2M} of wall wave 2 which are always smaller than $1.25h_0$, compared with the height according to (4) for untreated flow. The wave interference principle is applied in a simple manner, and results in an *unsymmetric* flow pattern much in contrast to the conventional engineering practise.

PHOTOGRAPHS

A selected series of photographs with two chute piers was prepared, one of which was not treated with the interference method and the other with the pier extension mounted, such that an immediate visual comparison of flow is possible.

Fig.4 shows tailwater views of flow with a Froude number of 4.2. Whereas practically no wave generation may be noted on the left side (in flow direction), the wall wave on the right channel side is impressive in height.

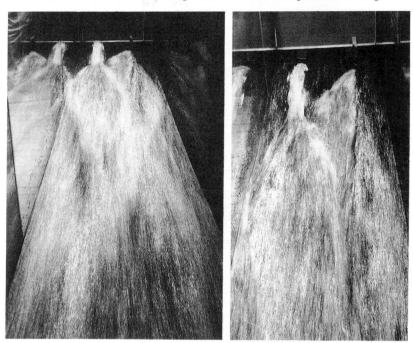

Fig.4   Tailwater views against pier flow with (right) and without (left) pier extension.

Side and plan views of the same flow configuration are seen in Fig.5. From the side, wave 1 looks large compared to the approach flow depth h_o whereas in plan view this difference cannot be really seen. However, the covering of waves is noted from the latter photo.

In Fig.6, a detailed side view reveals the ascending wave portions of untreated and treated piers. The difference is not large at this location because the wave covering occurs only beyond the primary wave.

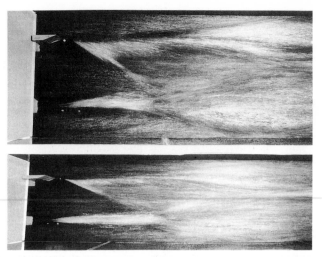

Fig.5   Chute flow without (right) and with (left) pier extension in plan view.

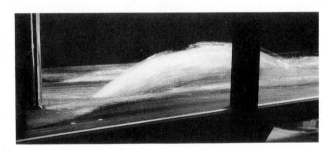

Fig.6   Side view to primary pier waves with untreated (front) and treated pier
(back).

CONCLUSIONS

Chute pier waves can be significantly reduced by the so-called pier extension.
A rational design procedure is presented, and the differences with the
untreated chute pier are illustrated both by experiments and photographs.

REFERENCES

- Golzé, A.Q. (1977). *Handbook of dam engineering*. Van Nostrand
  Reinhold: New York.
- Reinauer, R., Hager, W.H. (1994). Supercritical flow behind chute piers.
  Journal of Hydraulic Engineering **120**(11): 1292-1308.

1C12

DISCHARGE COEFFICIENT OF OBLIQUE SHARP CRESTED WEIR

S.M.BORGHEI
Assistant Professor of Civil Engineering
S U T , TEHRAN-IRAN

ABSTRACT
Standard rectangular sharp crested weir placed perpendicular to the flow is widely used as a main device for discharge measurement.Using the conventional weir equation and Rehbocks' formula for the discharge coefficient,the discharge depends on the length of the weir and the head on the weir.In order to increase the efficiency of the weir ,it can be used oblique;that is to increase the effective length and decrease the head.The result from comprehensive set of tests show that for the oblique weir, as the angle increases from 0° to 45°, C_d increases with increasing head,for 45° , C_d is constant and for higher angles, C_d decreases with increasing head.

INTRODUCTION
Sharp crested weir is one of the oldest flow measurement devices used in hydraulic structure and environmental projects as a simple,efficient and accurate equipment in the field and laboratory.Different geometric shapes are used, such as; triangular (or V-notch) [3], rectangular [6,8], cippoletti [2] and sutro [4].

For channels with limited width or depth and high discharge,rectangular weir across the width of the channel is used.In order to decrease the head of flow,the effective length can be increased by placing the weir oblique to the flow (Fig. 1.a). A comprehensive series of tests have been carried out for oblique weirs between 15 and 66 degrees.The results are presented in the form of graphs showing the relation between C_d and H/P for different angles.

STANDARD FORMULAS
The general formula for sharp crested rectangular weir is in the

conventional form of:

$$Q = \frac{2}{3} C_d L H \sqrt{2gH} \qquad (1)$$

where Q is the weir discharge, L is the weir length, H is the head on the crest (Fig. 1.b), g is the gravitational acceleration and C_d is the discharge coefficient. Rehbock [7] proposed an equation for C_d in the form of:

$$C_d = 0.611 + 0.075 \frac{H}{P} \qquad (2)$$

in which P is the weir height (Fig. 1.b) and H is large enough that surface tension does not effect the result.

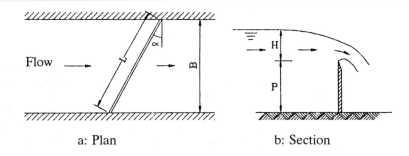

a: Plan   b: Section

Fig. 1. -Definition Sketch

Also the effect of the viscosity and surface tension is studied by Kindsvater and Carter [5] in their classic paper.They found that the effect of viscosity can be compensated by increasing the head by 1 mm and decreasing the weir width by 0.9 mm.Also the effect of weir contraction and larg values of H/P introduces new C_d values [2,4].

Aichel [1], without proposing a formula for C_d ,in oblique weir, introduced a mathematical and a graphical function to relate the discharge for the plane and oblique weir at different angles [2].

EXPERIMENTAL EQUIPMENT

The tests were conducted in a rectangular, horizontal, concrete channel, 71.4 cm width, 95 cm height and 800 cm long. Standard sharp crested weirs with 40 cm height and variable length (74 cm for 15° to 176 cm for 66°) were used.

Maximum discharge was over 90 lit/s and minimum discharge was chosen somehow that the head of water was never less that 4 cm, in order to avoid surface tension effect.

Table 1 shows the geometric and flow variables of the study.

$\alpha°$	Geometric Specification						
	15	30	45	49.5	54	59	66
L (cm)	74	82.5	101	110	121.5	139.5	176
	Flow Specification						
max Q (l/s)	83.15	77.63	91.07	90.12	86.07	88.56	91.11
min Q (l/s)	12.14	14.60	16.22	18.73	17.50	21.86	16.54

Table 1. Variables of Tests

ANALYSIS AND RESULTS

As it was expected the oblique weir increases the effective length of weir from B to L (or B/Cosα) and hence decreases the head H. Fig. 2 shows the variation of C_d with H/P in the form of straight lines with different slopes for different angles. As α increases from 0° (or plane) to 45°, C_d increases with increasing H/P, to a constant value for 45° angle irrespective of H/P values. However for $\alpha > 45°$, C_d decreases with increasing H/P; (ie for $\alpha < 45°$ the slope of the line C_d versus H/P is positive, For $\alpha = 45°$ the slope is zero and for $\alpha > 45$ the slope is negative).

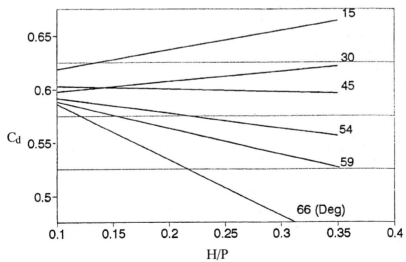

FIG. 2.- C_d for Different H/P and Oblique Angles (Eq. 2)

From Fig. 2 it is difficult to obtain exact formulas for C_d values, but it is obvious that such relationship as plane weir exist for oblique ones.

The best concluding result is for $\alpha = 45°$, where the effect of increasing H on C_d (which in the case of plane weir increases C_d) is compensated by the same amount due to oblique effect which changes streamline direction and hence results a constant C_d. The reason however is beyound the scope of this paper.

In equation 2, substituting B (channel width) for L (weir length), a new discharge coefficient is obtained as;

$$C_d{'} = \frac{C_d}{Cos\alpha} \qquad (3)$$

$C_d{'}$ versus H/P for 0°, 30°, 45°, 59° are shown in Fig.3. From this graph it can be seen that a 59° oblique weir, for small values of H/P, increases C_d almost twice and for higher values of H/P, 1.5 times the C_d values of plane weir.

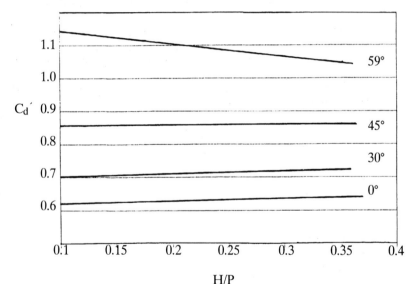

FIG. 3.- $C_d{'}$ for Different H/P and Oblique Angles (Eq. 3)

Since the aim of using oblique weir is to decrease H, therefore higher values of H/P than 1 is not recommended, and from another point of view, for higher values of 1 the oblique effect would be decreased.

CONCLUSION

Oblique weir can be used in place of plane weir for discharge measurement in channels with restricted width or height. Experimental results using 45° oblique weir instead of plane weir, show for H/P of 0.3 , C_d increases by 35% . This effect for 30° and 59° are 14% and 67% respectively .

ACKNOWLEDGEMENT

The laboratory work for this study was supported by the Water Research Center, affiliated to the Ministry of Energy, Tehran.

REFERENCES

1 Aichel,O.G.(1953)."Abflusszahlen fur schiefe wehre."(Discharge ratios for oblique weirs),Z.VDI 95. No. 1.
2 Bos, M. G. (1976). Discharge measurement structures. International Inst. for land Reclamation and Improvement, Wagemingen, The Netherlands.
3 Eli, R. N. (1986). "V - notch weir calibration using new parameters." J. Hyd. Engrg. Div., ASCE, Vol. 112, No. 4.
4 French, R. H. (1986). Open - Channel Hydraulics. Mc Grow-Hill Book Co. New York, N. Y.
5 Kindsvater, C. E., and Carter, R. W. (1957). "Discharge characteristic of rectangular thin plate weirs." J. Hyd. Engrg. Div., ASCE, Vol. 83, No. 6.
6 Ramamurthy, A. S., et al. (1987). "Flow over sharp - crested plate weir." J. Irrig. Drain. Div., ASCE, Vol. 113, No. 2.
7 Rehbock, T. (1929). "Generalized rectangular weir equation." J. Hyd. Engrg. div., ASCE, Vol. 114, No. 8.

NOTATIONS

The following symbols are used in this paper:

B = channel width
C_d = discharge coefficient for weir
$C_d{'}$ = $C_d/Cos\alpha$
H = head of water
L = length of weir
P = height of weir
Q = discharge
α = oblique angle

1C13

SCALING THE PERFORMANCE OF AERATORS IN A TUNNEL SPILLWAY

R.W.P. MAY, M. ESCARAMEIA
HR Wallingford, Wallingford, United Kingdom
I. KARAVOKYRIS
G. Karavokyris & Partners, Athens, Greece

1. INTRODUCTION

The Evinos Dam, now under construction in Greece, forms part of a scheme to divert water from the Evinos River, through a 29km long tunnel, to the existing Mornos Reservoir. This reservoir is the main source of water supply for Greater Athens and it is expected that the Evinos scheme will increase its annual yield by an average of $240 \times 10^6 m^3$. The Evinos Dam is a 120m high earthfill embankment with a central clay core. The spillway design discharge, corresponding to the 10 000 year storm, has been specified, after routing through the reservoir, as equal to $1600 m^3/s$.

The geology and topography of the site have led to a spillway design that minimises open excavation. The entrance section is a fan shaped frontal ogee with a 40m long crest followed by an 80m long transition section leading to a 10m diameter tunnel with a modified horseshoe cross-section. At its downstream end the tunnel is followed by a rectangular covered chute terminating in a flip bucket. At the design discharge the total drop between the reservoir level and the flip bucket equals 106m. Flow conditions have been studied in a general 1:80 scale model at HR Wallingford, UK. The study has shown the design to behave satisfactorily as an open channel spillway to a discharge of $2000 m^3/s$, with slug flow starting at about $2100 m^3/s$.

Maximum water velocities will reach values as high as 38m/s while the cavitation index has been calculated to fall to a minimum value of 0.20. Artificial aeration of the flow along the 100m of chute downstream of the tunnel is necessary to prevent possible cavitation damage. This paper describes the physical model study of the proposed aerators and air supply system which was carried out at HR Wallingford. The paper concentrates on the investigation of possible scale effects in the model.

Full details of the overall study are given in Ref.(1).

2. BRIEF DESCRIPTION OF AERATION MODEL

The aeration system was tested in a high-velocity tilting flume which is capable of producing a maximum flow velocity of about 15m/s. The aerator model reproduced one half of the chute cross-section at a scale of 1:30 and comprised a ramp aerator with vertical step and a vertical air shaft (see Fig.1). Measurements were made of: the water discharge by means of an acoustic flow meter; point water velocities using a fine-bore pitot tube connected to a pressure transducer; the water surface profile along the spillway with a point gauge; the air flow velocity in the vertical shaft using a hot wire anemometer; air pressures in the aerator step by means of manometers; the air cavity length with a conductivity probe; and point air concentrations by means of a void-fraction meter.

3. SCALING OF AIR DEMAND

Small models of aerators may underestimate the prototype air demand because the effects of surface tension and viscosity are not scaled correctly. However, laboratory research carried out at HR Wallingford on different sizes of ramp aerators (see Ref.(2)) showed that scale effects could be avoided if the model has: similar levels of flow turbulence to those of the prototype; an air supply system with the same non-dimensional head-loss characteristics; and water velocities greater than about 5-6m/s. The non-dimensional head-loss characteristic, E, of an air supply system can be defined by means of the following equation:

$$Q_a = EA_0 \, (2\Delta p/\rho_a)^{0.5} \tag{1}$$

where Q_a is the volumetric flow rate of air, A_0 is the outlet area of the duct, ρ_a is the air density, and Δp is the difference between atmospheric pressure and the average pressure in the cavity produced by the aerator.

Possible scale effects in a model can be evaluated by comparing the measured values of the air demand ratio, β (=air flow rate/water flow rate), with estimates based on prototype data. Pinto (Ref.(3)) obtained two prediction equations:

$$\beta = 0.47 \, (F_r - 4.5)^{0.59} \, (EA_0 \, / \, (Bd))^{0.60} \tag{2}$$

$$\beta = 0.29 \, (F_r - 1.0)^{0.62} \, (EA_0 \, / \, (Bd))^{0.59} \tag{3}$$

where B is the spillway width, d is the water depth and F_r is the Froude number of the flow (=$V/(gd)^{0.5}$). Eqn(2) applied to the Foz do Areia aerators which are similar in design to Evinos; Eqn(3) is a more general equation based on data from nine prototype aerators (including Foz do Areia). Analysis of the data used for Eqn(3) shows that, for each aerator, Q_a was related to the water velocity, V, by an equation of the form:

$$Q_a = k (V - V_0)$$ (4)

where V_0 is the velocity corresponding to the effective start of air entrainment. The value of k depends on the geometries of the aerator, spillway and air supply system. For a given aerator, the prototype data show that k and V_0 do not vary significantly with flow depth until the aerator cavity begins to collapse because of the imposed pressure. A similar type of linear relationship was also found in the HR laboratory research (see Ref.(2)). These tests indicated that V_0 was related to the vertical height, h, of the ramp above the line of the channel invert by:

$$V_0 = 7.4 (gh)^{0.5}$$ (5)

The equation also agrees well with nearly all the values of V_0 for the nine aerators. This suggests that the minimum entrainment velocity should scale satisfactorily in a Froudian model provided the three requirements mentioned previously are satisfied. Also, if model tests demonstrate the type of linear relationship given by Eqn(4), this is an indicator that scale effects are not significant in the model results.

4. TESTS AND RESULTS

Three types of test were carried out: tests where all the relevant quantities were scaled according to the Froude similarity law (Froudian tests); tests to investigate the sufficiency of the air supply system (collapse tests); and tests to investigate possible scale effects (non-Froudian tests). Collapse tests are important to determine whether there is any danger of the air cavity in the aerator collapsing due to large flow depths or lack of capacity in the air supply system. The four Froudian tests were therefore repeated with the inlet to the vertical air shaft reduced gradually in area so as to make the pressure in the air cavity more negative.This process was continued until the cavity collapsed. The non-Froudian tests were carried out to investigate any changes in the air demand characteristics at higher model velocities than those required for the Froudian tests. Having carried out a Froudian test with one of the selected discharges, the flow rate was then increased in steps while keeping the flow depth constant.

The tests showed that the measured operating pressures in the aerator cavity are well below the corresponding collapse pressures and therefore the air supply system has a satisfactory capacity. The non-Froudian tests were carried out up to velocities of the order of 12.5m/s. Fig.2 shows how the rate of air flow, Q_a, varied with the mean flow velocity, V, of the water upstream of the aerator for both Froudian and non-Froudian tests. The results demonstrate the same type of linear relationship found with the prototype and other laboratory data discussed in Section 3. The depth of flow (decreasing from test series 1 to 4) does not appear to affect this relationship.In Fig.2 it can be seen that the air demands in the Froudian tests (shown circled) lie below the trend of the other data and may have been subject to scale effects. These points were therefore not included when determining the best-fit line.

The predicted values of air demand ratio, β, for the aerator are given in Table 1. The first row of figures corresponds directly to the measurements made in the Froudian tests; the second row uses the best-fit line in Fig.2 to correct the air demands in the Froudian tests for possible scale effects; the last two rows give the values of β predicted by Pinto's Eqns.(2) and (3), which were derived from full-scale data. It can be seen that β obtained from the model are higher than given by Eqn (2), which was based on data for the Foz do Areia aerators, but are comparable with the values calculated from the more general Eqn (3). This suggests that the scale effects in the model were not very large.

Fig.3 shows longitudinal profiles of the air concentration ($C = \beta/(1+\beta)$) at a height above the bed of 2.5% of the water depth. Previous laboratory studies have suggested that a local air concentration of C=7-8% is the minimum necessary to ensure protection against cavitation damage. The possibility of scale effects due to relatively larger air bubbles in the model than in the prototype was also investigated. This was done by using a numerical advection-diffusion model named ADAM developed at HR for simulating 2-D air-water flows downstream of aerators (see Ref(4)). ADAM was applied to the case of Q=1600m³/s and, as expected, it predicts a relatively smaller rate of decrease of C than occurred in the model. Fig. 3 indicates that the local air concentrations are sufficient to prevent cavitation damage for a distance of at least 40m downstream of an aerator.

5. CONCLUSIONS

(1) Scale effects in models of aerators can be minimised by ensuring similarity in terms of flow turbulence and head-loss characteristics of the air supply system. The flow velocity in the model also needs to be greater than 5-6m/s.

(2) For a given aerator a linear relationship exists between the flow rate of air and the water flow velocity until the air cavity in the aerator starts to collapse.

6. REFERENCES

(1)   HR Report EX 2819 (1993). Evinos Dam. Part A: model tests of spillway; Part B: aeration studies.

(2)   May R.W.P., Brown P.M. and Willoughby I.R.(1991).Physical and numerical modelling of aerators for dam spillways. HR Wallingford, Report SR 278.

(3)   Pinto N.L. de S. (1991). Prototype aerator measurements. In "Air entrainment in free-surface flows", Ed. Wood I.R., A.A. Balkema, Rotterdam.

(4)   May R.W.P. and Escarameia M. (1992). Spacing of aerators for spillways: Development of a 2-D numerical model. HR Wallingford, Report SR 311.

7. ACKNOWLEDGEMENTS

The model studies of Evinos Dam were commissioned by the joint venture GR.IT.AU Evinos; the consultant was G. Karavokyris & Partners.

TABLE 1   PREDICTED AIR DEMANDS

Method of prediction	Tunnel discharge (m³/s)			
	500	1000	1600	2000
Froudian tests	0.249	0.194	0.130	0.110
Corrected (Fig 2)	0.352	0.222	0.158	0.134
Pinto (Eq (2))	0.151	0.110	0.0776	0.0655
Pinto (Eq (3))	0.212	0.159	0.125	0.112

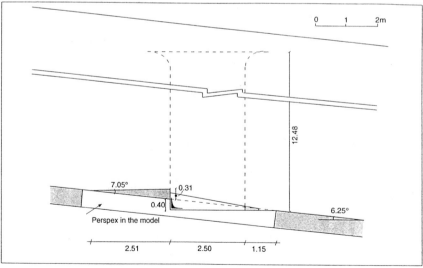

Fig.1 Longitudinal section of aerator tested

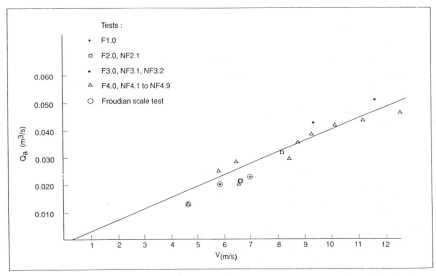

Fig.2 Relationship between water velocity and air discharge (model values)

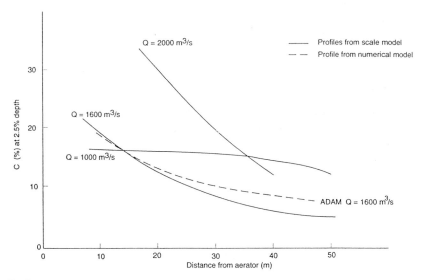

Fig.3 Longitudinal air concentration profiles at 2.5% of water depth

1C14

CHARACTERISTICS OF UNDULAR JUMPS IN RECTANGULAR CHANNELS

Iwao OHTSU , Youichi YASUDA , and Hiroshi GOTOU.
Dept. of Civ.Engrg.,Nihon Univ.,Coll.of Sci. and Tech.,
Kanda Surugadai 1-8, Chiyoda-ku, Tokyo, Japan , 101

1. INTRODUCTION

In the transition from supercritical to subcritical flow, water surface shows undulations for low supercritical-Froude numbers. This flow condition is called an undular jump.[1]

Several types of flow conditions have been observed in the course of the authors' experiments regarding the undular jump (Fig. 1).

Although the undular jump has been investigated by some researchers,[2],[3] the hydraulic conditions required to form each type of flow condition have not been studied. Also, the characteristics of each type of flow condition have not been clarified.

In this paper, the undular jump in rectangular horizontal channels has been investigated systematically. The flow conditions of undular jumps are classified. The non-dimensional factors governing each type of flow condition are represented as the supercritical Froude number F_1, the inflow condition (the state of the boundary-layer development of the supercritical flow)[4], the non-dimensional factor of the channel width including the effect of the lateral shock wave, and the Reynolds number.

2. DESCRIPTION OF FLOW CONDITIONS

The experiments were conducted under the wide ranges of supercritical depth h_1, channel width B, Froude number F_1 (= $v_1/\sqrt{gh_1}$; v_1 : average velocity at the toe of the undular jump), and Reynolds number Re (= q / ν; q = Q/B; Q : discharge) [2.00 cm $\leq$ h_1 $\leq$ 10.0 cm; B = 80, 40, and 20 cm; 1.0< F_1 < 3.0; 2.5 ×10^4 $\leq$ Re $\leq$ 1.4×10^5; and 2.0 $\leq$ B/h_1 $\leq$ 25].

For F_1 $\geq$ 1.2, the lateral shock wave is always observed at the first wave of the undulation (Fig. 1). If the cross point of the shock wave is located upstream of the first wave crest, the effect of the shock wave on the characteristics of the undular jump [e.g., flow conditon, wave height, wave length , etc.] is large. When the shock wave does not cross upstream of the first wave crest, i.e., L_s/L_w $\geq$ 1, the effect of the shock wave is negligibly small. Then, L_s is the horizontal length between the toe of the lateral shock wave and the cross point $[L_s = B/(2 \cdot \tan\theta) ; \theta :$ the angle of the shock wave to the side wall (Fig. 2)] and L_w is the horizontal length between the toe of the shock wave and the first wave crest of the undular jump.

For Re $< 6.5 \times 10^4$, the effect of the Reynolds number on the characteristics of the undular jump is large (Fig. 3 discussed in section 3). On the other hand, for Re $\geq 6.5 \times 10^4$, the characteristics of the undular jump are independent of the Reynolds number (Fig. 3).

For $L_s/L_w \geq 1$ and Re $\geq 6.5 \times 10^4$, the flow conditions of undular jumps depend on F_1 and the inflow condition δ/h_1 (δ :the thickness of the boundary layer at the toe of the jump), and can be classified as shown in Fig. 1.

<u>1) 1. 2 $\leq F_1 \leq$ 1. 7</u>

DEVELOPING INFLOW   When the boundary layer at the toe of the undular jump is developing (this inflow conditon is called the "developing inflow"), as shown in Fig. 1(a), stable undulations propagate far downstream. The flow is two-dimensional, and the undulations near the wall are nearly the same as in the case of the central plane of the channel. As F_1 increases, a reverse flow is occasionally observed near the bottom of the first wave crest.

DEVELOPED INFLOW   When the boundary layer is fully developed (this inflow conditon is called the "developed inflow"), stable undulations propagate far

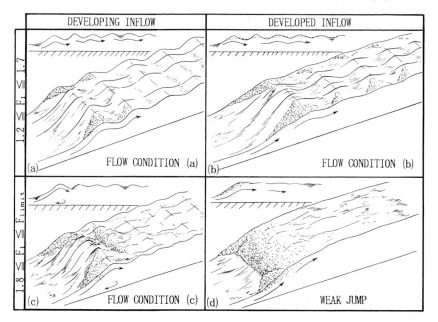

FIG.1 FLOW CONDITIONS OF UNDULAR JUMPS

downstream except for the flow near the wall [Fig. 1(b)]. The flow condition of this undular jump is almost two-dimensional, although the undulations near the wall are small and unstable. In the flow near the bottom, no reverse flow is observed at all.

2) $1.8 \leq F_1 \leq F_{limit}$

DEVELOPING INFLOW   A large stationary wave is formed as shown in Fig.1(c), and a reverse flow is always observed near the bottom of the first wave crest. The undulations downstream of the second wave are small and two-dimensional.   In addition, the range of F_1 for the formation of flow condition (c) depends on the state of the boundary-layer development (inflow condition). The upper limit of F_1 for the formation of (c) (i.e., F_{limit}) is empirically given by

$$F_{limit} = 2.80 - (\delta/h_1) \qquad [0.5 < \delta/h_1 < 1.0] \qquad (1)$$

DEVELOPED INFLOW   As shown in Fig.1(d), breaking (i.e., a surface roller) is always observed, and a weak jump is formed.   In this case, small undulations propagate far downstream because the main stream flows downstream along the water surface.

3. HYDRAULIC QUANTITIES GOVERNING UNDULAR JUMPS

Regarding the wave height of the first wave crest $h_{max.}$, the wave length between the first and second wave crests L, and the horizontal length between the toe of the lateral shock wave and the first wave crest L_w, arranging the experimental data in accordance with (2), Figs. 3 (a) $\sim$ (c) are obtained.

$$\frac{h_{max}}{h_1}, \frac{L}{h_2}, \frac{L_w}{h_1} = f(F_1, \frac{\delta}{h_1}, \frac{L_s}{L_w}, Re) \qquad (2)$$

where h_2 is the sequent depth of the jump {$h_2 = h_1 (\sqrt{8F_1^2 + 1} - 1)/2$, and L_s/L_w = $(B/h_1)/[2 \cdot \tan\theta \cdot (L_w/h_1)]$.

For $L_s/L_w \geq 1$ and $Re \geq 6.5 \times 10^4$, h_{max}/h_1 depends on F_1, and is independent of Re, L_s/L_w, and the inflow condition (δ/h_1) [Fig. 3 (a)]. The equation for h_{max}/h_1 is expressed by

$$\frac{h_{max}}{h_1} = 1.51 F_1 - 0.35 \qquad (3)$$
$$(1.2 \leq F_1 \leq 2.3, Re \geq 6.5 \times 10^4, L_s/L_w \geq 1)$$

Further, L/h_2 and L_w/h_1 depend on F_1 and the inflow condition [developing inflow or developed inflow] as shown in Figs. 3(b) and (c). The following equations for L/h_2 and L_w/h_1 are obtained.

For developing inflow,

$$\frac{L}{h_2} = \frac{1.14}{F_1 - 1} + 1.86, \quad \frac{L_w}{h_1} = 2.86 F_1 - 0.90 \qquad (4)$$
$$(1.2 \leq F_1 \leq 2.3, Re \geq 6.5 \times 10^4, L_s/L_w \geq 1)$$

For developed inflow,

$$\frac{L}{h_2} = \frac{0.90}{F_1 - 1} + 2.75, \quad \frac{L_w}{h_1} = 1.30 F_1 + 2.56 \qquad (5)$$
$$(1.2 \leq F_1 \leq 2.3, Re \geq 6.5 \times 10^4, L_s/L_w \geq 1)$$

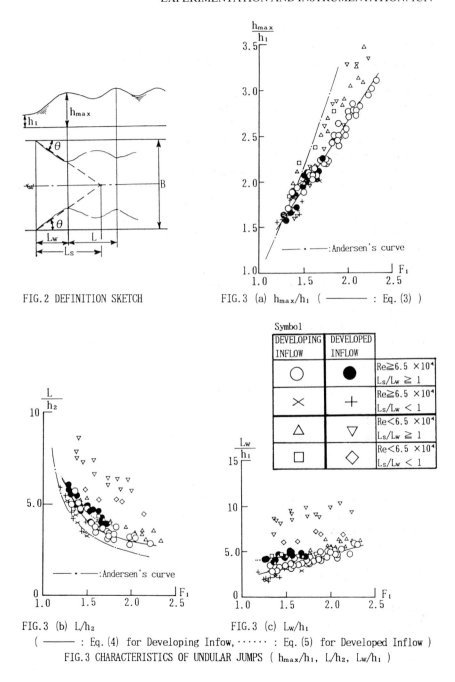

FIG.2 DEFINITION SKETCH

FIG.3 (a) h_{max}/h_1 (——— : Eq.(3))

FIG.3 (b) L/h_2

(——— : Eq.(4) for Developing Infow, ⋯⋯ : Eq.(5) for Developed Inflow)

FIG.3 (c) L_w/h_1

FIG.3 CHARACTERISTICS OF UNDULAR JUMPS (h_{max}/h_1, L/h_2, L_w/h_1)

For $L_S/L_w < 1$ and $Re < 6.5 \times 10^4$, the effects of L_S/L_w and Re on h_{max}/h_1, L/h_2, and L_w/h_1 are shown, moreover (Fig. 3).

Accordingly, for $L_S/L_w \geqq 1$ and $Re \geqq 6.5 \times 10^4$, the characteristics of un-dular jumps are generally governed by the Froude number F_1 and the inflow condition δ/h_1, and are independent of the Reynolds number Re and L_S/L_w.
In addition, the angle of the lateral shock wave θ for $Re \geqq 6.5 \times 10^4$ depends on F_1, and is empirically given by

$$\theta = 30.6 \ F_1^{0.65} \quad \text{(degree)} \quad (1.2 \leqq F_1 \leqq 2.3, Re \geqq 6.5 \times 10^4) \quad (6)$$

4. DYNAMICAL PRESSURE FIELDS IN UNDULAR JUMPS

Fig. 4 shows the dynamical pressure distribution (at the central plane of the channel) for the flow conditions (a) and (c). In Fig. 4, P_D is the dynami-cal pressure. The pressure was measured by using the pitot static tube.

FLOW CONDITION (a)   In this case, the dynamical pressure distribution is almost two-dimensional. As shown in Fig. 4(a), the positive and negative pres-sure fields alternate with each other, because stable undulations propagate far downstream. The maximum and minimum pressures occur near the bottom.
Regarding the magnitude and distribution pattern of the dynamical pressure down-stream of the first wave trough, the curvature of the stream line at the wave crest is larger than that at the wave trough, and the positive region is wider than the negative region. The absolute value of $P_{D \ min i.}$ in the negative region is larger than that of $P_{D \ max.}$ in the positive region.

FLOW CONDITION (b)   Except for the distribution near the wall, the dynam-ical pressure distribution is similar to that for flow condition (a). The magni-tudes of the pressure near the wall are almost hydrostatic, because the curva-ture of the stream line is small.

FLOW CONDITION (c),   As shown in Fig. 4 (b), the negative region in the first wave is wide. As for the pressure distribution downstream of the second wave, the pressure becomes hydrostatic in a short distance because the curvature of the stream line is small. Compared with flow condition (a), the positive and negative pressure fields do not alternate far downstream.

Accordingly, each type of flow condition shown in Fig.1 could be character-ized from the dynamical pressure distribution. Also, the validity of the classi-fication of the flow conditions was confirmed. In addition, each flow condition has also been characterized from the distributions of the mean velocity and the turbulent intensity, which were measured by using a one-dimensional Laser Doppler Velocity meter (LDV).[5]

CONCLUSION

The results obtained regarding the undular jump in rectangular channels are summarized below:
1. The factors governing the flow conditons of undular jumps can be represented as F_1, the inflow condition (δ/h_1), L_S/L_w, and Re.

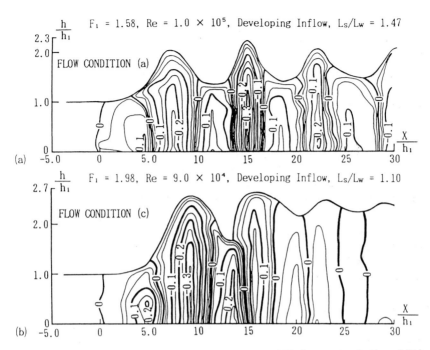

$F_1 = 1.58$, $Re = 1.0 \times 10^5$, Developing Inflow, $Ls/Lw = 1.47$

FLOW CONDITION (a)

$F_1 = 1.98$, $Re = 9.0 \times 10^4$, Developing Inflow, $Ls/Lw = 1.10$

FLOW CONDITION (c)

FIG. 4 EQUI-DYNAMICAL PRESSURE CONTOURS IN UNDULAR JUMP [———: $P_D/(\rho v_1^2/2)$]

2. For $Ls/Lw \geqq 1$ and $Re \geqq 6.5 \times 10^4$, the flow conditions of undular jumps depend on F_1 and the inflow conditon (δ/h_1), and can be classified as shown in Fig. 1.

3. Each type of flow condition can be characterized from the dynamical pressure distribution.

4. The characteristics of undular jumps (h_{max}/h_1, L/h_2, and L_w/h_1) have been expressed by (3), (4), and (5).

References

1. CHOW, V.T. (1959), "Open Channel Hydraulics", McGraw-Hill International, New York.

2. ANDERSEN, V.M. (1978), "Undular Hydraulic Jump", Jour. of Hydr. Div., ASCE, Vol. 104, No.HY8, pp.1185-1188.

3. CHANSON, H. (1993), "Characteristics of Undular Hydraulic Jumps", Res. Report, No. CE 146, Dept. of Civ. Engrg., Univ. of Queensland, Australia.

4. Ohtsu, I. and Yasuda, Y. (1994), "Characteristics of Supercritical Flow below Sluice Gtae", Jour. of Hydr. Engrg., ASCE, Vol.120, No.HY3, pp.332-346.

5. Ohtsu, I., Yasuda, Y., and Gotou, H. (1993), "A Few Experiments on Undular Hydraulic Jump", 48th Annual Meeting, JSCE, Japan, p. II-159 (in Japanese).

1C15

EXPANDING USBR STILLING BASIN III

Roger BREMEN,[1] and Willi H. HAGER,[2]

[1] Cons. Engr., Lombardi Ltd., Via R. Simen 19, CH-6648 Minusio,
Switzerland
[2] Sr.Res.Engr., VAW, ETH-Zentrum, CH-8092 Zurich, Switzerland

Based on extensive model tests, the hydraulic design for an expanding stilling basin is developed and its limitations are specified. Considerations regarding scour, tailwater waves and plunging flow, among others, are included.

INTRODUCTION

Stilling basins for low head hydraulic structures are normally provided with baffle elements (Mason 1982). The best known example of a typical baffle basin is the USBR basin III (Peterka 1983). It comprises chute blocks to corrugate the approach jet and lift a flow portion from the floor, baffle blocks as impact elements and a triangular end sill sloping 1:2 to deflect remaining bottom currents to the surface of the tailwater flow. Provided the unit discharge is limited to $20m^2s^{-1}$, basin III be considered as a minimum length structure (Hager 1992).

A disadvantage of the USBR basin III is the prismatic geometry, and a transition structure to the tailwater is needed. To overcome this deficiency and to understand the effect of a widening basin, the abruptly expanding stilling basin was developed. Based on detailed experiments, the hydraulic jump in an abrupt expansion on a horizontal smooth channel *without* appurtenances behaves always poorer than the classical hydraulic jump (i.e. a jump in a prismatic horizontal and smooth rectangular channel). Therefore, Bremen and Hager (1993) recommended expanding stilling basins only for forced hydraulic jumps. The expanding stilling basin may be considered as a generalisation of the USBR III basin. The limitations regarding maximum approach velocity, unit discharge and tailwater characteristics are identical, therefore.

OPTIMUM BASIN GEOMETRY

A stable hydraulic jump under variable tailwater level can only be obtained with a T-jump in an expanding horizontal channel, with the toe upstream from the expansion section (Bremen and Hager 1993). The T-jump is transitional between the unstable spatial jump occuring for a lower tailwater level, and the classical jump located entirely in the prismatic approach channel. Fig.1 shows a definition sketch including the jump front (index «1») located at position $+x_1$ relative to the expansion section, the approach flow depth h_1, the approach width b_1 and the approach velocity V_1, thus $F_1 = V_1/(gh_1)^{1/2}$ is the approach Froude number. The *baffle sill* (index «s») was found to be the optimum element to force the jump. It is located at position x_s from the expansion section, has a height s, and a width b_s. The tailwater depth is h_2 at location x_j, where the deaeration of air swarms is completed and thus the main turbulence action accomplished. The location of the radial standing wave downstream of

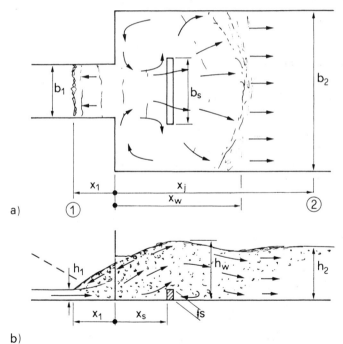

a)

b)

Fig.1 Abruptly Expanding Stilling Basin with Central Baffle Sill (a) Plan View, (b) Side View. (– – –) Sloping Approach Chute.

the sill is x_w, with the maximum wave height h_w. In actual designs, the approach channel upstream from the expansion section may be sloping in order to fix the toe location under variable tailwater level.

The performance of the expanding basin was rated with various criteria, involving (1) flow symmetry, (2) tailwater wave action, (3) corner vorticity as an index of flow divergence and (4) tailwater velocity distribution. Extended series of experiments involving tailwater channels 500 and 1500mm wide, width ratios $\beta=b_2/b_1=1.5$, 2, 3 and 5, and approach Froude numbers from 3 to 10 at approach flow depths between $h_1=25$ and 75mm were conducted (Bremen 1990). The location of the baffle sill was described with the position parameter $X_1=x_1/L_r^*$ where $L_r^* \cong 4.3 h_2^*$ is the roller length of the classical jump (superscript *) and its nondimensional height is $S=s/h_1$.

CENTRAL BAFFLE SILL

The sequent depth ratio $Y=h_2/h_1$ was found to be slightly influenced by the presence of the central baffle sill. For an expanding *free* hydraulic jump (index «j»), Bremen and Hager (1993) related the sequent depth ratio $Y_j=h_{2j}/h_1$ to the width ratio β and the position parameter X_1 as

$$\psi = \frac{Y^*-Y_j}{Y^*-1} = (1 - \beta^{-1/2})\,[1 - tanh(1.9X_1)] \,. \tag{1}$$

The function *tanh* is the tangenshyperbolic. The expansion effect vanishes for both $\beta=1$ (prismatic channel) or $X_1>1.4$ (classical jump in approach channel) and the asymptotic result of the classical Bélanger formula is $Y_j=Y^*$, with (Hager 1992)

$$Y^* = h_2^*/h_1 = \sqrt{2}\,\mathbf{F}_1 - (1/2). \tag{2}$$

Increasing either the expansion ratio β or approaching the toe position to the expansion section yields an increasing ψ-value, and thus a reduction of Y_j in Eq.(1). A smaller tailwater level is then required when compared to the classical hydraulic jump.

The effect of the central baffle sill on the sequent depths ratio $Y_s=h_{2s}/h_1$ is within $\pm 2\%$

$$Y_s/Y_j = 0.95. \tag{3}$$

Accordingly, the *tailwater level* of the forced jump in the expanding basin may be lowered by 5% as compared to the expanding free jump, according to detailed observations.

The *length of jump* L_{js} for the forced jump relative to the length $L_j^*=6h_2^*$ of the classical jump depends both on the expansion parameter $\psi(\beta, X_1)$ and F_1 as

$$L_{js}/L_j^* = (1 + \psi) (0.90 - 0.02F_1) . \qquad (4)$$

Compared to the expanding free jump, where the second paranthesis is equal to 1, a 20% length reduction occurs typically. No systematic effect of sill height and sill position was observed.

The *width of sill* b_s was optimised with regard to the standing swell at the sidewalls and the development of corner vorticity. The sill should extend over the approach channel width b_1, plus 25% of the expanding width (b_2-b_1), thus in total

$$b_s/b_1 = 1 + (1/4)(\beta-1). \qquad (5)$$

The *sill height* s has an optimum between underforcing with poor flow expansion, and overforcing with plunging flow behind the sill. The forcing of the jump increases when decreasing the relative sill position $X_s=x_s/L_r^*$ and with increasing the relative sill height S. Optimum sill action occurs for a linear combination of S and F_1 as

$$X_s = (3/4) (S/F_1)^{3/4} . \qquad (6)$$

The optimum sill height S thus increases both as the sill position parameter X_s and the Froude number F_1 increase. For USBR basin III the sill position is $x_s/h_2^*=0.8$ (i.e. increasing with F_1) and the sill height varies as $S=0.22F_1$. Therefore, a similar trend applies for the central baffle sill.

The corner vorticity is well-developed if the sill position x_s is equal to half the expansion width $(b_2-b_1)/2$. The *performance parameter* $X_p=2(x_s/b_1)/(\beta-1)$ should thus be between 0.8 and 1.2. Accordingly, the optimum sill geometry depends on the approach Froude number F_1, the expansion ratio β and the relative roller length L_r^*/b_1.

The *maximum wave height* h_w is almost independent of β and within $\pm 10\%$

$$\frac{h_w}{h_2} = 0.90 + \frac{1}{140}(15 - F_1)\frac{S}{X_s} . \tag{7}$$

For optimum basin flow, the value $h_w/h_{2s}=1.4$ is rarely exceeded. The location of the plunging point (Fig.1) does even not depend on S/X_s, and the data scatter ($\pm 15\%$) around the relation

$$\frac{x_w}{h_{2s}} = 5.6 - 0.2F_1 . \tag{8}$$

SCOUR CONSIDERATIONS

No detailed investigation on the scour behaviour was made, because Peterka has given detailed recommendations for the USBR III basin. The recommended end sill was inserted in the large test channel and its action was analysed with a 10mm uniform gravel. For a basin with an expansion ratio $\beta=3$, and for $F_1=7$, S=0.63, $X_s=0.20$, practically no erosion was visible after 15 minutes of operation. Only a –20mm erosion occured beyond the end sill in the axis domain, whereas large scour (–150mm) resulted for the equal experiment without the central baffle sill. A third run with a gradual expansion and a 1:1 expansion in plan view gave almost identical results as the abrupt expansion. From further tests no discernable effect of expansion angle in plan view was found provided the ratio is larger than 1:2 (26.5°), and $F_1>3$. For ease in construction, one may thus select any angle larger than indicated, and use the data previously refered to for the abrupt expansion. From scour experiments the *length of basin* L_B simply obtains

$$L_B = L_r^* . \tag{9}$$

The end sill is an essential part of the basin, and should be designed exactly according to Peterka (1983). His proposition is known to perform excellently, and it can thus directly be adopted for the modified design.

The *excavation volume* $V_s=b_2h_{2s}L_B$ of the expanding stilling basin is larger than for USBR basin III; no transition structure to the tailwater channel is needed, however, such that the overall length of the outlet structure is reduced. Economic considerations dictate the decision on the type of structure taken for a specific outlet works.

CONCLUSIONS

The modified USBR basin III involves an expanding dissipator geometry. This basin is suited for expansion ratios $1 < \beta \leq 5$, approach Froude numbers between 3 and 10, central sill heights $0.6 < S < 2$ and toe locations $X_1 \geq 0.2$. The central baffle sill is a significant component, whose optimum action was extensively tested in two model basins.

Although the excavation volume of the modified basin is larger than that of the original USBR basin III, the present design may be of interest when stilling basin and transition structure are combined. It involves a large expansion of flow and may be suitable for low head structures with a limited unit discharge. The experience gained with the popular USBR basin and the needs of a suitable transition structure are thus combined in the proposed design.

REFERENCES

- Bremen, R. (1990). Expanding Stilling Basin. Thesis 850, presented to the Swiss Federal Institut of Technology, EPFL: Lausanne.
- Bremen, R., and Hager, W.H. (1993). T-jump in abruptly expanding channel. *Journal Hydraulic Research* **31**(1): 61-78.
- Hager, W.H. (1992). *Energy Dissipators and Hydraulic Jump*. Kluwer Academic Publishers: Dordrecht, Boston.
- Mason, P.J. (1982). The choice of hydraulic energy dissipator for dam outlet works based on a survey of prototype usage. *Proc. Institution Civil Engineers* London, 72(5): 209-219. Discussions 74(2): 123-126.
- Peterka, A.J. (1983). *Hydraulic Design of Stilling Basins and Energy Dissipators. Engineering Monograph* 25, 7th printing. US Dep. Interior, Bureau of Reclamation: Denver.

1C16

THREE-DIMENSIONAL FEATURES OF
UNDULAR HYDRAULIC JUMPS

Hubert CHANSON
Dept. of Civil Engineering, The University of Queensland, Brisbane QLD 4072, Australia

1. INTRODUCTION

In open channels, an undular hydraulic jump is a stationary transition from super- to sub-critical flow. It is characterised by the development of regular and irregular free-surface undulations downstream of the jump. A recent study (CHANSON 1993) showed that the flow characteristics are functions of the upstream Froude number, the aspect ratio d_c/W and the inflow conditions. Further, the flow is three-dimensional (e.g. table 2).

With fully-developed inflows, five different flow regimes can be observed. Typical three-dimensional flow patterns include : lateral shock waves, free-surface recirculation, corner recirculation, roller (fig. 1). An important feature of undular jumps with fully-developed inflow is the lateral shock waves developing upstream of the first wave crest (fig. 1). The shock waves intersect slightly downstream of the top of the first wave. Further downstream, the crosswaves are reflected on the sidewalls and continue to propagate over long distances. CHANSON (1993) suggested a major flow redistribution between the start of the shock wave (SW) and the first wave crest (1C).

In this paper, new experiments are presented (table 1). They were performed in the 20-m long glass channel used by CHANSON (1993) and the upstream flows were a fully-developed boundary layer flows. For an aspect ratio $d_c/W = 0.286$, pressure and velocity distributions were recorded on the centreline ($z/W=0.5$), at $z/W = 0.25$ and $z/W = 0.05$ (i.e. near the sidewall) using a Pitot tube. For each position, the measurements were taken upstream of the jump (U/S), at the start of the shock wave (SW), at 1/3-distance between SW and the first wave crest (SW1), at 2/3-distance between SW and 1C (SW2), at the 1st wave crest (1C), 1st wave bottom (1B) and 2nd wave crest (2C) (see fig. 1).

The results provide new information on the three-dimensional flow redistribution immediately upstream of the first wave crest at undular hydraulic jumps with fully-developed inflow conditions.

2. PRESSURE DISTRIBUTIONS

Figure 2 presents dimensionless pressure distributions along the jump on the centreline (fig. 2A), at $z/W=0.25$ (fig. 2B) and near the sidewall (fig. 2C). The data are shown as $P/(\rho_w{}^*g^*d)$ versus y/d where P is the pressure and d is the local flow depth. At the upstream flow location (U/S), the pressure is hydrostatic. But along the jump, the data show explicitly that the pressure distributions are not hydrostatic.

On the channel centreline (fig. 2A), the pressure gradient $\partial P/\partial y$ is less than hydrostatic at the wave crests. And, at each wave trough, the pressure gradient is larger than hydrostatic as observed by CHANSON (1993).

All data indicate that, at each (x, z)-position located upstream of the lateral shock waves, the pressure gradient is larger than hydrostatic. And $\partial P/\partial y$ is less than hydrostatic downstream of the shockwaves : i.e., at $(SW1)_{z/W=0.05}$, $(SW2)_{z/W=0.05}$, $(SW2)_{z/W=0.05}$, $(1C)_{z/W=0.5}$, $(1C)_{z/W=0.25}$, $(1C)_{z/W=0.05}$.

Visual and photographic observations indicate that the free-surface is concave (i.e. curved upwards) upstream of the shockwaves and convex (i.e. $\partial^2 d/\partial x^2 < 0$) downstream of the crosswaves. With such streamline curvatures, the irrotational flow motion theory (e.g. ROUSE 1938, LIGGETT 1994) predicts a similar trend for the pressure distributions. But, at the wave crests, irrotational wave theories predict larger pressure gradient differences than those observed with undular jumps. Undular jump flows are stationary real-fluid flows and ideal-fluid flow theories cannot predict accurately their behaviour.

Fig. 1 - Flow pattern of undular hydraulic jump with fully-developed upstream flow

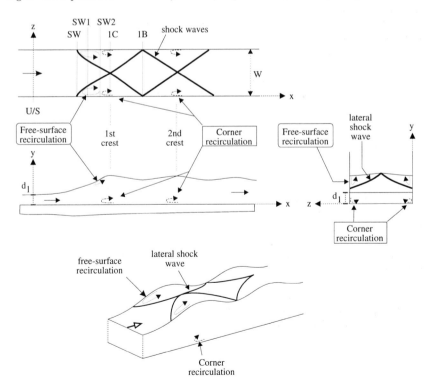

Fig. 2 - Pressure distributions on the centreline, at z/W = 0.25 and at z/W = 0.046

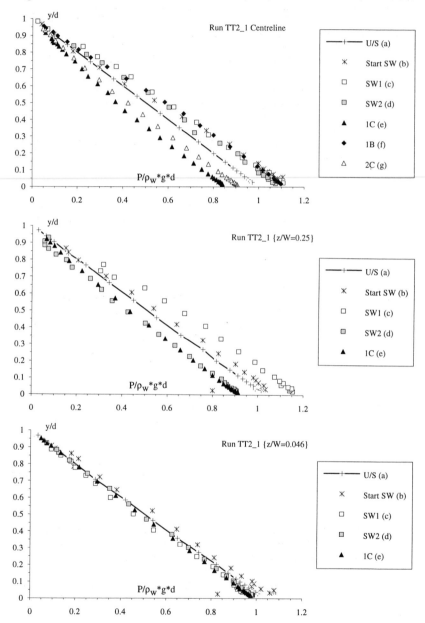

Table 1 - Experimental flow conditions

Run	q_w (m²/s)	Fr_1	x_1 (m)	Jump Type (CHANSON 1993)	Comments
TT2_1	0.06	1.35	10.38	C	Glass flume. W = 0.25 m.
TT2_2	0.06	1.14	9.0	A	Fully-developed upstream flows.

Table 2 - Free-surface profiles

Ref.	Run :	TT2_1			Run :	TT2_2		
	x (m)	d (m)			x (m)	d (m)		
		z/W=0.5	z/W=0.25	z/W=0.046		z/W=0.5	z/W=0.25	z/W=0.054
U/S	10.38	0.0587	0.0582	0.0575	9.0	0.0656	0.066	0.0664
SW	10.62	0.0618	0.0636	0.0646	10.0	0.0666	0.0662	0.067
SW1	10.69	0.0724	0.0800	0.0831	10.13	0.0674	0.0676	0.0718
SW2	10.75	0.0800	0.0911	0.0885	10.27	0.078	0.0773	0.0826
1C	10.82	0.1124	0.1006	0.0944	10.4.0	0.0968	0.090	0.088
1B	11.02	0.0859	0.0913	0.0959	10.60	0.0782	0.0845	0.0854
2C	11.22	0.116	0.1066	0.1043	10.85	0.0991	0.098	0.0954

3. VELOCITY DISTRIBUTIONS

Figure 3 presents the dimensionless velocity distributions at several cross-sections. All figures are plotted with the same scale.

Figure 3A (top left) shows a small velocity redistribution between the upstream flow (U/S) and the start of the shock waves (SW). Further, both figures 3B (top right) and 3C (bottom left) indicate a strong flow deceleration near the free-surface downstream of the shock waves : i.e., at $(SW1)_{z/W=0.05}$, $(SW2)_{z/W=0.05}$, $(SW2)_{z/W=0.05}$. Indeed, the shockwaves (also called oblique jumps) are associated with local energy dissipation and hence a loss of kinetic energy near the free-surface.

At the first wave crest (1C), a strong velocity deceleration is observed also next to the free-surface. Indeed, the shockwaves intersect at the first wave crest and a small "cockscomb" roller might take place. These processes enhance the energy dissipation at the free-surface and cause a local velocity reduction particularly on the centreline.

4. DISCUSSION

For the present series of experiments, the lateral shock waves occur for $Fr_1 > 1.2$ independently of the aspect ratio. Other researchers (FAWER 1937, IWASA 1955, IMAI and NAKAGAWA 1992) observed also lateral shock waves at undular hydraulic jumps. Visual observations indicate that the shock waves have a 'roller' form.

The importance of the lateral shock waves (or "Mach" waves) was highlighted by MONTES (1986) and CHANSON (1993). The crosswaves characterise a flow separation mechanism near the sidewalls. The sidewall boundary layers are subjected to an adverse pressure gradient (i.e. $\partial P/\partial x < 0$) and a flow separation is observed next to the free-surface (fig. 1). The lateral boundary layers force the apparition of critical conditions near the wall sooner than on the channel centreline. The shockwaves propagate then along the channel.

Fig. 3 - Velocity distributions at SW, SW1, SW2 and 1C

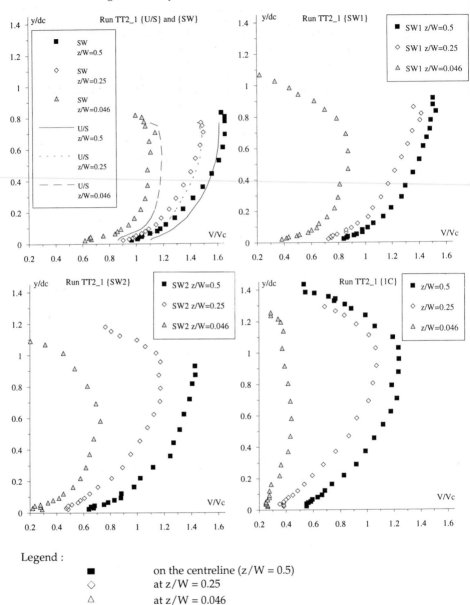

Legend :

■	on the centreline (z/W = 0.5)
◇	at z/W = 0.25
△	at z/W = 0.046

5. CONCLUSION

The three-dimensional characteristics of undular hydraulic jumps with fully-developed upstream flows have been investigated experimentally. The results are in agreement with earlier studies (FAWER 1937, CHANSON 1993). They confirm the importance of the lateral shock waves and their effects on the flow field.

Immediately upstream of the lateral shock waves, the free-surface is concave, and the pressure gradient is larger than hydrostatic. Downstream of the shock waves, the free-surface is convex, the pressure gradient is smaller than hydrostatic and a strong velocity deceleration is observed next to the free-surface.

ACKNOWLEDGMENTS

The author acknowledges the assistance of MM. B. TREVILYAN and L. TOOMBES, and the discussion with Dr J.S. MONTES (University of Tasmania, Australia).

REFERENCES

CHANSON, H. (1993), *Research Report No. CE146*, Dept. of Civil Engineering, University of Queensland, Australia, Nov..

FAWER, C. (1937), *Thesis*, Lausanne, Switzerland, Imprimerie La Concorde, 127 pages.

IMAI, S., and NAKAGAWA, T. (1992), *Acta Mechanica*, Vol. 93, pp. 191-203.

IWASA, Y. (1955), *Proc. 5th Japan Nat. Cong. for Applied Mechanics*, Paper II-14, pp. 315-319.

LIGGETT, J.A. (1994). "Fluid Mechanics." *McGraw-Hill*, New York, USA.

MONTES, J.S. (1986), *Proc. 9th Australasian Fluid Mech. Conf.*, Auckland, NZ, pp. 148-151.

ROUSE, H. (1938). "Fluid Mechanics for Hydraulic Engineers." *McGraw-Hill*, New York, USA.

LIST OF SYMBOLS

d   flow depth measured perpendicular to the channel bottom (m);

d_c critical flow depth(m) : in a rectangular channel $d_c = \sqrt[3]{q_w/g}$;

Fr   Froude number;

g   gravity acceleration : $g = 9.80$ m/s^2 in Brisbane, Australia;

P   pressure (Pa);

q_w discharge per unit width (m^2/s);

V   velocity (m/s);

V_c critical flow velocity (m/s);

W   channel width (0.25 m);

x   distance (m) along the flume (fig. 1);

y   distance (m) from bottom measured perpendicular to the channel bottom;

z   horizontal distance (m) from wall;

ρ_w water density (kg/m^3);

Subscript

c   critical flow conditions;

1   upstream flow conditions.

1C17

Self-Induced Sloshing Caused by Jet from Rectangular Tank Wall

M. Fukaya[*1], M. Baba[*1], K. Okamoto[*1] and H. Madarame[*1]

[*1] Nuclear Engineering Research Laboratory
University of Tokyo, Tokai-mura, Ibaraki, Japan

ABSTRACT

Self-induced sloshing caused by an interaction between the surface and flow was recently discovered by the authors. The flow pattern had a great effect on the sloshing growth. In this study, in order to inquire further into the effect of the flow pattern on the self-induced sloshing, an inlet jet angle and a size of rectangular test tank were changed from the previous study. The self-induced sloshing was observed with a certain flow pattern despite the difference of the inlet jet angle. The sloshing growth conditions were compared with the previous data. As a result, the governing parameter of the sloshing growth was founded to be the Strouhal number or the Froude number.

INTRODUCTION

Self-induced sloshing is one of the free surface oscillations which are caused by the flow itself without any external force. Okamoto et al.[1] recently discovered the self-induced sloshing in a rectangular tank having a horizontal jet. The free surface oscillated with the fundamental mode of the sloshing. The sloshing was observed in a certain water level and inlet velocity condition. The sloshing frequency was nearly equal to that of fluid without circulating flow in the tank. The same phenomenon was reported[2, 3] in the other different tank geometries.

Okamoto et al.[1] proposed the sloshing growth model based on the flow pattern transformation. Madarame et al.[4] explained the phenomenon by the superposition of the sloshing potential on that of circulating flow. Takizawa et al.[5] claimed that the sloshing was excited by the potential variation caused by the secondary flow. However, the growth mechanism has not been exhaustively explained, then further experimental and calculational studies are needed.

In this study, in order to evaluate the sloshing growth mechanism, geometrical effects on the sloshing was experimentally investigated. The sloshing growth mechanism was considered to have a great relation with the flow pattern in the tank. The flow pattern was thought to be affected by the inlet jet angle. Since the effect of the inlet jet angle on the sloshing was not experimented yet, the inlet jet angle was selected as the geometrical parameter.

EXPERIMENT

Figure 1 schematically shows the test tank. The thin rectangular tank had an inlet at the left side wall and an outlet at the bottom. The tank was a small scale model of that in the previous study[1]. Then, the effect of the tank size difference on the self-induced sloshing was able to be evaluated. In this study, the inlet jet angle(θ) and the inlet height(B) from the bottom were taken as geometrical parameters. The tank geometries were summarized as the Cases A ~ E in Table 1. The length of inlet channel(L_i) was over 6 times as large as the inlet hydraulic diameter.

Plane jet was injected into the test tank from the inlet channel. The inlet flow was supplied by a head tank which was pumped up from the dump tank. The water level in the head tank was maintained stable by overflow, therefore a pump vibration was not observed in the inlet jet. The flow rate was valve controlled, and measured by rotameters. The inlet velocity(V) was defined as an averaged velocity at the inlet, which was calculated dividing the flow rate by the inlet cross section. The water level(H) was defined as the mean level from the inlet center. Okamoto et al.[1] reported that the water level from the inlet center was more effective than that from the bottom. The H was controlled by an overflow tank gate which was connected to the downstream of the test tank. The water level data were measured using a condenser type level meter. The dominant frequency of the sloshing was obtained from the level data analyzing with Fast Fourier Transformation(FFT).

The tank geometries in the previous study[1] were also shown as the Cases PA ~ PC in Table 1.

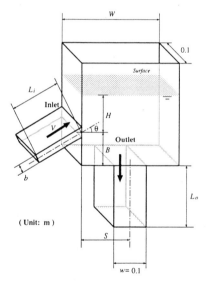

(Unit: m)

Figure 1. Test tank

Table 1   Geometrical parameters

Case	θ [deg]	B [m]	W [m]
A	0	0.10	0.30
B	30	0.10	0.30
C	45	0.10	0.30
D	30	0.05	0.30
E	30	0.15	0.30
PA	0	0.25	1.00
PB	0	0.45	1.00
PC	0	0.25	0.75

Case	b [m]	L_i/b	L_o/w	S/W
A-E	0.03	10	3	0.5
PA-PC	0.10	6	5	0.5

RESULTS AND DISCUSSION

Flow Pattern and Sloshing Region

In the stable free surface condition, there were two typical flow patterns in the test tank as shown in Fig. 2. At the low water level condition, the inlet jet went directly to the surface, resulting in a large clockwise circulating flow(Flow Pattern I). When the water level was high, twin circulations were formed at the both side of the jet showing a counterclockwise circulating flow in the upper region of the tank(Flow Pattern II).

Figures 3 and 4 show the flow pattern map of the Case B with respect to the inlet velocity and the water level. As shown in Figs. 3 and 4, a hysterisis of the flow pattern appeared. When the water level increased with a constant inlet velocity(Fig.3), the Flow Pattern I was observed until $H = 0.13$ [m]. Over the threshold water level, the Flow Pattern changed from I to II. Contrariwise, when the water level decreased(Fig.4), the Flow Pattern II was observed until $H \approx 0.08$ [m]. The hysterisis of the flow pattern remarkably appeared in the Cases B ~ E having a non-horizontal inlet jet. Such a hysterisis was not observed in the previous Cases PA ~ PC having a horizontal inlet jet.

In Figs. 3 and 4, the self-induced sloshing conditions were also plotted with triangles ($\triangle$ or $\blacktriangle$). The solid mark($\blacktriangle$) denoted a transitional condition. At the transitional condition, the sloshing was observed initially, however, the amplitude of the sloshing increased with time, resulting in the stable Flow Pattern I. In this study, the above condition was plotted with solid marks($\bullet$, $\blacktriangle$, $\blacksquare$, $\cdots$), and the solid marks region was called "Transitional Region".

As shown in Figs. 3 and 4, when the sloshing was observed, only the Flow Pattern II was formed. During the sloshing, the pathline of the inlet jet was winded as shown in Fig. 2.

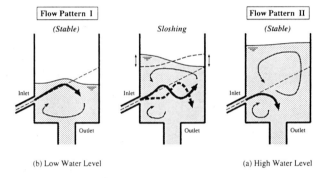

(b) Low Water Level   (a) High Water Level

Figure 2. Typical flow patterns

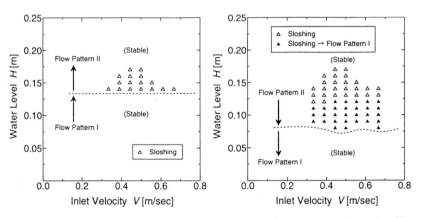

Figure 3. Sloshing region increasing H in the Case B

Figure 4. Sloshing region decreasing H in the Case B

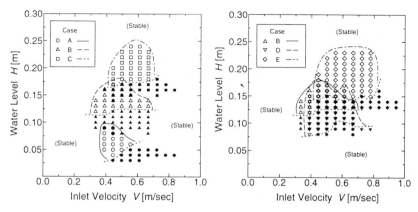

Figure 5. Effect of θ on sloshing region   Figure 6. Effect of B on sloshing region

The sloshing also occurred in the other tank geometries(Cases A,C~E). The flow pattern of the sloshing in all the Cases was the Flow Pattern II, while the sloshing was not observed with the Flow Pattern I. In the previous Cases PA~PC, the sloshing also occurred only with the Flow Pattern II. Then, the sloshing was considered to occur only with the Flow Pattern II.

Based on the the results of the Cases A~C, the effect of the inlet jet angle(θ) on the sloshing was investigated. The tank geometries of the Cases A~C were almost similar except the θ. Figure 5 shows the sloshing regions in the Cases A~C, which were mapped under the decreasing water level condition. The sloshing occurred only when the inlet velocity was 0.3 ~ 0.8 [m/s]. The water level of the sloshing region increased with increasing the inlet jet angle. Since the inlet jet more directly went to the surface in the larger θ geometry, the Flow Pattern I could be formed until the higher water level condition. The sloshing did not occur with the Flow Pattern I. Therefore, the sloshing region shifted to the higher water level condition with the larger θ geometry.

In the Cases B,D and E, the effect of the inlet height(B) on the sloshing was evaluated. The Cases B,D and E had almost similar tank geometries except the B. Figure 6 shows the sloshing regions of the Cases B,D and E. The inlet velocity for the sloshing region was 0.3 ~ 0.8 [m/s], while the water level for the sloshing region was different. In the previous study(1) having a horizontal inlet jet, the difference of the inlet height(B) had little effect on the water level condition for the sloshing. However, the inlet height could have a certain effect on the flow pattern with a non-horizontal inlet jet, resulting in the difference of the sloshing region. Therefore, in order to investigate the sloshing mechanism, the relation between the flow pattern and the sloshing should be more examined.

Strouhal Number

In order to evaluate the self-induced sloshing condition, the sloshing region map was rearranged with nondimensional parameters. The water level(H) was nondimensionalized by the tank width(W), i.e., H/W. The inlet velocity(V) was nondimensionalized by the sloshing frequency(f) and the tank width(W) as the following Strouhal Number(St).

$$St = \frac{fW}{V} \tag{1}$$

Figure 7 shows the rearranged sloshing regions in the Cases A~E and the previous Cases PA~PC. In this figure, the Transitional Regions were not shown. The sloshing regions distributed 0.1 ~ 0.8 in respect of the H/W. The Strouhal number for the sloshing in the Cases A~E distributed about 0.6 ~ 1.4 with the center value 1.0. The result nearly corresponded with the data of the previous Cases PA~PC. The V/f means the distance that the vertical fluctuation of the jet at the inlet travels in a sloshing period. When the Strouhal number was 1.0, the V/f was equal to the W. The sloshing could occur when the traveling distance was nearly equal to the tank width.

The Strouhal number for the sloshing was confirmed to be about 1.0 regardless of the tank geometries as the inlet jet angle and the tank size. Therefore, the Strouhal number was considered to be one of the governing parameters of the sloshing.

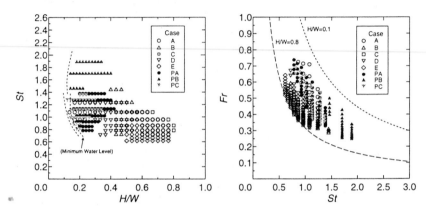

Figure 7. Strouhal Number   Figure 8. Froude Number

Froude Number

Froude number is the representative nondimensional parameter which describes the states of the free surface and flow. The self-induced sloshing was considered to be caused by the interaction between the surface and flow, therefore the sloshing condition was evaluated with the Froude number. Since the interaction between the surface and the jet was important, the water level difference between the surface and the jet, i.e., the H, was used as a representative length. Then, the Froude number(Fr) was defined as follows.

$$Fr = \frac{V}{\sqrt{gH}} \tag{2}$$

where g was the gravitational acceleration.

Figure 8 shows the sloshing regions without the Transitional Regions in respect of the Strouhal number and the Froude number. The sloshing occurred when the Froude number was 0.25 ~ 0.75. For the sloshing regions, the Froude number was in inverse proportion to the Strouhal number. The reason was explained as follows.

In the Cases A~E and the previous Cases PA~PC, the measured dominant frequencies were nearly equal to the theoretical sloshing frequencies without circulating flow. When the water level is high, the theoretical sloshing frequency is expressed as follows.

$$f = \frac{1}{2\pi}\sqrt{\frac{\pi g}{W}} \tag{3}$$

With substituting Eqs. (1) and (3) into Eq. (2), we got the following equation.

$$Fr = \frac{K}{St} \qquad K = \frac{1}{2\sqrt{(H/W)\pi}} \tag{4}$$

As shown in Fig. 7, the sloshing occurred when H/W was 0.1 ~ 0.8. Equation (4) explained that the Froude number had an inverse proportional relation to the Strouhal number at the sloshing as shown in Fig.8. Therefore, the Strouhal number and the Froude number were not independent and equivalent as the governing parameters of the sloshing.

CONCLUSION

The effect of the flow pattern on the self-induced sloshing was experimentally investigated, which was caused by jet from a rectangular tank wall.

(1) The self-induced sloshing occurred only when a certain flow pattern(Flow Pattern II) was formed. When another flow pattern(Flow Pattern I) was formed, the self-induced sloshing was not observed.

(2) The self-induced sloshing could occur despite the difference of the inlet jet angle.

(3) The governing parameter of the self-induced sloshing was the Strouhal number and/or the Froude number.

REFERENCES

(1) Okamoto, K., et al., Proc. IMechE, Flow Induced Vibrations, 1991-6, p.539.

(2) Amano, K., et al., Trans. of JSME (in Japanese), B, 57-538, (1991), p.1947.

(3) Eguchi Y., et al., Proc. Fall Meeting of the Atomic Energy Society of Japan (in Japanese), (1990), p.286.

(4) Madarame H., et al., ASME PV&P, Vol.232, (1992), p.5.

(5) Takizawa, A. and Kondo, S., Proc. Int. Conf. on Nonlinear Mathematical Problems in Industry, Iwaki, Japan, (1992).

1C18

THRESHOLD VELOCITY OF SAND MIXTURES

GUOLIANG YU
Currently at
School of Civil Engineering
The University of Birmingham

DUO FANG
State Key Hydraulics Laboratory
Sichuan Union University
Chengdu, Sichuan, China

ABSTRACT

A formula for the threshold velocity of sand mixtures is presented based on the following considerations. A small number of particles with the same size are at the threshold of motion. The local velocity acting on these particles is affected by nearby sands with the size d_{95}. The shade and exposure effects are taken into account by assuming that the river bed consists of uniform sands with the mean diameter of the mixture. A formula for threshold velocity for uniform bed load is also obtained as uniform bed load is a simple case of the mixture. Both formulae are tested against experimental and field data and good agreement achieved.

INTRODUCTION

Sediment threshold is very important for development and utilization of rivers and canals. Considerable research has been conducted and the topic continues to be one of basic active research (Hubbell and Sayre, 1964; Gesseler, 1976; Einstein, 1972; White, 1975, Yalin and Karahan, 1979; Du, 1963; Tang, 1963; Hua, 1965). Most of these however focus on uniform sands. Recent studies in sediment threshold motion of sand mixtures take into account the shade effect of bigger particles on smaller particles, and the exposure effect of smaller particles on bigger particles (Chang, 1988, Yu, 1991). Threshold motion of single-sized graded type of mixtures and bi-domal graded type of mixtures seem to obey different principles (Wu, 1984).

When considering threshold conditions or sediment transport rate, different view points should be considered. The shade effect is more important for sediment transport rate than the exposure effect. Conversely the exposure effect is more important for sediment incipient motion because of the lower probability associated with this, rather than the threshold of motion. It follows therefore that the local turbulent velocity is crucial to sediment threshold.

THRESHOLD VELOCITY OF SAND MIXTURES

The shade and exposure effects are taken into account by assuming that the river bed consists of uniform sands with the same mean diameter of the mixture. According to an analysis of the forces and momentum on a sand particle with a diameter, d, which lies on a channel bed with bed material mean diameter D, it has a momentum due to gravity as follows:

$$M_w = \frac{1}{2}(\gamma_s - \gamma)\alpha_1 d^3 \frac{dD}{d+D} \tag{1}$$

a momentum due to the drag force:

$$M_d = \frac{1}{2}C_d\alpha_2 \frac{\rho u_d^2}{2} \frac{d^3\sqrt{d^2+2dD}}{d+D} \tag{2}$$

a momentum due to the lift force:

$$M_l = \frac{1}{2}C_l\alpha_3 \frac{\rho u_d^2}{2} \frac{d^3 D}{d+D} \tag{3}$$

a momentum due to the cohesive force:

$$M_c = \kappa\varsigma_c d^2 \left(\frac{\gamma'}{\gamma'_c}\right)^{10} \frac{D}{d+D} \tag{4}$$

where γ_s is the specific weight of sand, γ is the specific weight of water, α_1 is volume coefficient of the sand, α_2 and α_3 are coefficients of area, u_d is the averaged velocity acting on the sand, C_d is the drag coefficient, C_l is the lift coefficient and $\kappa\varsigma_c$ is the cohesive coefficient, $\kappa\varsigma_c = 7.55*\pi*10^{-7} kg/m$. (Tang, 1963), γ' is the cavity ratio and γ'_c is the cavity ratio at static status. For a spherical sand particle, $\alpha_1 = \pi/6$, $\alpha_2 = \alpha_3 = \pi/4$. When the sand at the threshold of motion, it must satisfy the condition that all momentum reach equilibrium, i.e.,

$$M_d + M_l = M_w + M_c \tag{5}$$

Substituting Eqs. (1) to (4) into Eq. (5) and rewriting Eq. (5), gives the local threshold velocity of a mixture of sands as:

$$u_d = \sqrt{\frac{\dfrac{4}{3}\dfrac{\gamma_s-\gamma}{\gamma}gd + \dfrac{1.21\times10^{-5}}{\rho d}\left[\dfrac{\gamma'}{\gamma'_c}\right]^{10}}{C_l + C_d \dfrac{d}{D}\sqrt{1+2\dfrac{D}{d}}}} \tag{6}$$

When d=D, the bed material is uniform. The threshold velocity of uniform sands may therefore be written from Eq. (6) in the form

$$u_d = \sqrt{\frac{\dfrac{4}{3}\dfrac{\gamma_s - \gamma}{\gamma}gd + \dfrac{1.21 \times 10^{-5}}{\rho d}\left[\dfrac{\gamma'}{\gamma'_c}\right]^{10}}{C_l + \sqrt{3}C_d}} \tag{7}$$

For a mixture of bed material, when the sand size is less than the mean diameter, i.e., d<D, and because

$$\frac{d}{D}\sqrt{1 + 2\frac{D}{d}} < \sqrt{3},$$

it follows that:

$$u_d = \sqrt{\frac{\dfrac{4}{3}\dfrac{\gamma_s - \gamma}{\gamma}gd + \dfrac{1.21 \times 10^{-5}}{\rho d}\left[\dfrac{\gamma'}{\gamma'_c}\right]^{10}}{C_l + C_d\dfrac{d}{D}\sqrt{1 + 2\dfrac{D}{d}}}} > \sqrt{\frac{\dfrac{4}{3}\dfrac{\gamma_s - \gamma}{\gamma}gd + \dfrac{1.21 \times 10^{-5}}{\rho d}\left[\dfrac{\gamma'}{\gamma'_c}\right]^{10}}{C_l + \sqrt{3}C_d}}$$

That means that when a sand with a diameter d, where d<D, its threshold velocity in a mixed sand bed is larger than that in a uniform sand bed, i.e., threshold motion of those small particles in a mixture is more difficult than that in a uniform bed. Conversely, when the diameter of a sand is bigger than mean diameter, i.e., d>D the threshold velocity in a mixed sand bed is smaller than that in a uniform sand bed, i.e., threshold motion of those small particles in a mixture is easier than that in a uniform bed.

For a well graded mixture, the velocity acting on a sand particle is not only associated with the sand size d, the water depth and the depth-averaged velocity, but also with those sand sizes surrounding it. Although water separation following a bigger sand particle can have some shade effect, it also induces bursts and sweeps. It is the separation that makes the local velocity at many places nearby the same as that at the height of the particle size. According to our experimental observation, the threshold velocity acting on the sand may be represented by,

$$u_d = \frac{m}{m+1}U\left(\frac{d_{95}}{h}\right)^{1/m} \tag{8}$$

where d_{95} is the diameter which 95% of mixture sand are smaller, U is the depth-averaged velocity, h is the water depth and m is a coefficient. For natural rivers m=6 and for laboratory conditions without sand waves,

$$m = 4.7\left(\frac{h}{D}\right)^{0.06} \tag{9}$$

The drag coefficient, C_d=0.4 and the lift coefficient C_l=0.1. Therefore, the depth-averaged threshold velocity of a sand particle with size, d, in a mixed sand bed may be given by

$$U_c = \frac{m+1}{m}\left(\frac{h}{d_{95}}\right)^{1/m}\sqrt{\frac{\dfrac{40}{3}\dfrac{\gamma_s-\gamma}{\gamma}gd+\dfrac{1.21\times10^{-4}}{\rho d}\left[\dfrac{\gamma'}{\gamma'_c}\right]^{10}}{1+4\dfrac{d}{D}\sqrt{1+2\dfrac{D}{d}}}} \tag{10}$$

the depth-averaged threshold velocity of a uniform sand bed may be given by

$$U_c = \frac{m+1}{m}\left(\frac{h}{d}\right)^{1/m}\sqrt{\frac{8}{3}\frac{\gamma_s-\gamma}{\gamma}gd+\frac{1.51\times10^{-5}}{\rho d}\left[\frac{\gamma'}{\gamma'_c}\right]^{10}} \tag{11}$$

VERIFICATION USING EXPERIMENTAL AND FIELD DATA

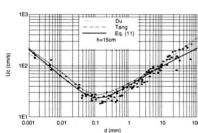

Fig. 1 Relationship between incipient velocity and sand size with cohesion

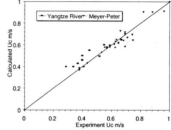

Fig. 2 Comparison between the calculated and experiment or field data.

As shown in Fig. 1, the suggested relation, Eq. (11), lies between the lines given by Tang and Du (Chien and Wan, 1986), when the water depth is 0.15m, $\gamma'/\gamma'_c=1.35$. For h/d<10, Eq. (8) needs to be revised, otherwise, the threshold velocity will be under predicted. Despite this a reasonable agreement was achieved.

Data used for the initial assessment on the uniform threshold relationship Eq.(11) include 15 field data (Han, 1988, listed below in Table 1) and experimental data (Meyer-Peter, 1948). Fig. 2 shows that the experimental and calculated values are well centralised on both sides of the perfect line of agreement.

Table 1 Field Data of Yangtze River

h (m)	d (mm)	U_{Field} (m/s)	$U_{Calculation}$ (m/s)	h (m)	d (mm)	U_{Field} (m/s)	$U_{Calculation}$ (m/s)
2.47	0.27	0.40	0.46	9.7	0.291	0.55	0.60
6.1	0.21	0.40	0.50	2.81	0.187	0.37	0.43
9.4	0.24	0.45	0.55	3.07	0.120	0.41	0.40
1.29	0.24	0.59	0.60	1.63	0.144	0.38	0.37

17.0	0.28	0.67	0.65	2.06	0.144	0.38	0.39
19.6	0.21	0.65	0.62	2.08	0.176	0.29	0.40
9.7	0.258	0.57	0.58	1.97	0.170	0.34	0.40
9.7	0.274	0.57	0.57	-	-	-	-

The experiments for well graded mixtures were carried out at the State Key Hydraulics Laboratory in Sichuan Union University, China. Two flumes were used for the experiments. One flume was 30 m long and 0.5 m wide. The other had a length of 68 m and width of 1.5 m. The slope and discharge were adjustable for both flumes. Their side walls were made of transparent glass, which allowed the convenience of visual observations. Water was supplied from a big sump with a constant head and circulated by a pump. The discharges were measured with a triangular weir installed beyond the entrance. An adjustable tailgate controlled the flow depth at the outlet end.

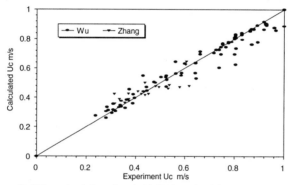

Fig. 3 Comparison between the calculated and experiment data.

In all some 102 sets of experiment were conducted to investigate the maximum threshold diameters and threshold velocities for well graded sand beds. The mixtures of 16 runs among them are bi-domal graded types. The characteristics of the data are summarized as follows:

d (mm)	D (mm)	D_{95} (mm)	h (cm)	U_c (m/s)
0.5~21.0	1.41~7.7	7.5~21.0	2.64~19.99	0.239~1.000

The results predicted by Eq. (11) are consistent with experiments and field data, as shown in Fig. 3.

CONCLUSIONS

Threshold conditions of which a small number of sands with the same size are at the threshold of motion are analytically investigated by momentum equilibrium. The local velocity acting on this kind of sand is significantly affected by nearby larger sand particles with the size d_{95}. The shade and exposure effects can be taken into account by assuming that the river bed consists of uniform sands with the mean diameter of the mixture. The threshold velocity formulae, i.e., Eqs. (10) for mixture bed load and (11) for uniform bed load, are obtained and both give good agreement with experimental and field data.

REFERENCES

Chien N., and Wan, Z.H., 1986, "Mechanics of sediment transport", *Press of Chinese Science.*

Chang, S.G., 1988, "On the threshold of non-uniform bed load", *Master thesis, Chengdu University of Science and Technology.*

Du, G.R., 1963, "On sediment threshold", *Chinese Journal of Hydraulics, No. 4, 1963.*

Gessler, J., 1976, "Stochastic aspects of threshold motion on river beds", *In stochastic approach to water resources, Vol. 2, by H W Shen (ed).*

Han, Q.W., 1982, "Threshold principle and velocity of sediment", *Chinese Sediment Research, No.6, 1982.*

Hua, G.X., 1965, "Threshold velocity of sediment", *Journal of Chengdu University of Science and Technology, No. 1, 1965.*

Hubbell, D.W. and Sayre, W.W.,1964," Sand transport study with radioactive tracers", *ASCE, HY 3, Vol.90.*

Meyer Peter, E. and Muller, R., 1948, "Formula for bed load transport", *Proc., 2nd. Meeting Intern. Assoc. Hyd. Res., Vol.6.*

Tang, C.B., 1965, "Threshold motion of cohesive sands", *Chinese Journal of Hydraulics, No. 4.*

White, W.R. et al., 1975, "Sediment transport theories; a review", *Proc. Ist. Civil Engrs., Vol. 59, Part 2.*

Wu, X.S., 1984, "Formation and threshold motion of wide-graded bed materials", *Master thesis, Chengdu University of Science and Technology.*

Yalin, M.S. and Karahan, E., 1979, "Threshold of sediment transport", *ASCE, HY11, Vol.105.*

Yu, G.L., 1991, "Two dimensional numerical modelling for alluvial rivers", *PhD Dissertation, Chengdu University of Science and Technology.*

1C19

SEDIMENT TRANSPORT IN UNSTEADY FLOW

W. H. Graf and T. Song
Laboratoire de Recherches Hydrauliques
École Polytechnique Fédérale de Lausanne, Switzerland

1. INTRODUCTION

A limited series of experiments were performed to study sediment transport under unsteady conditions in open-channel flow.

The relation developed by Graf and Acaroglu (1968) and extended by Graf and Suszka (1987) was used to calculate the intensity of sediment transport, Φ_A, with the intensity of shear, $\Psi_A = 1/\tau_*$:

$$\Phi_A = 10.4 \, K \left(\Psi_A\right)^{-1.5} \tag{1}$$

where:

$$
\begin{aligned}
K &= \Psi_A^{-1} &&\text{for } \Psi_A \leq 14.6 \\
K &= (1 - 0.045 \, \Psi_A)^{2.5} &&\text{for } 22.2 \geq \Psi_A > 14.6 \\
K &= 0 &&\text{for } \Psi_A > 22.2
\end{aligned}
$$

The respective intensities are defined as:

$$\Psi_A = \frac{(s_s - 1) \, d}{S_e \, R_{hb}} = \frac{(\gamma_s - \gamma) \, d}{\tau_o} \tag{2}$$

$$\Phi_A = \frac{C_s \, U \, R_{hb}}{\sqrt{(s_s - 1) \, g \, d^3}} = \frac{(q_s/q)U \, R_{hb}}{\sqrt{(s_s - 1) \, g \, d^3}} \tag{3}$$

where S_e is the energy slope (to be taken as the bottom slope $S_e = S_o$ in uniform flow), R_{hb} is the hydraulic radius of the bed and U is the average flow velocity. $C_s = q_s/q$ is the volumetric concentration of the sediments in the section; $q_s = g_s/\gamma_s$ is the unit sediment transport in volume, and g_s is the one in weight; q is the water flow in volume. The sediment is characterized by its density, $s_s = \gamma_s/\gamma$, and the equivalent particle diameter, $d=d_{50}$.

This relation, eq.1, will be tested firstly for its validity for the prediction of sediment transport in uniform flow and consequently for its use in non-uniform unsteady flow.

2. EXPERIMENTAL SET-UP

A recirculating tilting flume, being 17.8 m long and 0.6 m wide, was used (Fig.1). An *electromagnetic flowmeter*, installed in the supply conduit (underneath), measured the discharge passing through the channel. Six *ultrasonic limnimeters* were mounted to measure the flow depth during the passage of the unsteady flow hydrographes. An *Acoustic Doppler Velocity Profiler (ADVP)* was used to measure instantaneously the vertical velocity profiles (Song, Graf and Lemmin, 1994). This profiler, put into a streamline housing, was installed above the bed at such a level as to be during the passage of the hydrographes always in water. In this way, the inner region (y/D < 0.20) of the velocity profile was measured; the shear stress, τ_o, of the flow could thus be obtained. A *sediment trap*, consisting of the basket and strain gauge, was used to collect and to measure continuously the sediment transport.

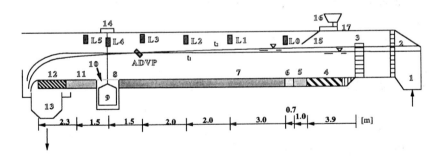

1 headbox; 2 grid; 3 straightening tubes; 4 rigid bed (concrete cubes); 5 rigid bed (sediment); 6 movable bed covered by wire netting; 7 measuring section with movable bed; 8 sediment trap; 9 basket; 10 grid of bars; 11 rigid bed (sediment); 12 rigid bed; 13 collection tank; 14 strain gauge; 15 plastic plate; 16 sediment-supply hopper; 17 sediment-transporting gutter; L1 to L5 limnimeters; ADVP Acoustic Doppler Velocity Profiler.

Fig.1 Experimental set-up.

3. EXPERIMENTS

Each hydrograph was repeated two times. Using a discrete Fourier transform, the mean flow velocity and the velocity fluctuation were obtained (Song and Graf, 1995).

The overall characteristic of a hydrograph was expressed by an unsteadiness parameter proposed by Graf and Suszka (1985):

$$\Gamma_{HG} = \frac{\Delta D}{\Delta T} \frac{1}{u_{b*}} \qquad (3)$$

where u_{b*} is the friction velocity of the base flow, $\Delta D = D_{max} - D_b$ is the difference between the maximum and the base flow depth and $\Delta T = \Delta T_R + \Delta T_D$ is the total time duration of the hydrograph.

Three different hydrographs were passed through the channel, whose hydraulic parameters are listed in Table 1.

Table 1 Summary of the flow parameters

No.	S_o	u_{b*}	D_b	ΔD	U	q	ΔT_R	ΔT_D	Γ_{HG}
(-)	(%)	(cm/s)	(cm)	(cm)	(cm/s)	(l/cm.s)	(s)	(s)	10^3
S75-944		9.34	15.5	5.7	95.7 - 116.1	1.5 - 2.5	40	48	6.1
S75-945	0.75	9.30	15.5	5.1	94.5 - 113.4	1.5 - 2.3	20	32	10.5
S75-946		9.30	15.5	3.8	93.2 - 110.4	1.5 - 2.1	12	22	16.1

S_o - bed slope; D_b - the water depth of the base flow; $\Delta D = D_{max} - D_b$ - difference between the maximum and the base flow depth; q - unit water discharge; U - cross-sectional average velocity; $Fr = U/\sqrt{gD}$ - Froude number; u_{b*} - the friction velocity of the base flow using Clauser's method; ΔT_R - time duration of the rising branch of a hydrograph; ΔT_D - time duration of the falling branch of a hydrograph; $\Delta T = \Delta T_R + \Delta T_D$ - the total time of a hydrograph; $\Gamma_{HG} = \Delta D/(\Delta T \, u_{b*})$ - an unsteadiness parameter.

For a given bottom slope, S_o, the evolution of the depth, $D(x, t)$, in distance and time was recorded. At the measuring section the vertical velocity distribution, $\bar{u}(y)$, was recorded with the ADVP instrument. With the data in the inner region (y/D < 0.20) the friction velocity, u_*, could be obtained, using the Clauser method. Only part of the outer region (y/D > 0.20) could be measured and if measurement was impossible, the data were artificially extended. The thus obtained entire profile was used to calculate the cross-sectional average velocity, $U = 1/D \int_0^D \bar{u} dy$. Subsequently the unit discharge was evaluated, q = UD.

The flume's bed consists of rather uniform gravel, having $d_{50} = 1.23$ cm, $d_{16} = 0.90$ cm and $d_{84} = 1.65$ cm. The specific density is $s_s = 2.75$.

The sediment transport, $g_s(t)$, - which was essentially a bed load -, was measured continuously with the strain gauge, as it accumulated in the measuring basket.

4. SEDIMENT TRANSPORT IN UNIFORM FLOW
Data for uniform flow have been obtained for $5+43 = 48$ experiments; they have been communicated by Song, Graf and Lemmin (1994). In addition, the data reported by Cao (1985) are also used. All the data are plotted on Fig.2, where it can be seen that they are in reasonable agreement with eq.1.

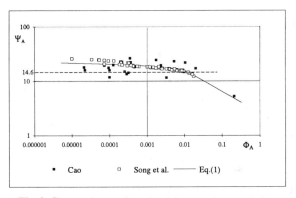

Fig.2 Comparison of eq.1 with experimental data.

5. SEDIMENT TRANSPORT IN UNSTEADY FLOW
In Figs.3 and 4 are plotted the sedimentograph, $g_s(t)$, as obtained from the measurements; also plotted are sedimentograph evaluated with eq.1. Two different types of evaluation were performed:

i) The Ψ_A-value, eq.2, was evaluated using a calculated bed-shear stress, $\tau_o = \gamma R_{hb} S_o$ or

$$\Psi_A = \frac{(\gamma_s - \gamma)\, d}{\tau_o} = \frac{(\gamma_s - \gamma)\, d}{\gamma R_{hb} S_o} \qquad (4)$$

where S_o is the bottom slope. While it would be more correct to use the slope of the energy-gradient line (i.e. the St. Venant equation) or

$$S_e = S_o - \left[\frac{\partial D}{\partial x} + \frac{1}{g} \frac{\partial U}{\partial t} + \frac{U}{g} \frac{\partial U}{\partial x} \right] \qquad (5)$$

this one is certainly difficult to determine. Nevertheless eq.1 was evaluated using this Ψ_A-value, eq.4, and the results are plotted in Fig.3. Obviously agreement with the measured values is not very good, notably in the falling branch of the hydrograph, where overprediction is evident.

ii) The Ψ_A-value, eq.2, was evaluated using the measured (i.e. calculated with the Clauser method) friction velocity, $u_* = \sqrt{\tau_o/\rho}$, or

$$\Psi_A = \frac{(\gamma_s - \gamma)\, d}{\tau_o} = \frac{(\gamma_s - \gamma)\, d}{\rho\, u_*^2} \qquad (6)$$

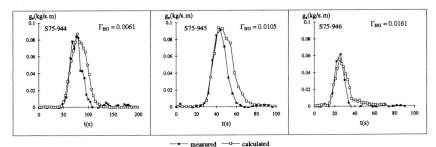

Fig.3 Sediment transport measured and calculated using $\tau_o = \gamma\, R_{hb}\, S_o$.

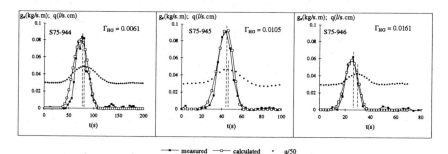

Fig.4 Sediment transport measured and calculated using the measured friction velocity, u_*.

In a previous communication (Graf and Song, 1995) we have shown that the measured friction velocity, u_*, is identical with the friction velocity calculated using the energy-gradient line, eq.5. Subsequently eq.1 was evaluated using this Ψ_A-value, eq.6, and the results are plotted in Fig.4. Obviously by using this more correct Ψ_A-value evaluation, agreement with the measured values is encouragingly good.

In Fig.4 is also plotted the discharge hydrograph, q(t). To be noticed is that the hydrograph has its peak later than the sedimentograph. (The later correlates directly with friction velocity which indeed arrives well before the peak of the hydrograph (see Graf and Altinakar, 1995, p.6))

6. CONCLUSIONS

Sediment transport, as bed-load, under unsteady flow condition was studied experimentally. It was shown that a sediment transport relation, such as eq.1, can be used to predict the sedimentograph, but the shear-stress under unsteady flow condition has to be used, either by evaluating eq.5 or by measuring directly the friction velocity.

7. REFERENCES

Cao, H. H. (1985). "Resistance hydaulique d'un lit de gravier mobile à pente raide; étude experimental." *Doctoral dissertation, No.589*, Ecole Polytechnique Fédérale, Lausanne, Switzerland.

Graf, W. H. and E. R. Acaroglu (1968). "Sediment Transport in Conveyance Systems (Part 1)." *Bull. Intern. Assoc. Sci. Hydr.*, XIIIe année, No.2.

Graf, W. H. and M. S. Altinakar (1995). *Hydraulique Fluviale, Tome 2.* Presses Polytechniques et Universitaires Romandes, Lausanne, Switzerland.

Graf, W. H. and L. Suszka (1985). "Unsteady Flow and its Effect on Sediment Transport." *XXI congress of IAHR*, Vol.3, Melbourne, Australia, 539-544.

Graf, W. H. and L. Suszka (1987). "Sediment Transport in Steep Channel." *J. Hydroscience and Hydr. Eng.*, JSCE, 5(1), 11-26.

Graf, W. H. and T. Song (1995)."Bed-Shear Stress in Non-Uniform and Unsteady Open-Channel Flows." (submitted for publication).

Song, T. and W. H. Graf (1995)."Velocity and Turbulence Distribution in Unsteady Open-Channel Flows." (submitted for publication).

Song, T., W. H. Graf and U. Lemmin (1994). "Uniform Flow in Open Channels with Movable Bed." *J. Hydr. Rech.*, IAHR, 32(6), 861-876.

1C20

Measuring river turbulence intensity with coarse-gravel bedload

G M SMART, G S CARTER
National Institute of Water & Atmospheric Research,
Christchurch, New Zealand

ABSTRACT
A robust, electronic velocity meter based on the Pitot tube, is described. The meter records at a rate sufficient to measure river turbulence. Typical data from a coarse gravel-bed river are presented. Limitations of the instrument are discussed.

BACKGROUND
New Zealand has many steep gravel bed rivers. Knowledge of hydraulic forces from these rivers on bridges, riverbeds and banks, is essential if the country's infrastructure is to be maintained and developed. Flows ranging up to 10,000 m^3/s (Buller and Haast rivers) and water velocities measured up to 5 m/s and estimated to range up to 7 m/s (Waiho River) make measurements of hydraulic data extremely difficult with conventional flow gauging equipment and impractical with most turbulence registering devices. The main problems are the high drag on the in-river sensors and the impacts and vibrations caused by gravel bed-material load. The difficulties of measuring turbulence during floods are compounded by floating logs and debris, usually heavy rain, poor visibility and precarious deployment positions for the researcher.

Quantitative measurement of river turbulence was first carried out in Holland in the 1950's with an electro-mechanical meter[1]. The projected area of the meter's sensor was 140 cm^2 which limited the sensitivity and prevented measurements in high velocity rivers. At about the same time the Americans measured laboratory turbulence using a Pitot tube coupled to a capacitance pressure transducer[2]. The advantages of using a Pitot tube for field measurements are that it is streamlined, strong and robust. The main disadvantage is that the velocity pressure head increases with the square of velocity so that accuracy is poor at velocities near zero. As Pitot tubes have proven reliable as airspeed indicators in aircraft, it was decided that a modification of this type of instrument should be suitable for use in rapid rivers bearing gravel.

INSTRUMENTATION

With the advent of micro electronics an amplifier can be located in a watertight housing, close to a miniature pressure transducer at the rear end of a Pitot tube. In 1990 a Pitot tube connected to pressure transducers was deployed in New Zealand rivers to measure water velocities[3]. Holes to detect hydrostatic pressure are located on the sides of the Pitot probe and connected to the reference side of a differential pressure transducer connected to the nose of the Pitot tube. In this way the transducer subtracts hydrostatic pressure from the nose stagnation pressure giving the true velocity head.

An absolute pressure transducer is connected to the side holes to indicate instrument depth. This device became known as a POEM (Pressure Operated Electronic Meter). The device proved suitable for measuring the velocity pressure head at very frequent intervals, thus providing an indication of the turbulence of flow. A more specialised prototype has now been developed, specifically for dynamic piezometry.

A Pitot tube, based on the British standard, is connected to a streamlined housing containing pressure transducers, preamp, multiplexer, A-D converter and line driver. Data is transmitted as 12 bit digital integers up a thin steel cable to a PC display and data logger. A tail is fitted at the rear of the electronics housing to point the instrument into the flow. This streamlined electronic Pitot tube has become known as a POTATo (Pressure Operated Turbulence Appraising Tool) and has the following characteristics:

> sampling frequency 56.25 Hz,
> diameter of Pitot sensor port 3 mm,
> Pitot tube and housing stainless steel,
> transducer response linear $\pm$ 0.5%,
> overall length 825 mm,
> diameter 50 mm

A schematic representation of the instrument is shown in figure 1.

The POTATo was tested below a constant velocity trolley in a current meter rating tank meeting the ISO 9002 standard. It measured the trolley velocity to within 2% when pulled through still water[4].

In order to determine the response time of the instrument an inflated balloon was connected to the nose of the Pitot tube and burst. The results, shown on figure 2, indicate that it takes around 0.07 seconds to register the assumed instantaneous drop in simulated velocity, from 2.5 m/s to 0.5 m/s. This range exceeds the turbulence intensity expected in natural rivers. The echo in impulse response evident when the Pitot is air filled, does not occur when the instrument is under water. A Fourier transform of the impulse response functions indicates that the 3dB bandwidth extends to 6Hz. As the actual bursting time of the balloon is not known, at this stage it can only be concluded that the instrument is sensitive to frequencies of at least 6Hz.

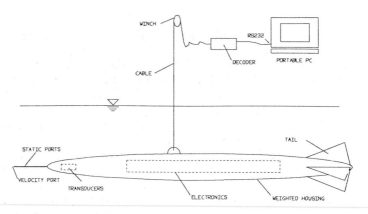

Figure 1. Schematic representation of POTATo

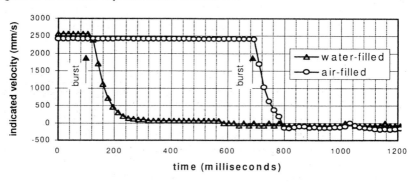

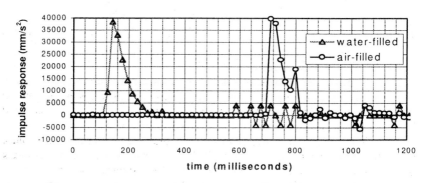

Figure 2.   Instrument response to step function simulated by balloon
                    burst.

FIELD MEASUREMENTS

The instrument has been tested on several rivers. The Waiho River in South Westland is fed by the Franz Joseph glacier. The valley slope at the gauging site is 0.012 and a Wolman sample of surface material gives a d_{50} of 95 mm and d_{90} of 250 mm. The mean annual flood is 1100 m³/s and bed movement occurs at flows as low as 100 m³/s. The data described below are of particular interest because measurements with a stable bed were recorded 12 hours prior to measurements with an active bed. At one side of the Waiho river the POTATo was slowly lowered from a bridge to the bed and raised again. The lowering winch was then moved some 3 m across the bridge and the exercise repeated until the other side of the river was reached.

Following measurement of turbulent velocity profiles at a flow of 46 m³/s a small flood occurred and the measurements were repeated with the flow at 214 m³/s, 12 hours later.

A typical vertical profile showing turbulent velocities and the depths at which they were recorded, is shown in figure 3. From these profiles mean velocity and relative turbulence intensity contour plots have been calculated (figure 4).

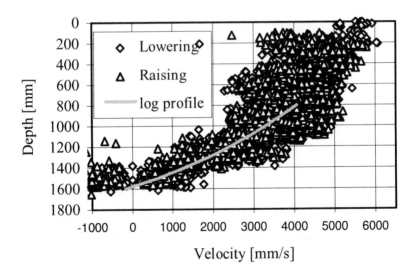

Figure 3.   Turbulent velocities and fitted log profile 63 m from left bank benchmark during 214 m³/s flood in Waiho river.

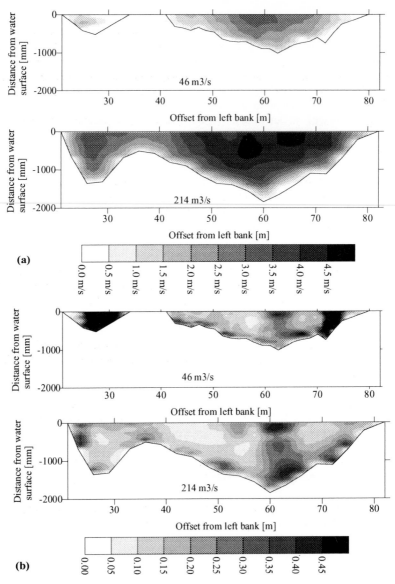

Figure 4.   Contours in the Waiho River of (a) velocity and (b) turbulence intensity scaled by mean velocity, measured without bedload (46 m³/s) and with bedload (214 m³/s).

DISCUSSION

The instrument collected turbulent velocity data under adverse conditions with a coarse gravel bedload and velocities up to 6m/s. The authors would be keen to hear of other attempts at this type of measurement. To the best of our knowledge, these measurements have not been previously attempted in an active bed, high velocity, alluvial channel and so the results must be interpreted with caution.

Researchers wishing to use this technique should bear in mind that the Pitot tube does not accurately measure velocities below about 0.3 m/s, including negative velocities.

CONCLUSIONS

The electronic Pitot tube responds to underwater velocity fluctuations up to frequencies of at least 6 Hz. The instrument does not accurately measure very low or negative velocities. The instrument performed satisfactorily in the field with river velocities up to 6 m/s and a coarse gravel bedload.

ACKNOWLEDGMENT

The field exercise was assisted by R D Henderson, W R Thompson and M J Duncan. This research was carried out under contract CO1414 from the Foundation for Research, Science and Technology (New Zealand).

REFERENCES

1. "Description of the Turbulence Meter Saturnus", Publication 10, code 42.10, Hydrometric Instruments Division, Hydraulics Laboratory Delft, March 1957.

2. "Turbulence in Civil Engineering: Measurement in free surface streams", A.T. Ippen and F. Raichlen, J.Hydraulics Div., ASCE, Vol. 83, No. HY5, October 1957.

3. "A P.O.E.M. on the Waiho (Electronic gauging of rivers)", G.M. Smart, NZ J. Hydrology Vol.30, No.1, 1991.

4. "River Turbulence Appraisal", G.M. Smart , in: Refined flow modelling and turbulence measurements. Presses Ponts et Chaussees, Paris, 1993, ISBN 2-85978-202-8.

1C21

CONSIDERATIONS FOR DEBRIS FLOW EXPERIMENTS USING A MOVING-BED FLUME

ALDO TAMBURRINO
Assistant Professor
Water Resources Center
Department of Civil Engineering
University of Chile
Santiago   -   CHILE

INTRODUCTION

It is well known that debris or mud flows can appear as trains of waves. There are some places where a pulsing motion of the debris or mud flow is the predominant flow mode (Zhaohui and Jignshen, 1993). Usually, the existence of periodic, permanent waves is explained as an instability phenomenon (Engelund and Wan, 1984; Liu and Mei, 1990; Lanzoni and Seminara, 1993; Ng and Mei, 1993; Wang, 1993), and the threshold condition to have roll waves is generally obtained by means of a linear analysis. Many models have been developed to determine the debris flow behavior, and most of them are based upon a constitutive equation representing the solid-water rheology. However, theory by itself is not enough to describe and explain the debris flow phenomena but it is necessary to have experimental and field data to improve our knowledge of this kind of flows.

One of the experimental facilities used to study debris flows is the moving-bed flume. It presents some advantages over traditional laboratory flumes. As inferred from its name, the main characteristic of the moving-bed flume is the moving bottom capability, usually a conveyor belt sliding over a plate. Taking advantage of the moving bottom, the frame of reference can be changed, making easier the measuring process. In this way, Hirano and Iwamoto (1981), Davies (1988) and Ling et al. (1990) simplified their measurement procedures setting the belt speed equal to the velocity of the front of the debris bore so that the position of the wave front remained stationary.

The objective of this paper is to present the relationship existing between the belt velocity and the debris flow characteristics in order to have permanent, periodic waves in the moving-bed flume. The solid-water mixture is modeled as a yield pseudo plastic fluid. From the equations, an estimation of the wave length can also be obtained, being another restriction of the experiment because it has to be compatible with the test section length. Finally, the relations to get the minimum slope of the flume in order to present waves are also indicated.

DERIVATION OF THE RELATIONSHIPS

Fig. 1 presents an idealized sketch of the train of waves to reproduce in the experimental facility. The belt speed is set up in such a way that the flow is stationary with respect to the laboratory frame of reference. Application of momentum theorem to the control volume indicated in Fig. 1b and assuming gradually varied flow between the sections (1) and (2) leads to:

$$\frac{dh}{dx} = \frac{sen(\theta) - \dfrac{\tau_0}{\gamma h}}{cos(\theta) - \dfrac{q^2}{gh^3}} \tag{1}$$

where τ_0 is the bottom shear stress and γ the specific weight of the solid-water mixture. The rest of the variables are defined in Fig. 1. The constitutive equation for the mixture is modeled as:

$$\tau = \tau_y + K \left(\frac{du}{dy}\right)^n \tag{2}$$

τ_y is the yield shear stress, K is a "viscosity" coefficient, and n is the flow index, smaller than 1. These parameters depend on the concentration, particle size distribution and chemical composition of the solids. u is the local velocity at the location y. The assumption of gradually varied flow allow us to get the velocity distribution:

$$\frac{u + U_b}{U + U_b} = \frac{1 - \left(1 - \dfrac{y}{h_0}\right)^{\frac{n+1}{n}}}{1 - \dfrac{n}{2n+1}\dfrac{h_0}{n}} \quad , \qquad 0 \leq y \leq h \tag{3}$$

$$\frac{u + U_b}{U + U_b} = \frac{1}{1 - \dfrac{n}{2n+1}\dfrac{h_0}{n}} \quad , \qquad h_0 \leq y \leq h \tag{4}$$

where U_b is the belt velocity and h_0 is the depth corresponding to the yield stress. U is the mean cross sectional velocity given by:

$$U = - U_b + \frac{n}{n+1} \left(\frac{\gamma sen(\theta)}{K}\right)^{\frac{1}{n}} \left(1 - \frac{n}{2n+1} \frac{h_0}{h}\right) h_0^{\frac{n+1}{n}} \tag{5}$$

If q is the discharge per unit width ($q = U\,h$), the relationship between h and h_0 is:

$$h_0^{\frac{n+1}{n}} \left(h - \frac{n}{2n+1}h_0\right) \frac{n}{n+1} \left(\frac{\gamma sen(\theta)}{K}\right)^{\frac{1}{n}} = q + U_b\,h \tag{6}$$

After evaluating Eq. 2 at $y=0$ to get the bottom shear stress, Eq. 1 can be written as:

$$\frac{dh}{dx} = \frac{sen(\theta) - \dfrac{\tau_y}{\gamma h} - \dfrac{K}{\gamma h} \left(\dfrac{n+1}{n} \dfrac{1}{h - \dfrac{n}{2n+1}h_0} \dfrac{q + U_b\,h}{h\,h_0}\right)^n}{cos(\theta) - \dfrac{q^2}{g\,h^3}} \tag{7}$$

or, replacing the numerator and denominator by A(h) and B(h), respectively:

$$\frac{dh}{dx} = \frac{A(h)}{B(h)} \tag{8}$$

In order to have a train of waves, it can be shown that Eq. 8 has to be undetermined in the form 0/0 (Brock, 1970; Ng and Mei, 1993), so

$$A(h) = 0 \quad , \quad B(h) = 0 \tag{9a, b}$$

Relation 9b corresponds to the critical condition, defining the value h_c

$$h_c = \left(\frac{q^2}{g\,cos(\theta)}\right)^{\frac{1}{3}} \tag{10}$$

Eq. 10 into condition 9a provides the belt velocity to have periodic, permanent waves:

$$U_b = - (g\,h_c\,cos(\theta))^{\frac{1}{2}} + \frac{n}{n+1} \left(h_c - \frac{n}{2n+1}h_0\right) h_0 \left(\frac{\gamma\,h_c\,sen(\theta) - \tau_y}{K}\right)^{\frac{1}{n}} \tag{11}$$

In this way, given the rheology of the mixture, the discharge q, and the channel inclination, the belt velocity can be determined using Eqs. 6, 10 and 11.

The wave length λ is obtained after applying the momentum theorem to the control volume defined by sections (1) and (2) in Fig. 1:

$$\frac{1}{2}\gamma h_1^2\cos(\theta) - \frac{1}{2}\gamma h_2^2\cos(\theta) + \gamma \operatorname{sen}(\theta) \int_{x_0}^{x_0+\lambda} h \, dx - \int_{x_0}^{x_0+\lambda} \tau_0 \, dx =$$

$$= \frac{\gamma}{g} q^2 \left(\frac{1}{h_2} - \frac{1}{h_1} \right) \tag{12}$$

The weight and force due to shear acting in the control volume defined between the sections (2) and (3) can be neglected because λ_f is much smaller than λ (see, for example, the records presented by Engelund and Zhaohui, 1984, or Hikida, 1990). As $h_1 = h_3$, the momentum theorem applied between these sections gives:

$$\frac{1}{2}\gamma h_2^2\cos(\theta) - \frac{1}{2}\gamma h_1^2\cos(\theta) = \frac{\gamma}{g} q^2 \left(\frac{1}{h_1} - \frac{1}{h_2} \right) \tag{13}$$

Eq. 13 is similar to the equation to compute the conjugate depths in the hydraulic jump and h_2 can be obtained:

$$\frac{h_2}{h_1} = \frac{1}{2} \left(\sqrt{1 + \frac{8 q^2}{g \cos(\theta) h_1^3}} - 1 \right) \tag{14}$$

Using Eq. 6 in the relationship for τ_0, it is possible to get:

$$\tau_0 = \tau_y + \gamma h_0 \operatorname{sen}(\theta) \tag{15}$$

Adding Eqs. 12 and 13:

$$\gamma \operatorname{sen}(\theta) \int_{h_1}^{h_2} h \frac{B}{A} dh = \int_{x_0}^{x_0+\lambda} \tau_0 \, dx \tag{16}$$

Thus, λ can be obtained using Eqs. 6, 7, 14 and 16. Because the complex structure of the equations involved, an iterative process is required to get λ. A suggestion of the iteration procedure is given in the flow diagram of Fig. 2.

Finally, it is possible to get the limit slope of the moving-bed flume in order to have a train of waves. Taking into account that the free surface profile between sections (1) and (2) is smooth, monotonic increasing with x (Ng and Mei, 1993), it is clear that waves will disappear if $h_1=h_2=h_c$. That means:

$$\frac{dh}{dx}\bigg|_{h=h_c} = 0 \tag{17}$$

As dh/dx is undetermined for $h=h_c$, it is necessary to apply L'Hôpital rule in Eq. 7 to impose the condition given by Eq. 17. The algebra involved is very cumbersome, and the final equation has to be solved numerically in order to get θ.

CONCLUSION
Some relationships to help to design debris flow experiments using a moving-bed flume have been developed in this paper. The conditions are restricted to those cases where trains of waves can be generated. Unfortunately, the relationships involved do not allow us to get directly U_b , λ and θ but a numerical method is needed to compute them. The complexity of the equations involved arises from the rheological model used to characterize the solid-water mixture.

ACKNOWLEDGMENTS
The author acknowledge the support given by Fondecyt by means of the Grant N° 1940545.

REFERENCES
Brock, R.R. (1970), "Periodic permanent roll waves", J. Hyd. Div., ASCE, Vol.96, N° HY12, pp. 2565-2580.

Davies, R.H. (1988), "Debris-flow surges -a laboratory investigation".Mitteilungen der Versuchsanstalt für Wasserbau, Hydrologie und Glaziologie. Vol. 96, Eidgenössische Technische Hochschule Zürich.

Engelund, F. and W. Zhaohui (1984), "Instability of hyperconcentrated flow", J. Hydr. Engng., ASCE, Vol. 110, N°3, pp. 219-233.

Hikida, M. (1990), "Field observation of roll-waves of debris flows", Proc. Int. Symp. on the Hydraulics and Hydrology of Arid Lands, San Diego, California, pp. 410-415.

Hirano, M. and M. Iwamoto (1981), "Experimental study on the grain sorting and the flow characteristics of a bore". Mem. Fac. Eng., Kyushu Univ. Vol. 41, pp. 193-202.

Lanzoni, S. and G. Seminara (1993), "Debris waves", XXV IAHR Congress, August 30 - Sept. 3, Tokyo, Japan, Vol. III, pp. 79-85.

Ling, Ch., Ch. Chen and Ch. Jan (1990), "Rheological properties of simulated debris flows in the laboratory environment", Proc. Int. Symp. on the Hydraulics and Hydrology of Arid Lands", San Diego, California, pp. 218-224.

Liu, K.F. and Ch.C. Mei (1990), "Waves in a fluid mud layer down an incline", Proc. Int. Symp. on the Hydraulics and Hydrology of Arid Lands, San Diego, California, pp. 403-409.

Ng, Ch. and Ch.C. Mei (1993), "Roll waves in mud flow",Proc. Conf. Hydr. Engng. '93, San Francisco, California, July 25-30, Vol. 2, pp. 1598-1603.

Wang, Z. (1993), "A study on debris flows surges",Proc. Conf. Hydr. Engng.'93, San Francisco, California, July 25-30, Vol. 2, pp. 1616-1621.

Zhaohui, W. and II. Jingshen (1993), "Analysis of pulsing phenomenon in viscous debris flow", Proc. Conf. Hydr. Engng. '93, San Francisco, California, July 25-30, Vol. 2, pp. 1610-1615.

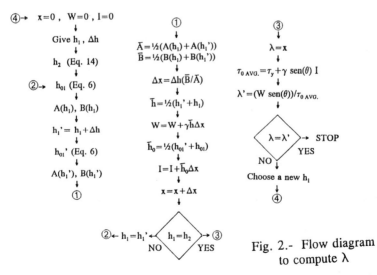

Fig. 1.- Sketch of the experimental facility and train of waves

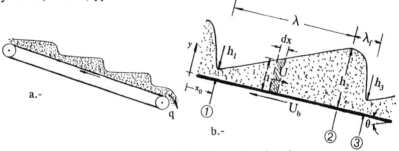

Fig. 2.- Flow diagram to compute λ

EXPERIMENTAL STUDY OF THE HUAIHE RIVER
FLOOD CONTROL MODEL

CHEN XIANPU   CHEN LIYE   LIANG BIN
(Huaihe River Water Resources Research Institute, Bengbu, China)

TAN PEIWEN   CHEN BIAO
(Huaihe River Planning & Design Institute, Bengbu, China)

The Huaihe river is located in east of China. The area of the Huaihe river is 20×104 km^2, and there are 1.04×10^8 poplications living in this region. In the reach from Huaibin to Zhengyangguan of the Huaihe river middle course, There are 10 inflow and a number of detention basins, and diversion channel. The meeting of the flood peak of tributaries is various. It is a complex flood control system. This region has frequently suffered flood disaster in history. The river regulation works will be a series of giant project, and expend enormous investment. It is necessary to take an unsteady flow river model for studying river regulation works and flood control. The length of test reach is 201 km. The layout of the Huaihe river flood control model is shown in Fig. 1. The horizontal model scale of $L_r = 1000$ and vertical model scale of $H_r = 100$ was taken. The area of model is 4000 m^2.

A computer system is employed to control model and collect experimental data. It can control 7 inflows which simulating 9 typical years flood process, a tail water level, 6 sluices, 11 outlets of breaking dike, and collect the water level of 37 points and velocity of 32 points. After ajusting model roughness and control system parameter, the prototype flood process was accurately repeated in model. The process of water level and discharge in model agree with prototype, as shown in Fig. 2 and Fig. 3. In the basices, the extended dike distance and Linhuaigang sluice storage plan are studing in this model. The results of model have provided scientific basis for the planning and design of river regulation works and flood control.

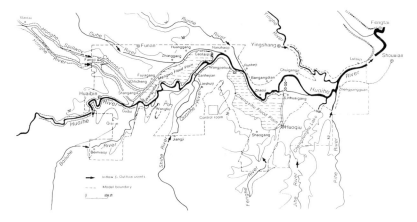

Fig 1.   Layout of the Huaihe river flood control model

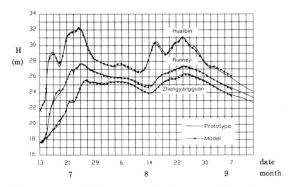

Fig. 2.   Comparison of the water level in model with prototype

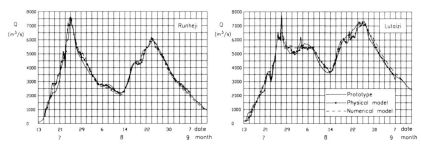

Fig. 3.   Comparison of the discharge process in model with prototype

499

1C23

The granulometry of the river bed gravelsand material by using a photo method.

VILIAM MACURA, Ass.Prof.
SILVIA KOHNOVÁ, Dipl.Eng.
Slovak Technical University
Department of Land and Water Resources Management
Radlinského 11, 813 68 Bratislava, SLOVAKIA

The character of the river bed gravelsand material is determined by the dynamic parameters of a river. Many characteristics of the river are coded in the river bed material. That's why the tasks of the river hydromechanic are determined by the particle-size analysis. The hand sieved and the machine sieved methods are well known . Using the sieved method in mountainous areas causes many problems due to the transport of samples to laboratories, or exhaustive sieving in the field. That's why a photo method was verified, which is the one method suitable to determine the granulometry, also in extreme conditions in mountainous areas. The photo-method which we verified and modified, has its origin in the V.V.Romanovskij and N.N. Kapitonov method.

THE BASICS OF THE PHOTO METHOD.

This method enables to determine the particle- size analysis of the upper active river bed layer, which is the most important from the hydraulic point of view. This layer is the interactive zone of the river. By using the photo method particle size analysis we presume the same relation between the weight percentage of individual fractions and the percentage of fraction area. The fraction area is determined by the photo of unattached river bed material.

THE PHOTO EVALUATION.

The area of individual fractions is fixed by a stencil - a simple model which consists of engraved circles on transparent material (organic glass). The diameters of these circles are determined by diameters of individual fractions in the same scale as the photo. A photo scale of 1:4 can be recommend .For better explanation we would consider a photographed sample 1x1 m.

To determine the particle areas with a diameter greater than 75 mm with the stencil is not recommended due to mistakes which could influence the accuracy of the particle size analysis. Area of fractions > 75 must be determined by using a planimeter. Fractions which are partially covered by the scale frame are put into a certain interval, accepting only that area projecting from the frame.
The area of elements smaller than 10 cm were defined as the difference of the whole photographed area, e.g. 1x1 m and the sum of area fractions ≥ 10 mm.

$$S_{f<10mm} = \left(10^6 - \sum S_{i \geq 10mm}\right).k_0 \quad \text{[mm}^2\text{]} \tag{1}$$

where : $S_{f<10\,mm}$ is the area of fractions with elements smaller a 10 mm,
$S_{f>10\,mm}$ is the area of fractions with elements greater a 10 mm,
k_0 is the coefficient determining the relation between the material weight of the volume and the specific weight of the material.

Because in the photo method all characteristics are planar, the coefficient k_0 has an area character too.

$$k_0 = \left(\frac{\rho_v}{\rho_m}\right)^{2/3} = 0.72 \tag{2}$$

where : ρ_m is the specific weight of the material (for mountain rivers in north Slovakia $\rho_m = 2630$ kg.m^{-3}),

ρ_v is the material weight of volume (for mountain rivers in north Slovakia $\rho_v = 1607$ kg.m^{-3}).

In rivers, where particle-size analysis was evaluated, the coefficient k_0 was equal 0,72. Because the area of fractions of d =10 mm is reduced by the air area between elements, it is important to reduce this area, by the whole area 1x1m. The resulting area S is :

$$S = 10^6 - \left[\left(10^6 - \sum S_{i \geq 10mm}\right).(1 - k_o)\right] \quad \text{[mm}^2\text{]} \tag{3}$$

d_e is determined analogicaly as by the sieved method,

$$d_e = \frac{\sum d_i.P_i}{100} \tag{4}$$

where d_i is the diameter of the fraction,

P_i is the percentage relation of fraction area to the whole sample area.

The representativnes of the photo method particle-size analysis to the sieved method was verified in rivers in Orava and Kysuce region and on the Danube river. Resulting from the analysis is , that comparison of both sieved, and photo method is influenced by the way of taking samples, the shape and the location of river bed material particles.

CONCLUSION:

Resulting from the measurements we can presuppose the sufficiency of the photo method granulometry of gravelsand mountain rivers and its ability to substitute the sieved method.

REFERENCES:

[1] JAIN,S.,KENNEDY,J.F.: The spectral evolution of sedimentary bed forms. J.of Fluid mechanics,1974, vol.63, N 2,p.301-314.
[2] MAREŠ,K: Zrnitosti složení povrchu hrubozrnného dna koryt horských a podhorských tokú. Vodní hospodářství,A37, 1987, 11, p.295.
[3] ROMANOVSKIJ,V.V.: Issledovanija i rasčety gidravličeskich parametrov nanosov. Leningrad 1973.

1C24

Numerical Computation of Slit-Type Bucket Supercritical Flow

LI,GUIFENL IU,QINGCHAO

China Institute of Water Resources
and Hydropower Resrarch(IWHR), Beijing, China

1.Introduction

Slit-type buckets(STB) is a new energy dissipator developed in recent years.The right bank spillway of Dongjiang Hydropower Project terminated in STB firstly in China. Thereafter,It was applied to spillway of other some big projects. As needed for engineering,a lot of works has been done on this type energy disspator.

Early application of this type disspator mainly depended on physical model tests.Recently some theoretical caculations and empirical analyses were conducted.2-D hydraulic compution models were developed for contracted supercritical flow.But the conventional coordinate system was used in this methods,as result,boundary conditions were dealt with approximately in compution.In addition,these methods can only be applied to cases with stright wall.According to those mentioned aboved,coordinate transformation,and 2-D equation under polar coordinate system are used in this paper for flow computation,improving the above methods.At last,some examples are caculated,the numerical model is proved to be stable,and more accurate results are obtained.

2.Compution of flow with curved wall and straight wall

According to the 2-D controling equations for open channel flow,and under OXYZ sestem,the velocity and depth of flow for any curved wall can be calculated.The boundery equation are complicable and the convergence is relatively poor because of the application of XOY system and approximate treatments on points along centerline.In order to improve the methods,this paper recomends the Eqs under polar system to deal with the same problems.According to this method the caculated results as shown in the Fig1.

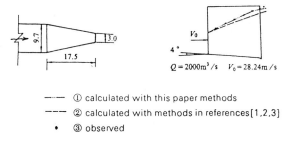

--·— ① calculated with this paper methods

——— ② calculated with methods in references[1,2,3]

• ③ observed

Fig.1 . water profile along the bucket wall applied at Longyingxia Gorge spillway

The observed data is obtained from experiments on STB,in which the spillway of Long Yangxia Gorge Project terminates.The spillway is 9.7m wide,and contracts to 3m at end the end of section.the deflecting angle of the STB is negative 4°.The cases for Q=2000m³/s are selected for compution.It can be seen from Fig.1.that the method developed in this paper has a lot improvements on previous,better agreement between results observed and caculated is obtained with the present method. But near to the end section water profile is under predication because the dissipation introduced with the recommended schemes is more than needed.Fig.2 shows the velocity distributions and developments of oblique waves.They are the caculated results for discharge Q=2000m³/s respectively.It can be seen that,the velocities near to walls are larger than those interior,therefore,when designing this type disspator,much attention should be given to the controls on abrupt wall turns and construction inregulations,to decrease or avoid cavitation destruction.

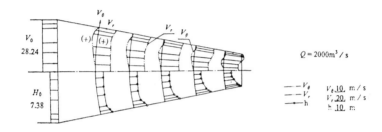

Fig.2.   velocity distribution and obliqne wave calculated with
the scheme in this paper.

3.Conclutions

This paper proposed a general methos to compute contructed supercritical flow.More attention was paid to the improvements on previous schemes.For this sake,the poler system was used in lieu of the conventional xoy system.Many caculated examples prove that the scheme is reliable,it can be applied to optimal design of STB to reduce the work on physical model tests.

1D1

DYNAMICS OF A SALTATING PARTICLE

ALLEN T. HJELMFELT, JR.
USDA/Agricultural Research Service, Columbia, MO USA

INTRODUCTION

When a fluid is transporting granular solids of greater density than the fluid, the motion of the larger grains takes place near the bed. Bagnold (1973) described the motion in this region as rolling or a succession of low jumps. Movement by the latter mechanism is termed "saltation." There is, however, disagreement concerning the nature of the movement. Abbot and Francis (1977) summarized their photographic records of flume studies as: "...previous ideas on saltation have always assumed, from eye observation only, that a grain on impacting a bed immediately starts a new trajectory. The photographic evidence shows otherwise... ." Nino et al.(1994) indicate that both the collision-rebound and collision-rolling modes were seen. The mode occurring depended mainly on the local configuration of the bed particles at the point of collision.

These disparate descriptions of the saltation process should be rectified. Numerical simulations will be analyzed using some of the tools of nonlinear dynamics to provide a possible explanation.

BASIC EQUATIONS

Any mathematical simulation requires idealization. In this simulation the most significant idealization is to restrict the motion to two dimensions (Fig. 1). A further simplifications include neglect bed particle interaction close to the bed, an that the fluid velocity distribution is unaffected by the saltating particle.

Several forces influence the trajectory a saltating particle. These include: particle weight, buoyant force, added mass, hydrodynamic drag and shear lift. Newtons second law can be expressed:

$$[\frac{\rho_p}{\rho_f} + \frac{1}{2}] \frac{du}{dt} = \frac{3}{4} C_d \frac{(U-u)}{d} \sqrt{(U-u)^2 + v^2} \tag{1}$$

$$[\frac{\rho_p}{\rho_f} + \frac{1}{2}] \frac{dv}{dt} = [\frac{\rho_p}{\rho_f} - 1]g - \frac{3}{4} C_d \frac{v}{d} \sqrt{(U-u)^2 + v^2} \tag{2}$$

The logarithmic velocity distribution for fluid flow near a rough wall was used and the drag coefficient for the particle was expressed by an equation due to Olson (1961). The particle trajectory was determined using a predictor-corrector method with variable time step. Impact of the particle with the bed was described through the coefficient of restitution and the impact and rebound were assumed to take place without sliding. The coefficient of restitution was taken to be 0.7, the particle diameter was set at 0.2 cm and the kinematic viscosity of the fluid at 0.01 cm sec^{-1}.

RESULTS

An example of the general response pattern for a saltating particle is shown in Fig. 2. The horizontal axis is the shear velocity which determines the magnitude of the fluid velocity. The vertical axis is the angular location on the bed element at which impact occurs (see Fig. 1). The computation process was: Motion was initiated by inserting a particle into the flowing stream. The particle trajectory was traced through 2000 impacts with the bed to escape start up transients. The trajectory was traced for an additional 150 impacts and the angular location of the impacts plotted as shown on Fig. 2. If the saltating particle strikes the bed at the same angular location in each of the last 150 impacts, all of the plotted points are coincident. If, however, the angular position changes, several points are plotted along a vertical line. The shear velocity is then incremented and the process repeated.

Large portions of this range yielded very confused patterns of motion(gray speckled blocks in Fig. 2). In these regions the particles jumped different distances on each consecutive hop and, at times, landed at nearly the same location after each hop. Reizes (1978) assumed that motion ceased if the same particle was struck more than eight times, but indicated that the number was insufficient as the stopping shear velocity was greater than the velocity for incipient motion. In this study the limit was increased to fifty and no problems were encountered. The numerous small jumps is interpreted here to be similar to the rolling motion described by several investigators.

Intervals of quite regular motion separate the blocks of confused motion in Fig. 2. These intervals of regular motion begin with saltation in which jump covers the same distance and with impact at the same relative location. Increasing the velocity, however, leads to a bifurcation after which the jumps alternate between two lengths that are nearly the same and impacts alternate between two slightly different relative locations. Further increases in shear showed stretching and folding, characteristics of chaotic motion. This simulation demonstrates that both the "regular" and "rolling and bouncing" motions described by various investigators are parts of the process called saltation.

REFERENCES

Abbot, J.E., and J.R.D. Francis. 1977. Saltating and suspended trajectories of solid grains in a water stream. Philos. Trans. R. Soc London, Ser. A, 284 pp 225-254

Bagnold, R.A. 1973. The nature of saltation and of "bed-load" transport in water. Proc. R. Soc. London. Ser. A. v332, pp 473-504.

Nino, Y., M. Garcia and L. Ayala. 1994. Gravel saltation, 1. Experiments. Water Resources Research, v30, n6, pp1907-1914.

Olson, R. M., 1961, *Essentials of Engineering Fluid Mechanics,* International Textbook Co., Scranton, Pennsylvania, 404p.

Reizes, J.A. 1978. Numerical study of continuous saltation. J. Hyd. Div., Am. Soc. Civil Engrs., v104, noHY9, pp1305-1321.

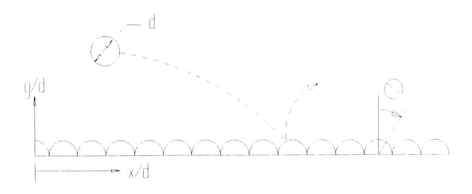

Fig. 1- Schematic of the idealized saltation problem.

velocity lead to additional bifurcations. Figure 3 is an expanded view of a portion of Fig. 2. Note the similarity between Figs. 2 and 3 and the widely displayed diagrams of orbital maps used to display the nonlinear dynamics of the logistic equation.

Motion can be shown in even more detail through use of Poincare diagrams. The particle velocity is plotted as the vertical coordinate and the vertical displacement as the horizontal coordinate. The diagram is formed by recognizing that a plot of the trajectory on the diagram, although displayed in 2-dimensions, actually shows points that represent a sequence of times. Thus it may be considered to also have a time coordinate perpendicular to the plane of the figure. In this 3-dimensional diagram the particle trajectory would appear as a line spiraling out of the page. One can pass planes through the t-axis, parallel to the y-v plane. Only points where the trajectory pierces the planes are selected and projected on to the y-v plane to form the Poincare diagram. Plane spacing based on travel distance was selected for this study. Fig. 4 is a series of diagrams for a shear velocity of 3.799cm/sec. The total jump length for all jumps is approximately 16 diameters. The spacings shown are 1, 8 and 15 diameters from impact. At two diameters all velocities are positive so the particles are all rising away from the bed, as would be expected. At eight diameters from impact all velocities are slightly negative indicating that the apex of the trajectory occurs slightly before the midpoint. At fifteen diameters all velocities are negative. The points all form a tight group that maintain a consistent relative orientation.

The shear velocity of 3.80 cm/sec represents a condition in the massive grey block. A series of Poincare surfaces for that shear velocity are shown in Fig. 5. The diagram for two diameters from impact shows both positive and negative velocities. Thus, some particles are already approaching the bed and will obviously not travel many more diameters. Progressing through the diagrams for 4, 8, 12 and 15 diameters shows that the swarm of points tend to stretch, fold back upon themselves, and finally coalesce into a tighter mass before impact. This stretching and folding is characteristic of chaotic motion. Chaotic motion was not, however, established conclusively in this study.

CONCLUSIONS

The two dimensional motion of a particle moving by saltation over a bed of identical particles has been investigated as a problem in nonlinear dynamics using numerical simulation. An iteration map was developed which showed the motion consists of regions both of regular continuous motion, and of less regular motion. More careful analysis of the less regular and confused regions was carried out using Poincare diagrams. The confused region

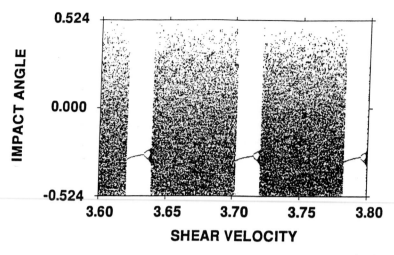

Fig. 2- Orbital map for shear velocities ranging between 3.60 and 3.80 cm/sec.

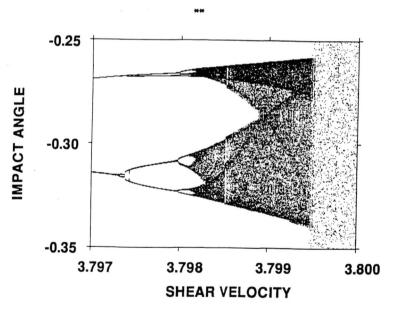

Fig. 3- Orbital map for shear velocities ranging between 3.797 and 3.80 cm/sec. Zooming in on the right hand edge of Fig. 2.

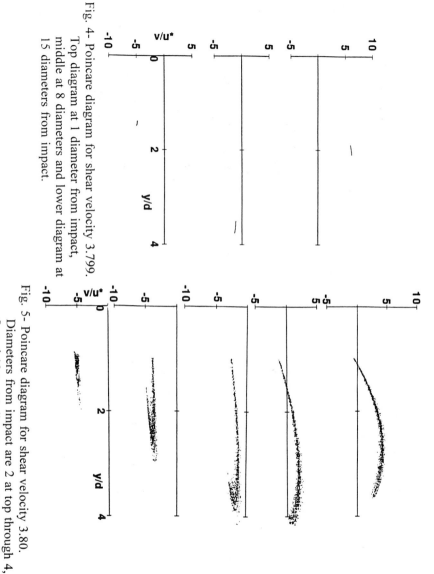

Fig. 4- Poincare diagram for shear velocity 3.799. Top diagram at 1 diameter from impact, middle at 8 diameters and lower diagram at 15 diameters from impact.

Fig. 5- Poincare diagram for shear velocity 3.80. Diameters from impact are 2 at top through 4, 8, and 12 to 15 diameters from impact at bottom.

1D2

On A Multivariate Model Of Channel Geometry

H. Q. Huang and Gerald C. Nanson
Department of Geography, University of Wollongong, Wollongong, NSW 2522, Australia

ABSTRACT: The channel geometry of alluvial streams is controlled by several factors in addition to flow discharge. An attempt to investigate the applicability of the multivariate model proposed by Huang and Warner (1995) demonstrates clearly that this model can satisfactorily quantify the influence of bank vegetation on channel geometry. Channel size is mainly determined by hydraulic conditions (discharge, roughness and slope), while channel shape (W/D) is largely controlled by bank materials and flow discharge. It appears that the regime theory developed in stable canals by engineers, and the minimum variance theory and the downstream hydraulic geometry used for natural streams by geomorphologists, are the simplified cases of this more complex multivariate model.

INTRODUCTION

The channel geometry of alluvial streams is usually described by a model known as 'hydraulic geometry' (Leopold and Maddock, 1953), which consists of simple power functions relating channel width (W), depth (D) and velocity (V) to water discharge (Q):

$$W = aQ^b; \qquad D = cQ^f; \qquad V = kQ^m \qquad (1)$$

Although investigations of this model in most geographical regions have proved successful, it has been found that besides flow discharge other factors such as channel slope, channel roughness, bank sediment composition and bank vegetation, also greatly effect channel geometry. It appears that the hydraulic geometry of alluvial channels is actually the product of a set of several controlling variables (e.g. Richards, 1982). In a study of the exact expressions of the multivariate controls of channel geometry, Huang and Warner (1995) obtained the following model:

$$W = C_W \cdot Q^{0.5} \cdot n^{0.355} \cdot S^{-0.156} \qquad (2)$$

$$D = C_D \cdot Q^{0.3} \cdot n^{0.383} \cdot S^{-0.206} \qquad (3)$$

$$V = C_V \cdot Q^{0.2} \cdot n^{-0.738} \cdot S^{0.362} \qquad (4)$$

in which n and S are the Manning's roughness coefficient and channel slope respectively, and the coefficients C_W, C_D and C_V were originally related to the critical tractive force for the movement of bank materials.

This multivariate model successfully explains the differences existing in hydraulic geometry relations for geographically widespread areas. Because this model was derived from numerous field observations found in the literature, it is believed to be a much more general expression of hydraulic geometry than the simple bivariate functions developed by Leopold and Maddock (1953). The purpose of this paper is to investigate the applicability of this multivariate model to three commonly involved practical cases: the quantification of the influence of bank materials on channel geometry; channel shape and its controlling factors; and the applicability of various channel geometry models to fluvial geomorphology.

Case 1: Quantification of the influence of bank vegetation

To examine the influence of bank materials on channel geometry (channel width (W), depth (D) and cross-sectional area (A)), Equations (2) to (4) are derived with the following linear forms:

$$W = C_W \cdot W'; \qquad D = C_D \cdot D'; \qquad A = C_A \cdot A' \qquad (5)$$

where $\quad W' = \dfrac{Q^{0.5} \cdot n^{0.355}}{S^{0.156}}; \qquad D' = \dfrac{Q^{0.3} \cdot n^{0.383}}{S^{0.206}}; \qquad A' = W' \cdot D' = \dfrac{Q^{0.8} \cdot n^{0.738}}{S^{0.362}}$

Field measurements of channel width (W), depth (D) and cross-sectional area (A) are plotted against their equivalent computed parameters (W', D' and A'). These should produce parallel lines for different bank types with the intercepts of log-log relationships yielding the values of C_W, C_D and C_A as indices of different bank strengths.

Hey and Thorne (1986) collected detailed field data on channel flow characteristics and channel boundary compositions from 62 locations on gravel rivers in the United Kingdom. By classifying vegetation cover into four categories, they found that bank vegetation exerted a significant influence on channel geometry. Their categories included:

A. Grassy banks with no trees or brushes;   B. 1-5% tree/shrub cover
C. 5-50% tree/shrub cover;   D. > 50% tree/shrub cover

and the following regression relations quantified this influence of bank vegetation on channel geometry:

$$W = 4.33 \ (3.33, \ 2.73, \ 2.34)Q^{0.5}; \qquad \text{Vegetation Type A (B, C, D)} \qquad (6)$$

$$D = 0.22Q^{0.37}d_{50}^{-0.11} ; \qquad\qquad \text{Vegetation Types A to D} \qquad (7)$$

where d_{50} is the average median sediment size. Equations (6) and (7) clearly indicate that bank vegetation only have a significant effect on channel width.

Figures 1 to 3 present the relations between W and W', D and D', and A and A' according to Equations (5) for the four groups of data. For channels with bank vegetation Type A, C_W varies between 5.0 and 7.0, with C_D between 0.3 and 0.4. For channels with bank vegetation Type B, C_W varies between 4.0 and 5.0, with C_D

around 0.4. For channels with bank vegetation Type C, C_W varies mainly between 3.0 and 4.0, with C_D between 0.4 and 0.5; For channels with bank vegetation Type D, C_W varies around 3.0, with C_D around 0.5. C_A varies in a very small range, from 1.5 to 2.0 for bank vegetation Types A to D.

These results demonstrate clearly that bank vegetation has a significant influence on channel width and depth, with width more responsive than depth, but that vegetation only exerts a limited effect on channel size. It can therefore be inferred from Equations (2) to (4) that channel size of alluvial streams is largely determined by hydraulic conditions (discharge, slope and roughness), while channel shape is determined not only by these hydraulic conditions but also by bank materials.

Considerable differences between Equations (6) and (7) and the results obtained from the the multivariate model are due to several problems. Equations (6) and (7) were obtained in ignorance of the influence of channel slope, using only the very simple case of channel roughness (d_{50}) and, most importantly, because of inadequate field data and the use of purely empirical methods.

Case 2: Channel shape and its controlling factors

According to Equations (2) and (3), channel shape is determined by the following relationship:

$$\frac{W}{D} = C_{W/D} \cdot Q^{0.2} \tag{8}$$

where
$$C_{W/D} = \frac{C_W}{C_D} \cdot n^{-0.028} \cdot S^{0.05} \tag{9}$$

As n and S have very small exponents (-0.028 and 0.05 respectively) in Equation (9), and C_W and C_D are determined by bank materials described in the above analyses, $C_{W/D}$ appears to be largely dependent on bank materials. Hence, channel shape is controlled largely by bank materials and water discharge (Equation (8)).

Equation (8) is very close to the following relationship obtained by Schumm (1968) in his study of the combined data from the Great Plains of the United States and the Murrumbidgee River in the southeastern Australia:

$$\frac{W}{D} = 21\frac{Q_{ma}^{0.18}}{M^{0.74}} \tag{10}$$

where Q_{ma} is the mean annual flood, and M is the channel silt-clay index defined to average silt-clay percentages in channel bed (S_b) and banks (S_c) by $M = [(S_b \cdot W) + 2(S_c \cdot D)]/(W + 2D)$.

When bank silt-clay content (B) was considered instead of using M, a close relation to Equation (8) was also obtained (Richards, 1982):

$$\frac{W}{D} = 800 \frac{Q_{ma}^{0.15}}{B^{1.20}} \tag{11}$$

According to Equation (8), Figures 4 and 5 present the relations for $C_{W/D}$ to bank silt-clay content B and channel silt-clay index M respectively on Schumm's (1960) data. The replacement of B with M does not improve this relation much. This suggests that it is the bank material rather than the total channel boundary material that has the greatest effect on channel shape, a notion consistent with the observations by other researchers (Richards, 1982).

Figure 6 shows the relationship represented by Equation (8) with Hey and Thorne's (1986) data. By comparison with Figures 1 to 3, it can be seen that bank materials have a much more significant effect on channel shape than on width, depth and cross-sectional area.

Case 3: Simplification of the multivariate model

Despite the multivariate complexity of channel geometry, the factors that control channel form might exhibit simple interrelationships downstream in some cases. For example, the power function relationships for channel slope (S) and the Manning's roughness (n) against discharge (Q) can take the simple form of:

$$S \propto Q^z; \qquad n \propto Q^y \qquad \text{(e.g. Leopold and Maddock, 1953)} \tag{12}$$

Channel banks are generally composed of coarse materials including boulders, cobbles and very coarse gravels in the upper catchment and become gradually finer downstream, although they still remain strong due to cohesive silt and clay and relatively dense vegetation. Hence, bank strength could be presumed to be relatively unchanged along the whole reach of some streams. Consequently, Equations (2) to (4) can be written as:

$$W \propto Q^{0.5+0.355y-0.156z} \propto Q^b \tag{13}$$

$$D \propto Q^{0.3+0.383y-0.206z} \propto Q^f \tag{14}$$

$$V \propto Q^{0.2-0.738y+0.362z} \propto Q^m \tag{15}$$

In terms of available models, Table I gives the values of the hydraulic geometry exponents both from the models and computed according to Equations (13) to (15). It can be noted that the agreement between the observed and computed values is nearly perfect for stable canals. This is understandable because in canals the relationships for channel slope and roughness to discharge are very well defined, and bank materials are relatively consistent (e.g. Blench, 1978; Huang and Warner, 1995). The surprising agreement between theoretical values from minimum variance theory and the values computed suggests that the theory represents a special case of the multivariate model (Table I). However, the agreement is quite poor for natural channels. This is possibly due to the fact that the simple power relationships for $Q - S$, and in particular for $Q - n$, generally do not exist (Leopold et al, 1964). Furthermore, as studied by Hey and Thorne (1986), bank materials also influence hydraulic geometry relations. Unfortunately, the natural streams from which the exponents listed in Table I were derived were studied in combination without considering the difference of bank strength existing between streams or even between cross-sections of the same stream (e.g. Leopold and Miller, 1956; Wolman, 1955). This means that the Leopold and Maddock's (1953) model of downstream hydraulic geometry is only a simplified case of the multivariate model.

Table I. Hydraulic geometry exponents observed according to Equations (1) and (12) and computed according to Equations (13) to (15)

Location of Streams	Observed exponents					Computed exponents		
	b	f	m	z	y	b	f	m
Stable canals[1]	0.5	0.333	0.167	-0.111	0.0	0.517	0.323	0.160
Minimum variance theory[2]	0.55	0.36	0.09	-0.74	-0.22	0.537	0.368	0.094
Ephemeral streams in semiarid, USA[3]	0.5	0.3	0.2	-0.95	-0.3	0.542	0.381	0.077
Brandywine Creek, Pennsylvania[4]	0.42	0.45	0.05	-1.07	-0.28	0.568	0.413	0.019

[1]Blench (1978) and Huang and Warner (1995) [2]Leopold and Langbein (1962) [3]Leopold and Miller (1956)
[4]Wolman (1955)

CONCLUSIONS

1. The multivariate model proposed by Huang and Warner (1995) can effectively quantify the influence of bank materials on channel geometry;

2. Channel size is mainly determined by hydraulic conditions (a combination of discharge, slope and roughness), while channel shape is largely controlled by a combination of bank materials and flow discharge;

3. The regime theory developed in stable canals by engineers, and the minimum variance theory and the downstream hydraulic geometry used for natural streams by geomorphologists, are the simplified cases of this multivariate model.

REFERENCES

Blench, T. 1978. *Regime behaviour of canals and rivers.* Ann Arbor, Michigan, University Microfilms International.
Hey, R. D. and Thorne, C. R. 1986. Stable channels with mobile gravel beds, *Journal of Hydraulic Engineering,* **112**, 671-689.
Huang, H. Q. and Warner, R. F. 1995. The multivariate controls of hydraulic geometry: a causal investigation in terms of boundary shear distribution. *Earth Surface Processes and Landforms,* in press.
Leopold, L. B. and Maddock, T. 1953. The hydraulic geometry of stream channels and some physiographic implications, *U. S. Geol. Survey Prof. Paper* **252**.
Leopold, L. B. and Miller, J. P. 1956. Ephemeral streams - Hydraulic factors and their relation to the drainage net, *U. S. Geol. Survey Prof. Paper* **282-A**.
Leopold, L. B. and Langbein, W. B. 1962. The concept of entropy in landscape evolution, *U. S. Geol. Survey Prof. Paper* **550A**.
Leopold, L. B., Wolman, M. G. and Miller, J. P. 1964. *Fluvial Processes in Geomorphology,* San Francisco, W. H. Freeman.
Richards, K. 1982. *Rivers: form and process in alluvial channels,* Methuen, London.
Schumm, S. A. 1960. The shape of alluvial channels in relation to sediment type, *U. S. Geological Survey Professional Paper* **352-B**, 17-30.

Schumm, S. A. 1968. River adjustment to altered hydrologic regimen - Murrumbidgee River and palaeochannels, Australia, *U. S. Geol. Survey Prof. Paper* **598**.

Wolman, M. G. 1955. The natural channel of Brandywine Creek, Pennsylvania, *U. S. Geol. Prof. Paper* **271**.

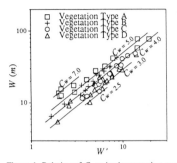

Figure 1. Relation of C_W to bank vegetation types

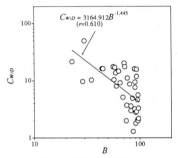

Figure 4. Relation of $C_{W/D}$ to bank silt-clay content B

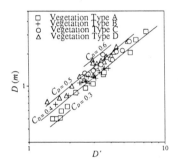

Figure 2. Relation of C_D to bank vegetation types

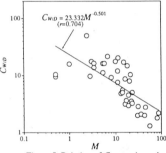

Figure 5. Relation of $C_{W/D}$ to channel silt-clay index M

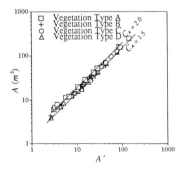

Figure 3. Relation of C_A to bank vegetation types

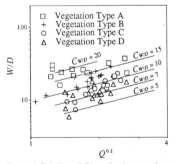

Figure 6. Relation of $C_{W/D}$ to bank vegetation types

1D3

DESIGN OF THRESHOLD CHANNELS

SHUYOU CAO   DONALD W KNIGHT
State Key Hydraulics Laboratory   *School of Civil Engineering*
Sichuan Union University   *The University of Birmingham*
Chengdu, Sichuan, China   *Birmingham, England, UK*

ABSTRACT
A new approach has been developed for designing the shape and dimensions of the cross section of a straight threshold channel based on the concepts of entropy and probability. The bank profile is a new type of simple parabolic curve. Bank equations are coupled with a frictional relationship to obtain a new design method. Channel dimensions and bank profiles predicted by this method are compared with those given by other design methods. The predicted channel dimensions are in reasonable agreement with laboratory experimental data.

INTRODUCTION
The cross-sectional shapes of stable alluvial channels are traditionally divided into two types (Henderson 1966). Type A contains two curved banks and a constant depth zone inserted between them. Type B is composed of only two curved banks, with the junction point of the two curves satisfying a continuity condition. It follows therefore that type B is a lower limit of a threshold channel. The concept of entropy, based upon probability theory instead of thermodynamic principles, has been applied in this paper to study the shape of a type B threshold channel cross section. Two boundary conditions are used to give the physical meaning of this entropy approach. One is that the dimensions and shapes are dependent on the discharge of flow and boundary sediment size or the angle of respose of the particles. Another is that the shape curve must satisfy the continuity condition at the junction point of the two bank profile curves.

BASIC EQUATIONS
The basic equations of this entropy design approach for threshold channels are obtained when the Lagrange multiplier in the equation for the cross-sectional shape of common regime channels approaches zero (Cao, 1995). The equation of boundary elevation above the bed at the channel centre is then

$$z = \frac{\mu}{2L} y^2 \tag{1}$$

where L = semi-width of channel water surface; y = lateral distance from channel centre; μ = tanφ = the submerged coefficient of static friction; φ = angle of internal friction for sediment. The depth at the channel centre is then given as

$$h_c = \mu L/2 \tag{2}$$

The aspect ratio for the channel is

$$B/h_c = 4/\mu \tag{3}$$

where B = 2L = top width of channel. Subtracting Eq. (1) from Eq. (2) gives the lateral distribution of depths, h(y), as

$$h(y) = \frac{\mu L}{2}\left[1 - \left(\frac{y}{L}\right)^2\right] \tag{4}$$

Integration of Eq. (4) gives the area of channel cross section as

$$A = \frac{2}{3}\mu L^2 \tag{5}$$

The wetted perimeter, P, is determined by arc integration of an element

$$dP = (dz^2 + dy^2)^{1/2} \tag{6}$$

where $z = \mu y^2/(2L)$. Hence

$$dP = (\mu/L)[y^2 + (L/\mu)^2]^{1/2} \, dy \tag{7}$$

Integration of Eq. (7) gives the wetted perimeter of a threshold channel as

$$P = \frac{L}{\mu}\left\{\sqrt{\mu^2 + 1} + \log\left[L\left(1 + \sqrt{1 + \frac{1}{\mu^2}}\right)\right] - \log\left(\frac{L}{\mu}\right)\right\} \tag{8}$$

Combining Eqs. (5) and (8) gives the equation for the hydraulic radius as

$$R = 2\mu L^2/(3P) \tag{9}$$

where P is given by Eq. (8). Table 1 lists values of h_c, A and B/h_c for three kinds of bank profile, in which Diplas's equation (1992) for B/h_c is given as

$$B/h_c = -16.1814\mu^3 + 44.3206\mu^2 - 43.5548\mu + 21.1496 \tag{10}$$

Table 1 shows that among the three approaches, Diplas (1992) gives the maximum values, USBR (see Henderson, 1966) gives the minimum values, and Cao & Knight gives values between them. However the results given by Cao & Knight's entropy approach are closer to Diplas's results than the USBR results.

DESIGN APPROACH

From an engineering point of view, the most important task of channel shape research is to provide design criteria for threshold channels. The variables considered in this problem should be the discharge, Q, the bed material characteristics of the channel boundary, d or μ, the streamwise water surface slope, S, the depth at the channel centre, h_c, and the top width of the channel, B. Among these variables, the former two are usually known. The latter three, as well as the bank shapes, have to be determined. There are now enough equations to design a threshold channel by a trial and error method if the discharge, Q, and the diameter of the boundary material, d, are known. The new design procedure is as follows:

1. Assume a depth at the centre of the channel, h_c;
2. Use Shields entrainment diagram to find a threshold Shields parameter for the given boundary sediment diameter. Then calculate the streamwise slope from

$$\Theta = \frac{\gamma h_c S}{d(\gamma_s - \gamma)} \tag{11}$$

3. Use an appropriate sediment resistance relationship, for example White et al's equation (1980), to calculate the average channel velocity;
4. Eq. (2) is then employed to determine the semi-width L;
5. Eq. (5) is then used to calculate the area of the cross section;
6. Multiply the cross section area obtained by step 5 by the velocity calculated by step 3 to obtain a calculated discharge, Q_c. If the calculated discharge equals that given, then the centre depth and semi-width of the channel are correctly obtained. Otherwise go back to step 1 to adjust h_c and then repeat steps 1 to 6.
7. Finally Eqs. (1) or (4) be used to determine the channel shape or depth distribution.

A FORTRAN program was developed to compute the procedure given above. A numerical example is given as Fig. 1, in which the boundary sediment diameter is $d_{50} = 0.8$ mm, $\mu = 0.5$, and the discharges, $Q = 0.05$ m^{3}s^{-1} respectively. The numerical results based on the entropy approach are compared with those developed by Diplas (1992) and USBR (Henderson, 1966). The three main parameters, S, B and h_c for five discharges are listed in Table 2. The shapes given by the entropy approach lie between those of Diplas and USBR approaches, but are closer to Diplas's values as shown in Fig, 1.

COMPARISON WITH LABORATORY DATA

Data related to channel cross section shape at threshold are difficult to find. However Stebbings's (1963) very small scale laboratory experiments can be used to test the top width and cross section area of the proposed entropy approach. A total of 34 sets of data in two series of experiments are used here. The dimensions of threshold channels for different flow discharges were obtained. The diameter of the channel boundary material was $d_{50} = 0.88$ mm with $\mu = 0.51$. During tests the centreline channel depth, h_c, was obtained from experiments. The aspect ratio, B/ h_c, the channel width, B, and the area of the cross section, A, were all calculated from Eqs. (3) and (5) respectively. The average value of the aspect ratio in all 34 data sets from Stebbings's experiments was 7.31, compared with 7.8 given by the entropy approach. The aspect ratio given by Diplas is 8.3, and by USBR is 6.16. The test results are shown in Fig. 2, which shows that the predicted values for the top width of the channel and the cross-sectional area (in cm^2) are in satisfactory agreement with the laboratory data. Fig. 3 shows the relationships between the discrepancy ratios, B_c/B, or A_c/A (calculated : measured), and the channel centre depths respectively. Fig. 3 shows that the accuracy of calculated values of B_c and A_c obviously depends upon the channel depth. The values of discrepancy ratios are much higher for h_c less than 2 cm than at larger depths. This phenomenon may be explained (Diplas, 1992; Van Burkalow, 1945; Wolman & Brush, 1961) by the fact that at shallow flow depths, the angle of repose of the bed material increases due to surface tension effects, resulting in narrower channels. In Stebbings's experiments surface tension effects became important for those experimental runs at flow depths smaller than 2 cm, for which the value of the channel width suddenly reduced. Assuming μ increases by 25% to $\mu = 0.64$ for depths smaller than 2 cm, then the values of the discrepancy rations are improved (Fig. 4).

CONCLUSIONS

A design approach for threshold type B channels has been proposed. The hydraulic geometry of channels is given by setting the Lagrange multiplier to zero in the entropy-maximisation principle. The predicted values of top width and area of the cross section agree well with Stebbings's (1963) data. The predicted dimensions and shapes are in between those predicted by Diplas (1992) and USBR (Henderson, 1966).

REFERENCES

Cao, S., 1995, "Regime theory and fluvial modelling with width adjustment function", *PhD thesis*, The University of Birmingham, England, UK.

Diplas, P., 1992, "Hydraulic geometry of threshold channels", *J. of Hyd. Engr.*, 118(4), 597-614.

Henderson, F. M., 1966, "Open channel flow", Macmillan, New York, 450-455.

Stebbings, J., 1963, "The shape of self-formed model alluvial channels", *Proc. Inst. Civ. Engrs.*, London, England, 25, 485-510.

Van Burkalow, A., 1945, "Angle of respose and angle of friction - an experimental study", *Geol. Soc. America Bull.*, 56(6), 669-708.

White, W. R., Paris, E., Bettess, R., 1980, "The frictional characteristics of alluvial stream: a new approach", *Proc. Inst. Engrs*, 2(69), 737-750.

Wolman, M. G., & Brush, L. M., 1961, "Factors controlling the size and shape of stream channels in coarse non-cohesive sands", *Paper 282-G USGS*, Washington D. C.

Author	Diplas	Cao & Knight	USBR
h_c	$(0.174\sim0.237)\mu B$ ($\mu=1\sim0.5$)	$0.25\mu B$	$0.32\mu B$
A	$0.69Bh_c$	$0.67Bh_c$	$0.64Bh_c$
B/h_c	Eq. (10)	$4/\mu$	π/μ
$B/h_c(\mu=0.5)$	8.42	8	6.28

Table 1. Comparison of three shape approaches

$Q\ (m^3s^{-1})$	Author	$S(\times10^4)$	B(m)	h_c(m)
0.05	Diplas	3.69	1.322	0.157
	Cao & Knight	4.66	1.249	0.156
	USBR	5.43	1.055	0.168
0.10	Diplas	2.67	1.827	0.217
	Cao & Knight	3.37	1.728	0.216
	USBR	3.95	1.452	0.230
0.15	Diplas	2.21	2.211	0.262
	Cao & Knight	2.78	2.091	0.261
	USBR	3.28	1.749	0.278
0.20	Diplas	1.93	2.530	0.300
	Cao & Knight	2.43	2.393	0.299
	USBR	2.87	1.996	0.317

Table 2 Three main parameters of threshold channels

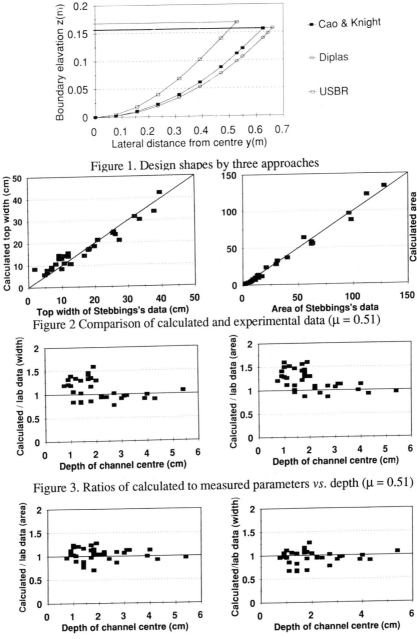

Figure 1. Design shapes by three approaches

Figure 2 Comparison of calculated and experimental data ($\mu = 0.51$)

Figure 3. Ratios of calculated to measured parameters *vs.* depth ($\mu = 0.51$)

Figure 4. Ratios of calculated to measured values *vs.* depth ($\mu=0.64$ for $h_c<2$ cm)

1D4

HIGHER ORDER MOMENTS OF DISCHARGE IN A STORAGE FUNCTION MODEL FOR FLOOD RUNOFF
–THE IMPACT OF MUTUALLY DEPENDENT RAINFALL INPUT–

M. KUDO[1], M. FUJITA[2] and G. TANAKA[2]
1) Head Office of Civil Department, JDC Corporation
Tokyo, 107, Japan
2) Department of Civil Engineering, Hokkaido University
Sapporo, 060, Japan

SUMMARY: The authors propose differential equations whose solutions give the first four central moments of discharge from a storage function model under the condition that rainfall input is a mutually dependent random variable. The validity of these equations are cross–checked by a simulation method.

1. INTRODUCTION

It is well known that runoff phenomena is more or less stochastic. Strictly speaking, basic equations for a runoff model are best described by random differential equations. For example, based on the conventional weather radar data, rainfall rate is expressed by empirical Z-R relations. The Z-R relations depend on storm characteristics such as their causes, and the constants included in the Z-R relation vary greatly. As a result, the estimated rainfall rate using radar leads to a noticeable error. It is considered that the Z-R relations give only the average rainfall rate. From the standpoint of runoff analysis, it is impossible to neglect deviations from the estimated rainfall rate by Z-R relations. The estimated rainfall rate must be understood in terms of probabilities.

This paper proposes a method which provides the high order moments of discharge under the impact of mutually dependent rainfall inputs.

2. BASIC THEORY

The runoff model considered here is the simplest storage function model described by Eq.(1) and Eq.(2).

$$\frac{dS}{dt} + q = r \qquad (1) \qquad S = Kq^P \quad K, P : const. \qquad (2)$$

$r(t)$: rainfall intensity(mm/hr), $q(t)$: discharge(mm/hr), $S(t)$: storage(mm)
If rainfall input $r(t)$ belongs to a random process, the output $q(t)$ and storage $S(t)$ are also described by a random process. Consider that $r(t)$, $q(t)$ and $S(t)$ consist of a mean and the deviation from its mean.

$$r = \bar{r} + \tilde{r} \ , \ E\{\tilde{r}\} = 0 \qquad (3) \qquad q = \bar{q} + \tilde{q} \ , \ E\{\tilde{q}\} = 0 \qquad (4) \qquad S = \bar{S} + \tilde{S} \ , \ E\{\tilde{S}\} = 0 \qquad (5)$$

Eq.(6) is obtained after eliminating q from Eq.(1) and Eq.(2).

$$\frac{dS}{dt} + \left(\frac{1}{K}\right)^m S^m = r \ , \ m = \frac{1}{P} \qquad (6)$$

Bras(1980) proposed the following equation in regard to the random variable with an exponent.

$$S^m = \alpha \bar{S} + \beta \tilde{S} \qquad (7)$$

$$\alpha = \overline{S}^{m-1}\left\{1 + \frac{1}{2}m(m-1)\frac{E(\tilde{S}^2)}{\overline{S}^2} + \frac{1}{6}m(m-1)(m-2)\frac{E(\tilde{S}^3)}{\overline{S}^3} + \cdots\right\} \tag{8}$$

$$\beta = \frac{\overline{S}^{m+1}}{E(\tilde{S}^2)}\left\{m\frac{E(\tilde{S}^2)}{\overline{S}^2} + \frac{1}{2}m(m-1)\frac{E(\tilde{S}^3)}{\overline{S}^3} + \frac{1}{6}m(m-1)(m-2)\frac{E(\tilde{S}^4)}{\overline{S}^4} + \cdots\right\} \tag{9}$$

Eq.(10) is derived by substituting Eq.(3), Eq.(5) and Eq.(7) into Eq.(6).

$$\frac{d(\overline{S}+\tilde{S})}{dt} + \left(\frac{1}{k}\right)^m(\alpha\overline{S}+\beta\tilde{S}) = \overline{r}+\tilde{r} \tag{10}$$

The expectation of Eq.(10) yields Eq.(11). Eq.(11) subtracted from Eq.(10) gives Eq.(12).

$$\frac{d\overline{S}}{dt} + \left(\frac{1}{K}\right)^m \alpha\overline{S} = \overline{r} \tag{11}$$

$$\frac{d\tilde{S}}{dt} + \left(\frac{1}{K}\right)^m \beta\tilde{S} = \tilde{r} \tag{12}$$

β is related to $\tilde{S}$ through the medium of the operator of expectation E as illustrated in Eq.(9). It is possible to solve Eq.(12) under the condition that its dependency is weak.

$$\tilde{S} = e^{-\int\left(\frac{1}{K}\right)^m\beta dt}\int_0^t \tilde{r}(\tau)e^{\int^\tau\left(\frac{1}{K}\right)^m\beta d\tau_1}d\tau \tag{13}$$

It is possible to calculate the second, third and fourth moment of $\tilde{s}$ by Eq.(13).

$$\sigma_s^2 = E(\tilde{S}^2) = e^{-2\int\left(\frac{1}{K}\right)^m\beta dt}\int_0^t\int_0^t E\{\tilde{r}(\tau_1)\tilde{r}(\tau_2)\}e^{\int^{\tau_1}\left(\frac{1}{K}\right)^m\beta d\tau_3+\int^{\tau_2}\left(\frac{1}{K}\right)^m\beta d\tau_4}d\tau_1 d\tau_2 \tag{14}$$

$$\mu_{S3} = E(\tilde{S}^3) = e^{-3\int\left(\frac{1}{K}\right)^m\beta dt}\int_0^t\int_0^t\int_0^t E\{\tilde{r}(\tau_1)\tilde{r}(\tau_2)\tilde{r}(\tau_3)\}e^{\int^{\tau_1}\left(\frac{1}{K}\right)^m\beta d\tau_4+\int^{\tau_2}\left(\frac{1}{K}\right)^m\beta d\tau_5+\int^{\tau_3}\left(\frac{1}{K}\right)^m\beta d\tau_6}$$
$$\times d\tau_1 d\tau_2 d\tau_3 \tag{15}$$

$$\mu_{S4} = E(\tilde{S}^4) = e^{-4\int\left(\frac{1}{K}\right)^m\beta dt}\int_0^t\int_0^t\int_0^t\int_0^t E\{\tilde{r}(\tau_1)\tilde{r}(\tau_2)\tilde{r}(\tau_3)\tilde{r}(\tau_4)\}$$
$$\times e^{\int^{\tau_1}\left(\frac{1}{K}\right)^m\beta d\tau_5+\int^{\tau_2}\left(\frac{1}{K}\right)^m\beta d\tau_6+\int^{\tau_3}\left(\frac{1}{K}\right)^m\beta d\tau_7+\int^{\tau_4}\left(\frac{1}{k}\right)^m\beta d\tau_8}d\tau_1 d\tau_2 d\tau_3 d\tau_4 \tag{16}$$

In order to calculate σ_s^2, μ_{S3} and μ_{S4}, the generalized moments such as $E\{\tilde{r}(\tau_1)\tilde{r}(\tau_2)\tilde{r}(\tau_3)\}$ must be known. It is obvious that $\tilde{r}$ in Eq.(12) shows continuous random variables. The continuous random variables of $r(t)$ may be transformed to discrete ones, $R(t)$ by Eq.(17). Let's define Eq.(18).

$$R(t) = \frac{1}{\Delta t}\int_{t-\Delta t}^t r(\tau)d\tau \tag{17}$$

$$R = \overline{R} + \tilde{R} \tag{18}$$

$\overline{R}$: mean, $\tilde{R}$: deviation from the mean

Eq.(19) is redefined by using Eq.(3) and Eq.(18).

$$\tilde{R}(t) = \frac{1}{\Delta t}\int_{t-\Delta t}^t \tilde{r}(\tau)d\tau \tag{19}$$

By Eq.(19), the following equations are obtained.

$$E\{\tilde{R}(\tau_1)\tilde{R}(\tau_2)\}=\frac{1}{\Delta t^2}\int\int E\{\tilde{r}(\tau_1)\tilde{r}(\tau_2)\}d\tau_1 d\tau_2 \tag{20}$$

$$E\{\tilde{R}(\tau_1)\tilde{R}(\tau_2)\tilde{R}(\tau_3)\}=\frac{1}{\Delta t^3}\int\int\int E\{\tilde{r}(\tau_1)\tilde{r}(\tau_2)\tilde{r}(\tau_3)\}d\tau_1 d\tau_2 d\tau_3 \tag{21}$$

$$E\{\tilde{R}(\tau_1)\tilde{R}(\tau_2)\tilde{R}(\tau_3)\tilde{R}(\tau_4)\}=\frac{1}{\Delta t^4}\int\int\int\int E\{\tilde{r}(\tau_1)\tilde{r}(\tau_2)\tilde{r}(\tau_3)r(\tau_4)\}d\tau_1 d\tau_2 d\tau_3 d\tau_4 \tag{22}$$

The autocorrelation function of the observed rainfall decays exponentially as shown in Fig.1. The autocorrelation function can be decided by

$$E\{\tilde{R}(t_1)\tilde{R}(t_2)\}=\sigma_R^2\rho^{t_1-t_2}\ ,\ (t_1\geq t_2)\quad \sigma_R^2:\ variance\ of\ \tilde{R}(t) \tag{23}$$

The theoretical autocorrelation function from the first-order autoregressive process satisfies Eq.(23). The AR(1) process is decided by

$$\tilde{R}(t)=\rho\tilde{R}(t-1)+N(t) \tag{24}$$

$N(t)$ consists of a sequence of uncorrelated random variables with mean zero and constant second, third and fourth moments, that is

$$E(N_t)=0\ ,\ E(N_t^2)=\sigma_N^2\ ,\ E(N_t^3)=\mu_{N3}\ ,\ E(N_t^4)=\mu_{N4} \tag{25}$$

The generalized third and fourth moment of $\tilde{R}(t)$ are derived by using Eq.(24).

$$E\{\tilde{R}(t_1)\tilde{R}(t_2)\tilde{R}(t_3)\}=\mu_{R3}\rho^{t_1+t_2-2t_3}\ ,\ (t_1\geq t_2\geq t_3) \tag{26}$$

$$E\{\tilde{R}(t_1)\tilde{R}(t_2)\tilde{R}(t_3)\tilde{R}(t_4)\}=(\mu_{R4}-3\sigma_R^4)\rho^{t_1+t_2+t_3-3t_4}+\sigma_R^4\{2\rho^{t_1+t_2-t_3-t_4}+\rho^{t_1-t_2+t_3-t_4}\}\ ,\ (t_1\geq t_2\geq t_3\geq t_4) \tag{27}$$

μ_{R3} and μ_{R4}: third and fourth moment

Next, the generalized moments of $\tilde{r}(t)$ must be determined under the condition that the left sides of Eq.(19) to Eq.(22) are known. The following equations are obtained.

$$E\{\tilde{r}(\tau_1)\tilde{r}(\tau_2)\}=e^{-\gamma(\tau_1-\tau_2)}\sigma_{r1}^2+\delta(\tau_1-\tau_2)\sigma_{r2}^2 \tag{28}$$

$$E\{\tilde{r}(\tau_1)\tilde{r}(\tau_2)\tilde{r}(\tau_3)\}=e^{-\gamma(\tau_1+\tau_2-2\tau_3)}\mu_{r31}+(\tau_2-\tau_3)e^{-\gamma(\tau_1-\tau_2)}\mu_{r32}+\delta(\tau_1-\tau_2)\delta(\tau_2-\tau_3)\mu_{r33} \tag{29}$$

$$
\begin{aligned}
E\{\tilde{r}(\tau_1)\tilde{r}(\tau_2)\tilde{r}(\tau_3)\tilde{r}(\tau_4)\}=&e^{-\gamma(\tau_1+\tau_2+\tau_3-3\tau_4)}\mu_{r41}+e^{-\gamma(\tau_1+\tau_2-\tau_3-\tau_4)}\mu_{r42}+e^{-\gamma(\tau_1-\tau_2+\tau_3-\tau_4)}\mu_{r43}\\
&+\{\delta(\tau_1-\tau_2)e^{-\gamma(\tau_3-\tau_4)}+\delta(\tau_2-\tau_3)e^{-\gamma(\tau_1-\tau_4)}+\delta(\tau_3-\tau_4)e^{-\gamma(\tau_1-\tau_2)}\}\mu_{r44}\\
&+\delta(\tau_3-\tau_4)e^{-\gamma(\tau_1+\tau_2-2\tau_3)}\mu_{r45}+\delta(\tau_1-\tau_2)\delta(\tau_3-\tau_4)\mu_{r46}\\
&+\delta(\tau_1-\tau_2)\delta(\tau_2-\tau_3)e^{-\gamma(\tau_1-\tau_4)}\mu_{r47}+\delta(\tau_2-\tau_3)\delta(\tau_3-\tau_4)e^{-\gamma(\tau_1-\tau_2)}\mu_{r48}\\
&+\delta(\tau_1-\tau_2)\delta(\tau_2-\tau_3)\delta(\tau_3-\tau_4)\mu_{r49}\ ,\ \tau_1\geq\tau_2\geq\tau_3\geq\tau_4
\end{aligned} \tag{30}
$$

Various coefficients included in Eq.(28) to Eq.(30) are listed below.

$$\gamma=-\log(\rho)/\Delta t\quad J_1=1-e^{-\gamma\Delta t}\quad J_2=1-e^{\gamma\Delta t}\quad J_3=1-e^{-2\gamma\Delta t}\quad J_4=1-e^{2\gamma\Delta t}\quad J_5=1-e^{-3\gamma\Delta t}\quad J_6=1-e^{3\gamma\Delta t}$$

$$J_7=e^{-\gamma\Delta t}-e^{\gamma\Delta t}+2\gamma\Delta t$$

$$\sigma_{r1}^2=-(\gamma\Delta t)^2\sigma_R^2/(J_1J_2)\qquad \sigma_{r2}^2=\Delta t J_7\sigma_R^2/(CJ_1J_2)$$

$$\mu_{r31}=2(\gamma\Delta t)^3\mu_{R3}/(J_2^2J_3)\qquad\qquad\qquad\qquad\qquad \mu_{r32}=-\gamma^2\Delta t^3e^{\gamma\Delta t}\mu_{R3}/(CJ_4)$$

$$\mu_{r33}=\Delta t^2\mu_{R3}[1+3e^{\gamma\Delta t}\{\gamma\Delta t(1+e^{2\gamma\Delta t})+J_4\}/(J_2^2J_4)]/C^2\qquad \mu_{r41}=-3(\gamma\Delta t)^4(\mu_{R4}-3\sigma_R^4)/(J_2^3J_5)$$

524

$$\mu_{r42}=2(\gamma\Delta t)^4\sigma_R^4/(J_1J_2)^2 \qquad \mu_{r43}=\mu_{r42}/2 \qquad \mu_{r44}=-\gamma^2\Delta t^3J_7\sigma_R^4/(CJ_1^2J_2^2)$$

$$\mu_{r45}=-2\gamma^3\Delta t^4(\mu_{R4}-3\sigma_R^4)/(Ce^{\gamma\Delta t}J_2J_5) \qquad \mu_{r46}=\Delta t^2\sigma_R^4\{1+4(\gamma\Delta t-J_1)(\gamma\Delta t+J_2)/(J_1J_2)^2\}/C^2$$

$$\mu_{r47}=\gamma^2\Delta t^4(\mu_{r471}+\mu_{r472}+\mu_{r473}+\mu_{r474}-3\sigma_R^4)/(C^2J_1J_2) \qquad \mu_{r471}=-6J_7\sigma_R^4/(J_1J_2^2)$$

$$\mu_{r472}=6\{2J_2+\gamma\Delta t(1+e^{\gamma\Delta t})\}\sigma_R^4/(J_1J_2^2)\qquad \mu_{r473}=-3(J_2+\gamma\Delta t e^{\gamma\Delta t})J_7\sigma_R^4/(\gamma\Delta tJ_1J_2^2)$$

$$\mu_{r474}=3(J_2+\gamma\Delta t)J_7\sigma_R^4/(\gamma\Delta tJ_1J_2^2)$$

$$\mu_{r48}=\gamma^2\Delta t^4(\mu_{r481}+\mu_{r482}+\mu_{r483}+\mu_{r484}+\mu_{r485}+\mu_{r486}-\mu_{R4})/(C^2J_1J_2)\qquad \mu_{r481}=3J_2(\mu_{R4}-3\sigma_R^4)/J_6$$

$$\mu_{r482}=-6(J_3-2\gamma\Delta t e^{-\gamma\Delta t})\sigma_R^4/(J_1^2J_2)\qquad \mu_{r483}=6\{2J_1-\gamma\Delta t(1+e^{-\gamma\Delta t})\}\sigma_R^4/(J_1J_2^2)$$

$$\mu_{r484}=3(J_1-\gamma\Delta t e^{-\gamma\Delta t})J_7\sigma_R^4/(\gamma\Delta tJ_1^2J_2)\qquad \mu_{r485}=-3(J_1-\gamma\Delta t)J_7\sigma_R^4/(\gamma\Delta tJ_1^2J_2)$$

$$\mu_{r486}=-3J_2^2(\mu_{R4}-3\sigma_R^4)/J_6=-\mu_{r48J}J_2$$

$$\mu_{r49}=-\Delta t^3(\mu_{r491}+\mu_{r492}+\mu_{r493}+2\mu_{r494}+\mu_{r495}+\mu_{r496}+\mu_{r497}+\mu_{r498}+\mu_{r499}-\mu_{R4})/C^3$$

$$\mu_{r491}=-2(6\gamma\Delta t+2e^{-3\gamma\Delta t}-9e^{-2\gamma\Delta t}+18e^{-\gamma\Delta t}-11)(\mu_{R4}-3\sigma_R^4)/(J_2^3J_5)$$

$$\mu_{r492}=12\{2\gamma\Delta t(2e^{-\gamma\Delta t}+1)-J_1(5+e^{-\gamma\Delta t})\}\sigma_R^4/(J_1J_2)^2$$

$$\mu_{r493}=-12\{\gamma\Delta t(2e^{-\gamma\Delta t}+4-\gamma\Delta t)-6J_1\}\sigma_R^4/(J_1J_2)^2\qquad \mu_{r494}=6\{\gamma\Delta t(2-\gamma\Delta t)-2J_1\}J_7\sigma_R^4/(\gamma\Delta tJ_1^2J_2^2)$$

$$\mu_{r495}=2\mu_{r483}J_7/(\gamma\Delta tJ_2)\qquad \mu_{r496}=6\{J_1(3-e^{-\gamma\Delta t})-2\gamma\Delta t\}(\mu_{R4}-3\sigma_R^4)/(e^{\gamma\Delta t}J_2J_5)\qquad \mu_{r497}=3\mu_{r46}C^2/\Delta t^2$$

$$\mu_{r498}=4C^2(\gamma\Delta t-J_1)\mu_{r47}/(\gamma^2\Delta t^4)\qquad \mu_{r499}=4C^2(\gamma\Delta t-J_1)\mu_{r48}/(\gamma^2\Delta t^4)$$

The second, third and fourth moment are given by Eq.(31) to Eq.(33).

$$\frac{d\sigma_S^2}{dt}+2\left(\frac{1}{K}\right)^m\beta\sigma_S^2=2\sigma_{r1}^2U_1(t)+C\sigma_{r2}^2 \tag{31}$$

$$\frac{d\mu_{S3}}{dt}+3\left(\frac{1}{k}\right)^m\beta\mu_{S3}=6\mu_{r31}U_2(t)+3C\mu_{r32}U_4(t)+C^2\mu_{r33} \tag{32}$$

$$\frac{d\mu_{S4}}{dt}+4\left(\frac{1}{K}\right)^m\beta\mu_{S4}=24\{\mu_{r41}U_5(t)+\mu_{r42}U_8(t)+\mu_{r43}U_{10}(t)\}+12C\mu_{r44}\{U_{11}(t)+U_{12}(t)+U_{13}(t)\}$$
$$+12C\mu_{r45}U_{15}(t)+6C^2\mu_{r46}U_{14}(t)+4C^2\{\mu_{r47}U_1(t)+\mu_{r48}U_{17}(t)\}+C^3\mu_{r49} \tag{33}$$

$$\frac{dU_1}{dt}+\left\{\gamma+\left(\frac{1}{K}\right)^m\beta\right\}U_1=1 \tag{34}$$

$$\frac{dU_2}{dt}+\left\{\gamma+2\left(\frac{1}{K}\right)^m\beta\right\}U_2=U_3(t) \tag{35}$$

$$\frac{dU_3}{dt}+\left\{2\gamma+\left(\frac{1}{K}\right)^m\beta\right\}U_3=1 \tag{36}$$

$$\frac{dU_4}{dt}+\left\{\gamma+2\left(\frac{1}{K}\right)^m\beta\right\}U_4=1 \tag{37}$$

$$\frac{dU_5}{dt}+\left\{\gamma+3\left(\frac{1}{K}\right)^m\beta\right\}U_5=U_6(t) \tag{38}$$

$$\frac{dU_6}{dt}+2\left\{\gamma+\left(\frac{1}{K}\right)^m\beta\right\}U_6=U_7(t) \tag{39}$$

$$\frac{dU_7}{dt}+\left\{3\gamma+\left(\frac{1}{K}\right)^m\beta\right\}U_7=1 \qquad (40)$$

$$\frac{dU_8}{dt}+\left\{\gamma+3\left(\frac{1}{K}\right)^m\beta\right\}U_8=U_9(t) \qquad (41)$$

$$\frac{dU_9}{dt}+2\left\{\gamma+\left(\frac{1}{K}\right)^m\beta\right\}U_9=U_1 \qquad (42)$$

$$\frac{dU_{10}}{dt}+\left\{\gamma+3\left(\frac{1}{K}\right)^m\beta\right\}U_{10}=U_{11}(t) \qquad (43)$$

$$\frac{dU_{11}}{dt}+2\left(\frac{1}{K}\right)^m\beta U_{11}=U_1 \qquad (44)$$

$$\frac{dU_{12}}{dt}+\left\{\gamma+3\left(\frac{1}{K}\right)^m\beta\right\}U_{12}=U_1 \qquad (45)$$

$$\frac{dU_{13}}{dt}+\left\{\gamma+3\left(\frac{1}{K}\right)^m\beta\right\}U_{13}=U_{14}(t) \qquad (46)$$

$$\frac{dU_{14}}{dt}+2\left(\frac{1}{K}\right)^m\beta U_{14}=1 \qquad (47)$$

$$\frac{dU_{15}}{dt}+\left\{\gamma+3\left(\frac{1}{K}\right)^m\beta\right\}U_{15}=U_{16}(t) \qquad (48)$$

$$\frac{dU_{16}}{dt}+2\left\{\gamma+\left(\frac{1}{K}\right)^m\beta\right\}U_{16}=1 \qquad (49)$$

$$\frac{dU_{17}}{dt}+\left\{\gamma+3\left(\frac{1}{K}\right)^m\beta\right\}U_{17}=1 \qquad (50)$$

The coefficient C is a constant with the dimension of time. The mean discharge and its deviation from the mean are derived by substituting Eq.(4) and Eq.(7) into Eq.(2).

$$\bar{q}=\left(\frac{1}{K}\right)^m\alpha\bar{S} \qquad (51) \qquad \tilde{q}=\left(\frac{1}{K}\right)^m\beta\tilde{S} \qquad (52)$$

The second, third and fourth moment of discharge are given by raising Eq.(52) to the power of two, three and four and by taking expectations.

$$\sigma_q^2=\left(\frac{1}{K}\right)^{2m}\beta^2\sigma_S^2 \qquad (53) \qquad \mu_{q3}=\left(\frac{1}{K}\right)^{3m}\beta^3\mu_{S3} \qquad (54) \qquad \mu_{q4}=\left(\frac{1}{K}\right)^{4m}\beta^4\mu_{S4} \qquad (55)$$

Consequently, the first four moments of discharge are given by solving Eq.(11) and Eq.(31) to Eq.(33) simultaneously.

3. EXAMINATION BASED ON SIMULATION METHOD

The adopted simulation method is carried out by directly solving Eq.(1) and Eq.(2) numerically as shown in Fig.2. The first four moments of discharge are calculated at each time interval. It is assumed that uncorrelated random variables in Eq.(24), $N(t)$ distribute exponentially. Their probability density function is expressed by

$$f(N)=\begin{cases}\lambda e^{-\lambda(N+1/\lambda)} & -1/\lambda\le N \quad \lambda: const. \\ 0 & elsewhere\end{cases} \qquad (56)$$

$$E(N)=0 \ , \ E(N^2)=1/\lambda^2=\sigma_N^2 \ , \ E(N^3)=2/\lambda^3=\mu_{N3} \ , \ E(N^4)=9/\lambda^4=\mu_{N4} \qquad (57)$$

The following parameters are set for simulation.

$$m=2 \ , \ K=5 \ , \ \Delta t=0.5(hr) \ , \ \bar{R}=\begin{cases}5(mm/hr) & 0\le t\le 20 \\ 0 & elsewhere\end{cases} \qquad (58)$$

On the other hand, the first four moments of $R(t)$ are given Eq.(24).

$$\sigma_R^2=\sigma_N^2/(1-\rho^2) \ , \ \mu_{R3}=\mu_{N3}/(1-\rho^3) \ , \ \mu_{R4}=\left(6\rho^2\sigma_N^2\sigma_R^2+\mu_{N4}\right)/(1-\rho^4) \qquad (59)$$

It is possible to calculate the theoretical first four moments of discharge by using Eq.(59). The solid and dashed lines in Fig.3 show the simulated and theoretical moments of discharge, respectively. Fig.3 denotes good agreement between them.

4. CONCLUSION

Fujita's previous work(1994) focused on stochastic response in which the rainfall input is an independent random variable. In this paper, the authors develop his work and propose new equations which can estimate the theoretical first four moments of discharge even if the rainfall input is a mutually dependent random variable.

REFERENCES

Bras, R.I. and Georgakakos K.P.: Real Time Nonlinear Filtering Techniques in Streamflow Forecasting, A statistical linearization approach, Third International Symposium on Stochastic Hydraulics, 95–105, 1980.

M. FUJITA, N. SHINOHARA, T. NAKAO and M. KUDO: Stochastic Response of Storage Function Model for Flood Runoff, Stochastic and Statistical Methods in Hydrology and Environmental Engineering, Vol. 2, 241–254, 1994.

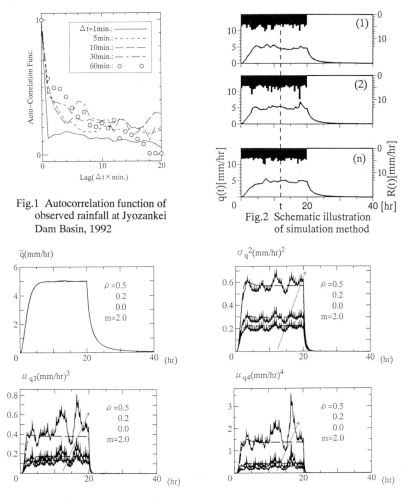

Fig.1 Autocorrelation function of observed rainfall at Jyozankei Dam Basin, 1992

Fig.2 Schematic illustration of simulation method

Fig.3 Comparison of results of simulated and computed

1D5

A Dynamic Model of Moisture and Heat Transfer in Soil-Plant-Atmosphere Continuum (SPAC)

H. P. HU, N. TAMAI, Y. KAWAHARA
Department of Civil Engineering, The University of Tokyo
Hongo 7-3-1, Bunkyo-ku, Tokyo 113, Japan

INTRODUCTION

The estimation of evapotranspiration from a vegetated landsurface has been increasingly drawn attention in recent decades since the evapotranspiration is a crucial process concerning with many natural phenomena, including water circulation, climate evolution, seed generating, plant growth, weather prediction and so on. In order to achieve this estimation it is essential to determine the fluxes of momentum, moisture, and heat at the land-atmosphere interface which is a boundary layer, containing vegetation and adjacent soil and air, saying Soil-Plant-Atmosphere Continuum (SPAC). For their own purposes many researchers have been involved in the study of exchange of energy and mass in the SPAC, but most of them pay more attention to independent submodels of soil, vegetation or atmosphere respectively by assuming remaining parts are in ideal or simple conditions. In recent years it is found that the coupled full-interacting system of the SPAC should be taken into consideration although the isolated and idealized analyses of the three components of the SPAC are also important and should be developed prior to the development of the theory of their coupled behavior (Milly, 1991).

Among the soil, vegetation, and atmosphere, vegetation is a more important regime than remaining two, since the presence of the vegetation greatly affects, and often largely controls the fluxes of mass, energy, and momentum at the landsurface. On the other hand, this greatly complicates the estimation of these fluxes as the number and complexity of the processes taking place in living organisms increase. Therefore, although many vegetation models have been developed, such as the Simple Biosphere model (SiB) (Sellers et al., 1986) and the Biosphere Atmosphere Transfer Scheme (BATS) (Dickinson et al., 1986), the applications of these models are still limited because of the complexity of the vegetation and a large amount of data requirements. Obviously, it is not possible to include all biophysical and biochemical processes into the vegetation model, but the essential processes must be involved into the model, otherwise the mechanisms of moisture and heat transfer in the SPAC could not be simulated correctly. The purpose of the present study is to establish a physically based, coupled, full-interacting SPAC model and its numerical simulation procedure to describe the moisture and heat transfer in soil, vegetation, and atmosphere. For the vegetation model, core of the SPAC, the number of the parameters which describe the physics and preserve the essential mechanisms are kept as low as possible. The present model is verified by field observation, meanwhile model sensitivity study is also performed and leads to a good understanding of the mechanisms of moisture and heat transfer in the SPAC.

MODEL STRUCTURE

SOIL MOISTURE AND HEAT

The mechanistic model for mass and heat flows in a partially saturated porous media developed by Witono and Bruckler (1991), both liquid water and water vapor flux being taken into account, are employed in this study, such that:

$$C_w \frac{\partial h}{\partial t} = \frac{\partial}{\partial z}(D_{hh}\frac{\partial h}{\partial z} + D_{hT}\frac{\partial T}{\partial z} + D_{l\,h}) - S_w \tag{1}$$

$$C\frac{\partial T}{\partial t} = \frac{\partial}{\partial z}(D_{TT}\frac{\partial T}{\partial z} - D_{Th}\frac{\partial h}{\partial z}) \tag{2}$$

where h is the pressure water head; T, soil temperature; D_{lh}, liquid water conductivity; S_w, intensity of root uptake; C_w, specific soil moisture capacity; C, volumetric heat capacity; D_{hh}, isothermal moisture conductivity; D_{hT}, thermal moisture diffusivity; D_{TT}, apparent thermal conductivity; D_{Th}, transport coefficient for heat flow.

INTERCEPTED WATER

The rainfall and dew water intercepted by the vegetation feed a storage of water within vegetation, being called intercepted reservoir. The intercepted water evaporates into the atmosphere at the potential evaporation rate from the fraction δ of the foliage covered with a film of water, whereas the water contained in the vegetation transpires from remaining part $(1-\delta)$ of the foliage. The equation of the water storage in the vegetation is set to:

$$\frac{\partial W_r}{\partial t} = veg \cdot P\text{-}E_r\text{-}R_r \tag{3}$$

where, W_r is the water content of the intercepted reservoir; veg, fraction of the vegetation; P, precipitation; E_r, evaporation from vegetation of the fraction δ. And R_r is the runoff of the intercepted reservoir, which occurs when W_r exceeds the maximum water content of the reservoir W_{rmax}.

SURFACE FLUXES

Following the basic equations of Noilhan and Planton (1989), the surface fluxes and surface resistance can be calculated by following equations.

1. Sensible heat flux, H

$$H=\rho_a C_p \frac{T_s-T_a}{r_a} \qquad (4)$$

where ρ_a is the density of the air; C_p, air specific heat; r_a, aerodynamic resistance; T_s and T_a, temperature of ground surface and air, respectively.

2. Evaporation from soil surface, E_g

$$E_g=(1-veg)\rho_a \frac{h_u q_{sat}(T_s)-q_a}{r_a} \qquad (5)$$

where $q_{sat}(T_s)$ is the saturated specific humidity at the temperature T_s; q_a, atmosphere specific humidity at reference height; h_u, relative humidity at the ground surface.

3. Evaporation from the δ fraction of the foliage covered by a water film, E_r

$$E_r=veg\rho_a\delta\frac{q_{sat}(T_s)-q_a}{r_a} \qquad \text{with} \qquad \delta=\left(\frac{W_r}{W_{rmax}}\right)^{\frac{2}{3}} \qquad (6)$$

4. Transpiration from remaining part $(1-\delta)$ of the foliage, E_{tr}

$$E_{tr}=veg\rho_a(1-\delta)\frac{q_{sat}(T_s)-q_a}{r_a+r_s} \qquad (7)$$

where r_s is the surface resistance. When $q_{sat}(T_s) \le q_a$, $\delta = 1$ in Eq.(6) and (7).

5. Aerodynamic resistance, r_a

The calculation of r_a is based on Monin-Obukhov similarity theory, hence:

$$r_a=\frac{1}{\kappa u_*}\left\{ln\left(\frac{z-z_d}{z_{oh}}\right)-\Psi_h(\xi)\right\} \qquad \text{with} \qquad u_*=\frac{\kappa u}{\left\{ln\left(\frac{z-z_d}{z_{om}}\right)-\Psi_m(\xi)\right\}} \qquad (8)$$

where κ is the von Karman constant; u_* the friction velocity; z, height of atmospheric measurement above ground; z_d, displacement height; z_{om} and z_{oh}, roughness heights for momentum and heat; ξ, dimensionless height; The determination of universal functions $\Psi_m(\xi)$ and $\Psi_h(\xi)$ can be referred to Hu et al. (1994).

6. Surface resistance, r_s

Surface resistance, r_s, depends on various governing factors, such as solar flux, leaf water potential, ambient temperature, ambient carbon dioxide, vegetation situation, and vapor pressure deficit. According to Dickinson et al. (1986), Noilhan and Planton (1989), the expression of r_s is taken to be:

$$r_s=\frac{r_{smin}}{LAI} \cdot \frac{1}{I_1 I_2 I_3 I_4} \qquad (9)$$

where r_{smin} is the minimum surface resistance, I_1, I_2, I_3, I_4 are fractional conductances.

SURFACE ENERGY BUDGET

At the ground surface, energy budget which links the models of soil, vegetation, and atmosphere into a SPAC model is written as:

$$R_n = G+H+lE \tag{10}$$

where R_n is the net radiation; G, heat flux; lE and H, latent and sensible heat flux respectively.

MODEL VALIDATION

In order to verify the present SPAC model and test its accuracy, field observation data obtained in a winter wheat field in Yucheng (36.98°N,116.65°E) in northern China (Wu, 1994) are utilized. The observation was started at 0700 local standard time (LST) in April 23 and finished at 0800 LST April 27, 1992. A lot of items were observed or measured, including soil temperature, soil moisture, evapotranspiration, vegetation parameters, and meteorological data. Meanwhile, parameters of the SPAC model are either specified in the laboratory, such as soil water conductivity, soil water characteristic curve, soil water capacity, soil water volumetric heat capacity, or determined by observation.

The SPAC model described previously was run with the observed initial and boundary conditions. The meteorological input data are net radiation, air temperature, water vapor pressure, and wind velocity, which are observed at 2 m above ground surface. Fig.1 shows comparison between the simulated evapotranspiration results and the observed data as a function of time. Considering the difficulty of the estimation of the evapotranspiration in such a complex situation, the calculation results are quite good, the biggest deviation being only about 20%. Fig.2 shows the comparison of the soil temperature at different depths in cases of simulation and observation, respectively. The simulated soil temperatures demonstrate

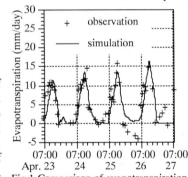

Fig.1 Comparison of evapotranspiration

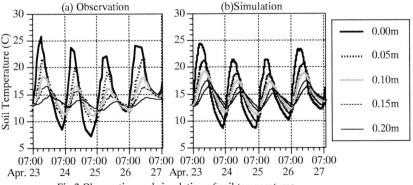

Fig.2 Observation and simulation of soil temperatures

fair agreement with the observations, the biggest difference being about 3 degrees. In general, good agreement is obtained between the simulated results and observed ones, thus the present model is verified.

SENSITIVITY STUDY AND DISCUSSION

A sensitivity study is performed in order not only to acquire a good understanding of the mechanisms of moisture and heat transfer in the SPAC, but also to examine the influential factors in an evapotranspiration process. The observation data obtained in the first 24 hours (from 0700 LST Apr. 23 to 0700 LST Apr. 24, 1992) during the field observation are chosen as the basic situation, which is utilized by all sensitivity study tests. Firstly, the sensitivity to the soil texture is carried out by varying soils from a coarse texture typical of sand, to a medium texture typical of loam, to a fine texture typical of clay, and the initial soil moistures are corresponded to a stable soil moisture profile for each soil with the ground water table at 2 m below the ground surface. It is found that the soil texture is not a sensitive factor. Although the surface moisture are very different for three kinds of soils (see Fig.3), the differences of evapotranspiration among them are very small, the biggest evapotranspiration rates being 13.44 , 12.85, 12.43 (mm/day) respectively for sand, loam, and clay. This conclusion is different from the study of Jacquemin et al. (1990). They got the conclusion that soil texture is a crucial influence factor for evaporation in the case of a bare ground with the surface moisture and average moisture for 1 m depth soil having 50% of saturated soil moisture.

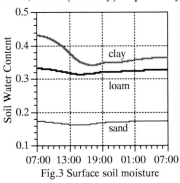

Fig.3 Surface soil moisture

Secondly, the sensitivity to soil moisture is examined by taking the initial surface soil moisture as 25% (dry), 50% (medium), and 75% (wet) of saturated soil moisture (0.4713, m^3/m^3) for loam, being 0.1183, 0.2365, and 0.3548 (m^3/m^3). Through the study, one can conclude that the soil moisture is a crucial influential factor. Fig.4 clearly shows that different soil moistures cause different evapotranspiration processes, especially for dry soil whose evaporation rate is much smaller than that of medium and wet soil. However, there is no obvious difference between medium and wet soil, because in these two cases there is enough water for evaporation, soil moisture

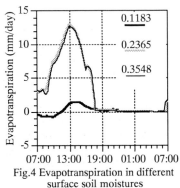

Fig.4 Evapotranspiration in different surface soil moistures

is no longer a limiting factor. At last, a synthetical analysis is carried out to investigate the sensitivity to a canopy, totally four parameters are examined, being leaf area index (LAI), vegetation cover fraction (veg), surface temperature (T_s), and minimum surface resistance (r_{smin}). Compared with soil moisture they four are not important influential factors, although they definitely affect the moisture and heat transfer in one or other ways.

CONCLUDING REMARKS

In this paper, a coupled full-interacting physical SPAC model to describe the dynamic aspects of the water and heat transfer in soil, vegetation, and atmosphere is developed. A vegetation model is employed to describe the vegetation as simply as possible, but preserving the essential characteristics of the botanical activities of vegetation. Transpiration is controlled by surface resistance, while the intercepted water of rainfall and dew evaporate at the potential rate. The required atmospheric input data are limited to only four, being net radiation, wind speed, vapor pressure, air temperature, which are all regularly observed at meteorological observation stations. Therefore, the present method can be easily applied in practical situations.

REFERENCES

[1]. Dickinson, R. E., A. Henderson-sellers, P.J. Kennedy, and M. F. Wilson, 1986, Biosphere-atmosphere transfer scheme (BATS) for the NCAR community climate model, Tech. Note NCAR/TN-275+STR, Natl. Cent. for Atmos. Res., Boulder, Colo., 1986.

[2]. Hu, Heping, N. Tamai, and Y. Kawahara, 1994, Moisture and heat transfer in unsaturated soil in relation to atmospheric conditions, Proceedings of 9th Congress of APD-IAHR, Singapore, Vol. 1, pp. 73-80.

[3]. Jacquemin B. and J. Noilhan, 1990, Sensitivity study and validation of a landsurface parameterization using the HAPEX-MOBILHY data set, Boundary-Layer Meteorol., Vol. 52, pp. 93-134.

[4]. Milly, P. C. D., 1991, Some current themes in physical hydrology of the land-atmosphere interface, Hydrological interactions between atmosphere, soil, and vegetation, IAHS publication, No. 204, pp. 3-10.

[5]. Noilhan, J. and S. Planton, 1989, A simple parameterization of landsurface process in meteorological models, Mon. Wea. Rev., Vol. 117, pp. 536-549.

[6]. Sellers, P. J., Y. Mintz, Y. C. Sud and A. Dalcher, 1986, The design of a simple biosphere model (SiB) for use within general circulation models, J. Atmos. Sci., Vol. 43, pp. 505-531.

[7]. Witono, H. and L. Bruckler, 1989, Use of remotely sensed soil moisture content as boundary conditions in soil-atmosphere water transport modeling: 1. field validation of a water flow model, Water Resour. Res., Vol. 25, pp. 2423-2435.

[8]. Wu, Qinglong, 1994, An investigation of numerical simulation for water movement and heat transfer under field evapotranspiration, Ph.D. thesis submitted to Tsinghua University, P. R. China.

1D6

A DYNAMIC APPROACH TO HYDROLOGY PROBLEMS: A STUDY OF DISASTROUS PLUVIOMETRIC EVENTS THROUGH WEATHER TYPE ANALYSIS

Alessandro Pezzoli
Department of Hydraulics – Turin Politecnico
Corso Duca degli Abruzzi, 24 – 10129 Turin, Italy

Abstract
Following a brief analysis of the relationships linking atmospheric circulation and precipitation, this report examines the most significant types of weather encountered in North Western Italy and, in relation to the catastrophic rainfall event which hit Piedmont on 5-6 November 1994, it shows that by means of the meteorological cartography supplied by the Bracknell (GB), Offenbach (D) and Rome (I) Centres it proved possible to obtain a reliable hydrometeorological forecast of this event from 36 to 48 hours in advance.

Résumé
Dans ce document, aprés une bréve analyse du rapport existant entre la circulation atmosphérique et les précipitations, on prend en compte les types de climat les plus significatifs dans l'Italie Nord-Occidentale. Par la suite, en relation avec l'événement pluviométrique catastrophique qui s'est produit au Piémont les 5-6 novembre 1994, on montrera comment il a été possible, en utilisant la cartographie météorologique fournie par les Centres de Bracknell (GB), Offenbach (D) et Rome (I), d'effectuer une prévision hydro-météoorologique fiable de l'événement avec une avance de 36-48 heures.

1. As is known for precipitation to occur, three basic stages are necessary, and namely:
i) the creation of saturation conditions. This thermodynamic condition, which is defined as the attainment of the dewpoint [6], is achieved through the cooling process associated with the ascending movement of moist air;
ii) a phase shift of the water content from vapour to liquid and/or solid state;
iii) the growth of the liquid droplets or ice crystals to the precipitation size.

Consequently, it is essential to investigate the stability (or instability) conditions of the air masses and the relative cooling mechanisms. The latter have been classified by Fletcher [3] as cyclonic, orographic or convective.

Cyclonic cooling can be either frontal or non-frontal; non-frontal cooling arises from the convergence and subsequent rising of air, as is observed in low pressure areas. Non-frontal precipitation of non tropical origin produces rainfall (or snowfall) of moderate intensity and rather long duration; this type of precipitation, in fact, may last from to 24 to 72 hours.

Frontal cyclonic cooling takes place instead when the air moves up along a frontal surface. When the front is warm, as a rule, the inclination of the frontal surface is modest, between 1:100 and 1:400. Because of this modest inclination, the clouds are mostly stratified, with a precipitation range (rain or drizzle of short duration) preceding the front to a depth of 150 ÷ 350 km.

The typical inclination of a cold front is much steeper, between 1:25 and 1:100. As a result, the air moves up and cools down rapidly thereby giving rise to tumultuous meteorological conditions, accompanied by the formation of cumulus type clouds with a marked vertical development, heavy rainfall of short duration, strong winds. The precipitation range following the front is 15 to 95 km deep and its width depends upon the physical difference between the two air masses meeting along the front.

Orographic cooling, on the other hand, takes place when moist winds rise along the mountain sides: the air expands as it is lifted, and it cools down as it meets the lower pressure areas higher up. Clouds are formed, first of all, and then, when the dew point is reached, precipitation occurs.

Convective cooling is produced when the vertical instability of moist air, as caused by the heating of the earth's surface, produces a convection current. Convective type precipitation phenomena are short lived, hardly ever lasting over an hour, but of great intensity.

2. Due to the special geographical position of the Italian peninsula, many parts of Northern Italy, and especially the Po Valley, are often hit by severe floods.

Obviously enough, such events are primarily associated with pressure conditions able to determine abundant rainfall over the central and northern regions of Italy.

From the pluviometric standpoint, the pressure configuration which is most commonly observed in connection with floods consists of a low pressure area over the western Mediterranean combined with very high "block type" pressure areas over the Balkan peninsula [5]. These conditions are quite frequent in Autumn, i.e. the season during which the highest annual amounts of rainfall are recorded and Alpine and Apennine river floods occur.

Among the factors associated with the morphology of the territory, orographic conditions play a fundamental role in giving rise to flooding in Northern Italy. On account of their geographical orientation, in fact, the Alpine and Apennine chains are exposed upwind to the air currents - moving in from the southern quadrants - which are always present in those meteorological situations that give rise to heavy and extended rainfall.

An investigation conducted by Giacobiello and Todisco [4] made it possible to select 72 flood related meteorological situations: 83% of them had well-defined characteristics, enabling these situations to be identified through three main synoptic patterns. A schematic representation of the topographic conditions of these three patterns at 500 hPa, as determined by working out, for each of them, the average geographical positions of the ridge and trough axes and those of the geopotential highs and lows, is illustrated in figs. 1a ÷ 1c, taken from the investigation mentioned above. These three patterns differ from one another on account of the different inclinations, relative to the meridians, of the axes of the troughs: from the north-east, the north and the north-west, respectively. They all share the following features:

i) out-flowing currents between south and south-west over Italy;
ii) the cyclonic curvature of the air currents in the area comprised between the western Mediterranean and north-western Africa;
iii) the presence of a vast ridge over eastern Europe and the Balkan peninsula.

The configurations exhibiting a trough from the north-west (see fig. 1a) are characterised, at 500 hPa, by a ridge from the south-west over the British Isles and a vast depression with its low over the western Mediterranean. In Northern Italy, the heaviest precipitation associated with this pattern occurs mainly in the Alps, the eastern Pre-Alps and the neighbouring areas. The pressure configuration exhibiting a trough from the north (see fig. 1b) is characterised by a large ridge over the Atlantic and a low on the North Sea, which is associated with a trough extending to the western Mediterranean where the secondary low is now generally absent. This situation is not as frequent as the other two and the heaviest rainfall nuclei occur over the Alps and the central-eastern Pre-Alps. The configuration involving a trough from the north-west (see fig. 1c) displays a pressure low near the north-western coast of France and a ridge of considerable width over eastern Europe. The precipitation associated with this pattern are generally very heavy and affect Piedmont and Lombardy more directly, as confirmed by Pangallo [5].

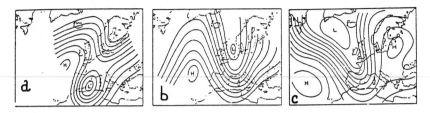

Fig. 1 - a) trough at 500 hPa from the north-east; b) trough at 500 hPa from the north; c) trough at 500 hPa from the north-west.

A mixed stochastic-deterministic model devised for the purpose of evaluating daily rainfall through an analysis of atmospheric circulation patterns, and hence of the relative high elevation cartography, has been developed by Bardossy and Plate [1]; furthermore, an accurate examination of Europe's typical atmospheric conditions has been worked out by Roth [7], who identified 18 different types of climate by correlating pressure conditions at ground level with those at 500 hPa.

Through such instruments it can be seen that the most dangerous pattern in connection with devastating rainfalls hitting the north-western regions of Italy (Piedmont, Liguria, Valle d'Aosta, Lombardia) is the one characterised by a low-pressure region over the North Atlantic with a trough over the Iberian peninsula (see fig. 2) and an anticyclone moving eastward and sometimes extending all the way to the Balkans. The great depression over the north Atlantic, acting as the centre of the action on the ocean fronts, generally displays two distinct pressure lows, of about 975 hPa. Over the Iberian peninsula, in lieu of an independent depression, we find a large trough, that is to say a pointed extension of the principal low-pressure region; in such cases, the 500 hPa chart is generally characterised by a trough with its axis arranged in the north-west direction (see fig. 1c).

In these circumstances, precipitation phenomena, in the form of rain, will occur in eastern France and north-western Italy. In the latter area, heavy rainfall is fostered by the so-called "stau" effect, i.e., the lifting of the air masses due to the Alpine barrier.

The trough over western Europe will evolve towards the east, thereby intensifying the flow of moist air from the south over Italy. The disturbance which can be observed over Gibraltar will also be conveyed in the north-eastern direction and will soon have an adverse effect on the atmospheric conditions over northern Italy and the Tyrrhenian coast; in most cases, this situation eventually gives rise to a depression over the western Mediterranean (see fig. 3), where the prevailing element of the situation is a well delimited low over the western Mediterranean with a principal low (about 1000 hPa) north-west of Sardinia moving towards the Gulf of Genoa, and a secondary low south-west of the Balearic Islands.

The air masses collect additional vapour as they move across the western

Mediterranean and give rise to sustained precipitation phenomena over northern Italy, which are further enhanced by the "stau" effect caused by the Alpine elevations. Another vast and deep depression is observed over the northern Atlantic, whilst the Azores anticyclone can be clearly identified off the Portuguese shore.

Fig. 2 - Meteorological chart at ground level: depression over the North Atlantic with a trough above Spain.

Fig. 3 - Meteorological chart at ground level: depression over the western Mediterranean.

These conditions give rise to heavy rainfall in western Italy and the northern part of the Tyrrhenian, as well as rain and storms over Corsica and Sardinia, i.e., along the cold front of the depression.

Bad weather will persist over central-northern Italy because of the anticyclonic block over Russia, and the cold front moving in from the west will make weather conditions even worse.

3. On 5 and 6 November 1994, a situation of the type described above caused a devastating flood in Piedmont, resulting in 65 dead and damage for some 1300 billion Italian Lire.

As we shall demonstrate below, by means of the meteorological cartography transmitted by radio by the European Meteorology Centres and received by the Meteorology Department of the Hydraulics Department of the Turin Politecnico [2], it was possible to foresee the event 36÷48 hours in advance, in its quantitative as well as qualitative terms. Forecast rainfall levels were then compared with the values actually measured by several pluviometric stations situated in the Piedmontese territory (Central Turin; Turin Piazza d'Armi; Migliorero Shelter–S. Gesso; Sella Balma–S. Ellero; Alpe Costapiana–S. Malone).

From the ground level forecast charts, issued 24, 48 and 72 hours before the event (see figs. 4d - 4e - 4f), it was possible to detect the presence of a massive trough (and relative low) extending over the entire Spanish peninsula; as mentioned above, this situation was to be rated as highly hazardous for north-western Italy.

Furthermore, the forecasts made available by elevation cartography at 500 hPa, 24, 48 and 72 hours before the event, showed a deep trough on the Iberian peninsula with its axis arranged in the north-west direction. Its presence, in line with the observations made earlier, suggested a pluviometric event of great severity, with rain falling all over Piedmont and even heavier rainfall, because of the orographic conditions, along the mountains with sides facing south/south-east on Saturday and those with sides facing south/south-west on Sunday (see figs. 4a - 4b - 4c).

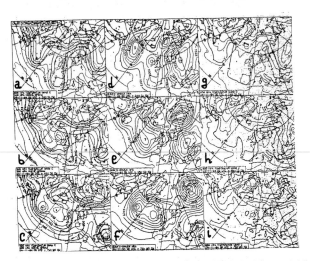

Fig. 4 - Forecast chart at 500 hPa: a) 05/11, 12:00 PM UTC; b) 06/11, 12:00 PM UTC; c) 07/11, 12:00 PM UTC. Ground level forecast charts: d) 05/11, 12:00 PM UTC; e) 06/11, 12:00 PM UTC; f) 07/11, 12:00 PM UTC. Temperature forecast charts at 850 hPa: g) 05/11, 12:00 UTC; h) 06/11, 12:00 PM UTC; i) 07/11, 12:00 PM UTC.

Moreover, forecast temperature charts at 850 hPa (see figs. 4g - 4h - 4i) gave high temperature values, with the thermal zero located at about 3000 m; this meant that the situation would be even more dramatic, since the precipitation would not be in the form of snow up to an elevation of $2500 \div 3000$ m.

Finally, another equally important indication came from the isohyet charts produced 48 hours in advance on the basis of complex mathematical models (see figs. 5a - 5b - 5c).

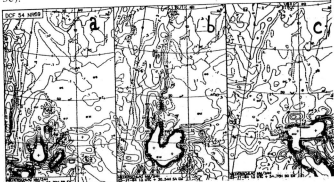

Fig. 5 - Isohyet forecast charts: a) 03/11, 6:00 PM UTC; b) 04/11, 6:00 PM UTC; c) 05/11, 6:00 PM UTC.

These charts, in fact, confirmed that the rain cell would extend over the entire Piedmont region and the minimum rainfall amounts in 24 hours would come to 50 mm on November 4, 150 mm on November 5 and 50 mm on November 6, such values being comparable with the amounts measured by rain gauges (see fig. 6).

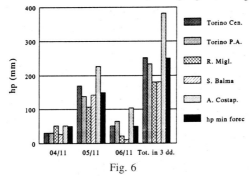

Fig. 6

These charts confirmed the severity of the event, since they indicated - 48 hours before the event - that the amount of rain falling over Piedmont alone, from Friday 4 through Sunday 6 November, would be, as a minimum, 250 mm. This enabled the local Civil Protection authorities to mobilise 2 days in advance and consequently, the Province of Turin, where immediate action was taken, was the only one among the areas hit by the flood were no fatal accidents were recorded along the provincial road network.

In conclusion, it can be stated that, provided it is used effectively, meteorological cartography can be an excellent tool to deal with civil protection and territorial management problems.

Acknowledgements

The Author wishes to his express his gratitude to the Meteomont Service of the Brigata Alpina Taurinense (Italian Army) that supplied measured rainfall data.

References

[1]   Bardossy A. e E.J. Plate (1991). Modelling daily rainfall using a semi–Markov representation of circulation pattern occurence. *Journal of Hydrology*, 122, 33–47.

[2]   Centri Meteorologici Internazionali (1994). Cartografia meteoidrologica. Raccolta del Dipartimento di Idraulica – Politecnico di Torino, 11, November 1994.

[3]   Flecter N.H. (1962). The physics of rainclouds. Cambridge University Press, New York.

[4]   Giacobello N. e G. Todisco (1979). Caratteristiche sinottiche di alcune situazioni alluvionali. *Rivista di Meteorologia Aeronautica*, 2.

[5]   Pangallo E. (1993). Caratteri meteorologici degli eventi alluvionali dell'autunno 1993 sulle Alpi Occidentali. *Nimbus*, 1, 2.

[6]   Pezzoli A. (1994). La previsione idrometeorologica nello studio delle reti idrauliche. *XXIV Convegno di Idraulica e Costruzioni Idrauliche*, Napoli, September 1994.

[7]   Roth G.D. (1993). Meteorologia. Arnoldo Mondadori Editore, Verona.

1D7

Fractal Interpolation on Simulation of Hydrologic Series

C.-L. LIU, T.-Y. LEE & S.-C. LIN
Professor, Ph.D. Candidate & Ph.D. Student
Dept. of Hydraulics & Ocean Eng'g., Cheng Kung University
Tainan, Taiwan 70101, ROC

ABSTRACT

By iterated function system (IFS), a new simulation of hydrologic series can be achieved. This method, different from the traditional one, is easy to deal with an irregular and complex data pattern. The result shows that it matches the observed data with enough interpolation points and the first fourth moments can be well preserved.

INTRODUCTION

For a dynamic system, the observation of time process is time series. Analyzing this series, the dynamic behavior can be realized. The traditional time series analysis uses only a single scale of time interval, so it can be regarded as Euclidean time series. On the other hand, if the different time scales are employed, then they show the statistical similarity, say the fractal time series [1,2]. Consider a function $y(x)$ to be discretized as $\{x_i, y_i | i = 1,2,3,...,N\}$, where N is the total data number. Traditionally a classical function, such as line, circle, sin, cos, and polynomials are utilized for data fitting. However if data set is smooth, the classical approach is good, but if it is complex and/or irregular, then the traditional method can not fit so well. For this reason a new alternative is employed with fractal interpolation technique. Its purpose is to fit the data of complexity and irregularity as close as possible. Therefore the fractal geometry provides the description for the qualities of change [3].

THEORETICAL DESCRIPTION

1. AFFINE TRANSFORMATION

Consider a point $p = p(x,y)$, and after scaling (resizing), rotation and translation, the new point $p' = p'(x',y')$. The relationship between these two coordinates can be described by affine transformation

$$A\begin{pmatrix} x \\ y \end{pmatrix} = \begin{pmatrix} x' \\ y' \end{pmatrix} = \begin{pmatrix} a & b \\ c & d \end{pmatrix}\begin{pmatrix} x \\ y \end{pmatrix} + \begin{pmatrix} e \\ f \end{pmatrix} \tag{1}$$

where A is the affine function. In the above equation six coefficients exist, namely a, b, c, d, e and f respectively. The first four elements are for scaling and rotation, the last two are for translation.

2. ITERATED FUNCTION SYSTEM (IFS)

The IFS was initially given by Barnsley [4,5], or called chaos game. The hydrologic data can be regarded as point set $\{(t_i, z_i) \mid i = 1, 2, ..., N\}$, where t is time (as x-axis) and z is observed hydrologic variable (as y-axis). Let $p_i = p(x_i, y_i)$ and deduce the IFS in general.

Assume x_i is an increasing function $x_1 < x_2 < x_3 < \cdots < x_N$ and no rotation with respect to y-axis, then $b = 0$. Eq.(1) has only 5 parameters remained to define the affine transformation. The form of IFS is

$$A_i\begin{pmatrix} x \\ y \end{pmatrix} = \begin{pmatrix} a_i & 0 \\ c_i & d_i \end{pmatrix}\begin{pmatrix} x \\ y \end{pmatrix} + \begin{pmatrix} e_i \\ f_i \end{pmatrix} \tag{2}$$

Let $p_1, ..., p_{i-1}, p_i, ..., p_N$ be a sequential set, the affine function $A_i (i = 2, 3, ..., N)$ has to concentrate their end points p_1 and p_N mapping to p_{i-1} to p_i. We have 5 parameters $(a_i, c_i, d_i, e_i, f_i)$ to define the affine transformation within p_{i-1} and p_i with only 4 equations provided. Let d_i be a constant, it follows

$$\begin{cases} a_i = \dfrac{x_i - x_{i-1}}{x_N - x_1} \\[2mm] c_i = \dfrac{y_i - y_{i-1}}{x_N - x_1} - d_i \cdot \dfrac{y_N - y_1}{x_N - x_1} \\[2mm] e_i = \dfrac{x_N x_{i-1} - x_1 x_i}{x_N - x_1} \qquad (i = 2, 3, \cdots, N) \\[2mm] f_i = \dfrac{x_N y_{i-1} - x_1 y_i}{x_N - x_1} - d_i \cdot \dfrac{x_N y_1 - x_1 y_N}{x_N - x_1} \end{cases} \tag{3}$$

If d_i is adjusted, the individual transformation will be changed.

3. FRACTAL DIMENSION IN TIME DOMAIN

To estimate the fractal dimension of a time series, the box-counting method is adopted. If the time scale is taken as x-axis, not consistent with y-axis in scale, its different scales will be overcome by the method of standardization (normalization) [6,7]. Let $x^* = |\max\{x_i\} - \min\{x_i\}|$ and $y^* = |\max\{y_i\} - \min\{y_i\}|$, thus the new data becomes $p_i^* = p(x_i^*, y_i^*) = p(x_i / x^*, y_i / y^*)$. The number $N_m(\varepsilon_m)$ covered by the square will be counted according to different length ε_m of box where $m = 0, 1, 2, \cdots, M$. When $m = 0$, the side length $\varepsilon_m = 1$ and box counting $N_m(\varepsilon_m) = 1$. M is the maximum iteration number for calculation.

Taking logarithm for $\{\varepsilon_m\}$ and $\{N_m(\varepsilon_m)\}$, the slope of the linear regression by least square method is fractal dimension D_F.

Besides, the parameter d_i is related to the fractal dimension of an attractor of IFS. According to Barnsley's theorem [5], if x_i values fall in equal increment and with $d_i = d$, we get

$$d = (N-1)^{D_F-2} \tag{4}$$

The d value can be obtained once the fractal dimension D_F is calculated. Substituting d in Eq.(3), the other affine parameters can be solved.

SIMULATION OF HYDROLOGIC SERIES

Based on the IFS from affine transformation and fractal theory, it is available to interpolate the data points, known as fractal interpolation. The random iteration algorithm is used for generating the random interpolating data. Theoretically there are infinite interpolating points showing the internal structure in detail for fractal object, in other words, the resolution is raised. As the interpolating points increase from few to many, the result is "virtual reality" in computer science while "fitting" or "simulation" in statistics.

The illustrated example is chosen from a reservoir catchment in northern Taiwan. For flow hydrograph of one storm event at the site, the fractal interpolation is used to simulate the hydrologic process. The time duration is 115 hours $(N = 115)$. Fig.1 shows the fractal dimension by means of box-counting method. The slope of the linear regression, or say the fractal dimension, is 1.3266 and the high correlation coefficient implies the fractality exists.

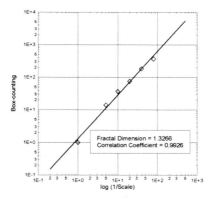

Fig.1 The diagram of fractal dimension by box-counting method

As mentioned in the above theory, the affine function $A_i (i = 2, 3, ..., 115)$ can be calculated for the hydrologic series. The different fractal interpolating points

542

can be adopted for the simulation of the original data through the IFS, and the results are illustrated in Fig.2. There are 8 cases, namely 20, 50, 100, 200, 500, 1000, 2000 and 5000 points. In this figure it is easy to find the more the interpolating points (the higher resolution), the better the fitting is. The simulated hydrograph will gradually match the observed hydrograph. Examining closely in Fig.3 to the statistical properties, the first fourth moments can be well preserved as the interpolating points increase to certain values, which really reflect the hydrologic characteristics. This performance can not be expected by traditional methods, say ARMA model et al., in particular for the preservations of the third and the fourth moments. The superior merits of fractal interpolation technique make itself high potential in application.

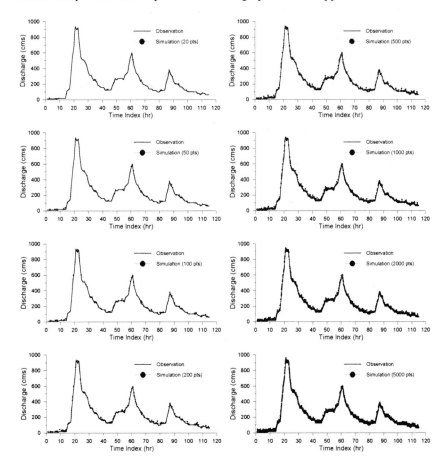

Fig.2 Simulated hydrograph on different interpolated data points

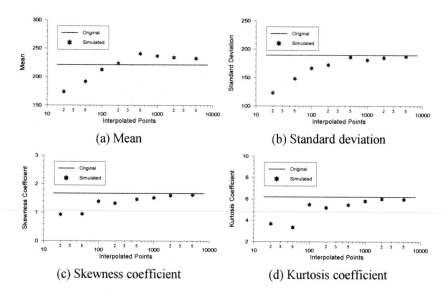

(a) Mean (b) Standard deviation

(c) Skewness coefficient (d) Kurtosis coefficient

Fig.3 Comparison of the first fourth moments on different interpolated data

In order to detect the sensitivity of fractal dimension in simulation, four "unreal" cases for fractal dimension are adopted, say 1.00, 1.80, 1.95 and 2.00. With these values the affine function A_i is calculated again $(N = 115)$. The simulation is shown in Fig.4. When $D_F = 1.00$ and $d = (115-1)^{-1} \approx 0.0088 \rightarrow 0$ nearly as linear interpolation, no much irregularity exists. When $D_F = 1.80$ (> 1.3266), the simulation deviates from the observation as shown in Fig.4(b). When $D_F = 1.95$, the deviation is obviously enlarged in Fig.4(c). When $D_F = 2.00$, the fitting is completely in vain. The oscillation of the fitting curve scatters in the whole plane and is shown in Fig.4(d). This phenomena tell us the fractal dimension up to 2 will be a plane, not a line anymore, in fractal geometry.

CONCLUSIONS

The fractal interpolation technique for fitting or simulating a hydrologic time series is proved possible and gives good performance. With increasing the interpolating points, the simulation will match the reality. If the number of interpolating points is greater than the population of original data of hydrologic series, the fitting series can be very well and preserve the first fourth moments. On contrary, the traditional methods seldom preserve the third moment and the fourth moment. The fractal dimension must be carefully estimated, especially as its value is close to 2.

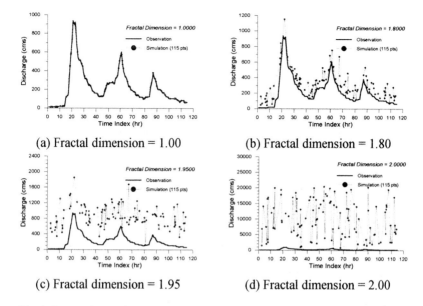

(a) Fractal dimension = 1.00

(b) Fractal dimension = 1.80

(c) Fractal dimension = 1.95

(d) Fractal dimension = 2.00

Fig.4 Comparison of discharge hydrograph on different fractal dimensions

REFERENCES

[1] Peters, E. E. (1991). *Chaos and Order in the Capital Markets: A New View of Cycles, Prices, and Market Volatility*, John Wiley & Sons, Inc., New York.

[2] Tsonis, A. A. (1992). *Chaos: From Theory to Applications*, Plenum Press, New York.

[3] Briggs, J. and F. D. Peat (1989). *Turbulent Mirror: An Illustrated Guide to Chaos Theory and the Science of Wholeness*, Harper & Row, Publishers, New York.

[4] Barnsley, M. F. (1986). Fractal Functions and Interpolation, Constructive Approximation, 2, pp. 303-332.

[5] Barnsley, M. F. (1993). *Fractals Everywhere*, Second Edition, Academic Press, Inc., Boston.

[6] Liebovitch, L. S. and T. Toth (1989). A Fast Algorithm to Determine Fractal Dimensions by Box Counting, Physics Letters A, 141(8,9), pp. 386-390.

[7] Chatterjee, S. and M. Yilmaz (1992). Use of Estimated Fractal Dimension in Model Identification for Time Series, Journal of Statistical Computation and Simulation, 41(3-4), pp. 129-142.

1D8

Modélisation des ressauts en ruissellement hydrologique quasi-tridimensionnel sur terrains quelconques

M. PIROTTON
Université de Liège, Institut du Génie Civil, Laboratoires L.H.C.N,
6, Quai Banning - B.4000 Liège - Belgique

INTRODUCTION

Comme l'ont encore dramatiquement prouvé les récentes crues hivernales, une prévision fiable de l'amplitude et de la distribution temporelle des débits, ainsi qu'une compréhension globale des processus de transformation hydrologique par un bassin d'un signal de pluies en un signal de débits à son exutoire, s'imposent comme fondements de toute gestion efficace des ressources hydrauliques, basée sur l'anticipation et respectueuse des populations aval.

En marge des modélisations hydrologiques traditionnelles qui convertissent, par une fonction mathématique paramétrée, un signal d'entrée en un signal de sortie, une alternative est examinée, basée sur la physique de l'écoulement à grande échelle de la lame d'eau sur le bassin. Les avantages d'un tel type d'approche découlent de l'interprétation physique inhérente à chaque paramètre utilisé.

Si les analyses théoriques et expérimentales conduisent à légitimer l'hypothèse cinématique pour le ruissellement hydrologique en fine lame, son application s'est longtemps cantonnée dans des applications unidimensionnelles, avec des topographies naturelles idéalisées en une succession de géométries élémentaires.

Cette communication démontre que sa généralisation à des topographiques digitalisées quelconques se heurte effectivement à d'innombrables difficultés, qui ne sont fiablement surmontées que par l'introduction, dans la solution, de discontinuités suggérées par une réflexion générale sur la signification physique des singularités d'écoulement.

APPLICATION DE LA THEORIE CINEMATIQUE A L'HYDROLOGIE

Lorsqu'on raisonne sur les équations de base de l'hydraulique et qu'on les particularise à une modélisation à grande échelle d'écoulements hydrologiques en très fine lame, deux paramètres significatifs émergent dont les valeurs relatives s'avèrent déterminantes dans le choix d'une approximation adéquate : le nombre de Froude f_0 ainsi que le nombre d'onde cinématique k_0, cité la première fois par Woolhizer et Liggett en 1967 et défini par :

$$k_o = \frac{l_o \, g \sin \theta}{u_o^2} = \frac{l_o \, tg \, \theta}{h_o \, f_o^2} \tag{1}$$

avec :

l_o : la longueur caractéristique d'écoulement
h_o : la profondeur de la lame à l'exutoire
u_o : la vitesse uniforme au même endroit
θ : l'angle local entre la surface d'écoulement et un plan horizontal
f_o : le nombre de Froude calculé avec ces valeurs caractéristiques

En s'appuyant sur l'expérience des hydrologistes, on peut conclure que l'écoulement sur des pentes naturelles situe généralement les couples de valeurs (f_o, k_o) dans des zones où l'approximation cinématique est licite. Par conséquent, le remplacement dans l'équation de continuité des composantes de la vitesse par leur forme explicite en fonction de la hauteur de lame rend le modèle quasi-tridimensionnel suivant :

$$\frac{\partial h}{\partial t} + \frac{\partial}{\partial x}(a'h^{m+1}\cos \theta_s) + \frac{\partial}{\partial y}(a'h^{m+1}\sin \theta_s) = \frac{\partial h}{\partial t} + \frac{\partial q_x}{\partial x} + \frac{\partial q_y}{\partial y} = (r-i)\cos \theta_z$$

$$\tag{2}$$

avec :

r, i   : respectivement les précipitations et la vitesse d'infiltration
$-\theta_z$: l'angle que fait la topographie locale avec l'axe z
$\cos \theta_s, \sin \theta_s$: les composantes dans le plan (x, y) de la tangente à une trajectoire quelconque C
a', m, c_f : des coefficients caractéristisant la topographie, sa couverture et le type d'écoulement au sein de la lame

La confrontation de la théorie et de l'expérimentation sur modèles réduits constitués de géométries élémentaires conforte le choix de l'approximation cinématique en révélant une structure majoritairement turbulente pour l'écoulement au sein de la lame (m =2/3). Remarquons immédiatement que par sa formulation non-linéaire, ce modèle réfute des conceptions classiques comme le temps de concentration considéré comme propriété intrinsèque d'un bassin, la présence d'isochrones et, plus généralement, les propriétés d'additivité liées à la théorie de l'hydrogramme unitaire. En sus, la non-linéarité implique de travailler avec des signaux de pluies brutes et de gérer simultanément les phénomènes d'infiltration puisque les quantités infiltrées contribuent localement et temporairement à gonfler la lame de surface et à modifier le champ de vitesse des volumes ruisselants qui parviennent jusqu'à l'exutoire.

Quoi qu'il en soit, l'excellente correspondance des résultats suggère d'étendre aux géométries naturelles les plus variées le cadre d'utilisation de cette théorie vérifiée sur des topographies simples. Cette généralisation ne peut cependant s'opérer sans une réflexion parallèle sur les conséquences des simplifications théoriques et la physique des processus.

Par opposition au caractère trivial de la formulation cinématique linéarisée qui décrit une simple translation, sans déformation, d'un signal à une célérité constante, le système non-linéaire introduit certains caractères essentiels des ondes hyperboliques avec apparition d'ondes de choc matérialisées par des sauts dans l'écoulement. Ces spécificités mettent localement en défaut l'hypothèse de la thèse cinématique identifiant pente de fond et de surface (dimensions spatiales de la discrétisation nettement supérieures à l'épaisseur de lame). Une telle situation peut être très simplement initiée en écoulement unidimensionnel par une pluie uniforme qui arrose la successsion des deux pentes reprises à la figure 1. Cette application, résolue théoriquement, conduit à des solutions triples de hauteurs d'eau en aval de la transition de pente, ce qui est physiquement inacceptable eu égard à la relation univoque qui lie hauteurs et débits.

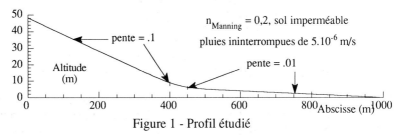

Figure 1 - Profil étudié

L'étude théorique avec le système d'équations complet établit qu'il est licite, à l'échelle des éléments utilisés, d'approximer à la fois de brusques transitions locales continues mais également d'éventuels ressauts à l'échelle de la fine lame fluide par ce que nous nommerons des ressauts simplifiés au sens de la théorie cinématique. Au-delà de ce constat primordial qui conforte le recours à l'approximation cinématique en conditions quelconques, se pose d'abord le problème théorique d'exprimer l'unique solution composite qui vérifie l'équation différentielle dans ses intervalles continus et satisfait à la condition appropriée de "saut" aux discontinuités.

Par ailleurs, cette solution, communément nommée solution faible du problème doit être adéquatement traitée numériquement par une méthode apte à gérer des ressauts en instationnaire. A défaut de travailler avec ces schémas spécifiques, les singularités de l'écoulement qui ni sont détectées ni analysées physiquement provoquent des approximations telles, sur les bilans volumiques notamment, que les résultats ne peuvent plus être fiablement interprétés.

Dans ce contexte a été élaboré un code qui procède par capture numérique des chocs pour travailler fiablement, sans aucune hypothèse simplificatrice, sur n'importe quelle topographie digitalisée. L'action sélective de la méthode conservative par éléments finis, développée avec des fonctions de pondération décentrées, est démontrée pour l'application élémentaire décrite à la figure 1.

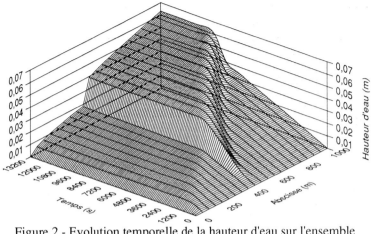

Figure 2 - Evolution temporelle de la hauteur d'eau sur l'ensemble
de la surface topographique, avec formation d'un ressaut

Toutes les lignes d'eau sur l'ensemble de la géométrie sont reprises dans le représentation tridimensionnelle de la figure 2. Elles confirment que la formation et la stabilisation d'un ressaut à la transition de pente sont parfaitement reproduites et conformes à la solution analytique, sans lissage exagéré ni onde parasite rémanente.

CONCLUSION
La méthode gère les hétérogénéités spatiales des propriétés évolutives des sols et des précipitations, grâce au développement d'éléments couvrants-découvrants. Des procédures annexes de maillage automatique et de détermination automatique des bassins versants ainsi que des outils graphiques complètent ce module hydrologique qui modélise la fine lame sur le bassin puis propage les hydrogrammes latéraux en réseaux de rus et de rivières [1]. Toute intervention humaine sur le bassin se répercute par une modification physiquement correspondante des paramètres de la modélisation, avec une quantification immédiate des impacts à l'exutoire, type d'utilisation prometteuse qui ouvre à l'approche hydrologique basée sur la physique de l'écoulement un bel avenir dans le domaine de l'aide à la gestion des ressources hydrauliques.

REFERENCE
[1] PIROTTON M., *Modélisation des discontinuités en écoulement instationnaire à surface libre Du ruissellement hydrologique en fine lame à la propagation d'ondes consécutives aux ruptures de barrages*, Thèse de doctorat, Université de Liège, 211 figures, 479 pages, Avril 1994.

1D9

Improved turbine control systems for hydropower plants obtained by
mathematical analysis.

Professor Hermod Brekke
The University of Trondheim
The Norwegian Institute of Technology
7034 Trondheim, Norway

INTRODUCTION

The purpose of this paper is to present methods for mathematical modelling and development of
stabilizing systems for control of Hydroelectric power plants with long high pressure tunnel
systems or traditional tunnels and long penstocks. The Structure Matrix Method (SMM) which
is the basic tool for stability analysis in the frequency domain by means of Laplace transformed
equations will be presented. The analysis includes a frictional damping term of the pressure
oscillations as function of the frequency and the flow amplitudes and the influence of the turbine
characteristics (Ref. [1]). The second part of the paper includes a presentation of a recently
developed pressure feedback system for a turbine governor, which in some cases can substitute
for a surge chamber. This system has been proven to work successfully in agreement with a
theoretical analysis made with the computer program STABANA based upon the SMM method.
A presentation of frequency response measurement compared with the computed stability analysis
for a power plant with a long penstock will be given.

THE BASIC THEORY

Allievis equations with a friction term added is the base for the theory. The equation for
equilibrium of forces and the equation of continuity yields:

$$\frac{\partial h}{\partial x} = \frac{Q_o}{gAH_o}(\frac{\partial q}{\partial t} + kq)$$

(1)

$$\frac{\partial q}{\partial t} = \frac{gAH_o}{a^2Q_o}\frac{\partial h}{\partial t} \qquad [h = \frac{\Delta H}{H_o}, \ q = \frac{\Delta Q}{Q_o}]$$

(2)

The Laplace transformed element matrix for a tunnel or pipe derived from Ellievis' equations may
be written as shown below for a pipe element with length = L, assuming that the relative flow is
positive out of the pipe at both ends and defining increasing pressure to be positive at both ends.

$$
\begin{bmatrix}
\dfrac{-s}{2h_w\,z\,\tanh\,(Lz/a)} & \dfrac{s}{2h_w\,z\,\sinh\,(Lz/a)} \\[2em]
\dfrac{s}{2H_w\,z\,\sinh\,(Lz/a)} & \dfrac{-s}{2h_w\,z\,\tanh\,(Lz/a)}
\end{bmatrix}
\begin{bmatrix} h_L \\[1em] h_R \end{bmatrix}
=
\begin{bmatrix} q_L \\[1em] q_R \end{bmatrix}
\tag{3}
$$

where

$$
K=(\tau_s+\tau_d)\pi D/(\rho Q_o q) \quad and \quad z=\sqrt{s^2+Ks} \quad and \quad s=j\omega
$$

τ_s and τ_d: are static and dynamic shear stresses respectively. Here τ_s is the stationary shear stress $\tau_s = f(\lambda)$ [λ= stationary friction factor], and $\tau_d = 0.5\rho(Q_o|q|/A)^2 f_d$ is the dynamic shear stress where $f_d=f(\omega,q,\lambda)$ (Ref. [1]).

Further element matrices for surge shafts and air accumulators yields:

$$
\begin{bmatrix} 1 & 0 \\ 0 & -sH_d A_{eqv}/A \end{bmatrix}
\begin{bmatrix} h_L \\ h_R \end{bmatrix}
=
\begin{bmatrix} q_L \\ q_R \end{bmatrix}
$$

Here $A_{eqv} = A_w$ = water level area in open surge shafts and V_o = volume of air at pressure H_{ao} mWC, $A_{eqv} = 1/(1/A_w + n\,H_{ao}/V_o)$.

In addition the element matrise for variation in cross sections and for T joints yields:
[This matrix can only be solved as a connection matrix in a global system].

$$
\begin{bmatrix} -1/K_{re} & 1/K_{re} \\ 1/K_{re} & -1/K_{re} \end{bmatrix}
\begin{bmatrix} h_L \\ h_R \end{bmatrix}
=
\begin{bmatrix} q_L \\ q_R \end{bmatrix}
\tag{4}
$$

For variation in cross section $K_{re} = Q_o\,Q_t/(gH_o)\;\zeta\;(1/A_{min}^2 - 1/A_{max}^2)$
For a T joint (se fig. 2) $K_{re} = K\,Q_o^2/(3qA_{shaft}^2\,H_o) + Q_t\,Q_o/(gA_{tun}^2\,H_o)$

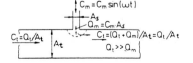

Figure 1. T joint

The element matrix for turbine governor and generator is based on the turbine characteristics, the governor and generator parameters and will be a 6 x 6 matrix which yields:

$$
\begin{bmatrix}
-B & -Q & 0 & -C & 0 & B \\
0 & D & E & F & 0 & 0 \\
0 & 0 & 1 & 0 & 0 & 0 \\
J & K & 0 & L & M & -J \\
0 & -T_a s & 0 & 0 & 1 & 0 \\
B & Q & 0 & C & 0 & -B
\end{bmatrix}
\begin{bmatrix}
h_R \\ n \\ P_i \\ y \\ P \\ h_L
\end{bmatrix}
=
\begin{bmatrix}
q_R \\ n_{ref} \\ P_{ref} \\ 0 \\ P_g \\ q_L
\end{bmatrix}
\tag{5}
$$

Here B, J, K, L and M are parameters determined from the turbine characteristics i.e. $\partial Q/\partial n$, $\partial \eta/\partial Q$, $\partial \eta/\partial n$ etc., see fig. 3. D, E, F are f(s) and are based on the turbine governor parameters (PID governor or a PI governor). The parameter T_a = generator inertia time constant and $s = j\omega$.

In fig. 2 a turbine characteristic diagram is shown. The most important parameter will be the slope of the constant guide vane curves i.e. $(\partial Q/\partial n)$ values which influences the parameters B, J and K.

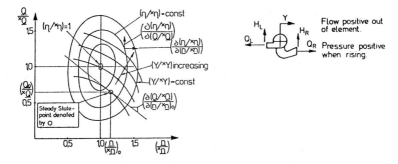

Fig. 2. Turbine characteristics diagram.

FREQUENCY RESPONSE TESTS

In order to verify the computed result, frequency response tests of the pressure/guide vane response have been carried out at various high head power plants in Norway.

For such testes a simplified turbine governor matrix has been established. This matrix contains only two of the parameters (B and C) from the matrix eq. (5) besides the parameter for the ratio (guide vane movement)/(guide vane reference) = Kq. The matrix equation yields

$$
\begin{bmatrix}
-B & -C & B \\
0 & 1/Kq & 0 \\
B & C & -B
\end{bmatrix}
\begin{bmatrix}
h_L \\ y \\ h_R
\end{bmatrix}
=
\begin{bmatrix}
q_L \\ y_{ref} \\ q_R
\end{bmatrix}
\tag{6}
$$

In fig. 3 the result from a frequency response test of a Pelton turbine power plant is shown. This power plant, Skjaak, in Norway has a very long (2355 m) penstock. The nominel technical data for the turbine is P = 31.5 MW, H_o = 638 m, n = 500 RPM and Q_o = 5.6 m³/sec.

The frequency response test was carried out at 26.13 MW, H_o = 665.7 m and Q_o = 4.44 m³/sek.

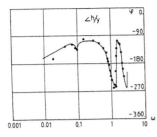
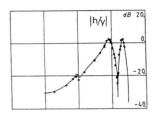

Fig. 3. Result from the pressure response test at Skjaak power plant. (Phase left, gain right).

The regulating stability at Skjaak power plant is poor with insufficient stability margins at full load (not shown in fig. 3).

IMPROVING GOVERNING STABILITY BY PRESSURE FEED BACK SYSTEM.
The principle of a pressure feed back system may be illustrated in the following simplified block diagram valid for a Pelton turbine where there is no influence from the speed on the flow in the turbine characteristic diagram, i.e. Q_n = 0 in fig. 4.

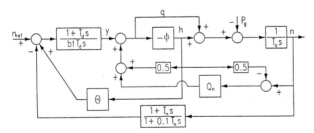

Fig. 4. Block diagram for a turbine governing system with a pressure feed back system.
(Influence from the efficiency not shown, see (Ref. [1])

In the block diagram $\phi = 2h_w$ tanh (L s/a)

$$\left(\frac{n}{n_{ref}}\right) = \frac{(T_d s + 1)(1 - \phi)}{b_t T_d s(1 + 0.5\phi(1 - 2((T_d s + 1)/(T_d s)\Theta))T_a^s} \tag{7}$$

553

By introducing $0.5(1-2((T_d s+1)/(T_d s)))\Theta = b$ then:

$$\frac{n}{n_{ref}} = \frac{(T_d s+1)(1-\phi)}{b_r T_d s(1+b\phi)T_d s} \tag{8}$$

In the block diagram and eq. (7) Θ represent the pressure feed back element which was first introduced by prof. Lein, (Ref [2]), denoted as a DT1 element: $\Theta(s)=KT_1 s/2(1+T_d s)$.

The author and his colleague Dr. Li Xin Xin introduced later an improved pressure feed back element based on the work by Dr. Lein. The expression for this element yields:

$$\Theta(B)=\frac{KT_1 s}{(T_1 s+1)(T_d s+1)(0.5s+1)} \tag{9}$$

It has been proven theoretically that by tuning such pressure feed back element an improved speed regulating could have been obtained for Skjaak power plant. In fig. 5a the complex plain plot of a Nyquist stability diagram open loop and closed loop is shown for alternatives without pressure feed back system, for a DT1 element with $T_1=Td$ and the improved optimized pressure feed back element.

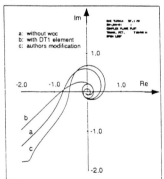
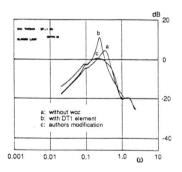

Fig. 5a. Nyquist diagrams open loop   Fig.5b. Close loop load-speed responses

Both the open loop and closed loop diagrams in fig. 5 show the advantage with an optimized improved pressure feed back system according to eq. (9) in the speed regulation compared with a pressure feed back signal with a non optimized DT1 element with $T_1 = T_d$ = integral time constant and the system without a pressure feed back system.

It can clearly be seen that an optimal pressure feed back system changes a poor regulating stability to be a fully acceptable stable system.

For Skjaak power plant the turbine governor was a mechanical governor and it was not convenient to implement a pressure feed back system for this plant at the time being. However, for Svartisen power plant which has a 350 MW Francis turbine operating at H_n = 560 m net head with n = 333.3 RPM a pressure feed back system has been installed saving a cost of 3 Mill US$ by not building an air accumulator surge chamber. This system is now in operation in Norway without a surge chamber, but still with a stable turbine governing.

CONCLUSION

By means of the presented advanced stability analysis it has been proven that regulating stability has been obtained in power plants with long tunnels or penstocks without surge chamber close to the turbine. The cost of a surge chamber may in many cases be saved by introducing a pressure feed back system on the turbine governor.

REFERENCES

1. Brekke, H. "A stability study on hydropower plant governing including the influence from a quasi nonlinear damping of oscillatory flow and fromt he turbine characteristics". Dr. Thesis, The Norwegian Institute of Technology, May 1984.

2. Lein, G. and Maurer, W. "Advance in control of hydropower plants". Will be published in book series for Hydraulic Machinery, Vol. 9. Gover.

1D10

Predicting The Caspian Sea Surface Water Level Using Artificial Neural Networks

MANOUCHEHR VAZIRI

Associate Professor of Civil Engineering, Sharif University
of Technology, Tehran, Iran

ABSTRACT

Fluctuations of the Caspian Sea mean monthly surface water level for the period of January 1986 to December 1993 were studied. A backpropagation artificial neural network was used to predict the time series data. The monthly prediction for the January to December 1994 presented a continuation of the Caspian Sea surface water level rise.

INTRODUCTION

The Caspian Sea is the largest lake in the world and is surrounded by Azerbaijan, Iran, Kazakhestan, Russia and Turkmenestan. The data for the mean yearly surface level for 1900 to 1977 has shown a decreasing trend, due to the occurrence of a dry period and a reduction of river inflows. The surface water level has risen 17.9 centimeters per year since 1977 which has resulted in tremendous costs to its surrounding countries. For the period 1977 to 1993, the mean yearly water surface level showed an average of -27.69 meters below the Baltic Sea surface water level and a standard deviation of 0.91. For the same period, the mean yearly water surface level showed a minimum of -28.95 meters in the year 1977 and a maximum of -26.33 meters for the year 1993. This rapid increase has resulted in an average overflowing of 60 meters into the coast, with tremendous costs and damages to the shore line infrastructures and therefore has given rise to efforts to predict the sea level (Davodi 1991).

Artificial neural network, ANN, has recently found application for predicting time series. In this study, to predict the Caspian Sea surface water level fluctuations, backpropagation ANN was used. An ANN is a parallel, distributed information processing system composed of many simple processing elements, which are interconnected via synaptic or weighted connections. Each processing element receives and processes weighted inputs from previous processing elements and transmits its output to the following set of processing elements through another set of weighted interconnections. An important mechanism for the neural network is how to adapt these connection weights so that the network processes the information properly. Of the many different ANN types, backpropagation ANNs have been applied to a wide variety of problems including time series analysis (Wasserman 1989, Williams 1994).

ANN MODELING

The key characteristics of a backpropagation ANN include: number of processing elements in each layer, number of layers, type of transfer function and learning rule. A basic backpropagation ANN consists of three layers, namely input layer, hidden layer and output layer. The backpropagation ANN gets its name from how it handles error. The function most commonly used for the error is the sum of the square of the difference between the actual and the desired output layer processing elements' output. Before deploying, the network needs to be trained and tested. Training consists of repeatedly presenting the examples to the network and updating the processing elements connection weights. Once the network is trained, it is tested by presenting to the network, examples not available during training and then evaluating the output errors.

For a one month ahead prediction of the Caspian Sea surface level, several one hidden and two hidden layer backpropagation ANNs were trained and tested. The selected ANN which had the smallest root mean square of error for the testing data was a basic three layer network. To predict one month ahead sea level, Y_t , the selected ANN would use the preceding 12 months sea level Y_{t-1} to Y_{t-12}. There were 12 processing elements in the input layer, taking in a sequence of the lagged Caspian Sea mean monthly surface level of Y_{t-1} to Y_{t-12} . There were 2 processing elements in the hidden layer. In the output layer, the one processing element provided the Caspian Sea mean monthly surface level Y_t as the ANN's output. The training data consisted of 72 samples of the current month and its preceding 12 months sea level for the period of January 1987 to December 1992. The testing data consisted of 12 months sea level which included the current month and its preceding 12 months of sea level for the period January to December 1993. A software called NeuralWorks Explorer was used to develop the aforesaid ANN. In this study, three types of transfer function, namely, hyperbolic tangent, sigmoid and sine were tried. For each transfer function, the training data was properly scaled by the software. The hyperbolic tangent was selected due to its superiority and its training convergence and testing results. The applied learning rule for the hidden and output layers was cumulative delta rule which accumulated the weight changes over several presentation of training examples and then applied to the weights. After 43518 iterations of randomly presenting the training data to the network the training converged to the root mean square error value of 0.025.

The trained backpropagation ANN was used to predict the mean surface water level for January to December 1993. The ANN predictions extended past seasonality. The superiority of the ANN was reflected in the mean surface water level prediction for the 12 months; the observed value was -26.33 meters and ANN predicted -26.36. The sum of square of errors for the 12 months for the ANN was 0.0332. The ANN predictions were very reasonable, an average error of 3 centimeters, when compared with the recorded levels. The application of ANN modeling required an understanding of the basic procedures for its development. Nevertheless, ANN was found to be a very powerful tool which did not require detail knowledge of statistics.

The testing data for January to December 1993 was used to enhance the trained ANN. The ANN was updated and retrained with the 84 samples of the current month and its preceding 12 months sea level for the period of January 1987 to December 1993. The retrained ANN was then used to predict the Caspian Sea mean monthly surface level for the period of January to December 1994. The predictions are presented in Table 1. The 1994 mean monthly sea level was -26.16.

The highest and lowest predicted sea surface levels of -26.05 and -26.31 were for the months of August and January respectively. The ANN predicted a continuation of the sea surface water level rise of 18 centimeters above the 1993 mean yearly surface water level.

TABLE 1. The Caspian Sea Mean Surface Water Level Predictions for 1994

Month	Prediction
January	-26.31
February	-26.30
March	-26.27
April	-26.20
May	-26.15
June	-26.06
July	-26.06
August	-26.05
September	-26.11
October	-26.16
November	-26.15
December	-26.15
Mean	-26.16

CONCLUSIONS

The artificial neural network modeling was found to be a useful tool for short term predictions of the Caspian Sea mean monthly surface water level. The developed backpropagation ANN's input and output consisted of the preceding 12 months and current month surface water levels respectively. The ANN had 2 processing units in its single hidden layer. The predictions of the ANN for 1994 confirmed the continuation of the sea level rise with highest and lowest levels for the months of August and January respectively. The ANN predicted a rise for 1994 of 18 centimeters above the 1993 mean yearly surface water level. The application of ANN in short term predictions could be incorporated with longer term computer simulation model predictions to enhance the control of the Caspian Sea surface water level.

ACKNOWLEDGMENT

The technical and financial supports by the Sharif University of Technology are gratefully acknowledged. Special thanks to the dearest person in my life, Dr. Lila Bahadori, for editing, input and support.

REFERENCES

Davodi, R. (1991). "Evaluation of the Caspian Sea surface water level." *Sharif University Civil Engineering Magazine*, No. 5, 12-15.

Wasserman, P. D. (1989). *Neural computing, theory and practice.* Van Nostrand Reinhold, New York.

Williams, T. P. (1994). "Predicting changes in construction indexes using neural networks." *ASCE Journal of Construction Engineering and Management,* Vol. 120, No. 2, 306-320.

1D11

A New Method for the Computation of Steady Open-Channel Trans-Critical Flow

I MACDONALD, M J BAINES, N K NICHOLS
University of Reading, Reading, UK
and
P G SAMUELS
HR Wallingford Ltd, Oxon, UK

ABSTRACT

A new method for computing the free surface profile for steady flow in an open channel is presented. The numerical scheme can solve trans-critical flows, automatically locating any hydraulic jumps or critical sections. Numerical results are given for a particular case, which are compared against the available exact solution.

1. INTRODUCTION

The use of numerical methods to compute the water profile and discharge for both unsteady and steady open channel flows is now very common in engineering work. The Saint Venant equations are almost always used to model such flows. A derivation of these equations can be found in Cunge, Holly and Verwey (1980). This paper restricts attention to the important case of steady flow. Apart from the intrinsic interest of the steady case, one reason why the computation of steady flows is so significant is that steady solutions are very often required as initial data for unsteady simulations. For steady flow the Saint Venant equations reduce to a single Ordinary Differential Equation (ODE), which, if the flow is wholly subcritical or wholly supercritical, can be efficiently integrated using a high accuracy numerical ODE solver. The situation becomes much more difficult if the flow is of mixed sub-supercritical type, i.e. has hydraulic jumps or critical sections. One approach for computing steady solutions is to apply an unsteady solver and proceed forward in time until all the transients have decayed and the solution has reached a steady state. This approach, however, can often be very inefficient. This paper presents a numerical scheme which was first applied to the steady flow problem in

MacDonald, Baines and Nichols (1994). The scheme is a shock capturing method (see Engquist and Osher (1981)). Shock capturing methods have previously been applied to the system of unsteady equations, but here are applied to the single steady equation. Since a hydraulic jump is in fact a steady shock, the shock capturing nature of the numerical scheme allows the scheme to locate automatically any hydraulic jumps.

2. The EQUATIONS

For a prismatic channel with no lateral inflow, the Saint Venant equations in conservation form are

$$\frac{\partial A}{\partial t} + \frac{\partial Q}{\partial x} = 0, \tag{1}$$

$$\frac{\partial Q}{\partial t} + \frac{\partial}{\partial x}\left(\frac{Q^2}{A} + I_1\right) = gA(S_0 - S_f). \tag{2}$$

Here x is the distance along the channel, t is time, $y(x,t)$ is the depth, $Q(x,t)$ is the discharge, $A(y)$ is the wetted area, and $I_1(y)$ is a pressure term given by

$$I_1 = g\int_0^y T(\eta)(y-\eta)d\eta, \tag{3}$$

where $T(y)$ is the channel width. $S_0(x) = -dz/dx$ is the bed slope, where $z(x)$ is the bed level above some horizontal datum. $S_f(y) = Q|Q|/K^2$ is the friction slope, where $K(y)$ is the conveyance. This work uses Manning's equation, $K = A^{5/3}/(nP^{2/3})$, where $P(y)$ is the wetted perimeter, n is the Manning friction coefficient and g is the acceleration due to gravity.

For steady flow we have $y=y(x)$ and $Q=Q(x)$, and so equations (1) and (2) reduce to

$$\frac{dQ}{dx} = 0, \tag{4}$$

$$\frac{d}{dx}\left(\frac{Q^2}{A} + I_1\right) = gA(S_0 - S_f). \tag{5}$$

At steady state the discharge Q must now be constant throughout any reach. From now on, Q will represent a known parameter with the value of the constant discharge, leaving only equation (5) to be solved for y. The convention is used that x is measured in the direction of the flow, and hence $Q \geq 0$.

3. THE NUMERICAL SCHEME

The functions $f(y)$ and $b(x,y)$ are defined as

$$f = -\left(\frac{Q^2}{A} + I_1\right), \quad b = gA(S_0 - S_f). \tag{6}$$

Equation (5) can now be written as

$$-\frac{df(y)}{dx} - b(x,y) = 0. \tag{7}$$

This ODE is discretised by a first order accurate finite difference scheme based on that of Engquist and Osher (1981). A reach of channel of length L is discretised with a uniform grid of spacing $\Delta x = L/N$. The numerical approximation to the solution on this grid will be given by $y_0, y_1, ..., y_N$. The difference scheme is defined by

$$(Ty)_i = -\frac{1}{\Delta x}\left[f_-(y_{i+1}) - f_-(y_i) + f_+(y_i) - f_+(y_{i-1}) \right] - b(i\Delta x, y_i), \tag{8}$$
$$i = 1, 2, ..., N-1.$$

If $d^2 f/dy^2 < 0$ and there is a critical depth $y_c > 0$ given by $df/dy(y_c) = 0$, then $f_{\pm}$ are given by

$$f_+(y) = f(\min\{y, y_c\}), \quad f_-(y) = f(\max\{y, y_c\}). \tag{9}$$

The assumptions on f hold for many common cross-sections, including rectangular and trapezoidal. More general expressions for $f_{\pm}$ are available for cross-sections that violate these conditions (see MacDonald, Baines and Nichols (1994)). The scheme is modified for the nodes at the ends of the channel as follows. If the depth is to be specified at inflow, say to y_{in}, then set $(Ty)_0 = y_0 - y_{in}$, otherwise set

$$(Ty)_0 = -\frac{1}{\Delta x}\left[f_-(y_1) - f_-(y_0) \right] - b(0, y_0). \tag{10}$$

Similarly, if the depth is to be specified at outflow, say to y_{out}, then set $(Ty)_N = y_N - y_{out}$, otherwise set

$$(Ty)_N = -\frac{1}{\Delta x}\left[f_+(y_N) - f_+(y_{N-1}) \right] - b(L, y_N). \tag{11}$$

The solution to the scheme is found by solving the system of nonlinear equations

$$(Ty)_i = 0, \qquad i = 0, 1, \ldots, N. \tag{12}$$

This system can be efficiently solved using Newton's method, since the Jacobian is tri-diagonal and hence easily inverted. Special care is required to ensure that the depths remain positive after each iteration.

Details of the method described in this paper can be found in MacDonald, Baines and Nichols (1994) with some theoretical results. These results apply to many useful cases. The main restriction on the theory is that it requires the bed slope to be positive, i.e. the flow must be always going downhill. The cross-sectional shape is also slightly restricted; however the theory applies to many common shapes, including rectangular and trapezoidal. The main result shows that the scheme converges to the correct physical solution in the limit $\Delta x \to 0$. By a physical solution it is meant that any hydraulic jump satisfies the correct jump condition ($\Delta f = 0$ in the previous notation) and has no increase in total energy across the jump. Some stability results are also available; the numerical solution is shown to be uniformly bounded in Δx.

4. RESULTS

In MacDonald, Baines, Nichols and Samuels (1995) a method is given for constructing test problems with known exact solutions. Figure 1 shows numerical results for the scheme presented in this paper for a test problem constructed by this method. Results are shown for three different grid spacings against the exact solution. The bed level for this problem is shown in Figure 2 with the computed surface level against the exact surface level. The solution has two features, a hydraulic jump and a critical section. It is observed that the numerical method successfully locates both these features, with the resolution of the jump improving as the grid is refined. In this paper only one example is shown, but the scheme has been successfully applied to many other examples, some of which can be found in MacDonald, Baines and Nichols (1994).

5. CONCLUSIONS

A numerical scheme has been presented that can efficiently solve for steady flow, regardless of the type of flow. The ability of the scheme to automatically locate hydraulic jumps has been demonstrated by an example. Although the scheme can currently only be applied to prismatic channels, extension to non-prismatic and eventually natural channels is envisaged.

REFERENCES

[1]   J A Cunge, F M Holly and A Verwey (1980). *Practical Aspects of Computational River Hydraulics*. Pitman, London, England.

[2]   B Engquist and S Osher (1981). *One Sided Difference Approximations for Nonlinear Conservation Laws*. Mathematics of Computation, **36**, 321-351.

[3]   I MacDonald, M J Baines and N K Nichols (1994). *Analysis and Computation of Steady Open Channel Flow using a Singular Perturbation Problem*. Numerical Analysis Report 7/94, Department of Mathematics, University of Reading.

[4]   I MacDonald, M J Baines, N K Nichols and P G Samuels (1995). *Steady Open Channel Test Problems with Analytic Solutions*. (submitted for publication).

ACKNOWLEDGMENTS

This work was funded by the Engineering and Physical Sciences Research Council, UK and HR Wallingford Ltd., UK.

FIGURES

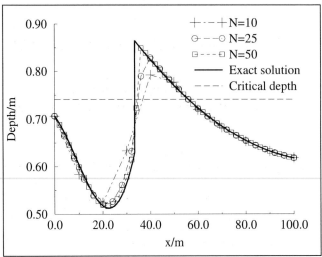

Figure 1: Numerical solutions for several grid spacings against the exact solution.

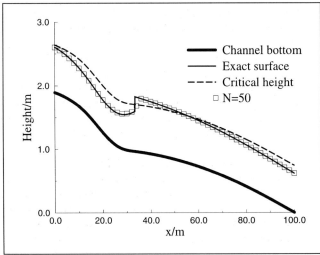

Figure 2: Computed surface level against exact surface level.

1D12

UNCERTAINTY IN FLOOD LEVEL PREDICTION

Dr P G SAMUELS FIMA MICE MIWEM
Rivers Group, HR Wallingford, Howbery Park, Wallingford, UK

1 SUMMARY
Computational hydraulic modelling is not a precise science although the fluid mechanics of open channel flow were formulated over 100 years ago as the St Venant Equations. The accuracy of computational models constructed using these equations is not known, neither is the amount of data required to deliver a given precision from the modelling. This paper draws together the results of several investigations into a framework for assessing the accuracy of flood level predictions. Some guidelines are suggested for the spacing of cross-sections and the need for good model calibration data is highlighted.

2 MOTIVATION
The National Rivers Authority (NRA) advises local planning authorities in England and Wales of land at risk of flooding. This is often taken to be the area likely to be inundated in the 1 in 100 year event, although the local standard of flood defence may vary with the riparian land-use. The NRA makes frequent use of 1-D computational models to estimate flood levels. The work reported in this paper was carried out to answer specific questions on the amount of data and type of modelling needed to delineate the flood plain of any watercourse in NRA Thames Region. It built upon research commissioned at HR Wallingford by the Ministry of Agriculture Fisheries and Food (MAFF) on river flood protection.

3 SOURCES OF UNCERTAINTY
There are three categories of uncertainty which are those in the hydrological estimates, those in the flow to level correlation and those in the determining the flood limit from the local flood level. This paper does not consider all the categories of uncertainty listed above but concentrates on the likely accuracy of the flood level prediction for a given flood flow. The uncertainties in each category are independent and are capable of separate refinement. The uncertainty in the flow to level correlation through hydraulic modelling depends upon the survey

accuracy, the topographic discretisation, numerical modelling errors, quantity of calibration data and the quality of the model calibration. Frequently a model will be calibrated for moderate flood conditions and then the model extended to higher discharges.

4 A FRAMEWORK FOR THE ASSESSMENT OF UNCERTAINTY

To date a full mathematical assessment of the uncertainty in modelling has not proved possible and so this paper draws mainly upon the results of numerical experiments. Burnham & Davies (1990) examined the effects on flood level of topographic survey accuracy and the reliability of model calibration for steady flow modelling of the 100 year flood condition in many US rivers. They assumed that the models themselves had no discretisation or numerical errors. Defalque et al (1993) examined the effects of coarsening steady flow models of several UK rivers and introducing random error into the topographic survey. They considered several flows from half bank full to five times bank full (which in the UK will exceed the 100 year flood flow in almost all rivers). Yung & Kung (1994) reported an investigation of topographic data and roughness uncertainties on the propagation of a dam break flood wave with an unsteady flow model which was much more severe than the normal flood discharges in the valley. The measure of uncertainty which has been used by both Burnham & Davies and Defalque et al is the mean magnitude of the error at any location and this will be used in the current paper. Burnham & Davies also identify the magnitude of the maximum error.

The computations of Yung and Kung showed an attenuation of the discharge of the flood wave along the 30 km reach studied, by approximately 40% of the upstream value. However, the spread in discharge due to the data uncertainties was an order of magnitude smaller (4% at the downstream end). Thus since both processes depend upon the curvature of the flood discharge hydrograph it may be concluded that the influence of topographic and roughness error on the magnitude of the flood flow in a natural river should be small. The principal case in which this conclusion might be questioned is when the uncertainties in water level or flow allow a different flow pattern to be established. Nevertheless we will proceed upon the assumption of steady flow in the river and apply the same criteria to unsteady flow models. The framework for the analysis is to assume that the potential errors from the different processes are independent and thus can be combined as a RMS (root mean square) value. This choice is pragmatic rather than theoretical but intuitively it seems implausible that all the sources of error would act purely additively on the total uncertainty. Previous studies will provide formulae for the contributions to the overall uncertainty with some *ad-hoc* adjustments as described below.

5 TOPOGRAPHIC AND ROUGHNESS UNCERTAINTY

The work of Burnham & Davies (1990) provides the basis for assessing the combined uncertainty E_{tr} associated with the topographic survey and the roughness values for the river channel and the flood plain. Burnham & Davies present several equations for measurements in Imperial units (as used in the US) with dimensional constants. These have been converted for metric units below.

Conventional Field Survey of Flood Plain Topography

$$E_{tr} = 0.12 \ HD^{0.6} \ S^{0.11} \ (5Nr)^{0.65} \tag{1}$$

$$E_{max} = 1.7 \ (E_{tr})^{0.8} \tag{2}$$

Aerial survey providing spot levels on section lines

$$E_{tr} = 0.12 \ HD^{0.6} \ S^{0.11} \ (5Nr + \ Sn)^{0.65} \quad if \ Nr > 0 \tag{3}$$

$$E_{tr} = 1.5 \ S^{0.49} \ Sn^{0.83} \quad if \ Nr = 0 \tag{4}$$

$$E_{max} = 1.7 \ (E_{tr})^{0.8} \tag{5}$$

Flood plain topography determined from contour maps

$$E_{tr} = 0.63 \ HD^{0.35} \ S^{0.13} \ (Nr + \ Sn) \quad if \ Nr > 0 \tag{6}$$

$$E_{tr} = 1.4 \ S^{0.23} \ Sn^{1.18} \quad if \ Nr = 0 \tag{7}$$

$$E_{max} = 2.1 \ (E_{tr})^{0.8} \tag{8}$$

E_{tr} is the expected mean absolute error (m) in the water level and E_{max} is the maximum expected error (m) in the water level. S is the river slope (m/m) and HD is the hydraulic mean depth flood flow condition. These equations involve two parameters which quantify the uncertainties, the calibration reliability number Nr and the survey reliability number Sn. Burnham & Davies describe Sn in detail and Table 1 gives this in metric units.

Table 1 Survey Accuracy Number Sn

Contour Interval	Spot Accuracy	Sn
250 mm	50 mm	0.08
500 mm	100 mm	0.17
1000 mm	250 mm	0.33
2000 mm	500 mm	0.66

However they give little guidance on assigning values to Nr except that Nr = 0 corresponds to perfect knowledge and Nr = 1 to best professional judgement ie using the US Soil Conservation Service method or a set of photographs and river descriptions both of which may be found in French (1987). The key factor in setting Nr must be the density of the calibration data and this may be measured by the mean drop in water level between successive pairs of gauge boards or recorders assuming that they are reasonably well distributed in the reach. A suggested relationship is tabulated below with linear variation between the tabulated values.

Table 2 Calibration Reliability Number Nr

Mean Level Drop (m)	Reliability Number Nr
less than 1.0	0
1.0 +	0.1
4.0 +	0.4
10 +	0.7
15 +	0.8
20 +	0.9
no calibration data	1

6 UNCERTAINTIES DUE TO CALIBRATION QUALITY

The quality of the calibration should be assessed as the mean magnitude of the difference between the simulated and observed water levels. It would seem wise to allow the uncertainty associated with calibration quality to increase if the model is used beyond the discharge of the calibration event. A pragmatic adjustment would be to multiply by the ratio of the benchmark flood discharge Q_T to the highest calibration flood discharge Q_C in such cases. Thus the calibration uncertainty may be estimated by:

$$E_c = (1/N)\{ \Sigma \ |(H_{model} - H_{obs})|\} \ max[\{Q_T/Q_C\}, 1.0] \qquad (9)$$

where H_{model} is the maximum flood water level predicted in the model, H_{obs} is the observed water level and N is the number of observation points in the sum.

7 DISCRETISATION AND APPROXIMATION ERRORS

Most commercial models are second order accurate in space but this limiting behaviour may not be achieved chiefly because the river cross section geometry may be non-uniform between sections and the hydraulic properties of adjacent sections may show substantial differences. Samuels (1990) gives some guidelines on cross section location from an analysis of the approximations to key terms in the flow equations. In addition a new limit on the ratio of conveyance between adjacent sections can be derived as follows. The average value of the square of the conveyance, K, is needed to estimate the friction slope between adjacent sections. Common formulae are the arithmetic mean, the geometric mean, and various harmonic means. Two examples are set out below.

Arithmetic mean AM
$$K^2 = \{ 0.5 (K_1 + K_2) \}^2 \qquad (10)$$

Second Harmonic Mean H2M
$$K^2 = \{ 0.5 ([1/K_1]^2 + [1/K_2]^2) \}^{-1} \qquad (11)$$

For $K_1 = 1$ and $K_2 = 1.1, 1.2, 1.25, 1.5$, the value of AM is always the largest and

H2M always the smallest of the means. The percentage difference between these largest and smallest estimates is 0.7%, 2.5%, 3.8% and 12.8% for the cases quoted. Hence, whilst the formulae will give the same value in the limit of small variations between sections, their performance in practice will be different for large changes in hydraulic properties between cross-sections. This analysis suggests that the ratio of conveyance at successive cross-sections should be in the range of about 0.8 to 1.25 for the error in the friction slope averaging to be less than 4% (the limit suggested in Samuels (1990)).

Defalque *et al* (1993) undertook a series of numerical experiments in which they coarsened and compared the results of previously calibrated models. The cross-section spacing was increased and the mean magnitude of the relative changes MARE in the model results were calculated for six flows. The results gave the surprisingly simple relationship between MARE and the section spacing DX,

$$MARE = 0.1 \, DX \, / \, L \tag{12}$$

where L is the Backwater Length and is given by Samuels (1989) as $L = 0.7 \, D/S$. In the calculation of MARE the actual differences in water level were normalised (divided) by the bankfull depth of flow D. We may deduce that

$$dy = 0.14 \, DX \, S \tag{13}$$

where dy is the mean magnitude of the water level error. Using Equation 13, Table 3 gives the section spacing against slope which will keep the potential error from discretisation error in the range $0.01 \, m < dy < 0.03 \, m$.

Table 3 Recommended Cross-section Spacing

Slope (1 in ...)	Section Spacing (m)	Comment
300 - 1000	75	or 100 m for a round number
1000 - 3000	200	
3000 - 10000	500	
10000+	1000	or greater

These results justify the historic good practice of engineers based upon experience and common sense. When computing the overall accuracy of the modelling the contribution from the discretisation error should be based upon the MARE defined above. Thus we set the discretisation uncertainty E_d as

$$E_d = 0.1 \, D \, DX \, / \, L \tag{14}$$

8 COMBINATION OF LEVEL UNCERTAINTIES FROM ALL SOURCES

The total uncertainty E_{total} in the simulated water level for an approximate 100 year flood may now be compounded from the factors described above. As described in Section 4, the individual conributions are squared and summed, thus

$$E_{total} = \{(E_{tr})^2 + (E_c)^2 + (E_d)^2\}^{0.5} \qquad (15)$$

In the spirit of the work of Burnham & Davies (1990) the maximum error might be expected to be larger than this estimate of the mean error magnitude. If the survey and calibration errors dominate as suggested by Burnham & Davies, then the maximum error might be estimated by

$$E_{max} = 1.7 \, (E_{total})^{0.8} \qquad (16)$$

9 CONCLUDING REMARKS

Several important factors may be deduced from the previous studies referenced in this paper. It is of utmost importance to gather good calibration data and without this the benefits of high accuracy topographic survey cannot be realised. This conclusion is supported by the work of Burnham & Davies, Defalque *et al* and Yung & Kung. Also their work shows that, given good calibration data, the computed water levels have a smaller uncertainty than that quoted for the flood plain topography. Defalque *et al* showed that the uncertainty in the modelling is greatest for just out of bank floods. Table 3 provides new guidelines for cross-section spacing for 1-D models based on numerical experiment and an over-riding condition is that the ratio of conveyance between sections should lie in the range (0.8,1.25). The procedures outlined in this paper still require validation through pilot testing.

10 ACKNOWLEDGEMENTS AND DISCLAIMER

The support of MAFF and the NRA through a research and a consultancy commission respectively is gratefully acknowledged. The opinions expressed in this paper do not constitute the official view of either MAFF or the NRA.

11 REFERENCES

1   Burnham M W and Davis D W (1990), *Effects of data errors on computed steady-flow*, Jnl Hyd Eng, ASCE, 116(7), pp 914, 929.

2   Defalque A, Wark J B and Walmsley N (1993), *Data density in 1-D river models*, Report SR 353, HR Wallingford, March

3   French R H (1987), *Open Channel Hydraulics*, McGraw Hill

4   Samuels P G (1989), *Backwater Lengths in Rivers*, Proc ICE (Lond), Vol 87, Pt2, December

5   Samuels P G (1990), *Cross-section location in 1-D models*, Proc 1st Int Conf on River Flood Hydraulics, Ed White, J Wiley, pp339, 350

6   Yang X L and Kung C S (1994), *Parameter uncertainty in dam-break flood modelling*, Proc 2nd Int Conf on River Flood Hydraulics, ed White & Watts, J Wiley, pp 357, 370.

1D13

Long wave equations for waterways curved in plan

J.D. FENTON and G.V. NALDER
Mechanical Engineering Dept, Monash University, Clayton, Vic 3168, Australia
and Department of Mathematics and Statistics, The University of Waikato,
Te Whare Wānanga o Waikato, Private Bag, Hamilton, New Zealand

INTRODUCTION

The St Venant equations hold a central place in the theory of the propagation of waves and floods in open channels. Their one-dimensional simplicity is in keeping with the approximately known flow and geometry of the problem, however the effects of curvature of the stream have generally been neglected. The present paper is an attempt to incorporate the effects of channel curvature on the propagation of floods and long waves, retaining the relative simplicity of the one-dimensional equations. The effects of stream curvature are expressed only by the offset of the centroid of the cross section and the middle of the stream relative to the local radius of curvature. The essential structure of the equations is the same as the traditional form, and the usual methods may be modified relatively simply to solve them. An expression is given for the speed of long waves as modified by curvature effects in arbitrary channels. It is seen that in real rivers floods travel faster due to the effects of curvature than they would in the same channel straightened out fictitiously for computational purposes, which may have important practical consequences.

CURVILINEAR COORDINATE SYSTEM

Consider a waterway of arbitrary cross-section, which meanders such that it is curved in plan as shown in Figure 1. Let there be some arbitrary curve along the river, where the distance along the curve is denoted by s, and at any point on this curve let there be a local orthogonal curvilinear co-ordinate system (s, n, z), where n is horizontal and transverse to s, and z is vertically upwards. The local radius of curvature of the reference axis is r, such that in an elemental increment ds it turns through an angle $d\theta$, then $ds = r\,d\theta$. An elemental unit of volume at (s, n, z) is shown in plan; its component lengths are $(r - n)/r\,ds$ (by proportionality), dn, and dz, but it is more convenient to introduce the curvature $\kappa = 1/r$, so that the elemental dimensions are $(1 - \kappa n)ds$, dn, and dz.

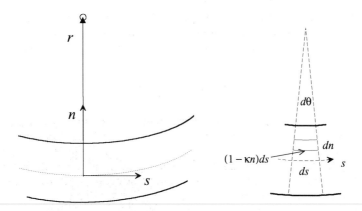

Figure 1. Orthogonal curvilinear coordinate system, physical dimensions, and control volume

The approach in this work parallels that of Fenton (1995) for straight channels. Consider a control volume as shown in plan in the figure, consisting of a slice of the channel bounded by the two vertical planes perpendicular to the local s axis at s and $s+ds$. The control volume made up of this elemental section is continued into the air such that the lateral boundaries are the river banks, the upper boundary being arbitrary but never intersected by the water. The control volume thus contains liquid of constant density ρ in the lower part and air of negligible density in the upper part, whose motion does not contribute to mass and momentum exchange. This choice of control volume is important. If the upper surface of the liquid had been chosen as the upper boundary, complicated manipulations including the kinematic boundary condition on the free surface would have to be performed.

Although the choice of position of the s axis is arbitrary, it will usually be reasonable to choose it so that it follows the course of the river, possibly chosen to be the midpoint of the surface of the river at a particular stage, as taken from cross-sections or aerial photographs, or chosen to be the path followed by the deepest part or talweg.

MASS CONSERVATION EQUATION

The Mass Conservation equation in integral form [Streeter and Wylie, 1981, #3.3] is:

$$\frac{\partial}{\partial t} \int_{CV} \rho \, dV + \int_{CS} \rho \, \mathbf{u} \cdot d\mathbf{S} = 0, \tag{1}$$

where t is time, $\mathbf{u}$ is the velocity vector, and $d\mathbf{S}$ is a vector representing an area element of the control surface, with direction normal to and directed outwards from the control surface. It can be shown that the first term can be written

$$\rho\, ds \int_{n_R}^{n_L} (1 - \kappa n)\frac{\partial \eta}{\partial t}\, dn, \tag{2}$$

where $z = \eta(s, n, t)$ is the elevation of the free surface, and $n = n_L(s, t)$ and $n = n_R(s, t)$ are the coordinates of the water's edge at the left and right banks.

Considering the second term in equation (1), it is easily shown that the net mass leaving the control volume across the two vertical faces is $\rho\, \partial Q/\partial s\, ds$. If there is inflow from rainfall, groundwater or tributaries entering the channel at a volume rate of q per unit length, the extra mass leaving the control volume is $-\rho q\, ds$. Combining the individual contributions to equation (1), and as surface elevation may vary across the section, we use the cross-sectional area of the flow A instead, giving the Mass Conservation Equation:

$$(1 - \kappa n_m)\frac{\partial A}{\partial t} + \frac{\partial Q}{\partial s} = q + O\left((\kappa n_L)^2, (\kappa n_R)^2\right), \tag{3}$$

where n_m is the transverse distance of the midpoint of the river surface from the streamwise co-ordinate. For a straight river, $\kappa = 0$, and the usual expression is obtained. The terms which we have ignored, shown on the right side, are of the order of the square of the curvature times the distance from the reference axis to the bank. For the theory to be accurate the curvature and radius of curvature should then satisfy $(\kappa n_L)^2 = (n_L/r)^2 \ll 1$ and $(\kappa n_R)^2 = (n_R/r)^2 \ll 1$, such that the theory will be most accurate for streams where the width of the river is small compared with the radius of curvature. From now the order of accuracy will not be shown explicitly. Throughout this work, all mathematical operations will performed to give expansions in κn to first order. This level of approximation is similar to the usual one for the conventional formulation of the St Venant equations for channels presumed to be straight, where the approximation made is (depth / disturbance length)$^2 \ll 1$, which will also later be made in this work.

MOMENTUM CONSERVATION EQUATION

Now consider the integral form of the Momentum Conservation Equation [Streeter and Wylie, 1981, #3.3] :

$$\frac{\partial}{\partial t} \int_{CV} \rho\, \mathbf{u}\, dV + \int_{CS} \rho\, \mathbf{u}\, \mathbf{u}.d\mathbf{S} = \mathbf{F}, \tag{4}$$

where $\mathbf{F}$ is the force acting on the fluid in the control volume, including both surface and body forces. The contribution of pressure to the surface forces can be written as $-\int_{CS} p \, d\mathbf{S}$, where p is the pressure, the negative sign showing how the local force acts in the direction opposite to the outward normal. This term, in its form shown here, is difficult to evaluate for non-prismatic and/or curved channels, as the pressure and the non-constant unit vector have to be integrated over all the submerged faces of the control surface. A considerably simpler derivation is obtained if the term is evaluated using Gauss' Divergence Theorem [Milne-Thomson, 1968, #2.61], which states that the net pressure force on a closed surface equals the volume integral of the pressure gradient throughout the region enclosed by that surface. It is much easier to evaluate the volume integral of the pressure gradient than the surface area integral.

Combining the individual contributions, making the hydrostatic approximation for the pressure, and collecting all terms gives the momentum equation expressed in terms of integrals over the cross-sectional area:

$$\frac{\partial}{\partial t} \int_A (1 - \kappa n) u \, dA + \frac{\partial}{\partial s} \int_A u^2 \, dA + g \int_A \frac{\partial \eta}{\partial s} dA + g S_f A (1 - \kappa \bar{n}) = q u_q, \qquad (5)$$

in which g is gravitational acceleration, S_f is the friction slope, $\bar{n}$ is the transverse distance of the centroid of the cross-section from the streamwise coordinate, and u_q is the velocity of inflow before mixing.

Now the three integrals in this expression will be approximated for practical implementation. It can be shown, to within the accuracy of the derivation of the Mass Conservation Equation, where the effects of curvature are included to first order, that the contribution of the time derivative term is

$$(1 - \kappa \bar{n}) \frac{\partial Q}{\partial t} + \frac{Q}{A} (\kappa \bar{n} - \kappa n_m) \frac{\partial A}{\partial t}. \qquad (6)$$

The next integral can be written

$$\int_A u^2 dA = \frac{Q^2}{A} + \text{Neglected terms in second and higher derivatives of velocity.}$$

This is a surprising result, that the approximation is exact for linear vertical and horizontal shear flows. Traditionally it has been believed that setting the integral equal to Q^2/A is exact only if flow velocity is constant over the section. The term becomes, after differentiation:

$$\frac{\partial}{\partial s} \int_A u^2 dA \approx \frac{\partial}{\partial s} \left(\frac{Q^2}{A} \right) = 2 \frac{Q}{A} \frac{\partial Q}{\partial s} - \frac{Q^2}{A^2} \frac{\partial A}{\partial s}. \qquad (7)$$

The remaining term, $g \int_A \partial\eta/\partial s \, dA$ is a rather complicated expression to obtain, and the details will not be shown here. Expressing it in terms of the integrated quantities and collecting the other integral contributions to equation (5), equations (6) and (7), gives the Momentum Equation:

$$(1 - \kappa\bar{n})\frac{\partial Q}{\partial t} + \frac{Q(\kappa\bar{n} - \kappa n_m)}{A}\frac{\partial A}{\partial t} + \frac{2Q}{A}(1 + \kappa n_m - \kappa\bar{n})\frac{\partial Q}{\partial s}$$

$$+\left(\frac{gA}{B} - \frac{Q^2}{A^2}(1 + 2\kappa(n_m - \bar{n}))\right)\frac{\partial A}{\partial s} + \frac{Q^2 \kappa'(n_m - \bar{n})}{A} - gA\bar{S} + gAS_f(1 - \kappa\bar{n}) - qu_q = 0,$$

where $\bar{S}$ is the mean downstream slope of the channel, and $\kappa' = d\kappa/ds$. It is more convenient to have just one time derivative in each equation, and so eliminating $\partial A/\partial t$ using the Mass Conservation Equation (3), gives:

$$(1 - \kappa\bar{n})\frac{\partial Q}{\partial t} + \frac{Q}{A}(2 + 3(\kappa n_m - \kappa\bar{n}))\frac{\partial Q}{\partial s} + \left(\frac{gA}{B} - \frac{Q^2}{A^2}(1 + 2(\kappa n_m - \kappa\bar{n}))\right)\frac{\partial A}{\partial s}$$

$$= \frac{Q^2(\kappa'\bar{n} - \kappa' n_m)}{A} + gA\bar{S} - gAS_f(1 - \kappa\bar{n}) + q\left(u_q + \frac{Q}{A}(\kappa n_m - \kappa\bar{n})\right), \qquad (8)$$

where partial derivative terms have been retained on the left side while forcing terms have been taken to the right. As the friction slope S_f and inflow q are only approximately known, there is room for some simplification by dropping the curvature components in those terms.

DISCUSSION AND CONCLUSIONS

Equations (3) and (8) are the long wave equations which govern the propagation of floods and long waves in curved waterways, a pair of partial differential equations, which, provided boundary and initial conditions are specified, may be solved numerically. Importantly, the structure of the equations is the same as that of the conventional St Venant equations. The effects of curvature appear in them simply by the presence of the terms κn_m and $\kappa\bar{n}$ (and similar terms involving the derivative κ') in the coefficients and in the forcing terms on the right. These terms are, respectively, the ratio of the offset of the midpoint of the surface to the radius of curvature, and a similar ratio involving the offset of the centroid of the water cross section.

The two partial differential equations can be written as ordinary differential equations in a characteristic formulation, from which it can be deduced that c, the speed of propagation of disturbances relative to still water, is given by

$$c = \sqrt{\frac{gA}{B}} \left(1 + \frac{1}{2}\kappa n_m + \frac{1}{2}\kappa \bar{n} \right). \tag{9}$$

It is well known that C, the speed of long waves in a straight channel, is given by $C = \sqrt{gA/B}$, where A/B is capable of simple interpretation as the mean depth. It is perhaps surprising that the modification for a curved channel is so simple. However, it is not so simple to put a physical interpretation on the length scale $\sqrt{A/B} \times \left(1 + \frac{1}{2}\kappa n_m + \frac{1}{2}\kappa \bar{n} \right)$ in this equation. Unfortunately, the simple interpretation that the square of the wave speed is given by $g \times$ Mean Depth does not seem to hold in this case where the channel is curved in plan.

For real rivers, which tend to be shallower on the inside of bends and steeper on the outside, the location of the streamwise axis will usually be determined by normal flows in the lower part of the bed. As the water level rises in a flood, the surface spreads out more over the shallower inside bank causing both n_m and $\bar{n}$ to increase. This remains true for both positive and negative curvature: for typical rivers both n values move in the direction of the centre of curvature and both κn_m and $\kappa \bar{n}$ are positive, whether the river curves to left or right.

This has important implications for flood prediction, as the speed of propagation of floods and waves in such real rivers is then greater than that given by the straight channel approximation. Even if all the computational details of the present method were not to be included in operational models, there should be at least some allowance made for this effect.

REFERENCES

Fenton, J.D. 1995 "On the St Venant long wave equations", Paper submitted for publication.

Milne-Thomson, L.M. 1968 *Theoretical Hydrodynamics* (Fifth Edn.), Macmillan: London.

Streeter, V.L., and E.B. Wylie 1981 *Fluid Mechanics* (First SI Metric Edn.), McGraw-Hill Ryerson: Toronto.

1D14

THE THREE-DIMENSIONAL EFFECTS ANALYSIS
BASED ON GEOMETRIZED NAVIER-STOKES EQUATIONS

VYSOTSKY L.I.
Doctor of Sciences, Saratov technical university
Saratov, Russia

SUMMARY
The 3-D effects in laminar steady flows are reduced in the end to changing of kinematic and geometrical parameters. The latter are characterized by radii of curvature and stream lines' divergence/convergence angles. It is offerred to execute the direct analysis of influence of mentioned parameters based on 3-D geometrized Navier-Stokes equations suggested by the author.

THE GENERALIZED EQUATIONS
Three-dimentional laminar liquid streams are described by Navier-Stokes and continuity equations. At $\rho_l = \text{const}$ they are:

a) in vector form

$$\vec{a} - \frac{1}{\rho_l}\text{grad } p + \vec{T} = \frac{d\vec{u}}{dt}; \quad \text{div } \vec{u} = 0; \tag{1}$$

b) in tensor form (in covariant components)

$$-\frac{\partial}{\partial q_i}(\Pi_{(m)} + P) + \nu g^{km}(u_{i;\,m})_{;\,k} = \frac{\partial u_i}{\partial t} + u^k u_{i;\,k}; \tag{2}$$

c)
$$\left(\rho_l\right)_{,\,t} + \left(\rho_l u^i\right)_{,\,i} = 0; \tag{3}$$

d) dissipative function (without taking of second viscosity into account):

$$\Phi = \mu\left\{\frac{\partial u^i}{\partial q^i}\frac{\partial u^j}{\partial q^i} + g^{ki}G_{mj}\left[\frac{\partial u^j}{\partial q^i}\frac{\partial u^m}{\partial q^k} + u^p\left(u^m\Gamma^j_{in} + \frac{\partial u^j}{\partial q^i}\right)\Gamma^m_{kp} + \right.$$

$$+ \frac{\partial u^m}{\partial q^k} u^n \Gamma_{in}^j + u^r u^s \Gamma_{jr}^i \Gamma_{is}^j + 2u^r \frac{\partial u^i}{\partial q^i} \Gamma_{ir}^j \right\} . \tag{4}$$

THE WORKING EQUATIONS

The main equations and expressions can be transformed to various coordinate systems. For purposes, mentioned in report's title, it is convenient to use the system of coordinates, entered in [1] and connected with stream line in (l, y_1, z^*). Both the system and appropriate geometrical parameters are shown in Fig.1.

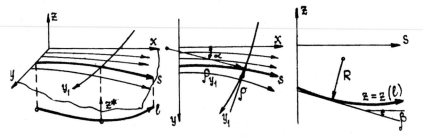

Fig.1. System of coordinates and geometrical parameters.

After executing of all necessary transformation of initial equations, (enough combersome procedure) they can be put in a form

$$-\frac{\partial}{\partial l}\left(\Pi_{(m)} + P\right) + T_l = \frac{\partial u}{\partial t} + \frac{\partial}{\partial l}\left(\frac{u^2}{2}\right); \tag{5}$$

$$-\frac{\partial}{\partial y_1}\left(\Pi_{(m)} + P\right) + T_{y_1} = \frac{\partial u_{y_1}}{\partial t} + \frac{u^2 \cos^2 \beta}{\rho}; \tag{6}$$

$$-\frac{\partial}{\partial z}\left(\Pi_{(m)} + P\right) + T_z = \frac{\partial u_z}{\partial t} + \frac{\partial}{\partial l}\left(\frac{u^2}{2}\right)\sin \beta + \frac{u^2 \cos \beta}{R}; \tag{7}$$

$$\frac{\partial u}{\partial l} + u\left(\frac{\cos \beta}{\rho_{y_1}} + \frac{1}{\cos \beta}\frac{\partial \beta}{\partial z} - \frac{\tan \beta}{R}\right) = 0; \tag{8}$$

where covariant components of viscous term $\vec{T}$ defined as $T_i = \nu g^{km}(u_{i;\,m})_{;\,k}$ are equal to

$$\frac{T_l}{\nu} = \frac{1}{\cos^2 \beta}\left(\frac{\partial^2 u}{\partial l^2} + \frac{\partial^2 u}{\partial z^2}\right) + \frac{\partial^2 u}{\partial y_1^2} - \frac{\partial^2 u}{\partial l \partial z}\frac{2\tan \beta}{\cos \beta} + \frac{\partial u}{\partial l}\left(\frac{\tan \beta}{R\cos^2 \beta} + \frac{1}{\rho_{y_1}\cos \beta}\right) -$$

$$-\frac{\partial u}{\partial y_1}\frac{1}{\rho} - \frac{\partial u}{\partial z}\left(\frac{1}{R\cos^3\beta} + \frac{\tan\beta}{\rho y_1}\right) + u\left[\frac{\sin 2\beta}{\rho}\frac{\partial \alpha}{\partial z} - \frac{1}{R^2\cos^2\beta} - \right.$$

$$\left. - \cos^2\beta\left(\frac{1}{\rho^2} + \frac{1}{\rho_{y_1}^2}\right) - \left(\frac{\partial \alpha}{\partial z}\right)^2 - \left(\frac{\partial \beta}{\partial y_1}\right)^2 - \frac{1}{\cos^2\beta}\left(\frac{\partial \beta}{\partial z}\right)^2\right]; \quad (9)$$

$$\frac{T_{y_1}}{\nu} = 2\left[\frac{\partial u}{\partial l}\left(\frac{1}{\rho} - \tan\beta\frac{\partial \alpha}{\partial z}\right) + \frac{\partial u}{\partial y_1}\frac{\cos\beta}{\rho y_1} + \frac{\partial u}{\partial z}\left(\frac{1}{\cos\beta}\frac{\partial \alpha}{\partial z} - \frac{\sin\beta}{\rho}\right)\right] +$$

$$+u\left[\frac{2\sin\beta}{\rho^2}\frac{\partial \rho}{\partial z} - \frac{1}{\rho^2}\frac{\partial \rho}{\partial l} - \frac{\cos\beta}{\rho_{y_1}^2}\frac{\partial \rho_{y_1}}{\partial y_1} - \frac{2\tan\beta}{R\rho} + \frac{4\tan\beta\sin\beta}{\rho}\frac{\partial \beta}{\partial z} - \right.$$

$$\left. \frac{\sin\beta}{\rho y_1}\left(2\frac{\partial \beta}{\partial \rho y_1} + \frac{\partial \alpha}{\partial z}\right) + \frac{\tan^2\beta - 1}{R}\frac{\partial \alpha}{\partial z}; \quad (10)\right.$$

$$\frac{T_z}{\nu} = \left(\frac{\partial^2 u}{\partial l^2} + \frac{\partial^2 u}{\partial z^2}\right)\frac{\tan\beta}{\cos\beta} + \frac{\partial^2 u}{\partial y_1^2}\sin\beta - \frac{\partial^2 u}{\partial l\partial z}2\tan^2\beta +$$

$$+\frac{\partial u}{\partial l}\left(\frac{\cos^2\beta + 1}{R\cos^3\beta} + \frac{\tan\beta}{\rho y_1} - 2\tan\beta\frac{\partial \beta}{\partial z}\right) + \frac{\partial u}{\partial y_1}\left(2\cos\beta\frac{\partial \beta}{\partial y_1} - \frac{\sin\beta}{\rho}\right) +$$

$$+\frac{\partial u}{\partial z}\left[\frac{2}{\cos\beta}\frac{\partial \beta}{\partial z} - \frac{\tan\beta\sin\beta}{\rho y_1} - \frac{\tan\beta(2\cos^2\beta + 1)}{R\cos^3\beta}\right] +$$

$$u\left[\frac{2\tan\beta}{R^2}\frac{\partial R}{\partial z} - \frac{1}{R^2\cos\beta}\frac{\partial R}{\partial l} + \frac{1}{R\rho y_1} + \frac{\tan^2\beta - 1}{R}\frac{\partial \beta}{\partial z} - \frac{\sin\beta}{\rho_{y_1}}\frac{\partial \beta}{\partial z} - \frac{\cos\beta}{\rho_{y_1}}\frac{\partial \beta}{\partial y_1} + \right.$$

$$\left. + \cos\beta\frac{\partial^2 \beta}{\partial y_1^2} + \frac{1}{\cos\beta}\frac{\partial^2 \beta}{\partial z^2} - \sin\beta\left(\frac{\partial \beta}{\partial y_1}\right)^2 - \frac{\tan\beta}{\cos\beta}\left(\frac{\partial \beta}{\partial z}\right)^2\right]. \quad (11)$$

Dissipative function Φ, after untangling of expression (4) in coordinates (l, y_1, z^*) is the follow:

$$\Phi = \mu\left\{\frac{\partial}{\partial y_1}\left(\frac{u^2}{2}\right)\frac{\cos^2\beta}{\rho} - \frac{\partial}{\partial l}\left(\frac{u^2}{2}\right)\frac{\tan^2\beta}{R} + \frac{\partial}{\partial z}\left(\frac{u^2}{2}\right)\frac{2}{R\cos\beta} + \right.$$

$$\left. + \left(\frac{\partial u}{\partial l}\right)^2\frac{\cos^2\beta + 1}{\cos^2\beta} + \left(\frac{\partial u}{\partial y_1}\right)^2 + \left(\frac{\partial u}{\partial z}\right)^2\frac{1}{\cos^2\beta} - \frac{\partial u}{\partial l}\frac{\partial u}{\partial z}\frac{\tan\beta}{\cos\beta} + \right.$$

$$+u^2\left[\frac{2\cos^2\beta}{\rho_{y_1}^2} - \frac{1}{R^2} - \frac{4\tan\beta}{R\cos\beta}\frac{\partial\beta}{\partial z} + \frac{2}{\cos^2\beta}\left(\frac{\partial\beta}{\partial z}\right)^2 + \left(\frac{\partial\beta}{\partial y_1}\right)^2 + \left(\frac{\partial\alpha}{\partial z}\right)^2 + \right.$$

$$\left.\left.+2\frac{\partial\alpha}{\partial z}\frac{\partial\beta}{\partial y_1}\right]\right\} \tag{12}$$

THE ANALYSIS

As far as in received geometrized Navier-Stokes' equations include the inner geometric parameters in explicit form, they can be used for direct analysis of degree of their influence while solving of the most diverse problems. For example, by means of integration of equation (5) at steady motion along l one can receive the Bernoulli integral (if $\Pi_{(m)} = gz$ and $P = p/\rho$) as follows:

$$B_1 = B_2 + h_l, \tag{13}$$

where $B = z + p/\rho g + u^2/2$ is Bernoulli three-term and $P = p/\rho$, $h_l = -\frac{1}{g}\int_0^l T_l dl$. Obviously, h_l represents the change of specific mechanical energy along stream line caused by influence of viscosity. Also obviously, that $h_l = h_f + h_m$ where h_f represents the dissipatized mechanical energy, and h_m is the transferred mechanical energy. They are expressed as follows:

$$h_f = \frac{1}{\rho_l g}\int_0^l \frac{\Phi}{u}dl; \qquad h_m = -\frac{1}{g}\int_0^l \left(T_l + \frac{\Phi}{\rho_l u}\right)dl. \tag{14}$$

As Φ and T_l contain the kinematic and geometrical parameters

$$\left(\frac{\partial^2 u}{\partial l^2}, \frac{\partial^2 u}{\partial y_1^2}, \frac{\partial^2 u}{\partial z^2}, \frac{\partial^2 u}{\partial l\partial z}, \frac{\partial u}{\partial l}, \frac{\partial u}{\partial y_1}, \frac{\partial u}{\partial z}, u, \rho^2, \rho_{y_1}^2, R^2, R\rho, \rho, \rho_{y_1}, R, \alpha, \beta, etc\right)$$

in explicit form,it is easy to analyse the expressions' simplyfication in various private cases ,when it is possible to neglect obviously by influence of some factors. Let's demonstrate it on some elementary examples. 1. Let the radii of curvature R, ρ and ρ_{y_1} are high enought to one may neglect the parameters including the appropriative curvatures to the the first and second power. Then

$$\frac{T_l}{\nu} = \frac{1}{\cos^2 \beta} \left(\frac{\partial^2 u}{\partial l^2} + \frac{\partial^2 u}{\partial z^2} \right) + \frac{\partial^2 u}{\partial y_1^2} - \frac{\partial^2 u}{\partial l \partial z} \frac{2 \tan \beta}{\cos \beta} - u \left[\left(\frac{\partial \alpha}{\partial z} \right)^2 + \left(\frac{\partial \beta}{\partial y_1} \right)^2 + \right.$$

$$\left. + \frac{1}{\cos^2 \beta} \left(\frac{\partial \beta}{\partial z} \right)^2 \right]; \tag{15}$$

$$\Phi = \mu \left\{ \frac{1}{\cos^2 \beta} \left[\left(\frac{\partial u}{\partial l} \right)^2 (\cos^2 \beta + 1) + \left(\frac{\partial u}{\partial z} \right)^2 \right] + \left(\frac{\partial u}{\partial y_1} \right)^2 - \frac{\partial u}{\partial l} \frac{\partial u}{\partial z} \frac{\tan \beta}{\cos \beta} + \right.$$

$$\left. + u^2 \left[\frac{2}{\cos^2 \beta} \left(\frac{\partial \beta}{\partial z} \right)^2 + \left(\frac{\partial \beta}{\partial y_1} \right)^2 + \left(\frac{\partial \alpha}{\partial z} \right)^2 + 2 \frac{\partial \alpha}{\partial z} \frac{\partial \beta}{\partial y_1} \right] \right\}. \tag{16}$$

It means particularly that if streamlines of the flow are almost linear, then the dissipative function doesn't include gradients of squared velocity.

2. If it is possible to neglect the influence, of $\partial \beta / \partial y_1$, $\partial \alpha / \partial z$, $\partial \beta / \partial z$, in addition , expressions (15) and (16) may be further simplified as

$$\frac{T_l}{\nu} = \frac{1}{\cos^2 \beta} \left(\frac{\partial^2 u}{\partial l^2} + \frac{\partial^2 u}{\partial z^2} \right) + \frac{\partial^2 u}{\partial y_1^2} - \frac{\partial^2 u}{\partial l \partial z} \frac{2 \tan \beta}{\cos \beta}; \tag{17}$$

$$\Phi = \mu \left\{ \frac{1}{\cos^2 \beta} \left[\left(\frac{\partial u}{\partial l} \right)^2 (\cos^2 \beta + 1) + \left(\frac{\partial u}{\partial z} \right)^2 \right] + \left(\frac{\partial u}{\partial y_1} \right)^2 - \frac{\partial u}{\partial l} \frac{\partial u}{\partial z} \frac{\tan \beta}{\cos \beta} \right\}. \tag{18}$$

Here the terms containing the first and second degrees of velocity disappear.

3. If besides it is possible to suggest $\partial u / \partial l \ll \partial u / \partial y_1, \partial u / \partial z$, then

$$\frac{T_l}{\nu} = \frac{1}{\cos^2 \beta} \frac{\partial^2 u}{\partial z^2} + \frac{\partial^2 u}{\partial y_1^2} - \frac{\partial^2 u}{\partial l \partial z} \frac{2 \tan \beta}{\cos \beta}; \quad \Phi = \mu \left[\frac{1}{\cos^2 \beta} \left(\frac{\partial u}{\partial z} \right)^2 + \left(\frac{\partial u}{\partial y_1} \right)^2 \right]. \tag{19}$$

4. If besides $\partial u / \partial y_1 \ll \partial u / \partial z$, then

$$\frac{T_l}{\nu} = \frac{1}{\cos^2 \beta} \frac{\partial^2 u}{\partial z^2} - \frac{\partial^2 u}{\partial l \partial z} \frac{2 \tan \beta}{\cos \beta}; \qquad \Phi = \frac{\mu}{\cos^2 \beta} \left(\frac{\partial u}{\partial z} \right)^2. \tag{20}$$

5. If, in addition, $\beta \approx 0$ then $T_l/\nu = \partial^2 u/\partial z^2$; $\quad \Phi = \mu(\partial u/\partial z)^2$ (21) Similarly other possible cases may be easy analyzed, when it is possible, for example, to suggest the flow flat and vertical one, or flat and horizontal one, or cylindrical (vertical or horizontal) one etc. Untangling the dependence for h_m one can receive it as

$$
h_m = -\frac{\nu}{g} \int_0^l \left\{ \frac{1}{\cos^2 \beta} \left(\frac{\partial^2 u}{\partial l^2} + \frac{\partial^2 u}{\partial z^2} \right) + \frac{\partial^2 u}{\partial y_1^2} - \frac{\partial^2 u}{\partial l \partial z} \frac{2 \tan \beta}{\cos \beta} + \frac{\partial u}{\partial l} \left(\frac{1}{\rho y_1 \cos \beta} + \right. \right.
$$

$$
+ \frac{\tan^3 \beta}{R} - \frac{\partial u}{\partial y_1} \frac{\sin \beta}{\rho} + \frac{\partial u}{\partial z} \frac{1 - \tan^2 \beta}{R \cos \beta} + \frac{1}{u} \left[\left(\frac{\partial u}{\partial l} \right)^2 \frac{\cos^2 \beta + 1}{\cos^2 \beta} + \right.
$$

$$
+ \left(\frac{\partial u}{\partial y_1} \right)^2 \frac{1}{\cos^2 \beta} - \frac{\partial u}{\partial l} \frac{\partial u}{\partial z} \frac{\tan \beta}{\cos \beta} + u \left(\frac{\sin 2\beta}{\rho} \frac{\partial \alpha}{\partial z} - \frac{4 \tan \beta}{R \cos \beta} \frac{\partial \beta}{\partial z} - \frac{\cos^2 \beta + 1}{R^2 \cos^2 \beta} + \right.
$$

$$
+ \cos^2 \beta \left(\frac{1}{\rho y_1} - \frac{1}{\rho^2} \right) + \frac{1}{\cos^2 \beta} \left(\frac{\partial \beta}{\partial z} \right)^2 + 2 \frac{\partial \alpha}{\partial z} \frac{\partial \beta}{\partial y_1} \right] \right\} dl . \tag{22}
$$

It will be noted, that the expressions for h_l, h_f and h_m are aplicable to particular stream line that makes them extremely convenient at using at numerical simulation of filamented flows.

CONCLUSION

The geometrized form of Navier-Stokes equations and expressions for components of viscous term $\vec{T}$, dissipative function, change of specific mechanical energy and etc., offered in report,are rather convenient for both direct analysis of their dependencies on local geometrical and kinematic parameters, and justified simplification of offered expressions and their using at numerical solving of problems.

THE MAIN DESIGNATIONS

T_i is components of viscous term $\vec{T} = \nu \Delta \vec{u}$; q_i is curvelinear coordinate; g^{km} is contravariant component of metric tensor; G_{mj} is algebraic addition; , is private differentiation; ; is contravariant derivative .

REFERENCES

1. Vysotsky L.I. Supercritical flows on spillways controlling. Moscow.: Energoatomizdat. 1991. 240 P.

1D15

ENTROPY AND OPEN-CHANNEL FLOW PROPERTIES

CHAO-LIN CHIU AND WEIXIA JIN
University of Pittsburgh, Pittsburgh, PA 15261, USA

INTRODUCTION

A channel section has propensity to establish and maintain an equilibrium or quasi-equilibrium state that corresponds to a value of Shannon's entropy (Chiu and Said 1995). In natural channels, such a state is probably established by the channel-forming "bankfull discharge" (Leopold and Wolman 1970). Under the Shannon's entropy concept, the intuitive notion of the tendency for entropy to increase is interpreted probabilistically into a framework amenable to logical analysis. The concept can be applied in hydraulic engineering to complement the rational hydrodynamics which can deal with only the observable aspect of a physical process in the space-time framework.

ENTROPY AND VELOCITY DISTRIBUTION

Chiu (1989) derived a system of velocity distribution equations by maximizing the Shannon's entropy (Shannon 1949), subject to the constraints of the conservation laws governing the transport of mass, momentum and energy. The simplest of the equations derived is

$$u = \frac{u_{max}}{M} \ln\left[1 + (e^M - 1) \frac{\xi - \xi_0}{\xi_{max} - \xi_0}\right] \tag{1}$$

in which u = velocity at ξ; ξ = independent variable with which u develops such that each value of ξ corresponds to a value of u; ξ_{max}= maximum value of ξ where the maximum velocity u_{max} occurs; and ξ_0 = minimum value of ξ, which occurs at the channel bed where u is zero. M is called "entropy parameter" since it is the parameter of both the probability distribution $p(u/u_{max})$ corresponding to (1) and its entropy that can be derived as

$$H = -\int_0^1 p\left(\frac{u}{u_{max}}\right)\ln p\left(\frac{u}{u_{max}}\right)d\left(\frac{u}{u_{max}}\right) = 1+\ln\frac{e^{M}-1}{M} - \frac{Me^{M}}{e^{M}-1} \qquad (2)$$

A smaller value of M means a more uniform probability distribution and a larger entropy value. As a velocity-distribution equation, (1) is flexible in the sense that, by selecting a suitable form of ξ, it can describe one- or two-dimensional velocity distribution in an open channel. For example, for a wide channel, $\xi = y/D$; and, for an axially symmetrical flow in a circular pipe, $\xi = 1- (r/R)^2$ (Chiu et al 1993). According to (1), the ratio of the mean velocity $\bar{u}$ to the maximum velocity u_{max} is

$$\frac{\bar{u}}{u_{max}} = \frac{e^{M}}{e^{M}-1} - \frac{1}{M} = \phi \qquad (3)$$

which is also a function of M and, hence, the entropy. In Fig. 1, u_{max} is plotted against $\bar{u}$ in a wide rage of $\bar{u}$ and, hence, the discharge in the four channels indicated. It shows a stable linear relationship which, along with (3), supports the theory of constant values of ϕ, M and entropy. This is further supported by Fig. 2 which shows a similar result obtained from experiments in a circular conduit, with the slope, discharge, and water depth adjusted in wide ranges (Tsujimoto et al 1993). The practical way to determine M and entropy of a channel section is to use the relation shown in (3) and Figs. 1 and 2. The constant M value of a channel section has been used as a basis for an efficient method of velocity measurements to determine the discharge (Chiu and Said 1995). Fig. 3 is based on (1) and (3), and shows that a channel section of a smaller M value and, hence, a larger entropy value has a less uniform velocity distribution. It also indicates that the location of $\bar{u}$ depends on M. This is an important concept to use in flow measurements in rivers and streams. When the discharge and, hence, $\bar{u}$ fluctuate the channel adjusts u_{max} and its location, to keep ϕ and M constant as shown in Fig. 4.

ENTROPY AND ENERGY COEFFICIENT

By using the probability density function p(u) corresponding to (1), the mean values of u^2 and u^3 can be obtained, which in turn give the momentum and energy coefficients as functions of M (Chiu 1991). Fig. 5 shows the relation of M to the energy coefficient α given by the following equation that can be derived:

$$\alpha = \frac{(e^{M}-1)^2[e^{M}(M^3-3M^2+6M-6)+6]}{[e^{M}(M-1)+1]^3} \qquad (4)$$

which indicates that a channel section of a smaller M and a larger entropy has a larger value α. Therefore, for a given value of mean velocity $\bar{u}$, the mean rate of transport of kinetic energy $\alpha \bar{u}^2/2g$ through a channel section increases with the entropy. The method to determine α from the M values of channel sections, as shown by Fig. 5, should facilitate using the energy equation to compute the water surface profile and related transport processes in rivers and streams.

ENTROPY AND SEDIMENT CONCENTRATION

A simple differential equation governing the sediment concentration along a vertical under a steady state is:

$$-\varepsilon_s \frac{dC}{dy} = v_s C \qquad (5)$$

in which ε_s = diffusion coefficient for sediment transfer; C = sediment concentration; and v_s = settling velocity of sediment particle. ε_s is often estimated as $\beta \varepsilon_m$ where β is a coefficient that can be estimated by existing methods (Householder and Goldschmidt 1969); and ε_m is the diffusion coefficient for momentum transfer that can be obtained as

$$\epsilon_m = \frac{\tau}{\rho} \left(\frac{du}{dy} \right)^{-1} \qquad (6)$$

in which ρ = fluid density; du/dy = velocity gradient along the y-axis; and τ = shear stress that can be obtained approximately as $\tau_0(1-y/D)$ where τ_0 is the shear stress at $y = 0$. If u_{max} occurs on the water surface and the velocity gradient du/dy is obtained by (1) with ξ/ξ_{max} equal to y/D, the solution of (5) yields

$$\frac{C}{C_0} = \left[\frac{1 - \dfrac{y}{D}}{1 + (e^M - 1) \dfrac{y}{D}} \right]^{\lambda'} \qquad (7)$$

in which C_0 is the sediment concentration at $y = 0$; and

$$\lambda' = \frac{v_s u_{max}(1 - e^{-M})}{\beta u_*^2 M} = \frac{v_s \bar{u}(1 - e^{-M})}{\beta u_*^2 M \, \phi(M)} = \lambda G(M) \qquad (8)$$

in which

$$\lambda = \frac{v_s \overline{u}}{\beta u_*^2} \tag{9}$$

$$G(M) = \frac{1 - e^{-M}}{M \, \phi(M)} \tag{10}$$

G(M) represents the entropy effect. Under a given set of flow and channel conditions, an increase in λ means an increase in the sediment size. Eq. (7) appears similar to the well-known Rouse equation; however, the sediment concentration given by it remains finite at the channel bed while that by the Rouse equation does not. It also separates the entropy from other factors, and, hence, enables linking the sediment concentration to the entropy or the overall characteristics of a channel section. Fig. 6 shows the variation of sediment concentration with M, given by (7) at $\lambda = 1/4$. It indicates that a channel section of a smaller M value and, hence, a larger entropy value has a less uniform distribution of sediment concentration on a vertical. Fig. 6 and similar figures computed for a wide range of λ also indicate that, at a given value of λ, the location at which the mean sediment concentration occurs depends on M and, hence, the entropy. The dependence on M increases with λ. This concept can be used as a basis to facilitate measuring the mean sediment concentration, in a similar way the mean velocity is located according to the M value.

SUMMARY AND CONCLUSION

The entropy of a channel section has been linked to the velocity distribution, energy and momentum coefficients, and distribution of sediment concentration. It is advantageous to determine the entropy value of a channel section as represented by M or ϕ value, since it can be used as a basis to facilitate flow measurements and characterization.

REFERENCES

Bridge, J. S., and Jarvis, J. (1985). "Flow and sediment transport data, River South Esk, Glen Glova, Scotland." An unpublished report, Dept. of Geological Sciences, State Univ. of New York, Binghamton, NY 13901.

Chiu, C.-L. (1989). "Velocity distribution in open-channel flow." J. Hydr. Engrg., ASCE, 115 (5), 576-594.

Chiu, C.-L. (1991). "Application of entropy concept in open-channel flow study." J. Hydr. Engrg., ASCE, 117(5), 615-628.

Chiu, C.-L., Lin, G.-F., and Lu, J.-M. (1993). "Application of probability and entropy concepts in pipe-Flow study." J. Hydr. Engrg., ASCE, Vol. 119(6), 742-756.

Chiu, C.-L., and Said, C. A. A. (1995). "Maximum and mean velocities and entropy in open-channel flow." J. Hydr. Engrg., ASCE, 121(1), 26-35.

Culbertson, J. K., Scott, C. H., and Bennett, J. P. (1971). "Summary of alluvial-channel data from Rio Grande Conveyance Channel, New Mexico, 1965-69." Open File Report, U. S. Geological Survey, Water Resources Division, Aug., 1971.

Guo, Z.-R. (1990). Personal Communication, Southeast China Environmental Science Institute, Yuancun, China.

Guy, H. P., Simons, D. B., and Richardson, E. V. (1966). "Summary of alluvial channel data from flume experiments, 1956-61." Geological Survey Professional Paper 462-I, U.S. Govt. Printing Ofc., Washington, D. C.

Householder, M. K., and Goldschmidt, V. W. (1969). "Turbulent diffusion and Schmidt Number of Particles." J. Engrg. Mech., ASCE, 95(6), 1345-1367.

Leopold, L., and Wolman, G. (1970). "River channel patterns." Chapter 7, Rivers and river terraces, ed. by G. H. Dury, Praeger Publishers, New York, 197-237.

Shannon, C. E. (1948). "A mathematical theory of communication." The Bell System Tech. J., 27 (Oct.), 623-656.

Tsujimoto, T., Okada, T., and Motohashi, K. (1993). "Experimental study on velocity distribution of flow with free surface in a circular conduit." KHL Progress Rept., Hydraulic Lab., Kanazawa Univ., Kanazawa, Japan, December, 61-76.

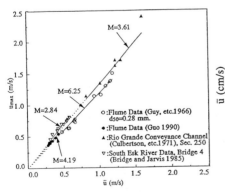

FIG. 1. Relation between Maximum and Mean Velocities

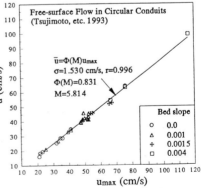

FIG. 2. Relation of u_{max} to $\bar{u}$ in Circular Conduit

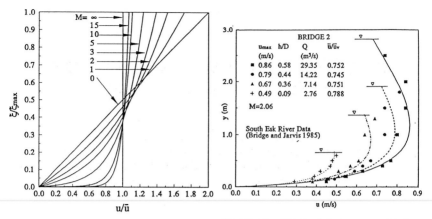

FIG. 3. Relation of M to Velocity Distribution

FIG. 4. Variation of Velocity Distribution at a Channel Section

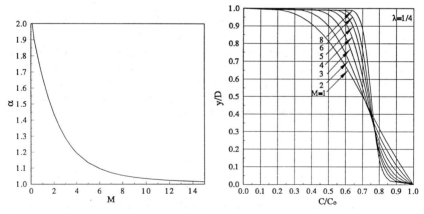

FIG. 5. Relation of M to Energy Coefficient

FIG. 6. Relation of M to Distribution of Sediment Concentration

1D16

EVALUATION OF FLOW VARIATION BY AQUACULTURE FACILITIES USING A NUMERICAL MODEL

LEE, J.S.[*], PARK, I.H.[**] and LEE, M.O.[***]

[*] Dept. of Civil Eng., National Fisheries Univ. of Pusan, Pusan, Korea
[**] Dept. of Ocean Eng., National Fisheries Univ. of Pusan, Pusan, Korea
[***] Dept. of Ocean Civil Eng., National Yosu Fisheries Univ., Yosu, Korea

ABSTRACT

Recently an enhancement of the productivity in coastal culture ground become to the important subjects, and much efforts have been conducted in coastal hydraulic problems to get some physical processes such as sea water exchange or conveyance. Numerical experiments on the flow variation due to the hanging culture facilities are carried out. The experiment conditions are determined from the oyster culture grounds located in Kamak bay, Korea. In the evaluation of the flow field, an additive pressure drag force is considered. The pressure drag is expressed only as a drag coefficient, cross section area of facility and facility density. The velocity decay rate in the numerical model is comparable with the field measurement data. The horizontal velocity distribution is affected sensitively by the density and layout of the facilities.

INTRODUCTION

Aquaculture facilities located in coastal or inner bay causes a flow resistance which affects to the conveyance of sea water. Therefore, the flow resistance is an important factor in determination of the optimum density and arrangement to evaluate a flow variation due to the aquaculture facilities. Nakamura(1991) developed a formula for the flow resistance considered a drag coefficient C_D, and estimated a flow decay rate by the aquaculture facilities in channel.

The Kamak bay located in southern sea of the Korean peninsula is very famous as a hanging oyster culture ground. The oyster production of the bay possesses about 30% of the total production of Korea. However, the recent production in the bay is decreasing in spite of increased aquaculture areas. Lee(1993) pointed out that the flow structure and sea water conveyance directly affect to the survival rate of oyster. According to the flow measurement in Kamak bay by authors in 1994, the current velocity in the hanging culture facilities decreases about 30 % than the natural condition.

However, it is very difficult to investigate the time and spatial variation around aquaculture facilities using by field observation or laboratory experiment. In this study, a numerical model is used for the evaluation of horizontal flow variation due to the aquaculture facilities.

NUMERICAL MODEL

In this study, the flow model, DIVAST, developed by Falconer(1986) is used, the governing equation is based on the depth integrated time dependent shallow water equation. The model, DIVAST, evaluates the bottom friction force as a quadratic law including the Chezy drag coefficient C.

$$C = -18.0 \log (k / 12h) \tag{1}$$

where, h is the total water depth, and k is the bed roughness. However, a pressure drag force should be considered when a body is set on the flow field. The pressure drag force F_P for the unit hanging oyster culture facility can be expressed by eq.(2).

$$F_P = C_{DP} \cdot A_P \cdot \frac{\rho}{2} \cdot | \vec{V} | \cdot \vec{V} \tag{2}$$

where, A_P is the cross section area of unit facility, ρ is the sea water density, $\vec{V}$ is the velocity vector and C_{DP} is the drag coefficient depends on the shape of the unit facility. Now, to calculate the total pressure drag forces in each grid point $(F_T)_{i,j}$ a facility density per unit area $(n)_{i,j}$ should be considered in the model.

$$(F_T)_{i,j} = \Sigma F_P = C_{DP} \cdot \frac{A_P}{(A)_{i,j}} \cdot \frac{\rho}{2} \cdot | \vec{V} | \cdot \vec{V} \cdot (n)_{i,j} \tag{3}$$

where, $(A)_{i,j}$ is the flow section area of each grid point($\Delta S \times h$).

FIELD OBSERVATIONS

Fig.1 shows a standard hanging oyster culture system of 1 set used in Kamak bay. The system has the facility density about 20 set/ha. To investigate the velocity decay rate, the tidal current is observed around hanging oyster culture grounds, where the bottom is almost flat, and the water depth is 5 m. The current velocity is measured simultaneously at 2 points by the self-recording electromagnetic current meter(ACM-III, ALEC Co.) during 25 hours in spring tide. One current meter was set across the hang oyster culture facilities from another one at a distance of 100 m, and the velocity was measured at 2 m below of sea surface every one minute. Fig.2 shows a variation of velocity differences between upstream and downstream point during flood tide. The velocity differences have a value within 5 cm/sec, where the maximum velocity decay rate becomes to 0.5.

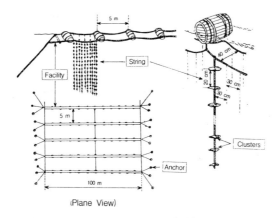

Fig.1 Hanging oyster culture facility system.

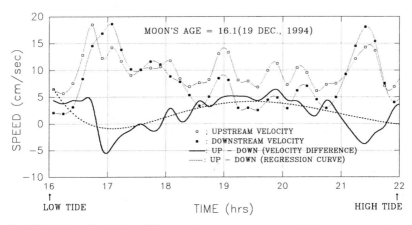

Fig.2 Variation of velocity differences between upstream and downstream point in the field.

NUMERICAL EXPERIMENTS

The calculation domain has a 600 *m* width and 600 *m* length and it has a constant water depth, 5 *m*. In the numerical model, the grid interval is 10 *m*, then other physical coefficients have normal values. If we assume the cluster shape of Fig.1 as a cube, the drag coefficient has the common value 1.0 (Munson *et al.*, 1994).

On the other hand, the 3 kinds of facility density are used in the numerical experiment. Also, to investigate the effect of facility arrangement, the 3 types of layout are considered. Table 1 denotes the numerical experiment cases. All of the cases are carried out under the same water level difference $\Delta\eta$ = 5.9 cm in both open boundaries of the channel, which has 9.83×10^{-5} hydraulic grade. After the run time of 3 hours, a steady uniform current field is made in the calculation domain.

The magnitude of velocities considered only bed roughness is u_{10} = 118 cm/sec when k = 10 mm, u_{50} = 101 cm/sec when k = 50 mm and u_{100} = 92 cm/sec when k = 100 mm. Above velocities coincide well with the Manning's resistance formula.

Table 1. Experiment conditions for the flow resistance by the aquaculture facilities

CASE	Bed Roughness	Drag Coeff.	Facility Density	Facility layout
I-1			$10 \times 10^4 ea/ha$	1ha×1ea (located at centre)
I-2			$20 \times 10^4 ea/ha$	
I-3			$30 \times 10^4 ea/ha$	
II-1	50 mm	1.0		1ha×3ea (in longitudinally)
II-2			$20 \times 10^4 ea/ha$	1ha×3ea (in crossly)
II-3				1ha×9ea (in equally)

RESULTS AND DISCUSSIONS

EFFECTS OF FACILITY DENSITY

Fig.3 shows a spatial variation for $(u)_{i,j}/(u_{50})_{i,j}$ according to the various facility density. In Fig.3, the flow patterns are weakly affected by the facility density, however, the velocity decays up to 68 % at the rear of the facility in CASE I-3. Above results show a similar pattern of the flow measurement data in the field.

EFFECTS OF FACILITY ARRANGEMENT

Fig.4 and Fig.5 show a spatial variation of flow field $(u)_{i,j}/(u_{50})_{i,j}$ and $(v)_{i,j}/(u_{50})_{i,j}$ due to the facility arrangement. A complex flow pattern occurs by the facility layout. In Fig.4, the maximum velocity decay rate shows 52 % at the rear of facilities in the left and right figure, while it has 75 % in the middle of figure. In Fig.5, the transverse flow component $(v)_{i,j}/(u_{50})_{i,j}$ varies within ±4 % as a bilateral symmetry.

Fig.6 denotes a water level variation $(\eta)_{i,j=32}/(\eta_{50})_{i,j=32}$ along the centre line in case of CASE II-1, II-2 and II-3, where the water level variation corresponds well to the facility locations. On the other hand, the maximum water level difference $\Delta\eta$ in both side of the channel occurred in CASE II-3. It means that the larger facility area increase, the more flow resistance increase.

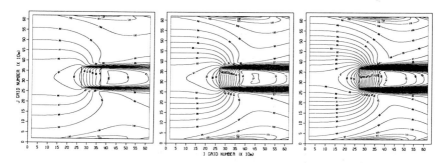

Fig.3 Spatial variation of $(u)_{i,j}/(u_{50})_{i,j}$ due to the facility density.

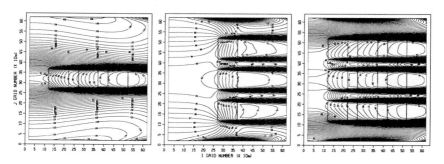

Fig.4 Spatial variation of $(u)_{i,j}/(u_{50})_{i,j}$ according to the facility arrangement

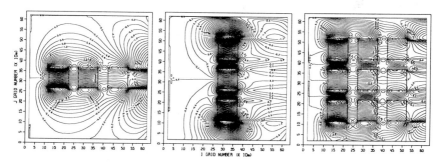

Fig.5 Spatial variation of $(v)_{i,j}/(u_{50})_{i,j}$ same as Fig.4.

From the above results, we can see that the density and layout of the aquaculture facilities sensitively affects to the magnitude and horizontal distribution of the flow field. The decrease of sea water conveyance in the aquaculture facilities would directly affects to the survival rate of oyster *etc.*.

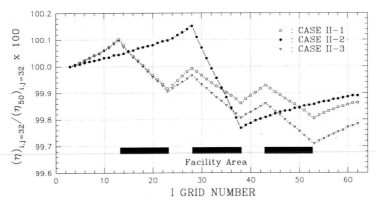

Fig.6 Water level variation $(\eta)_{i,j=32}/(\eta_{50})_{i,j=32}$ due to the facility lay out along the centre line.

Therefore, for the appropriate management of the aquaculture ground, it is necessary to consider a flow resistance due to the density and layout of aquaculture facilities for the prediction of flow structure and sea water conveyance.

ACKNOWLEDGEMENT

This study was performed by the research fund of the RCOID(Research Centre for Ocean Industrial Developments), an Engineering Research Centre designated by KOSEF. Also the Authors wish to express their sincere thanks to Prof. G.H. Lee and Prof. D.S. Kim of National Yosu Fisheries Univ. for providing the current meters.

REFERENCES

Falconer, R.A., 1986. A Two-dimensional Mathematical Model Study on the Nitrate Levels in an Inland Natural Basin, Proc. Inter. Conf. Water Quality Modellings in the Inland Natural Environment, BHRA, Fluid Engineering, Bournmouth, Paper J1, pp.325-344.

Lee, G.H., 1993. Fisheries Oceanographical Studies on the Production of the Farming Oyster in Kamak Bay, The Ph.D. Thesis, National Fisheries Univ. of Pusan, Korea, 180p.

Munson, B.R, Young, D.F, Okishi, T.H., 1994. Fundamentals of Fluid Mechanics, 2nd Ed., John Wiley & Sons, Inc., pp.612-613.

Nakamura, M., 1991. Fisheries Civil Engineering, Kogyo Sisa, Japan, pp.514-516.

1D17

SAFETY EVALUATION ON NUCLEAR POWER PLANT SITE DUE TO POTENTIAL DAM BREACH

Choi, Chul Soon[1], Yi, Sung Myeon[2] and Han, Kun Yeun[3]

1),2) Civil Engr. of Korea Power Eng. Co., INC., P.O. Box 631, Kang Nam, Seoul, 135-606, KOREA
3) Prof. of Civil Eng., Kyungpook Nat'l Univ., Daegu, 702-701, KOREA

1. INTRODUCTION

The damage of nuclear power plants resulting from certain hydrologic events, especially upstream dam failure due to its overtopping, often results in the loss of human life and causes extensive property loss. In addition, the abnormal operation of nuclear power plant may lead to severe threats such as the leakage of radioactive material.

Therefore the nuclear power plant shall be designed to withstand the worst possible hydrologic events. Thus the site for nuclear power plant(hereinafter Site) shall be selected and planned to acquire the safety of power plant during its operation duration, usually 40 years.

This technical study is to estimate the flood stage due to the potential earthdam failure and evaluate the safety of Site located downstream from the dam.

2. DAM-BREAK FLOOD ROUTING MODEL

The National Weather Service DAMBRK Model(Fread 1988) was used to perform all analyses. DAMBRK is a popular and widely used state -of-the-art model applied for dam-break and/or inundation analyses.

DAMBRK is a computer model used to develop a breach outflow hydrograph and route downstream flood wave. This model is composed of two functions. First, one routes an inflow hydrograph through the reservoir, develops the breach patterns and compute the outlet discharge through the dam breach as outflow hydrograph type. Second, the one routes the combined outflow discharge through the downstream channel. This scheme is performed using the complete one-dimensional Saint-Venant equations of unsteady flow. Those equations are solved by means of a nonlinear weighted four-point implicit finite-difference method.

DAMBRK can carry out either subcritical, supercritical or mixed dynamic flow routing. The downstream flow can be modeled as either supercritical or subcritical flow. A specified reach immediately downstream from the dam can be modeled as supercritical flow with the remaining channel modeled as subcritical, or a mixed flow option can be used which allows DAMBRK to model situations where flows change between subcritical and supercritical.

3. FIELD DATA COLLECTION ANALYSIS

Channel Description

The first reach of channel downstream from dam has an average slope of 0.010 in high-gradient, which decreases to about 0.004 for next middle reach of stream in the Bugu stream. The last reach of channel adjacent to the coastal area shows the average slope of 0.002. The total channel length to be routed is equal to 5.63km and the Site for nuclear power plant is located beside the channel bank 5.13km~5.34km downstream from the Bugu dam. The left bank of stream is used as paved local road to access the nuclear power plant. Also the right side of stream is widely used as farm field. The stream mouth connected to the East Sea is formed with the sand dune transported from the sea wave activities. The channel bottom elevation is varies from EL.39.0 to EL.1.8m and that of Site is EL.3.5m.

The n-values are estimated on the basis of standard hydraulic texts in reference with the field survey. The modified n-values for first reach of stream in high-gradient was appiled to compute the anticipated supercritical flow as subcritical flow. This procedure can reduce the painstaking of computation and simplify the model application.(Triesta, 1992)

The schematic outline of Bugu stream and the Site is presented in Figure 1.

Dam and Site Description
The Bugu reservoir plays role of supplying the industrial and domestic water to be required for plant operation. The detailed description of dam features are given as Table 1.

The Site is located about 5.2km downstream from the Bugu dam and also adjacientto the East Sea for the good sppliance of cooling water. Its elevation was graded as EL.10.0m by land datum level. Presently, 2 Units are being operated to generate power and another two Units have been constructed since 1990. Two future Units will be designed in a few years.

4. RESULT AND ANALYSIS

Breach Scenario
The PMP(Probable Maximum Precipitation) fall down the headwaters of Bugu reservoir. The inflow rate may exceed the outflow rate through the uncontrolled spillway and so the water level can rises to the crest of earth-fill dam. The inflow flood(PMF) can overtop the dam crest. The downstrean embankment of Bugu dam may undergo the break occurred from the erosion effect.

By the assumption on the above dam breach scenario, the breach parameters of Bugu dam were discussed in detail in Table 2.

The inflow discharge due to PMP was obtained from the Korean standard. The peak inflow floods were estimated as $410m^3$/sec, $378m^3$/sec, $352m^3$/sec corresponding to each failure mode, respectively.

The breach outflow hydrographs was obtained with respect to each failure case and hydrographs on extreme failure cases were shown as Figure 2.

Flood Routing
A total of 58 cross sections was located at unequal distance and time step size was selected as to satisfy the recommendation of concerned code. The estuary boundary condition was adopted as the maximum measured tidal level, EL. 3.3m.

The Manning n-values used vary from 0.08 to 0.035 based on the Site survey and the weighting factor in Preissmann scheme was used as 0.6.

The computed stage profiles of flood waves in Bugu stream after Bugu dam break were shown as Figure 3, particularly extreme flooding cases.

Conclusively the flood stage at the Site for nuclear power plant was obtained as Table 2.

5. CONCLUSION

The design concept for nuclear power plant shall require the safety against the potential extreme hydrologic accidents. Therefore this attempt was performed to establish the procedure of flood routing due to extreme hydrologic event such as dam failure occurring from PMP.

The flood wave transition can be simulated and evaluated for the various combinations of the breach parameters with the PMF(Probable Maximum Flood) condition. The applied flood routing model has yields reasonable result for the siting of nuclear power plant and the Site has been evaluated to be safe from the Bugu dam failure under the extreme environmental condition.

As a result, the range of outflow peaks and flood stages downstream from the dam can be determined for regulatory and disaster prevention measures which are necessary for the safety of nuclear power plant site.

REFERENCES

1. Fread, D. L., DAMBRK: The NWS Dam-Break Flood Forecasting Model, Office of Hydrology, National Weather Service(NWS), 1988, pp. 1~37.
2. Han, Kun Yeun et al., " A Forecasting Model for the Floodwave Propagation from the Hypothetical Earth Dam-Break", Proc. of KSCE, Vol.6, No.4, 1986, pp. 69~78.
3. Han, Kun Yeun et al., " An Analysis of Outflow Hydrograph Resulting from an Earth Dam-Break ", Proc. of KSCE, Vol.5, No.2, 1985, pp. 41-50.
4. Hunt, B., "Dam-Break Solution", J. of HY Div., ASCE, Vol. 110, No. 6, June, 1984, pp. 675~686.
5. Katopodes, N. D. and Schamber, D. R., "Applicability of Dam-Break Flood Wave Models", J. of HY Div., ASCE, Vol. 109, No. 5, May, 1983, pp. 702~721.
6. MacDonald, T. C., and Jennifer, L. M., "Breaching Characteristing of Dam Failures", J. of HY Div., ASCE, Vol. 110, No. 5, May, 1984, pp. 567~586.
7. Ponce, V. M., and Tsivoglou, A. J., "Modeling Gradual Dam Breaches", J. of HY Div., ASCE, Vol. 107, No. 7, July, 1981, pp. 829~838.

8. Singh, K.P. and Snorrason, A.,"Sensitivity of Outflow Peaks and Flood Stages to the Selection of Dam Breach Parameters and Simulation Model", J. of Hydrology, Vol.68, 1984, pp. 295~310.

9. Triesta, D.J.,"Evaluation of Supercritical/Subcritical Flow in High Gradient Channel", J. of Hyd. Eng., ASCE, Vol.118, No.8, 1992, pp.1107~1118.

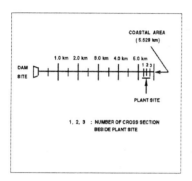

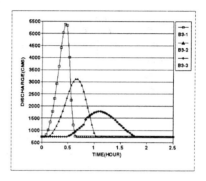

Figure 1. Schematic outline of Bugu stream

Figure 2. Breach outflow hydrograph of Bugu dam(B3 case)

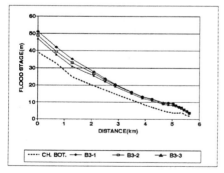

Figure 3. Peak flood stage profiles(B3 case)

Table 1. Characteristics of Bugu reservoir

Drainage Area (km^2)	11.86
Dam	
Type	Earth-fill
Elevation, Top of Dam (m)	68.0
Elevation of Stream Bed (m)	39.0
Length (m)	240.0
Freeboard(m)	3.5
Reservoir	
Elevation, High Water Level (m)	64.5
Area, HWL (ha)	34.5
Capacity, HWL (ha-m)	390.0
Spillway	
Type	Uncontrolled Ogee-crested Side Channel
Elevation, Spillway Crest (m)	64.5
Length of Crest (m)	53.0

Table 2. Evaluation result of flood stage at Site due to Bugu dam breach

CASE ID	Inflow Flood (CMS)	Breach Parameter		Outflow Peak (CMS)	Peak Discharge at Plant Site (CMS)			Max.Elevation at Plant Site (m)		
		BB (m)	TF (hrs)		# 1	# 2	# 3	# 1	# 2	# 3
B1-1	410	29 m	0.5 hrs	4628	1293	1273	1254	8.27	7.36	6.70
B1-2	378	29 m	1.0 hrs	2806	1081	1065	1050	7.95	7.14	6.54
B1-3	352	29 m	2.0 hrs	1634	822	812	809	7.46	6.73	6.17
B2-1	410	58 m	0.5 hrs	5402	1313	1292	1271	8.30	7.37	6.72
B2-2	378	58 m	1.0 hrs	3000	1140	1123	1106	8.04	7.20	6.58
B2-3	352	58 m	2.0 hrs	1724	829	818	811	7.51	6.79	6.19
B3-1	410	87 m	0.5 hrs	5648	1332	1310	1288	8.33	7.38	6.74
B3-2	378	87 m	1.0 hrs	3111	1198	1180	1162	8.13	7.26	6.62
B3-3	352	87 m	2.0 hrs	1769	835	824	813	7.54	6.84	6.20

Note) # 1 : 5.14 km point, # 2 : 5.24 km point # 3 : 5.34 km point
BB; final breach width TF; failiture duration time

1D18

Theoretical Analysis on Self-excited vibration of long span shell roller gate

KUNIHIRO OGIHARA
Toyo University
2100, kujirai, Nakanodai, Kawagoe Saitama,JAPAN
HIROJI NAKAGAWA
Kyoto University
Yoshidahonmachi Kyoto, JAPAN
SACHIHIKO UEDA
Ishikawajima Heavy Industry
Ootemach Chiyodaku, Tokyo, JAPAN

PREFACE

For needs of the discharge control by shell roller gate in the river, the operation under the small gate opening is coming up now. But it brings the vibration problem in such operation. Therefore it is necessary to make clear the origins of vibration and to make develop the cure measure for this vibration. This paper is the summary of the

Photo.1. Waves in upstream under vibration

results from the research work which has been done under the task committee of Japan Hydraulic Gate and Penstock Association. Self-excited vibrations of long span shell roller gate have been observed both in the field and in the model under the small gate opening (Photo. 1). But under which conditions of the self-excited vibrations arise, and what is the origins of vibration have been not so clear that the operation of gate system is determined except the operation under small gate opening. Namely the gate does not stop under the small gate opening in long shell roller gate. This paper shows the results of theoretical analysis on this self-excited vibration by taking account the results of field test and the model test which is made by the elastic similarity model.

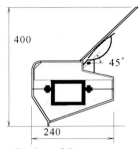

Fig.1 Section of Gate   (in mm)

OUTLINE OF SELF-EXCITED VIBRATION
The scale of model gate is 1/12 for the standard long span shell roller gate of prototype and it is satisfied with the similitude of Froude and the similitude of elasticity. Three type beams are used in model test, which have the different rigidity, namely ratio of deflection and span length are 1/600, 1/800 and 1/1200. The section of model gate is shown in Fig. 1 and the relations between double amplitude, gate opening and upstream water depth are shown

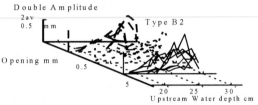

Fig.2 Double Amplitude of Vibration

in Fig. 2. The scale in Fig. 1 is for the prototype; gate height is 4m, the depth of gate is 2.4m and span length is 40m. In the model gate, there is the elastic beam in the center of gate section for adjusting the deflection of beam. The self-excited vibrations of model gate, are observed under 5 mm of gate opening and over 20 cm of upstream depth. The flows of down stream under the vibration, are from the free flow to the slightly submerged flow. The records of vibration are shown in Fig. 3, such as the variation of water depth in upstream (H1), the vertical deflection of middle point of beam, the vertical deflection of beam at the support point and the acceleration of the vertical vibration at the middle point beam.

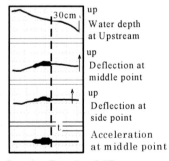

C series Opening 3.75mm
Fig.3 Records of Vibration

There can be seen the record of vibration in three records except the record of water depth of upstream.

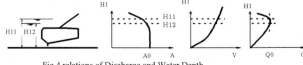

Fig.4 relations of Discharge and Water Depth

Under which conditions for the self-excited vibration arise, can be determined by the maximum discharge condition which is derived by taking account the deflection of beam. Because the recorded line of upstream water depth is changed from the decreasing condition to the increasing condition at the point of self-excited vibration arises in Fig. 3. This condition is explained by the relation as shown in Fig. 4, that is, the velocity under the gate increases by the rising the water depth of upstream, but the area of water flow under the gate decreases by the increase of upstream water depth. So the water discharge is derived by the multiply of the area and velocity, has the critical point which the discharge becomes the maximum. When the water depth of upstream comes over this critical point, the vibrations occur as shown as black wide record in Fig. 3 in any rigidity of beam. Therefore this critical condition can be applied to the same type of gate.

THEORETICAL MODEL FOR SELF-EXCITED VIBRATION

BASIC EQUATION FOR THE VIBRATION

The basic equation for the self-excited vibration for this phenomenon is derived as shown Eq. (1). by taking account the motion of deflection of beam. The external forces of this equation can be determined by the motion of beam and flow condition. When the external forces are not acted on this, it becomes the free vibration and the natural angular frequencies are written as equation (2).

$$EI_y \frac{\partial^4 y}{\partial z^4} + m \frac{\partial^2 y}{\partial t^2} + R \frac{\partial y}{\partial t} = f_y(z,t), \quad EI_x \frac{\partial^4 x}{\partial z^4} + m \frac{\partial^2 x}{\partial t^2} + R \frac{\partial x}{\partial t} = f_x(z,t) \quad (1)$$

$$\omega_{yn} = \frac{n^2 \pi^2}{L^2} \sqrt{\frac{EI_y}{m}}, \quad \omega_{xn} = \frac{n^2 \pi^2}{L^2} \sqrt{\frac{EIx}{m}} \quad (2)$$

EXTERNAL FORCE FOR SELF-EXCITED VIBRATION

Many external forces for this vibration are considered, but two basic forces are considered here, those are derived by the gate motion themselves. When considering the self-excited vibration, the forces produced from the gate motion are important. The first one is the inertia force, namely related to the added mass on the gate, and it can be expressed by the acceleration term. The second is the force by the velocity of gate motion which is related to the water flow and its own motion. The first one can be written as equation (3).

$$f_{y1} = -\rho V C_{my} \frac{\partial^2 y}{\partial t^2}, \quad f_{x1} = -\rho V C_{mx} \frac{\partial^2 x}{\partial t^2} \quad (3)$$

The second one is derived by the momentum relations when the gate motion make the flow change under the small gate opening. When the gate moves y at the gate opening a and discharge q in unit width, the mass change in unit time by this motion can be shown as Eq.4.

$$\Delta M = \frac{\rho q}{a} y \quad (4). \qquad \Delta F = \frac{\rho q}{a} y \frac{dy}{dt} \qquad (5)$$

This change has been derived by the velocity by/dt and it makes the force by the momentum change as shown equation (5). The forces for vertical and horizontal direction are given by multiplying the coefficient Cy and Cx.

$$f_{x2} = C_x \frac{\rho q}{a} y \frac{dy}{dt}, \quad f_{y2} = -C_y \frac{\rho q}{a} y \frac{dy}{dt} \qquad (6)$$

Another momentum force is derived by the motion of gate itself, is written as follows.

$$f_{y3} = -\left[\rho b \frac{dy}{dt} + \rho b_1 \frac{dH_1}{dt}\right]\frac{dy}{dt} = \rho(b + b_1 \sin\alpha)\left(\frac{dy}{dt}\right)^2 \quad (7), \quad f_{x3} = -\left[\rho H_1 \frac{dx}{dt}\right]\frac{dx}{dt} \qquad (8)$$

The equation (7) has the two terms which show the effect of mass change by gate motion. The first one is derived from the gate motion itself but the second one is derived from the change of water level by the gate motion. This is determined by the curve between discharge and water depth as shown Fig.5 , and the angle of α is defined as shown in same figure by dH_1/da. The total external forces are written as equation (9) from the equations (3) to (8).

$$f_{y3} = -\rho V C_{my} \frac{\partial^2 y}{\partial t^2} + \left[\rho(b + b_1 \sin\alpha)\left(\frac{dy}{dt}\right) - C_y \frac{\rho q}{a} y\right]\left(\frac{dy}{dt}\right),$$

$$f_{x3} = -\rho V C_{mx} \frac{\partial^2 x}{\partial t^2} + C_x \frac{\rho q}{a} y \frac{dy}{dt} - \left[\rho H_1 \frac{dx}{dt}\right]\frac{dx}{dt} \qquad (9)$$

UNSTEADY CONDITION FOR THE VIBRATION

The unsteady condition can be determined by whether the sum of the works in one cycle of vibration are plus or minus. Now gate motion gives as equation (10), and its frequency is nearly same as natural frequency of vertical bending motion of gate vibration system. And also the frequency of horizontal motion is assumed to be same as vertical one in this self-excited vibration from the results of model test. When the total works of this vibration is plus, the energy is supplied to the gate vibration system continuously, and the vibration becomes larger. The works of external forces of vibration for the one cycle of motion for each direction are derived as equation (11) and (12) respectively.

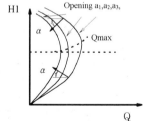

Fig.5 Limiti condition and angle α

$$y = a_y \sin\omega t \sin\left(\frac{2\pi z}{L}\right), \quad x = a_x \sin(\omega t - \varphi)\sin\left(\frac{2\pi z}{L}\right), \quad \omega = \omega_{yn} = \omega_{xn} \qquad (10)$$

$$\frac{2a_y}{3\pi}\left[-2\rho\left(b+b_1\sin\alpha\right)-C_y\frac{\rho q}{a\omega}\right]>\frac{\left(m+m_y'\right)\gamma_y}{\omega}\ ,\ 2\gamma_y=\sqrt{\frac{R}{m+m_y'}}\ ,m_y'=\rho\,C_{my}V \qquad (11)$$

$$\frac{2a_x}{3\pi}\left[-2\rho\,H_1+C_x\frac{\rho q}{a\omega}\left(\frac{a_y}{a_x}\right)^2\cos\varphi\right]+\frac{2\rho\,gH_1}{\omega^2}\left(\frac{a_y}{a_x}\right)\sin\varphi>\frac{\left(m+m_x'\right)\gamma_x}{\omega}\ , \qquad (12)$$

$$2\gamma_x=\sqrt{\frac{R}{m+m_x'}}\ ,m_x'=\rho\,C_{mx}V \qquad (13)$$

The most effective terms can be selected by applying the dimensions of model gate for these equations. Its result shows that the first term in equation (11) and third term in equation (12) are most effective for the self-excited vibration. So finally the unsteady conditions are given as follows by using the damping ratio h.

$$\sin\alpha<-h\frac{3\pi\left(m+m'_y\right)}{4\rho\,a_y b_1}-\frac{b}{b_1}\ ,\ \frac{2\rho\,gH_1}{\left(m+m'_x\right)\omega^2}\frac{a_y}{a_x}\sin\varphi>h\ ,\quad h=\frac{\gamma_y}{\omega} \qquad (14)$$

The first equation of (14) gives the unsteady condition for the vertical motion, and it gives $\sin\alpha=\dfrac{dH_1}{da}<0$. This means that the zone over the limit condition of maximum discharge in Fig.5 is corresponding to this equation. The second equation gives the condition for horizontal motion. It is related to many factors such as the vertical motion such as amplitude a_y and phase $\sin\varphi$, the horizontal motion such as amplitude a_x and angular frequency ω and water depth of upstream H_1. The following conditions give the unsteady condition for horizontal motion: the values a_y and $\sin\varphi$ are larger in first term, the values a_x and ω are smaller in second term and the value H_1 is larger in third term. This means that the horizontal vibration strongly depends on the vertical vibration.

COMPARISONS BETWEEN THE THEORY AND RESULTS OF EXPERIMENTS

The comparison between the result of theory and experimental results is shown in Fig.6 in the case of the ratio of deflection and span length is 1/800. The curve in this figure is determined by the limit condition of maximum discharge condition taking account of the deflection of beam. In this figure the black points are the cases of the self-excited vibration arise and the circle points are the case of non-vibrations. This result shows the good correlation between the theory and the

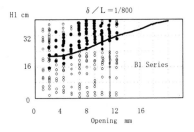

Fig.6 Comparison between experiments and theory

experimental results.

CONCLUSION AND ACKNOWLEDGMENT

The unsteady condition of vibration in the bending motion of long span roller gate, can be derived by theoretical analysis and its results gives good correlation between the results of model tests. This phenomenon has been observed and measured by several methods such as the measurements by accelerometer and pressure gauge, and the record of video cameras both in model test and in fields test. The results from these measurements give the good correlation to the results from the theory. This work has been done under the project research of Japan Hydraulic Gate and Penstock Association (JHGPA) since 1990 with the corporations of Ministry of International Trade and Industry, Ministry of Construction, and Ministry of Agriculture, Forestry and Fisheries. The results are taking account to the operation system and design works for the long span shell roller gates in futures. The authors make the appreciation for the members of three Ministries and the members of JHGPA to the corporations.

REFERENCES

(1) K. Ogihara, H. Nakagawa and S. Ueda, Self-excited vibration of long span shell roller gate by three dimensional experimental model, 24th IAHR Congress. pp. D433-D439, 1992, 9

(2) K. Ogihara, H. Nakagawa and S. Ueda, Unsteady condition for the self-excited vibration of long span shell roller gate, The research reports of Toyo University, vol.28, pp63-68, 1992,   in Japanese

(3) K. Ogihara, H. Nakagawa and S. Ueda, Theoretical analysis on the conditions of self-excited vibration in shell roller gate, 9th APD-IAHR Congress. vol.2, pp.125-131, August 24-26, 1994

(4) K. Ogihara, H. Nakagawa and S. Ueda, Theoretical model for self-excited vibration of shell type roller gate, Transaction of JSCE, No.503/II-29, pp69-78, 1994, 11, in Japanese

1D19

RUN OFF CHARACTERISTICS OF A CATCHMENT IN THE SOUTH-WESTERN PART OF SRI LANKA

DR. SHAHANE DE COSTA
Penta - Ocean Construction Co., Ltd.,   Sri Lanka

ASSOC. PROF. Y. TSUNEMATSU AND DR. T. MISHIMA
University of Hiroshima,   Japan

(1) INTRODUCTION   - The yeilding pattern of water from season to season and its variation within a river system from upstream to downstream was looked into for a Sri Lanakan catchment, namely the Kaluganga basin $(2,600 km^2)$, in order to forecast the discharge. Kaluganga, begins from the central hilly areas (1500m M.S.L.) of Sri Lanka and flows to the south - western coast.

(2) HYDROLOGICAL DATA   - The gauging points being Dela, Ratnapura, Kukulegama ,Ellagawa and Putupaula from upstream to downstream of the Kaluganga river system are as in FIG -1. Six sets of data, namely every other month from the year 1988 was selected from all five gauging points for analysis.

(3) METHOD OF ANALYSIS   - The discharge was seperated into two components, namely ground water flow and subsurface flow using the filter separation method. The unit graphs of each component for all six sets of data were obtained using the AR method. From the unit graphs obtained it was observed that, while there is no appreciable variation from upstream to downstream of the river system, that there exist a clear pattern in the variation of the unit graphs from season to season. From this it was concluded that the catchment consists of two types of unit graphs as in FIG-2. This is due to the storage condition of the soil strata and reflects the temporal variation of yeildability.

(4) FORECASTING OF DISCHARGE   - The rainfall is separate to two components using eq. (1) . The discharge is calculated using eq. (2) and the above mentioned unit graph pattern.
$XX (N) = XN (N) * RR*XL (N)$,   For subsurface flow.   (1)
$XX (N) = XN (N) * RR* (1 - XL(N)$,   For ground water flow.

XN : rainfall (mm/h), RR : runoff ratio.
XL $= Y * XN^\beta / (1.0 mm/h)$,
  $Y = 0.5$, $\beta = 1$, $0 < XL < 0.95$
YY (N) $=$ h(N) * XX (N-K) , h(N) : unit graph values. (2)

FIG - 3 indicates the calculated and observed discharge for the year 1988 for Dela. As observed from FIG-4 the degree of accuracy of the calculated hydrograph is comparatively high. This model was applied for the years 1986 and 1984, and the same results were obtained.

(5) CONCLUSION - A model is developed to forecast the discharge of the Kaluganga basin for long time periods with sufficient accuracy. It was concluded that this basin has 2 distinct yeilding patterns which correspond to the month in question. Therefore the variation of yeildability must be incorporated in runoff forecasting.

REFERENCE : (1) Irrigation Dept. Sri Lanka, Kaluganga river basin.
      (2) Meteorological Dept. Sri Lanka, Rainfall data of North
        - Western province.

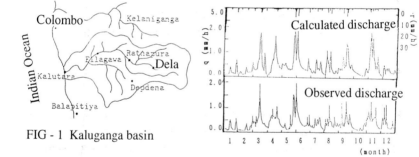

FIG - 1 Kaluganga basin

FIG - 3 Observed and Calculated discharge

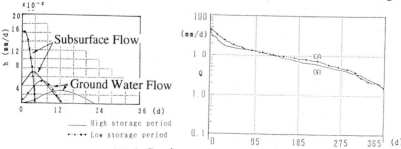

FIG - 2 Two types of Unit Graphs FIG - 4 Observed and Calculated discharge
                  in decending order.

1D20

TO PROBLEM OF BOUNDARY CONDITIONS
ON SPATIALLY CURVED SOLID SURFACE

VYSOTSKY L.I.
Doctor of Sciences, Saratov technical university
Saratov, Russia

SUMMARY
In the report results received while formulating boundary conditions on
spatially curved solid surfaces using specialized non-orthogonal curvelinear
coordinate system [1] are presented.

THE INITIAL EQUATIONS
Basing on three-dimentional Navier-Stokes tensor equations (in covariant
components):

$$-\frac{\partial}{\partial q_i}(\Pi_{(m)} + P) + \nu g^{km}(u_{i;\,m})_{;\,k} = \frac{\partial u_i}{\partial t} + u^k u_{i;\,k} \tag{1}$$

in l, y_1, z^* coordinate system, entered in [1], the expressions for covariant
(and contravariant) components of vorticity, angular deformation speeds
, tangential stresses, "viscous" terms etc. were received. It is convenient
to use the results at decision of spacial problems in coordinates (ω, ψ) and
other cases. Some of received results are adduced below.

THE BOUNDARY CONDITIONS OF SOME FLOW PARAMETERS
COVARIANT COMPONENTS OF VORTICITY

$$(\omega_l)_S = 0; \qquad (\omega_{y_1})_S = \frac{1}{2}\left(\frac{\partial u}{\partial z}\frac{1}{\cos\beta}\right)_S \tag{2}$$

$$(\omega_Z)_S = -\frac{1}{2}\left(\frac{\partial u}{\partial y_1}\cos\beta\right)_S. \tag{3}$$

COVARIANT COMPONENTS OF VISCOUS TERM $\vec{T}$:

$$(T_l)_S = \nu \left[\frac{\partial^2 u}{\partial z^2} \frac{1}{\cos^2\beta} + \frac{\partial^2 u}{\partial y_1^2} - \frac{\partial^2 u}{\partial l \partial z} \frac{2tg\beta}{\cos\beta} - \frac{\partial u}{\partial y_1} \frac{1}{\rho} - \frac{\partial u}{\partial z} \left(\frac{1}{Rcos^3\beta} + \frac{tg\beta}{\rho y_1} \right) \right]_S ;$$

$$(4)$$

$$(T_{y_1})_S = 2\nu \left[\frac{\partial u}{\partial y_1} \frac{\cos\beta}{\rho y_1} + \frac{\partial u}{\partial z} \left(\frac{1}{\cos\beta} \frac{\partial \alpha}{\partial z} - \frac{\sin\beta}{\rho} \right) \right]_S ; \qquad (5)$$

$$(T_z)_S = \nu \left\{ \frac{\partial^2 u}{\partial z^2} \frac{tg\beta}{\cos\beta} + \frac{\partial^2 u}{\partial y_1^2} \sin\beta - \frac{\partial^2 u}{\partial l \partial z} 2tg^2\beta + \frac{\partial u}{\partial y_1} \left(2\cos\beta \frac{\partial \beta}{\partial y_1} - \frac{\sin\beta}{\rho} \right) + \right.$$

$$\left. + \frac{\partial u}{\partial z} \left[\frac{2}{\cos\beta} \frac{\partial \beta}{\partial z} - \frac{tg\beta \sin\beta}{\rho y_1} - \frac{tg\beta(2\cos^2\beta + 1)}{Rcos^3\beta} \right] \right\}_S . \qquad (6)$$

SPEED DEFORMATION TENSOR

$$(e^{ll})_S = -\left(\frac{tg\beta}{\cos\beta} \frac{\partial u}{\partial z} \right)_S ; (e^{ly_1})_S = \frac{1}{2} \left(\frac{\partial u}{\partial y_1} \right)_S ; (e^{lz})_S = \frac{1}{2} \left(\frac{1}{\cos^2\beta} \frac{\partial u}{\partial z} \right)_S ;$$

$$(7)$$

$$(e^{y_1 y_1})_S = (e^{zz})_S = (e^{y_1 z})_S = 0; \qquad (\tau^{ij})_S = 2\mu(e^{ij})_S. \qquad (8)$$

BERNOULLI EQUATION FOR STREAMLINE ON SOLID WALL:

$$\Pi_{(m)} + P + \int_{l_0}^{l} \left(T_l \right)_S dl = \text{constant}, \qquad (9)$$

where $\left(T_l \right)_S$ is defined by (4).

NOTATIONS

e_{ij}, ω_i, T_i are covariant components of angular deformation speed, vorticity and viscous term $\vec{T} = \nu \Delta \vec{u}$; l, y_1, z^* is curvilinear coordinates system; $\Pi_{(m)}$ is mass force potential; ν is kinematic viscosity; P is pressure function; g^{km} is contravariant metric coefficients.

REFERENCES.

1. Vysotsky L.I. Supercritical flows on the spillways controlling. Moscow.: Energoatomizdat. 1991. 240 p.

1D21

Problèmes dynamiques dans le projet et le fonctionement des usines hydroélectriques et des stations de pompage

Prof. M. POPESCU et Prof. D. ARSENIE

Faculté d'Ingéniérie , Université de Constantza , Bd. Mamaia Nr 124,8700
Constantza , Roumanie

Cette communication s'appuie sur l'expérience de plus de 25 années , dans la réalisation de certaines recherches théoriques et études hydrauliques effectives afin de faire le projet d'un grand nombre d'usines hydroélectriques et stations de pompage de Roumanie à référence directe aux suivants problèmes dynamiques : *le coup de bélier , stabilité hydraulique et résonance hydraulique* .

1. En partant d'un résultat obtenu par les auteurs dans le cadre du laboratoire d'hydraulique de l'Institut de Recherches Hydrotechniques de Bucarest , en ce qui concerne la conception et la réalisation d'un nouveau type de dispositif , constitué par des pièces fixes , capable de réaliser la résistance hydraulique asymétrique à un haut degré d'asymétrie , on a conçu un système efficace de protection au *coup de bélier* des stations de pompage du type "réservoir d'air prévu à un dispositif spécial de résistance asymétrique , qui est capable de bien réduire , ou dans certaines conditions d'éliminer même complétement les surpressions causées par les régimes hydrauliques transitoires du système hydraulique en question .

Ce nouveau système de protection au coup de bélier a été appliqué pour le projet de plus de 20 stations de pompage de Roumanie , qui ont été réalisées et présentent un bon comportement dans l'exploitation [2] , [4] .

2. Les études pour la *stabilité hydraulique* de fonctionnement de certaines usines hydroélectriques sont parmi les plus anciennes dans le domaine de l'ingéniérie hydraulique .

De nouveaux éléments ont été générés par l'utilisation à grande échelle dans la pratique d'ingéniérie de certains schémas hydrauliques complexes , [3] .
Dans telles situations les méthodes classiques d'analyse se combinent avec des traitements qualitatifs qui s'appuient sur la fonction Liapunov , en même temps avec de procédures numériques et études expérimentales , [4] , [6] .

Des études théoriques plus récentes ont indiqué qu'aux certaines usines hydroélectriques , dans des conditions de fonctionnement aux pouvoirs réduits , où les pertes de charge hydrauliques peuvent être considérées linéaires , le manque de stabilité surgit comme une règle et on doit le combattre par la surveillance dans l'exploitation et l'évitement de telles situations , [4] , [5] .

3. L'expérience de certains systémes hydrauliques appartenant aux usines hydroélectriques et aux stations de pompage a mis en evidence la présence de certains phénomènes dangereux , parmi lesquels sur premier plan se situe la *résonance hydraulique* .

L'étude par modèles mathématiques des phénomènes de résonance a démontré que certains systèmes sont plus exposés à de tels comportements .

Ayant comme point de départ ce genre d'analyses , on a introduit dans ce rapport la notion de *sensibilité à la résonance* et on a recommandé que outre les études théoriques et les simulations numériques , il faut introduire dans l'exploitation l'appareillage de surveillance des régimes de pressions , dans les systémes estimés comme sensibles , [1] , [4] .

Bibliographie

1. Popescu M. , Halanay A. :
Resonance sensitivity of hydropower and pumping station , Water Power and Dam Construction, England , No.9 , 1984 .
2. Popescu M.
Highly efficient protection systems against waterhammer for pumping stations , International Conference on the hydraulics of pumping stations , Manchester , England , 17-19 sept. , 1985 .
3. Popescu M.
Computation of hydraulic transients in complex hydro-schemes , Water Power and Dam Construction , England , No. 9 , 1986 .
4. Popescu M. , Arsenie D.
Méthodes de calcul hydraulique pour des usines hydroélectriques et des stations de pompage , Edition Technique , Bucarest , 1987 (en roumain) .
5. Halanay A. , Popescu M.
Une propriété arithmétiques dans l'analyse du comportement d'un système hydraulique comprenant une chambre d'équilibre avec éntranglement , C.R.Acad. Sci.Paris , t 305 , Sérié II , p1227-1230 , 1987 .
6. Popescu M. , Stanuca A.
The high-head power plant Lotru in Roumanie . Hydraulic transients , L'Energia Electtrica , Italy, No. 7-8 , 1988 .

1D22

Some pecularities of transporting capacity of the non-uniform flow and their application to the river-bed formation problems .

Sanoyan V.G., E.T.Djrbashian, H.V.Tokmajian
Armenian Research Institute of Water Problems and Hydraulic Engineering
Yerevan, Armenia

It's irrefutable that in the formation of alluvial river-bed the transporting capacity of the flow has considerable contribution. Existing dependences unfit for the use for river bed flows, cause they apply to the simpliest kind of motion - uniform, while river-bed flow is essentially non-uniform and non-stationary.

As a result of the oretical and experimental researches held in the Armenian Scientific-Research Institute of Water Problems and Hydraulic Engineering a regularity in the change of the transporting capa-city of the fiow in the general case of motion was possible to establish. In accordance to this regularity some of the accepted conceptions about alluvium transportation are rejected and with its help a number of formerly unknown and practically important features of the suspention-carrying flow are discovered, which allow to substantiate existence ofdifferent kinds of river-bed forms. The mentioned regularity jointly with other single-dimentional equations derived by us alow to get quantitative solutions of the problems connected with forecasting of the influence of the hydrotechnical constructions on deformation of the river-bed.

The problems connected with forecasting of the stable forms of the alluvial river-beds (in statistical meaning) are of special interest.

The main difficulty in this case is that the boundary conditions of the problem are unknown i.e. they are searchable themselves. That's why solution of this problem by only hydraulical approach is impossible. Researches showed good resultitaveness of combination of two approaches: hydraulical and morphological. It allowed to substantiate the curvative movement of the river flows in the stabilized river-beds and pull out a new conception which can explain cause of meandering and delta formation of river-beds. It is proved that:

- At free motion of the flow in the washed away enviroment after long interaction between the flow and the river-bed the latter from alternative forms gets such a form which has the greatest resistance to the flow (i.e. the principle of minimal dissipation of energy is rejected).
- Speed of energy dissipation is the same in all the sections of the flow.

To the shown conditions satisfy flows when meandering takes place, the flow divides into branches and also at ribbon-bed collateral types of river-bed forms. In the latter two cases the quantity of suspended alluviums and the slope of the location should be small.

Analys of the achieved results shows that the axis of the meandering flow represents by itself a periodical curve like sinusoid and the river-bed consists of located round that curve and following each other periodically repeated expanding and narrowing elements. Expanding areas are almost rectilinear with little slope of the bed; depth and velocity of the flow here is also small. The narrowing areas, on the contrary, are more curvative and have greater slope of the bed. Depth and velocity of the flow are also great.

Above stated ideas have been rendered when defining river-bed's parameters (width of the river-bed, depth of the flow, stride and amplitude of the meandres) of rivers falling into lake Sevan (Armenia). As a result formulas are deduced which express dependence of the mentioned parameters upon liquid and solid expenditures of the flow and average-sized alluvium carried by the flow. Comparison of the results of the calculations with natural data gave satisfactory results.

Index of authors

Good cites "overwhelming" evidence that "the Empire's problems were not ones of economic failure but of economic success."[11]

By 1901, when Schumpeter began his legal studies at the University of Vienna, it had become one of Europe's great research centers in mathematics, medicine, psychology, physics, philosophy, and economics faculties. While German economics was dominated by a "historical school"—led by Gustav Schmoller at the University of Berlin—that despised abstraction and worshipped the imperial state, Carl Menger had established Vienna as the ideological and intellectual antipode of Berlin and the Continental leader in theoretical economics.

Law occupied a loftier status and connoted a more liberal education in German-speaking universities than in English or American ones. Along with canon and Roman law, Schumpeter took courses in history, philosophy, and economics. Schumpeter soon decided that economics, especially theoretical, interested him more than the law. Menger was now too old and infirm to lecture, but the intellectual battle he had long waged against the historical school was now led by two brilliant disciples, Eugen von Böhm-Bawerk and Friedrich von Wieser. Schumpeter attended their seminars where he distinguished himself from older classmates such as Ludwig von Mises, a prominent liberal; and Otto Bauer and Rudolf Hilferding, two of Europe's leading Marxists, by his "cool, scientific detachment" and "playful" manner.[12] In his final year at university, the twenty-two-year-old succeeded in publishing no fewer than three articles in Böhm-Bawerk's statistical monthly. By the time he was awarded his doctor of laws degree in early 1906, he had identified himself as a firm supporter of modern economy theory, or "English economics" as it was known in Berlin despite its well-known Austrian, French, and American contributors. His first postgraduate publication was a long and provocative essay, "On the Mathematical Method in Theoretical Economics."

Having declared his colors, so to speak, Schumpeter embarked on the intellectual "grand tour" so beloved by university graduates in the German-speaking world. Nursing an unspoken ambition to reconcile the warring schools of economic thought and perhaps eventually to breach the Continent's most important university, he spent the spring term at the University of Berlin, getting to know the chief representatives of the German Historical School. In the summer, he spent several weeks in Paris, where he heard the mathematician Henri Poincaré lecture on physics. His ultimate destination was England, the country he admired as "the apotheosis of the civilization of capitalism" and whose economists he had studied exhaustively.[13]

Arriving in London in early fall, Schumpeter proceeded to live the strangely double life for which his education had prepared him. His public persona was that of a gregarious, slightly flamboyant, pleasure-loving Continental aristocrat. Finding English manners, customs, and institutions "completely congenial," he affected the routines of London's fashionable set. He rented a flat on Princes Square, near Hyde Park. He ordered his suits on Savile Row. He kept his own hunter for daily rides on Rotten Row. He spent evenings at plays and dinners, weekends at house parties in the country.

His other, equally elegant persona divided his waking hours between the austere and deliberately plebeian precincts of the London School of Economics, and the British Museum's hushed, high-ceilinged reading room, where he made a point of working at the same table where the overweight, sloppily attired Karl Marx had composed *Das Kapital*. Convinced that truly original thinkers had their best ideas before turning thirty, and intent on reaching the first rung of his projected academic career as quickly as possible, the twenty-four year old Schumpeter was racing against a self-imposed deadline.

Before leaving Vienna, he had outlined two books that he intended to write. In the first he would introduce "English" or theoretical economics to a hostile, ill-informed German audience. The second would be reserved for a groundbreaking contribution that he confidently expected to revolutionize economic theory. Like most intellectuals of his generation, Schumpeter was fascinated by the implications for society of Darwin's theory of natural selection. Was it not ironic, he thought, that while constant change was a hallmark of modern times, economic theory ignored the process that was making the economy more productive, specialized, and complex. Economic evolution was elusive, like "certain natural processes" to which Marcel Proust would allude in *Swann's Way* that were "so gradual that . . . even if we are able to distinguish successively, each of the different states, we are still spared the actual sensation of change."[14] Economists made do with the assumption that the economy merely cloned itself year after year, becoming marginally bigger over time, but remaining essentially unchanged in all other particulars. True, if the economist wished to analyze how a small change in one economic variable affected all the others, "static" theory fit reality like a well-tailored suit. But existing theory fit badly or not at all if the change in question was large, or the timeframe too long to safely ignore structural changes in technology, the labor force, or institutions. And, contrary to the claims of German economists, economic history could do no better. Science, unlike history, was general. History concerned itself with what did happen, science with what could or could not happen under specific circumstances. That's what made science an instrument of mastery. If economics was to be science it too had to be general.

What was needed was a theory of economic development and the new university graduate intended to produce it. Schumpeter's ambition was to replace static with dynamic economic theory, just as Darwin had swept aside traditional with evolutionary biology. As he would observe years later, he thought that his idea was "exactly the same as the idea . . . of Karl Marx," who had also had a "vision of economic evolution as a distinct process generated by the economic system itself."[15]

On at least one occasion, Schumpeter took the train to Cambridge to seek Alfred Marshall's advice. Now sixty-five and in poor health, Marshall was still recovering from a confrontation over Britain's free trade policy with Joseph Chamberlain, the present colonial secretary, and was on the verge of relinquishing his Cambridge chair. Nonetheless, he gave the brash young man breakfast at Balliol Croft and listened tolerantly as he described his plans for constructing a theory of economic evolution.

As Schumpeter was well aware, such a theory had been one of the older man's unfulfilled dreams. Although Marshall had borrowed tools from physics to analyze the interplay of supply and demand in individual markets, he had always insisted that economic phenomena more closely resembled biological than mechanical processes, and criticized earlier economists for assuming that institutions, technology, and human behavior were fixed. Indeed, he introduced the latest edition of his *Principles of Economics* with the claim that "The Mecca of an economist lies in economic biology."[16] Nonetheless, Marshall had stopped short of developing a theory of economic development as Schumpeter proposed to do. Evidently the English Oracle expressed some skepticism in the course of the hour-long conversation because Schumpeter told him in parting that their exchange had cast him as "an indiscreet lover bent on an adventurous marriage and you a benevolent old uncle trying to persuade me to desist." Marshall had replied good-humoredly, "And this is as it should be. For if there is anything to it, the uncle will preach in vain."[17]

Possibly Schumpeter had been hinting that he was about to embark on adventure of a more personal nature. He was having an affair with a woman twelve years his senior. Gladys Ricarde-Seaver was English, upper class, and "stunningly beautiful," the daughter of a "high Anglican church official" who had grown up in a spacious villa in the shadow of St. Peter's Cathedral near Harrow. Although biographers have been able to agree on very little else about her, including her age, public records strongly suggest that she was one of Webb's "glorious spinsters[s]" living the "British museum life." Thirty-six years old and never married, Ricarde-Seaver probably met Schumpeter at the LSE, which catered to women like her who were interested in feminism, social reform, and the popular Fabian cause of eugenics. The decision to marry was sudden, and neither bride nor groom seems to have anticipated parental approval, if indeed they notified their parents in advance. Gladys's brother was the only witness at the civil ceremony in Piccadilly. While Anglophilia and the prospect of an aristocratic alliance might have accounted for Schumpeter's impulsive decision, a pregnancy scare would seem to be a plausible motive. Later Schumpeter hinted to friends that Ricarde-Seaver had taken advantage of his youthful naïveté and when Ricarde-Seaver died in 1933 she left what was by then a considerable estate to a birth control society.[18]

By breaking his private rule against wedlock before the end of the "sacred decade," Schumpeter had forced the issue of how he would support himself. A fragment of a novel found among his papers after his death revolves around an Austrian aristocrat who marries "an English girl with a great pedigree and absolutely no money," a clue that Gladys's income was, at least at the time, too modest to support them both.[19] Obtaining a professorship in Austria would necessarily be a tortuous and uncertain affair. He briefly toyed with the idea of joining the London bar, but that too would have taken years.

This was the age in which enterprising young men with expensive tastes, limited incomes, and brides to support went east to make their fortunes. Possibly Ricarde-Seaver suggested that Cairo might offer more moneymaking opportunities than London or Vienna for someone with legal training but no experience. The Englishwoman in Schumpeter's unfinished novel "had connections which she resolutely exploited for her darling," and various Ricarde-Seavers were involved in large-scale business ventures from North America to North Africa. One uncle, for example, had been a close associate of Cecil Rhodes and the first prominent railroad engineer to support Rhodes's scheme for building a "Cape to Cairo" transcontinental railroad.

However the decision was made, the newlyweds wasted no time after exchanging their vows and headed south to Egypt with the swallows of winter.

Travel let Edwardians see that currents of change were unsettling the whole world. In a rapidly shrinking globe, even ancient civilizations such as Egypt's were not immune. For someone who had hitherto thought of economic development as a European phenomenon, Egypt was bound to challenge the notion, not so much of limits to growth, but of limits on *who* could grow. Had Schumpeter not gone to Cairo he might very well have merited the economic historian W. W. Rostow's unfair appellation as "a rather parochial economist of the advanced industrial world."[20]

Hard as it is to imagine today, Egypt was the China of the turn of the twentieth century. Anthony Trollope had visited Cairo on post office business in 1859. In *The Bertrams,* composed on the way home, he commented drily:

> Men and women, or I should rather say ladies and gentlemen, used long
> ago, when they gave signs of weakness about the chest, to be sent to the
> south of Devonshire; after that, Madeira came into fashion; but now they
> are all dispatched to Grand Cairo. Cairo has grown to be so near home
> that it will soon cease to be beneficial.[21]

Napoléon I's slaughter of the Mamluks in 1798 began the conquest of Egypt by the West, but Egypt's transformation from an Ottoman fiefdom into a British dependency was mostly the work of entrepreneurs, bankers, and lawyers in the second half of the 19th century.

The American Civil War and the resulting cotton famine turned Cairo into a Klondike on the Nile. Egypt's ruler, the khedive Ismail Pasha, seized the opportunity to turn the whole country into a giant state-owned cotton plantation. As British trade with India grew, he saw a way to exploit that too, hence, the construction of the Suez Canal. Colossal amounts of foreign capital, mostly in the form of loans, flowed into Egypt. For Rosa Luxemburg, the Polish revolutionary, Egypt was a microcosm of the "madness" of modern imperialism:

> One loan followed hard on the other, the interest on old loans was de-
> frayed by new loans, and capital borrowed from the British and French
> paid for the large orders placed with British and French industrial
> capital. While the whole of Europe sighed and shrugged its shoulders at
> Ismail's crazy economy, European capital was in fact doing business in
> Egypt on a unique and fantastic scale—an incredible modern version
> of the biblical legend about the fat kine which remains unparalleled in
> capitalist history.[22]

Inevitably, the debts piled up to complete Suez, and a host of other grandiose projects proved unsustainable. Within six years, the khedive was bankrupt, forced to sell his 44 percent stake in the canal and obliged to let his government be placed in what was essentially receivership. Had he invested more cautiously and avoided debt, some historians speculate, Egypt might have entered in the twentieth century as another, if smaller, Japan.

A period of de facto British rule commenced in 1883. Evelyn Baring, the 1st Earl of Cromer, scion of the banking family and one of the great imperialists of the age, was installed as the power behind the khedive's throne. Baring's top priority was restoring Egypt's solvency. He placed British officials at the helm of Egyptian bureaucracies, paid interest on the debt, balanced the budget, and spent what money remained on irrigation and infrastructure. An Anglo-French agreement reached in 1904 extended British rule indefinitely, sparking another, even more spectacular investment boom. Not much bigger than Holland, Egypt attracted as much British capital as India. Within three years, the nominal value of Egyptian equities had grown fivefold, and more than 150 companies with £43 million of capital had been formed. Lord Rathmore, a director of the Bank of Egypt, described the speculative mania that overtook investors: "People were apparently mad; I do not know what other word to use; they seemed to think that every company that came out was worth double its value before it had even started business."[23]

Nonetheless, the flood of foreign capital was transforming Egypt's feudal economy. Historian Niall Ferguson argues that while old empires extracted tribute, modern ones injected capital and promoted economic growth. In 1900 Egyptian manufacturing had consisted of two salt factories, two textile mills, two breweries, and a cigarette factory. Sugar refining had been by far the most important industry, employing 20,000 workers. By 1907, brand-new industries such as ginning and pressing cotton, cottonseed oil, and soap manufacture employed 380,000 workers. Wages rose along with cotton prices and Sultan Hussein Kamel, who had succeeded his father as khedive, marveled at the rapidity with which Egyptians were acquiring European culture: "I have seen in our factories the most intricate machines handled by Egyptians."[24]

Egypt's foreign colony—expats as well as Jews, Copts, and Greeks who had settled there hundreds of years earlier—helped make Egypt "almost the most cosmopolitan country in the world." Cairo swarmed with fortune hunters, bankers, brokers, and entrepreneurs who invested in tourism, railways, banking, sugar, and, of course, cotton. Thomas Cook and Son tamed the Nile and provided English tourists with a "little piece of the West floating along the African river." John Aird & Company completed the Aswan High Dam in 1902. Cecil Rhodes promoted his dream of a transcontinental African railway. Not all the entrepreneurship was for profit. J. P. Morgan, an acknowledged "Egyptomaniac," was only one of several American millionaires—including John D. Rockefeller, the founder of Standard Oil—who was financing archeological digs along the Nile.

Egypt became the poster child for the new imperialism. Speaking to a Liberal Party club in London after his retirement, Baring boasted, "History, so far as I am aware, records no other instance of so sudden a leap from poverty and misery to affluence and material well-being as that which has taken place in Egypt." [25] Over-Baring, as he was known, was, of course, a self-interested promoter. Yet as hostile a critic of British imperialism as Luxemburg did not contradict him.

When William Jennings Bryan, the three-time Democratic presidential candidate, stopped in Cairo on his way back from India in 1906, he found the first sight of the city disconcertingly, even disappointingly modern. Instead of crumbling stones and "picturesque Oriental wonders," he found bright lights, electric trams, automobiles, hydraulic bridges designed by Alexandre-Gustave Eiffel, bottled water, and as many high rises as minarets. It was no harder to buy a cold Bass Ale or a copy of the *Daily Mail* than in New York or London. The business district, with its towering glass and cast-iron department stores, pharaonic luxury hotels, multitude of banks, and telephone and telegraph offices, gave Cairo the appearance of a European city. Its pastel-colored belle époque apartment buildings, wide boulevards, and outdoor cafés reminded Bryan of Paris. [26]

Nile cruises were a particular favorite among affluent honeymoon couples, but Schumpeter arrived in Cairo with more pressing matters on his mind than holding hands with Gladys on the deck of a Cook's steamer. As they made their way to Egypt, by train to Marseilles, steamboat to Alexandria, and again train to Cairo, news of the global financial crisis had followed them. In each capital that experienced a violent stock market collapse, the wave of bank runs and bankruptcies was called by a different name. Many businessmen assumed that the trouble in their community was the worst and that its causes were primarily local. In fact, identical symptoms erupted in half a dozen countries both before and after the panic in New York. The links in a chain that stretched around the globe were popping.

In Cairo the trouble began when Sir Douglas Fox & Partners, the British engineering firm that had built the first leg of Rhodes's transcontinental railroad, had tried to obtain a concession to build a "funicular railway from the base to the summit of the pyramid of Cheops." Perhaps the gods of the underworld took offense, wrote economic historian Alexander Noyes, or else investors saw the proposal as a sign of how crazy speculation had become.[27] In any case, the Egyptian stock market plunged. Brokers and businessmen dismissed the decline as temporary. Within a month a fancy dress ball drew such a huge "rollicking, pleasure-seeking and picturesque crowd" that the dance floor was too crowded for dancing. But in April the market crashed a second time and, this time, kept going down. As the *Economist* reported from London:

> Piles of shares were waiting to be sold, though the market was so satiated with paper that the offer of threescore shares in any security sent down quotations whole points. It was equally difficult at one time to buy. It was well known that a number of small houses were tottering, and when the crisis became most acute, one of these firms suspended payments.[28]

This time full-scale panic erupted. In a matter of weeks, nearly one-quarter of the value of the companies listed on the Cairo exchange had vanished like a mirage. The effect on the real estate boom was immediate. The "great staircase of unsound values" constructed with borrowed money came tumbling down. In May, rumored difficulties at several Cairo banks triggered a bank run. "The total diminution in Egyptian interests, other than Governments, appears to represent about a billion dollars since the completion of the Assouan Dam," reported the *New York Times*'s correspondent grimly.[29] It hardly helped matters that the political situation had suddenly become, in the words of a senior British diplomat, "simply damnable," nationalist agitation having turned "intensely virulent."[30]

Baring and other British officials tried to put the best face on the situation. Repeating the conventional mantra that depressions were the economic equivalent of fasting after a binge, they insisted that the "crisis must in the end be extremely beneficial to Egypt and Egyptian finance, as it will purge the financial arteries of much that is unhealthy."[31] But when credit dried up completely, the Bank of England was forced to send an "instantaneous shipment of $3,000,000 gold." A leading Egyptian expressed familiar regrets when he admitted, "We have been working beyond our means, by using capital which was not ours."[32]

The Egyptian crash was, however, part of a worldwide phenomenon, just as Cairo was a link in a chain that stretched from San Francisco to Santiago, London to Bombay, New York to Hamburg and Tokyo. The chain had been forged not just by ships, railroads, and telegraph cables, but by bills, notes, bank transfers, and gold and the boom that had appeared unique to Cairenes had in fact been almost universal. As a London banker observed after the fact, "beginning about the middle of 1905, a strain on the whole world's capital supply and credit facilities set in, which increased at so portentous a rate during the next two years that long before October, 1907, thoughtful men in many widely separated markets were discussing, with serious apprehension, what was to be the result."[33] The event that triggered the chain reaction had taken place on the far side of the world. Besides practically leveling the city, the great San Francisco earthquake and fire of 1906 resulted in enormous claims to insurance companies in London. As insurers were forced to sell pounds in order to buy dollars to settle them, the pound started to fall in terms of its gold price. To stem the outflow of gold, the Bank of England raised its discount rate to 6 percent in October 1906. The result was a credit squeeze for borrowers.

Under the gold standard, when England sneezed, the United States caught cold. The New York stock market plummeted in March 1907, and in May economic activity began to decline. The recession set the stage for the last and worst banking panic—the panic of 1907—that focused on the New York trusts. The resulting credit freeze forced thousands of banks and businesses around the United States into bankruptcy. A severe economic downturn continued for more than a year, and business conditions did not fully recover until 1910. In England, and on the Continent, the slump was even deeper and longer. In Cairo, on the other hand, the Panic of 1907 was only a pause.

One week after departing from Paddington Station, Schumpeter and his bride were sitting on the elegant terrace of the legendary Shepheard's Hotel overlooking bustling Al Kamel street, flicking flyswatters, listening to "a hundred different propositions from guides and traders,"[34] and imbibing "the peculiar colonial atmosphere of Cairo" along with their drinks.[35] Young and beautiful, they blended perfectly in a cosmopolitan scene where, as the London *Traveler* explained, "Americans, Britishers, Germans, Russians are assorted with Japanese, Indians, Australians, South Africans, well-to-do, well-dressed and handsome specimens of what we call civilization."[36]

The collapse of stock and real estate prices had left a mountain of civil lawsuits in its wake. Schumpeter joined an Italian law firm and was soon representing European businessmen before Egypt's peculiar Mixed Court, a relic of Ottoman administration. The court building overlooked Atabael-Khadra, where all the electric trams converged. Cairo's noisiest square was filled with "the raucous shouting of the pedlars, the rattling of the water-carriers' tiny brass trays, the blowing of motor car trumpets and the ringing of tram bells . . . the uproar heightened by the voices of men and women in passionate controversy."[37]

Schumpeter found that his legal practice, though lucrative, hardly required all his time. Instead of going straight to the country club when he left the court, he often ducked into a favorite coffeehouse—for Cairo, like Vienna, was a city of coffeehouses. These male-only retreats served as chess parlors, offices, literary salons, and, increasingly, headquarters for

Islamic fundamentalists and anti-imperialist conspirators. Sipping Turkish coffee and puffing on a hookah that was passed around in the same manner as in Vienna, Schumpeter composed quickly and neatly, his pen flying over the sheets of paper.

"German economics does not really know what 'pure' economics is all about," the twenty-four-year-old author opined. Schumpeter intended his book to be a plea to critics, especially German economists, "to understand not fight; learn, not criticize; analyze and find what is correct . . . not just simply accept or dismiss" economic theory, and a rebuttal of the view, popular in German universities, that "English" or theoretical economics was a dying discipline.[38] True, "Economics, like mechanics, gives us a stationary system, not like biology, a narrative of evolution."[39] It could shed no light on the dynamic process that had transformed, first Britain, then France, Germany, and Austro-Hungary, and now Egypt. But this gap in economic theory was an argument for constructing a new, dynamic theory, not for abandoning theoretical economics.

Schumpeter closed his book's final chapter on the future of economics by posing two questions: Could one prove the existence of *economic* development in the sense that growth could be traced to *economic* rather than demographic, political, or other external causes? Was it possible to devise a plausible narrative of economic evolution under the assumption that existing social arrangements—capitalism and democracy—would persist? His provisional answer to both was a strong affirmative.

Schumpeter had only just sent his 600-page manuscript off to a publisher in Germany in early March 1908, when the sirocco began blowing from the south and he was stricken with Malta fever, a debilitating, often fatal bacterial infection. The omnipresent dust, furnace-like heat, and the threat of complications convinced him that it was time to return to London. He had achieved his twin objectives in coming to Cairo. Besides finishing his first book, he was now, if not rich, at least back in funds. His law practice had prospered and he had been lucky, having ingratiated himself with one of the khedive's daughters and become her investment manager. Having succeeded in doubling the rental income from her estates, as he later recalled, and overseen the reorganization of a sugar refinery, he had earned a substantial sum.[40] By October 1908, he was back in London recuperating at his brother-in-law's house, plotting his return to Vienna.

He had recovered sufficiently by February 1909 to deliver his "Habilitation" lecture on *The Nature and Essence of Economic Theory* at the University of Vienna. His performance on the podium won rave reviews and the title of Privatdozent. The reviews of his book were more mixed, although even the critics were impressed. To his chagrin, his alma mater did not make him an offer. Instead of a prestigious appointment in one of Europe's great capitals, he had to accept an associate professorship in a remote outpost of the Empire not dissimilar to his place of birth.

Cernowitz was a polyglot town of transients, and its university was new and undistinguished. The inhabitants were divided among German Protestants, Jews who spoke German, and Romanian Catholics, most of whom had arrived relatively recently and many of whom were itching to move on to Vienna, Paris, or New York. Partly because few had deep roots there, no single ethnic or religious group dominated the rest, engaged in proselytizing, or had much impulse to do anything but mind their stores and businesses and take a stroll through the city park on Sunday. Schumpeter expressed his resentment by being unfaithful to Gladys, snubbing his colleagues, and thumbing his nose at decorum. He scandalized the faculty by sauntering into meetings in his jodhpurs. On one occasion, he challenged the university librarian to a duel.

Reflecting on the years he spent in the Bern patent office from 1902 to 1909, Albert Einstein observed that the solitude and monotony of his country life stimulated his "creative mind." He recommended similar periods of enforced isolation to other scholars bent on producing works of genius, by taking up temporary occupation as a lighthouse keeper, for example. It gave one time to think one's thoughts and, of course, to write them down. It also cut down on the distracting buzz of other people's ideas.

Cernowitz proved to be Schumpeter's lighthouse. In the two years he spent there, he distilled what he had absorbed, observed, imagined, and thought between the ages of twenty-four and twenty-six when he was living abroad, and poured the resulting combination into *The Theory of Economic Development.*

For Schumpeter, the process of development not only implied that the economy was getting bigger, but also that its structure was evolving, its workers becoming more productive, its industries more specialized, its financial system more sophisticated. He took for granted that the aim of all production was "the satisfaction of wants,"[41] and that a rising living standard was the result of development. But development was not the "mere growth of population or wealth." A nation with a rapidly growing population could produce more output without delivering higher average wages or consumption. Predatory empires like ancient Egypt's could enrich themselves at the expense of weaker powers without achieving ever higher levels of productivity. New, sparsely populated territories could be opulent without having developed a capacity for specialization or a high degree of interdependence.

A nation's ability to provide its citizens with a high standard of living depended first and foremost on its productive power, which enabled the economy to produce more and more with the same resources, like the porridge pot in the Grimm brothers' fairy tale. Output per worker had doubled or tripled in his own lifetime after more or less stagnating for nearly two thousand years between the births of Christ and Victoria. In the same spirit as Mark Twain had ventured in 1897 that "the world has moved farther ahead since the Queen was born than it moved in all the rest of the two thousand put together,"[42] Schumpeter treated economic development as a fact rather than a theoretical possibility. By contrast, Malthus and Mill had

> lived at the threshold of the most spectacular economic developments ever witnessed. Vast possibilities matured into realities under their very eyes. Nevertheless, they saw nothing but cramped economies, struggling with ever-decreasing success for their daily bread. They were convinced that technological improvement . . . would . . . fail to counteract the fateful law of decreasing returns . . . and that a stationary state . . . was near at hand.[43]

It was no longer possible to dispute, as it had been possible in 1848 or even in 1867, that the living standards of ordinary people had improved. In the rich countries, consumption of food as well as clothing, tobacco, meat, and sugar had risen sharply. Better nutrition was reflected in demographic trends. Infant mortality began to plummet after 1845, life expectancy at birth to rise after 1860, and average height, which fell between 1820 and 1870, to increase after 1870. The twin blights of homelessness and begging began to disappear. "The capitalist process, not by coincidence but by virtue of its mechanism, progressively raises the standard of life of the masses," Schumpeter wrote. Even the normally cautious Alfred Marshall had insisted in 1907 that "The Law of Diminishing Return is almost inoperative . . . just now." [44]

If development were being driven mainly by globalization, as Marx had hypothesized, and local conditions mattered little, average living standards should have become more, not less, similar. But anyone who had recently lived in Cairo as well as in London, Cernowitz, and Vienna had to be struck by the stunning differences in the level and rate of economic development of different countries. In 1820, the average standard of living in the world's richest country—still Holland—was roughly three and a half times that of the poorest nations of Africa and Asia. By 1910, however, the lead of the richest over the poorest had grown to more than eight-fold. [45] These differences in living standards primarily reflected differences in productive power rather than in territory, natural resources, or populations. For any given amount of capital and labor, the most efficient economies could produce many times as much as the least efficient ones. [46] What's more, the productivity of some economies was growing several times as fast as others. So the question was not just what process could increase productive power by multiples in the course of two or three generations, but why the process operated so much faster in some countries than others.

The traditional answer would have been that a nation's development depended on its resources. Schumpeter took the opposite view. What mattered was not what a nation had, but what it did with what it had. He identified three *local* elements of "industrial and commercial life" that drove the process: innovation, entrepreneurs, and credit. The distinctive feature of capitalism, he believed, was "incessant innovation," the famous "perennial gale of creative destruction."[47] Marx too had observed that "The bourgeoisie cannot exist without constantly revolutionizing the instruments of production" but he had had in mind primarily factory automation.[48] Schumpeter took a broader view. "Innovation," by which he meant the profitable application of new ideas rather than invention per se, could involve many types of change, he pointed out: a new product, production process, supply source, market, or type of organization.

Marshall, whose motto was that nature took no leaps, had stressed continual incremental improvements by managers, and skilled workers that accumulated over time.[49] Schumpeter stressed innovative leaps that were dramatic, disruptive, and discontinuous. "Add successively as many mail coaches as you please, you will never get a railway thereby," he insisted, ". . . the essence of economic development consists in a different employment of existing services of labor and land."[50] But new technologies alone could not explain why some economies were developing and others were not, since new machines and methods could be, and were, transferred globally. Marx had explicitly ruled out any role for the individual in his economic drama. Webb once complained that Marx's "automaton owner" was driven by forces over which he had no control, and into which he had no insight, blindly pursuing "profit without even being conscious of the existence of any desire to be satisfied."[51] Schumpeter focused on the human element. For him, development depended primarily on entrepreneurship. He shared the obsession of late nineteenth-century German culture with leadership. Having heard Sidney Webb expound on the Fabian theory of hereditary genius as the cause of unequal incomes, he had also become interested in the work of Francis Galton, Darwin's cousin, and Karl Pearson, a professor at the LSE, on hereditary genius and the role of elites.

The central character in Schumpeter's narrative was the visionary leader. The entrepreneur's function was "to revolutionize the pattern of production by exploiting an invention or, more generally, an untried technological possibility."[52] This might mean new products like cars or telephones, new processes like the cyanide method for extracting South African gold, new organizations such as the trust, new markets such as Egypt for rail cars and cotton ginning machinery, or new sources of supply such as India for cotton. In contrast to Marx's automaton capitalist or Marshall's owner-engineer, the entrepreneur distinguished himself by a willingness to "destroy old patterns of thought and action" and redeploy existing resources in new ways. Innovation meant overcoming obstacles, inertia, and resistance. Exceptional abilities and exceptional men were required. "Carrying out a new plan and acting according to a customary one are things as different as making a road and walking along it," Schumpeter wrote.[53]

His entrepreneurs are motivated less by the love of money than by a dynastic urge—"the drive and will to found a private empire," as well as the urge to dominate, fight, and earn others' respect. Finally there was "the joy of creating, getting things done, or simply exercising one's energy and ingenuity."[54] While Marx had cast the bourgeois as a parasite whose activities would ultimately destroy society, Schumpeter took up and developed Friedrich von Wieser's notion that "growth was the result of the "heroic intervention of individual men [who] appear as leaders toward new economic shores."[55] He never grew tired of pointing out "the creative role of the business class that the majority of the most 'bourgeois' economists so persistently overlooked." Science and technology were not independent forces, he insisted, but "products of the bourgeois culture" just like "business performance itself."[56] Though many did realize great fortunes, entrepreneurs did more to eliminate poverty than any government or charity.

Despite his energy, vision, and domineering nature, the entrepreneur could thrive only in certain environments. Property rights, free trade, and stable currencies were all important, but the key to his survival was cheap and abundant credit. To carry out his plans, Schumpeter argued, the entrepreneur had to divert land, labor, and machines from their existing uses to his projects. His enablers are "bankers and other financial middlemen who mobilize savings, evaluate projects, manage risk, monitor managers, acquire facilities and otherwise redirected resources from old to new channels."[57] True, the financial sector's peculiar dependence on confidence and trust made it vulnerable to panics and crashes. But without well-functioning credit markets and a robust banking system, an economy would be deprived of the low interest rates and abundant credit needed for innovation. What distinguished successful economies was not the absence of crises and slumps, but, as Irving Fisher also stressed, the fact that they more than made up lost ground during investment booms.

The highest interest rates in the world were found in the poorest countries. As the economic historian David Landes writes, "In these 'underdeveloped' countries, where the civilizing rays of capitalism had not yet exercised their mysterious faculties of enlightenment, there were few banks but many money-lenders, little investment but much hoarding, no credit but much usury."[58] In Egypt, entrepreneurs faced huge obstacles because of the backward state of local banking and the primitive facilities of credit and exchange. Interest rates were twice or three times as much as in the West. The best securities paid interest of 12 to 20 percent. The poor peasant, meanwhile, paid 5 to 6 percent a month.

Having shown that economic theory had been designed for a "system essentially without development," Schumpeter succeeded in formulating a new one for a body in motion, while building on existing theory. He showed how an economy could produce more with the same resources while evolving a new, more specialized structure. What's more, his theory implied that any nation could do it. By emphasizing the local business environment instead of natural resources, Schumpeter's theory suggested that nations made their own destinies. Governments that wished to see their citizens prosper should give up territorial ambitions and focus on fostering a favorable business climate—strong property rights, stable prices, free trade, moderate taxes, and consistent regulation—for entrepreneurs at home. There were no intrinsic limits to growth. Human wants were infinite. Rising incomes and new desires provided just as much opportunity for profitable ventures as opening up new territories. As long as trade was possible, innovation could offset the constraints of population, territory, and resources. It was a beguiling, romantic, even a heroic narrative. His was an equal opportunity, optimistic, and, not coincidentally, unwarlike formula for economic success.

Schumpeter completed the manuscript of *The Theory of Economic Development* in May 1911. By then he was back in Vienna, staying at his mother's apartment, and waiting to hear whether he had been chosen to fill an empty chair at the University of Graz. The city, where he had spent part of his childhood, was a pleasant provincial town and its university was not especially distinguished. Nonetheless, it had the advantage of being just two and a half hours by train from the capital. The faculty proved to be anything but amenable, deemed his work "barren, abstract and formalistic," and voted to appoint another candidate. Only intervention at the education ministry by his mentor Böhm-Bawerk succeeded in overturning the decision, fulfilling Schumpeter's ambition to become the empire's youngest full professor at the age of twenty-eight.

When Schumpeter started teaching in Graz in fall of 1911, he got a chilly reception from students, who boycotted his classes, as well as from his new colleagues. But it was nowhere as frigid as that accorded to his magnum opus, which appeared that fall. It was, he later testified, "Met with universal hostility."[59] Even Böhm-Bawerk was sharply critical, so much so that he devoted sixty pages to an attack on Schumpeter's book the following year. More discouraging, his mentor was one of the few economists to review it at all.

When Schumpeter got an invitation to spend the academic year 1913–1914 as the first Austrian exchange professor at Columbia University, he accepted with alacrity. Gladys, however, made it clear that she did not intend to accompany him. Their marriage had soured soon after it had begun, perhaps because an English feminist and Fabian and a Viennese prince were hardly likely to be compatible or simply because both were, at least on Schumpeter's testimony, promiscuous. Accepting that the marriage had been a mistake, he made no effort to dissuade her. In August 1913, he sailed alone from Liverpool on the RMS *Lusitania*, while Gladys resumed her former life in London.

Schumpeter's sabbatical was a great success. He adored New York and was adored in turn by Americans who were charmed by his sparkling conversation and astonished by his personal habits, such as taking an hour every day to complete his toilette. One colleague at Columbia called his inaugural lecture at the university "a remarkable performance . . . and . . . very unusual—both brilliant and profound."[60]

The triumph was crowned by a note from the president of Columbia informing him that the trustees had voted to give him an honorary degree. Lecture invitations from Princeton, Harvard, and other universities poured in. Irving Fisher invited him to New Haven for Thanksgiving. At dinner they talked about the possibility of a European war. Like the English politician Norman Angell, Fisher was convinced that economic integration made the possibility remote. Many nations were now so dependent on foreign capital, he said, that they could no longer afford to misbehave. Schumpeter listened skeptically.

Before leaving the United States, he could not resist touring the country by train as Marshall had done. He did not return to Vienna until August 1914.

Act II

FEAR

War of the Worlds

The world will seek the greatest possible salvage out of the wreck.

—*Irving Fisher, 1918*[1]

"Sidney had refused to believe in the probability of war among the great European powers," Beatrice Webb later commented in the margin next to her original diary entry for July 31, 1914.[2] Investors didn't see the Great War coming either, judging by the stock and bond markets that hovered near all-time highs. Surely war was economic suicide and therefore unthinkable. One week after Germany overran Belgium, George Bernard Shaw predicted in the *New Statesman,* his venture with the Webbs, that the war would be over in a few weeks. As Webb observed in early August, the war seemed like "a terrible nightmare sweeping over all classes, no one able to realize how the disaster came about."[3]

For Webb, the war was "a blank and dismal time." Her political stock had been falling since before 1914 and continued to decline. She hoped that she and Sidney would be given some important role in the Liberal-led wartime coalition government of David Lloyd George. No offer materialized, and the Webbs eventually went off on a somewhat desultory world tour. When Beatrice was finally appointed to a commission to study the gap between women's and men's wages in 1918, she regretted accepting the assignment almost immediately. "I am not the least interested in the subject," she complained. She was too preoccupied by fears of "the kind of world we shall live in when the peace has come."[4]

. . .

A Cambridge don, civil servant, speculator, and art patron, John Maynard Keynes was rather homely and quite rude. He made up for these short-comings with cleverness, a charming voice, and efficiency in practical matters. His best friends, who invariably called him Maynard, were the artists, writers, and critics known collectively as "the Bloomsbury Group." They counted on him to buy their paintings, advise them on real estate, and invest their trust funds—all while they debated whether he was or wasn't a hopeless philistine.

When Britain declared war in August 1914, Bloomsbury immediately agreed with Shaw that war was madness and would "benefit only a few capitalists."[5] Keynes pledged to seek conscientious objector status, but then dismayed his friends by changing his mind. He upset them further by accepting the invitation of the then Chancellor of the Exchequer, David Lloyd George, to serve on the Treasury staff. Keynes argued that while the war was undoubtedly bad, his presence in the government would make it less so.

The Treasury's task was not only to achieve "maximum slaughter for minimum expense" but also to finance the war without debauching the world's safest currency or jeopardizing Britain's supremacy as the world's banker.[6] As the war dragged on, Britain loaned massive sums to her Euro-pean allies and was obliged to borrow even more massive sums from the United States. Because the loans were so colossal, writes Keynes's biographer Robert Skidelsky, the "headache of inter-allied debt . . . became the chief source of irritation, misunderstandings and almost constant squab-bling within the Alliance." Within a few months, Keynes became "the go-to official for inter-Allied (read American) loans."[7] Communication within Whitehall was by memorandum, and Keynes was a wizard with a pen. His energy, confidence, and nerve never seemed to flag.

An episode toward the end of the war captures Keynes's uncanny abil-ity to stay focused on the big picture. In the early spring of 1918, the Ger-mans caught the Allies off guard and smashed through the western front. Tens of thousands of German troops were soon encamped a few miles from the Arc de Triomphe, and Paris was being shelled day and night. The

Big Bertha howitzers were sowing terror as far away as London. If Paris were to fall, anxious Britons reasoned, the Germans could move them to the channel beaches and bombard the southern counties.

Keynes was too tantalized by a suggestion from one of his Bloomsbury friends to indulge in such anxious speculation. The painter and critic Roger Fry had given him a heads-up about an extraordinary collection of modernist paintings that was about to go on sale. Edgar Degas, who had been an art dealer before he devoted himself full-time to painting, had amassed hundreds of works by Manet, Corot, Ingres, Delacroix, and other contemporary artists over his long career, rarely parting with a single canvas. This treasure trove was to be auctioned off at the Galerie Roland in Paris on March 26 and 27.

Recognizing a chance to save a piece of the civilization he loved and for which his country was fighting, Keynes did not hesitate. He immediately contacted Charles Holmes, director of the National Gallery, to urge him to lobby the war cabinet for a £20,000 war chest. Judging that his superiors at the Treasury would react with disapproval to such extravagance at a time of national sacrifice, Robert Skidelsky relates that Keynes recast the scheme as an insurance policy against default: "Under our agreements with the French Treasury we are entitled to set off British government expenditure in France against our loans to them," he began his memorandum to the Chancellor. At that point in the war, France owed Britain such gargantuan sums that the likelihood of collecting interest on the debt, never mind the principal, looked exceedingly remote. How much better, Keynes argued, to collect "priceless pictures than dubious French bonds."[8]

A few days later, Keynes sent the painter Duncan Grant, his former— and Vanessa Bell's current—lover, a triumphant telegram: "Money secured for pictures."[9] Meanwhile, he had contrived to get himself and Holmes invited to an Allied conference in Paris. They crossed the channel escorted by "destroyers and a silver airship watching overhead" and traveled to Paris by train.[10] To avoid tipping off French dealers and inquisitive British reporters, Holmes disguised himself with a false mustache and whiskers, and both he and Keynes used aliases. The ruse was so successful that at the close of the auction two days later, Keynes wrote gleefully to his mother, "I bought myself four pictures and the nation upwards of twenty." [11]

In fact, he came home with one of Cézanne's still lifes of apples and two Delacroix, while Sir Charles Holmes returned to the National Gallery with twenty-seven drawings and paintings, including a Gauguin still life and Manet's *Woman with a Cat.* Prices had slumped under the threat of German occupation, and Keynes was especially pleased that Holmes had had to use only half of his budget. Of his return from France, Vanessa Bell wrote to Roger Fry that "Maynard came back suddenly and unexpectedly late last night having been dropped at the bottom of the lane . . . and said he had left a Cezanne by the roadside! Duncan rushed off to get it." [12]

An Anglophile and constitutional monarchist, Joseph Schumpeter was horrified when Austria and Germany entered the war as allies. When he got a draft notice in December 1914, he promptly applied for a permanent exemption on the grounds that he was the only professor of economics at the University of Graz. According to his biographer Robert Lorie Allen, he hoped for an advisory role in the government and spent as much time as possible in Vienna cultivating politicians of all parties. (He was equally promiscuous in his private life, Gladys having announced her intention of remaining in England for good, even though she had not consented to a formal divorce.) But Schumpeter proved too radical for his own party, the conservative Christian Socials, and too conservative for the Socialists. The longer the war dragged on, the more frustrated he became at being "totally cut off from any possibility of being effective."

Having opposed the war, Schumpeter lobbied the emperor and his advisors to make a separate peace with the Allies—which, in fact, Franz Joseph nearly did—as well as to seek a postwar alliance with England. On the eve of surrender, Schumpeter was waging a personal campaign on two fronts: against the increasingly popular notion of a postwar economic and political union or "Anschluss" with Germany, and against the increasingly fatalist attitude on the part of the Austrian middle class toward the future of democracy and private enterprise in Europe. He spent the final year of the war anticipating the problems that the Austrian government would face after the war.

Six months before the armistice, at a public lecture at the University of Vienna, Schumpeter proposed a blueprint for postwar economic recovery. Like Keynes, he was an optimist. In *The Crisis of the Tax State*, based on his lecture, he argued against the inevitability of Socialism and predicted that the capitalist welfare state, which he called the "*Steuerstaat*" or "tax state," would survive the war. The crisis he foresaw would result not from the triumph of Socialism, he argued, but rather from a gap between the voters' expectations and their willingness to pay taxes. The major challenge for democratic governments would be avoiding chronic budget deficits and inflation.

Even young men who had experienced the triumph of "brutality, cruelty and mendacity" over civilization assumed that civilization would eventually reassert itself.[13] On August 31, 1918, hundreds of soldiers waited on a train platform in the Alpine resort where the emperor had declared war on Serbia four years earlier. A remarkable-looking man—short, taut, radiating nervous energy, with gaunt features, graying hair, and cold blue eyes wearing a K.u.K. uniform—strode through the crowd toward a skinny young corporal. "Aren't you a Hayek?" he asked him. "Aren't you a Wittgenstein?" the other shot back.[14]

The Hayeks and the Wittgensteins were among Vienna's leading families. The former were senior civil servants and academics; the latter, wealthy industrialists and art collectors. Friedrich von Hayek and Ludwig Wittgenstein were cousins, although the latter was old enough to be Hayek's uncle. They had never exchanged more than a few words at family gatherings, but both had volunteered within a few weeks of each other, having welcomed the war partly out of the hope that facing death would make them better men. Both had endured months of semistarvation, grossly inadequate clothing, nonexistent shelter, influenza, malaria, and intensifying ethnic tensions. Both had taken part in the disastrous Piave offensive, the final desperate gesture of the hopeless Austro-Hungarian army. They had witnessed their comrades in arms wading through the mosquito-infested salt-water marshes, rifles held over their heads, until they fell. Unlike one hundred thousand other members of the K.u.K. army, they had survived.

Hayek was anxious to get home to Vienna to learn whether his application to the air force had been accepted. Wittgenstein had taken leave from his unit in order to see a publisher who had expressed an interest in a manuscript he was carrying in his rucksack, the *Tractatus Logico-Philosophicus,* soon to be recognized as one of the most important works of twentieth-century philosophy. When the train for Vienna came, the two men got into the same compartment, and as it rolled eastward through the night, they talked.

Wittgenstein went on and on about Karl Kraus, whose antiwar journal, *Die Fackel,* satirized the mendacious Austrian media and emphasized "the duty of genius" to seek and tell the truth. Hayek was troubled by Wittgenstein's gloom about the future, but at the same time was profoundly impressed by his "radical passion for truthfulness."[15] When they arrived in Vienna, he and Wittgenstein went their separate ways. In another world war, Hayek would fulfill his own duty to the truth by writing *The Road to Serfdom.*

The best and brightest of the generation that had been too young to fight was Frank Ramsey, a protégé of Maynard Keynes. Like Keynes, Ramsey came from an old Cambridge family. His father was the head of a college and his younger brother the future archbishop of Canterbury. An awkward but brilliant great bear of a boy, Ramsey helped to translate Wittgenstein's *Tractatus* when he was sixteen. At age nineteen he wrote a criticism of Keynes's thesis on probability so devastating that Keynes gave up any notion of a mathematic career. He was drafted to revise the *Principia Mathematica,* Bertrand Russell and Alfred North Whitehead's prewar attempt to reduce all of mathematics to a few logical principles.

Ramsey was eleven when war was declared, and the war radicalized him, as it did many of the boys at his school. He upset his headmaster by threatening to switch from mathematics to economics, which he considered more likely to make the world a better place. Instead of specializing exclusively in mathematics, philosophy, or economics, Ramsey contributed original ideas to all three disciplines. He published only two papers in the *Economic Journal* before his tragic death from a botched operation at age twenty-six, but both became classics as Keynes accurately predicted.

A free spirit with a passion for literature, psychoanalysis, and the many women who adored him, Ramsey personified Keynes's attitude that whatever the limitations of formal logic, imaginative solutions could be found for social problems. Even as an undergraduate, he was indifferent to the notion, seemingly proven by the world war, that vast, impersonal forces beyond human control would determine society's future. In a talk delivered at a meeting of the Apostles, the Cambridge secret society to which Keynes and Russell had also belonged as undergraduates, Ramsey announced that the "the vastness of the heavens" did not intimidate him. "The stars may be large, but they cannot think or love; and these are qualities which impress me far more than size does," he said, adding, "My picture of the world is drawn in perspective, and not like a model to scale. The foreground is occupied by human beings and the stars are all as small as three-penny bits." [16]

Appalled by the colossal waste of human life and capital in the war, Irving Fisher redoubled his efforts on behalf of public health and a postwar League of Peace. Between 1914 and 1918, he helped to found the Life Extension Institute, whose purpose was to promote best practices in individual hygiene and so further "the preservation of health . . . and increase vitality" [17]; coauthored a best seller, *How to Live*, on what today would be called wellness; and started campaigning in earnest for alcohol prohibition. Even as he argued for America to enter the war against Germany, he deplored the negative "eugenic" effects of sending the cream of the younger generation to be slaughtered and maimed on the battlefield. He became president of a labor group that lobbied for safety legislation, automatic cost-of-living wage increases, and universal health insurance. Oddly, the war seemed to strengthen rather than weaken his belief in modern science and man's improvability.

Before it was over, though, he suffered a grievous blow that might have caused a less confident man to question whether his certitude was justified. In the late spring of 1918, after months of nagging anxiety, he was forced to confront the agonizing possibility that his twenty-four-year-old daughter Margaret, to whom he was extremely close and who had recently become engaged, might be incurably insane. Shortly after her fiancé received his officer's commission, she began to talk incessantly about strange portents, God and immortality, and her conviction that her fiancé would be killed.[18] When it became obvious that she had begun hearing voices, and her behavior became more and more bizarre, Fisher took her to the Bloomingdale Asylum in upper Manhattan. The diagnosis, dementia praecox, was devastating. Unable to accept that Margaret was unlikely to recover, he tapped his contacts in the medical community in search of a more hopeful prognosis.

He soon discovered Henry Cotton, the medical director of the New Jersey State Hospital at Trenton, who had been reporting extraordinary success in treating schizophrenia. A prominent psychiatrist and medical reformer, Cotton was convinced that mental illnesses were due to "focal infections." What distinguished him from researchers with similar views was his willingness to apply his theory to his patients by aggressive removal of infected teeth, tonsils, colons, and reproductive organs. He claimed that he had completely cured hundreds of hopeless cases since the start of the war.

This is what Fisher, a parent desperate to save his child and a true believer in modern medical miracles, wanted to hear. Elated at having found someone who promised a cure, Fisher had his daughter transferred to Cotton's care in March 1919. When the doctor reported the presence of "pure colon bacillus," Fisher assented to his recommendations for treatment. Cotton had two of Margaret's wisdom teeth removed immediately. When she remained suspicious, delusional, apathetic, and perplexed, he had her cervix removed. Before and after the surgery, she was inoculated repeatedly with her own streptococcus bacilli, the last time in September. Later, Cotton had to admit to a Princeton audience that Patient Number Twenty-four was a "treatment failure." Margaret died of septicemia on November 19, 1919, at the age of twenty-five.[19]

Fisher was devastated. Yet he never questioned the wisdom of Cotton's "treatment" or his conclusion that the origin of Margaret's psychosis, and the indirect cause of her death, was her parents' failure to deal in a timely fashion with her impacted wisdom teeth and her tendency to constipation. Nor was his boundless faith in medical science shaken. If anything, Fisher's campaigning grew more frenetic. He told himself over and over that some good would come out of the twin calamities of Margaret's death and the war. In his mind, the two became inextricably linked. He predicted that society would enter "a period of life conservation" and use science to extend life and improve health: "The war has for a time withdrawn much of the world and destroyed and maimed a large part of that which it has withdrawn," he said. "The world will seek the greatest possible salvage out of the wreck." [20]

The wreck was beyond calculation: 8.5 million dead, 8 million permanently disabled, mostly young men. Ninety percent of the Austro-Hungarian army and nearly three-fourths of the French army were killed, wounded, imprisoned, or missing. "The toll of the war for our family is three killed and four others wounded, two seriously injured, out of a total of seventeen nephews and nephews in law in khaki," Webb recorded. "Every day one meets saddened women, with haggard faces and lethargic movements, and one dare not ask after husband or son." [21]

World War I destroyed globalization, disrupted economic growth, severed physical, financial, and trade links, bankrupted governments and businesses, and led weak or populist regimes to rely on desperate measures that were supposed to head off revolutions but just as often hastened them. When the war was over, the victors as well as the vanquished were crippled by colossal debts and subjected to vicious attacks of inflation and deflation. Poverty, hunger, and disease, those Malthusian scourges, once again seemed to have the upper hand. In London and Paris as well as Berlin and Vienna, citizens of the great capitals of Europe were forced to realize that they and their nations were now a great deal poorer. Virginia Woolf could not stop brooding about the war and its devastating effects. In her novel *The Voyage Out*, published in 1915, a sheltered West End matron discovers that "after all it is the ordinary thing to be poor and that London is a city of innumerable poor people." In *Mrs. Dalloway*, a decade later, "the war was over" except for victims like the suicidal working-class veteran Septimus Smith and the impoverished and crazed Socialist Doris Kilman, who continue to suffer five years on. In *To the Lighthouse*, Mrs. Ramsay and her family are haunted by the threat of tuberculosis, another of the war's legacies.

The war had dealt a blow to the legitimacy of private property, free markets, and democracy while providing an impetus to violent revolutionary movements from Moscow to Munich. "The people are everywhere rejoicing," observed Beatrice Webb anxiously on Armistice Day. "Thrones are everywhere crashing and men of property are secretly trembling."[22] From their respective positions in Austria and England, Joseph Schumpeter and Maynard Keynes tried to convince their countrymen that political healing would depend on economic recovery, as would any dampening of dangerous revolutionary ardor. Reviving the world economy would require the Allies to draw political boundaries that made economic sense, they argued; and, more important, to give up the fantasy that exacting reparations from the losers would make up their own losses. Both men pleaded for stabilizing national currencies, restoring the flow of credit, and eliminating trade barriers.

Bertrand Russell, the philosopher, was among the many Western intellectuals who were convinced that "The Great War showed that something is wrong with our civilization."[23] His initial reaction to news of the Bolshevik Revolution was cautious optimism. He was prepared to believe that if not the promised land, Soviet Russia was at least a grand futurist experiment. But unlike many others who let themselves be swept away by their own hopes and fears, he made up his mind to reserve judgment until he had a chance to examine the new society that the revolutionaries claimed they were building—firsthand.

Chapter VI

The Last Days of Mankind:
Schumpeter in Vienna

The hour of socialism has not yet struck.

—*Joseph Schumpeter, 1918*[1]

If Austria was a pathetic ruin . . . there was plenty of material,
I thought, with which to rebuild the ruins.

—*Francis Oppenheimer, British Treasury Representative, 1919*[2]

When the armistice was announced on November 11, 1918, Webb reported, London exploded in "a pandemonium of noise." In Paris, there was "a wild celebration" until dawn. Even Berlin was "elated," its citizens glad to be rid of the war and the dynasty that had dragged them into it.[3] Of the four great European capitals, only Vienna was silent. An enormous gray crowd gathered in the Ringstrasse in front of the parliament building. A few soldiers stripped the imperial eagle from their uniforms and forced others to do the same. A mile or so away in the Berggasse, Sigmund Freud sat in his study and jotted in his pocket diary: "End of war." Significantly, he avoided the word *peace*.[4]

The disintegration of the multinational Austro-Hungarian Empire had been a fait accompli for weeks. A city with a population as large as Berlin's, Vienna suddenly found itself the capital of a "mutilated and impoverished republic" of 6 million, one-tenth the size of the old empire. After a final session of the imperial parliament in which legislators hurled inkwells and briefcases at speakers' heads, Czechoslovakia, Hungary, and Yugoslavia had seceded, taking many German-speaking lands with them. As a result, Austria's eastern and northern frontiers now lay just beyond Vienna's outer suburbs.[5] And Austria's new neighbors challenged even these borders and constantly threatened to invade. Meanwhile, Austria was in no position to defend itself or issue counterthreats. By November 12, after the emperor and his family had slipped into exile and the new republican government was formally proclaimed, the 4-million-man Austro-Hungarian army had completely dissolved. In the interval between the armistice offer in November and its signing a few days later, hundreds of thousands of soldiers had been imprisoned in Italian POW camps. Most wouldn't find their way home for several years.

The revolutionary firestorm that had been ignited by defeat and hunger in St. Petersburg in February 1917 was now spreading west to Budapest, Berlin, and Vienna. Two Marxists dominated Austria's provisional government. Most observers had taken the inevitability of a Communist putsch for granted since January 1918. The week after New Year's, militants at the Daimler-Benz works had struck to protest a halving of the flour ration. Half a million men and women who had been drafted by the imperial authorities to work in munitions plants had abandoned their factories. Rumors were circulating of an imminent uprising in Hungary and revolution in Germany.

The city was bracing for the return of Austria-Hungary's defeated troops. In *The Last Days of Mankind,* Karl Kraus, the antiwar satirist, warned that embittered, half-starved, armed hordes would turn Austria into a battlefield. "The war . . . will be child's play compared to the peace that will not break out."[6] Hundreds of thousands of men, including the nineteen-year-old Friedrich Hayek, had abandoned their units on the Venetian plain and joined a "hungry, disorganized and undisciplined" mass exodus to the north. Along the way, they exchanged military horses, cars, and artillery for food and looted shops or set them ablaze. By early November, the mass was trying to squeeze through the single, narrow exit from Italy over the Brenner Pass and into Innsbruck. Gun-wielding soldiers commandeered trains. "Roofs, platforms, buffers, steps of carriages, even the engines themselves, swarmed with soldiers," reported one correspondent. "Seen from a distance each train looked like a madly rushing swarm of bees."[7] Hundreds were thrown to their deaths when the trains roared into tunnels und under bridges, and their bodies littered the embankments on either side of the tracks.

Determined to avoid Austria's destruction by "bloody anarchy," the civil servants of the now defunct empire simply tried to keep the trains moving. At one point, a British businessman reported, the Trieste-Vienna line was carrying off between seventy thousand and one hundred thousand men every twenty minutes or so. Worried about anarchy and Communist takeover, the bureaucrats set up depots at the outskirts of Vienna where soldiers turned in their arms before they entered the city. Inside the city, the police continued to report for duty. After an incident in which some Red Guards "liberated" some food and weapons depots, the Social Democratic government hastily recruited unemployed factory workers into a peoples' militia. Thanks to such measures, as well as to the overwhelming desire of Hungarian, Czech, and Yugoslav soldiers to get home as fast as possible, Vienna remained relatively calm.

The returning soldiers found a city under siege. In this most middle-class of European cities, there was hardly any food or fuel. Virtually from the moment the new republic was announced, no more manufactured goods left Vienna and no shipments of beef, milk, potatoes, or coal arrived. Not since 1683, when the city was briefly encircled by the Ottoman Turks, had Vienna been so cut off from the outside world. Travel to Munich, Zurich, or nearby Budapest became difficult, if not impossible. Mail service was fitful. Telegrams took two or three weeks to reach their destinations, if they ever did. Packages arrived without their contents or not at all. "Don't feed the customs officials or railway workers," Freud warned relatives in England.[8]

It hardly needs saying that a city of 2 million must buy food from elsewhere in order to eat. Before the war, Vienna and the Alpine provinces had relied on imports from the non-German-speaking parts of the empire for nearly all of their potatoes, milk, and butter, one-third of their flour, and two-thirds of their meat.[9] But Hungary had suspended exports to Austria in the middle of the war. Now Austria's other new neighbors— mainly Czechoslovakia and Yugoslavia—imposed blockades. As the British high commissioner put it, "For hundreds of years trade has followed certain channels and lines of communication have developed accordingly. These channels and lines have been suddenly blocked . . . The result is districts that are starving close to districts having a superfluity of food-stuffs."[10]

Austria also had plenty of arms, salt, timber, and manufactures to sell; Czechoslovakia had sugar, potatoes, vegetables, and coal; Hungary and Yugoslavia had milk. But despite efforts by the provisional government to work out barter deals with the new states, nationalist politics and anxiety over the latter's own potential shortages prevented the exchanges from taking place.

That was not all. The Allies announced that their wartime blockade of Germany would continue until the Central Powers had signed the peace terms proposed by the victorious Entente. That meant that the only country still willing to sell food to Austria had none to sell. Herbert Hoover, who had come to Europe on a fact-finding mission for the American government, commented bitterly, "The peacemakers had done about their best to make [Austria] a foodless nation."[11]

Making matters immeasurably worse, Austria's rural provinces created an informal blockade of Vienna. Several were, indeed, threatening to pursue union with Germany or Switzerland. In the farm districts, the war had wrecked domestic agriculture. The men were not at home to plant fields. Livestock, the chief source of fertilizer, was slaughtered to supply the military. And the government's policy of forcing farmers to sell food at controlled prices led to less planting and more hoarding. As food shortages worsened, especially in the war's final year, rural districts began to act independently by enacting local export bans on food, passing laws prohibiting tourists from visiting, and organizing stop-and-search operations to prevent anyone from taking food out of the district.

The new government had inherited crushing war debts but no gold reserves with which to buy food for its citizens. The governments of

Hungary and Czechoslovakia had seized the last of the gold deposited in the central bank. Hoover arrived in Paris in mid-December to set up a program for reviving food trade and, if necessary, delivering food aid. He was shocked by the state of Austrian finances: "The citizenry that paid the taxes to pay the army and the bureaucracy had seceded. The state—which paid the salaries of the army, the railway workers—was bankrupt."[12]

Serious food shortages had cropped up almost as soon as the war began. As early as 1915, Vienna's airy white rolls had been replaced by leaden "*Kriegsbrot*," and "meatless" weeks had become routine as early as 1915. Everything became ersatz: not only was bread made "of anything but flour," wrote Stefan Zweig, an Austrian journalist and novelist, but "coffee was a decoction of roasted barley, beer yellow water, chocolate colored sand."[13] The government's requisitioning and distribution efforts drove a growing share of what food supplies there were underground into the black market. And now, despite the end of hostilities, the city's supplies continued to dwindle. Ludwig von Mises, Austria's leading business economist, recalled that "at no time during the first nine months of the Armistice did Vienna have a supply of food for more than eight or nine days."[14] Government depots, the only legal source of food, had absurdly small amounts of pickled cabbage and "war bread" to hand out to housewives who had stood on a queue for hours. The bread ration was six ounces a week per person, less than one-fourth of average prewar consumption. The meat ration had fallen to 10 percent of the prewar level. There was no milk ration for children over the age of one. But one expert estimated that average daily calorie consumption had fallen to little more than a thousand calories—not enough to sustain life for more than a few weeks.

Crowds on the street were wan and listless, while children looked three years younger than they were. "Now we are really eating ourselves up," Freud wrote to a friend. "All four years of the war were a joke compared to the bitter gravity of these months, and surely the next ones, too."[15] Franz Kafka, a clerk in an insurance office, wrote a short story, "Ein Hungerkünstler," about the art of starving. Tuberculosis was common in middle-class neighborhoods, where it had virtually disappeared before the war. By early 1920, the prewar burial rate of forty to fifty corpses per day would climb to two thousand per day. Felix Salten, an Austrian theater critic and novelist best known for *Bambi* (1923), remembered "hearing of lions, of panthers, of elephants, of giraffes that died in the drawn-out agony of starvation in their cages in the zoological gardens, people shrugged their shoulders. How many human beings lay in their beds in the throes of death, and dragged on, emaciated, wracked by suffering, to the bitter end."[16]

The city itself began to waste away. The population was displaying the classic symptoms of starvation—lassitude, indifference, and passivity, alternating with bouts of mania. Despite the influx of demobilized soldiers, imperial civil servants, and several thousand Jewish refugees fleeing pogroms in the east, the city's population, which had expanded rapidly during the early 1900s boom, shrank by several hundred thousand people. Like a body deprived of food that begins to absorb its own muscle, the whole country began living off its accumulated possessions. At one point, the government announced that Austria was willing to pawn "anything"— castles, palaces, shooting boxes, game preserves, and chateaus owned by the Hapsburgs.[17]

The misery of the food blockade was multiplied by a "cold blockade." One week after the armistice, there was no coal for heating apartments and just one week's worth of supplies for cooking. The weekly fuel ration consisted of one twenty-five-watt lightbulb per apartment, one candle, and a little more than a cup of fuel. Even for middle-class households, baths and laundry became unaffordable luxuries. Schools, which had already been closed due to the influenza pandemic, now announced Kaltferien, or cold vacations. Stores had to close by four in the afternoon. Cafés were required to kick out their patrons before nine o'clock. People chopped up their doors, stripped bark off the trees, and cut down trees in city parks. Entire tracts of the Vienna Woods were denuded. The telephone poles and trees that had lined Vienna's elegant boulevards disappeared. Wooden crosses vanished from cemeteries. A visitor wrote, "The whole life of Vienna is haunted by this lack of fuel."[18]

Any number of maddening Catch-22s cropped up, as the Social Democratic *Die Arbeiter-Zeitung* pointed out: "People need wood because they have no coal, but no wood can be shipped because there's no coal for the locomotives."[19] According to the historian Charles Gulik, the Austrian Republic had inherited 30 percent of the former empire's factory workers, 20 percent of its steam-generating capacity, and only 1 percent of its coal supplies. No fuel meant that factories; blast furnaces; bakeries; brick, lime, and cement works; and power plants had to be shut down, choking off industrial production, homebuilding, and power generation. Half of Vienna's sixteen industrial concerns with more than one thousand workers shut down for good. In the city that had been a pioneer of electrification, blackouts became routine, even on Christmas Day. Tram service, which depended on electricity, had to be suspended. Railroad traffic was restricted to freight trains carrying food shipments. In turn, the energy shortage, the falloff in arms production, and demobilization continued to swell the ranks of the unemployed.

On Christmas Eve 1918, just before midnight, Thomas Cuninghame, the official British representative in the former Hapsburg Empire, drove up one of Vienna's stateliest streets, Mariahilferstrasse. "There was not a soul stirring and hardly a light in the streets," he wrote in his diary. "The beautiful old city had become 'Die Tote Stadt.'"[20] William Beveridge was mobbed on Boxing Day by desperate housewives at a market "twittering around us like ghosts in Hades, saying that they wanted food."[21] One of the great European capitals seemed to be on the verge of death.

Joseph Schumpeter's latest ambition—to become commerce secretary in the monarchy's last cabinet—had come to naught a few weeks before the armistice. Since then he had been cooling his heels in Graz, halfheartedly preparing his lectures for the spring term. With the first national election imminent and the Social Democrats and right-wing Christian Social Party expected to form a coalition government, he put feelers out to the Left about the possibility of being appointed finance minister. A Burkean liberal who favored maximum individual freedom and minimum government intervention, he was on generally good terms with Socialists. The two Social Democrats who were temporarily in charge of the country were old friends from his university days. Otto Bauer, a middle-class Jew with pan-German sympathies, was the party leader and interim foreign secretary. Karl Renner, the bluff, portly eighteenth child of a Moravian peasant, was chancellor. Though both were Marxists, their politics had more in common with Fabianism than with Bolshevism. Nonetheless, another man got the job.

Early in the New Year, a fresh political opportunity presented itself. Another university pal, a German Socialist who would soon become the Weimar Republic's first finance minister, wrote to Schumpeter from Berlin with an interesting proposal. Would he join a group of Socialist luminaries assembled that December to advise the new German government on the transition to Socialism; in particular, the possible nationalization of the coal industry?

Strange as it might seem, the Socialist politicians now in charge of the well-being of 60 million citizens had never given any serious thought to how a socialist economy might operate. Marx had expressly forbidden his followers to indulge in what he dismissed as utopian "fantasizing." The leading German Marxist would admit only that such leaps of the imagination might be "a good mental exercise." [22] But the growing radicalization of German workers had forced the issue. Since the armistice in November, there had been mutinies and strikes, extortionist wage demands, physical intimidation, and "spontaneous expropriations" of firms by their employees. The German working class had sacrificed for more than four years. Now they wanted to be compensated. The leaders of the Left parties had been promising for years to wrest control from employers and put workers in charge. Yet now that they were in power, they realized that no government could survive unless it could revive production. The commission's job was to suggest a way out of the dilemma.

Schumpeter accepted the invitation with alacrity. The shortest road to Vienna might lie through Berlin, he saw. With Socialists in power in both Austria and Germany, the likelihood that the two German-speaking states would merge was growing, and Bauer, the Austrian foreign secretary, was a member of the commission. Besides, Schumpeter expected the commission to adopt a gradualist view. Later, Schumpeter justified his willingness to get involved in a Socialist project by saying, "If someone insists on committing suicide, it's better for a doctor to be present."[23] But at the time, most investors, bankers, and industrialists did not expect the commission to propose any such thing. Eduard Bernstein, a prominent German Socialist on the panel, had recently warned, "We cannot seize the wealth of the rich people for then the whole system of production would become paralyzed."[24] Bauer, who wanted to replace boards of directors with representatives of management, workers, and consumers, had stressed that socialized and private enterprises would function side by side "for generations."[25]

Schumpeter lost no time in applying for a leave of absence from the University of Graz, which the provost promptly granted. The journey to Berlin took four days instead of the usual two, but when he arrived in the Prussian capital, he was in a city that no one could possibly describe as "dead" even in these desperate times.

Berlin, January 1919: The city had been spared Allied occupation and had emerged from the war structurally intact, if shabby, woefully short of supplies, and expensive. But each wave of demobilized soldiers—embittered, excitable, and by now addicted to violence—threatened to overwhelm the city. Any stray spark could start a conflagration.

The explosion occurred between Christmas and New Year's. The Communist Spartacists called a general strike, and a full-scale civil war ensued. Mass demonstrations stretched from Alexanderplatz to the Reichstag. Trains were paralyzed, banks barricaded, the university closed, stores boarded up. Spartacists seized virtually every factory, power plant, government building, newspaper, and telegraph office. Tanks rumbled through the streets, and, after the revolutionaries rained grenades and machine-gun fire on government troops, the German chancellor authorized the use of flamethrowers and field artillery. Terrified citizens jammed the train stations in a vain attempt to flee. Berlin's most famous citizen, Albert Einstein, had already withdrawn to Zurich: "Glorious reading about events in Berlin here, under sunny skies, eating chocolate," he wrote to a friend.[26] Schumpeter, however, relished being in the thick of things.

The Socialist brain trust had been meeting for several weeks in the basement of the Reichsbank, which had been spared a takeover by alert civil servants who barricaded the building. Despite the chaos and blood-shed, the commission continued to conduct its proceedings like a university seminar, calmly and deliberately considering alternatives ranging from nationalization to laissez-faire, and addressing practical questions such as how enterprises could be socialized without undermining efficiency gains or innovation.

Schumpeter adopted the same supercilious, cynical attitude he had displayed in Böhm-Bawerk's seminar at the University of Vienna. "I have no idea whether socialism is possible, but if it is, one has to be consistent," he would say flippantly. "In any case it would be an interesting experiment to try it once."[27] He treated the issue of nationalization as a technical one, a banker recalled: "'If one wanted, at the end of a war, to socialize large firms, one had to proceed in a certain way.'"[28]

In the end, as expected, the commission rejected both laissez-faire and Soviet-style state ownership and opted for a combination of public ownership and private management. When the report was drafted, two liberal commission members refused to sign it and issued a minority report. Schumpeter, however, added his name to the majority report. His stint on the commission paid off exactly as he had hoped. Impressed by his cooperative attitude and technical expertise, Hilferding urged Bauer to consider Schumpeter for the position of finance minister of Austria, and by the time the commission report was published on February 15—the day before the Austrian parliamentary elections—the Viennese press was already hinting that Schumpeter would be asked to join the government. Two weeks later, Bauer was back in the German capital for four days of secret Anschluss talks with Weimar's foreign minister, Ulrich von Brockdorff-Rantzau. (Union with Germany was Bauer's top priority. He had already enlisted Robert Musil, the writer, who was "officially charged with the task of indexing newspaper clips . . . In reality, assigned to promote union with Germany in various newspapers."[29]) At that point, "Schumpeter was in a hurry to leave," recalled another commissioner.[30] On the eve of another general strike and bloody uprising, he left Berlin in Bauer's company.

. . .

The post of finance minister was one of two or three in the new coalition government so unrewarding that no career politicians could be found to take them.

How was one to prevent the currency of a bankrupt nation from falling, buy food from abroad without gold or dollars, or cobble together a budget when every variable from borders to reparations was being decided by the Allies in Paris? As soon as Renner floated his name, Schumpeter's own party, the conservative Christian Socials, hastily agreed. They did not necessarily trust him. In an anti-Semitic culture, the party of landowners and aristocrats considered Schumpeter a Judenfreund because he associated with Rothschilds and other Jewish bankers and businessmen. He had also shown a woeful lack of party loyalty by blocking the promotion of the Christian Social whip, who was on the faculty at Graz. But the Socialists considered Schumpeter, who was regarded as "something of a genius in the economic sciences," the right man to take charge of the republic's shaky finances.[31] Since Austria's fortunes depended on the Allies, Renner and Bauer reasoned, his pro-Western sentiments and early opposition to the war were welcome assets, as was his experience living abroad, his American honorary degree, and his fluency in English and French.

Pamphleteers on the left and right instantly labeled Schumpeter as an opportunist: "How delightful it must be to have three souls in one," liberal, conservative, and Socialist, *Die Morgen* began. Karl Kraus called him an "exchange professor in his convictions."[32] But Schumpeter's desire to join the government was hardly discreditable. If the fledgling Austrian Republic failed to deliver bread along with peace, democracy was doomed. He had an economic recovery plan. For Schumpeter, becoming finance minister in a time of revolution was an opportunity to save his country from ruin.

In some ways, the new finance minister and the Viennese housewife faced similar challenges. To pay for food and fuel for her family, Anna Eisenmenger, whose extraordinary diary provides a window onto day-to-day life in those catastrophic times, had three options: to earn, to borrow, or to sell her belongings. To keep the trains running, the militia on patrol, and the soup kitchens open, Schumpeter had the same choices. To survive, Eisenmenger and her family wound up applying for pensions, renting out rooms, working for an American relief organization, and, as a last resort, selling off Dr. Eisenmenger's precious stash of prewar cigars. Schumpeter could collect taxes, cajole bankers into buying government bonds, dip into the country's reserves of cash and gold, if these existed, and, in a pinch, sell assets belonging to the state.

Of course, if an individual had to buy things from abroad—or even take a trip to next-door Geneva—he had to get his hands on foreign currencies. If he had a Swiss bank account he could tap it, as did Max von Neumann, a banker in Budapest who took his family, including his son John, into temporary exile after Béla Kun's Communist coup. If he had no such reserves, the foreign currency had to be earned or borrowed. Sigmund Freud and his fellow psychoanalysts took in English patients, such as James Strachey and his wife, Alix, who paid him in pounds. Eisenmenger borrowed dollars from a cousin in America. Most of the time, the individual had to use krone to buy pounds or dollars.

Because so much of what Austria needed to stay alive had to be imported, the Austrian finance minister had to find foreign currencies or gold with which to buy it. If he couldn't, he had to arrange a foreign loan or hope for a gift. But his principal job was to defend the krone's value vis-à-vis other currencies. Every uptick in the exchange value of the krone meant that Austria could pay less for coal or pork. Every downward movement meant that Austria had to pay more. That is why housewives stood outside the windows of currency shops waiting for the latest report on the krone's value with "constricted chests." For the finance minister, the value of the currency was of even greater consequence, because he was also responsible for the government's budget. Every decline in the krone's value caused the government's deficit to rise. The finance minister's most important single task was to prevent the currency from collapsing. Ultimately it was a confidence game. People took one's money if they believed that they could settle their debts with it. What gave them this confidence was, of course, the knowledge that they could settle their debts with it. So every finance minister had to be bullish on his currency, and if he had no gold or foreign currency reserves, he had to use air to keep it afloat.

Schumpeter, the youngest finance minister in Austria's history, was delivering his maiden speech in a gilded marble palazzo tucked in a narrow lane—Heaven's Gate—in the city center. Pacing back and forth, waving his hands, enunciating in his best Theresianum accent, he alternated between playfulness and passion. Success in modern politics, he was aware, depended on a leader's ability to "fascinate," "impress," and "engage" the public. And economic stabilization required "a popular government and a credible leader of brilliance, will, power and words that nations can trust."[33] In the gloomy, frigid hall full of black-coated officials, he radiated energy, optimism, and hope.

The war had saddled all the combatants, including England and France, with unprecedented debts, but Austria's case was extreme. The imperial government had not dared to raise taxes during the war. In 1919, as a result, tax receipts covered only two-thirds of government spending. The government owed huge interest payments on its war debt, a disproportionate share of which was inherited by the new Austrian Republic. It had also promised relief for the unemployed, principally the cost of sustaining the militia. It had to pay civil servants, including thousands who flocked to Vienna from outposts of the old empire. Finally, it had to provide food subsidies to cover the difference between the price paid by the government and that charged to consumers. The old imperial government had assumed that the lion's share of the huge debts they were accumulating would be paid by the losers. This, of course, only postponed the day of reckoning.

Most Austrians could imagine only two alternatives now: to be adopted by Germany or to become a permanent ward of the Entente. Otto Bauer was an enthusiastic supporter of Anschluss with Germany. He saw nothing wrong, either, with a little inflation, regarding it as "a means of animating industry and raising the standard of life of the workers."[34] Bankers and industrialists leaned toward an alliance with the Entente. They shared

the fondest wishes of British Treasury officials, in particular those of Maynard Keynes, that "Austria will never be allowed to go under. The Entente
will put her finances straight. A large loan in sterling is all that is needed." [35]

Schumpeter took a different view. He believed that a shrunken Austria
had the means to recover economically. His deepest conviction was that
nations' resources matter less than what they did with what they had. As
long as entrepreneurs were allowed to create new enterprises, the financial
system was functioning efficiently, and there were not too many barriers
to trade, society could regenerate itself. He rejected the popular assumption that economic viability depended on vast territories, huge populations, and natural resources. In an extraordinary essay on the sociology
of imperialism written in 1919 with Germany in mind, he described how
ancient Egypt's military-industrial complex impoverished the empire by
chronic warfare: "Created by wars that required it, the machine now created the wars it required." [36] England became the richest country before
she acquired an empire. Switzerland, whose per capita income rivaled
that of Britain, was no bigger than Scotland. And before the war, Vienna
had been the most important financial, transportation, and trading center in Central Europe. As long as the Allies or her neighbors did nothing
to prevent Austria from trading freely or its government from restoring
her creditworthiness, he saw no reason why Vienna could not resume her
prewar economic role and once again earn a good living, provided no insuperable obstacles were placed in her way. "People so often say that German Austria is not viable," Schumpeter admitted. But, he added forcefully,
"I believe in our future . . . One must not think that a country in order to
survive economically must possess all essential raw materials within its
own frontiers . . . The neighboring countries cannot exist without us or
without our financial mediation." [37]

To be sure, the nation had to deal with its massive war debt. The historian Niall Ferguson points out that there are five, and only five, ways
to ease such burdens: de jure repudiation, as practiced by Lenin in 1918
and Hitler in 1938, and varying degrees of de facto repudiation involving
changing the repayment terms, lowering the value of the money in which
the debt is repaid (inflation), or achieving such rapid economic growth

that income rises faster than interest payments. The most respectable, of course, is simply to pay it off.

Reflecting his faith that Austria could help itself, Schumpeter told his audience that he strongly favored the last option. It was the fastest way to restore investor confidence in Austria's creditworthiness and revive production. But no postwar government would get away with raising taxes on farmers and the middle class to compensate wealthy bondholders. Jacking up income taxes would also discourage investment just when the economy desperately needed an infusion of fresh capital. Schumpeter's preferred solution was to force the rich to shoulder Austria's war debt by levying a steep, one-time tax on *property*. In effect, he wanted to pay wealthy bondholders off with their own money by seizing a big chunk of their liquid assets, including cash, bonds, and stock.

The genius of Schumpeter's plan, which reflected the priorities he had set out in his theoretical treatise *The Crisis of the Tax State*, was that while the ownership of business enterprises, farms, and other property would be reshuffled, it would remain in private hands. Taxing existing property rather than future income had the further advantage of not discouraging investors from making fresh capital available for investment or businessmen from expanding production. To reduce the risk the government would inflate its way out of debt, Schumpeter also proposed creating a central bank that, like the Bank of England, was independent of the Treasury. At the same time, he favored stabilizing the krone at its current value rather than its prewar parity. These measures would bolster the confidence of foreign investors, on whom Schumpeter pinned his hopes, and ensure that Austrian investments would be bargains for them.

Schumpeter's recovery program required two conditions in order to work: peace terms that did not impose insuperable obstacles to a renewal of trade, and a sustained effort to raise enough taxes to cover the government's spending. "At the moment we cannot get any credit even abroad because foreigners have no faith in our future," he told his staff. Eliminating or even dramatically reducing the government's deficit would require heroic measures, he admitted. He favored sin taxes on "conspicuous consumption" of such proletarian indulgences as beer and tobacco, as well as sales taxes on "luxury foods, luxury entertainment, luxury textiles, luxury stores, servants, luxury clothing."[38] It was not a plan designed to win friends on either the right or the left. His own party was dead set against a tax on property, especially if it included farms. The Socialists considered the notion of taxing beer a hilarious example of Schumpeter's political cluelessness.

By the third day of Schumpeter's tenure as finance minister, the krone was in free fall. Communist guerrillas, led by a former Austro-Hungarian army corporal who had been trained and armed by Moscow, were riding around Budapest in open trucks festooned with red flags. A Red Guard of demobilized Austrian soldiers immediately set off for the Hungarian capital to express solidarity. The Bolshevik victory was widely interpreted as Hungary's having thrown herself into Moscow's arms rather than submit to the Entente. It prompted Lloyd George, the British prime minister and something of a hawk on reparations, to issue a warning to the peace conference. While the Entente, no less "tired, bleeding and broken" than the losers, were intent on making the Germans and their allies pay for reconstruction, Lenin's disciples were busy trying to seduce Germans with promises of "a fresh start," that is, as Lloyd George explained, a chance "to free the German people from indebtedness to the Allies and indebtedness to their own richer classes" that had lent the Reich the resources with which to fight the war. If the Allies insisted on imposing overly harsh terms on Germany, the inevitable result would be "Spartacists from the Urals to the Rhine."[39]

As if on cue, Lloyd George's gloomy prophecy proceeded to unfold. On April 7, in Munich, a band of anarchists declared a Bavarian Soviet Republic. Within a week the homegrown mob was replaced by professional revolutionaries—Russian émigrés with ties to the Internationale—who promptly unleashed a reign of terror. A Russian document captured in a police raid suggested that Lenin's army was poised to march into Germany via Poland to join the insurrectionists. The word in Paris was that Vienna, now flanked by two red capitals, would be the next domino to fall. In a Parliamentary debate on whether or not to leave British troops in Russia to help defeat the Bolsheviks, Winston Churchill warned that "Bolshevism is a great evil, but then it has arisen out of great social evils." Six weeks later, the British cavalry officer in Russia, General Briggs, wrote to Churchill arguing for British support: "starvation means Bolshevism." [40]

Béla Kun's emissaries did appear in working-class districts of Vienna to make dramatic pledges to supply food to the proletarians—but not the bourgeoisie—in the future Soviet Republic of Austria. They painted fantastic pictures of life in Budapest, of prices in first-class hotels now on a par with those of rough taverns, workers' families living like royalty in confiscated palaces, and social equality between the bourgeoisie and the proletarians. In his memoir, Bauer recalled:

> As soon as [Béla Kun] realized that we had no intention of [proclaim-ing the Austrian Soviet] he embarked upon a campaign against us. The Hungarian Embassy in Vienna became a centre of agitation. Large sup-plies of money came from Hungary to the Communist Party of Austria, which not only served to strengthen its propaganda, but which was also expended for the purpose of bribing trusted individuals among the workers and soldiers. The communist propaganda sought to persuade the workers that there were large supplies of food in Hungary which were sufficient to meet all the requirements of Austria. [41]

To counter such propaganda, Herbert Hoover sent cables from his Paris headquarters at 51 Avenue Montaigne urging his deputies to plas-ter Vienna's city walls with fliers warning that "any disturbance of public order will render food shipments impossible and will bring Vienna face to face with absolute famine." [42] Meanwhile, he stepped up relief operations in a race against Communism and death. In Vienna, the government ordered half a company of the Socialist militia, the Volkswehr, to take up residence in the courtyard of 7 Herrengasse, where the cabinet held its meetings.

Fears of a coup may explain a curious incident related by the food minister, Hans Loewenfeld-Russ. Apparently, Schumpeter had telephoned him on the last day of March, invited himself to dinner, and asked him to invite Ludwig Paul, the transportation minister, as well. As soon as the three men were alone, Schumpeter asked the others abruptly whether, in the event of a coup, they would be willing to join the new Bolshevik government with him. "Not even in my dreams," retorted Paul sharply.[43] Loewenfeld-Russ nodded angrily in agreement. Schumpeter immediately backtracked, chiming in that he too wouldn't dream of joining such a government.

When Loewenfeld-Russ demanded that Schumpeter explain why he had wanted a private meeting and had asked such a peculiar question, Schumpeter replied that he was merely sounding out the only two cabinet members besides himself who had been appointed not by the chancellor but by a political party.

He was probably telling the truth. At around that time, Cuninghame reported that one of his sources had submitted a "long circumstantial report . . . detailed plan, worked out by the Socialist Party, for establishing a Socialist form of government." According to Cuninghame's informant, this government was supposed to be a decoy, "Soviet in appearance rather than reality."[44] Supposedly Renner and other moderates in the cabinet were ready to adopt this ruse, although the more left-leaning members like Bauer refused to have anything to do with it. Cuninghame was instructed by the British Foreign Office to inform the Austrian defense minister that a Bolshevik government, phony or genuine, would mean a suspension of food aid and a resumption of arms shipments to Poland, which was harassing Austria with territorial demands.

As the Austrian republic hovered between life and death, the cabinet was in permanent session. Sessions typically started hours after the end of the normal work day. Around the time that the opera let out, the fifteen ministers and their undersecretaries converged by car or on foot on the Palais Modena at 7 Herrengasse, one of the finest streets in Vienna, dating from the late Middle Ages. Anxious, sleep-deprived men trudged past the disheveled Volkswehr guards camped in the courtyard and climbed the imposing stairway to the once brilliantly lit, elegant rooms where emperors had met with their councilors and where Karl Renner had set up his chancellery. They had to take care to avoid tripping over the machine guns haphazardly positioned at the windows and were forced by the cold and damp to keep their heavy overcoats on. The meetings usually lasted well past midnight, and the prime minister sometimes sent out to a nearby restaurant for a meager meal and a glass of beer to keep them all going.

On April 17, the ministers had barely begun to tackle the "unimaginably long" agenda when thousands of gaunt and ragged men with "pinched and yellow" faces came marching along the Ringstrasse, now strewn with litter, past luxury high rises with boarded-up windows, to gather in front of parliament a few blocks away. Most were unemployed factory workers and demobilized soldiers, many with missing limbs or other visible war wounds. Scattered among them was a small cadre of armed Communist Party members and foreign agitators. After a few hours, they succeeded in whipping the crowd into a sufficient frenzy to storm the parliament building and, once inside, set it ablaze. When shooting started, the Volkswehr rushed in. In retaking the building, the "peoples' militia" shot some fifty demonstrators dead and wounded several hundred, or so the first reports claimed.

One episode shocked the public even more than the attempted putsch. At the height of the fighting, a horse was shot out from under a policeman in the street in front of the parliament. As the animal lay dead in the street, a hungry mob tore it to pieces and carried off hunks of bloody meat. For ordinary Viennese, who adored the emperor's white show horses the way Americans loved boxing champions, the incident seemed to prove that civilization was reverting inexorably to barbarism. No one could have been more appalled than the republic's newly installed finance minister, who, even in these desperate times, kept several thoroughbreds.

In Budapest, the opinion on the street was that revolution was imminent in Vienna, but by midafternoon the insurrection had petered out. Friedrich Adler, the recently released assassin of the monarchy's penultimate prime minister and a popular Socialist politician, had arrived to urge calm. The Communist leaders themselves could not agree on whether to proclaim a Soviet republic. The next day, the heads of the workers' councils declined to call a general strike. Ellis Ashmead-Bartlett, the *Daily Telegraph*'s war correspondent, rushed to the Austrian capital from Budapest. "Instead of finding Vienna in flames, I found the town absolutely quiet." [45]

Hotel Sacher, across from the opera and famous for its voluptuous chocolate cake, was the preferred trysting place for Vienna's diplomats, spies, and counterrevolutionaries. Madame Sacher was said to be an ardent monarchist. Schumpeter lunched there often. On May 2, Sir Thomas Cuninghame discovered Schumpeter in one of the "salles privees" at the rear of the hotel, huddled with four other men, including Ellis Ashmead-Bartlett, the British correspondent who had broken the story of the slaughter at the battle of Gallipoli. Halfway through their meal, Cuninghame, who had apparently heard that Ashmead-Bartlett was in town, joined them.

Ashmead-Bartlett was trying to raise money on behalf of about 150 Hungarian officers who were hanging around Vienna, terrified of being deported on the one hand and eager to organize a counterrevolution against Bela Kun on the other. Their trouble was the complete lack of money or credit with which to hire as much as a train, beyond that extended by the sympathetic Madame Sacher. Von Neumann, the Budapest banker, was in Vienna to help in the fund-raising effort. Other rich sympathizers were afraid to lend them cash for fear that such a loan would reach the ears of Austria's Socialist government. Louis Rothschild, on whom the plotters had pinned their hopes, was changing his conditions daily. Finally, Cuninghame had suggested that Ashmead-Bartlett see Schumpeter, known to the British as a bitter opponent of union with Germany, which Britain also opposed.

The journalist was impressed by Schumpeter's intelligence, lively manner, and flawless command of English. He noted approvingly that Schumpeter was not yet forty and betrayed none of the caution typical of Treasury men. "We discussed the future of Austria," he recalled. Schumpeter had immediately declared himself in favor of a constitutional monarchy such as Britain's and agreed that "the only way to eliminate the Red danger from Vienna was to drive the Soviet Government out of Hungary." After saying that he would gladly advance the revolutionaries money from the Treasury were it not for the need to account for every kroner to parliament, he offered to assure Rothschild that if he lent the money, the Treasury would look the other way. "This was good news," said Ashmead-Bartlett, "as it did away with Louis Rothschild's main objection . . . namely his fear of being asked awkward questions by the Austrian government."

As it turned out, when the monarchists seized the Hungarian embassy in Vienna on May 4, they unearthed a large cache of the money—reportedly 135 million krone and 300,000 thousand Swiss francs—earmarked for fomenting revolution in Vienna. So, just as the negotiations with Rothschild were winding up, Schumpeter sent his secretary to inform the banker, "There is no need for him to advance any money, as it has been raised elsewhere."[46] When Béla Kun tried to retrieve his war chest and get the aristocratic officers extradited, Schumpeter intervened on their behalf. Before the matter went any further, the Béla Kun government was overthrown by the right-wing Admiral Miklós Horthy and his followers.

Over the next few weeks, the Austrian government went on a spending spree. The Socialists dominated the coalition government formed in March 1919 because they were the only ones who could control the unemployed, soldiers, workers' councils, and radicals. Arguing that the large conservative peasant majority would not permit a Socialist revolution and that any putsch would result in Allied intervention, Bauer pressed for a variety of social welfare measures. Aware that they might have only a narrow window in which to act, the Socialists succeeded in laying the groundwork of the Austrian welfare state in a few short weeks. In Vienna alone, sixty thousand war invalids, dependents of POWs, and officials of the former empire and their families qualified for relief. Within a year, one-sixth of the population would be on welfare and not producing any salable goods.

Meanwhile, Schumpeter's efforts to gather support for his tax proposals stalled. No loans were forthcoming from the Allies. Reserves of gold and hard currencies were minuscule. The government had little choice but to finance its deficit by printing more money.

The government looked for ways to shift the burden onto business via what Bauer called a "far reaching encroachment on the rights of private enterprise, originally conceived as emergency regulation for a few months." In May the cabinet passed a decree requiring large companies to boost employment by 20 percent. It was soon followed by others compelling employers to recognize trade unions and give workers paid holidays, and forbidding layoffs without government approval. Not surprisingly, the law resulted in a sharp decline in productivity, complaints about *"Arbeitsunlust"* or absenteeism, and a further slump in tax receipts.

Nonetheless, the Renner cabinet forged ahead with socialization. In mid-May, Otto Bauer announced a program of partial nationalization of mining, iron foundries, power stations, forests, and timber. Schumpeter objected that as long as the government administration got its finances in order and stabilized the krone, business owners would once again invest and expand. After alienating conservatives by proposing that the rich bear the burden of the war debt, he alienated his Social Democratic colleagues by claiming that socializing private enterprises would make it impossible to attract foreign investors and smother any recovery.

Schumpeter's Social Democratic colleagues trusted him no more than his own party did and called him "vain," "conceited," and "affected" behind his back. The other ministers wore shabby clothes and shoes with holes in their soles. Schumpeter dressed like an English banker or diplomat. The cut of his Savile Row suit was impeccable. The silken handkerchief that he tucked under his heavy gold watch was snowy. Newspaper caricatures invariably depicted him in jodhpurs, high boots, and a Homburg. He carried a riding crop under his arm as if to suggest that he intended to whip his ministry, the cabinet, or the whole country into shape. The other ministers lived in modest flats with their frumpy wives. Separated from Gladys, apparently for good, Schumpeter flaunted his lavish bachelor lifestyle. He rented a suite at the posh Hotel Astoria around the corner from the ministry, an apartment on Strudlhofgasse, and half of a palais belonging to a count, where he threw teas and dinners for the likes of the Rothschilds, Wittgensteins, and other plutocrats, as well as foreign diplomats, journalists, and politicians. He often pulled up to the ministry in an ostentatious horse-drawn carriage. He ate in the best restaurants, drank the finest French champagne, and often had a call girl or two on his arm or sitting beside him in his carriage. It was a manner of living far beyond a cabinet officer's pay grade, and it was obvious that Schumpeter was running up debts to his wealthy friends. Even his old mentor Friedrich von Wieser got the impression that Schumpeter did "not really care about the general misery" or suspected that "as soon as his vanity is no longer satisfied . . . he will retire."[47] Schumpeter made matters worse by pretending to be indifferent to criticism. He would tell reporters, "Do you think that I want to remain Minister of a State that goes bankrupt?"[48]

Anschluss, which Otto Bauer regarded as Austria's only chance for economic revival and Schumpeter was the only cabinet member to oppose, was another major source of friction. In late May, the local correspondent for *Le Temps* discovered Le Docteur Schumpeter, seated at his desk in the yellow ballroom. The sumptuous baroque palace in the heart of a starving city, the lavish gold leaf decorating the Treasury's empty vaults, and the floor-to-ceiling frescoes glorifying Austria's past military triumphs amid anarchy and defeat struck the reporter as highly ironic. And then to come upon Schumpeter, the "bourgeois whipping boy" in the cabinet, sitting "at the feet" of a portrait of Ferdinand I was too much![49] *Le Temps* readers would have gotten the joke: the famously feeble-minded and sexually impotent Austrian emperor had been forced to abdicate in 1848, another year of revolution. Such a fate seemed likely for the similarly weak and helpless Austrian Republic and its minister, even one as strong-willed, brilliant, and notoriously libertine as Le Docteur Schumpeter.

The Austrian government was trying to influence the terms of a peace treaty that would be dictated by the Entente by waging a propaganda war for "the right to union with Germany."[50] When the *Le Temps* interview appeared quoting Schumpeter, Bauer accused him at the next cabinet meeting of secretly lobbying the French and English to forbid the merger. As Bauer complained in his memoir, "The French statesmen were able to answer us that the leading men of Austria, the bankers and the industrial magnates daily assured the Entente diplomats at Vienna that Austria did not need union and could get along pretty well by herself, provided the peace conditions were relatively favorable."[51]

Bauer's accusation was largely true. Schumpeter had been giving anti-Anschluss speeches for weeks. He had also proposed the idea of a monetary union with France to Henry Allize, head of the French Military Mission to Vienna. His contention that the new Austrian state could avoid bankruptcy was based on an expectation that France would give a higher priority to establishing a non-German-dominated common market in Central Europe. As late as the end of June, Schumpeter declared publicly that he was hopeful that the Entente would ensure an "equitable distribution of the burden" of war debts and would not insist on the confiscation of Austrian assets in Czechoslovakia, Hungary, and Yugoslavia. As he put it, "In the case of Germany, the peace terms were drafted to check recovery; in the case of Austria, they must encourage it." [52]

At the end of May, Schumpeter once again attacked the Anschluss policy in a "sensational" interview with *Neues 8 Uhr Blatt,* warning, "Our safety lies in our peaceful intercourse with all states, and especially with our immediate neighbors." [53] Bauer wrote him a furious letter, but instead of heeding his warning, Schumpeter tried to engineer a secret side deal with the British. He gave Francis Oppenheimer, Keynes's emissary to Vienna, the draft of a "secret" plan involving Allied control of Austria's finances and central bank in exchange for long-term loans. As Oppenheimer reported in a cable to his boss warmly supporting Schumpeter's plan, the Austrian finance minister

> does not share the general opinion that Anschluss to Germany was Austria's only salvation. He wanted, if possible, a strong Allied Finance Commission to take charge of Austria on the lines of the British financial administration in Egypt, but whatever form the control might take, it would have to safeguard Austria's amour propre. He insisted that a single currency throughout the successor states, with Vienna remaining the banker of them all, was perhaps the most important item in the program of Austria's recovery.

He added, "Suffice it to say that it was a rare and fortunate privilege to have had to deal with such a genial, open-minded expert." [54] The two continued to meet frequently. Among other things, Schumpeter was actively trying to help the British acquire the Austrian companies that controlled shipping on the Danube. As Oppenheimer had informed Keynes, "Dr. Schumpeter had agreed to facilitate the transfer of this company, possibly of the other three as well, into British ownership on exceptional terms for cash, and he promised to maintain for us a first refusal until we had either accepted or declined the offer."[55] Naturally, nothing in Vienna stayed secret for long. "Schumpeter carries on with his intrigues," Bauer wrote to Renner. "I shall do nothing for the time being, but after the conclusion of the peace treaty it will be inevitable to force his resignation."[56]

Almost as soon as the Allied treaty terms had been presented to the Germans on May 7 in Versailles, the Austrian delegation, with the prime minister Karl Renner at its head, left Vienna for France. On June 2, 1919, after spending two weeks cooling their heels in the old royal chateau in Saint-Germaine-en-Lay, relishing the French food and wine, they learned the Entente's terms for Austria. "It was a terrible document," Otto Bauer recalled. Large chunks of German-speaking Austria were parceled out to the Czechs, Yugoslavs, and Italians. "Equally harsh were the economic provisions . . . They were simply a copy of the German Peace Treaty."[57] The draft Treaty of Saint-Germaine acknowledged that the Austro-Hungarian Empire had broken up but penalized only Austria for its crimes. Three million German-speaking Austrians were to live under Czech rule. The private property of Austrian citizens was to be confiscated. Austria's government was to pay reparations for thirty years. The coup de grâce, at least from Bauer's point of view: union with Germany was expressly forbidden.

In Vienna, the reaction was shock mingled with disbelief. Schumpeter told a reporter that "the Allies' motivation can only be to destroy German Austria."[58] On June 30, he said, "It is not easy to kill a people. In general it is impossible. But here we have one of the few cases in which it is possible . . . fiscal collapse inevitably brings with it social collapse."[59] When the foreign exchange market issued its verdict on the treaty, the krone collapsed once more. As Friedrich Wieser told a London conference on relief and reconstruction at which Keynes was present a few months later, the currency markets have declared thereby that they do not consider the Austrian Republic, with boundaries as fixed by the Peace Treaty, and with the burdens laid upon her therein, as capable of life. The Austrian who loves his country will do everything in order to keep her alive. But it is not surprising that the outside world, to whom her existence is a matter of indifference, has declared that she is incapable of life.[60]

By treating Austria as harshly as Germany, the Allies not only destroyed the viability of the new state but also shredded what little was left of Schumpeter's credibility. He was forced to admit that his political judgment had been naïve. He was, as he confided to his diary, a man without an intuitive feel for political reality, "a man without any antennae."[61]

Schumpeter's political demise was agonizingly drawn out. At the July 15 cabinet meeting, Bauer hurled yet another accusation at him, this time of sabotaging the "socialization" of basic resource industries by conniving to deliver a major Austrian mining and timber concern into the hands of Italy's Fiat company, making it impossible for the government to take it over. Schumpeter tried vainly to defend himself by portraying a series of transactions with a foreign exchange trader named Kola as an attempt to raise gold and hard currencies with which to defend the krone.[62] Two weeks later, Schumpeter had the humiliating task of defending the government's plan to sell or mortgage several of the nation's "immortal works of art," including the emperor's prized Gobelin tapestries. There was no other way to raise the requisite foreign currencies for buying food abroad, he argued, warning dryly, "This process cannot be often repeated." He begged the lawmakers one final time to pass his budget: "The greatest problem of the State would be to get through the next three years without Government bankruptcy and without the issue of new notes," he pleaded, knowing that his arguments were falling on deaf ears.[63] It was Schumpeter's last appearance in front of parliament.

In mid-October, utterly isolated and constantly ridiculed in the media, Schumpeter was finally dismissed. The manner and circumstances were so brutal that one liberal newspaper accused Renner of character assassination. Nor was that the end, for several of his actions as finance minister resulted in investigations that continued for months. Felix Somary, the banker, observed that "Schumpeter made light of everything" and attributed his cool manner to his training at the Theresianum, "where the students learned to cultivate self-control and under no circumstances to show emotion. One should master the rules of the game of all parties and ideologies, but avoid commitments."[64] Inwardly, though, Schumpeter was shattered. He was convinced that he lacked "the quality of leadership,"[65] and his public humiliation was made more painful by his mother's disappointed hopes. The fact that subsequent Allied stabilization programs for Austria were modeled on his or that the government that sacked him was judged to be "incapable of governing the country" failed to ease the sting of failure. When asked about his experiences, he rarely said more than "I held the minister-ship in a time of revolution, and it was no pleasure, I may assure you."[66]

In November, when Wieser returned from London, his acquaintances were still talking about Schumpeter's fall. The older man observed, "It seems that Schumpeter is utterly ruined in the eyes of all the Parties. Even the young economists, who regarded him as their leader, have written him off. No one has any more expectations of him."[67] His former admirers had sold him short. After two terms at the University of Graz, where he licked his wounds, Schumpeter did what many former public servants do. He joined the private sector.

His timing was impeccable. The destruction of Austria's hopes for the future coincided with a stock market boom and deal-making frenzy. As one observer recalled,

> Stock quotations began to adapt themselves from day to day to the falling value of money. The capitalists sought to preserve their capital from depreciation by investing it in securities and bills . . . The Stock Exchange speculated upon a continuous fall in the Krone. The Krone's exchange

value vis a vis other currencies fell faster than its internal purchasing power. As a result, Austrian prices were far below the level of world market prices and large profits could be realized by exporting Austrian products.[68]

In a last-minute expression of appreciation, parliament had awarded Schumpeter a golden parachute in the form of a banking license, and by 1921 Schumpeter had parlayed it into the presidency of a small but old, highly respected bank. He had run through his savings and borrowed heavily to live far beyond the means of a professor and politician. Now he needed to make money.

Chapter VII

Europe Is Dying: Keynes at Versailles

Expert opinion is being ignored.

Keynes has been too splendid about the Austrian treaty. He is going to fight. He says he's going to resign.

—*Francis Oppenheimer, 1919*[1]

Vienna was not unique. In January 1919, famine and pestilence raged from St. Petersburg to Istanbul. To the Britons and Americans who came to Europe to survey the damage, the whole continent seemed to be on the verge of dying. After a ten-hour drive from the coast to Lille, in eastern France, one observer wrote in his diary that he could recall seeing no "human being not connected with the Army ... or any animal ... or really any living thing except rank grass or any inhabited house." In Ypres, Belgium, where some of the worst fighting had taken place, "the colors of the bricks and stone are mellowed; grass and moss are beginning to grow over the ruins."[2]

Eight weeks after the signing of the armistice, the restoration of peaceful conditions had proved impossible. The blockade was still in effect. The Allies dared not give up their most effective weapon against Germany too soon. Fighting involving hundreds of thousands of troops continued as dozens of small wars erupted. Pogroms, expulsions, and mass murder were under way. Eight and a half million men had lost their lives. Nearly as many were left physically disabled or psychologically maimed. An entire generation of children in Central Europe—the Kriegskinder, or children of war—was growing up underfed and undersized.

In the aftermath of the war, the "universal age" and its economic achievements seemed as unreal as a dream. In addition to the staggering loss of life and property, the prewar channels of trade and credit were in ruins. Everywhere new barriers to exports and imports were springing up. Those who possessed something to sell were often reluctant to part with it for paper currencies issued by bankrupt governments; a large share of trade reverted to barter. Winners and losers alike had mortgaged themselves to the hilt to fight the costliest war in history, exhausting not only their reserves but also their limited powers of taxation. As late as 1916, France, Germany, and Russia had no income tax. Now there was no credit to feed the population, fuel the furnaces, repair the damaged factories, or finance renewed trade. The threat of bankruptcy as much as the thirst for revenge was making shaky governments determined to make someone else foot the bill.

"The economic mechanism of Europe is jammed," wrote David Lloyd George, Britain's wartime prime minister, to Woodrow Wilson.[3] Everything depended on economic revival, but the heads of victorious states gathering in Paris seemed incapable of paying attention. This, in any case, was the gloomy view from the third floor of the magnificent Hotel Majestic near the Place d'Etoile, where the British delegation was housed and where Maynard Keynes, the rising Treasury star, was dashing off a letter to Vanessa Bell, assuring her that she would "really be amused by the amazing complications of psychology and personality and intrigue which make such magnificent sport of the impending catastrophe of Europe."[4]

Keynes had arrived at the Hotel Majestic on the tenth of January, the wettest, most depressing month of the year. President Woodrow Wilson had been in Paris for a month, prime minister Lloyd George was not due for another day. The city, which had managed to avoid falling to the kaiser's troops despite heavy bombardment, was now an occupied zone. American Express branches had sprouted like chanterelles. Giant British printing presses were groaning in the Champs de Mars. Black sedans carrying diplomats and drab military vehicles clogged the streets while young men and women wearing the uniforms of some twenty-seven countries jammed the sidewalks. The whole world was in Paris, it seemed.

A mini-Whitehall and a mini–White House constituted themselves on the Seine. Winston Churchill, Britain's minister of munitions, accompanied as always by his faithful secretary, Eddie Marsh, shuttled back and forth between the two. On the far end of the Champs-Elysees were President Woodrow Wilson and a team of advisors that included Bernard Baruch, the financier; John Foster Dulles, counsel to the American team; and Felix Frankfurter, former assistant to the secretary of war. The delegations brought their own fleets of cars and airplanes, set up their own telephone and telegraph networks, and operated their own trains.

Keynes was not a member of Lloyd George's inner circle. Accordingly, while the prime minister and his mistress, Frances Stevenson, were installed in a luxury flat, Keynes was bunking in the Hotel Majestic with the rest of the British delegation. The hotel, which had been undergoing preparations since shortly after the armistice, had its own physician, a chaperone for the female staff, and a security detail of Scotland Yard detectives who were supposed to forestall leaks. As a result, it was easy to get out of, but "extremely difficult to get in," recalled Harold Nicolson, a British diplomat in the delegation who was married to the writer Vita Sackville-West and was an old friend of Keynes's. The hotel was "staffed from attic to cellar with bright British domestics from our own provincial hotels. The food, in consequence, was of the Anglo-Swiss variety," Nicolson added.[5] Oddly, no one thought of replacing the French staff in the Hotel Astoria next door, where the British delegation had its offices and kept its confidential maps and papers.

The most extraordinary people floated in and out of the Majestic. Ho Chi Minh, the future leader of the Viet Cong, washed dishes in the kitchen. T. E. Lawrence, aka Lawrence of Arabia, was often in the lobby, as were Jean Cocteau, the playwright; and Marcel Proust, "white, unshaven, grubby, slip-faced, wearing a fur coat and white kid gloves." Nicolson described their encounters:

He asks me questions. Will I please tell him how the Committees work? I say, "Well, we generally meet at 10:00, there are secretaries behind . . ." "No, no, you're going too much too fast. Start over. You took the official car. You got out at the Quai d'Orsay. You climbed the stairs. You entered the room. And, then? What happened? Be precise, my dear, precise!" So I tell him everything. The sham cordiality of it all; the handshakes; the maps; the rustle of papers; the tea in the next room; the macaroons. He listens enthralled, interrupting from time to time—"Be more precise, my dear man, don't go too fast."[6]

Journalists outnumbered diplomats. Frederick Maurice, a former major general, was in Paris on assignment for the London *Daily News*. He had nearly brought down the government by accusing the prime minister of lying to Parliament about British troop strength late in the war. His favorite child, Nancy, was in Paris too, the brand-new secretary of the married, middle-aged, conservative Major General Edward Louis Spears, who would one day become her husband—one of countless young female assistants whose khaki uniforms inspired several mildly salacious chansons. Her younger sister, Joan, a precocious fifteen-year-old student at St. Paul's Girls School in London, who showed no sign of becoming one of the century's most famous economic thinkers, would have given anything to be in Paris too. She had to content herself with Nancy's patchy, pompous bulletins to their mother.

Maynard Keynes was considered "one of the most influential men behind the scenes" in Paris. Even his critics acknowledged that he was "clear headed, self-confident, with an unerring memory."[7] He complained, with justification, of overwork, but his dinner companions wryly noted his "unsurpassable digestion" and capacity for champagne. At thirty-six, Keynes was still as thin and lanky as an undergraduate. His upturned nose and fleshy lips had earned him the nickname Snout in his school days and he had the hungry look of someone who was, as Lady Ottoline Morrell, one of Bertrand Russell's lovers, remarked disparagingly, "greedy for work, fame, influence, domination, admiration."[8] Keynes's arrogance could be breathtaking, his manners appalling, his dress sloppy. Yet his luminous eyes, animated features, and aura of confidence made him attractive. Men and women alike found his silky, melodious voice irresistible.

Born in 1883, the same year as Joseph Schumpeter, Keynes was the favorite son of a highly successful, close-knit Cambridge academic family, at home with, and in some cases related by marriage to, other intellectual dynasties, including the Darwins, Ramseys, Maurices, Stephens, and Stracheys. Neville Keynes, his father, was a lecturer in moral philosophy and a close friend of Alfred Marshall. His mother, Florence, who became mayor of Cambridge in 1932, was active in local politics and philanthropy. They were bright, attentive, affectionate parents to Keynes and his two younger siblings.

Recognized as a genius in adolescence, Keynes was groomed to be a Cambridge fellow virtually from the cradle. Neville Keynes encouraged his gifted son to pursue mathematics. After graduating with honors from Eton in 1902 and obtaining a top score on the Cambridge entrance examination, Keynes entered one of the oldest Cambridge colleges, King's, on a scholarship. The publication of the philosopher G. E. Moore's *Principia Ethica* at the end of Keynes's freshman year was the great event of his undergraduate career, all the more so because Moore was a member of the Apostles, which served as a link between generations of Cambridge intellectuals. *Principia Ethica* was concerned with defining a good life. The Victorian preoccupation with striving, moneymaking, and obeying rules was the target. Rejecting the utilitarian values and do-good moralism of Alfred Marshall's generation, as well as its sexual mores, Moore espoused a kind of radical individualism and aestheticism tempered by the Golden Rule. "Nothing mattered except states of mind, our own and other people's, of course, but chiefly our own," Keynes recalled in 1938. "These states of mind were not associated with action or achievement or with consequences. They consisted in timeless, passionate states of contemplation and communion." [9]

Such reflections give no hint of Keynes's devotion to boating, riding, tennis, and especially golf, his passion for public debate, and his commitment to the Liberal Party or the prestigious social or intellectual student

societies that he was invited to join or lead. His college career showed
Keynes to be a natural leader as well as a brilliant intellect. Though rarely
in bed before three in the morning, he graduated with first-class honors
on the eve of his twenty-first birthday. Assuming that he would follow in
Neville's footsteps, he spent a year studying for the mathematics Tripos. By
1905, the queen of sciences was significantly harder to subdue than when
Marshall had scored his second place. Keynes's twelfth-place finish was
hardly embarrassing, but it was not good enough to win a fellowship at
King's. To give an idea of the competition, the great number theorist G. H.
Hardy, best known as the author of *The Mathematician's Apology,* was still
waiting for a university lectureship after finishing in fourth place on the
Tripos in 1900.

Keynes escaped to hike in the Alps with a copy of Marshall's *Principles
of Economics.* He was back in Cambridge in the fall, sufficiently intrigued
to attend Marshall's lectures while preparing to take the civil service examination. "Marshall is continually pestering me to turn professional
Economist and writes flattering remarks on my papers," Keynes wrote to
his close friend Lytton Strachey. "Do you think there is anything in it? I
doubt it." [10]

Nonetheless, the subject grew on Keynes, and he began to think that
he might like to "manage a railway or organize a Trust, or at least swindle
the investing public." [11] With an academic appointment off the table,
Keynes set his sights on the Treasury. But a second-place finish in the
civil service examination resulted in a temporary exile to the India Office,
where he was assigned to work involving the Indian rupee. Unlike Cecily
in Oscar Wilde's *The Importance of Being Earnest,* Keynes found the rupee
more fascinating than not. He took the view that the currency—any currency, actually—was a clue to the state of a nation's economy and, since
countries were connected to one another through trade and investment,
the world's economy.

Everybody was willing to accept British pounds in return for goods
or services, but not everyone was willing to take rupees. The value of
money—whether the giant millstones favored by ancient Micronesians,
gold coins, or entries on a bank balance sheet—depended strictly on peo-

ple's willingness to accept it. So a nation's currency must reflect the world's confidence in its economic prospects, solvency, and willingness to make good on its promises. In that sense, a currency was like a pulse, a vital sign that could signal anything from disease or injury to a momentary rush of excitement or fear. The challenge for a doctor was to find the cause of a racing pulse before the patient went into shock or made him look foolish by hopping off the gurney in apparently perfect health. If the patient was thousands of miles away and more details about his condition were impossible to obtain, the challenge was that much greater. With his nimble mind, knack for spotting connections, and gift for synthesis, Keynes not only relished such conundrums but proved to be a natural diagnostician.

Keynes dispatched his official duties vis-à-vis the rupee with such ease that he had ample time in the office to write a treatise on probability, which he hoped would win him the elusive college fellowship. It also left him with free evenings and weekends in which to cultivate his social life. He lived in London, where he rented a flat at 46 Gordon Square in a fashionably louche part of town. His upstairs neighbors were the beautiful, intimidating, wildly talented Stephen sisters, the future Vanessa Bell and Virginia Woolf. Keynes got on especially well with Vanessa, a painter who loved to gossip and talk dirty. Keynes's sex diary, as meticulously detailed as the notebooks in which he recorded his expenses and golf scores, indicates that his love life blossomed too. In contrast to the period from 1903 to 1905, when the number of his sex partners was "nil," by 1911 there were eight, and in 1913 they peaked at nine. Among them were lovers and lifelong friends Duncan Grant, Lytton Strachey, and J. T. Sheppard, the openly gay provost of King's.[12] Still, he rarely missed Sunday lunch in Cambridge with the large Keynes clan.

For most of his twenties, Keynes was Britain's resident expert on obscure currencies. Thinking about currencies got him into the habit of thinking about economies holistically instead of focusing on "trade" or "labor" or "industry" in isolation, and taught him how to draw salient conclusions from a handful of indicators. It also gave him a feeling for which government actions exerted systemic effects, like those of the moon on tides, instead of effects only on a particular industry or group. By 1908,

however, he had quit the India Office. Arthur Pigou, the successor to Alfred Marshall's chair at Cambridge, and Keynes's father offered to support
him for up to a year while he finished his treatise. When in 1909 the completed work failed to win him the coveted King's fellowship—a license,
essentially, to take on paying students and dine at high table—Marshall
personally financed an economics lectureship at Cambridge for Keynes. At
that point, the fellows of King's College elected him as one of their own.

In his first communication to his parents after arriving at King's College
as a freshman, the eighteen-year-old Keynes had announced, "I've taken a
good look around the place and come to the conclusion that it's pretty inefficient." [13] As his biographer Robert Skidelsky points out, the institutions
would vary over his life, but never Keynes's view of them or, indeed, of the
world as he found it. They were badly run and in need of more competent
management. Though given to fits of "ungovernable anger," [14] especially
when confronted by stupidity, Keynes was on the whole more exasperated
than outraged, more impatient than self-righteous. He parted company
with his Bloomsbury friends in that he had none of the artist's disdain
for worldly success or people in power. Like Winston Churchill, who confessed to his wife that even when "everything tends towards catastrophe
and collapse. I am interested, geared up and happy," [15] Keynes was more
invigorated than depressed by the world's problems and could not repress
his impulse to make the bad slightly less so or the good a little better.

His response to the war was a characteristic blend of patriotism, op-
portunism, and pragmatism. When England declared war on Germany
in August 1914, he had not known what to think. An incorrigible optimist, he shared the general view that the fighting would be over in a few
months, if not weeks. The first time that the Chancellor of the Exchequer,
David Lloyd George, had asked for his advice was before the fighting broke
out. He spent a day trying to convince Lloyd George not to cave in to pressure from City bankers to suspend gold convertibility of the pound until
absolutely necessary. Keynes's optimism evidently exceeded that of City
bankers.

He was formally drafted by the Treasury in January 1915 and assigned to war finance. The military draft, introduced in 1916 for males ages eighteen to forty-one, raised the personal stakes considerably, for Keynes had become part of the British war machine. At least a half dozen of his closest friends and former lovers were pacifists who had resolved not to fight. They constantly pressed him to end his complicity in a war that he professed to despise. On one occasion he found a note from Strachey on his dinner plate: "Dear Keynes, Why are you still at Treasury? Yours, Lytton." [16] As long as he was on the Treasury staff, Keynes was not actually at risk of being drafted, since the military exempted men who were "engaged in work of national importance." Under intense pressure from his friends to take a stand against the war, Keynes constantly threatened to quit, and in February 1916 he alarmed his parents by going so far as to apply formally for conscientious objector status. In his application, he made it clear that he objected to the coercion of the draft rather than to the war, that is, on libertarian rather than pacifist grounds. After he informed the draft board that he would be too busy at the Treasury to attend his hearing, his application was rejected and he pursued the matter no further. His friends eventually forgave him, especially after he began using his Whitehall connections to protect them when he could. Still, most of Keynes's biographers prior to Skidelsky had considered the episode so potentially damaging to his reputation that they had covered it up, just as they had avoided mentioning his homosexuality.

Keynes's job was to help the Treasury borrow dollars from the Americans on the cheapest possible terms while lending pounds to the French and Britain's other Continental allies on the most lucrative ones, all while protecting the foreign exchange value of the pound sterling. Rounding up scarce currencies such as Spanish pesetas in emergencies was another of his duties, one that gave him hands-on experience as a foreign exchange trader and left him addicted to that risky but thrilling game of betting on one currency's rise, another's fall. Ultimately, writes Skidelsky, all matters of war finance—and many related to postwar finance—passed through Keynes's hands.

Toward the end of the war, as popular hopes that its staggering cost could be recouped from Germany soared, Keynes was increasingly drawn into the vexing debate over reparations. Lloyd George, who had become prime minister of a wartime coalition government at the end of 1916, asked the Treasury to estimate how much the Germans could pay. He had taken it for granted that "the Treasury experts naturally had their mind primarily set on securing some source of revenue which would reduce the crushing burden of taxation by the payment of interest on our gigantic war debt for the next two generations." [17] But other considerations weighed on Keynes, who was ultimately assigned to draft the Treasury position paper. When he delivered his report on reparations to the incoming Treasury chief, Austen Chamberlain, Joseph's son, on the heels of the December 14, 1918, general election, it was a bombshell.

An Allied reparations commission headed by a former governor of New York, Charles Evans Hughes, had already recommended that Germany should be made to pay $40 billion, or roughly one-third of the Allies' war expenditures. Keynes concluded that the most that could be extracted from Germany was £3 billion or $15 billion, less than the amount Britain and France owed the United States. Pointing out that the Allied commission's figure was twice the estimated prewar value of Germany's gold reserves, securities, ships, raw material inventories, factories, and machinery, Keynes warned that placing too high a figure on reparations would ultimately undermine British economic interests by increasing the risk that Germany would ultimately repudiate her debt.

The report caused a furor. Most Britons felt that since the Germans had started the war, they should bear its cost. After all, as Lloyd George pointed out, the burden had to be carried by someone. Prewar tax revenues would not have sufficed to pay *interest* on the war debt. The French national debt had multiplied tenfold; the British, fourfold since 1914. If the Germans didn't pay, then innocent Britons and Frenchmen would have to shoulder higher taxes in order to pay it down. One reason that the British electorate was so passionate on the subject was that nearly 40 percent of the British population owned government securities. British business was also in the pro-reparations camp. They wanted German firms, not British industry, to be taxed in order to repay the debt.

Keynes refused to backtrack, insisting that his £3 billion figure was probably too high. In the furious row that erupted inside the Treasury, he stood his ground, consistently representing the low end of estimates. Lloyd George started referring to him as "the Puck of economics," after Shakespeare's mischief maker and speaker of the immortal phrase "Lord, what fools these mortals be!"[18]

While journalists, politicians, and the public were fixated on the amount Germany should pay, Keynes drew attention to *how* the indemnity might be collected. The easiest method was the oldest and the one that Germany had planned to use to extract reparations from Britain, France, and Belgium had she prevailed on the western front. It was the method proposed by the Hughes commission and involved stripping Germany of her portable public and private property, from stock certificates and gold reserves to ships and machinery. Keynes favored the second alternative, which was to leave Germany's existing wealth more or less intact, supply her with raw material, and levy an annual tribute on her future export earnings. "Having thus nursed her back to a condition of high productivity," Keynes explained, the Allies could "compel her to exploit this product under conditions of servitude for a long period of years."[19]

According to Skidelsky, Keynes went to Paris with twin objectives that were not easily reconciled; namely, reviving the European economy without damaging British export prospects. Two conditions were essential for his strategy to work: a relatively low German indemnity and the willingness of Americans to forgive Britain's war debt. That was the only way Germany could avoid running huge trade surpluses—that is, export more than she imported in order to earn pounds or francs—and that Britain could avoid head-to-head competition with the German export juggernaut. Keynes refused to be daunted by the fact that neither part of his plan was remotely acceptable to the American, French, and British public, something their elected representatives could not possibly ignore.

Ten days after the German surrender, Keynes boasted to his mother, "I have been put in principal charge of financial matters for the Peace Conference."[20] This was an overstatement. His formal role at the conference involved relief, not the political tar baby of reparations. His immediate brief was to help Herbert Hoover work out the financial arrangements required for Europe's transition from war to peace, especially the provision of food.

The armistice called for continuing the blockade of Germany and Austria but permitted exceptions for needed food and medicine. The French had subsequently placed a lien on Germany's remaining gold, hard currencies, and other liquid assets, arguing that they had to be set aside for reparations. With her accounts frozen, Germany could buy no food and faced slow starvation. Keynes was determined to overcome the obstacles placed by the French.

Within a few days of his arrival in France, Keynes was on his way to an "extraordinary adventure" in occupied Germany. He had been asked to join a team of American and French financial experts in Trier, the ancient town on the Mosel River at the intersection of France, Germany, and Luxembourg, where Karl Marx had grown up. Adjacent to the headquarters of Marshal Ferdinand Foch of France and currently occupied by the US Army, Trier had been chosen as the site for renegotiating the November armistice. Though curious "whether the children's ribs would be sticking out," the Allied experts had hardly ventured from the train for three entire days except for a little shopping spree to buy wartime scrip, paper clothing, and other souvenirs.[21] A bridge foursome had formed the first night, and Keynes played more or less around the clock.

Keynes's mission involved finance as well as food. Like Hoover, he was appalled at the blockade and, like President Wilson, convinced that "so long as hunger continued to gnaw, the foundations of government would continue to crumble."[22] He was in Trier ostensibly to find a way to get food trains rolling into Germany. Typical of all negotiations during the Peace Conference, however, things were not quite that simple. As an entirely separate matter, the Allies were determined to get hold of Germany's merchant fleet, now anchored off the city of Hamburg, but were at somewhat of a loss about how to accomplish the takeover. They had not stipulated the surrender of the ships in the armistice, and sending the navy to seize them seemed politically unwise. So it occurred to Allied leaders that the food crisis might provide a convenient opportunity for getting the Germans to strike a deal. Keynes's job was to convince them that "ships against food was . . . a reasonable bargain." As he later admitted, there was an element of bluff involved, not to speak of the difficulty of making it clear to "bewildered, cowed, nerve-shattered and even hungry" financiers "to comprehend how the ground really lay."

In Trier, Keynes watched curiously as the German financiers, dressed like undertakers, approached their train. They walked "stiffly and uneasily," lifting their feet "like men in a photograph or a movie." After they climbed into the railway carriage, they did not extend their hands but only bowed stiffly. They were a sad lot "with drawn, dejected faces and tired staring eyes, like men who had been hammered on the Stock Exchange." [23]

The head of the Reichsbank looked like "an old, broken umbrella." The "sly Corps type" from the Foreign Office had a "face cut to pieces with dueling." The spokesman for the German team was a third figure, "a very small man, exquisitely clean, very well and neatly dressed, with a high stiff collar which seemed cleaner and whiter than an ordinary collar" and with "eyes gleaming straight at us, with extraordinary sorrow in them, yet like an honest animal at bay." This was Carl Melchior, a Jewish banker from Hamburg. He was a liberal, a critic of submarine warfare, and a partner of the banker Max Warburg who had extensive connections in the United States.

Keynes spoke first and asked whether everyone understood English. In his memoir, Max Warburg described Keynes's face as an expressionless mask but said that his voice and his phrasing of questions conveyed sympathy. When it was Melchior's turn to speak, he did so in "moving, persuasive, almost perfect English." The banker used ingenious arguments to plead for a loan, while Keynes endeavored "clearly and coldly" to communicate the idea that a loan was politically out of the question.[24] They managed to agree that the Germans would immediately hand over £5 million in gold and hard currencies in exchange for milk and butter, but that was all.

When Keynes next met the Germans a month later, again in Trier, a stalemate had developed on the question of ships for food. The Germans were determined to hold on to their ships as long as possible, because they saw them as their best bargaining chip in the upcoming peace negotiations and were determined not to give them up without a quid pro quo. What's more, the Germans were under the impression that the United States would be willing to advance them the funds needed to purchase the first few installments of food—a sizeable chunk of which was to consist of surplus American pork.

By the end of the second meeting, the Germans had declared that large-scale food imports couldn't be financed without a loan. If an Allied loan should prove politically impossible, as Keynes had warned would likely be the case, then they would not deliver the ships. If the negotiations broke down and Germany couldn't get food, no one could prevent "the flooding of Bolshevism over the whole of Europe." [25] The negotiators had reached an impasse. Nothing could be done except by the Big Four—the leaders of the United States, England, France, and Italy—but the Big Four were busy arguing over the number of members of the Brazil delegation and hearing proposals from "Copts, Armenians, Slovaks and Zionists." T. E. Lawrence, ostensibly the interpreter for Emir Faisal of Saudi Arabia, took advantage of the Emir's decision to quote several passages from the Koran to propose a scheme for Arab self-rule in the old Ottoman territories. [26]

The next meeting between Keynes and Melchior took place at the beginning of March in Spa, in Belgium, at the former headquarters of the German high command amid hills covered in black pines "far also from the starved cities and growling mob." [27] But the talks again went nowhere, and Keynes was feeling desperate that two months had passed since the first meeting in Trier with no progress toward freeing up gold to pay for food. He sensed that Melchior might feel the same way and asked for permission to sound him out. Getting past Melchior's sullen clerks, Keynes caught him alone and, quivering with excitement, asked if they could speak privately. He recalled:

> Melchior wondered what I wanted . . . I tried to convey to him what I
> was feeling, how we believed his prognostications of pessimism, how we
> were impressed, not less than he, with the urgency of starting food sup-

plies, how personally I believed that my Government and the American
Government were really determined that the food should come, but that
. . . if they, the Germans, adhered to their attitude of the morning a fatal
delay was inevitable; that they must make up their minds to the handing
over the ships.[28]

Melchior promised that he would do his best but held out little hope.
"German honor and organization and morality were crumbling; he saw
no light anywhere; he expected Germany to collapse and civilization to
grow dim; we must do what we could; but dark forces were passing over
us."[29] Keynes's meetings with Melchior confirmed his own pessimism
about the war's devastating consequences, and, not surprisingly given the
uprisings in Berlin and elsewhere in Germany, he shared Melchior's fears
that Germany would succumb to Bolshevism if the treaty terms were too
onerous.

By the next evening, it was obvious that Melchior's effort had come
to nothing and that the new German government in Weimar was digging
in its heels. At times, Keynes seemed more anxious than the Germans
about the threat of revolution and the glacial pace of the negotiations. He
couldn't be sure whether Germany's food supplies were really as depleted
as the British thought. Convinced that a dramatic gesture was required
to break the logjam, Keynes proposed a public rupture and convinced the
team to order its train back to Paris in the middle of the night so that the
Germans would awaken in the morning to find them gone. Back in Paris,
Keynes found that his ploy had succeeded in capturing the attention of the
Big Four. As Lord Riddell, newspaper baron and Britain's wartime press
secretary, wrote in his diary on March 8, 1919:

The Council decided to victual the Germans, provided they hand over
their ships and pay for the food in bills of exchange on other countries,
goods or gold. The French strongly opposed this. LG said to me afterwards that the French are acting very foolishly, and will, if they are not
careful, drive the Germans into Bolshevism. He told me that he had
made a violent attack on Klotz, the French Minister of Finance, in which

he said that if a Bolshevist state is formed in Germany, three statues will be erected—one to Lenin, one to Trotsky, and the third to Klotz. Klotz made no reply ... The Americans are pleased ... All the commercial people, British and American, favor abolishing the blockade and urge an early settlement with Germany so that the world may again get to work.[30]

Four days later, Keynes was on a train bound for Trier in the company of the British admiral Rosslyn Wemyss, whom the Big Four had deputized to deliver the ultimatum to the Germans. The French had succeeded on one point: the Germans had to agree unconditionally to hand over the ships before they were to be told about the food. "D'you think you could see to it that they don't make any unnecessary trouble?" the admiral asked Keynes. So Keynes once again sought out Melchior in private and told him that if the Germans declared their willingness unconditionally, there would be a quid pro quo. "Can you assure me von Braun will do this?" Keynes asked, referring to the head of the German delegation. Melchior paused for a moment before "he looked at me again with his solemn eyes. 'Yes,' he replied, 'there shall be no difficulty about that.'" The next day, everyone stuck to their script: "All was settled now and the food trains started for Germany."[31]

With considerably less difficulty, Keynes also convinced the Allies to approve a loan to pay for British food shipments to Austria by early 1919. After this small triumph, Keynes had the Germans installed at the Château de Villette outside Paris. A plan was afoot to collect financiers from many countries to discuss reconstruction. As it turned out, Keynes visited the chateau only once or twice. Not long after the Germans moved in, the Peace Conference turned away from the issue of reconstruction and became terminally entangled in the matter of reparations.

"The subject of reparations caused more trouble, contention, hard feeling, and delay at the Paris Peace Conference than any other point in the Treaty," Thomas Lamont, the US Treasury representative, wrote afterward.[32] Harold Nicolson remarked that while the conference was often portrayed as a duel between the forces of darkness and light—Wilson versus Georges

Clemenceau of France, Carthaginian versus Wilsonian peace, Keynes versus Klotz—in fact, it was "not so much a duel as a general melee." [33]

The Allies were at loggerheads among themselves. President Wilson opposed saddling Germans with the entire cost of the war. It was reasonable to demand that Germany pay for the damage inflicted by her troops, he argued, but that was all. Then there was the tricky question of what share of the levy on Germany each of the victors would be entitled to and how long. When Lloyd George suggested that payments cease after thirty years, Clemenceau said they should extend for a thousand years if necessary. As late as March 1919, the Allies could not agree over the issue. The French were calling for £25 billion while the United States refused to countenance any figure over £5 or £6 billion. The official British figure was £11 billion. In early March, Keynes finally suggested leaving the total amount of reparations to be paid out of the treaty. That solution was ultimately adopted.

Lloyd George, annoyed by constant press leaks, suggested that the Big Four meet privately. Thus, the second half of the Peace Conference, from mid-March to mid-May, took place in Woodrow Wilson's "tiny study." At first alone except for one interpreter, the heads of state of the United States, UK, France, and Italy—Woodrow Wilson, David Lloyd George, Georges Clemenceau, and Vittorio Orlando—sat around a fire in overstuffed chairs, recalled Nicolson, "with maps spread out on the floor which they were sometimes obliged to crawl about on their hands and knees to study, the big Four managed to hammer out what was, in effect, the penultimate version of the settlement." [34]

April turned out to be the cruelest month. As the weather turned warm, the formerly festive atmosphere in Paris suddenly became frenetic. The reservations that many participants had had about holding the conference in Paris were confirmed: bedbugs, medieval plumbing, and price gouging were the least of it. The press had turned vituperative. "The constant clamour of their newspapers, the stridency of their personal attacks, increased in volume," observed British diplomat Harold Nicholson. "The cumulative effect of all this shouting outside the very doors of the Conference produced a nervous and as such unwholesome effect." [35] Lloyd

George had to face down a rebellion in Parliament, where conservatives feared he was not being tough enough on Germany. Clemenceau became the bête noir of the French press, which was convinced that he was being outmaneuvered by the English and Americans. Orlando left the conference. And Woodrow Wilson became terribly ill with either food poisoning or influenza. By May the fights among the four grew so bitter that on one occasion Wilson was forced to intervene physically between Lloyd George and Clemenceau.

The conference within the conference not only froze out the representatives of the smaller countries but left experts like Keynes in the cold too. The Big Four made far-reaching economic decisions on the fly with hardly any input. President Wilson considered the British plea for debt forgiveness for a few minutes before summarily rejecting it. The British prime minister consulted Keynes when he wanted "to wriggle out of his commitments," observed Lloyd George's biographer, but "he never thought of taking his advice."[36] After twelve-hour days climbing in and out of drafty cars and racing from one overheated room to another, Keynes often had dinner with Jan Smuts, a South African member of the British war cabinet and strong proponent of the League of Nations and reconciliation with Germany.

> Poor Keynes often sits with me at night after a good dinner and we rail
> against the world and the coming flood . . . Then we laugh, and behind
> the laughter is Hoover's horrible picture of thirty million people who
> must die unless there is some great intervention. But then again we think
> things are never really as bad as that; and something will turn up, and
> the worst will never be.[37]

On May 7, 1919, Herbert Hoover, who was solid and square and struck most Europeans as gratuitously pugnacious in his day-to-day dealings, was heading down the Champs-Elysees before dawn. The streetlights still glimmered, and the avenue was deserted. He walked slowly and kept his head down, like a pugilist after a losing match. He did not expect to meet anyone he knew. Except for a few ascetic French generals, the delegates at the Peace Conference liked to linger over their *Times* of London and English marmalade at breakfast. So he was surprised to see two familiar figures in bowlers crossing the boulevard in his direction. Keynes and Smuts were talking animatedly, heads together and seemingly oblivious to his presence. What were those two doing out at this hour?

When they got close enough to recognize him, they too gave little starts of surprise. It dawned on all three at once: each had been up since at least four in the morning, when the freshly printed draft treaty was delivered to their rooms by messenger. None had seen the treaty in its entirety before, although Keynes had read, with mounting dismay, parts of the draft as early as May 4. Despite their insider's knowledge and Keynes's and Smuts's cynicism about the proceedings, they were shocked. Each had been driven out of doors by anger, disbelief, and awful premonitions. After this flash of telepathy, Hoover, Keynes, and Smuts all began to talk at once. As Hoover recalled, "We agreed that it was terrible." [38]

Within two weeks, the incorrigible optimist had moved out of his room at the Majestic, rented an apartment with cook and valet on the edge of the Bois de Boulogne, and crawled into bed, too depressed to get up except when the PM summoned. By May 14, feeling like "an accomplice in all this wickedness and folly," Keynes had made up his mind to resign. "The Peace is outrageous and impossible and can bring nothing but misfortune," he wrote to his mother, Duncan Grant, and others. [39]

Keynes's final intervention was a protest against "murdering Vienna." [40] The negotiations over Austria had been postponed until the German terms were settled. He was getting regular reports from Francis Oppenheimer, the Treasury's emissary on the scene, who had been in constant touch with Joseph Schumpeter, who, in turn, was supplying data on Austrian assets, tax revenues, and the like. On May 29, Keynes sent Lloyd George a memorandum pleading the case that Austria should pay no reparations. On the thirtieth, he attended a meeting of the Austrian reparations commission and won a major concession, getting a demand for 10 billion gold crowns in reparations dropped. Quoting ghastly statistics about children dying of TB and malnourishment, he was partly successful in modifying a demand from the French that Austria surrender her milk cows.

Keynes agreed with one Viennese newspaper's harsh criticism of the treaty:

> Never has the substance of a treaty of peace so grossly betrayed the intentions which were said to have guided its construction as is the case with this Treaty ... in which every provision is permeated with ruthlessness and pitilessness, in which no breath of human sympathy can be detected, which flies in the face of everything which binds man to man, which is a crime against humanity itself, against a suffering and tortured people.[41]

Although he must have known that it was a lost cause, Keynes continued to plead with Bernard Baruch for the US Treasury to endorse "my Grand Scheme for putting everyone on their legs."[42] Lloyd George called a special meeting of the British delegation and promised to refuse the services of the British Army to advance into Germany, or the services of the British Navy to enforce the blockade of Germany in order to obtain eleventh-hour changes in the treaty. But, as Keynes predicted in a letter to his mother, it really was too late for grand gestures. The French were furious, and President Wilson, who should have been sympathetic, had grown increasingly suspicious of British intentions. He vetoed Lloyd George's proposal as peremptorily as he had rejected Keynes's proposal for debt forgiveness a month earlier. Lloyd George did not press the case further, possibly because of an intelligence report that the German cabinet had already secretly decided to sign the treaty. Nonetheless, he predicted gloomily, "We shall have to do the whole thing over again in twenty five years at three times the cost."[43]

By the time the Germans actually signed the Treaty of Versailles on June 28, Keynes had been back in England for nearly a month. He had retreated to Charleston, Virginia and Vanessa Stephen's house in the country, where he spent long hours weeding furiously to distract himself. He had dashed off a letter of resignation to Austen Chamberlain, the Chancellor of the Exchequer, on June 5. On the same day, he had also written to Lloyd George, "The battle is lost. I leave the twins [the judge Lord Sumner and financier Lord Cunliffe, head of the British reparations commission] to gloat over the devastation of Europe and to taste what remains for the British taxpayer."[44]

Austin Robinson, son of an Anglican minister, war pilot, and Cambridge undergraduate, dated his "conversion to the faith of the economists" to October 1919, when he attended one of Keynes's last lectures for the term.[45] Keynes read from his half-finished manuscript about the peace treaty in front of a large audience. Robinson was incredibly moved by "the very obvious depth of his dedication to the problems of the world and his hatred of failure to avert foreseeable disaster."[46] For Robinson's generation, which wanted to put the war behind it by doing something to heal its wounds, Keynes's argument that getting the economics right was essential for preventing future wars was a genuine revelation. He was intrigued by Keynes's conviction that ideas mattered as much as, if not more than, competing economic and political interests.

Keynes had begun writing almost as soon as he returned to Cambridge from Paris. He plucked his theme for *The Economic Consequences of the Peace* from a clever remark by Jan Smuts's mistress: "Mrs. Gillett, referring to the Anti–Corn Law League, had reminded Smuts that economic reform had preceded franchise reform in the nineteenth century, and that 'now it seems as though in the same way the political and territorial questions won't be solved 'til the economic world is righted.'" Smuts reported this remark to Keynes, who said "how true it was and he had never thought of it that way."[47] Margot Asquith, the amusing wife of the former prime minister, had suggested to Keynes that he include portraits of the major personalities. In August, Macmillan's in London agreed to publish the book, although Keynes had to agree to pay the printing costs. Felix Frankfurter, with whom he had become friendly in Paris, arranged for an American edition.

Keynes blasted the treaty as a rank betrayal by the older generation of political leaders. Not only had the Big Four done nothing to restore the prewar European economy, but they had not seriously considered the need to do so. They had simply assumed that the restoration of broken ties and rebuilding would happen spontaneously.

The Treaty includes no provisions for the economic rehabilitation of Europe, nothing to make the defeated Central Empires into good neighbors, nothing to stabilize the new States of Europe, nothing to reclaim Russia; nor does it promote in any way a compact of economic solidarity amongst the Allies themselves; no arrangement was reached at Paris for restoring the disordered finances of France and Italy, or to adjust the systems of the Old World and the New. . . .

It is an extraordinary fact that the fundamental economic problems of a Europe starving and disintegrating before their eyes was the one question in which it was impossible to arouse the interest of the Four. Reparation was their main excursion into the economic field, and they settled it as a problem of theology, of politics, of electoral chicanery, from every point of view except that of the economic future of the states whose destiny they were handling.

A Carthaginian treaty, "if it is carried into effect, must impair yet further, when it might have restored, the delicate, complicated organization, already shaken and broken by war, through which alone the European peoples can employ themselves and live."[48]

The Economic Consequences of the Peace is extraordinarily gloomy, prompting Leonard Woolf to nickname its author Keynessandra. "In continental Europe the earth heaves and no one but is aware of the rumblings," Keynes writes. "There it is not just a matter of extravagance or 'labor troubles'; but of life and death, of starvation and existence, and of the fearful convulsions of a dying civilization." Part of Keynes's gloom stems from his sense that it was "not only the war that has made Europe poorer." Looking backward, Keynes now saw the prewar prosperity as a fool's paradise.

We assume some of the most peculiar and temporary of our late advantages as natural, permanent, and to be depended on, and we lay our plans accordingly. On this sandy and false foundation we scheme for social improvement and dress our political platforms, pursue our animosities and particular ambitions, and feel ourselves with enough margin in hand to foster, not assuage, civil conflict in the European family.

Living standards could not have continued to rise much longer, he maintained. The prosperity of Europe had been based not on the "ingenious mechanism" of competition, an environment friendly to entrepreneurs and ample finance, but rather on a happy historical accident that had temporarily removed certain limits to growth. Thanks to the large exportable surplus of foodstuffs in America, Europe had been able to feed herself cheaply.

The trouble was, Keynes wrote, that American grain *couldn't* stay cheap when US consumption caught up with supply. He reprised an argument by Arthur Jevons, a gifted contemporary of Marshall's, who predicted in 1870 that dwindling coal supplies would choke off England's economic growth. Instead of fuel, the binding constraint for Keynes was wheat. There might be no shortage of wheat in the world as a whole, he acknowledged. But to call forth more supply in the future, he argued, would require England to offer a higher real price. In short, the law of diminishing returns would at last reassert itself, requiring Europe to offer more and more other goods and services to obtain the same amount of bread.

Keynes's bleak economic forecast turned out to be too pessimistic. In the short run, Europe's economy recovered in spite of the war's devastation and the flaws in the treaty. In the long run—starting in the Great Depression and continuing past the end of the twentieth century—food became cheaper, not more expensive, absolutely as well as relative to wages. Keynes's political prediction—that "vengeance . . . will not limp" and that "nothing can then delay for very long that final civil war between the forces of reaction and the despairing convulsions of revolution"—was far more prescient.

World War I and its aftermath set Keynes's intellectual priorities and shaped his thinking about the economy, Skidelsky argues. Henry Wick-

ham Steed, editor of the *Times* of London, characterized Keynes's ideas as
a "revolt of economics against politics."[49] Keynes was asserting the importance of something about which generals and prime ministers were only
superficially familiar: how the modern world made its living, and that the
ability to make a living was a prerequisite if not a guarantor of peace.

Keynes appreciated how specialized the global, especially the European, economy had become, how dependent each part was on the others,
how subject to psychological shifts, and, consequently, how easily a breakdown in one could spread to the rest. Keynes had not yet identified policy
levers—instruments of mastery—that would let governments exert more
control over their economy's course. But he was beginning to think in
terms of an "economics of the whole," and of the consequences of government action and inaction.

The war had deepened his distrust of conventional wisdom and had
disabused him of any notion that progress was automatic. It was, all in
all, a brutal lesson in the destructive powers of governments that willfully
ignored economic realities. The Victorian economic miracle had produced
the rapid growth of productive power and a dramatic rise in living standards. But the miracle had depended on certain government actions—
spreading free trade, enabling the gold standard, upholding the rule of
law—as well as untrammeled competition. Having absorbed that lesson,
Keynes could not conceive how government could ignore its responsibility
to restore prosperity.

In mid-October, Keynes was on the Continent for an international bankers' conference. "There has never been as a big a business transaction as
the peace treaty," Melchior's partner Max Warburg had remarked.[50] Now
his brother Paul, the American financier, hoped to organize commercial
credits, financed mostly by American banks, so that Germany could import raw materials. On a whim, Keynes had invited Melchior by telegram
to meet him. Three days later, the two were strolling along Amsterdam's
canals in the rain talking freely for the first time and marveling at how
"extraordinary [it was] to meet without barriers."[51]

After resigning, the German delegation in protest before the signing of the peace treaty and twice turning down offers to become Weimar's finance minister, Melchior had gone back to his Hamburg bank. He told Keynes that the German president had betrayed Germany's intention to sign the treaty to a British agent ahead of time. Melchior was certain that the tip had led Lloyd George to abandon his efforts to modify the treaty. After lunch, Keynes invited Melchior and Warburg back to his hotel room and read aloud his chapter on President Wilson. Keynes had portrayed the American leader as having raised the world's hopes only to disappoint them:

> With what curiosity, anxiety, and hope we sought a glimpse of the features and bearing of the man of destiny who, coming from the West, was to bring healing to the wounds of the ancient parent of his civilization and lay for us the foundations of the future.
>
> The disillusion was so complete, that some of those who had trusted most hardly dared speak of it. Could it be true? they asked of those who returned from Paris. Was the Treaty really as bad as it seemed? What had happened to the President? What weakness or what misfortune had led to so extraordinary, so unlooked-for a betrayal?

Wilson could preach ringing sermons on his Fourteen Points, but lacked

> that dominating intellectual equipment which would have been necessary to cope with the subtle and dangerous spellbinders whom a tremendous clash of forces and personalities had brought to the top as triumphant masters in the swift game of give and take, face to face in Council.[52]

Warburg, who despised the president, giggled as Keynes read these lines, but Melchior listened solemnly and looked as if he was about to cry.

When the bankers held their meeting, Keynes urged them to support a reduction in reparations, cancellation of Allied war debts, and an international loan for Germany. He and Warburg drafted an appeal to the League of Nations and got a dozen of the conference participants to sign. Thus, the first of many attempts to revise Versailles was drafted before the ink on the treaty had dried.

With his daily quota of a thousand words "fit for the printer" seven days a week, Keynes had piled up sixty thousand words by October. As chapters were finished, he read or sent them to various people, including his mother, Frances, and Lytton Strachey. The whole publishing industry seemed dedicated to churning out books about the peace treaty. Keynes's book was the first out of the gate, appearing two weeks before Christmas. By Easter, a hundred thousand copies had been sold in England and the United States. Keynes's "reparation" to Bloomsbury for having abetted the war was graciously accepted. Lytton Strachey, whose *Eminent Victorians* had been the literary sensation of 1918, called Keynes's argument "crushing" and predicted that "nobody could ignore it."[53] Though grumbling that Keynes had been very indiscreet, Austen Chamberlain confessed to his wife that the book was "brilliantly written" and had given him "malicious pleasure."[54] All reviewers were lavish in their praise of Keynes's style and many were persuaded that it would be impossible for Germany to comply with the treaty.

Keynes's book brought a simmering controversy to the boiling point. Some of the critics argued that Germany could afford to pay much more than Keynes said. Others called him politically clueless. Among the less flattering suggestions were that he was a "dehumanized intellectual" for his lack of partisanship. Predictably, the Tory attack on Keynes questioned his loyalty and suggested that perhaps he deserved an iron cross. A. J. P. Taylor, the historian, succinctly, and not unfairly, summarized the message of *The Economic Consequences of the Peace* as "Precautions should be taken against German grievances, not against German aggression."[55] Captain Paul Mantoux, the Big Four's translator, attacked the book on the grounds that Keynes "had never been present at one of [the Council of Four's] meetings."[56] But the most common criticism was simply that Keynes had missed the point. Wickham Steed, editor of the *Times* of London, noted that

If the war taught us one lesson above all others it was that the calculations of economists, bankers, and financial statesmen who preached the
impossibility of war because it would not pay were perilous nonsense.
Germany went to war because she made it pay in 1870–1 and believed
she could make it pay again.[57]

American reviewers suspected Keynes of advancing British interests
under the guise of altruism toward Europe. Thorstein Veblen, the sociologist, scolded him for getting Woodrow Wilson "all wrong."[58] On the
first anniversary of the treaty, the *New York Times* called *The Economic
Consequences of the Peace* "a very angry book" and claimed that "insofar as
American opinion has changed, it is into distrust of all Europe and a desire to break away from entanglement from abroad."[59] Bernard Baruch expressed the administration's position when he claimed that Keynes wanted
that "America shall pay instead of Germany."[60]

Some historians now accept that Keynes's criticism of President Wilson was unfair, and consider his condemnation of the French too partisan.
If anything, they contend, British claims for reparations were less justifiable than those of the French. On the other hand, Margaret MacMillan's
Paris 1919: Six Months That Changed the World and other recent histories
of the Peace Conference show that Keynes's view that the Allies had bla-
tantly violated their contract with Germany, and should have allowed the
losers to negotiate some elements of the peace, is widely accepted now.
And few disagree with Keynes's principal point—that no peace based on
such shaky economic foundations could possibly last.

Not surprisingly, *Economic Consequences* turned Keynes into a hero in
Vienna and Berlin. Excerpts, translations, and new editions poured forth
from the presses. Given that no ceiling on the amount of reparations was
fixed in the treaty, the view that Keynes had not only voiced the German
case but was in a position to influence opinion made perfect sense. Joseph
Schumpeter, the former Austrian finance minister, called the book "a masterpiece."[61]

Chapter VIII

The Joyless Street:
Schumpeter and Hayek in Vienna

The alternating boom and bust is the form economic
development takes in the era of capitalism.

—*Joseph Schumpeter*[1]

The 1920s are almost always viewed in a rearview mirror and judged solely
as a preamble to, if not the cause of, the Great Depression, the rise of Fascism, and the triumph of Bolshevism. For the West, it is supposed to have
been a time of decadence, delusions, fake prosperity, and false beliefs. But
seen through the eyes of four individuals—Joseph Schumpeter, Friedrich
Hayek, John Maynard Keynes, and Irving Fisher—it was as inventive, exciting, and genuinely progressive an era as any in the last century.

Keynes and Fisher became economic oracles. They prospered financially. More important, they created new intellectual wealth. The violent
inflations and deflations that followed World War I convinced them that
free markets and democracy could not long survive such pathologies and
focused their minds on systemic causes. Like the doctor in Molière's *Le
Malade Imaginaire,* they shifted their attention from individual parts of
the economic body to its circulatory system. They concluded that inflation and deflation, seemingly polar opposites, were symptoms of the same
underlying disease, and that the system of money and credit creation was
both its source and its transmission mechanism.

Solving the immediate problem of how to revive the interlocking parts of the world economy, some in extremis, required a new framework. Fisher and Keynes nourished the hope that the violent booms and depressions could be avoided. They no longer believed, as Alfred Marshall had, that booms and depressions resulted from random external shocks or, as Karl Marx had, that they were intrinsic to the market economy. Instead of acts of nature, extreme gyrations were man-made disasters capable of being averted. Fisher, Keynes, and Hayek searched for instruments of mastery, confident that these existed and could be made to work—even if the Englishman and the American were prepared to trust the discretion of civil servants while the Austrian, product of a more tragic national history, insisted that governments be bound by rules. Only Schumpeter could be described as fatalistic, as much because of temperament and personal tragedy as because of intellectual conviction.

When Schumpeter was driven from office in the fall of 1919, Austria's fiscal crisis was entering an acute stage. Faced with a ballooning deficit and too fearful of public unrest to impose austerity measures, the cash-strapped government of Karl Renner printed more and more paper money to pay its bills. Ludwig von Mises, the president of the Chamber of Commerce, described the "heavy drone" of the central bank's printing presses: "They ran incessantly, day and night . . . [Meanwhile] a large number of industrial enterprises were idle; others were working part time; only the printing presses stamping out notes were operating at full speed."[2] The more kronen the government issued, the less one krone could buy. Vienna's police chief complained that "every issue reduces the value of the krone."[3] The effect on the krone's exchange value was immediate, and because Austria imported so much of her daily requirements, the plunging exchange rate sent domestic prices spiraling upward. Ironically, the Social Democrats initially welcomed inflation as an economic stimulant, not suspecting how soon it would end in prostration—and political ruin—like any other episode of mania.

Initially, easy credit and rising prices seemed to jolt the paralyzed economy back to some semblance of life. Investment, exports, and employment perked up as inflation slashed the real cost of borrowing and the falling krone let Austrian exporters undercut foreign rivals. But eventually Austria's trading partners began imposing tariffs on her exports, businesses had difficulty restocking inventories, and unemployment began to swell again.

Meanwhile, inflation went from a trot to a canter to a wild gallop. Despite constant renegotiation of union contracts, a worker who had earned 50 kronen a week before the war earned about 400 kronen at the end of 1919. But his new salary let him buy just one-quarter as much food, coal, or clothing as his old one. Instead of an eightfold pay raise, he had actually suffered a 75 percent pay cut. Within a year he would need more than eight weeks' pay to buy a cheap suit and a pair of boots.[4] Civil servants and pensioners found that their weekly incomes afforded them no more than a couple of eggs or loaves of bread. That was just the beginning. At one point, Freud, who was considering moving to Berlin, complained, "One can no longer live here and foreigners needing analysis no longer want to come."[5] By October 1921, prices were rising by more than 50 percent every month on average, marking the beginning of a hyperinflation. By October 1922, the price level was two hundred times higher than a year earlier.

Inflation wiped out the entire savings of the middle class and undermined its faith in democratic government argues historian Niall Ferguson. "All of us have lost 19/20ths of what we possessed in cash," Freud wrote to a friend.[6] Worthless bank notes, like ersatz food and paper suits, created a universal feeling of being cheated. In Stefan Zweig's "The Invisible Collection," a blind art connoisseur believes that his portfolio of old master drawings is intact. In reality, his family has bartered them away and substituted blank sheets of vellum for the missing works. Anna Eisenmenger, the diarist, wrote of feeling betrayed as she surveyed her "remaining 1,000 kronen notes; lying by the side of my food [ration] cards in the writing table drawer . . . Will not they perhaps share the fate of the unredeemed food cards, if the State fails to keep the promise made in the inscription on every note?"[7] As confidence in the krone collapsed, daily commerce reverted to barter. Many peasants and shopkeepers refused to accept cash. For the middle classes, it meant trading a piano for a sack of flour, fifty prewar cigars for four pounds of pork and ten pounds of lard, a gold watch chain or, in Freud's case, a journal article for a few sacks of potatoes.

Until the shelves were bare, one could buy anything in a Viennese shop, including its entire contents, for a few pounds or dollars. *La Peine des Hommes: Les Chercheurs D'Or,* a novel published in 1920 by the French journalist Pierre Hamp, portrays an inner city populated by carpetbaggers who flock to Vienna like vultures. Salzbach, the hero, accuses them of turning misery into gold.[8] Bigger bargains were found in Austria's farmland, mines, railways, ships, power plants, factories, and banks. As the krone depreciated, these too became available at fire sale prices if the buyer paid in pounds, dollars, or another "hard" currency. Foreign takeovers stoked popular anger, one reason that the Kola affair involving Fiat's purchase of Alpine Montan, the iron concern, continued to dog Schumpeter long after he left office.

While war veterans hung around outside dozens of restaurants inside the Ring waiting for scraps, a new class of millionaires drank champagne and dined on delicacies "equal in quality and quantity to that obtained in London."[9] The stark contrast between the newly rich and the newly poor that had disgusted the young Adolf Hitler before the war grew more extreme. Panhandlers, beggars, and refugees seemed to be everywhere. Popular resentment focused on black marketeers, war profiteers, foreigners, and especially Jews. Every surge in food prices was followed by demonstrations against the rising cost of living and outbursts of violence. In December 1921, a huge crowd smashed shop windows, attacked hotels, and looted food shops. One visitor to Vienna wrote to his wife that "hand in hand with the exasperation caused by the continual rise in prices goes a feeling of intense resentment and hate against all those who have made money out of Austria's misfortune, the 'Schiebers,' speculators on the exchange market, and their like 'who are mostly Jews.'"[10]

Inflation turned the old Vienna into a funhouse of inverted, topsyturvy values. In *The Joyless Street,* Georg Pabst's 1925 film starring Greta Garbo, senior civil servants huddle in dark, unheated apartments, neighbors spy on one another, housewives break the law, girls from good homes become prostitutes, and sober citizens turn into manic stock speculators.

Gilt-edged stock certificates became the inflation hedge of choice. People who had never invested in anything but government bonds suddenly poured what was left of their cash into the stock market, where huge profits were being made.

In her diary, Anna Eisenmenger reports a conversation with her bank manager that captures the helplessness of the middle class in the face of the speculative fever that was attacking the whole population:

> "If you had bought Swiss francs when I suggested, you would not now have lost three-fourths of your fortune."
>
> "Lost?" I exclaimed in horror. "Why, don't you think the krone will recover again?"
>
> "Recover?" He said with a laugh . . . Our krone will go to the devil, that's certain."
>
> "Come into my room for a moment . . ." There he began to explain to me that the monarchy was compelled to issue war loans and that the subscription to these loans was often compulsory. This was done because the State had already used up its gold reserves and had no money left for carrying on the War. With the money from the war loans the War was continued, but there was practically no cover for the notes at present in circulation.
>
> "Just test the promise made on this 20 kronen note and try to get, say, 20 silver kronen in exchange for it," he said, holding out a 20 kronen note . . . "Now you will understand me when I tell you that at the present time it is well to possess a house or [land] or shares in an industry or mine or something else of the sort, but not to possess any money, at least no Austrian or German money. Do you understand what I mean?"
>
> "Yes, but mine are government securities. Surely there can't be anything safer than that."
>
> "But, my dear lady, where is the State which guaranteed those Securities to you? It is dead."[11]

The conversation ended with the bank manager advising Madame Eisenmenger to put her money in stocks. Like countless other Viennese, she did.

• • •

Although his political career was apparently over and he had been forced to return to his university post in Graz, Schumpeter still had friends in high places. To compensate him for his ignominious sacking, conservatives in parliament awarded him a banking charter the following year. The charter was his to sell, use, or sock away. Since there were fewer than two dozen investment banks in Vienna and many banking partnerships were desperate to raise capital by selling shares to the public, a license to start a bank had many potential takers. Parliament's gift to Schumpeter proved a golden parachute of considerable value.

On July 23, 1921, Schumpeter was elected president of Vienna's oldest investment bank, M. L. Biedermann, on the day it went public. He was twenty-nine years old. In return for the use of the charter and his signature on banknotes and such, Schumpeter got a magnificent office, an annual salary of 100,000 kronen (about $250,000 in today's dollars), and enough stock to make him the bank's second-largest shareholder. The biggest perk was a practically unlimited line of credit to invest on his account.

His timing was perfect. The League of Nations was finally cobbling together a rescue package for Austria that bore a striking resemblance to the stillborn Schumpeter plan of 1919. In return for an emergency loan, the government promised to embrace fiscal and monetary discipline by creating a new central bank that would be barred from financing the government's deficits by buying Treasury bills, by balancing its budget by firing a hundred thousand civil servants and closing tax loopholes, and, when Austria's foreign debt had shrunk to a specified level, by returning to the gold standard. The rumor of the impending deal and the simultaneous announcement that the Allied Reparation Commission was renouncing claims on Austria sufficed to halt the krone's decline and brought inflation down from 1,000 percent to 20 percent even before the protocols were signed in August.[12]

The speculative fever did not abate, but instead shifted to gilt-edged stocks. As businesses issued shares instead of taking out loans at higher real interest rates, banks gobbled up the new stock certificates. Soon banks were the largest shareholders in Austrian business. According to the historian C. A. Macartney,

> Austrian banks—a very few conservative concerns of established reputation always excepted—did not confine their investments by considerations of safety, things went on just as merrily on these shares as they had on the exchange. Industry itself, including the most reputable, had become speculative. It passed largely into the hands of the banks, its shares were bandied about and used for the most improbable purposes.[13]

As expected, Schumpeter left the banking to Biedermann's capable longtime chief and became, in effect, a money manager and venture capitalist. He promptly took large stakes in several enterprises, in some cases with a partner who was an acquaintance from the Theresianum. Within months, he was a director of the Kauffman Bank, a porcelain works, and a chemical subsidiary of a German multinational.[14]

The frenzy of deal making, buying, and selling was intoxicating. Schumpeter may have dressed like a bank president, but, as the Viennese press observed snidely, his lifestyle was as extravagant as a lord's. He still owed large personal debts and even larger tax arrears. He had given up his hotel suite and his half of the Palais, but he threw lavish dinners at his apartment and spent prodigiously, whether on his mistresses, horses, or clothes. He was as careless of his reputation as he was of his money. In response to a business associate's warning about appearing in public with prostitutes, "he rode up and down . . . a main boulevard in the inner city . . . with an attractive blond prostitute on one knee and a brunette on the other."[15]

At the beginning of 1924, Schumpeter considered his financial affairs to be "perfectly in order,"[16] his Biedermann line of credit being covered by gilt-edged securities. Then came the spectacular collapse of the Vienna stock exchange on May 9, 1924. Between breakfast and dinner, three-fourths of the value of the "highly marketable securities" that constituted Schumpeter's personal collateral had disappeared in a puff of smoke.[17]

In the frantic days that followed, he was forced to dump the best of his remaining stocks into a falling market. Thanks to a wrong-way bet against the French franc, the Biedermann bank had sustained huge foreign-currency losses. To raise cash, the bank's directors, Schumpeter included, had to sell large numbers of Biedermann shares to a subsidiary of the Bank of England. Over the summer, several of his companies failed, forcing him as a director to compensate their shareholders. His Theresianum partner turned out to be, if not a crook, then guilty of shady dealings, and Schumpeter was named in several lawsuits as well as a criminal inquiry that dragged on for years.

The combination of personal insolvency and an unsavory business associate was too much for Biedermann's British investors, who insisted that Schumpeter resign. By the time he did so in September 1924, amid accusations in the media that he had used his Biedermann connections to do favors for a government minister, nothing was left of his millions. The bank directors granted Schumpeter a severance equal to a year's salary, but his debts were far larger and he had no prospect of recouping his losses. The financial crisis triggered a prolonged recession. Several big banks and hundreds of industrial and commercial firms failed. Ultimately, Biedermann was liquidated, although, amazingly, the investors were all repaid. In the depths of the slump, Ludwig von Mises motioned another economist to his office window and, pointing down to the Ringstrasse, that symbol of Vienna's liberal age, said gloomily, "Maybe grass will grow here because our civilization will end."[18]

If Schumpeter's enemies judged him harshly, they didn't condemn

that connotes

since the start of the war; his entire thirties, more or less. At forty-one, he found more to regret than to look forward to.

The black mood didn't last. The need to defend himself and to find a way to make money energized him. And at the end of his annus horribilis he found a reason to smile again. Like most Don Juans, Schumpeter had been infatuated—even obsessed—countless times, but he had never been truly in love. Annie Reisinger disarmed him because she was young, working class, and vulnerable. She was the twenty-one-year-old daughter of the concierge of his mother's apartment building. He had known her since she was a baby. When she was eighteen, he had begun a mild flirtation with her, only to be rebuffed. She had been more frightened by his reputation as a womanizer than by the fact that he was a public figure twice her age. He had run into her again on Christmas Day when she paid his mother a visit. She was prettier, more womanly, and more self-possessed than he remembered. Jaded as he was, he found her cheerful good nature and lack of intellectual pretention refreshing.

He was lonely and sore. She was recovering from an unhappy affair with a married man. They were both on the rebound. Schumpeter made a project of their romance. He courted her daily. He swept her off to operas, balls, restaurants, and weekends in the country. He showered her with flowers and expensive trinkets. When he asked her to marry him, he went down on his knees.

Appalled as his mother may have been at the prospect of a shopgirl as a daughter-in-law, she bit her tongue. A man who was as notorious as he was penniless was hardly in a position to make the brilliant match she had craved for him. Besides, he was still legally married to his first wife. Schumpeter had not seen Gladys, who had since resumed the use of her maiden name, since they parted in 1913. It isn't clear whether she had refused to consider a divorce or he simply hadn't bothered to ask for one. What is plain is that they were still legally married and that Gladys, if she wished, could have prevented his remarriage or sued him for bigamy. Lu██d Vi██████ eralized the divorce laws, and ████████████████ to marry Annie. Annie overcame her own and her parents' misgiving██ went along.

Meanwhile, Schumpeter's friends were looking for ways to rescue his career. Despite his misadventures in politics and banking, his reputation as a brilliant economic theorist had survived. True, he had made enemies in Vienna and Berlin who would block his appointment at either of those cities' universities, but many others abroad, including the University of

Tokyo, were eager to recruit him. Ultimately, the University of Bonn, the first university out of which Karl Marx had flunked, offered him a chair in public finance. "Schumpeter is a genius," began a letter from one of his supporters to the cultural ministry in Berlin. German universities were completely cut off from contemporary economics, the writer pointed out, and Schumpeter would turn Bonn from a backwater into an important center of theoretical economics.

"Bonn conquered!" Schumpeter telegraphed his fiancée triumphantly in October 1925 when he heard that he had beaten out his Viennese rival, von Mises. Somewhat to his own surprise, he was eager to go. Although his chair was in public finance, he had been promised that he could lecture on pure theory. In early November he and Annie were married in the presence of just two witnesses before embarking on a leisurely tour of luxury spas in northern Italy. They arrived in Bonn just before the start of the spring term.

Schumpeter and his wife soon became the most glamorous couple in Bonn. In a typically grandiose gesture, Schumpeter rented an imposing stucco mansion overlooking the Rhine where Kaiser Wilhelm had lived as a student. By the time Annie attended her first faculty tea, Schumpeter had invented a new identity for her. Instead of a super's daughter who had worked as a bank teller in Vienna and an au pair in Paris, he presented her as the pampered offspring of a prominent Viennese family who had been educated at an expensive French finishing school. Schumpeter's crushing debts forced him to moonlight as a journalist and public speaker, but everyone who knew him agreed that he was happier than he had been in years. Among other things, Annie was pregnant with their first child.

The idyll was not to last. The sudden, unexpected death of his mother in mid-June was a severe blow. For most of his life, she had been "the great human factor," and he often spoke of "his unconditional attachment to her, his unbounded confidence in her."[19] Two weeks after Schumpeter returned from her funeral in Vienna, he suffered a second horrible loss, witnessing Annie's "terrible death" during childbirth.[20] Their baby boy lived less than four hours.

The Biedermann fiasco and the loss of the only two people to whom he felt close scarred Schumpeter permanently. He would need more than a decade just to pay off his debts, and he never got an opportunity to rebuild his fortune. Half a dozen years later, Schumpeter wrote in a letter from Singapore:

> There is no real liberation. I can't get rid of bad memories and premonitions ... mistakes, failures, hardships, etc. and the year 1924, never stand out so clearly before my eyes as when I am on a beautiful boat, seemingly comfortably safe on a still Ocean. And the feeling of decadence, intellectual and physical often condenses into direct forebodings of death.[21]

Yet on the future of capitalism, Schumpeter remained remarkably sanguine. Israel Kirzner, the economist, observed that the questions that drove research on business cycles in the 1920s were: Can capitalism work?[22] Can an economy with private property and free markets survive? Karl Marx had believed that panics and slumps were generated by the economic system and would eventually destroy it. Alfred Marshall had taken the opposite view, attributing recessions to random shocks that originated outside the economy. Schumpeter stood Marx on his head by viewing the business cycle as intrinsic but essentially benign. As "commonly, prosperity is associated with social well-being, and recession with a falling standard of life. In our picture they are not, and there is even an implication to the contrary."[23]

Despite frequent crises and depressions since 1848, he pointed out, production and living standards had risen by multiples. Growth had occurred in spurts because innovations were not "evenly distributed in time ... but appear, if at all, discontinuously in groups or swarms."[24] Innovation bred imitators, another burst of investment, and secondary rounds of innovation. Then investment subsided and consumer goods flooded the market, driving down prices and pushing costs higher. The squeeze on profits resulted in recession.

Constant dislocations were the downside to innovation, rising productivity, and higher living standards. In Schumpeter's theory of economic

In Jane Austen's lifetime, "nine parts of all mankind" were doomed to dire poverty and life-long drudgery. A generation later, Charles Dickens was convinced that "we are moving in a right direction towards some superior condition of society."

3

Henry Mayhew, the first investigative journalist, wanted to learn whether the wages and standard of life of London's poor could improve. In a time of cholera, he scoured London's back streets and alleyways for facts and created an extraordinary portrait of life and labor in the capital of the world. But the answer to his question eluded him.

4

5   6

For Friedrich Engels (*left*), Victorian London was a modern-day Rome and Judgment Day, both inevitable and imminent.

His friend and dependent Karl Marx (*right*) promised to reveal modern society's Law of Motion, but suffered from writer's block. He never learned English or visited a single factory while composing *Das Kapital*.

7   8

A mathematician and missionary manqué from London's lower middle class, Alfred Marshall had an overriding aim to "put man into the saddle," and his deepest conviction was that a proletariat was not a necessity of nature. He and his Cambridge-educated bride, Mary Paley, set out to turn economics into a compass to guide mankind out of poverty.

9

10

Beatrice Potter was born into Britain's governing class but was torn between conflicting desires: to pursue a career as a social investigator or to become the wife of a powerful man, namely the charismatic and domineering Joseph Chamberlain.

11

12

(left): She found a perfect partner in Sydney Webb, the clever son of a London hairdresser, and together they invented the idea of the welfare state and the "think tank." *(right):* The former Tory and new "thunderer on the Left," Winston Spencer Churchill, picked Beatrice's brain.

13

14

The greatest *American* economic thinker of the last century was a Yankee tinkerer, teeto-taler, and TB survivor. Trained in mathematics but desiring "contact with the living age," Irving Fisher invented the rolodex, the consumer price index, and the economic forecast. By the 1920s, Fisher *(bottom, left)* was America's economic oracle, wellness guru, and stock picker, his celebrity rivaling that of Alexander Graham Bell *(right)*.

In a postgraduate year in London, Joseph Alois Schumpeter rode, fenced, dressed, and talked like one of the Viennese aristocrats he wished to be taken for. He spent most of his time at the British Museum writing a book criticizing economic theory for ignoring how the economy evolved over time.

After marrying impulsively, Schumpeter rushed off to Egypt, the miracle economy of the Belle Epoque, to make his fortune as a lawyer and money manager. In Cairo, he found inspiration for his greatest work, *The Theory of Economic Development.*

15

16

Friedrich von Hayek got interested in how markets and modern economics functioned in the trenches during World War I as a corporal in the Austro-Hungarian army. In the Second World War, Hayek obeyed Wittenstein's injunction by writing *The Road of Serfdom*, a devastating attack on command-and-control economies.

17

18

Datum *Sj. 379, am 30. Juni 1918.*

Unterschrift des Besitzers

Hayek's cousin, Ludwig Wittgenstein, an aviation engineer turned philosopher, impressed upon young Hayek that the duty of genius was to tell uncomfortable truths and to speak of the unspeakable.

19

World War I wrecked the foundations of the nineteenth-century economic miracle and bankrupted the governments of victors and vanquished alike, leaving in its wake famine, hyperinflation, and a revolutionary firestorm that spread from the Urals to the Rhine.

20

As the finance minister of a mutilated, penniless, and starving nation, Schumpeter (*standing, third from left*) tried to convince Austrians that they could recover economically without throwing themselves into the arms of either red Russia or resentful Germany.

John Maynard Keynes (*center*), the clever, ambitious, self-confident heir to one of England's intellectual dynasties, defined the good life as the one available to a London gentleman on the eve of World War I. He is shown here with Bloomsbury pals Bertrand Russell, the philosopher (*left*), and Lytton Strachey, the biographer (*right*).

Keynes collected artists and writers as well as, thanks to his speculative genius, art. The great love of his youth was Duncan Grant, the painter (*left*) who, like most of Keynes's other bohemian friends, refused to serve in World War I and urged Keynes to apply for conscientious-objector status.

Keynes became the point man in Britain's wartime Treasury for loans from the United States to France and other allies and played a supporting role at the 1919 peace conference. Favoring debt forgiveness among the victors and modest reparations for the losers, he resigned in protest after the Big Four refused to make postwar economic recovery in Europe a priority in the Versailles peace treaty.

◄ 23

In 1923, Hayek spent a postdoctoral year in New York City, where he met Irving Fisher and wrote a withering critique of monetary reformers who claimed that central banks could tame the business cycle by managing the money supply. He doubted that forecasters could anticipate the economy's ups and downs ahead of time sufficiently reliably to serve as guides to policymakers.

24

25   26

The postwar slump drew Joan Robinson, the dreamy but driven daughter of a general, to economics, a war-hero husband, and England's celebrity economist, John Maynard Keynes. Self-confident, articulate, uninhibited as a writer, she broke into Keynes's all-male inner circle of disciples, developing a theory of how the rise of big business might lead to an unwelcome combination of higher prices and lower employment. She enlisted the help of her gifted but neurotic lover, Richard Kahn, who served as a go-between with the great man.

To the surprise and disapproval of his Bloomsbury friends, Keynes married the Russian ballerina Lydia Lopokova, a member of Sergei Diaghilev's itinerant Ballets Russes. Her riotous sense of humor, fractured English, and lack of intellectual pretension made her the love of his life.

27

Irving Fisher (*left*) and Joseph Schumpeter (*right*), shown in New Haven in 1932, embraced opposing prescriptions for how to fight the Great Depression but joined forces to promote the use of mathematics in economics.

28

29

Months before D-Day, Franklin Delano Roosevelt called on the Allies to avoid the mistakes made after World War I by focusing on postwar economic recovery.

30

31

Young Milton Friedman (with wife, Rose), one of the legions of young Keynesian supporters of the New Deal, played a key role in Secretary Henry Morgenthau's wartime Treasury, which was, for all practical purposes, run by the brilliant but devious Harry Dexter White. Keynes (*right*) and White were the principal architects of the Bretton Woods monetary system that paved the way for postwar economic recovery in the West. A Soviet agent of influence and spy, White was taken by complete surprise when Stalin refused to join.

Paul Anthony Samuelson was the most influential American Keynesian of the post–World War II era. With a world view shaped by the collapse of the farm belt, the Florida land bubble, and the Great Depression, he modernized economics with mathematics, Keynes's theories, and numerous original ideas of his own. Postwar generations of Americans, including John F. Kennedy, imbibed the new economics through his textbook and *Newsweek* column, and he is regarded as the guiding spirit behind the 1963 Kennedy tax cut.

32

33

In the 1950s, Joan Robinson, the most famous of Keynes's English disciples, repudiated her brilliant early work and became one of Stalin and Mao's trophy intellectuals. She was a harsh critic of American leadership in mainstream economics. She is shown here, partially hidden, in Beijing in July 1953 with Dr. Chi Chao-ting, Roland Berger, and Harold Spencer at the signing of the first "Icebreaker" trade deal.

34

35

Robinson urged her protégé, Amartya Sen, who came to Trinity College, Cambridge, from Calcutta in 1953, to give up "that ethics rubbish." Democracy and the people's welfare were luxuries that poor countries could not afford, she insisted. Sen ignored her advice to work on famines, economic justice, and the problem of translating individual into social choices.

development, booms were followed by busts—"perennial gales of creative destruction"—but the economy was inherently stable. If the system
was in jeopardy, the threat originated in politics. Marx and Engels saw
recessions as signs of failure and sources of instability. Schumpeter took
the opposite view. Since the cycle produced development, depressions
were healthy, a way to drive out inefficient firms and force companies to
trim costs and rationalize their operations. The death of firms and industries was as inevitable as the death of human beings. Nothing lasted,
Schumpeter observed: "No therapy can permanently obstruct the great
economic and social process by which businesses, individual positions,
forms of life, cultural value and ideals sink in the social scale and finally
disappear." But death also made room for new life. Growth required
managerial talent, labor, and other resources to be shifted from old to new
industries. Thus if nations wanted progress, they had to accept slumps.
Like it or not, he liked to say, "the pattern of boom and bust is the form
economic development takes in the era of capitalism." [25]

Innovations as huge as electricity or as small as toothbrushes are
"primarily responsible" for the recurrent "prosperities" that revolutionize the economic organism and the recurrent "recessions" that are due to
the "disequilibrating impact of the new products or methods." Recessions
produced enormous suffering—rising unemployment, declining wages,
losses, and bankruptcies—but didn't last long. "The phenomena felt to be
unpleasant are temporary," Schumpeter wrote, while "the stream of goods
is enriched, production is partly reorganized, costs of production are
diminished, and what at first appears as entrepreneurial profit finally increases the permanent real incomes of the other classes." [26] He insisted that
constant change was a requirement for economic stability, just as motion
is needed to keep a bicycle upright.

In Bonn, he had plunged into two separate books, cultivated a group
of bright young students, written dozens of newspaper columns, and
spent hundreds of hours on the lecture circuit speaking to German business groups. He had justified his compulsive activity as necessary to pay
off his staggering debt but had used it as an anesthetic. The diary into
which he emptied his battered heart every night was little more than a

catalog of regrets and self-recrimination. Since his mother's funeral, he had never once returned to Vienna.

In the fall of 1927—two years after the deaths of his mother and Annie—Schumpeter accepted an invitation to teach at Harvard and came to the United States for a second time. He was perhaps not as enamored as he had been in 1912, but he was stunned by American opulence, energy, and optimism. A few financial experts were warning of a speculative bubble in the stock market. In an essay written in the spring of 1928, he readily acknowledged that the boom might well be followed by a plunge in stock prices and a period of falling output and high unemployment. But he concluded, "The instabilities, which arise from the process of innovation, tend to right themselves, and do not go on accumulating." Thus, he explained, capitalism was "economically stable, and even gaining in stability."[27]

A tall, dark-haired, slightly shabby young man who vaguely resembled Leon Trotsky sat in the main reading room of the New York Public Library examining yellowed copies of the *New York Times*. He was looking for stories from the final months of the war about the Austro-Hungarian army. Again and again, the blue eyes behind wire-rimmed spectacles widened in surprise. How surreal to come halfway around the world only to learn that everything you thought you knew about an episode in your own life was a fiction.

As cynical as all Viennese were about the Austrian press, Friedrich von Hayek, Ludwig Wittgenstein's cousin and a former corporal in the k.u.k. army, was shocked. Until now he had believed, as the Austrian media had claimed, that the Piave offensive was a bold strategic gamble that failed because of various blunders. It was clear from the *Times* account, however, that American and British war correspondents had unanimously regarded the defeat of the Austro-Hungarian forces as absolutely certain *weeks* before the offensive. In other words, one hundred thousand lives, of which Hayek's might easily have been one, had been squandered for a lie.

In August 1918, Hayek had been swept along by a disintegrating army in chaotic retreat from Italy across the Alps. When he finally reached

Vienna, he had given up an earlier dream of becoming a diplomat and enrolled as a law student at the University of Vienna. He later attributed his interest in social sciences to the war and especially to the experience of serving in a multinational army. How could a society harmonize competing desires and interests without relying on military-style coercion? How could individuals with different languages, cultures, and educations communicate and agree on common actions? The dysfunctional command of the Austro-Hungarian army obviously had not found the answers, but the conduct of trench warfare had left Hayek with time to read. Among the books he read and reread were two turgid volumes on political economy.

At the University of Vienna, which hardly functioned for lack of coal, light, or food, Hayek became best friends with another veteran, a law student named Herbert Furth, who had been gravely wounded at Piave. Furth was the son of a Vienna city councilman and Austria's first suffragette. He introduced Hayek to a sophisticated crowd of left-wing students from assimilated, relatively affluent Jewish families. Furth and his friends hung out at the Café Landtmann across from the Rathaus park and argued about Marxism and psychoanalysis. The sons of lawyers, academics, and businessmen, they struck Hayek as considerably more self-confident and cosmopolitan than other young men of his age. He recalled that "what went on in the intellectual world of France and England was to them nearly as familiar as what happened in the German-speaking world." Through them, he discovered Bertrand Russell and H. G. Wells, Proust and Croce, and imbibed "that genuine devotion to things of the spirit need not mean being impractical in the art of getting on in life."[28]

After the war, student politics at the University of Vienna was dominated by virulent Catholic nationalism and violent Communism. Hayek and Furth, who considered themselves Fabian Socialists, found both repellant. Eager to create an alternative, they organized a mildly Socialist organization, the Democratic Students Association, in their first term.

Hayek attended lectures by Friedrich von Wieser, the economist who had served as the monarchy's last finance minister and was Austria's most effective international spokesman. He read the work of Austrian economic thinkers such as Carl Menger and Eugen von Böhm-Bawerk. But as one might expect in a city of ten thousand coffeehouses, a severe apartment shortage, and a surplus of underemployed intellectuals, Hayek's most important education took place in cafés among his peers. In their third year, Hayek and Furth organized a fortnightly seminar that they called the *Geist-Kreis* in jest. *Geist* can refer to the Holy Ghost or to the secular, even demonic, spirits that are channeled at séances. The group's twenty-odd members discussed cultural topics from plays to logical positivism and included the economists Oskar Morgenstern, Gottfried Haberler, and Fritz Machlup, the philosopher Erich Voegelin, and the mathematician Karl Menger (Carl's son), as well as historians, art historians, musicologists, and literary critics.

Hayek completed his doctor of laws degree in the spring of 1922, at the height of the hyperinflation. He immediately took a day job as a minor civil servant in the war claims settlement bureau. Like Einstein's sinecure at the Swiss patent office, Hayek's job was sufficiently undemanding to allow him to earn a second doctorate in political science. A friend of his commented that having one's salary rise from 5,000 kronen to 1 million kronen in the space of nine months, as Hayek's did, was "apt to shape a person's mind." [29] That is probably an exaggeration, but it is safe to say his exploding paycheck and its shrinking purchasing power drew Hayek's attention to the role of money, just as falling asleep in his desk chair at the Bern patent office had drawn Einstein's to the theory of special relativity. Although Hayek preferred collecting old books to investing in stock certificates, he began to daydream about someday becoming president of Austria's central bank.

Another development caught Hayek's attention. The Bolsheviks' "lightning socialization" [30] of 1919 and the Renner government's threat to nationalize key industries had made the most urgent questions for left-wing Viennese intellectuals: Could Socialism work? Could it deliver the goods? Was planning feasible? The German sociologist Max Weber had already weighed in with a blistering "No." [31] Foreign Minister Otto Bauer and Joseph Schumpeter both said "Yes," the affirmative in the latter case being limited to "the right circumstances." [32]

At that point, Hayek's employer and mentor, the liberal economist Ludwig von Mises, propelled the debate to a new intellectual level. In a provocative essay, "Die Gemeinwirtschaft," or "The Collective Economy," Mises essentially reframed the argument by shifting its focus to information. His premise was that an economy was like a computer, a machine for solving a mathematical problem. He argued that a centrally planned economy lacked the necessary data to reduce the number of unknowns to the number of equations and, therefore, lacked the means to calculate prices that brought supply and demand into balance.

Mises allowed that planners could draw up a list of consumer goods and services. But, he asked, then what? How will the authorities assure themselves that the value of, say, an automobile to consumers will equal or exceed the value of the labor, steel, rubber, and other resources that must be sacrificed to produce it? How will they know that the car will be worth more to consumers than the bus that could have been made with those same resources?

To make such calculations in a market economy, Mises said, individual businesses and consumers use price data. Take the question of whether the cost of making the car is more or less than the amount consumers are willing to spend on it. To figure the cost, add up the hours of labor, pounds of steel and rubber, marketing, distribution, and other inputs, multiply by their prices, and add everything up. To figure the value the consumer places on it, take the selling price and multiply by one for one car. Does it make sense to produce cars? If your cost is less than your revenue, you can keep on making them. If it costs more to make them than people are willing to pay, you'll have to reconsider.

The trouble with substituting planners for markets, Mises argued, is that without markets there are no longer market prices to use for making your calculation. Can't you just make some up? Sure, but if no one was producing for, or buying in, markets, they wouldn't be *market* prices. They wouldn't reflect the subjective preferences of the consumers who are demanding a good or the calculations of the businesses deciding whether to supply the good—in real time too. They wouldn't give you the information you need to make a rational decision. You'd have no way of knowing whether you were making the most of your resources or squandering them heedlessly.

The debate over socialization and von Mises's notion of markets as calculators and transmitters of information made an enormous impression on Hayek. It inspired him to write a paper on government rent control. For many families, the severe shortage of housing, another residual of the war, was becoming as pressing a problem as the lack of food or jobs. In 1922, the Social Democrats decided to fix rents at four times the prewar level. Since the consumer price index had risen 110-fold since January 1921, the city council was unwittingly decreeing a virtual rent holiday. As a strategy for ending the housing shortage, it was bizarre. As soon as the controls took effect, new construction ceased, existing buildings became more dilapidated, and overcrowding and homelessness grew worse. Intended to protect the poor, the policy blunder kept people from moving, created more inequality, and reduced the savings available for investment.[33]

Hayek jumped at a chance to spend the academic year 1923–24 at New York University in Greenwich Village as a research assistant to Jeremiah Whipple Jenks, a currency expert on the Allied Reparation Commission whose manners and appearance had confirmed Beatrice Webb's prejudices about Americans. When Hayek arrived in New York with only a few dollars in his pocket, he was aghast to learn that Jenks had gone off to Cornell, where he held a second professorship.

Jenks returned in the nick of time to save Hayek from having to work as a dishwasher at a Sixth Avenue diner. Apart from collecting data for Jenks, Hayek took classes at NYU, started writing a book on how capital goods such as machinery and factory buildings were priced, and finished a long article analyzing the performance of the decade-old Federal Reserve. He met with Irving Fisher, for whom Schumpeter had supplied a letter of introduction. He also crashed the lectures by Wesley C. Mitchell and John Bates Clark at Columbia, the leading center of American research on business cycles.

Hayek's main motive for coming to New York was to learn as much as he could about American thinking about booms and depressions. He was less interested in pursuing the abstract question of whether capitalism could work than in finding out whether economic forecasting could. Was it possible to predict how output or prices would behave six months or a year hence—that is, accurately enough to allow monetary authorities to head off incipient inflations or deflations ahead of time? They were not purely academic questions on Hayek's part. Von Mises, who had recommended him to Jenks, had been talking to Hayek about starting a program of business cycle research and producing economic forecasts at the Vienna Chamber of Commerce.

Hayek would have welcomed the chance to stay in New York for a second year, but by the time the Rockefeller Foundation informed him that its trustees had renewed his grant, he was already sailing back to Europe. By the end of May 1924, he was back in Vienna at his dull job at the reparations office, unhappy and depressed. Before leaving, he had fallen in love with his cousin Helene Bitterlich, who worked in the office as a secretary. He had almost asked her to become engaged to him before he left for New York, but in the end had not. He was furious with himself. In his absence she had married another man.

His spirits were lifted by an invitation from Mises to join his private seminar, "the most important center of economic discussion in Vienna and perhaps in Continental Europe." Along with a dozen or so former members of the *Geist-Kreis,* the group included the economist Steffi Braun, the philosophers Felix Kaufman, Alfred Schutz, and Fritz Schreier, and the historian Friedrich Engel-Janosi. The first paper Hayek delivered at the seminar was on his analysis of rent control in Vienna.

Von Mises had been trying to obtain a position for Hayek at the Chamber of Commerce. When he failed, he raised enough money to create an independent forecasting institute and put Hayek in charge of it. The Austrian Institute for Business Cycle Research was modeled on the academic and private American organizations that Hayek had visited in the United States, and Hayek became its first director. Thus, at thirty, he found himself running a research institute with ties to similar organizations abroad and publishing a monthly forecast for an international audience—though his entire staff consisted of two typists and one clerk.

In 1928, Hayek submitted the book he had started to write in New York, *Monetary Theory and the Business Cycle,* as his "habilitation" to the

University of Vienna. Lionel Robbins, a young working-class Liberal at the London School of Economics who was looking for intellectual allies, happened to attend Hayek's trial lecture on "The 'Paradox' of Saving" and was so impressed that he asked whether he had any interest in coming to London. Robbins also expressed interest in the institute's latest forecast. In his April 1929 newsletter, Hayek noted corporate borrowing was growing faster than production in the United States and warned of "unpleasant consequences." That observation led Robbins subsequently to credit his protégé with prophesying the 1929 stock market crash. In fact, Hayek's alarm was transitory. In his October 1929 newsletter, he reassured readers that neither "a sudden breakdown of the New York Stock Exchange" nor a "pronounced" economic crisis were imminent.[34]

Chapter IX

Immaterial Devices of the Mind:
Keynes and Fisher in the 1920s

The world is gradually awakening to the fact of its own
improvability. Political economy is no longer the dismal science.

—*Irving Fisher, 1908*[1]

We should be led to control and reduce the so-called "business
cycle."

—*Irving Fisher, 1925*[2]

The Great War had postponed the need for Keynes to settle on a career. At
one time, he had thought he wanted to run a railroad, but railroads were
no longer as glamorous as before the war. That high ground was now occupied by finance. The business of borrowing, lending, and insuring had
been transformed by floating currencies, huge war debts, the urgent need
for credit, and the vexing issue of reparations. Once a staid if mysterious
backwater, finance had become the fastest-growing industry—or, in the
eyes of skeptics, a giant casino.

Oswald "Foxy" Falk, a stockbroker friend whom Keynes had brought
to the Treasury during the war, introduced him to the City, London's
Wall Street. Within a year Keynes found himself chairman of an insurance company. He knew nothing about insurance or the desirability of
diversifying an investment portfolio. A life insurance company "ought to
have only one investment and it should be changed every day,"[3] he opined

at his first board meeting. That Fisher's notion of a trade-off between an investment's risk and its rate of return had not occurred to Keynes is a sign of how novel it was. Like so many ideas that seem too obvious to require discovery, the idea that putting all one's eggs into a single basket was risky was generally as little understood as Einstein's theory of relativity.

Keynes by no means limited himself to running the insurance company. The collapse of the global gold standard, with its fixed exchange rates—something like a single world currency—during the war and its replacement by floating exchange rates had created a foreign exchange speculator's paradise. When his speculation in francs, dollars, and pounds prospered, as in the fall of 1919 and the spring of 1920, Keynes was able to buy paintings by Seurat, Picasso, Matisse, Renoir, and Cézanne. "The affair is of course risky but Falk and I, seeing that our reputations depend on it, intend to exercise a good deal of caution," Keynes assured his father, who, like several of his son's Bloomsbury trust-fund friends, had blithely handed over several thousand pounds for him to manage. Perhaps the son's next thought—"Win or lose this high stakes gambling amuses me"— should have set off warning bells.[4]

In this expansive frame of mind, Keynes took Vanessa Bell and Duncan Grant on a whirlwind tour of the Continent in the spring of 1920. They paid a visit to the American art historian and promoter of Renaissance painters Bernard Berenson. At Berenson's Florentine villa, I Tatti, Keynes and Grant each pretended, to their own but not their host's great amusement, to be the other. But mostly they went shopping. Even Keynes, who tended to be a tightwad where trivial sums were concerned, bought seventeen pairs of leather gloves. In March, around the time Joseph Schumpeter was poised to embark on his own high-stakes gambling spree in Vienna, Keynes had decided to go long in dollars on behalf of his syndicate. Prices were rising faster in Britain—and even more so in Europe—than in the United States, he reasoned, so the pound would be sure to weaken against the dollar. His logic was perfectly sound, his timing, not so much. No sooner had he returned to London than the franc, mark, and lire began, perversely, *appreciating* against the dollar. By the time fundamentals once again prevailed, Keynes was wiped out. Through some reverse alchemy his £14,000 of profits had turned into a loss of more than £13,000. Astonishingly, his investors' confidence in his genius did not falter. His father and friends were convinced that Keynes would soon recoup his and their losses, and his broker agreed to reopen his account if he could put up £7,000. Even more amazingly, these remarkable acts of faith proved to be justified. By the end of 1924, Keynes was a wealthy man.

After his success as a best-selling author, Keynes had turned to journalism to help finance the lifestyle to which he was becoming accustomed. He wrote for the *Manchester Guardian* and Lord Beaverbrook's' London *Evening Standard,* and the American *New Republic.* According to his biographer Robert Skidelsky, Keynes's career in print supplied one-third of his income during the 1920s and culminated in his becoming publisher of the left-wing political weekly founded by the Webbs and G. B. Shaw, the *New Statesman.* Peter Clarke, another of Keynes's biographers, observed that launching "assaults of thoughts upon the unthinking" seemed to bring out the remarkable range of Keynes's talents.[5]

In 1922, his topic of choice was money and banking. Before World War I, monetary economics had been more or less an American obsession. But Irving Fisher, virtually the only American economic theorist taken seriously in Cambridge, had convinced Keynes that money had a far more powerful effect on the "real" economy than accepted theory allowed.[6] As early as 1913, a couple of years after he and Fisher met at George V's coronation, in a speech to a group of businessmen in London, Keynes was echoing Fisher's view that the key to booms and depressions was "the creation and destruction of credit."[7] The economic disorders that followed the war seemed to bear out Fisher's argument.

In 1923, Keynes was so excited by the new ideas that he distilled what he had been thinking and writing about in *A Tract on Monetary Reform:*

> The fluctuations in the value of money since 1914 have been on a scale so great as to constitute, with all that they involve, one of the most significant events in the economic history of the modern world. The fluctuation of the standard, whether gold, silver or paper, has not only been of unprecedented violence, but has been visited on a society of which

the economic organization is more dependent than that of any earlier epoch on the assumption that the standard of value would be moderately stable.

He tried to show that inflations and deflations made it difficult for investors and businessmen to calculate the effects of decisions and, to a much greater degree than the public appreciated, distorted decisions to save or invest. But he also took pains to convey a more general point on which he and Fisher were of one mind: "We must free ourselves from the deep distrust which exists against allowing the regulation of the standard of value to be the subject of deliberate decision. We can no longer afford to leave [things to nature]." The evil of inflation was that it redistributed existing wealth arbitrarily, pitting one group of citizens against another and, ultimately, undermining democracy. The evil of deflation was that it retarded the creation of new wealth by destroying jobs and incomes.

It is not necessary that we should weigh one evil against the other. It is easier to agree that both are evils to be shunned. The individualistic capitalism of today, precisely because it entrusts saving to the individual investor and production to the individual employer, presumes a stable measuring rod of value and cannot be efficient—perhaps cannot survive—without one.

Again and again, Keynes stressed his main message, namely, that there was a remedy: "The remedy would lie . . . in so controlling the standard of value that, whenever something occurred which, left to itself, would create an expectation of a change in the general level of prices, the controlling authority should take steps to counteract this expectation." And failure to make money the "subject of deliberate decision" would leave a dangerous vacuum in which "a host of popular remedies . . . which remedies themselves—subsidies, price and rent fixing, profiteer hunting, and excess profits duties—eventually became not the least part of the evils."

The most famous of Keynes's phrases—"In the long run we are all dead"—appears in the *Tract* in the following context: "This long run is a misleading guide to current affairs. In the long run we are all dead. Economists set themselves too easy, too useless a task if in tempestuous seasons they can only tell us that when the storm is long past the ocean is flat again."[8] Later, Schumpeter and other critics interpreted Keynes's flippant phrase to mean that he was indifferent to the inflationary consequences of short-term monetary or fiscal stimulus. But it is clear from the passage that he was attacking the belief that inflation and deflation would cure themselves without active management. His point was, nations had to make deliberate choices between two desirable but incompatible goals. He borrowed the idea from Fisher, whom he called "the pioneer of price stability as against exchange stability."[9] In a world in which capital flowed freely across borders, countries had to choose between stable prices for their imports and exports, on the one hand, and stable prices for their domestically produced goods and services, on the other. They couldn't have both. They had to choose. Keynes left no doubt as to which choice he favored. Domestic price stability was of paramount importance in avoiding socially disruptive transfers of wealth and high unemployment.

World War I had wrecked the gold standard. Since 1875 the British government had guaranteed that £6 could be exchanged at the Bank of England for one troy ounce of gold, and it was the bank's job to see to it that the supply of pounds grew no faster or slower than the rate required to maintain that parity. When other countries pegged their currencies to gold, the effect was, of course, to fix the rate of exchange between all "hard" or gold-standard currencies. For example, since the US government determined that $30 could be exchanged for one troy ounce of gold, £1 equaled $5. In other words, as the economist Paul Krugman has observed, the nineteenth-century gold standard operated almost like a single world currency regulated by the Bank of England.

When the war broke out, one combatant after the other went off gold in order to buy arms and feed their armies. After the war, the holy grail of British politicians and their Chancellors of the Exchequer was the earliest possible return to the gold standard. No politician was a stronger supporter of reinstating the prewar gold standard than Winston Churchill, who had rejoined the Conservative Party and had been appointed Chancellor of the Exchequer by Stanley Baldwin, the leader of the new Tory government.

On March 17, 1925, Keynes attended a fateful dinner with Churchill at which he tried to convince the chancellor that the pound would be grossly overvalued at the prewar parity. While a strong pound would be a boon to Britain's financial industry, it would cripple the old export industries— textiles and coal especially—and result in mass unemployment. This was an argument that he and Irving Fisher had long been making in the press. Keynes did not succeed. As Churchill said afterward, referring to a 1918 campaign promise: "This isn't an economic matter; it is a political decision." [10]

"The Economic Consequences of Mr. Churchill"—as Keynes called a pamphlet he wrote a few months later—were more or less precisely what he, Fisher, and other opponents had predicted. In anticipation of the new policy to raise the foreign exchange value of the pound by 10 percent, the Bank of England had raised its discount rate from 4 percent to 5 percent, a full point above the New York rate, in December 1924. The purpose was to stimulate demand for the pound by attracting short-term American funds to London. As higher interest rates choked off the flow of new credit and the strong pound dampened demand for exports, Britain's heavy industry cratered while unemployment in England's north soared. Keynes blamed the slump on Churchill's failure to take his advice.

Here it is necessary to backtrack slightly. As Keynes succeeded in working out how he was going to make a living and where he would spend his energies, he began to think more about how he wanted to live. He was in his late thirties. Something was missing. For much of 1921 and 1922, he had considered himself "married" to Sebastian Sprott, one of the beautiful undergraduates he met while lecturing at Cambridge. He had also had other affairs. Not only did none of these attachments match the intensity of his relationship with Duncan Grant a decade earlier, but they also intensified his dissatisfaction. They were a reminder that for a variety of reasons, including that homosexuality was both illegal and socially unacceptable, such relationships could never provide him with a partner with whom he could share his rich, varied, and increasingly public life.

Keynes had always been happy in the bosom of his own family. His old Bloomsbury friends were mostly married, living with someone, setting up households, having children. They more or less expected him to do the same, but his choice—a Russian ballerina with a voluptuous body and a droll sense of humor but no obvious intellectual interests—first amazed, then horrified them. Keynes met Lydia Lopokova, who danced comic roles, on an opening night of the Ballets Russes. Their passionate affair commenced in May 1921 when he found an excuse to put her up in the Bloomsbury apartment above his own, belonging to the as yet unsuspecting Vanessa Bell. Four years later, on August 3, 1925, they married in London amid great fanfare and with huge crowds gathered outside. Before the wedding, Keynes purchased a country estate, Tilton, in Surrey, where he strode around in tweeds, inspecting hogs and wheat and behaving like a country squire.

He spent his honeymoon in Russia as a guest at his in-laws' in Saint Petersburg—now named Leningrad—and subsequently as a guest of the Soviet government in Moscow. Along with several other Cambridge dons, he represented the university at the bicentennial of the Russian Academy of Sciences. Keynes's VIP schedule included visits to the economic planning ministry and the state bank, *Hamlet* in Russian, the ballet, and endless banquets. As he wrote to Virginia Woolf, his hosts "embarrassed him with a medal set in diamonds." When he and Lydia turned up at Woolf's house in Surrey after the trip, she found that Keynes had traded his country squire tweeds for an embroidered Tolstoy shirt and Astrakhan fur cap. Afterward she summarized Keynes's impressions of Russia for the benefit of their mutual friends:

> Spies everywhere, no liberty of speech, greed for money eradicated, people living in common . . . ballet respected, best show of Cézanne and Matisse in existence. Endless processions of communists in top

hats, prices exorbitant yet champagne produced, & the finest cooking in
Europe, banquets beginning at 8:30 and going on until 2:30 . . . then the
immense luxury of the old Imperial trains; feeding off the Tsar's plate.

As usual, he displayed his journalist's verve for telling detail, false
notes, and delicious contradictions, but he also used his analytical prow-
ess to distinguish appearance from reality. The other VIPs left Moscow
incredibly impressed by the relatively well-fed, clothed, and housed Soviet
worker, who, apparently, never had to fear unemployment as his Western
counterparts did. But Keynes could explain to *New Republic* readers that
the Soviet economic "miracle" was a Potemkin village. The typical urban
worker did indeed live better than before the war. Indeed, he lived "at a
standard of life that is higher than its output justifies," Keynes reported.
But the other six in seven Soviet citizens were small farmers who were
being exploited even more ruthlessly than under the tsar:

> The Communist Government is able to pamper (comparatively speak-
> ing) the proletarian worker who is of course its especial care, by exploit-
> ing the peasant . . . The official method of exploiting the peasant is not
> so much by taxation—though the land tax is an important item in the
> budget—as by price policy.

Moscow could pay urban workers two or three times what a peasant
earned by the simple expedient of forcing peasants to sell their crops to
the government at prices far below those of the world market. The result
was not just to lower the living standards of the majority of Russians but
also to wreck the economy. Farm output, "the real wealth of the country,"
was falling, farm income was drying up, and an uncontrolled rural exodus
was under way. Moscow and Saint Petersburg were full of homeless illegals
and had unemployment rates closer to 20 or 25 percent than the official
zero. "The real income of the Russian peasant is not much more than half
of what it used to be, whilst the Russian industrial worker suffers overcrowding and unemployment as never before," Keynes concluded.[11]
Though he advised his Soviet hosts to reverse their ruinous poli-

cies, he conceded that the Soviet economy was not "so inefficient as to be unable to survive," albeit "at a low level of efficiency" and low living standards. And he did not contradict the prediction of Grigory Zinoviev, Stalin's second in command, that ten years hence "the standard of living will be higher in Russia than it was before the war, but in all other countries lower,"[12] although only because he had qualms about the West. Perhaps because his in-laws in Saint Petersburg were being persecuted or, more likely, because he was appalled to a greater degree by inefficiency, ugliness, and stupidity than by cruelty, he dismissed the notion that Soviet Russia held the key to the West's salvation:

> How can I adopt a creed which preferring the mud to the fish exalts the boorish proletariat above the bourgeois and intelligentsia and who, whatever their faults, are the quality of life and surely carry the seeds of all human advancement? Even if we need a religion, how can we find it in the turbid rubbish of the Red bookshops?

Displaying his Bloomsbury prejudices, he blamed the "mud" and "rubbish" on "Some beastliness in the Russian nature—or in the Russian and Jewish natures when, as now, they are allied together."[13] When the editor of the *New Republic* asked Keynes to remove the offending sentence for the sake of American readers, he refused.

In late 1925 and early 1926, Keynes was momentarily distracted from monetary issues. Along with the whole country, he was mesmerized by an ugly conflict between coal barons and miners, and the threat of a nationwide strike. The first victim of the stronger pound had been Britain's decaying coal industry, already saddled with excess capacity, outdated technology, high costs, and inept management. After a standoff between owners and unions over pay cuts, the Conservative government had tried to buy time by subsidizing the miners' pay. But when the subsidies were due to expire, the standoff remained and a strike loomed. Keynes's friends in the Liberal Party did not believe, as Conservatives did, that a strike would be the first step toward revolution. Nonetheless, they supported the government, insisting that such an action would be illegal, unconstitutional, and an assault on democracy. Keynes, who sympathized with the miners who were not to blame for Churchill's decision, weighed in with proposals for a compromise. In return for the unions' taking a modest pay cut, and the owners' shutting down their least efficient pits, the government would continue the subsidies. Everyone would win.

It was wishful thinking. The ten-day general strike of May 1926 was a bust. The miners stayed out another six months, until starvation forced them back into the pits on the very terms they had rejected. Meanwhile, however, the Liberal Party had split in two. Keynes wound up siding with his old nemesis Lloyd George, who attacked the government's hard-line response, and against his old friends in the Party. Among Keynes's new friends was Beatrice Webb, whom he met for lunch several times. She attributed his siding with the miners to his recent marriage:

> Hitherto he had not attracted me—brilliant, supercilious, and not sufficiently patient for sociological discovery even if he had the heart for it, I should have said. But . . . I think his love marriage with that fascinating little Russian dancer has awakened his emotional sympathies with poverty and suffering.[14]

Keynes's antipathy to the herd—whether wealthy bankers, trade unions, proletarian culture, or ostentatious patriotism—made him ill suited for politics, Webb thought astutely, although she thought he might be valuable as a cabinet minister.

In September, Keynes was in Berlin giving an informal status report on the general strike as well a formal lecture on "The End of Laissez-Faire." At the University of Berlin, large and excited crowds gave him a warm welcome not usually extended to Englishmen. His attack on the Versailles peace treaty, condemnation of the French seizure of the Ruhr, and support for reparations reductions and foreign loans packages made him a hero there. The latest and most important, the Dawes Plan, had slashed Germany's reparations bill and opened the floodgates to an enormous surge of foreign, mostly American, loans. Awash in money, a magnet for immigrants and foreign visitors, Weimar was in its golden age. Keynes found the atmosphere in Germany's Babylon almost giddy.

He saw his old friend Carl Melchior, who had also gotten married in the interim, again and met Albert Einstein for the first and only time. His reaction to them was tinged with Bloomsbury-ish disgust with money and paranoia that German culture was being endangered by an alien one. "[Einstein] was a Jew . . . and my dear Melchior is a Jew too," he reflected.

> Yet if I lived there, I felt I might turn anti-Semite. For the poor Prussian is too slow and heavy on his legs for the other kind of Jews, the ones who are not imps but serving devils, with small horns, pitch forks, and oily tails . . . It is not agreeable to see a civilization under the ugly thumbs of its impure Jews who have all the money and the power and the brains. I vote rather for the plump Hausfraus and thick fingered Wandering Birds.[15]

His peculiar momentary identification, more propitiatory than sympathetic, with the slow, heavy, and thick masses as opposed to the clever devils that he preferred, reflects his fear of the mob, a theme that he expressed in less objectionable language in his formal talk on "The Death of Laissez-Faire." Governments of democracies risk violence if they are foolish enough to leave the economic circumstances of their citizens to chance.

Keynes continued to lecture at Cambridge throughout the 1920s. One undergraduate recalled him as "more like a stockbroker than a don, a city man who spent long weekends in the country."[16] Nonetheless, his glamour and fame attracted large crowds at his lectures. On Monday nights an invitation-only political economy club met in his rooms at Kings and attracted clever undergraduates and ambitious dons.

"Let us be up and doing, using our idle resources to increase our wealth," Keynes told an assembly of Liberal Party politicians on March 27, 1928.

"When every man and every factory is busy, then will be the time to say that we can afford nothing further."[17] At the time of the general strike, Keynes had assumed that new theories about controlling the business cycle, packaged as a solution to Britain's unemployment problem, might provide an alternative to the high tariffs advocated by the Right and the exorbitant taxes advocated by the Left. Lloyd George, his new ally, had been actively plotting a political comeback and hunting for a new philosophy. Keynes briefly considered running as the Liberal candidate for Cambridge University, but rejected the idea after a few days of agonizing. Instead he became the architect of policies on which Lloyd George campaigned in the spring of 1929. In other words, the germs of Keynes's *General Theory* grew in the Petri dish of a political campaign.

Keynes regarded instability, not inequality, as the great threat to capitalism. The arbitrary windfalls and losses—unrelated to hard work, thrift, or good ideas—not the gap between rich and poor, were what he meant by inequity. "The most violent interferences with stability and with justice, to which the nineteenth century submitted . . . , were precisely those which were brought about by changes in the price level," he wrote, echoing Irving Fisher. So the "first and most important step . . . is to establish a new monetary system."[18] Unlike Webb, Keynes rejected the politics of class war. He was too much of an elitist. Labor "put on an appearance of being against anyone who is more successful, more skilful, and more industrious, more thrifty than the average," he groused. "It is a class party, and the class is not my class . . . I can be influenced by what seems to me to be justice and good sense; but the class war will find me on the side of the educated bourgeoisie."[19]

Lloyd George, whom Keynes had reviled in 1919 as "the devil incarnate," had been forced out of office in 1922 for bartering favors for campaign contributions, womanizing, and a host of other ethical lapses. Yet the "Welsh wizard" retained his hold on the Liberal Party and on Keynes. Essentially unemployed for most of the 1920s, he turned his estate, Churt, into an economic think tank, pouring his energies, time, and a Party fund that he controlled into producing a Liberal program. Now he was planning a comeback on the basis of a plan to fight unemployment. Keynes was the campaign's chief economist.

After 1919, unemployment in Britain never dropped below 1 million, inching up year after year until it reached 10 percent in 1929. At that point Britain had yet to fully recover from World War I. The volume of British exports shrank even as world trade was expanding. In 1913, Britain had been the world's top exporter; by 1929, she had slipped to second place behind the United States.[20] The workshop of the world consisted largely of the old smokestack industries—coal, iron, and steel, textiles, shipbuilding—at a time when the consumers of the world wanted more oil, chemicals, cars, movies, and other products of new industries. Moreover, the national averages hid a sharp cleavage between the prosperous south of England and the chronically depressed industrial north, reviving the old notion, reminiscent of the Hungry Forties in the previous century, of England as two nations estranged from each other; one rich, and the other poor.

On September 25, 1927, Keynes was one of fourteen professors summoned to Churt by Lloyd George for an intimate gathering of "a few trying to lay the foundations of a new radicalism."[21] Keynes coauthored Lloyd George's inquiry "Britain's Industrial Future," backed with £10,000 of the latter's money. The report finally appeared in early February 1928 and quickly took on the moniker of "The Yellow Book," after its yellow cover. Though Keynes wrote to H. G. Wells that he hoped never again "to be embroiled in cooperative authorship on this scale," he conceded that the white paper was a "pretty serious effort to make a list of things in the politico-industrial sphere which are practicable and sensible."[22]

The report gave Keynes his first opportunity to learn something about industrial as opposed to financial companies. He told Liberal candidates that the trend toward bigness in business was driven not just by technology and finance but also by the threat of unsold inventories. Big business had evolved naturally and had to be accepted as such. It was not quite the warm endorsement of giant corporations offered by Schumpeter, but it was distinctly un-Socialist.

"We Can Conquer Unemployment" was the Liberal slogan in the campaign of 1929. On March 1, Lloyd George made a dramatic pledge to reduce unemployment to "normal" proportions within a year.[23] The heart of his program was a huge deficit-financed public works program intended to jump-start the economy. Higher growth was supposed to generate the tax revenue to pay for roads, sewers, telephone lines, electric transmission, and new housing while unemployment insurance would be used to pay workers. Keynes weighed in with a pamphlet titled "Can the Liberal Pledge Be Carried Out?" less than three weeks later. After the Treasury fired back that public works jobs would merely replace private ones, he followed it up with a second, "Can Lloyd George Do It?"

> The fact that many workpeople who are now unemployed would be receiving wages instead of unemployment pay would mean an increase in effective purchasing power which would give a general stimulus to trade. Moreover, the greater trade activity would make for further trade activity; for the forces of prosperity, like those of a trade depression, work with a cumulative effect.[24]

This, Skidelsky points out, was the germ of the idea of a multiplier. Developed two years later by one of Keynes's beautiful young men, Richard Kahn, the idea is that increasing government spending by $1 will generate more than $1 of private spending, since the initial increase in consumption by recipients leads to more hiring and income and another, if smaller, increase in spending, and so on.

Confident as ever before the May 30 general election, Keynes made a bet that the Liberals would win one hundred seats. In fact, they won just fifty-nine, effectively ending Lloyd George's political career, and Keynes had to pay out £160, only partly offset by a £10 bet he collected from Winston Churchill. The campaign also forced him to rewrite large swaths of his *Treatise on Money*. The summer of 1929 was idyllic, taken up with his manuscript, the filming of a five-minute ballet scene for one of the first British "talkies," *Dark Red Roses*, tennis, and a meeting with the government's point man for public works, Oswald Mosley, a rising star in the

Labour Party who would become a Fascist in the 1930s. The only source of irritation was the sorry outcome of Keynes's commodity speculation. He had been long on rubber, corn, cotton, and tin in 1928 when the markets suddenly turned on him and he was forced to liquidate part of his stock portfolio to cover his losses.

Irving Fisher bought his first gasoline-powered car in 1916. The last and most luxurious of the Fishers' electric models, a superdeluxe enclosed Detroit, had had to be driven to a garage every night for recharging and couldn't go faster than twenty-five miles per hour. Now Fisher, who logged thousands of railroad miles every year, hit the road in a brand-new gas guzzler, a Dodge. The highways between New York and Boston were still mostly unpaved, rutted, and dotted with potholes that could swallow a wheel or worse, but for Fisher the new car "opened up almost unlimited vistas."[25] Throughout the 1920s, Fisher got a new car every two years or so, trading up and up as his and the country's fortunes prospered. By the end of the decade, in addition to a Lincoln, he owned a La Salle convertible and a brand-new Stearns-Knight, America's answer to the English Rolls-Royce. And, like Jay Gatsby, he employed an Irish chauffeur.

By 1929, one in five American families had a car. As Fisher had predicted in 1914, the war left the United States with the biggest and strongest economy in the world. Unlike for Britain and France,

> the First World War had not been a cause of unalloyed economic loss; it had on occasions brought economic and social advantages. What is more it had demonstrated to all the combatant powers that it lay in the hands of government to formulate strategic and economic policies which could to some extent determine whether or not a war would be economically a cause of gain or loss; they were not the hopeless prisoners of circumstance.[26]

Thanks to wartime production and exports to the United Kingdom and Europe, the United States overtook Britain in annual output by 1918.[27]

Instead of collapsing, as in Germany or Austria, or being choked off by monetary authorities, as in Britain, America's recovery from the postwar recession started in 1921 and kept going. There were two recessions in the middle of the decade, each a little over a year long, but they were so mild that most Americans—farmers excepted—were unaware of them. For the entire period from 1921 to 1929, the economy expanded at an average rate of 4 percent a year while unemployment averaged less than 5 percent. In 1929, the economy was 40 percent bigger and per capita income 20 percent higher than in 1921, a remarkable performance for any country in any decade and rarely equaled since.[28]

But the averages hardly convey the convulsive changes that new forms of energy brought in their train. They inaugurated a new way of living. The modern era of the car, the suburban house, California, oil, the telephone, daily newspapers, stock quotes, refrigerators and fans and electric lighting, radio and movies, working women and smaller families, declining union membership, and shopping centers took over. The hitherto unknown concept of retirement took hold among men who had reached sixty. "Scientific management" and "Taylorism" became new corporate buzzwords after Louis Brandeis argued successfully that railroads did not have to increase rates in order to pay higher wages, as long as they organized work according to principles pioneered by Frederick Winslow Taylor. RCA and AT&T were the Microsoft and Google of the day. Meanwhile, the old economy of farms, coal mines, woolen mills, and shoe factories—those great sources of American wealth in the nineteenth century—slipped into senescence.

The steamship, railroad, and telegraph had exploded limits on mobility and communication for Alfred Marshall's generation. The automobile and telephone did the same for Fisher's, but in a way that individualized travel and long-distance interaction. Fisher thrilled at his escape from timetables, just as Beatrice Webb gloried in going miles without a chauffeur when she got her first bicycle. Mass production made possible mass ownership of the car, radio, telephone, fan, refrigerator, and prefabricated house, and these, in turn, made life in the suburbs attractive and affordable. Consumers were getting their hands on instruments of mastery that let them turn the dials, flip the switches, and get in the driver's seat.

While Webb absolutely refused to drive a car, and Geoffrey Keynes once called his brother an "anti-motoring motorphobe, giber of all forms of motoring,"[29] Fisher personified America's love affair with cars and also with gadgetry of all kinds. He ordered two wireless sets in March 1922 after giving his first radio speech. It was, he wrote to his son, "perhaps the largest audience I ever addressed." He told "an audience I couldn't see or hear or quite believe existed" that the newly inaugurated transatlantic broadcasts made "the whole world a neighborhood."[30] Not long after a twenty-five-year-old US Airmail pilot named Lindbergh flew a single-engine monoplane nonstop from Long Island to Paris in 1927, Fisher, who was in Paris, took advantage of the new transatlantic telephone service by arranging to have a nine-minute conference call with his wife in Rhode Island, his mother in New Jersey, and his son-in-law in Ohio. Irving Jr. recalled that Fisher "kept his eye on the second hand of his watch."[31] By then, Fisher was handling most of his business correspondence by telephone, doing most of his writing by Dictaphone, and, when he was in a hurry, which was almost always, dictating directly to a typist seated in front of an Olivetti. His home office had long since swallowed up the entire third floor of his New Haven mansion, with filing cabinets and typing tables spilling into hallways and stairwells. His staff included eight to ten "assorted females" who used telephones with glass mouthpieces and typed to the hum of an ozone machine installed to invigorate the office's atmosphere.

Fisher was spending most of his time crusading for the League of Nations, immigration restrictions, environmental conservation, and public health reforms, including universal insurance. He lived by the same precepts. Virtually the whole top floor of Fisher's house was devoted to a home gym, Fisher's "garage for keeping his personal engine in top form." Along with health food and vitamins, he was a sucker for exercise equipment. The gym was crammed with Indian clubs, dumbbells, weight-lifting devices, a rowing machine, an electric cabinet, a sun lamp; a vibrating chaise that his children claimed looked like an electric chair and "an outlandish mechanism for administering all-over rhythmic massage." [32] By 1929, Fisher had a full-time personal physician and a trainer on his payroll.

Again and again, Fisher argued that history was a bad guide to human potential. In a 1926 speech before a public health group,[33] Fisher argued that human beings had no more reached the limit of longevity than they had the limit of consumption. The true limit, he argued, was one hundred. He pointed out that by 1931, the life expectancy of an English boy would be nearly twenty years longer than in 1871.[34] Equally important, seven in ten people were healthy enough to enjoy life and do a day's work. At the end of the war, by contrast, six out of nine had ranged from "infirm" to "physical wrecks" to invalids "with a precarious hold on life." [35] He predicted—accurately as it turned out—that the average life span would increase from fifty-eight to eighty-two by 2000.[36]

Fisher's faith in the improvability of man and the limitless possibilities of science and free enterprise grew in tandem with the twenties boom:

> The world is gradually awakening to the fact of its own improvability. Political economy is no longer the "dismal science," teaching that starvation wages are inevitable from the Malthusian growth of population, but is now seriously and hopefully grappling with the problems of abolition of poverty. In like manner hygiene, the youngest of the biological studies, has repudiated the outworn doctrine that mortality is fatality, and must exact a regular and inevitable sacrifice at its present rate year after year. Instead of this fatalistic creed we now have the assurance of Pasteur that "It is within the power of man to rid himself of every parasitic disease." [37]

Fisher became one of the founders and first president of the American Eugenics Society. Eugenics—the application of genetics to marriage, health, and immigration practices—was by no means only a Fabian cause. Selective breeding of human beings has of course been practiced by most societies in varying forms from Spartan infanticide to the arcane mating rituals of the British aristocracy. In the late Victorian era medical and scientific advances and the spirit of reform endowed eugenics with its name and immense popularity. One of Richard Potter's closest friends,

Charles Darwin's cousin Francis Galton, is regarded as the father of the field. Major Leonard Darwin, Charles Darwin's son, established the International Eugenics Society in 1911. Beatrice and Sidney Webb and, indeed, most prominent Fabians, including G. B. Shaw and H. G. Wells, were enthusiastic eugenicists. Keynes, who served as vice president and board member of the British Eugenics Society as well as treasurer of its Cambridge branch, considered eugenics "the most important, significant and, I would add, genuine branch of sociology."[38] Eugenics was a bipartisan cause. Conservatives such as Arthur Balfour, the Conservative Prime Minister from 1902 to 1905; Winston Churchill; Lord Beveridge, architect of the post-WWII British welfare state; the writers Leonard Woolf and Virginia Woolf; and feminists Victoria Woodhull and Margaret Sanger were all enthusiastic eugenicists.

To be fair, eugenics hardly meant in 1910 or 1920 what it came to signify in the 1970s after it was discredited by association with the Nazi genocide and Jim Crow. The "general spirit" of the first international congress in London in 1912, which Fisher attended, was "conservative."[39] He and Keynes were libertarians, and Fisher in particular was an antiracist who was committed to "eliminating . . . race, prejudice, as well as other antisocial prejudices, such as underlie the Ku Klux Klan."[40] That said, Fisher and the American Eugenics Society were major forces behind the 1924 immigration law aimed not only at, as Fisher put it, "the immigration of the extremely unfit such as formerly were dumped into our population out of the public institutions of Europe"[41] but at radically reducing all immigration from southern and eastern Europe.

Fisher had focused on the evil effects of inflation and deflation on debtors and creditors, the arbitrary redistribution of wealth they caused, and the "vicious remedies" that governments adopted at the behest of victims but that "like the remedies of primitive medicine, they are often not only futile but harmful."[42] He had not yet linked fluctuations in the price level to ups and downs in employment and output, far less assigned them a primary role. Indeed, his *Principles of Economics,* published in 1911, has no index entries for *boom, depression,* or *unemployment.*

The brief but steep recession of 1920–21 focused Fisher's attention on what the government could do to fight unemployment. In 1895, the US federal government had neither the means nor the responsibility to manage the overall level of economic activity. It was small relative to the economy. Taxes were a means of financing government activities, mainly military, and tariffs were a way to aid specific industries. Money creation was left to the banks, and, under the nineteenth-century gold standard, its pace was strictly governed by the rate of growth of the world's gold supply.

Now the United States had a central bank—the Federal Reserve, created in 1913—and more discretionary power to influence the level of economic activity by encouraging or discouraging money creation and lending. The severity of the downturn convinced Fisher that in attempting to roll back wartime inflation, the Fed had slammed the brakes too hard and too long. The widespread distress among farmers—reminiscent of the 1890s—and factory workers convinced him that the greatest evil associated with unstable prices was their effect on output and employment. That chain of causation stretching from money creation to job creation became the focus of Fisher's research during the twenties.

Fisher's concerns were gradually shifting to booms and busts, and the role that money played in the economy's stability or volatility. He suspected that fluctuation in the supply of money and credit not only caused inflation and deflation but also accounted for the ups and downs of economic activity and employment. He was becoming convinced that better monetary management could lead to a "lessening of cyclical fluctuations." [43]

In addition to a steady stream of academic articles, Fisher spent more and more of his time writing for newspapers. Like Keynes and the Webbs, he knew that his best shot at selling his ideas to government policy makers was indirectly and as an outsider. In article after article, he did his best to convince the public that inflation and unemployment had a common monetary cause. He admitted that any connection between the banking system and "a matter as intensely human as the unemployment program" would strike most people as far-fetched. True, commentators had recognized the link between a general decline in the average price level and a rise in unemployment in the severe postwar recession in the United States and Britain. Likewise, inflations were associated with upswings in production and hiring. Yet theories of the "business cycle"—the alternation of boom and bust in output and employment—typically bore no mention of changes in the price level, and other researchers could find no correlation between prices and employment.

As Fisher discovered, other forecasters had missed the empirical link between prices and employment. They had confounded the level of prices with changes in the level—the distinction that had come to him in a flash in the Swiss Alps—a mistake comparable to mixing up the rate at which water flows into a bathtub with the depth of the water in the tub. As Fisher put it, other analysts had "missed the clear distinction between high prices and rising prices and likewise, between low prices and falling prices. In short they scrutinized the price level but not its rate of change."[44] One reason for the confusion was that there were no good gauges of how fast the average level of prices in the economy was changing. Fisher devoted most of the 1920s to developing and publishing accurate price measures that could be used to forecast economic activity and let the public keep track of changes in the dollar's purchasing power.

Fisher was convinced that once the causes of economic cycles were correctly identified, forecasters would be able to "predict business conditions in a truly scientific manner . . . much as we forecast the weather." In 1926 he wrote that "monetary theory ought, for instance, to help us analyze and predict the price level." He assumed that once the central bank could forecast prices accurately, it could forestall anticipated price swings and, hence, eliminate or at least moderate booms and depressions. For Fisher, means typically dictated ends. "We should be led to control and reduce the so-called business cycle" instead of ascribing "a sort of fatalistic nature" to depressions and booms, he argued.[45]

In short, by the mid-1920s, Fisher had added business cycles to the list of economic ailments that, far from being untreatable, were shortly to give way to modern cures: "The idea that it's inevitable and unpredictable is entirely false. On the contrary, the causes are well known, in the main, and we know now in large degree to prevent the intensity of these alternate chills and fevers of business."[46] He attributed his confidence to the apparent success with which the Federal Reserve was already achieving a "rough stabilization of the dollar," citing the central bank's efforts to prevent periods of speculation. "We have in our power, as a means of substantially preventing unemployment, the stabilization of the purchasing power of the dollar, pound, lira, mark, crown and many other monetary units."[47] Like Keynes, he insisted that a stable currency was primarily a societal issue. "If our vast credit superstructure is to be kept from periodically falling about our ears," he wrote, "we must regard banking as something more than a private business. It is a great public service."[48]

In a 1925 piece for the Battle Creek Sanitarium's health newsletter, Fisher explained "Why I would rather be a Sanitarium Employee than a Millionaire."[49] Yet while there were many things he valued more than money, he had always secretly wanted someday to become his wife's equal in financial terms. The first of his inventions to achieve commercial potential was the product of his impatience. Having to thumb through boxes of dog-eared index cards drove him mad with frustration, so he fashioned an ingenious device that held the cards in place and kept them visible to the user. Fisher tried to convince a dozen office equipment manufacturers that his nifty gadget was the perfect solution to the increasingly voluminous record-keeping requirements of modern business and that companies would leap at any product that let them organize and store records more efficiently.

Initially, the Rolodex suffered the same fate as many other inventions: the inventor was forced to go into the business himself using his own, or rather his wife's, money. Fisher set up a tiny factory in New Haven with a staff consisting of his brother, a carpenter, and a helper. The firm's capital consisted of a loan of $35,000 from Margaret. A year after the war, Index Visible needed a three-story factory to house its operation, as well as a sales office in the New York Times Building on Nassau Street in downtown Manhattan. Fisher's first big client was New York Telephone, which helped push the company into the black in 1925. Seizing the moment, Fisher engineered a merger with his chief competitor to form the nucleus of Remington Rand. Having by this time poured a total of $148,000 into his start-up, he swapped Index Visible's common stock for $660,000 in cash, a bundle of preferred stocks, bonds, options, and dividends, and a seat on the board of the new entity, Rand Kardex. Afterward, he confessed to his son that paying his own way had been one of his "suppressed desires ever since I was married . . . Inventing offered the one chance I saw of making money without a great sacrifice of time."[50] At fifty, Fisher realized his dream and became a millionaire many times over.

Meanwhile, economic forecasting was really taking off. The boom created a market for economic forecasts. Fisher began writing a syndicated economic outlook column. He also began publishing a weekly Purchasing Power of Money index, one of several price measures that the US government eventually adopted. Before long, he had set up the Index Number Institute and was mailing wholesale price data to dozens of newspapers from its headquarters in his home office at 460 Prospect Street in New Haven. After the sale of Index Visible, Fisher moved his forecasting and data operation into the New York Times Building and his indexes and charts started to appear in the *Philadelphia Inquirer,* the *Journal of Commerce,* the *Minneapolis Journal,* the *Hartford Courant,* and other newspapers.

Always keen to apply his ideas in the real world, Fisher had begun indexing his office workers' pay to inflation during the war. He was probably the first employer to ever grant an explicit, annual, automatic "cost of living" adjustment. Ironically, the experience taught him that indexing was not a practical solution to the problems created by inflation and deflation. He explained:

> As long as the cost of living was getting higher, the Index Visible employees welcomed the swelling contents of their "high Cost of Living" pay envelopes. They thought their wages were increasing, though it was carefully explained to them that their real wages were merely standing still. But as soon as the cost of living fell they resented the "reduction" in wages.[51]

Fisher cited his employees' reaction as proof of the omnipresent "money illusion." He also hazarded a guess that Wall Street traders were as prone as typists to be misled by the false perception that their own currency's value was steady while the price of goods and services, or other currencies, bobbed up and down. A total return on a stock of 10 percent might look like a terrific investment. But if inflation was 11 percent, the investor would actually be losing money. Fisher bet that investors and unions would pay for a yardstick that enabled them to figure their "real" rates of return or whether a pay offer was a "real" increase or not.

Interest in monetary stabilization had led Fisher to an interest in index numbers, and now led to an interest in studying stock returns. The US stock market collapsed in 1921 when the Federal Reserve raised interest rates to quash wartime inflation, but share prices rebounded sharply the following year. By mid-1929, stock prices were three times higher in nominal terms than in 1921 and roughly nineteen times as high as after-tax corporate profits.[52] Fisher's Remington Rand stock had appreciated tenfold in real terms between 1925 and 1929.

As early as 1911, Fisher had argued that a diversified portfolio of stocks was a better long-term investment than bonds. The value of bonds reflected only the government's ability to repay its debt and its willingness to resist inflation. Stocks, on the other hand, could capture the effects of private sector productivity gains on profits and, hence, had far more upside potential. As the twenties boom continued, Fisher grew more and more bullish. By 1927, he had become the New Economy's most prominent promoter and was borrowing hundreds of thousands of dollars to invest on margin. He had a few scares. Once, when he returned from a trip to Paris and Rome that fall, his personal secretary was waiting for him at the New York dock. A plunge in the market had forced her to use $100,000 in his agent's account to repay short-term bank loans. Within a month, however, Fisher was urging Irving Jr. to "risk half of your present holdings by borrowing on it as collateral and using proceeds of loan for buying more. Six months or a year later you could probably sell at substantial advance and then diversify."[53]

In August 1929, unemployment was 3 percent. The tempo of innovation had picked up after the war. More patents had been filed in the previous ten years than in the previous century. Not surprisingly, an economic commission appointed by Herbert Hoover, the newly elected president and former head of the American effort to avert starvation in Europe after World War I, concluded, "Our situation is fortunate. Our momentum is remarkable." [54] When bearish investors such as Roger Babson warned that stock prices had risen too far too fast, Fisher countered that they were not out of line with corporate profits. On another occasion, he listed reasons why corporate profits were likely to keep growing: mergers were increasing scale economies and lowering production costs; companies were spending more on R&D; recycling was on the increase; management was becoming more scientific; automobiles and better roads would increase business efficiency; and the growth of business unionism presaged less industrial strife.

By 1929, Fisher was a director of Remington Rand, an investor in a half dozen start-ups, and the head of a successful forecasting service. Meanwhile, he spent most of that year revising his 1907 masterpiece, *The Theory of Interest*. Reflecting on one of the most spectacular bull markets in the history of the US stock market, Fisher attributed the surge in stock prices to an explosion of innovation since the war and the resulting growth in profitable investment opportunities. He delivered his manuscript in September and immediately began work on a book about stocks. He was scheduled to address a group of loan officers at the Hotel Taft in New Haven on October 29. Two weeks earlier, the *New York Times* reported, Professor Irving Fisher of Yale University had confidently told members of the Purchasing Agents' Association that stock prices had reached "what looks like a permanently high plateau." [55]

Chapter X

Magneto Trouble: Keynes and Fisher in the Great Depression

Men and women all over the world were seriously contemplating and frankly discussing the possibility that the Western system of society might break down and cease to work.

—*Arnold J. Toynbee, 1931*[1]

Keynes spent the first half hour of every day in bed in London reading the financial pages and talking to his broker and other City contacts on the phone. But his daily research turned up no early warnings of the American stock crash of October 1929. The King's College endowment, which he managed as bursar, plunged by one-third, and his personal portfolio fared even worse. The trouble, explains Robert Skidelsky, wasn't that Keynes owned much American stock. Rather, he had gone long in rubber, cotton, tin, and corn in the expectation that the American boom would drive commodity prices higher, and he had done so by borrowing on a ten-to-one margin. When commodity prices began to weaken in 1928, Keynes was forced to sell most of his stock in a falling market to cover his commodity positions. By the end of 1929, his net worth had plummeted from £44,000 to less than £8,000.[2] The experience converted Keynes into a value investor, convincing him that "the right method in investment is to put fairly large sums into enterprises which one thinks one knows something about and in the management of which one thoroughly believes."[3]

In the face of financial calamity and misplaced hopes, Keynes was his usual optimistic self. He was certain that the American monetary authorities would inaugurate "an epoch of cheap money" to head off a severe recession.[4] Three lunches with the new Labour prime minister, Ramsay MacDonald, whose party had soundly defeated both the Tory incumbents and Keynes's own Liberal candidate, Lloyd George, in the general election of May 1929, convinced Keynes that the new government would reject what Churchill had called "orthodox Treasury dogma."[5]

The Treasury's traditional cure for financial crises was to reassert fiscal rectitude by balancing the government's books while the Bank of England raised interest rates to defend the gold value of the pound sterling. Restoring business and investor confidence, the reasoning went, was the shortest route to recovery. Any attempt by government to act as employer of last resort would merely result in less hiring by private employers. As Winston Churchill, the outgoing Tory Chancellor of the Exchequer, reiterated before Parliament, "Whatever might be the political and social advantages, very little additional employment and no permanent additional employment can in fact, and as a general rule be created by State borrowing and State expenditure."[6] Keynes was confident that the Labourites would embrace Liberal proposals for public works spending and lower interest rates, their effect on the government deficit and the gold value of sterling be damned. An invitation the following July to chair MacDonald's Economic Advisory Council, the prime minister's "economic general staff," confirmed his upbeat expectations.[7] "I'm back in favor again," he crowed in a note to Lydia.[8]

Keynes was certain that easier money would stabilize the economy. Unemployment might ratchet up for several months, he wrote in a *Times* of London column, but as long as interest rates fell even faster than prices, business investment would bounce back and commodity prices and farm incomes would recover. He also had faith in the activism of the new president, Herbert Hoover, in contrast to Calvin Coolidge's passivity. Hoover had appointed an energetic Federal Reserve chairman, Eugene Isaac Meyer, the future publisher of the *Washington Post*, and had announced a program to fast-forward federal construction projects. The successful former mining executive and European food aid czar was inviting business bigwigs to the White House for brainstorming sessions. A few weeks after the stock market crash, his treasury secretary, Andrew Mellon, had gone to Congress to ask for a 1 percent tax cut for corporations and individuals.[9] And, as always, Keynes was confident enough to back his forecasts with cash. By September 1930, reports Skidelsky, he was once again buying up large amounts of American and Indian cotton.

Demand for Keynes's opinions soared, and he used his newspaper columns, radio talks, and newsreel interviews to promote monetary activism to fight the slump. In December 1930, he wrote a long piece for the *Nation* that began: "The world has been slow to realize that we are living this year in the shadow of one of the greatest economic catastrophes of modern history." To dispel resignation, he used every public forum to dismiss the popular narrative that cast booms and busts as episodes in a morality play. He vigorously denied the notion that recessions were the inevitable punishment and welcome correctives for extravagance, imprudence, and greed. Instead, Keynes told his readers, "We have involved ourselves in a colossal muddle, having blundered in the control of a delicate machine, the working of which we do not understand."[10]

The problem, in other words, was a technical one. For Keynes, depressions, like car wrecks, were the result of accidents and policy blunders. They involved permanent losses in output that, like time, could never be recouped and were not restoratives but simply a waste. Bad harvests, hurricanes, wars, and other bolts from the blue did sometimes trigger downturns, but the origin of most recessions was bad or erratic decisions by economic policy makers. In principle, that meant that downturns could be minimized or prevented altogether. Keynes was especially eager to rebut the notion that booms, rather than depressions, were the problem. As he put it a few years later, echoing one of Schumpeter's sentiments from *The Theory of Economic Development*, "The right remedy for the trade cycle is not to be found in abolishing booms and thus keeping us permanently in a semi-slump, but in abolishing slumps and thus keeping us permanently in a quasi boom."[11] He insisted that, contrary to the accusations of moralists, the slump meant that past economic gains had been phantasmagorical. Referring to the investment boom of the twenties, he wrote that "the other was not a dream. This is a nightmare, which will pass away with the morning. For the resources of nature and men's devices are just as fertile and productive as they ever were . . . We were not previously deceived."[12]

The economy was suffering from a mechanical breakdown for which there was a (relatively) easy fix. In one column he wrote that there was nothing more profoundly wrong with the economic engine than a case of "magneto" or starter trouble.[13] Prices had fallen so much that farmers and businessmen couldn't sell their products for what it cost to produce them. Hence, they had no choice but to slash production and investment, setting off another round of unemployment and causing prices to fall still further. To break the vicious circle, all the monetary authorities had to do was to lower interest rates by creating more money until business could raise prices and found it worthwhile to begin investing again. He was convinced that easier money would head off anything worse than a garden-variety recession.

Keynes used automotive analogies to make the point that, as Skidelsky put it, immense catastrophes could have trivial causes and trivial solutions. To many ears, however, his message sounded counterintuitive, even flippant. While the eminent mathematician and Marxist G. H. Hardy was ridiculing the notion of mechanical solutions to deep scientific problems— "It is only the very unsophisticated outsider who imagines that mathematicians make discoveries by turning the handle of some miraculous machine"—Keynes was reassuring his readers that once the problem was correctly diagnosed, there was a solution—if only the authorities had the *conviction* to act:

Resolute action by the Federal Reserve Banks of the United States, the Bank of France, and the Bank of England might do much more than most people, mistaking symptoms or aggravating circumstances for the disease itself, will readily believe . . . I am convinced that Great Britain and the United States, like-minded and acting together, could start the machine again within a reasonable time; if, that is to say, they were energized by a confident conviction as to what was wrong. For it is chiefly the lack of this conviction which to-day is paralyzing the hands of authority on both sides of the Channel and of the Atlantic.

The lack of conviction was partly, or even mainly, intellectual. Keynes attributed the magnitude of the catastrophe to the fact that "there is no example in modern history of so great and rapid a fall of prices from a normal figure as has occurred in the past year." [15] As Keynes knew, old theories could not be refuted with facts alone. New theories were required. To add ballast to his editorializing, Keynes hurried his two-volume *Treatise on Money* into print, finishing the preface in mid-September 1930.

The focus of the *Treatise* was the possibility of controlling the business cycle by stabilizing prices. When investment exceeded saving, the result was inflation. When the reverse was true, the results were a falling price level, slumping output, and rising unemployment—a recession, in other words. Thus, depressions could be cured by encouraging spending and discouraging saving, exactly the opposite of the medicine that traditionalists such as Churchill extolled. "For the engine which drives Enterprise is not Thrift, but profit," he argued, asking rhetorically, "Were the Seven Wonders of the World built by thrift? I doubt it." [16]

His upbeat message was that if deflation was driving farmers, miners, and businessmen to slash output, the authorities possessed the cure. In his 1921 book *Stabilizing the Dollar,* Irving Fisher had argued that the central bank could control the quantity of money and credit by manipulating the interest rate. By raising rates when inflation threatened and lowering them when deflation loomed, the central bank could restrain or encourage investment, depending on whether it wished to stimulate or slow economic activity. And by controlling investment, the monetary authorities could keep it in line with saving, and prices in line with costs. This is what Keynes believed in 1931, when he was still confident that concerted action to lower interest rates would end the slump.

As Skidelsky observes, Keynes failed to appreciate the economic orthodoxy of Socialist politicians. Even though high unemployment had dominated public concerns for at least nine years, Labour still had no program of its own for attacking it. Beatrice Webb was an exception. A vocal critic of the Treasury view, she had criticized "Treasury book-keeping" and annual budget balancing in her controversial 1909 *Minority Report*.[17] In boom times, she had argued, government ought to raise taxes on the rich and create a surplus. In bad times, it should fund public works even if it meant running a budget deficit. But by 1930 she had become convinced that unemployment was intrinsic to capitalism. Ignoring the fact that unemployment in the United States had averaged less than 5 percent for most of the 1920s, she had concluded that it could not be eliminated until private industry had been nationalized.[18]

Most members of the Labour cabinet hewed as steadfastly to the Treasury view as had Winston Churchill. One minister wrote to the prime minister, "The captain and officers of a great ship has run aground on a falling tide; no human endeavor will get the ship afloat until in the course of nature the tide again begins to flow." MacDonald replied that the "letter expresses exactly my own frame of mind."[19] Cutting benefits and raising taxes seemed more prudent than embracing the radical stimulus measures advocated by Keynes and Fisher.

At the end of 1930, Keynes's advisory council of economists came up with a hodgepodge of conventional and radical policies: cut the unemployment benefit, adopt a 10 percent tariff on imports, and implement "a big public works program" to create jobs for the unemployed.[20] They explicitly rejected the view that any additions to government payrolls would merely displace private employment. "We do not accept the view that the undertaking of such work must necessarily cause any important diversion of employment in ordinary industry."[21] But the Labour cabinet, in which Sydney Webb served as colonial minister, adopted only the first measure and rejected tariffs and public works.

By early 1931, reports Skidelsky, Keynes's finances were so strained that he tried to sell his two best paintings, including Matisse's *Deshabille*.[22] He found no buyers at his minimum asking price.

In the summer of 1929 Irving Fisher had not only splurged on his Stearns-Knight but also watched with satisfaction as a crew of workmen finished a lavish renovation of his and Maggie's New Haven house. The best thing about it, he told his son, was that he, not his wife, was footing the bill.

At sixty, Fisher looked fitter and more distinguished than ever, with his thick white hair, trim figure, and a thoughtful gaze that gave no hint that he was blind in one eye. He had borrowed heavily to take advantage of options on Remington Rand stock that came his way as part of his sale of Index Visible. Four years after the sale of Index Visible to Rand, the value of his stock portfolio had multiplied tenfold. His Index Number Institute, still housed in the New York Times Building, had inaugurated a subscriber service for stock indices. Fisher wrote a syndicated weekly column for investors that appeared in newspapers around the country every Monday. In the public's eye, he was identified not only with Prohibition and the wellness craze, but also with the stock market boom and New Era optimism on the economy.

As questions about the durability of the bull market accumulated in 1929, he dismissed the dire warnings of professional stock market bears such as Roger Babson by pointing to the remarkable combination of low inflation and rapid economic growth that had characterized the decade. "We have witnessed probably the greatest expansion in history, within any similar period of time, of the real income of a people,"[23] he wrote. In mid-October, according to the *New York Times,* Fisher had predicted that the stock market was poised to go "a good deal higher within a few months."[24]

After the crash, Fisher was by no means convinced that a recession was inevitable. In January 1930 he wrote:

> The fall of paper values was largely a transfer of wealth, not a destruction of physical wealth . . . Physical plans are unimpaired . . . The redistribution of corporate ownership was confined to a very small percentage of the population, and consequently will have little effect upon the purchasing power of the great mass of consumers.[25]

His competitor the Harvard Economic Society agreed that a repeat of the severe 1920–21 recession was not in the cards. Days after the crash, the Harvard forecasters informed their subscribers, "We believe that the present recession both for stocks and business, is not a precursor to a business depression."[26]

Fisher wasted little time bemoaning his losses and instead focused his attention on producing a postmortem of the crash. He wrote much of *The Stock Market Crash—and After* in November and December 1929. He defended his optimism that stock prices would recover by pointing out that they were now only eleven times earnings, below their long-run historical average, and "too low a ratio in view of the expectations of a faster rate of earnings in the future." He rejected the popular explanation that the inflated stock prices were to blame, arguing that "between two thirds and three fourths of the rise in the stock market between 1926 and September, 1929 was justified" by earnings and productivity gains, a conclusion that some recent analyses confirm. At the same time, he explained how investors like him had been lured by a combination of low interest rates and high returns to take on too much debt: "When new inventions give an opportunity to make more than the current rate of interest there is always a tendency to borrow at low rates to make a higher rate from investment." Instead of artificially high stock prices, the problem was excessive borrowing:

> Investors found themselves confronted on the one hand by wonderful opportunities to make money and on the other a low rate for loans. They could borrow at much less than they expected to make. In short, both the bull market and the crash are largely explained by the unsound financing of sound prospects.[27]

Fisher continued to predict a stock market recovery and to deny that the crash had made a depression inevitable. He pointed out that economic activity had begun to decline before the stock market crash and predicted a typical recession. As long as businesses did not succumb to doom and gloom by scaling back production and firing employees, he insisted, the real economy would weather the storm. Month after month for the next year, Fisher maintained that an upturn was around the corner. Like Keynes, he had confidence in Hoover's competence and resolve.

For several months, Fisher's optimism looked plausible. By April 1930, the stock market was back to the level it had reached in early 1929. Prices were not falling as fast, and unemployment was not rising as rapidly, as in 1921. Indeed, as late as June 1930, the unemployment rate was 8 percent. In 1921, it had been 12 percent. Interest rates were extremely low. But as Milton Friedman and Anna Schwartz observe in their magisterial *A Monetary History of the United States, 1867–1960,* instead of the anticipated recovery there was a palpable "change in the character of the contraction."[28]

A further plunge in industrial prices offset any benefit to borrowers from lower interest rates. Billions in assets evaporated in a wave of bank failures in the fall of 1930 and the summer of 1931. Even when Fisher was finally forced to admit the severity of the depression, he insisted that the market and the economy were both bottoming. His optimism, overconfidence, and stubbornness betrayed him, and, like so many others who kept hoping that the tide would turn, he hung on to his stock. Had he adopted Herbert Hoover's cautious formula and paid off his bank loans while Remington Rand stock was climbing to $58 a share in 1928 and 1929, Fisher would still have been a millionaire eight to ten times over. Even if he had sold his stock one year after the crash, he would still have been comfortable. In late 1930 Remington Rand was selling for $28 a share. By 1933, it would fall to $1 a share. By April 1931 Fisher's net worth had shrunk to a little more than $1 million. In August he was forced to shutter the Index Number Institute and disband his staff of economists and statisticians. As if this was not devastating enough, the IRS sued him for $75,000 in back taxes related to sales of Remington Rand stock in 1927 and 1928. He was forced to turn to his sister-in-law Caroline Hazard, the retired president of Wellesley, who eventually turned over the management of the loan to a committee consisting of her lawyer and two nephews.

Public recrimination and ridicule added to the stress and humiliation of financial ruin. The former president of the American Economic Association attacked Fisher in the *New York Times* for "always insisting that all was well and talking of prosperity, a new era and increased efficiency of production as justification of the high stock prices."[29] The paper also reported that "Secretary Mellon, former President Coolidge and Professor

Irving Fisher of Yale were named yesterday as the individuals most responsible for 'continuing and extending the mania'" of speculation which preceded the Wall Street crash.[30] When the CEO of a company in which he had invested heavily was indicted for fraud, Fisher sued. The publicity tarnished his reputation further. His son recalled hearing two strangers discuss the lurid details of the case, which were being reported daily in the *New York Times.* "Gosh, he's supposed to know all the answers, and look how he got burned."[31]

Instead of running its course, the economic slide accelerated and spread across the globe. US industrial output plunged to less than half its 1929 level, and unemployment shot up to 16 percent. The tone of commentary turned panicky: by midyear, newspapers were referring to "the Great Depression."[32] Fisher confessed that "the most important economic event in the lifetime of all of us here" would be "an enigma" for years to come.[33] He and Keynes had both been blindsided, but Fisher had lost his credibility with the public as well.

Keynes and Fisher both spent the first week of July 1931 in the drought-stricken Midwest. Two dozen monetary experts were meeting at the University of Chicago to discuss the government's response to what was now being called the Great Depression. Keynes praised the Hoover administration for cutting taxes and signing off on a raft of building projects, among them the Hoover Dam. He complimented the Federal Reserve for cutting interest rates to record lows to prevent deflation. "The depression must be fought by price-raising, not price-cutting," he told reporters.[34] He was still convinced that lowering interest rates would suffice to end the recession, but prudent enough to recognize that not putting all eggs in one basket— "attacking the problem on a broad front, trying simultaneously every plausible means"[35]—made economic as well as political sense in a situation that no one had foreseen.

Keynes chaired a roundtable that took up the question "Is it possible for Governments and Central Banks to do anything on purpose to remedy Unemployment?"[36] Typical midwestern fiscal conservatives, the Chicago economics faculty nonetheless supported the Hoover administration's policy of more government spending and easier money. Keynes was not the only one to have the insight that shortfalls in demand—the means and desire of households and businesses to spend—caused recessions and that the solution was for the government to make up for them. Indeed, the Chicagoans were decidedly more enthusiastic than was Keynes about Hoover's public works program and business lending initiative. Keynes had less confidence in the organizational capability of American as opposed to British civil servants.

After he returned to London, Keynes lent his name to the Labour government's *Report of the Committee on Finance and Industry,* by Lord Hugh Macmillan, proposing that Britain, the United States, and France join in a concerted effort to expand credit by a number of means, including canceling war debts, making emergency loans, and removing obstacles to trade. Labour's attempt to restore confidence in the pound by proposing £70 million of spending cuts plus £70 million of tax increases was to no avail. By August 1931, the Labour government had split over the policies proposed by the Economic Advisory Board, and Ramsay MacDonald had resigned as prime minister. A few weeks later, the collapse of the largest Austrian bank, Kreditanstalt, triggered a financial crisis on the Continent and a run on the British pound as European investors frantically raised cash by withdrawing sterling from their London accounts. The Bank of England responded by more than doubling the discount rate to 6 percent.

On September 21, Britain finally took the step that Keynes and Fisher had been advocating all along: devaluing the pound by 30 percent and suspending gold payments. Rather than leaving interest rates at their September high to prevent a further outflow of gold and hard currency reserves and defend the gold value of the pound—a measure that would have forced another round of investment and job cuts—the Bank of England lowered the rate from 6 percent to 2 percent in the first half of 1932.[37] In a congratulatory telegram to Prime Minister MacDonald on "the breaking of the gold standard," Fisher assured him that the step was "not something to be ashamed of."[38]

Keynes was reassuring. Vanessa Bell wrote to her sister Virginia Woolf in October, after she and Duncan Grant went to see a movie in London:

Suddenly Maynard appeared on the screen enormously big . . . blinking at the lights & speaking rather nervously & told the world that everything was now going to be all right. England had been rescued by fate from an almost hopeless situation, the pound would not collapse, prices would not rise very much, trade would recover, no one need fear anything. In this weather one can almost believe it.[39]

It was too late for the Labour government. The general election in October resulted in a landslide for Tories and Liberals. Ramsay MacDonald retained his prime ministership, but Conservatives once again controlled domestic economic policy.

Despite his financial straits, damaged reputation, and advancing age, the sixty-five-year-old Fisher seemed more energized than depressed by the economic calamity. In 1932 he published an extraordinary number of scientific papers and newspaper pieces. He bombarded the Hoover administration and the Federal Reserve with advice and organized other economists to do the same. His chief objective was to convince President Hoover to take the United States off the gold standard, if not de jure then de facto by having the Federal Reserve do nothing to prevent the foreign exchange value of the dollar from falling. He met with the bankers at the Federal Reserve to urge them to adopt an aggressive program to buy bonds from the banks and the public in order to pump money into the banking system. To his frustration, the "Federal Reserve men thought it would be 'safer' if they waited!" as he later complained. "That waiting, in my opinion, cost the country the major part of the depression."[40]

In January 1932, Fisher attended a second meeting of monetary experts at the University of Chicago. This time, he organized a telegram urging the president to permit the federal budget deficit to rise, pump reserves into the crippled banking system, slash tariffs, and cancel interallied debts. Thirty-two prominent economists from Chicago, Wisconsin, and Harvard universities signed the statement, in which Fisher pointed out that Sweden, Japan, and Britain were recovering after going off gold the previous year. The signatories reflected the extent to which Fisher and Keynes's view of the crisis, with its emphasis on its global nature, monetary causes, forecasts of its future course, and the need for concerted monetary intervention, had gained adherents. On the other hand, theirs was still a minority view. In the same month, two Harvard instructors, Harry Dexter White and Lauchlin Currie, had issued a similar manifesto. Calling the depression "an international calamity," they insisted that the government do more than aid the victims and focus on preventing the slump from worsening:

> With the reparations problem involved, economic distress throughout Europe on the increase, with the progressive mal-distribution of gold reserves, the growing loss of confidence in banks, the mounting trade barriers, disorders in Spain, India, and China, the outlook for recovery in the near future is not encouraging . . . In view of . . . the failure on the part of the government to adopt other than palliative measures there devolves upon the economist the responsibility of recommending a course of action which will hasten the approach of recovery.

Calling for massive public spending, the Harvard dissidents referred derisively to "economists who believe that the course of the depression cannot be checked, that political and economic changes are beyond human control." [41] At Harvard apparently these included the entire senior faculty. The third Harvard signature on the manifesto was also that of an instructor.

By 1932 the depth and global nature of the depression was becoming clear, and Herbert Hoover was on his way to becoming the "most hated man in America." Bombarded with conflicting advice, the president adopted a grab bag of inconsistent policies to combat rising unemployment. Under attack for cutting taxes and raising spending while the budget deficit continued to widen, Hoover reversed course, raising taxes and cutting spending. Bankers, businessmen, and, indeed, the economics community refused to support such unconventional measures. After a meeting with a Treasury undersecretary, Fisher wrote to Maggie, "I told him he and Hoover should choose *some* way and go right after it!" [42]

In truth, there was no consensus anywhere about what government should do. Most governments reacted to falling prices, production, and tax revenues by trying to balance their budgets. The effect of tax increases and spending cuts was to make the slump worse and to trigger further price declines. Banking panics created huge liabilities for governments. Thus, as the economic historian Harold James points out, the action of governments, especially Washington, helped to spread the deflation and depression and made the Great Depression truly global.

Any hope that 1932 would be like 1923, when the American economy had roared back after the steep 1920–21 recession, was soon quashed. Instead of recovering, the economy's slide accelerated. By 1933, stocks were trading at one-fifth of their 1929 values while retail prices had plunged by 30 percent. National output and income had shrunk by one-third. An extraordinary 25 percent of the labor force was out of work. Death by suicide was up sharply, as one might expect. One of the few bright notes was that Americans on the whole were getting healthier and living longer and had a lower chance of dying before their time. Apparently, the 1920s prosperity, with its plethora of opportunities to work and consume, had not been a wholly unalloyed blessing.

By the time Keynes and the American journalist Walter Lippmann conducted their first transatlantic broadcast in real time in July 1933, Franklin Delano Roosevelt was in the White House. Lippmann concluded the broadcast with an observation calculated to win over his interviewer:

> It may be that at the present stage of human knowledge we are not equipped to understand a crisis which is so great and so novel . . . Nowhere in the whole world has there been a prophet of whom it can be said that his teachings were comprehensive and prompt and sufficient . . . It is also a crisis of the human understanding, and our deepest failures have not been failures arising from malevolence but from miscalculation.[43]

Most economic historians agree that not only did no one predict the Great Depression on the basis of any previous depression, but no one could have predicted it on the basis of any existing theory.[44] In retrospect, modern scholars put the primary blame on mistakes by the Federal Reserve, the collapse in confidence and spending by consumers and business, and the wave of selling into falling markets by increasingly panicky investors. But, as David Fettig at the Federal Reserve Bank of Minneapolis has observed,

> In the end, if the Great Depression is, indeed, a story, it has all the trappings of a mystery that is loaded with suspects and difficult to solve, even when we know the ending; the kind we read again and again, and each time come up with another explanation. At least for now.[45]

For those of a scientific bent, being spectacularly wrong is often the most powerful stimulus to fresh thinking. By late 1932, it had become clear that Keynes and Fisher's theory that price stability was a sufficient condition for economic stability—that is, full employment—was flawed or, at the very least, missing some crucial variable. Neither had a truly satisfactory explanation for the magnitude of the economy's collapse between 1929 and 1933. Without a compelling theory that accounted for the crisis, no government would have the confidence to take strong, consistent action. Thus, both men were driven to examine their earlier assumptions and to look for forces that they had overlooked or misunderstood.

Fisher thought he had discovered the missing variable: debt. He first proposed a new theory to explain the magnitude of the economic collapse by emphasizing the toxic interaction of excessive debt and rapid deflation at a meeting of economists in New Orleans. "Over-investment and over-speculation are often important," he told them, "but they would have far less serious results were they not conducted with borrowed money."[46] Public and private debt levels had exploded since World War I, not only in the United States but worldwide.[47] American households took on debt to buy cars, appliances, and houses while European governments still owed gargantuan sums from the war.

The initial fall in stock prices was enough to rattle the confidence of heavily indebted businesses and households and overextended banks, who rushed to liquidate their debts and shore up their balance sheets. This led to an initial wave of distress selling—"selling not because the price is high enough to suit you, which is the normal characteristic of selling, but because the price is so low it frightens you"[48]—and further declines in stock prices, which in turn caused bank deposits to contract. As the supply of money shrank, prices began to slide downward across the board.

Deflation, as a fall in the overall price level, should, in principle, raise real incomes by increasing the purchasing power of a given nominal wage. As the prices of everything from gasoline to shoes fall, a given wage buys more. In his 1911 book *The Purchasing Power of Money*, Fisher had shown that falling prices could also depress income. The real value of a $1,000 loan is $1,000 divided by the average price level. If prices decline, the real value of the debt increases, impoverishing debtors and enriching creditors. A second effect follows from the redistribution of income from debtors to creditors. Debtors tend to spend more and save less of their income than creditors, one reason they took out a loan in the first place. Thus, their spending falls by more than creditors' spending increases.

If everyone expected prices to fall in the future, Fisher argued, companies would become reluctant to borrow to invest in new factories and equipment, because they would have to repay the banks later in more valuable dollars. As businesses slashed investment plans, spending on capital goods would fall, as would the incomes of capital goods producers and workers. As income slipped, the demand for money and the nominal rate of interest would both fall. But the nominal rate of interest would fall less than the price level, so the real interest rate would wind up higher. In both cases, falling prices would lead to lower production and higher unemployment.

Fisher's point was that the effort of businesses to get rid of debts actually resulted in increasing the debt burdens in real terms, a dramatic instance of actions that were beneficial for an individual but harmful in the aggregate. Even businesses that were debt free would find themselves in trouble as the prices they could charge for their products fell faster than the costs of labor and raw materials. The squeeze on their profits would inevitably result in layoffs and production cuts. The rational attempt by banks and individuals to solve their own difficulties by slashing their debt, he emphasized, had the perverse effect of making things worse.

Fisher had already concluded that the immediate cause of the crisis was "the collapse of the credit system under the weight of these debts."[49] Between 1929 and 1933, three banking panics wiped out billions in business, farm, and personal assets—equal to one-third of the nation's money supply. Yet the Federal Reserve began raising interest rates in the fall of 1931 and did nothing to shore up the banking system, on the grounds that weeding out unfit banks was laying the groundwork for recovery. Fisher blamed lingering war debts, the Smoot-Hawley tariff, and the absence of a strong leader at the Federal Reserve. Benjamin Strong, who as president of the New York Federal Reserve Bank had dominated the Federal Reserve, had possessed a deep knowledge of banking and close ties with the head of the Bank of England. His death at the end of 1928, Fisher was certain, had deprived the relatively untested American central bank of strong leadership—and credibility overseas—just when such leadership was most needed. He told a reporter that "the effect of the economic crisis could have been mitigated 'by at least 90 percent' if the Federal Reserve banks had followed the stabilization policy of former Governor Benjamin Strong of the New York Bank."[50]

Nonetheless, Fisher's optimism that greater understanding would ultimately make preventing and mitigating depressions possible was undimmed:

> The main conclusion of this book is that depressions are, for the most part, preventable and that their prevention requires a definite policy in which the Federal Reserve System must play an important role. No time should be lost in grappling with the practical measures necessary to free the world from such needless suffering as it has endured since 1929.[51]

Judging by newspaper headlines of the early 1930s, popular wisdom viewed economics through a biblical lens: recessions were the wages of sin. When good times lasted too long, businesses and individuals threw caution to the wind and behaved badly. Recessions—periods when output, employment, and income contract instead of expand—occurred when private businesses and households unwound past excesses, wrote off bad investments, and behaved with restraint once again. Recessions, in this view, were regrettable but necessary correctives, like a detox program for a drunk. When they occurred, the government had to prevent business and consumer confidence from being damaged further by balancing the budget and guarding against excessively easy monetary policy. That, of course, is the platform on which Franklin Delano Roosevelt campaigned.

The Brain Trust around FDR was a group of campaign advisors from Columbia University that included Adolph Berle, a law professor and expert on corporate governance; Rexford Tugwell, an agricultural economist; and Marriner Eccles, a millionaire Western banker. Its members distrusted economic radicals such as Keynes and Fisher almost as much as did the British Labourites, considering them inflationists hardly better than William Jennings Bryan and the silverites of the 1890s. This was unfair. Fisher and Keynes advocated that the Treasury and the central bank stop targeting the gold exchange rate and instead target the overall price level. They wanted, in other words, the monetary authorities in the major economies to let their foreign exchange rates depreciate while preventing deflation of domestic prices. For the Brain Trusters, this was a distinction without a difference. Tugwell recalled, "We were at heart believers in sound money."[52] In their way, Roosevelt's advisors were as conservative on money matters and as wedded to the Treasury view as was the British Labour Party.

David Kennedy describes FDR's own brain as "a teeming curiosity shop continuously restocked with randomly acquired intellectual oddments ... open to all number and manner of impressions, facts, theories, nostrums, and personalities ... particularly inflation-preaching monetary heretics like Yale's Professor Irving Fisher."[53] Tugwell recalled, "All the old schemes for cheapening money were apparently still alive, and there were many new ones. The Governor [FDR] wanted to know all about them. We shuddered and got him the information."[54]

Inflation's appeal was political. Two-thirds of the Democratic Party consisted of southern and western farmers who were caught between debt and declining crop prices and were hostile to gold. On the other hand, the prospect of inflation inspired more dread among bankers and businessmen than one would suppose in a year when the average price level had dropped by more than 10 percent and one-third of the nation's banks had defaulted. Memories of the violent inflations during and after World War I and the deflations that had been necessary to cure them were still too fresh to ignore. FDR was especially hostile to international cooperation to fight the depression.

Unused to thinking like mathematicians, the Brains Trusters found the notion that huge disturbances might have trivial causes counterintuitive. FDR's economic advisors were more inclined to blame the depression on traditional Democratic nemeses: income inequality, monopolies, and, as had Fisher, the Smoot-Hawley tariff. FDR himself was intrigued by popular theories of overproduction and underconsumption that blamed the depression on either too much wealth or too much poverty. In a speech in May 1932 at Atlanta's Oglethorpe University, the candidate decried the "haphazardness" and "gigantic waste" in the American economy, along with the "superfluous duplication of productive facilities," and called for thinking "less about the producer and more about the consumer." He also predicted that the American economy was nearing its limits and that "our physical economic plant will not expand in the future at the same rate at which it has expanded in the past."[55]

David Kennedy observes that FDR's speech at the Commonwealth Club of San Francisco on September 23, 1932, reflected the "eclecticism and fluidity" of the candidate's views:

A *mere* builder of more industrial plants, a creator of more railroad systems, an organizer of more corporations is as likely to be a danger as a help. The day of the great promoter or the financial Titan, to whom we granted everything if only he would build, or develop, is over."

Extraordinary as it sounds, at a time when one-third of the nation was destitute, FDR blamed the depression on too *much* rather than too little production:

It is the soberer, less dramatic business of administering resources and plants already in hand, of seeking to reestablish foreign markets for our surplus production, of meeting the problem of under-consumption, of adjusting production to consumption, of distributing wealth and products more equitably.[56]

Naturally, FDR's advisors had their own political agendas too. Berle promoted the notion that the economic crisis had created a unique window for enacting major social reforms. Kennedy points out that the economic recovery program on which FDR campaigned "was difficult to distinguish from many of the measures that Hoover, even if somewhat grudgingly, had already adopted: aid for agriculture, promotion of industrial cooperation [price fixing], loans to business, support for the banks, and a balanced budget."[57] The first budget bill FDR sent to Congress cut the federal budget far more than Hoover had dared.

Keynes and Fisher both considered the candidate's emphasis on social welfare reforms before the economy had been stabilized wrongheaded and risky. A few weeks before FDR's inaugural, Keynes sent the president a letter warning against mixing long-run reforms with the recovery program and advocating "open market operations to reduce the long term rate of interest."[58] Fisher urged FDR to announce a retreat from the gold standard on inauguration day, arguing that it "would reverse the present deflation overnight and would set us on the path toward new peaks of prosperity."[59] At the end of 1933, Keynes wrote an open letter to FDR, published in the *New York Times*, to reiterate his argument. "Even wise and necessary reform may . . . impede and complicate recovery. For it will upset the confidence of the business world and weaken their existing motives to action."[60] Fisher shared Keynes's reservations about the New Deal:

> It's all a strange mixture. I'm against the restriction of acreage and production but much in favor of reflation. Apparently FDR thinks of them as similar—merely two ways of raising prices! But one changes the monetary unit to restore it to normal, while the other spells scarce food and clothing when many are starving and half naked.[61]

The single exception to the continuation of Hoover's policies was a very large one: FDR's decision to abandon the gold standard. This was the step that Keynes and Fisher had been urging in one form or another since the 1929 crash. In practical terms, going off gold meant that the Federal Reserve would not push up interest rates to prevent the dollar's exchange rate against the pound and other foreign currencies from falling. The first beneficiaries would be farmers and miners, since a cheaper dollar meant that their grain and ore became more competitive abroad, and then businesses and households that borrowed to buy houses or make capital improvements.

After Roosevelt announced that the United States would go off gold on April 19, 1933, Keynes praised the president for being "magnificently right." Fisher once again let his hopes rise. He wrote to Maggie: "Now I am sure—so far as we ever can be sure of anything—that we are going to snap out of this depression fast." [62] This time, Fisher's economic forecast proved prescient. The US economy hit bottom within a month of Roosevelt's inauguration, marking the beginning of a recovery. On the other hand, Fisher's hope that his personal finances could be mended was not realized. Going hat in hand to his sister-in-law was the least of the humiliations he was to suffer. Had Yale University not agreed to buy his New Haven home and let him live there rent free, he would have been evicted from it. The Fishers' summer cottage at the shore was handed over to Caroline Hazard, who forgave the rest of the debt in her will. Without income from dividends, Fisher had to support himself with directors' fees.

Keynes met FDR for the first time at 5:15 in the evening on May 28, 1934. After days of dawn-to-dusk meetings with cabinet members, Brain Trusters, NRA bureaucrats, and other officials, he had finally gotten in to chat with the president for an hour. Afterward he reported to Felix Frankfurter, now an advisor to FDR, that he had told him that if the government increased federal stimulus spending from $300 million a month to $400 million a month, the United States would have a satisfactory recovery. [63] The president said that he had a "grand talk with Keynes and liked him immensely" but complained that he talked like a "mathematician." [64] The next day, the *New York Times* ran another open letter from Keynes to the president praising the New Deal and calling for deficit spending to the tune of 8 percent of GDP. "This, he promised, might directly or indirectly, increase the national income by at least three or four times this amount . . . Most people greatly underestimate the effect of a given emergency expenditure, because they overlook the Multiplier—the cumulative effect of increased individual incomes, because the expenditure of these incomes improves the incomes of a further set of recipients and so on.[65]

The next evening, Keynes attended a dinner at New York's New School for Social Research with Fisher and Schumpeter.[66] In his talk he spelled out his theory of deficit-financed public works spending, including his notion that the cumulative effect of $1 of such spending could be much greater than $1. Whereas Fisher never departed from his conviction that the Great Depression was the result of monetary blunders, that "of all things tried, monetary policies have succeeded most," and that "the only sure and rapid recovery is through monetary means," Keynes had clearly suffered a crisis of faith in the potency of monetary stimulus.[67] Fisher listened in bemused silence. "His paper was interesting but to me—and I think to everyone else—rather obscure and unconvincing," he wrote to Maggie afterward. "He was very skillful in answering questions and objections but seemed to get nowhere."[68]

As the Great Depression dragged on, Keynes's faith in the effectiveness of monetary policy ebbed further. By the time A Treatise on Money appeared, he was beginning to pose a theory of the causes of unemployment. Cambridge undergraduates were his first audience. The nub of the new theory was that, as he put it in an article published in the American Economic Review in December 1933, "circumstances can arise, and have recently arisen, when neither control of the short-rate of interest nor control of the long-rate will be effective, with the result that direct stimulation of investment by government is a necessary means."[69]

In the severe depression, prices fell even faster than interest rates. So reductions in nominal rates did not prevent real rates from climbing. Once nominal rates fell to zero, there was nothing further that the central bank could do to make borrowing cheaper or to ease debt burdens and thus to end the depression—with incalculable political consequences, what

Keynes called The Liquidity Trap. As he had once observed, "The inability
of the interest rate to fall has brought down empires."[70] Once monetary
policy was rendered ineffectual, the only option for shoring up demand
was getting money into the hands of those who could spend it.

> All past teaching has . . . been either irrelevant, or else positively injuri-
> ous. We have not only failed to understand the economic order under
> which we live, but we have misunderstood it to the extent of adopting
> practices which operate most harshly to our detriment, so that we are
> tempted to cure ills arising out of our misunderstanding by resort to
> further destruction in the form of revolution.[71]

Keynes finished the first draft of *A General Theory of Employment, Interest
and Money* in 1934 after returning from the United States. He began circulating the manuscript in early 1935. To George Bernard Shaw, he wrote
that he believed he was "writing a book on economic theory which will
largely revolutionize—not I suppose at once but in the course of the next
ten years—the way the world thinks about economic problems."[72]

The prime innovation in the *General Theory* was to show that in severe depressions, monetary policy would not work. Economists grounded
in classical models were like

> Euclidean geometers in a non-Euclidean world who, discovering that in
> experience straight lines apparently parallel often meet, rebuke the lines
> for not keeping straight as the only remedy for the unfortunate colli-
> sions that are occurring. Yet, in truth, there is no remedy except to throw
> over the axiom of parallels and to work out a non-Euclidean geometry.
> Something similar is required today in economics.

His innovation has sometimes been misunderstood. It was not that
governments should spend more in bad times or run deficits in a slack
economy. Beatrice Webb, Winston Churchill, and Herbert Hoover had all
embraced deficit spending before Keynes. It was also not that wise behavior on the part of an individual can be self-defeating if everyone behaves

similarly. Nor that the classical proposition that excess supply or insufficient demand for labor could always be cured by a fall in wages or interest rates:

> As many of us were forced by the logic of events to realize, the economics of the system as a whole differs profoundly from the economics of the individual; that what is economically wise behavior on the part of a single individual may on occasion be suicidal if engaged in by all individuals collectively; that the income of the nation is but the counterpart of the expenditures of the nation. If we all restrict our expenditures, this means restricting our incomes, which in turn is followed by a further restriction in expenditures.[73]

As Herbert Stein, the economist, pointed out, Keynes asked a very different question from the one posed by Hayek and Schumpeter. In explaining depressions in terms of the preceding booms, the Austrians were trying to figure out how the economy had gotten there. Keynes was less interested in the genesis of slumps than in the more basic puzzle of how high unemployment and slack capacity could persist for long in a free market economy with unrestricted competition.

Not only should unemployment be temporary under standard economic assumptions, but, by and large, it had been. In Fisher's hydraulic machine—as well as in economic models in the heads of Marx, Marshall, and Schumpeter—a bad harvest, a war, a strike, an innovation, or some other "shock" could produce a temporary imbalance between supply and demand that, if large enough relative to the size of the economy, could result in unemployment. But, in that event, competition among workers and among lenders should drive down pay and interest rates until it was once again profitable to hire and invest.

Say's law, which stated that supply creates its own demand, was already considered out of date by the mid-nineteenth century. Based on the truism that every purchase creates an equivalent income, the law presumed that income was earned solely so that it could be spent. But saving was, of course, also an important motive, and even in the Victorian era the saving of working-class households was significant. As soon as the possibility of spending less than one earned was acknowledged, Say's law became obsolete.

What Keynes did, writes Skidelsky, was essentially to avert his eyes from market-clearing equilibrium. Instead he let money flows (such as income) functionally determine other money flows (such as consumption). The denial of supply-demand equilibrium is what Schumpeter simply could not stomach. Thus what made the *General Theory* so radical was Keynes's proof that it was *possible* for a free market economy to settle into states in which workers and machines remained idle for prolonged periods of time—that there were depressions that, unlike the garden-variety ones, were not brief and didn't end of their own accord as a result of falling prices and interest rates, or, at an extreme, that free market economies tended naturally to stagnate even when there were idle workers and machines available. In such depressions, unfreezing credit flows through monetary policy didn't provide a sufficient stimulus, because even zero-percent interest rates could not tempt businesses to borrow while prices were falling and there was reason to think that demand would recover. The only way to revive business confidence and get the private sector spending again was by cutting taxes and letting businesses and individuals keep more of their income so that they could spend it. Or, better yet, having the government spend more money directly, since that would guarantee that 100 percent of it would be spent rather than saved. If the private sector couldn't or wouldn't spend, then the government had to do it. For Keynes, the government had to be prepared to act as the spender of last resort, just as the central bank acted as the lender of last resort.

James Tobin has pointed out that Fisher came close to producing the elements of a general theory in his 1930 book *The Theory of Interest*. He had a theory of investment and savings, as well as how production and prices are determined in the short run. In *Booms and Depressions,* in 1932, he introduced the role of debt in self-reinforcing slumps. But, unlike Keynes, Fisher never combined these separate components into a single unified model that showed how interest rates, the price level, output, and, therefore, employment were determined.

As often happens with novel doctrines, most of the measures urged by Fisher and Keynes, except for the abandonment of gold, were not adopted in either the UK or the United States. Still, in England, the worst was over by August 1932, when the economy began slowly expanding. By 1937, Japan's economy had been growing for a half dozen years. In Germany, where the economic collapse was as bad as in the United States, unemployment had virtually disappeared by 1936. Keynes noted the bitter irony of Nazi Germany and Fascist Italy achieving full employment by engaging in massive deficit spending, repudiating their foreign debts, and letting their currencies depreciate. The same was true of Imperial Japan. Of course, the goal of these governments was to wage war and to pay off their debts by exploiting their victims.

In the United States, however, the depression had come roaring back with a vengeance in 1937—largely, it seems, because of blunders by the administration and especially by the Federal Reserve. In 1936, after three years of recovery, FDR raised taxes and scaled back spending on New Deal programs such as the WPA. A onetime bonus payment for World War I veterans in June 1936 briefly pumped up the federal deficit, but federal spending fell sharply thereafter. Meanwhile, the Social Security Act of 1935 created a payroll tax that began in 1937. Together these two ill-timed actions brought the federal budget into virtual balance by late 1937.

Early in the Depression the Federal Reserve had remained passive in the face of a traumatized banking system and credit markets. The Banking Act of 1935 gave the Fed authority to change reserve requirements. Between August 1936 and May 1937, the Federal Reserve, worried about growing excess reserves and inflationary pressures, abruptly doubled reserve requirements. As excess reserves fell, so did the stock of money. From May 1937 to June 1938, the US economy contracted by one-fifth, industrial output plummeted by one-third, and unemployment, which had fallen to 10 percent, jumped back to 13 percent. The official rate, which excluded temporary government jobs, rose from 15 percent to nearly 20 percent. The stock market plunged too, completing Irving Fisher's financial ruin.

Keynes, who invested heavily in depressed American stock in 1936 and hung on after the 1937 crash, recouped his losses and more. But his heart failed him. Keynes collapsed at his office in London and was diagnosed with a potentially fatal heart condition. He dropped out of public life, seemingly for good. Irving Fisher continued to speak and write but never established the rapport with the Roosevelt administration that he had enjoyed with the Hoover administration. His public reputation was as battered as his stock portfolio.

Hayek's and Schumpeter's predictions that doing nothing would lead to a recovery did not pan out either, and both wound up intellectually isolated and increasingly disheartened by the economic decline and the growing political extremism in Germany and Austria.

But no economist there or anywhere else had a satisfactory theory in the early 1930s to explain the cascading global crisis. In the absence of such a theory, English economists quickly divided into two rival camps: an interventionist group led by Keynes and the "Cambridge Circus," which included Keynes's communist disciples Piero Sraffa, Joan Robinson, and Richard Kahn, and, on the other side of the divide, a group of young "liberals" at the London School of Economics led by the thirty-year-old Lionel Robbins. One of the few prominent British economists who was the son of a miner or had strong intellectual ties to Continental economics, Robbins had spent considerable time in Vienna with Ludwig von Mises and his circle. Not only did Robbins find Mises's arguments in the debate over Socialism's viability compelling, but he also shared Mises's dismay over the seemingly inexorable trend toward government intervention in the economies of England and America.

Robbins resented the dominance of Cambridge and Keynes in English economics and regarded Keynes, with whom he clashed bitterly on protectionism while serving on Ramsey MacDonald's Economic Advisory Board, as a political opportunist and intellectual bully. Ironically, Robbins's ambition was to turn the London School of Economics, founded and patronized by Fabians, into the liberal counterweight to Cambridge collectivism. In search of potential political allies, Robbins had spotted Hayek, the thirty-one-year-old Austrian protégé of von Mises, and invited him to come to LSE to give a series of lectures in January 1931. Hayek, who was running his business cycle research institute in Vienna and working on a massive history of monetary policy, had impressed Robbins by correctly predicting the collapse of the American boom back in the spring of 1929, when other pundits were issuing sunny forecasts: "The boom will collapse within the next few months." [74] Hayek later recalled that he had said that there was "no hope of a recovery in Europe until interest rates fell, and interest rates would not fall until the American boom collapses, which I said was likely to happen within the next few months." [75]

Mises and Hayek had developed a theory blaming depressions on excessive money creation and overly low interest rates in the preceding boom that led to a massive misallocation of capital—or, as Robbins put it, "inappropriate investments fostered by wrong expectations." [76] Hayek thought the theory explained the Great Depression, which he argued was "due to monetary mismanagement and State intervention operating in a milieu in which the essential strength of capitalism had already been sapped by war and by policy." [77]

If it was true that overinvestment during the boom—not underinvestment in the recession, as Keynes contended—was to blame for the slump, then what was needed was simply "time to effect a permanent cure by the slow process of adapting the structure of production"—in other words, waiting until excess capacity was absorbed or written down and new investment was once again called for. "The creation of artificial demand," Hayek argued, would do nothing to undo the misallocation of capital and therefore would only lead to another burst of inflation and another downturn, as it had in 1921, when Austria suffered a hyperinflation.

Hayek's LSE lectures were "a sensation," according to Robbins. "At once difficult and exciting . . . they conveyed such an impression of learning and analytical invention." William Beveridge, the director of the LSE and the acknowledged father of the English welfare state, was so impressed by the "tall, powerful, reserved" Austrian that he promptly offered him a vacant professorial chair. Hayek had written a stinging review of Keynes's *Treatise on Money* and had engaged in a high-profile debate with Keynes and his disciples. Hayek's grave expression, courtly manners, and reserve that hinted at some private sorrow appealed to his English audience. His enigmatic expression, fearlessness, and ascetic refusal to prescribe easy cures reminded them of his cousin Ludwig Wittgenstein. Hayek had found credible new arguments for the traditional liberal policies of sound money, free trade, and respect for property rights and the view that recessions heal themselves.

Lionel Robbins's 1934 book *The Great Depression* was a skillful application of Hayekian theory to the boom and bust of the interwar period. (Decades later, in his 1971 *Autobiography of an Economist,* Robbins recanted, confessing that he would "willingly see it forgotten." [78]) Hayek supported Robbins's public campaign to counter Keynes's proposals. He was one of the cosigners, with Robbins and other LSE professors, supporting a balanced budget policy in 1932. [79]

Hayek's star did not glitter for long. By 1935, Beatrice Webb said of "Robbins and Co"—the "Co." being Hayek—that "they and their credo are side-tracked, without influence or even relevance to the present state of the world." [80] She was right. By the time Keynes's *General Theory* appeared the following year, the debate was over, and the economics profession had swung solidly to the Keynesian view, which, according to one of Hayek's friends, "fitted the times of deflation and mass unemployment better than Hayek's monetary temperance." [81]

At the time, Hayek was left less embattled than entirely eclipsed. Speculating about Hayek's failure to attack the *General Theory* in print, Bruce Caldwell, editor of Hayek's collected works, hazards a guess that Hayek was simply not invited to review it. Harsh criticisms of the early Hayek— from his adversaries, his erstwhile defenders, and his political allies—are prevalent in the literature. Keynes referred to his 1931 work *Prices and Production* as a "frightful muddle," [82] and Milton Friedman described himself as "an enormous admirer of Hayek, but *not for his economics.*" [83] Before long, Hayek's exchanges with Keynes were confined to their common passion for antiquarian books.

After three stints as a visiting professor, Schumpeter moved to Harvard for good in 1932. His reasons for leaving Germany had less to do with the rise of left- and right-wing political extremism (the Nazis fared poorly in the 1932 election) than with his failure to obtain a chair in Berlin and his desire to avoid marrying his longtime mistress, Mia Stöckel. Germany had been for him a place of exile, irrevocably associated with the greatest disappointments and tragedies of his life, including the deaths of his beloved second wife, Annie, and his mother.

A severe blow was the publication of Keynes's *Treatise on Money*, which convinced Schumpeter, who had been working on his own book on monetary origins of the business cycle, that his project was "useless." He told one of his students, "The only thing left for me to do now is to throw the money manuscript away."[84] His reaction suggests that his own ideas coincided with those of Keynes and Fisher and that he realized he had little to add. Otherwise, Schumpeter surely would have welcomed a chance to criticize Keynes's theory and contrast his own.

Schumpeter's depression was deepened by the stunning collapse of the German economy after Black Thursday. As American investors rushed to liquidate their foreign holdings and American merchants slashed their imports of German grain, German industrial production fell by 40 percent and unemployment shot to more than 30 percent.[85] The depression in Germany was even deeper than in the United States—deeper, in fact, than in any other major economy.

Twenty years before in the midst of another global economic crisis, Schumpeter and Keynes had advocated similar responses. Now Schumpeter defined his position in opposition to that of Keynes. At the annual meeting of the American Economic Association in December 1930, Schumpeter had attracted a spurt of media attention when he suggested that no politically palatable cure for the depression existed.[86] Joseph Dorfman, the historian of economics, attributed that response to Schumpeter's "somber outlook," which struck many in the United States as "a useful counterbalance to the characteristic optimism of the Anglo-American tradition."[87]

Schumpeter's insistence that monetary expansion was bound to fail intensified over time. It is a bit of a puzzle, especially in light of his praise of Japan's decision to abandon the gold standard in 1931. To be sure,

Schumpeter's theory of the business cycle emphasized causes other than monetary ones far more than Keynes's or Fisher's, particularly the consequences of new technologies, chemical and mechanical, that were revolutionizing farming. Schumpeter also believed that "creative destruction" of obsolete firms or industries was a precondition for long-term growth of productivity and living standards. But had he believed these things any less in 1919? His extreme fatalism struck at least some of his students and colleagues as new.

Schumpeter took part in efforts to find jobs for Jewish economists who were being persecuted by the new Hitler administration. He formed "a Committee [with the American economist Wesley Clair Mitchell] to take care of some of those German scientists who are now being removed from their chairs by the present government on account of their Hebrew race or faith." In a letter written shortly after Hitler became chancellor in Germany's coalition government in March 1933—but before the creation of the Nazi dictatorship—Schumpeter expressed his growing sense of isolation and unhappiness:

> In order to avoid what would be a very natural misunderstanding allow me to state that I am a German citizen but not a Jew or of Jewish descent. Nor am I a thorough exponent of the present German government, the actions of which look somewhat differently to one who has had the experience of the regime which preceded it. My conservative convictions make it impossible for me to share the well-nigh unanimous condemnation the Hitler Ministry meets with in the world at large. It is merely from a sense of duty towards men who have been my colleagues that I am trying to organize some help for them which would enable them to carry on quiet scientific work in this country should the necessity arise.[88]

Schumpeter must have imbibed some of the new attitudes that Hayek brought to the LSE when he gave a series of lectures there on the depression. By the time he arrived at Harvard for good a year later, he was asserting that economists had no business giving advice, although, as his student Paul Samuelson remarked sardonically, "He was always giving advice." He organized an informal seminar with like-minded colleagues, the "Seven Wise Men," who met once a week. The group, which included Wassily Leontief, the Russian-born mathematical economist, eventually published a laissez-faire manifesto attacking the New Deal.

> Recovery is sound only if it does come of itself. For any revival which is merely due to artificial stimulus leaves part of the work of depressions undone and adds, to an undigested element of maladjustment, new maladjustment of its own which has to be liquidated in turn, thus threatening business with another crisis ahead. Particularly, our story provides a presumption against remedial measures which work through money and credit. For the trouble is fundamentally not with money and credit, and policies of this class are particularly apt to keep up, and add to, maladjustment, and to produce additional trouble in the future.[89]

When Keynes's *General Theory* appeared, Schumpeter, who had earlier been on the most cordial terms with Keynes and sympathetic to his views, wrote a singularly splenetic review: "The advice (everybody knows what it is Mr. Keynes advises) may be good. For the England of today it possibly is. That vision may be entitled to the compliment that it expresses forcefully the attitude of a decaying civilization."[90]

Chapter XI

Experiments: Webb and
Robinson in the 1930s

The Soviet Union presents a blazing contrast to the rest of
Europe.

—*Walter Duranty,* New York Times, *July 20, 1931*[1]

Two experiments on a large scale are actually going on in the
world of today—American Capitalism and Russian Communism.

—*Beatrice Webb, April 1932*[2]

The apparent helplessness of Western governments in the face of a global
economic calamity seemed to confirm the thesis of the Webbs' 1923 book
The Decay of Capitalist Civilization. Interpreting Labour's stunning electoral defeat as more of "a victory for the American and British financiers"
than a repudiation of the government's shaky response to the slump, Beatrice Webb lost what little remained of her faith in the Fabians' "inevitability of gradualism."[3] Initially hostile to the Bolshevik regime, she now saw
the Soviet Union as the only nation that was "increasing the material resources and improving the health and education of its people." Somewhat
impulsively, she decided to make this "new social order" the subject of her
and Sidney's next magnum opus.[4]

One week after the general election that evicted Sidney from his
cabinet post on October 27, 1931, the seventy-eight-year-old Beatrice was
asking herself, "How shall we spend our old age?"[5] She wondered if she

was strong enough to travel to Russia to collect material, even though her reason for wanting to go was merely to lend "vividness" to her account.[6] She had already made up her mind that the Soviet experiment was working just as surely as the Western one was not, declaring that "without doubt we are on the side of Russia."[7] Before their departure aboard the Russian steamer *Smolny,* she had "summarized the immense book that she and Sidney were to write on their return."[8]

Stalin had not foreseen the worldwide depression any more than Keynes or Fisher had, but he seized the opportunity to recruit Western sympathizers and allies. Prominent fellow travelers were even more prized than more ordinary party members, and extraordinary efforts were expended to cultivate them. A phalanx of official guides, interpreters, and drivers met the Webbs in Leningrad and whisked them off on a strenuous two-month tour of factories, farms, schools, and clinics to inspect what Webb now referred to as a "new civilization."[9]

In London, the dinner invitations, political consultations, and newspaper interviews had dried up after Labour was ousted. In Russia, "We seem to be a new type of royalty," Beatrice observed with pleasure.[10] We now know that while the Webbs were being ferried about in limousines and special trains, Stalin was transforming the Ukraine into a giant concentration camp. Moscow had been selling grain to the West in return for machinery, but the collapse of world grain prices meant having to double the tonnage for export. The Soviet dictator, who was so economically illiterate that he once, after a shortage of small coins developed, had several dozen bank cashiers shot, demanded one-half of the harvest for export. The inevitable famine ultimately claimed at least six million lives, onequarter of a rural population that was already decimated by forced collectivization.

Once back in England, Webb added her voice to the denials issuing forth from Moscow. She was relying on the testimony of Western correspondents in Moscow like the *New York Times*'s Walter Duranty, who insisted, "There is no famine or starvation, nor is there likely to be."[11] But Duranty had not strayed from the capital and was merely echoing the government's disavowals. After Malcolm Muggeridge, a *Manchester Guardian* correspondent who was married to her niece, went to the Ukraine to see for himself, Webb refused to believe his shattering description of starving peasants and official abuse. She dismissed her nephew's reports as "hysterical" and suggested that Soviet Communism had become the innocent target of "poor Malcolm's complexes" and "a well of hatred in [his] nature." Beatrice invited Ivan Maisky, the new Soviet ambassador, and his wife over for the weekend and was "comforted" by their assurances that there was no famine.[12] In *Soviet Communism: A New Civilization,* which was published in 1935, she insisted that "what the Soviet Union was faced with, from 1929 onward, was not a famine but a widespread general strike of the peasantry, in resistance to the policy of collectivization."[13]

Bertrand Russell, who was critical of the Webbs for their "worship of the state" and "undue tolerance of Mussolini and Hitler," was even more appalled by their "rather absurd adulation" of the Soviet government.[14] The historian Robert Conquest faults their naïve faith in official statistics, inclination to depreciate anecdote, and ignorance of history: "They had no background of knowledge, let alone 'feel' for the great slave empires of antiquity, the millenarian sects of the 16th century, the conquerors of medieval Asia."[15] But Keynes probably put his finger on the real source of Webb's infatuation with the Soviet Union when he called Communism a religion "with an appeal to the ascetic in us."[16] In her eighth decade, Webb had found a new faith. As Muggeridge complained, "You couldn't change her mind with facts."[17]

Although Keynes had an "unmitigated contempt for the official Labour Party,"[18] he was an old-fashioned liberal like Russell. He bracketed the Soviet Union with Fascist Germany and loathed Stalin, predicting in 1937 that "an eventual agreement between him and Germany [is] by no means out of the question, if it should happen to suit him."[19] On being asked to contribute to a Festschrift for Webb's eightieth birthday, he replied that "the only sentence which came to my mind spontaneously was that 'Mrs. Webb, not being a Soviet politician, has managed to survive to the age of eighty.'"[20]

Keynes was inclined to see young Communists and fellow travelers in his circle at Cambridge as amateurs whose fanaticism was a harmless eccentricity or a passing phase. He didn't see why ideology should get in the way of friendship or research and, if anything, he admired their idealism and courage. In 1939, he ventured that "there is no one in politics today worth six pence outside the ranks of Liberals except the post war generation of intellectual communists under thirty five." Though deluded, they were "magnificent material," too good to waste.[21]

Joan Robinson, who would become the most famous of Keynes's Cambridge disciples, was almost certainly one of the "intellectual communists" that Keynes had in mind when he wrote that these members of the younger generation were "the nearest thing we now have to the typical nervous non-conformist English gentleman who went to the Crusades, made the Reformation, fought the Great Rebellion, won us our civil and religious liberties and humanized the working classes last century."[22] Robinson's commanding manner, zeal, and combative instincts were bred in the bone. Born Joan Violet Maurice, she came from a long line of military officers, university dons, civil servants, and dissenters. Her mother, the indomitable and perpetually youthful Lady Helen Marsh, was the beneficiary of a trust created by Parliament in 1812 after the assassination of her ancestor the British prime minister Spencer Perceval. Her great-grandfather F. D. Maurice, a famous university radical whom Alfred Marshall had known in the Grote Club, gave up his Cambridge chair rather than agreeing "to believe in eternal damnation."[23] Her father, Major General Frederick Maurice, sacrificed his military career when he publicly accused Prime Minister Lloyd George of lying in World War I, then went on to become a war correspondent, a military historian, the head of two London colleges, and the author of nineteen books. Robinson's maternal uncle, Eddie Marsh, was Winston Churchill's longtime private secretary. He devoted his free time to writing bad poetry and promoting the work of comely young writers and artists, among them Rupert Brooke, Siegfried Sassoon, and Duncan Grant. Robinson's family, her husband, Austin, said, was "a trifle frightening."[24]

Like Webb, Robinson had to reinvent herself. Despite her impressive pedigree, cavernous family mansion, and posh private schooling, she was being groomed to support a husband's career rather than to pursue one of her own. But at fourteen, she was already dreamy, bookish, and introverted. The world of her imagination seemed more vivid than the world around her. She wrote constantly: essays, stories, poetry. She wanted an audience badly enough to declaim her poems at Poet's Corner in Hyde Park.

The Maurice affair, which occupied Parliament in 1918, was a source of pride as well as pain. Even by Edwardian standards, Major General Maurice was an aloof and distant father. All emotion, he believed, was selfish. When he was forced out of the army, he wrote to his children that he was "persuaded that I am doing what is right, and once that is so, nothing else matters to a man," adding that this was what Christ meant when he directed his followers to forsake their parents and *children* for his sake. His son-in-law Austin Robinson observed, "He no more noticed anything irrelevant to his immediate preoccupation than shadows reflected on a wall."[25] On one occasion, Joan's sister Nancy was behind their father on a ski trail when she slipped on a bridge and wound up hanging upside down over the gorge. A passing ski instructor had to rescue her.

Despite her family's numerous connections to Cambridge, Robinson was the only one of the four Maurice sisters to go to university. Higher education was still considered superfluous for an upper-class English girl. And her father's forced retirement may have made it unaffordable had Robinson, as single-minded as her father when she wanted something, not won a teacher's scholarship. She enrolled at Girton College, the oldest women's college at Cambridge, whose mock-medieval architecture and remoteness from the men's colleges prompted the philosopher C. S. Lewis to compare it to the Castle of Otranto in Horace Walpole's Gothic novel.[26]

As a student at St. Paul's School for Girls during the painful and prolonged recession of 1920–21, Robinson had done volunteer work at a London settlement house. When she went up to Cambridge in the late summer of 1922, the downturn was entering its third year. With unemployment in the double digits and the subject of heated political debate, Robinson decided that she would give up history, her favorite subject at St. Paul's, and take up economics instead. As one of her biographers, Marjorie Turner, observed, poverty and unemployment were blemishes of the society in which she and her family occupied a privileged position, and she felt compelled to understand them.

Cambridge in the 1920s may have seemed like a lush suburb of Bloomsbury, where T. S. Eliot, Roger Fry, G. E. Moore, and John Maynard Keynes wandered about, but female undergraduates were forbidden many of its fruits. Countless rules limited intellectual intercourse with resident geniuses, whether dons or students. The one that forbade them to wear gowns to lectures like the men and required them to wear dresses and hats instead was only one of many daily reminders of their inferior status. When Bertrand Russell was scheduled to lecture at Newnham College, the second-oldest women's college, the panicked authorities first threatened to rescind the invitation, then issued an injunction forbidding any young lady "to accompany him from lecture room to door."[27] Robinson and other female students of Arthur Pigou, an eminent economist who held Alfred Marshall's former chair, could only deliver their essays to the porter's lodge, whereas his male students could bring them directly to his rooms, where they might easily be invited to stay and chat. The Union, where the undergraduate Keynes had sharpened his debating skills against those of future prime ministers, was off-limits to women, except for the upstairs gallery. So was the Cambridge Conversazione Society, aka the Apostles, where the mathematician Frank Ramsey, who was exactly Robinson's age, came to the attention of his future mentors Keynes and Russell. Keynes's own nursery for future stars, the Monday Political Economy Club, was open—by invitation only—to male but not female undergraduates.

Instead of having one of Cambridge's Olympians as a tutor, Robinson was assigned a smartly dressed daughter of a New York perfume manufacturer. Still in her twenties, Marjorie Tappan had studied economics at Columbia University—although there is no record there of her having received a doctorate, as she claimed—before working for the American economics team at the Paris peace talks for two years. Robinson detested her. Whether her resentment was due to the fact that Tappan was a rich American whose family was in "trade" or simply the fact that Tappan was not one of the luminaries is hard to know. The only thing that seems to have rubbed off on Robinson was Tappan's habit of using a long cigarette holder when she smoked, and gesturing with it while talking with her students.

Robinson attended the lectures of Pigou on economic theory and the less frequent ones of Keynes on current economic issues, but her undergraduate papers gave little hint of her future. "Beauty and the Beast," delivered at the Marshall Society in her third year at Cambridge, was a charming pastiche that proved that she could write and had a firm grasp of Alfred Marshall's *Principles of Economics*. But compared with the problems some of her male peers were solving, it was sophomoric. At twenty-one, Keynes's protégé Frank Ramsey had published a devastating paper on Keynes's probability theory, a forceful critique of Wittgenstein's *Tractatus*, and an article for Keynes's *Economic Journal* showing that a wildly popular economic panacea, the Douglas social credit scheme, was based on a faulty premise.

Despite some early successes, Robinson's undergraduate career ended in tears. She took part one of the economic Tripos in 1924, and part two the following year. Her second-class results on both, which dashed any hope of a college fellowship, were "a great disappointment" to her.[28] Years later she was still fretting over "being so badly educated."[29] Mortified, she moved back home to London, where she spent the fall and winter in a "wretched state" living in a "grubby room" in London's East End and working in a government housing office.[30] She was so miserable that she asked her father to investigate various possibilities in America, among them a scholarship to Harvard's sister college, Radcliffe. But in the spring, she decided to opt for the traditional solution to a female career quandary. On the eve of the General Strike of May 1926, Robinson was in Paris with her sister Nancy shopping for wedding clothes.

Her fiancé was a clean-cut thirty-two-year-old Cambridge don. The son of an impecunious parson, Austin Robinson was a decorated World War I seaplane pilot who was so electrified by Keynes's 1919 lectures on the Versailles peace treaty that he switched from classics to economics. Bright, efficient, and incredibly hardworking, he was invited by Keynes to join the Monday-night Political Economy Club, got a first-class degree in economics, and was elected fellow of Corpus Christi College. By Joan's second year, he was giving lectures on monetary economics. They did not become a couple until Joan left Cambridge for London.

While Austin was besotted, Joan was cooler, refusing him the first time he proposed. He was handsome, intelligent, upright, respected, kind—and seemingly unthreatened by her expressed desire for some money-making occupation. Yet, against the bold canvas of her imagined future life, he lacked color. A dozen years later, when Stevie Smith, one of her many literary acquaintances, invited her to suggest a plot for a novel, Robinson proposed one about a girl who was torn between two lovers, one of whom was a conventional young man with a good job who promised to provide her with an "orthodox life" that she "tries to force herself to want." [31] It was an unpromising start for a marriage.

"I want desperately to stay in Cambridge," Austin confessed to her after they became engaged. [32] But despite Keynes's patronage, his prospects of a salaried university position there or anywhere else in England were far from good. There were simply no academic openings. When Joan learned, from a friend's father, that the old maharajah of Gwalior, India—an Anglophile who insisted on naming his children George and May and importing tutors from Cambridge—had died, leaving behind a ten-year-old heir in urgent need of instruction, she pressed Austin to apply for the post. While they waited for job opportunities to open up at home, she pointed out to Austin he would be earning several times as much as any lecturer in England.

The newlyweds wound up spending the first two years of their married life in an ancient Indian city with "broad streets, beautifully carved balconies, doors and latticed windows, mosques and temples, old and new Palaces," [33] on the main line between Delhi and Bombay. Although Joan was close to her family, it was delightful for the couple to be on their own. Life in Gwalior involved riding at dawn with the lancers and the boy maharajah, lessons in Hindi at lunchtime, tennis, newspapers, and cocktails at the club before dinner. With a personal staff of a dozen servants, including five gardeners, Joan felt free to teach an economics course at a local secondary school. She worked on a paper that Austin was asked to write on

India's likely future contribution to British tax revenues. Meanwhile, she thought about how she might best help her husband secure a permanent lectureship at Cambridge University, and also about what work she might do. Dorothy Garratt had teased her that if she had not married a minister's son she would "probably be scrubbing lavatory seats in a leper colony or embroidering chasubles for curates."[34] At one point, she thought about opening an import business in Indian handicrafts.

With her husband's three-year tutoring contract due to expire at the end of 1928, Robinson went back to Cambridge on her own that July. Her idea was to personally deliver the report they had written together and to use her connections to pave the way for Austin's return. She was, and would always be, an enterprising and persistent networker. Less than two years later, by May 1930, Austin had his appointment as a permanent, full-time university lecturer. Until then, while Austin wrote his first book, they lived on their considerable savings. Only after Austin's future was assured, say her biographers, did she begin to focus seriously on her own career.

India and marriage had restored her intellectual self-confidence, and Austin gave her access to the university community. Her husband's success and his friendship with luminaries such as Keynes were gratifying. Lacking a college fellowship and a first-class degree, she paid a £5 fee to get a master's diploma and let it be known that she was available to coach undergraduates for a modest wage. She could not help being aware that she was still on the outside, an onlooker instead of a participant in the intellectual feast. High table, fellows' rooms, and clubs were all off-limits to her because of her sex.

Everything changed in the months after the American stock market crash. Two developments were crucial.

While she was waiting for Austin to get his appointment in the 1929–30 academic year, Robinson attended a seminar where she learned about a theoretical challenge that had preoccupied some of Keynes's Cambridge disciples. The seminar had been organized by Piero Sraffa, a brilliant but neurotic autodidact, economist, and communist who fled Mussolini's Italy in 1927. He had gotten Keynes's attention with an article calling for a revamping of economic theory to reflect the monopoly elements of modern business: the rise of giant corporations, branding, and advertising. Economists assumed competitive markets with large numbers of buyers and sellers selling identical products. Under such circumstances, no single firm could influence the price at which it sold its output, any more than a farmer could influence the price of wheat or a miner the price of silver. But modern businesses mimicked monopolies and spent large sums to influence prices. That invalidated, Sraffa argued, the principle rationale for competition, namely that a free market economy produced maximum output at minimum cost and opened the door to government intervention. What was needed was a theory. He and several others were already working on various approaches.

Robinson also befriended Keynes's "favorite pupil" Richard Kahn, a beautiful, dark-eyed Orthodox Jew who became her ally and helpmeet. Kahn was so gifted that Keynes enlisted him to help him revise A *Treatise on Money* even though he had had less than a year's formal training in economics. It was thrilling to interact with men whose intellect she could worship because it was superior to her own.[35] She started telling Austin that he was a mere plow horse while Sraffa was a tiger, and she was willing to overlook Kahn's immaturity, narcissism, and dysfunction. She was becoming aware of a bigger game, and now she wanted to become a player.

Austin suggested a topic for Sraffa's challenge when he, Joan, and Kahn had lunch one day. With the help and support of her lover from mid-1930 to early 1933, she took up the challenge. She and Kahn developed a theory to show how advertising, branding, and product innovation caused firms in seemingly competitive industries—that is, in industries with lots of buyers and sellers and no barriers to entry—to behave like monopolies. Instead of minimizing prices to consumers and maximizing output and employment, they used their market power to gouge consumers and earn extraordinary profits, depressing employment and lowering wages. In the context of the Great Depression, Robinson saw herself working out an explanation of how, even under ideal circumstances, the free market economy tended toward long-run unemployment, excess industrial capacity, and stagnation.

As Robinson's confidence grew, so did her ambition. In March 1931

Robinson informed Kahn, "I am now toying with the idea of producing a complete book with all this stuff . . . It is not I who am bringing out this book. It is a syndicate of you A + me." [36] Like a general commanding her army, she assigned tasks; Austin would write the introduction, Kahn would pose problems and write the mathematical appendix, and she would draft the book. Six months later, Robinson asked Dennis Robertson, a highly regarded collaborator of Keynes and an expert on the theory of the firm, to write a preface. She told him that she had written five chapters and sketched out another ten. As Aslanbeigui and Oakes observe, Robinson was "clearly planning to publish under her name alone." [37] For the next year and a half, Robinson and Kahn worked intensively on the book, which Robinson soon renamed her "nightmare."

Meanwhile, her collaboration with Kahn let her join Keynes's inner circle. In the first half of 1931, Keynes was grappling with criticisms of *A Treatise on Money*—especially those of Hayek—and working out some of the ideas that would mature into *The General Theory*. Between January and May a group of young Cambridge economists, including Sraffa, Kahn, and Austin, that called itself "the Circus" acted as Keynes's sounding board. Joan attended the weekly meetings and started to send notes to Keynes via Kahn. "Keynes seemed to play the role of God in a morality play," another participant recalled. "He dominated the play but rarely appeared himself on the stage. Kahn was the Messenger Angel who brought messages and problems from Keynes to the Circus and who went back to Heaven with the result of our deliberations." [38] For Robinson, it was an extraordinary opportunity to gain access to Keynes's latest thinking as he tried to understand the worst economic crisis in modern history, as well as to hone her own analytic powers.

Whether her new status was helpful in winning her first formal, if temporary, university teaching post is unclear. In any case, she was appointed junior assistant lecturer. One of her students that year remembered Joan as "young, vigorous and beautiful." He described her lectures: "She addressed us in abstruse terms . . . I understood little but sat spellbound." [39]

Despite the new demands on her time, her own manuscript was nearly complete by October 1932. At that point, any hesitation on her part to claim ownership had disappeared, write her biographers.[40] Husband, wife, and lover seemed to communicate by Cambridge's five-times-a-day post the way modern couples exchange e-mails. Robinson dashed off a triumphant note to Austin:

> I have found out what my book is about. It was quite a sudden revelation which I only had yesterday. What I have been and gone and done is what Piero said must be done, in his famous article. I have rewritten the whole theory of value beginning with the firm as a monopolist. I used to think I was providing tools for some genius in the future to use, and all the time I have done the job myself.[41]

Hitherto she had regarded herself strictly as a teacher. "I used to feel 'I must tell these people what economists think'—now I really feel I am an economist and I can tell them what I think myself."[42] She told Kahn that "AR" would find her "a Changed Woman. I have recovered my self-respect." She left no doubt, however, that she now considered herself first among equals, the original thinker, the guiding genius: "You and Kahn and I have been teaching each other economics intensively these two years. But it was I who saw the great light and it is *my* book." It is hard to miss the note of glee at having beaten the boys.

Meanwhile, Kahn was falling in love with Robinson. By 1931 they were having an on-again, off-again affair, greatly alarming Keynes, who feared for his protégé's career, and made Joan nervous lest a scandal spoil her own imminent academic success. Austin had gone to Africa for six months, and Joan insisted that Kahn also leave Cambridge to cure himself of "lovesickness." He decided to visit America for a year. Alone, under great stress, and feeling that she was on the verge of a breakdown, Joan worked feverishly to finish her book. While she was proofreading her manuscript, Kahn was at the University of Chicago promoting the book, convincing a doctoral student and future Soviet spy, Frank Coe, to incorporate her as-yet-unpublished analysis into his thesis. Then Kahn delivered a bombshell. Edward Chamberlin, a young Harvard professor, was about to publish a book, *The Theory of Monopolistic Competition,* that overlapped with hers and would precede hers by at least six months. In February, Kahn visited Harvard, where he arranged to deliver a lecture one day before Chamberlin's book was released. When he claimed that Robinson's theory and analytical techniques were superior, Chamberlin, who was in the audience, failed to deliver an effective rebuttal. "I feel a viscious [*sic*] pleasure at hearing that Chamberlin is no good," Robinson wrote to Kahn on March 2, 1933, in response to his account of the confrontation. She added that she would "just put in a note" in her preface that she had known nothing of his work. She considered asking Keynes to let her review Chamberlin's book for the *Economic Journal* but realized that "on second thought that would be bad" and that she could "deal with him sometime after I am out."[43]

To Joan's disappointment, Keynes "was not much interested in the theory of imperfect competition" and refused to be convinced that monopoly was a major cause of periodic shortfalls in effective demand.[44] After warning his publishers at Macmillan that they probably wouldn't find the book exciting, Keynes nonetheless urged them to publish it. *The Economics of Imperfect Competition* appeared in the fall of 1933. Robinson's book was an instant critical success, garnering numerous respectful, and even some superlative, reviews. Schumpeter, who had already called her "one of our best men,"[45] responded instantly to Kahn's suggestion that he promote her new book. In his review, Schumpeter praised Robinson for "genuine originality" and concluded that the book gave her "a claim, certainly to a leading, perhaps to the first place" among economic theorists in this area, placing her ahead of Kahn and Sraffa as well as Chamberlin.[46]

Robinson had an enormous edge over Sraffa and Kahn. Both suffered from severe writer's block, and Sraffa was so disabled by severe anxiety that he could not deliver lectures. She, on the other hand, was a superb speaker and writer who, once she found that she had something to say, was one of the most prolific in her discipline. As soon as the final correction had been made on her manuscript, she plunged into a series of articles and reviews.

Less than a year after *Imperfect Competition* was published, Joan gave birth to her first child. "How well you do things," gushed her friend Dorothy Garratt in May 1934. "A discovery in economics and a baby girl."[47] Robinson was elated by her *succès d'estime*. That September, when Kahn went to Tilton to work on Keynes's new book, she wrote to him, asking cheekily, "Would Maynard like me to write him a preface for the new work showing in what respects his ideas have altered?"[48] Given that most of her interactions with Keynes had been via Kahn or by letter, her suggestion was nervy, even more so considering that Kahn was the only member of the Circus who had made an original contribution (the multiplier) to Keynes's new theory. Nonetheless, there was no doubt that she had clearly won Keynes's respect as an economist. A few years later, he acknowledged that Robinson was "without a doubt within the first half dozen" economists at Cambridge, a group that included Pigou, Sraffa, Kahn, and Keynes himself.[49]

Andrew Boyle, the Scottish journalist who in 1979 exposed Anthony Blunt as the fourth member of the notorious Cambridge Five Soviet spy ring, claims that Robinson was a founding member of the first Communist cell at Cambridge. It was supposed to have been organized by Maurice Dobb, the economics lecturer who would soon recruit his student and later spy Kim Philby in 1931.[50] But Boyle, who corresponded with Robinson, gives no sources. Geoffrey Harcourt, who knew Robinson toward the end of her life, dates her infatuation with Stalin—her "radicalization" as he puts it—to 1936.[51]

That year Robinson's views were unquestionably in a violent state of flux. When she reviewed John Strachey's *The Theory and Practice of Socialism* in mid-1936, she was critical of Strachey's argument that Soviet-style central planning was the cure for the Great Depression. She did not call Strachey's logic "an insult to my intelligence," as Keynes had, but she took him to task for conflating flaws in mainstream economic theory with fatal defects in the economic system. "We cannot be recommended to overthrow merely because its economists have talked nonsense about it," she quipped.[52]

Six months later, Robinson seems to have had a conversion. She described capitalism as "a system which allows effective demand to fall off amongst a population which is overcrowded and underfed, which meets unemployment with schemes to restrict output and can offer no help to distressed areas except orders for armaments." Marxist dogma might be "over simple," she admitted, but at least it didn't stifle "simple common sense." Indeed, she regarded Marxism as an effective vaccine "against the sophistications of laissez faire economics." [53]

In May 1936, her friends the Garratts had introduced Joan to a couple who were visiting Cambridge from Aleppo, Syria, the setting for Agatha Christie's best seller *Murder on the Orient Express.* Dora Collingwood was an English landscape painter whose father was a noted archaeologist, artist, and secretary to John Ruskin, the art historian. Her husband, Ernest Altounyan, was an Anglo-Armenian doctor. Dorothy Garratt described him as "a very strange but attractive person, living at an emotional level which made me feel very suburban." He was in his mid-forties, myopic, graying, but with "a good forehead and nose," [54] an insinuating voice, and all sorts of romantic friends, including Arthur Ransome, the children's author, and T. E. Lawrence, aka Lawrence of Arabia. The latter had recently died in a motorcycle crash, and Altounyan told Robinson that he had hoped to find a publisher for his epic poem celebrating their friendship. Robinson offered to read it and send it to her uncle. Altounyan was terribly impressed and grateful. They began exchanging notes. By the end of the month, he was writing confidentially that she was "by far the loveliest thing that has happened to me in England" and that meeting her had been "intoxicating." [55]

Altounyan loved to dance, and one of his daughters said that "he tried to live his whole life as a kind of dance and got depressed and frustrated if he was prevented from doing so." He was also bipolar. From Aleppo, he began writing Robinson long, rambling love letters. Robinson, meanwhile, was proofreading the poem. Eddie Marsh, Keynes, and a dozen other literary friends pronounced it dreadful, but Robinson persisted until she finally badgered the editor of Cambridge University Press into publishing it.

On March 12 of the following year, one month before the publication of Altounyan's poem, Joan boarded the Orient Express at Victoria

Station. Two months pregnant and traveling alone, she resembled Mary Debenham in Christie's novel: "There was a kind of cool efficiency in the way she was eating her breakfast and in the way she called to the attendant to bring her more coffee, which bespoke a knowledge of the world and of traveling . . . She was, he judged, the kind of young woman who could take care of herself with perfect ease wherever she went. She had poise and efficiency."[56] Joan was reunited with Altounyan in Aleppo before proceeding to Jaffa and Tiberius, in Palestine.

She saw him alone a second time on April 14, on her return trip. By then, seeing him surrounded by an untidy, unhappy household, she may have begun to feel that her lover's appeal had been largely of her own imagining. Reviews were very scarce for his poem, *Ornament of Honor*. The *Palestine Post* hailed him a "Tennyson minor."[57] When she returned to Cambridge, the old ménage à trois with Austin and Richard Kahn again became the emotional pivot in her life. "In another age she would have been on a camel, riding through the desert," Frank Hahn, an economist, once observed. "A part of her personality was simply upper-class refusal to go with the herd; a need to distinguish herself from the herd."[58]

One year after her second baby was born and weeks after Hitler seized Czechoslovakia, Robinson suffered a serious attack of mania. She spent many months in a sanitarium. By the time she got out of the hospital, Austin had been assigned to war work in Whitehall, and a physical separation ensued. One by one her colleagues were enlisted. Eventually, Kahn too had to leave Cambridge. He was eventually posted to Cairo and wound up spending most of the war there. Robinson stayed behind in Cambridge.

Chapter XII

The Economists' War:
Keynes and Friedman at the Treasury

In War we move back from the Age of Plenty to the Age of
Scarcity.

—*John Maynard Keynes, 1940*[1]

The outbreak of war gave Hayek and Keynes an opportunity to make
peace. Both had hoped that war could be avoided, but neither had any illusions about accepting Hitler's "peace" offers. Both hoped and believed that
the United States would enter the war. Otherwise, by the time Germany
collapsed, Hayek said, "the civilization of Europe will be destroyed."[2] Both
saw the war as a defense, not only of Britain, but of the eighteenth-century
Enlightenment. At a benefit performance at the Cambridge Arts Theatre
to raise money for refugees in December 1940, Keynes told the audience
that there were a thousand Germans in Cambridge. There were, he said,
now "two Germanys":

> The presence here of Germany in exile is . . . a sign that this is a war not
> between nationalities and imperialism, but between two opposed ways
> of life . . . Our object in this mad, unavoidable struggle is not to conquer
> Germany, but to convert her, to bring her back within the historic fold
> of Western civilization of which the institutional foundations are . . . the
> Christian Ethic, the Scientific Spirit and the Rule of Law. It is only on
> these foundations that the personal life can be lived.[3]

By the time the blitz began in the summer of 1940, Keynes and Hayek had been exchanging notes for months about the evacuation of the London School of Economics to Cambridge, help for Jewish academics fleeing Nazi-controlled Europe, and efforts to win the release of foreign colleagues as "enemy aliens" in the panicky weeks after the fall of France in June 1940. In October Keynes had arranged rooms and high table privileges at King's College for Hayek. On the long weekends that Keynes continued to spend in Cambridge, they frequented G. David, the antiquarian book dealer around the corner from the Cambridge Arts Theatre, and exchanged historical tidbits.

More surprisingly, the war put Hayek and Keynes on the same side of the economic policy debate. For most of the 1930s, Hayek had dismissed Keynes's proposals to fight the Great Depression with easier money and deficit spending as "inflation propaganda" and once referred to his rival privately as a "public enemy."[4] But by 1939, Hayek was praising Keynes in newspaper articles. Much to the chagrin of some of his left-wing friends and disciples, the war had turned Keynes into an inflation hawk.

What happened? Circumstances had changed. Britain had practically dissolved its army and air force after World War I, so playing catch-up with Hitler's Germany required massive increases in government spending starting in 1937. Partly out of fear that raising taxes would aggravate unemployment, which still hovered at around 9 percent, and partly because rearmament was unpopular, the government of Prime Minister Neville Chamberlain opted not to raise taxes, instead issuing IOUs to the public in the form of bonds. Thus, even before war was declared, Britain's national debt had climbed to dizzying heights. The first war budget, published in September 1939, projected a deficit of £1 billion, or a stunning 25 percent of Britain's annual national income.

Massive deficit spending had a dramatic effect. The economy boomed, especially in the south of England, where ports and bases were being expanded and arms factories were being built. This was, belatedly, the cure that Keynes had advocated in 1933, and it seemed to vindicate the *General Theory*.

One might have expected Keynes to be pleased that the Treasury, having stubbornly resisted his advice in the late 1920s and early 1930s, had finally turned "Keynesian." Instead, writes Skidelsky, he was increasingly worried and disapproving. By running up a huge debt and then printing money to hold down interest rates, the government was sowing the seeds of future inflation. Now that war was certain, things could only get worse. Keynes denied that he was having a change of heart. It was circumstances that had changed. In 1933, the unemployment rate was 15 percent; in 1939 it was below 4 percent and falling, and industrialists were complaining of shortages of skilled mechanics and engineers. Keynes invented the economics of plenty to address a massive shortfall of demand in a depression. Now he was applying the same logic to the opposite condition: namely, an excess of demand during a war.

After World War I, the consequence of inflationary war finance and huge debt burdens had been economic and political chaos. He unveiled the "Keynes Plan" in two articles in the *Times* of London in mid-November 1939.[5] To plug the gap of £400 million to £500 million between spending and tax receipts, he proposed a wartime levy on income. The twist was that the money was to be refunded after the war, enabling Keynes, like Schumpeter in 1919, to call his tax "forced savings." Skidelsky points out that the articles, published a few months later as *How to Pay for the War,* illustrated "Keynes's conception of the budget as instrument of economic policy."[6] One of the warmest endorsements came from Hayek, who seconded Keynes's proposals in a column in the *Spectator* and followed up with a note: "It is reassuring to know that we agree so completely on the economics of scarcity, even if we differ on when it applies."[7]

As Keynes well knew, he was living on borrowed time. A massive heart attack in 1937 had forced him into premature retirement at Tilton. Two years of being cared for by Lydia, a German miracle drug, and Germany's mad dreams of world conquest gave him the opportunity for a third and final act.

On the eve of the Battle of Britain, Hitler's attempt to destroy the RAF on the ground, Keynes was back at the Treasury with "no routine duties and no office hours" but with "a sort of roving commission plus membership of various high up committees."[8] The prime minister, Winston

Churchill, Britain's last lion, paid scant attention to how the war against Hitler would be financed, and even less to postwar economic arrangements. These became the bailiwick of Keynes, who took on the role of Churchill's de facto Chancellor of the Exchequer during World War II. When Keynes wrote his impassioned outburst against the Versailles Treaty in 1919, he had warned that "vengeance, I dare predict, will not limp" if the victors insisted on impoverishing the vanquished. After he was proved tragically right, Skidelsky writes, Keynes's "one overriding aim" was for the Allies "to do better than last time."[9]

After the stunning collapse of France left Britain to face the German juggernaut alone, the Treasury's, and therefore Keynes's, obsession became to raise money to keep fighting. While Hitler's tactic of serial conquest did not require Germany to place her economy on a total war footing, Britain did not have the luxury of waging a limited war. As the aggressor, Hitler could decide when to attack, and his strategy of Blitzkrieg was self-financing insofar as despoiling its victims paid the military bill. Britain's choices boiled down to two: One was accepting Hitler's offer of "peace," which meant sharing France's ignoble fate. While Keynes's old political mentor, Lloyd George, was prepared to become King George's Marshal Pétain and the Left was holding peace vigils, that option was a nonstarter with the British electorate. The other choice was to throw fiscal prudence to the wind and wage total war regardless of the postwar consequences. Although Keynes had no doubt that the latter was the correct choice, he never stopped racking his brains for clever ways to soften any negative consequences. Once again he was "interested, geared up, and happy."[10] As he wrote to a friend, "Well here am I, like a recurring decimal, doing very similar work in the same place for a similar emergency."[11]

From August 1940, Keynes spent as many as eighteen hours a day at his desk and, quite often, in the Treasury's deep cellar. Like Hayek, who had insisted on staying in London and commuting to Cambridge during the first phase of the bombing, he ignored the danger, dismissed the possibility of a German invasion, and hoped that his books and pictures would survive. Now that he was an insider with access to "all the innermost secrets" as well as to the chancellor, whose office was next to his own, he had a far bigger hand in shaping British financial policy than during World War I. Insider, yes, but still an iconoclast. Neither middle age, celebrity status, nor bad heart had dimmed his impatience with the inefficiency of the King's College freshmen or the fury expressed in *The Economic Consequences of the Peace*. "To the carpenter with a hammer, everything looks like a nail," goes the old saw. To Keynes, everything looked like a problem that he was better qualified to solve than those who were assigned the responsibility. He meddled in matters from tariffs to beer taxes, often getting the facts wrong and ruffling feathers. He once sent Richard Kahn, now posted in Egypt, a plan for reorganizing Cairo's entire transportation system.

As in the previous war, Keynes's job became to loosen American purse strings. In early May 1941, before America's entrance into the war and at the height of a bitter controversy with the United States over providing navy destroyer escorts for arms shipments to Britain, Keynes spent eleven weeks in Washington, D.C., as Britain's envoy. It was his third visit to the United States—such visits being "considered in the nature of a serious illness to be followed by convalescence" [12]—but this time he eschewed his favorite mode of transatlantic conveyance, the *Queen Mary,* and flew on Pan American's Atlantic Clipper. With German U-boats roaming the North Atlantic sinking British ships at the rate of sixty a month, flying was safer, although not necessarily much faster due to erratic schedules. Stepping onto the tarmac at La Guardia airport, where reporters waited, Keynes first fantasized aloud about a daily shuttle between London and New York, then took aim at American isolationists.

A German victory would mean that America's ties to the Old World would be severed permanently, he pointed out. "The American economy could not function at all on its present basis. It doesn't bear thinking about." Not everyone appreciated his lecture. The arch-isolationist Senator Burton Wheeler of Montana sneered, "The American people resent the fact that these foreigners are trying to involve us in the war by giving us free advice on how to run our country when they've made such a miserable failure with their own." [13] By "miserable failure," the senator meant that Britain couldn't pay her bills. Having converted her economy for total war, Britain was forced to pay for imported matériel with hard currencies, even as her ability to earn hard currencies by exporting evaporated. When Lord Lothian, Britain's ambassador, came right out and said, "Well, boys, Britain's broke. It's your money we want," the US Treasury refused to believe that the British Empire could possibly be short of gold.[14]

Yet American antipathy to sacrificing lives and treasure in fratricidal European wars was so strong after World War I that the United States embraced unilateral disarmament in a period when Germany, Russia, and, belatedly the British and French, were all rearming. Although the United States maintained the world's largest navy, its army was "a tiny skeleton force" of two hundred thousand men, and its entire air force consisted of 150 fighter planes. In 1940, the United States was spending less than 2 percent of its annual income on defense, and all arms sales to foreign governments were restricted by law. The Johnson Act of 1934 was aimed specifically at Britain. It prohibited arms sales to any country that had defaulted on its World War I debts.

The fall of France and the near destruction of the British expeditionary force at Dunkirk in June 1940 provoked a sharp reappraisal. Even in an election year, it was no longer possible to deny that Germany—especially in alliance with the Soviet Union—posed a serious potential threat to the United States. Hitler, who had a huge program to build navy destroyers and aircraft and was badgering the *Caudillo de España,* Francisco Franco, to allow German bases in western Spain, clearly had America in his sights. Congress quickly approved some $4 billion in munitions spending and set a target of 2 million "men under arms" for the end of 1941.

Nonetheless, rearmament was described as being strictly for "hemisphere defense."[15] The overwhelming majority of voters were convinced that Britain could not avoid defeat. Ironically, the historian Alan Milward points out, the American decision to rearm made that dismal prospect somewhat more likely. Britain had ordered some $2.4 billion in munitions from American defense contractors—enough ships, planes, and trucks to keep defense plants busy for several years. Now it risked having its orders bumped by American orders.

Lend-Lease was FDR's inspired strategy for keeping the United States out of the war while keeping Britain in it. Unlike his ambassador in London, Joseph Kennedy, and many of his closest advisors, the president thought that with adequate support from the United States, Britain could and would prevail. Churchill's "we will never surrender" oration during the Dunkirk evacuation convinced him that "there would be no negotiations between London and Berlin" of the kind that antiwar groups from the Communist Party to the America First Committee, two members of the British war cabinet, and Ambassador Kennedy, were demanding.[16]

Arming the British was already reviving the American economy and driving down unemployment. The only trouble was that the flow of arms could not continue on America's cash-and-carry terms, since Britain could no longer earn dollars by exporting—as Churchill explained to the president in his "begging letter," waiting until FDR's reelection in November 1940 to send it.[17] Roosevelt's response was delivered at a press conference where he told reporters that "the best immediate defense of the United States is the success of Great Britain in defending herself."[18] He was not above reminding Americans of the economic benefits of supplying Britain. He illustrated the point with a parable: If your neighbor's house was on fire and you had a water hose, you wouldn't try to sell it to him; you would lend him the hose and tell him to give it back when he had put out the fire. "What I am trying to do . . . is get rid of the silly, foolish old dollar sign," he said.[19] The United States would send Britain whatever arms and supplies she needed, paid for by American taxpayers, in exchange for Britain's promise to repay in kind when the war was won. In a radio "fireside chat" on December 29, the night that German bombers reduced London's financial district to rubble, he declared "We must be the great arsenal of democracy."[20]

The proposal required congressional approval because Roosevelt was asking for an initial appropriation of $7 billion. Opponents argued that Lend-Lease would inevitably drag America into the war by provoking a German attack. Others raised the specter that arms sent to Britain would pass to the Nazis after Britain's inevitable defeat. But the president prevailed and Congress approved the measure, with an amendment that forbade the navy from sending its ships into the war zone, on March 10, 1941.

Churchill hailed Lend-Lease as "the most un-sordid act in the history of any nation," and, indeed, the new arrangement signaled the start of a

$50 billion procession of ships, planes, and food from American factories and farms and a suspension of the traditional American practice of treating loans to allies as strictly business. But, of course, there were strings, and Keynes was determined to loosen them.

The dispute that broke out between Britain and the United States exactly one day after the White House sent the Lend-Lease bill to Congress turned on the fact that the law would cover only orders placed after the bill went into effect, not those placed before. Churchill maintained that down payments on past orders "had already denuded our resources."[21] When he complained that "we are not only to be skinned, but flayed to the bone," he was referring to one particularly onerous condition.[22] To prove that she truly needed help, Britain was supposed to exhaust all of her dollar reserves before tapping Lend-Lease—in effect, to pay for the construction of the American plants that were going to be producing arms for Britain. That meant handing over the country's dwindling gold supplies. The United States actually sent a destroyer to Cape Town, South Africa, to pick up $50 million in bullion that London had placed there for safekeeping. Britain was also required to sell stock in American companies and American subsidiaries of British corporations into a weak market. In the weeks before the passage of Lend-Lease, the British Treasury representative in New York, who was liquidating Britain's stock portfolio at the rate of $10 million a week, detected a jockeying for postwar commercial advantage.

The ever optimistic Keynes was convinced that the United States would never stand by while Britain became another Vichy France, but he failed to appreciate how committed Americans were to staying out of the war. Lend-Lease, of course, was designed to reconcile those goals. Quite apart from his election promise, "I have said this before, but I shall say it again and again and again: Your boys are not going to be sent into any foreign wars,"[23] FDR repeatedly assured Congress that the United States would not fight unless attacked. His critics on the left and the right accused him of secretly maneuvering to create provocations, but recent evidence shows that, until Pearl Harbor, the president continued to hope that he could avoid entering the war. "The time may be coming when the Germans and the Japs will do some fool thing that would put us in," he told his aides. "That's the only real danger of our getting in, is that their foot will slip."[24] One clear sign that the president meant what he said was that when Keynes arrived in Washington, the United States was monitoring encryptions from the Enigma machine, provided by the British in April, not to hunt down German submarines but to *avoid* them.[25]

Keynes accused the United States of "treating us worse than we have ever ourselves thought it proper to treat the humblest and least responsible Balkan country" and argued that Britain had to fight to keep "enough assets to leave us capable of independent action."[26] The point was to limit Britain's dependence on Lend-Lease and thus American control over the British balance of payments. Keynes went to Washington as the chancellor's personal envoy in order to try to work out better financing for Britain's pre-Lend-Lease orders. His target was to replenish Britain's reserves up to $600 million. Building up cash reserves under the cover of Lend-Lease was precisely what the Americans were on guard against.

His initial meeting with Roosevelt's Treasury secretary, Henry Morgenthau, was a disaster. Keynes's condescending professorial manner irritated the secretary. His proposal to the US Treasury to refund $700 million of the previous down payments already made on existing orders ran afoul of the president's assurance to Congress that Lend-Lease would apply only to future orders. Keynes saw Roosevelt twice, the second time in 1941 after Germany broke its pact with Stalin and invaded the Soviet Union. He managed to raise a loan enabling Britain to postpone the sale of its assets at distressed prices by offering prime British properties as collateral and agreeing to a hefty interest rate.

Keynesianism really took hold in the first year or two of the war. The huge deficit-financed military buildup had accomplished what earlier efforts to fight the Great Depression never did; mopping up the huge residue of unemployment left at the end of the 1930s. After the apparent failure of monetary policy to restore full employment, this struck young economists as a convincing demonstration, in the eyes of young economists, that the economy worked the way Keynes said it did in the *General Theory*. By

1941, self-identified Keynesians were scattered around the wartime bureaucracy in Washington like raisins in a cake.

A forecasting coup early in the war gave the young Keynesians in the government bureaucracy instant credibility. Most of the businessmen who were consulting for the government's War Production Board were convinced that the economy's productive capacity was "very limited" and skeptical that the output of weapons and matériel could be ramped up as quickly as the president wanted. The Keynesians at the Office of Price Administration disagreed. On one of Keynes's trips to Washington, they had asked their leader his opinion. Keynes displayed his knack for coming up with quick and dirty estimates from just a few facts. "Well, how much was 1929 real output over 1914?" he asked. "Well, that was a fifteen year period and it's been twelve years since 1929, so let's take 12/15ths of that increment . . . I think that would be a reasonable goal."[27] The OPA forecasters thought so too. Keynes was reasoning that because World War I through the twenties was a long period of low average unemployment, it was a good indicator of how fast the economy *could* grow when demand wasn't depressed. His and their forecasts proved to be remarkably accurate. As one OPA staffer said, "The Keynesian wing of the U.S. civil service had been vindicated."[28]

By 1941, Keynesians dominated three New Deal agencies: the Agricultural Adjustment Administration, the Bureau of the Budget, and the National Resources Planning Board. There was also a group at the Treasury. A few had risen high enough in the Roosevelt administration to influence economic policy. They included John Kenneth Galbraith, deputy chief of the Office of Price Administration; Marriner S. Eccles, chairman of the Federal Reserve; Lauchlin Currie, one of FDR's six administrative assistants; and Harry Dexter White, Treasury Secretary Henry Morgenthau's de facto chief of staff. If old adversaries now found they could make common cause with Keynes, some of his most fanatical fans in Washington were dismayed. At a dinner at the Curries' house, a number of the younger men tried to convince Keynes that the "Keynes Plan" was the wrong prescription for the United States. The official unemployment rate was still in the double digits, and some industries were still saddled with unused capacity. Spending cuts, tax increases, and other austerity measures would only aggravate these and might abort the recovery long before the economy approached full employment. Keynes happened to be right, and in any case he was not swayed. Still, he allowed, "The younger Civil Servants and advisers strike me as exceptionally capable and vigorous." He found, however, "the very gritty Jewish type perhaps a little too prominent." [29]

John Kenneth Galbraith, a farm boy from Canada who looked and sounded like an English lord, liked to say that Keynes's ideas came to Washington via Harvard.[30] But it would be more accurate to say that they had also come by way of the University of Wisconsin, Columbia, the City University of New York, MIT, Yale, and, more often than not, the University of Chicago.

Milton Friedman, a recent Ph.D. from Chicago, did not attend the dinner with Keynes at Lauchlin Currie's house, but in 1941 the future leader of the anti-Keynesian monetarist revival of the Reagan years was nonetheless one of the brightest young Keynesians in the Treasury. And, as it happens, he did more than most to make Keynesianism practically feasible in the United States.

The son of blue-collar Hungarian Jewish immigrants who settled in Brooklyn in the 1890s, Friedman was born just before World War I. He grew up over his parents' store on Main Street in Rahway, a gritty New Jersey factory town on the railway line between New York and Philadelphia, whose main claim to fame was that George Merck had moved his chemical plant there in 1903. Friedman grew up witnessing his parents struggle unsuccessfully to make a go of one business after another, including an ice cream parlor. His mother essentially supported the family, but his father was the one who died of angina at age forty-nine, when Friedman was fifteen. In high school, he read *This Side of Paradise*, F. Scott Fitzgerald's coming-of-age novel about Princeton. Amory Blaine, the protagonist, has "personality, charm, magnetism, poise, the power of dominating all contemporary males, the gift of fascinating all women." If Friedman's being less than five feet tall, wearing glasses, and being poor meant that the resemblance was not perfect, he could at least cultivate the trait Blaine valued most: "Mentally—Complete, unquestioned superiority." [31]

In Friedman's world that meant becoming an actuary. The high school debating champion went to Rutgers, not Princeton, intending to do just that. The Great Depression and a young instructor and future Federal Reserve chairman, Arthur Burns, lured him from accounting to economics. To keep his own economy afloat, the undergraduate sold fireworks, coached other students for exams, and wrote headlines for the student paper. When he graduated in 1932, he took a cross-country road trip before enrolling that fall at the University of Chicago, where the faculty were "cynical, realistic and negative" about reform, yet reformers at heart, and where being a lower-class Jew wasn't a bar to admission.[32] By the end of his first year, he had met Rose Director, the younger sister of one of his professors, taken her to the Chicago World's Fair, and fallen in love.

Three years later, when he had finished his coursework and depleted his savings, the New Deal was "a lifesaver."[33] All during the summer of 1935, he had waited in vain for an offer of a lectureship. Not only was the number of academic openings negligible, but anti-Semitism made the likelihood of his landing one remote. If one of his professors hadn't gotten him a research post in Washington, he might well have abandoned his chosen career and returned to accounting. But his enthusiasm for the New Deal was real—Rose's conservative brother attested to Friedman's "very strong New Deal leanings"[34]—and he went to assist at the "birth of a new order" that promised social change of all kinds.[35]

His new employer, the National Resources Committee, was one of the dozen or so "planning agencies" created during the first Roosevelt administration. "Planning" was then enjoying a great vogue. Proposals for setting agricultural production targets, prices in a host of industries, and minimum wages had their roots not in Stalinist economic doctrine but in the platforms of British Fabians and Labourites. In practice, however, the New Deal planners mostly engaged in constructing national income accounts and forecasting future output and employment. John Maynard Keynes had been badgering the governments of Britain and the United States to create a system of national income accounts analogous to a corporation's annual income statement. Without reliable measures of how much output an economy produced every year, how much income it generated in the form of wages, profits, interest, and rents, or how much and on what households, businesses, and government spent, the government and businesses were operating in the dark. There was no way to detect imbalances between output and demand or to gauge their magnitude. With only desk calculators, constructing national income accounts was an agonizingly labor-intensive, time-consuming project. Thus was born a huge public works program for economics graduate students. Herbert Stein, one of Friedman's Chicago classmates, once estimated that the number of economists in Washington had shot up from a mere one hundred in 1930 to five thousand by 1938.[36]

Friedman was put to work assembling the first large database on consumers and their purchases. Although the labor involved was purely statistical, he later credited the experience for some of his best work, including his "permanent income hypothesis," cited when he won the Nobel Prize in 1976. It explains, among other things, why consumers typically spend a smaller fraction of one-time tax cuts or other windfalls than of permanent tax cuts or other ongoing additions to income.

Two years later, as the far from completed economic recovery that began in 1933 went into reverse, Friedman left Washington for New York and the National Bureau of Economic Research. There he joined a team assembled by Simon Kuznets, a Columbia professor, who was constructing the first complete set of national income accounts for the United States. In addition to filling gaps in the data, Friedman's job was to create detailed estimates of the income of self-employed professionals.

In the course of his research, he was dismayed to discover that despite the huge influx of émigré Jewish physicians after Hitler came to power in 1933, the number of medical licenses had not increased in the intervening five years. Furious at the power of professional groups to prevent outsiders from entering their disciplines, he wrote a scathing indictment of licensing. He bore the brunt of power himself when his study's publication was delayed for three years by a member of the NBER's board of directors with ties to the pharmaceutical industry. Meanwhile, he wondered why he bothered. "The world is going to pieces . . . and we sit worrying about means and standard deviations and professional income," he wrote to his fiancée, the younger sister of Aaron Director, in 1938. "But what the hell else can we do?"[37]

That summer, he married Rose Director, as peppery, energetic, and conservative as her brother. When Friedman returned to Washington the second time, in the fall of 1941, he had finished his doctorate and survived a hellish first academic job at the University of Wisconsin, where sentiment was overwhelmingly pro-neutrality and anti-Semitic. The young couple consoled themselves with the thought that, sooner or later, the United States would have to enter the war. By the time Hitler attacked his Soviet ally, the Friedmans were overjoyed to be going to Washington, where there would be important war work for them to do. Over the summer, Friedman had coauthored a paper, "Taxing to Prevent Inflation," with a public finance professor on the Columbia faculty who recruited him to join the Tax Research Division at the Treasury. In Friedman's first stint in Washington he had come there as a statistician. Now he was poised to play a more influential role in shaping policy.

After Dunkirk, in light of the growing likelihood that the United States would be drawn into the war, the Roosevelt administration had become preoccupied with paying for it. The US economy was already being converted to aid for European allies, and a larger bill was coming due for the military buildup that was under way. One unwelcome side effect of shifting the economy to a wartime footing was that inflation had reappeared. Between 1940 and 1941, consumer prices jumped 5 percent, the largest one-year increase since 1920. While hardly scary by present standards, the surge was enough to revive unpleasant memories of post–World War I inflation and cost-of-living protests, along with the severe recession that followed and was seen as a direct consequence.

During World War I, tax revenues had covered two-thirds of Washington's costs. The rest was financed by issuing bonds. A reasonable inference would be that the government was borrowing to close the gap between revenues and spending—reasonable, but wrong. Most of the "borrowing" was a disguised form of printing more money. The newly created Federal Reserve had urged its member commercial banks to lend their customers money with which to buy war bonds. To increase their reserves commensurately, the banks, in turn, had borrowed from the central bank by discounting the loans at the Federal Reserve—i.e. borrowing from the Federal Reserve on the security of loans for which government bonds served as collateral. As a result, while . . . currency and deposits at the Federal Reserve . . . increased by $2.5 billion . . . only about a tenth of that represented direct purchases of government securities; the remainder consisted of credit extended to member banks.[38]

The result of the massive expansion of the money supply was a surge of inflation. For farmers, miners, and real estate developers inflation had meant a giddy extension of the wartime boom. But when the Federal Reserve jacked up interest rates, wholesale prices plunged by 44 percent and the boom turned into a nasty slump. The political fallout swept Republican Warren Harding, who campaigned on the slogan "A Return to Normalcy," into the White House. For the officials at the Democratic Treasury, how to avoid a repetition of this disaster was Topic A after World War II.

By the time the Friedmans moved into their apartment near Dupont Circle, a short walk from the Treasury building, the Treasury secretary's bulldog assistant, Harry Dexter White, was growling that things weren't going very well. "It's getting away from you," he hissed at Galbraith after one meeting about the inflation problem. "You must get moving."[39] The secretary had already ordered the Tax Division to prepare a restructuring of the federal tax system. Virtually all of the debate in Washington about fighting inflation concerned the relative effectiveness of wage and price controls versus taxation. Ultimately, the Roosevelt administration embraced both.

Selective price controls to avert "price-spiraling, rising costs of living, profiteering and inflation" had already been in effect since April 1941, and the OPA had been created to administer them.[40] After Bernard Baruch told a Congressional committee, "I do not believe in piecemeal price fixing. I think you have first to put a ceiling over the whole price structure, including wages, rents and farm prices . . . and then adjust separate price schedules upward or downward, if necessary,"[41] the Office of Price Administration was granted sweeping powers to set prices and wages in most industries.

The Treasury and the OPA had initially been at odds over the tax estimates, since one of Baruch's arguments for more authority over business was that granting it would reduce the need for tax hikes. But once the General Maximum Price Regulation was enacted in 1942, the two agencies were able to agree on taxes. Friedman's first major assignment was to estimate how much taxes had to be raised to contain inflation.

On May 7, 1942, during his first appearance before a congressional committee, Friedman proposed $8.7 billion of additional taxes as "the smallest amount that is at all consistent with successful prevention of inflation."[42] Following Keynes's reasoning in defense of the 1940 Keynes Plan, Friedman pointed out that with government demand and household income soaring, it was essential to restrict consumer spending to prevent a situation in which more money was chasing a fixed output of consumer goods. As Friedman told the committee a trifle pompously, "Taxation is the most important of those measures; unless it is used quickly and severely, the other measures alone will be unable to prevent inflation." Among those other, less potent measures, Friedman listed "price control and rationing, control of consumers' credit, reduction in governmental spending, and war bond campaigns."[43] Nowhere did Friedman mention monetary policy. Looking back on his wartime work in 1953, Friedman attributed the oversight to "the Keynesian temper of the times,"[44] but Friedman numbered among Keynes's American disciples and would do so until the late 1940s.

True to his Keynesian convictions, Friedman was inclined to view the income tax as "more effective in preventing an inflationary price rise and . . . a better distribution of the cost of the war" than a sales tax, which of course was regressive.[45] That summer, he helped develop a proposal for a consumption tax, largely as a measure to avoid raising income tax rates. White, who was very taken with the idea of taxing spending instead of income, proposed combining the consumption tax with Keynes's suggestion for compulsory savings accounts that couldn't be tapped until after the war. After a stormy Treasury meeting that ended with a vote of sixteen to one against the plan, Morgenthau decided to back White and took the proposal to Congress anyway. It was dead on arrival. This contretemps was Friedman's first exposure to the challenges of getting legislation enacted, writing speeches for his superiors, and, eventually, trooping up to Capitol Hill to testify before congressional committees.

Without a doubt, the key to any tax plan was tax *collection*. This is where Friedman placed his stamp on government forever. Before 1942, income taxes were due on the previous year's income in four quarterly installments. It was the taxpayer's responsibility to come up with the money when it was due. That posed no problems, either to taxpayers or to the tax collectors, as long as tax rates were low and only a small fraction of the population paid them. In 1939, fewer than 4 million returns were filed, and the total collected was less than $1 billion, roughly 4 percent of taxable income. The Friedmans' income placed them among the top 2 percent of American households, but their tax bill was just $119, less than 2 percent of their taxable income. They had no trouble paying the whole amount on March 15, the deadline for filing federal taxes before 1955. Under the planned overhaul, their tax bill would be something like $1,704, or 23 percent of their taxable income. It was obvious that if the Treasury wanted to collect more taxes, it would have to find a way to collect the income at the time it was earned, not a year later.

The solution was tax withholding at the source. The Treasury collected taxes from employers when they paid their workers' wages. Recipients of other kinds of income—interest, dividends, money earned by the self-employed—were required to pay taxes quarterly on income earned that year based on advance estimates of liability by the recipient. A major departure from German and British practice, which had relied on collecting taxes at the source for years, was that payments would be treated as tentative and subject to adjustment later. The only serious opposition came from the IRS, which envisioned "an almost insuperable burden" for tax collectors, opposition overcome by having IRS officials visit businesses to study payroll practices so that the mechanics of withholding could be designed with those practices in mind.[46]

Friedman was back on Capitol Hill. This time he got a lesson in getting to the point and keeping things simple. When he started to answer a question by Texas senator Tom Connally, he cleared his throat and said, "There are three reasons. First . . ." Connally cut him off. "Young man, one good reason is enough," said the senator, who was wearing his trademark flowing black neckpiece in place of the usual bow tie.[47] The Treasury secretary, a man of "meager intellectual capacity" in Friedman's opinion, always insisted that his aides explain problems in terms that a high school student like "my daughter Joan" could grasp—even after Joan went off to college.[48]

In his weekly dispatch from the British Embassy, Isaiah Berlin, the historian of ideas, called it "a tax bill of unprecedented dimensions" and reported that the new law was expected to raise $7.6 billion.[49] On the twenty-second of August, he wrote excitedly that "the tax bill will affect more citizens than any ever passed by Congress."[50] For the first time, the United States had a broad-based income tax. A family of four with an income of $3,000 owed no tax in 1939 but owed $275 in 1944; a $5,000 family's taxes went from $48 to $755; a $10,000 family's from $343 to $2,245. Income taxes collected in 1939 equaled little more than 1 percent of personal income; by 1945, the figure had jumped to just over 11 percent. Morgenthau sent the withholding proposal to Congress in early 1942, and the Current Tax Payment Act of 1943 was introduced in the Senate on March 3, 1942.

The most enduring effect of Friedman's wartime efforts was to create "an enormously powerful revenue-raising machine."[51] That machine was so powerful, Herbert Stein pointed out, that revenues would rise faster than GDP for decades after the war because of the interaction between economic growth and progressive tax rates. As incomes rose, more and more taxpayers would be pushed into the higher tax brackets. That dynamic ensured that postwar administrations could keep raising spending while cutting tax rates occasionally without running huge deficits. What's more, withholding rendered taxation far less painful.

It was now possible to manipulate taxes to stabilize the economy. Before the war, Stein observed, taxes were too small a share of national income to leave much scope for stimulating or restraining the economy. More important, large swings in tax collections became automatic. When the economy slumped, tax revenues fell; when it rebounded, the opposite happened. Thus Keynesian stimulus became automatic in recessions, Keynesian restraint in booms. The irony was that Friedman, the future patron saint of low taxes and small government in the Reagan years, made it possible.

Chapter XIII

Exile:
Schumpeter and Hayek in World War II

> While history runs its course it is not history to us. It leads us
> into an unknown land and but rarely can we get a glimpse of
> what lies ahead.
>
> —*Friedrich Hayek,* The Road to Serfdom, *1944*[1]

For Keynes and many of his disciples who were called to serve their countries, the war was a time of intense engagement, extraordinary intellectual challenge, and unprecedented influence. For Schumpeter and Hayek, the Second World War was a time of enforced inactivity, isolation, and exile. They were out of favor intellectually. As émigrés, they were not asked to join the war effort. They were left behind in shells of universities populated by the old, disabled, alien, and female. They could not rejoice in the inevitable Allied victory without also mourning the suffering and devastation on the enemy's side.

As eyewitnesses to—and victims of—the collapse of the Austro-Hungarian Empire after World War I, they could imagine possibilities that those who had come of age in the United States or Britain did not and could not. Keynes was not only determined that the Allies would not make the same mistakes after the war as in 1919 but also confident that his voice would be heard and his viewpoint would prevail. Fifty-six when Britain declared war on the Axis, he was in a position to influence governments and public opinion in ways that he could not at thirty-six. He was the leader of a revolution in economic thought with many adherents, Churchill's de facto Chancellor of the Exchequer, the chief British financial negotiator in Washington, and one of the architects of the postwar monetary system.

Schumpeter was haunted by a sense of personal failure, depressed by the catastrophe engulfing Europe and Japan, and alienated by prowar fervor. He grew increasingly isolated from his colleagues and students at Harvard. He did not bother to hide his bitterness over the fact that Americans condemned Germany and Japan categorically while embracing the Soviet Union as an ally. As a result, he attracted the attention of the FBI, which investigated him for more than two years.

For Schumpeter, the political triumph of left- and right-wing Socialist parties in Europe after World War I had proved that economic success alone was no guarantee of a society's survival. Capitalism and democracy were an unstable mixture he believed. Successful businessmen would conspire with politicians to bar the entry of new rivals, government bureaucrats would stifle innovation with taxes and regulation, and hostile intellectuals would attack capitalism's moral flaws while singing the praises of totalitarian regimes and even, on occasion, secretly or openly providing aid and comfort to the West's sworn enemies. His fear that bourgeois society was spawning its own gravediggers, as Marx had predicted, had hardened into certainty.

Instead of joining the war effort like other Austrian expatriates in the United States, the fifty-six-year-old Schumpeter poured his premonitions into the book that most shows him as the ultimate ironist that he was. Published in 1942, when faith in free enterprise was dwindling in the West, *Capitalism, Socialism and Democracy* was an encomium disguised as a funeral oration that challenged Keynes's conclusion that capitalism was innately failure-prone. Whatever its shortcomings—financial crises, depressions, social strife—it was in capitalism's nature to deliver the goods to that "nine parts of mankind" who had been enslaved and impoverished throughout human history. "The capitalist engine is first and last an engine of mass production," Schumpeter asserted confidently at a time when American GDP had barely recovered from the Great Depression.[2]

Thanks to that engine, he wrote in an oft-quoted passage, modern working girls could afford stockings that were once too costly for any women, even queens a century earlier. If the United States economy were to grow as fast in the half century after 1928 as in the half century before, he observed in what turned out to be a gross underestimate, the US economy would be 2.7 times bigger in 1978 than in 1928. He wasn't predicting that outcome—the opposite, as it turns out—only impressing his readers with the power of the "ingenious mechanism."

Having argued that competition was an ingenious social contrivance for harnessing creative genius and raising living standards, Schumpeter promptly prophesied the system's demise. To his own rhetorical question "Can capitalism survive?" he replied, "No. I don't think it can."[3] The entrepreneur, the creative force in capitalism's success, was under attack, as was the ideology of economic liberalism, not only in the Soviet Union but in the West. As one reviewer commented, he "predicted the triumph of socialism but proceeded to deliver one of the most passionate defenses of capitalism as an economic system ever written."[4]

Doubtless, the feeling that opportunities for extraordinary individuals were shrinking reflected Schumpeter's middle age and depressive tendencies. He was haunted by thoughts of death and fears that he had become little more than an anachronism himself. At Harvard, his ideas were increasingly regarded as quaint, just like his courtly manner and flowery speech. "A new economics" was needed, he wrote in his diary, but he did not feel up to creating it. In a statement of unconscious irony, he added, "I do not carry weight."[5]

When Friedrich von Hayek and his family had moved to London in the fall of 1931, he had expected to return to Vienna. Within two years, he recognized that his exile would likely be permanent. For several years, Hayek found himself at the head of the liberal economic camp in his adopted country. By the time he became a British subject in 1938, however, his disciples had deserted him. As John Hicks, a prominent Keynesian, recalled in 1967, "It is hardly remembered that there was a time when the new theories of Hayek were the principal rival of the new theories of Keynes."[6]

Hayek's sense of intellectual isolation was compounded by the gloomy developments in Austria. Well before Hitler marched into Vienna and declared *Anschluss* in 1938, Hayek's old associates—including Ludwig von Mises, who had been fired from a university post—had begun to drift abroad to escape growing anti-Semitism. In 1935, he wrote to Fritz Machlup, a Jewish member of his old *Geist-Kreis* seminar, who had informed him of his decision to stay in America permanently. As a Jew, Machlup had had little choice. Hayek agreed, but added that "the mass emigration of intellectuals from Vienna and especially the demise of our school of economic thought pains me deeply."[7] And the following year: "The speed of the intellectual surrender and the corruption of politics (to say nothing of finances) is shattering."[8]

Days after Hitler's troops marched into Vienna to cheering crowds, Hayek was making the rounds of his former *Geist-Kreis* friends, who told him horror stories about Gestapo arrests, firings, and harassments. That year he applied for and got British citizenship. He attacked the Nazi regime in print and condemned anti-Semitism. He became involved in efforts to help Jewish colleagues on the Continent to emigrate.

Hayek's unhappy marriage compounded his misery. He had been pressing his wife for a divorce that she refused. What's more, he had never stopped loving Helene. He had seen her in August 1939, just before the news of the Stalin-Hitler pact signaled the inevitability of war and the impossibility of meeting again until it was over.

By the time war finally came, Hayek's isolation had turned into virtual seclusion. Barely forty, a full decade younger than Keynes, he felt old. He had, among other things, completely lost his hearing in one ear. His deafness epitomized how cut off he had become both from his old world and his adopted one. He had stayed in London during the first six weeks of the blitz to show his loyalty to Britain and his indifference to danger, but eventually he was forced to follow the LSE—shrunken now to a few dozen female students and himself—to Cambridge, where the school remained for the duration of the war. His wife and children went to stay in the country, his old ally Lionel Robbins went to Whitehall, and one by one his other colleagues disappeared to do war work.

The Road to Serfdom was Hayek's contribution to the Allied war effort. He called it "a duty I must not evade."[9] For a few short weeks after war was declared, he had high hopes and expected an assignment in the propaganda ministry. He peppered Lord Macmillan, head of the ministry, with memoranda suggesting possible strategies for German broadcasts— "I am free and anxious to put my capacities to the best use which, after very careful consideration, I believe would be in connection with propaganda work."[10] But it soon became obvious that, because he was foreign-born, he would be shut out of war work. Bitterly, he resigned himself to running the LSE's shrinking economics department, more or less by himself.

Hurt and frustrated, he toyed with the idea of joining his friends in America. "I . . . resent this complete seclusion,"[11] he wrote to Machlup. Still, when Machlup echoed his thought in a subsequent letter, he bristled at any suggestion of jumping ship. "I have given up all thought of going . . . while I am wanted here in any way. It is after all my duty."[12] When the New School offered him a temporary professorship in 1940, he cabled terse, almost haughty regrets.[13] Later, he wrote to another friend, "I envy you a little your chance of doing something connected with the war—when it is all over I shall probably be the only economist who has no such opportunity whatever and shall, *nolens, volens,* have been the purest of pure theoreticians."[14] As always, when faced with disappointment, he shifted his focus to the future. "I seem very early to have lost the capacity quietly to enjoy the present, and what made life interesting to me were my plans for the future—satisfaction consisted largely in having done what I had planned to do, and mortification mainly that I had not carried out my plans."[15]

Paradoxically, the next three years became some of the most productive of his life. "I have . . . done more work this summer than ever before in a similar period."[16] At one point—while the bombs were falling—he was working on no fewer than three different books. Pretty soon he was also filling the pages of *Economica,* LSE's journal, practically single-handedly. "So far the bombing attack is an abject failure," he wrote when he arrived in Cambridge. "What drove me out of London was simply the discomfort of an empty house and of frequent journeys."[17] Still, as a precaution, he mailed chapters of his new book to friends in America for "safe-keeping."

In January 1941, Hayek alluded explicitly for the first time to his ambition to write a book aimed, like Keynes's *The Economic Consequences of the Peace,* at a mass audience: "I am mainly concerned with an enlarged and somewhat more popular exposition of the themes of my *Freedom and the Economic System* which, if I finish it, may come out as a sixpenny Penguin volume."[18] He owed the book to his fellow man: "Since I can do nothing to help winning the war my concern is largely for the more distant future, and although my views in this respect are as pessimistic as can be—much more so than about the war itself—I am doing what little I can to open peoples' eyes."[19]

He worked on *The Road to Serfdom* for two and a half years, from New Year's 1941 to June 1943. "I am a frightfully slow worker," he complained at one point, "and with my interests, as they are at the moment, divided between so many different fields, I shall have to live very long to carry out what I should like to do at the moment."[20]

Hayek began *The Road to Serfdom* by invoking history—its relevance to the present—and his own history of living in two cultures:

> While history runs its course, it is not history to us. It leads us into an unknown land . . . It would be different if it were given to us to live a second time . . . Yet although history never quite repeats itself, and just because no development is inevitable, we can in a measure learn from the past to avoid a repetition of the same process.

Addressing the reader directly, Hayek describes a powerful sense of déjà vu. The drift toward collectivism in England reminded him of Vienna in the aftermath of World War I. "The following pages are the product of an experience as near as possible to twice living through the same period—or at least twice watching a very similar evolution of ideas." He expressed the conviction shared by earlier European observers of English society, from Engels and Marx to Schumpeter, that by moving from one country to another, one may sometimes twice watch similar phases of intellectual development. The senses have then become particularly acute. When one hears for a second time opinions expressed or measures advocated which one has first met twenty or twenty-five years ago, they assume a new meaning . . . They suggest, if not the necessity, at least the probability, that developments will take a similar course.[21]

What opinions, what measures, what works did he have in mind? Of recent books, one was surely Adolf Hitler's *Mein Kampf,* which appeared in an unabridged English edition for the first time in 1939. Another was doubtless the Webbs' 1936 paean to central planning, *Soviet Communism: a New Civilization,* which Hayek reviewed for the Sunday *Times.* Although it was politically far removed from either, he was doubtless also thinking of Keynes's *General Theory.*

Hayek's book was a defense of markets and competition couched in terms of the modern information economy:

We must look at the price system as such a mechanism for communicating information if we want to understand its real function . . . The most significant fact about this system is the economy of knowledge with which it operates, or how little the individual participants need to know in order to be able to take the right action.[22]

It was also a warning. Herbert Spencer had been the first to caution that infringements of economic freedom would lead to infringements of political freedom. Hayek's mentor, Ludwig von Mises, had identified the welfare state as a Trojan horse, "merely a method for transforming the market economy step by step into socialism . . . What emerges is the system of all-round planning, that is, socialism of the type which the German Hindenburg plan was aiming at in the First World War." But Hayek was by no means advocating laissez-faire. In fact, he repudiated benign economic neglect quite explicitly:

There is, finally, the supremely important problem of combating general fluctuations of economic activity and the recurrent waves of large-scale unemployment which accompany them. This is, of course, one of the gravest and most pressing problems of our time. But, though its solution will require much planning in the good sense, it does not—or at least need not—require that special kind of planning which according to its advocates is to replace the market. Many economists hope, indeed, that the ultimate remedy may be found in the field of monetary policy, which would involve nothing incompatible even with nineteenth-century liberalism. Others, it is true, believe that real success can be expected only from the skilled timing of public works undertaken on a very large scale. This might lead to much more serious restrictions of the competitive sphere, and, in experimenting in this direction, we shall have carefully to watch our step if we are to avoid making all economic activity progressively more dependent on the direction and volume of government expenditure.[23]

Later he would tell American audiences in a speech that "You must cease to argue for and against government activity as such . . . We cannot seriously argue that the government ought to do nothing."[24]

Early in 1943, Machlup sent around several chapters to American publishers. The first responses were not encouraging:

Frankly we are doubtful of the sale which we could secure for it, and I personally cannot but feel that Professor Hayek is a little outside the stream of much of present day thought, both here and in England . . . If, however, the book is published by someone else and becomes a bestseller in the non-fiction field, just put it down to one of those mistakes in judgment which we all make.[25]

Harper's dismissed it as "labored" and "overwritten."[26]

In June 1943, Hayek finally signed a contract with Routledge for publication in the UK. It was not until February 1944, shortly before the book was to appear in England, that Hayek heard that the University of Chicago Press had decided to accept it.

Act III

CONFIDENCE

Nothing to Fear

On January 11, 1944, FDR had been in bed for days with the flu. Exhausted from meetings of the Big Three in Cairo and Tehran, suffering from hypertension, hypertensive heart disease, cardiac failure (left ventricle), and acute bronchitis, any of which could kill him, he was too weak to make his customary trip to Capitol Hill to deliver his annual State of the Union message.[1] Knowing that the newspapers couldn't print the full text of his speech, which he had sent by messenger to Congress, he insisted on speaking directly to the American people in a "fireside chat" over the radio. The D-day landing in Normandy was months away and the United States was locked in a life-and-death struggle in the Pacific, but the president urged the country to look beyond the war: "It is our duty now to lay the plans and determine the strategy for the winning of a lasting peace."[2]

Again and again, the president hammered home his theme that the foundation for a lasting peace was not the defeat of gangster regimes alone but also rising living standards. Economic security was the supreme responsibility of democratic governments. He was determined not to repeat the mistakes made by the Allies after World War I that he believed had helped lead to the current war. Maintaining that the welfare state and individual liberty went hand in hand, he warned, "People who are hungry and out of a job are the stuff of which dictatorships are made." Roosevelt called on Congress to support postwar economic recovery at home and abroad.

His major domestic proposal was for an "economic Bill of Rights"—
namely, government guarantees of jobs, health care, and old age pensions.[3]

The most radical speech of FDR's presidency, says his biographer
James MacGregor Burns, "fell with a dull thud into the half-empty chamber."[4] Congress had a Republican majority, and the president's references
to joblessness and hunger did not seem to resonate with millions of Americans gathered around the radio. When Keynes arrived in Washington a
few months later, he found that "on this continent the war is a time of immense prosperity for everyone."[5] Not only had the war years shaped up as
the best of times, but 60 percent of the population told pollsters that they
were "satisfied with the way things were *before* the war."[6]

The war itself was responsible. Even before 1939, intensifying fears of
war had resulted in a huge influx of gold into the United States as investors in Europe and Asia sought a safe haven for their savings. As a result,
American banks were flush and interest rates remained near zero. And,
since 1939, spending by the federal government had soared from 5 percent
of GDP to nearly 50 percent, much faster than tax revenues, despite the
dramatic increases in income and profits taxes and the imposition of the
new Social Security payroll tax. This was deficit spending on a scale that
dwarfed the anti-Depression fiscal policies of the first Roosevelt administration.

The combination of massive deficit spending and the accidental
monetary stimulus from abroad sparked a boom. With 11 million men
and women in uniform, and factories, mines, and farms operating flat
out, the official jobless rate had plunged from 15 percent at the end of
1939—11 percent counting "temporary" government workers—to well
below 2 percent at the end of 1943. Thanks to the tight labor market, factory pay was up 30 percent after inflation. And after four years of war, the
average American household was consuming more, not less, than in 1939.

The United States was supplying planes, ships, and tanks by cranking
up production, not by tightening belts. The economy's annual output, or
GDP, was growing at a nearly 14 percent annual rate, three times as fast as
in the "wild twenties," when, the president said sourly, "this Nation went
for a joy ride on a roller coaster which ended in a tragic crash."[7] To be

sure, Americans couldn't have new cars, refrigerators, or houses, but they were so confident the dollar would retain its prewar value that they were willing to save nearly one-quarter of their pay to buy them after the war. Nor could they go on their beloved road trips. But they could buy more clothes, food, alcohol, cigarettes, and magazines, listen to more radio and records, and see more movies and ball games. The contrast with Britain, where per capita consumption was down by 20 percent, was extraordinary. As readers of Elizabeth Jane Howard's *Cazelet Chronicles* novels know, the life of English citizens was complicated for years by shortages of housing, clothes, coal, gasoline, and many foodstuffs. Nor did austerity end with V-E Day. As late as 1946, the Labour government was forced to consider, in secret, whether to impose bread rationing. The last of the controls wasn't lifted until 1954.

Although it was clear that the American economic system was hardly on its last legs, the president and his advisors feared that wartime prosperity couldn't last. Among the "economic truths that have become accepted as self-evident," FDR implied in his speech, was that a *new* New Deal would be needed to prevent the Great Depression from roaring back when the troops came home. "A return to the so-called normalcy of the 1920s" after the war would mean that "we shall have yielded to the spirit of Fascism here at home,"[8] he warned melodramatically.

The president's position reflected only one side of a hot debate between Keynesians and anti-Keynesians. The more upbeat the public and businessmen became about postwar prospects, the more American disciples of Keynes worried that the economy would sink into another slump. Public spending would plunge with demobilization. Alvin Hansen, an advisor to the Federal Reserve who was sometimes called "the American Keynes," foresaw "a postwar collapse: demobilization of armies, shutdowns in defense industries, unemployment, deflation, bankruptcy, hard times."[9] Paul Samuelson, a consultant for the main postwar planning agency, warned the administration not to become complacent about unemployment. "Before the war we had not solved it, and nothing that has happened since assures that it will not rise again." They had little faith that business and consumers would pick up the slack. As Samuelson put it, "If a man goes without an automobile for 6 years, he does not then have a demand for six automobiles."[10] Having concluded from the 1930s that business was too timid to invest and that monetary policy was a poor weapon with which to fight recessions, the Keynesians were convinced that the only solution was to slow the cuts in public spending by slowing demobilization and by beefing up spending on infrastructure.

Anti-Keynesians were also worried about stagnation, but of a different type and coming from a different quarter. Schumpeter was worried about the prospects for economic growth—in the longer run. He feared that the economy would no longer produce gains in productivity and living standards—not because of inadequate demand but because of government policy. In an article published in 1943, he agreed that "everybody is afraid of a postwar slump" but argued that the popular fears were overblown: "Viewed as a purely economic problem, the task [of reconstruction] might well turn out to be much easier than most people believe . . . But in any case, the wants of impoverished households will be so urgent and so calculable that any postwar slump that may be unavoidable would speedily give way to a reconstruction boom. Capitalist methods have proved equal to much more difficult tasks."[11]

The real threat to postwar growth, he believed, came from antibusiness policies enshrined in the New Deal. He and Hayek feared that governments would continue their wartime management of production and distribution—including price and wage controls, deficit financing, and high taxes—after the war was won. Such efforts, intended to avert stagnation, might well produce that very result. Schumpeter called it "capitalism in the oxygen tent."[12] Hayek was less concerned about the possible loss of dynamism than he was about the loss of freedom. While the president cautioned that a "return to normalcy" would be tantamount to a victory for Fascism, Hayek warned that continuing wartime management of production and distribution would ultimately result in a radical restriction of economic and political rights as well. Their fears proved more realistic for the UK and Europe than for the United States, where virtually all the wartime agencies were dismantled starting in 1945.

• • •

Apart from military victory, FDR's top priority was not to repeat the mistakes made by the Allies after World War I that he believed had helped lead to the current war. He pointed to the talks among the Big Three about postwar financial, trade, and political arrangements that were already underway by January 1944 as an example of doing better than last time. Attacking the "ostrich isolationism" of "unseeing moles" who regarded the parleys with suspicion, he lashed out at those who viewed prosperity in the rest of the world as a threat to American economic interests. In Tehran he had extracted Stalin's commitment to a new League of Nations. The "one supreme objective of the future" was collective security, the president insisted, including "economic security, social security, moral security" for the "family of Nations." After bringing the aggressors under military control, "a decent standard of living for all individual men and women and children in all Nations" was essential for peace. "Freedom from fear is eternally linked with freedom from want." [13]

There was no disagreement between Keynesians and anti-Keynesians on the need for international cooperation. On this issue, they had seen eye to eye since 1919. Few believed that a favorable global economic environment would arise spontaneously. The bilateral trading blocs of the interwar era were designed to allow the Soviet Union and Nazi Germany to break away from the world economy. Even Hayek, who by virtue of experience and temperament was more skeptical of the potential for positive government intervention, was convinced that democracies were capable of more competence than they had shown a generation before. That governments had to actively plan and cooperate to ensure the revival of world trade, the resolution of war debts, and the stabilization of currencies this time reflected a consensus.

From Europe, however, FDR's upbeat vision of one world in which the major powers were focused on economic growth instead of expansionist aggression seemed far too rosy. On March 9, 1944, Gunnar Myrdal, the head of one of Sweden's postwar planning commissions, delivered a considerably darker prognosis. The young economist had spent the early part of the war traveling through the American South reporting his classic study of race relations, *An American Dilemma: The Negro Problem and*

Modern Democracy, before returning to his native Sweden—which had retained its status as a nonbelligerent nation despite having supplied the German war machine—in 1942.

Myrdal saw the future through a much darker glass. He feared that autarky, economic stagnation, and militarism—the very pathologies that had helped to breed the second global conflagration within a single generation—had not been defeated, despite four years of unprecedented effort, sacrifice, and suffering. The dream of a single world community—the United Nations—bound together by trade, convertible currencies, and international law was a dangerous illusion, he argued. Dismissing the "overoptimism" of American economists, he predicted that the present wartime boom would turn into a depression more severe than the Great Depression and mass unemployment. A depression in the United States would necessarily have repercussions for the whole world, especially for Sweden and other countries that depended on exports to pay for the imports they needed to survive as modern economies. Inevitably, economic chaos would produce an epidemic of strikes and civil unrest and fuel nationalist rivalries—just as similar economic conditions had done before the war. A general trend of militarism and autarky[14] similar to that which had prevailed in the interwar period would continue. In particular, the world would inevitably break up into three great competing empires—Russian, British, and American—as the conflicting economic and political interests of the Big Three displaced the Allies' common goal of defeating the Axis. In Myrdal's global dystopia, the new imperialism would not only be oppressive but also inherently unstable.

This, of course, is the world of *1984.* George Orwell, who completed his dystopian novel in 1948, pictured a world carved up into three empires—Oceania, Eurasia, and Eastasia—that are locked in a permanent Cold War. Too evenly matched to win or lose, the superpowers use external threats to justify totalitarian rule and economic stagnation. The hero—an everyman named Winston Smith, who "displays flashes of Churchillian courage"—learns that the "splitting up of the world into three great superstates was an event which could be and indeed was foreseen before the middle of the twentieth century."[15]

Ironically, one who viewed the nightmare with satisfaction rather than fear was Stalin. FDR had returned from Tehran convinced that the Allied leaders shared a common interest, once the enemy was defeated, in creating a framework within which all countries could focus on economic growth. He had assured Americans that "All our allies have learned by experience—bitter experience that real development will not be possible if they are to be diverted from their purpose by repeated wars—or even threats of war." [16]

In reality, Stalin was convinced that his capitalist allies were inherently incapable of cooperating for long and that once their common enemy was defeated, the drive for profits would soon set the United States and Britain at each other's throats. In his mind, Anglo-American war was an "inevitability." [17] In that case, he could extract aid and territory from his allies while waiting for the coming crisis to provoke a war and drive their citizens into pseudo-political parties whose first loyalty was to Moscow.

Why did he ignore abundant evidence to the contrary? According to John Lewis Gaddis, the foremost American historian of the Cold War, Stalin was a genuine captive of Lenin's primitive economic theory—a theory based on a false analogy between economic competition and warfare. Instead of FDR's belief that growth in one country would benefit rather than harm its trading partners, Stalin was convinced that trade, like war, was a zero-sum game in which one side's gain was the other's loss. Indeed, Lenin had believed that war was merely a more aggressive form of economic competition.

In *The General Theory,* Keynes had expressed his belief that ideas matter: "Madmen in authority, who hear voices in the air, are distilling their frenzy from some academic scribbler of a few years back." [18] Thanks in no small part to the ideas of Keynes, Hayek, and their followers, those in authority were neither mad nor in the thrall of barbaric relics. They were determined to avert such nightmares.

Chapter XIV

Past and Future: Keynes at Bretton Woods

> Economic diseases are highly communicable. It follows therefore that the economic health of every country is a proper matter of concern to all its neighbors, near and distant.
>
> —*FDR, message to delegates at Breton Woods*[1]

Keynes described his and Lydia's crossing on the *Queen Mary* in mid-June 1944, barely two weeks before the international monetary conference at Bretton Woods, New Hampshire, as "a most peaceful and also a most busy time."[2] Traveling with Friedrich von Hayek's, and now his, close friend Lionel Robbins and a dozen other British officials, Keynes presided over no fewer than thirteen shipboard meetings and had a major hand in writing two "boat drafts" on the two major institutions that were to administer postwar monetary arrangements: the International Monetary Fund and the World Bank.[3] In his spare time, he lounged in his deck chair devouring books. Along with a new edition of Plato's *Republic* and a life of his favorite essayist, Thomas Babington Macaulay, he read Hayek's *The Road to Serfdom*.

In contrast to his more doctrinaire disciples, Keynes was a genius capable of holding two opposing truths in his mind: "Morally and philosophically," he wrote in a long letter to Hayek, "I find myself in agreement with virtually the whole of it; and not only in agreement but deeply moved agreement." Hayek might not have succeeded in drawing "the line between freedom and planning satisfactorily,"[4] and therefore might not be a useful guide through the "middle way" of actual policy making, but he was articulating values that Keynes considered essential "for living a good life."[5] Robbins mused that Keynes, "so radical in outlook in matters purely intellectual, in matters of culture he is a true Burkean conservative."[6]

Keynes went on to say that Hayek was too quick to dismiss the possibility that some planning was compatible with freedom, particularly if the planning was done by those who shared their values: "Dangerous acts can be done safely in a community which thinks and feels rightly which would be the way to hell if they were executed by those who think and feel wrongly."[7] He meant that a war economy run by Churchill or FDR was unlikely to lead to a totalitarian state, even though ones run by Stalin and Hitler had.

Keynes and Lydia were whisked to the White Mountains of New Hampshire by private train. The Mount Washington Hotel in Bretton Woods was a turn-of-the-century grande dame meant to recall grande dames such as the Hotel Majestic in Paris, where Keynes had been at the end of the last war—only with 350 rooms, private baths, a ballroom, an indoor pool, and a palm court with Tiffany windows. But the slightly seedy resort, long past her prime, was hardly prepared for the onslaught of 730 delegates from forty-four Allied countries. "The taps run all day, the windows do not close or open, the pipes mend and un-mend and nobody can get anywhere," Lydia wrote to her mother-in-law. They were installed in an enormous suite next to Secretary of the Treasury Henry Morgenthau. Unlike the voyage over, the conference was "a madhouse," Lydia observed, "with most people working more than humanly possible."[8]

FDR had issued the invitations to the conference, and Morgenthau acted as titular host, but the principal architects, planners, and parliamentarians were his aide Harry Dexter White and Keynes. The principals came with different ideas, divergent interests, and, in many cases, hidden agendas. The hotel was crawling with spies. Delegates had no authority to bind their governments. But the organizers recognized that they must guarantee an economic recovery and that no recovery could take place without cooperation. The framers shared the determination expressed by FDR in his State of the Union address not to repeat the mistakes made after World War I, and to take a global, multilateral, "United Nations" approach. The very fact of the conference reflected a radical redefinition— and enlargement—of government's responsibilities. Just as Washington, London, and Paris now accepted responsibility for keeping employment high at home, virtually every government in the West accepted some measure of responsibility for keeping employment high in their trading partners' economies too.

The precise features of the new order reflected a shared view of what had gone wrong the last time and a conviction that getting things right had ramifications that were more than economic. FDR, Churchill, and Keynes and his American disciples believed that economic pathologies— inflation and unemployment—had produced Fascism and fatally weakened many democracies. They believed with equal conviction that the breakup of the pre–World War I global economy—produced by the frantic beggar-thy-neighbor attempts of each nation to insulate itself from the worldwide economic crisis—and the accompanying decline in world trade were partly responsible for world war. Economic rivalry might lead to war. As Cordell Hull, the American secretary of state, put it: "[Un]hampered trade dovetailed with peace; high tariffs, trade barriers and unfair economic competition with war . . . If we could get a freer flow of trade . . . so that one country would not be deadly jealous of another and the living standards of all countries might rise . . . we might have a reasonable chance of lasting peace."[9]

The great innovation of the 1920s and 1930s—the economics of the whole developed by Fisher, Keynes, and, to a lesser extent, Schumpeter and Hayek—taught that what was good for one nation might easily be bad for all. Devaluing one's currency, erecting trade barriers, and clamping controls on capital outflows might be effective for reducing balance-of-payment deficits, stopping the outflow of gold, and pumping up government revenues. But if everybody adopted the same tactics, the eventual result would be universal impoverishment and unemployment.

In the 1930s, world trade fell by half, and trading continued mostly within currency blocs like the pound sterling bloc within the British Empire, the Soviet sphere, and the bilateral trading bloc designed by Hitler's economics minister, Dr. Hjalmar Schacht. It was now commonly acknowledged that keeping free enterprise functioning globally required the visible hand of government. In a way, biographer Robert Skidelsky points out, the new arrangement devised by White and Keynes was Keynesianism applied globally.

The purpose of the Bretton Woods conference was to revive world trade and stabilize currencies and to deal with war debts and frozen credit markets. The war had left much of the world dramatically poorer, and countries had to be able to earn their way back to prosperity. In the broadest sense, salvage meant rebuilding and reconstruction, moving back toward pre-1913 globalization, but without reviving the pre–World War I assumption that the economic machinery worked automatically. For the West, it meant learning from the past in order to avoid the mistakes of the interwar era—the very lesson that Marxists claimed capitalists could not learn—and restoring lost moral and material credibility. Economic stability was a key to political stability, and economic growth was a necessary if not sufficient condition for the long-run survival of the West. Modern societies could not survive if the ingenious mechanism malfunctioned or broke down, any more than great cities could survive without electricity or trains.

Unlike the British thinkers who championed free trade in the 1840s, neither Keynes nor Fisher (nor Schumpeter nor Hayek) believed in an automatic tendency toward peace and progress such as had been cheerily assumed by so many in the Belle Epoque. Governments had to intervene; international cooperation was required. No system was spontaneously generated or self-regulating, as had been taken for granted before 1914. It would require the sole remaining superpower in the West and the once powerful but now humbled European empires to create one. The alternative was unthinkable. White suggested that failure would lead once more to war: "The absence of a high degree of economic collaboration among the leading nations will . . . inevitably result in economic warfare that will be but the prelude and instigator of military warfare on an even vaster scale." [10]

In other words, White and Keynes shared the fears of George Orwell, Gunnar Myrdal, Schumpeter, Hayek, and many others, but they were neither slaves to economic determinism nor radically distrustful of government. They were not prepared to believe that governments could not now be convinced to avert both depression and war by establishing a common framework for cooperation. They believed that democratic governments could learn from past mistakes and rejected both the Marxist notion of historical necessity and the traditional presumption of Great Power rivalry. They certainly did not share Stalin's conviction that war was part of capitalism's DNA.

The real test, of course, was not only whether the West could learn from history but whether, with the help of its ingenious mechanism, the West would draw the *right* lessons.

In 1944 England was fighting for her life at any cost, even if it meant losing much of its empire, cooperating with the Soviet Union, and playing second fiddle to an increasingly assertive United States. All British visions of the postwar world, except for a tiny band of Communists, had in common the overwhelming priority of keeping the Americans engaged in Europe.

At the end of World War I, the United States was already the world's biggest and richest economy, but not a superpower. At the end of World War II, it was the sole superpower. As successive American administrations were to learn, greater wealth and power turned out to involve more rather than less interdependence. At the end of World War I, Woodrow Wilson's argument for America's continued engagement in European affairs fell on deaf ears. In 1944 the argument that the world had to be made safe for the United States no longer seemed far-fetched. Pearl Harbor had shattered, once and for all, the American illusion that two oceans could protect it from foreign security threats.

According to historian John Gaddis, Roosevelt's wartime priorities were supporting the Allies, since the United States could not defeat Japan and Nazi Germany alone; winning their cooperation in shaping postwar settlements, since no lasting peace was possible without Soviet participation; insuring a multilateral approach to security; and preventing another Great Depression. Finally, because the United States was a democracy and politicians had to defer to popular opinion, Roosevelt was committed to convincing the American people that a return to prewar isolationism was unthinkable.

At the Hotel Majestic in Paris in 1919, Keynes had been one of hundreds of technical advisors with little hope of being heard and still less hope of shaping the outcome. At the Mount Washington Hotel in 1944, he was a pooh-bah, to use Lydia's favorite expression. The Allies had learned from experience. They now assumed that peace depended on economic revival. In 1918 that presumption had been shared by only a few— Schumpeter, Keynes, and Fisher among them—but hardly by the leaders of the victorious nations or their electorates.

Britain's bankrupt status and financial dependence on the United States meant that the Americans would largely determine the outcome while putting on an appearance of cooperation. On their side, although Treasury Secretary Morgenthau was nominally in charge, Harry Dexter White, his deputy, was the only one "who knows the complete matter" and "who can prevent a vote on anything he doesn't want voted on."[11] White orchestrated everything from press conferences to having transcripts typed and distributed.

Keynes, typically, took little trouble to disguise the fact that he was ramming his views down the throats of the banking committee that he chaired. Morgenthau had to go around to Keynes's suite and "ask him would he please go slow and talk louder and have his papers in better arrangement."[12] Skidelsky points out that if Keynes was not inclusive, he was at least efficient, and that his haste in rushing through the agenda reflected exhaustion and a growing determination to get away as soon as possible. Keynes gave the final speech at the banquet, his arrival prompting the entire meeting to stand up, silently, until he made his way to the dais and sat down.

. . .

"The Soviet Union is a coming country, Britain is a going country," Harry White told Keynes at one point in their long and difficult negotiations.[13] As Skidelsky points out, Keynes was sometimes puzzled by White's obsession with Russia and often outraged by his hostility to Britain. What he did not apparently suspect was that his most influential American disciples—and, more often than not, adversaries at the negotiating table—were passing government secrets to the Soviet Union and helping the Soviets to spy on him and other delegates. Among the gaggle of government economists White brought with him to Bretton Woods were a dozen or more employees of the Treasury's Monetary Research Division who were members of the "Silvermaster ring" of agents for the KGB.

The wartime alliance, Soviet heroism and sacrifice in defeating Germany, and the role of European Communists in the resistance all explain why the first revelations that the Soviets had set up a large-scale espionage operation seemed at first incredible, and later so shocking. Most disturbing was the Soviets' reliance on a fifth column of American citizens so reminiscent of the highly successful Nazi strategy of relying on a network of sympathizers in Europe. The newly burnished image of the Soviet Union explains not only why FDR and Truman were slow to accept that World War II would be followed by a Cold War, but also what now seems unfathomable: how some of the brightest and best were willing to serve as spies, agents of influence, and apologists for a foreign regime and why most, apparently, had no regrets. They did what they did for the sake of "mankind."

Even in the depths of the Great Depression, the Communist Party of the United States of America (CPUSA) never remotely attained the status of a mass political movement, much less an independent one. Its membership peaked in 1944 at eighty thousand or so, the overwhelming majority of members drifted away in less than a year, and it exerted scant influence beyond a few neighborhoods in the Bay Area, Boston, and New York and a handful of trade unions. The spies were sometimes poor or economically insecure, often first in their family to attend university. Many smarted under casual anti-Semitism and snobbery. The rise of Hitler and Franco, with their explicitly anti-intellectual and militaristic threat to civilization, gave the Party some cachet at universities. Fighting the Great Depression became a political movement, like Civil Rights in the fifties and sixties. Just as physicists at the Manhattan Project saw themselves as part of the war effort, cranking out forecasts at Treasury was part of the fight to defeat fascism.

In the 1930s, Lauchlin Currie had been an instructor at Harvard and had coauthored several prostimulus and New Deal manifestos with his best friend, Harry Dexter White. In 1939 he become one of six administrative assistants on the president's staff and was soon advising Roosevelt on momentous matters such as mobilizing the economy for war, the wartime budget, and Lend-Lease for China. Currie organized the Flying Tigers. He practically ran Lend-Lease for China and was closely involved in US-British and US-Russian loan negotiations, and in the talks leading up to the Bretton Woods conference. Compelling evidence from multiple independent sources shows that Currie and White were not innocent victims of dirty anti–New Deal politics, and certainly not of McCarthyism. The charges against them were made by two independent sources and corroborated by Soviet cables captured and encrypted by the US government long before Senator Joseph McCarthy made his sensational claims, and confirmed decades later in material from the KGB archives.

The charge against Currie was that he, and possibly at the behest of the president, pressured the OSS to return purloined Soviet cipher traffic and to suspend encryption operations. The evidence against Harry Dexter White was particularly damning. According to two of his biographers, David Rees and R. Bruce Craig, Whittaker Chambers, a *Time* magazine editor and former agent of the GRU, Russia's foreign intelligence service, who went to the assistant secretary of state in 1939 with the names of other Soviet agents, volunteered that White and Currie were agents. Chambers produced copies of a Treasury document that White had given him to deliver to the GRU (Chief Intelligence Directorate). His charges were corroborated independently by at least two other ex-agents. One Venona cable, dated November 1944, concerns an offer to White's wife, delivered by Nathan Gregory Silvermaster, to help with college tuition for the Whites' daughter. Two other cables document unauthorized discussions between White and a KGB general, Vitaly Pavlov, including one in 1941 over lunch in a Washington restaurant.

Although Moscow valued them as spies, Currie's and White's real importance was as agents of influence. Occupying positions of great sensitivity, reach, and authority, they took actions and promoted measures that may or may not have served the interests of their government but were definitely intended to advance those of the Soviet Union. Ironically, they were as clueless about Soviet intentions as the most naïve American politician. Unlike FDR and Truman, whose views shifted sharply after the Yalta conference in 1945, these calculating, hard-nosed, duplicitous men reacted with the shocked incomprehension of jilted lovers when Stalin made fools of them.

The generation that came to economics during or as a direct consequence of the Great Depression seized the message of *The General Theory of Employment, Interest and Money* like drowning men lunging toward a lifeline. Keynes was their hero and they were his disciples—intellectual disciples, that is; the "Keynesian" label did not imply support for Keynes's policy proposals, much less his politics. Some were political conservatives. Some, particularly in Europe, were Socialists. Most fell within the spectrum defined by mainstream parties. That some rose to positions of power and influence and used those positions to pursue hidden agendas out of loyalty to a totalitarian regime says a great deal about them and their times, but very little about Keynesian ideas, much less about Keynes the man—except perhaps that, like everyone else, he could not imagine how such smart men could be so stupid or so bad.

Chapter XV

The Road from Serfdom:
Hayek and the German Miracle

> It cannot be said too often—at any rate, it is not being said nearly
> often enough—that collectivism is not inherently democratic,
> but, on the contrary, gives to a tyrannical minority such powers
> as the Spanish Inquisitors never dreamed of. . . .
>
> Since the vast majority of people would far rather have State
> regimentation than slumps and unemployment, the drift towards
> collectivism is bound to continue if popular opinion has any say
> in the matter.
>
> —*George Orwell, review of* The Road to Serfdom, *1944*[1]

On March 31, 1945, Isaiah Berlin reported in his weekly dispatch from
Washington that "*The Reader's Digest,* which in effect is the voice of Big
Business, has printed a digest of Professor Hayek's notorious work" and
that "the imminent arrival of the Professor himself is eagerly anticipated
by the anti–Bretton Woods party, who expect him to act as the heavy artillery."[2]

Hayek's transatlantic crossing "by slow convoy" in stormy March
weather was considerably less agreeable than Keynes's the previous June,
but when he stepped onto the pier in New York, he was greeted by flashing
cameras and a mob of reporters. Three thousand people turned up at New
York University for his first lecture, and for the next six weeks—four weeks
longer than he had originally planned to stay in the United States—his

schedule of speeches, radio broadcasts, and press interviews was so tightly packed that he could hardly manage a brief late-night rendezvous with his old friend Fritz Machlup, who had been faithfully sending Hayek food parcels with Spam, nuts, prunes, rice, and the like since 1943.

"The voice of Big Business" had turned Hayek into an instant celebrity. A front-page *New York Times* book review by *Newsweek* writer Henry Hazlitt did the rest. The sensational success of *The Road to Serfdom* was partly timing. In the spring of 1945, the recently concluded Yalta conference and the imminent defeat of Nazi Germany by the Red Army had focused American public opinion on postwar settlements and, in particular, on future US-Soviet relations. Among the issues before Congress were a trade bill, a huge loan to the British, and, of course, ratification of the global monetary agreement endorsed at Bretton Woods the previous July—all administration initiatives to which Republicans were strongly opposed. Although most of the references in Hayek's book were to Nazi Germany, not Stalin's Soviet Union, his antistatist message resonated with opponents of the New Deal. As Berlin predicted, American conservatives were ecstatic and rushed to embrace the Viennese professor. But Hayek proved to be an unreliable poster boy. In his next dispatch, Berlin reported with some amusement that the American Bankers Association was wavering in its opposition to the Bretton Woods Treaty, thanks "curiously enough" to a Professor Hayek, who, "at a meeting of influential New York bankers attended by both Winthrop Aldrich and various Morgan partners as well as by Mr. Herbert Hoover and others, argued passionately in favor of Bretton Woods."[3]

A month later, Berlin gloated, "Professor Friedrich von Hayek, upon whom the economic Tories in this country placed so much hope, founded upon the Professor's indubitably anti–New Deal views, has proved a most embarrassing ally to them since his passion for free trade makes him no less hostile to tariffs and monopolies."[4]

Unbeknownst to his Republican sponsors, Hayek had begun to warm to FDR before the war. "I suppose Roosevelt knows what he does," he wrote to Machlup, confessing that Roosevelt's 1938 "Message to Congress on the Concentration of Economic Power" had prompted "a considerable revision of my views about him."[5] Hayek was by no means intimidated by the discomfiture and embarrassment of his supporters. On his final night in Washington, Albert Hawkes, a Republican senator from New Jersey, gave a dinner for him. Bored and disappointed by Hayek's abstract argument and dry delivery, another senator rose to demand his opinion on a piece of pending trade legislation. Hayek responded icily, "Gentlemen, if you have any comprehension of my philosophy at all, you must know that the one thing I stand for above all else is free trade throughout the world. The reciprocal trade program is intended to expend world trade, and so naturally I would be for such a measure."

The *Washington Post* columnist Marquis Childs, who was among the guests, reported gleefully that "the temperature in the room went down at least ten degrees since the Republican Party had decided to take a stand against the extension of the trade program." The cooling-off process continued when, a little later, Hayek repeated that while he did not like many of the features of the Bretton Woods Monetary Agreement, he was in favor of it. The alternative to such an agreement, he said, was "too grim to contemplate."[6]

Congress gave its approval to the Bretton Woods Treaty in July. The British Parliament waited until December, refusing its imprimatur until after Washington finally gave the green light to an $8.8 billion loan to Britain that Keynes had given so much to obtain. The choice between autarky versus globalization, free trade versus protection, had been made. The Russians shocked the Roosevelt administration—and its own moles—by refusing to ratify the treaty. George Kennan, the diplomat and architect of the Truman Doctrine, recalled:

> Nowhere in Washington had the hopes entertained for post-war collaboration with Russia been more elaborate, more naïve, or more tenaciously (one might almost say ferociously) pursued than in the Treasury Department. Now, at long last, with the incomprehensible unwillingness of Moscow to adhere to the Bank and the Fund, the dream seemed to be shattered, and the Department of State passed onto the embassy, in tones of bland innocence, the anguished cry of bewilderment that had floated

over the roof of the White House from the Treasury Department on the other side. How did one explain such behavior on the part of the Soviet government? What lay behind it?[7]

In contrast to Churchill, FDR and Truman had viewed Stalin very much as Neville Chamberlain had viewed Hitler prior to 1938: as a ruler with legitimate grievances and limited aims who would make deals and live up to them if handled properly. They took big-power rivalry and commercial conflicts for granted but assumed that the United States and the Soviets had a common interest in ensuring that those conflicts took place within a cooperative framework. But the view that Stalin could be negotiated with began to crumble even before FDR's death of a cerebral hemorrhage on April 12, 1945, two weeks after Hayek's arrival in America. The dictator's abrupt refusal to join the IMF and the World Bank was one of the decisions that led to a radical reappraisal, beginning with the famous "Long Telegram" sent by George Kennan, the number-two minister in the US Embassy in Moscow, to the secretary of state in February 1946, describing a Soviet Union that did, indeed, resemble the totalitarian empires of George Orwell's imagination.

Keynes and Hayek never fully resolved their long-running debate over how much and what kind of government intervention in the economy is compatible with a free society. Nonetheless, Keynes endorsed *The Road to Serfdom* and nominated Hayek, rather than his disciple Joan Robinson, for membership in the British Academy. When Keynes's heart finally gave out on April 21, 1946, Hayek wrote to Lydia that Keynes was "the one really great man I ever knew, and for whom I had unbounded admiration."[8]

By early 1947, Keynes's hopeful vision of one world was half in ruins. One after the other, Poland, Hungary, and Romania fell to Soviet domination. Churchill had delivered his Iron Curtain speech. Truman announced that the United States would "support free peoples who are resisting attempted subjugation by armed minorities or by outside pressures . . . primarily through economic and financial aid which is essential to economic stability and orderly political processes."[9]

Hayek had avoided returning to Vienna at the end of the war. His closest friends were either dead or in exile abroad. After the Yalta conference, Stalin had postponed the Red Army's assault on Berlin to grab what he saw as a valuable bargaining chip. After heavy aerial bombardment and fierce street-to-street combat, Vienna fell to the Russians. Some of her finest buildings were reduced to rubble. Water, electricity, and gas lines were destroyed. Abandoned by the police and other local authorities, defenseless residents were terrorized by criminal bands. The Soviet assault forces showed some restraint toward the civilian population, but the second wave of troops to arrive in the city engaged in a six-week frenzy of rape, looting, and violence.

During the war Hayek had dreamed of re-creating his old *Geist-Kreis* on the Continent as a way of demonstrating that the ideals of the European Enlightenment were still alive: "The old liberal who adheres to a traditional creed merely out of tradition ... is not of much use for our purpose. What we need are people who have faced the arguments from the other side, who have struggled with them and fought themselves through to a position from which they can both critically meet the objection against it and justify their views."[10] On his second visit to the United States, the conservative Volker Fund offered to sponsor a conference to found a community of like-minded liberals. Hayek convened the first meeting of the Mont Pelerin Society in Switzerland on a hill overlooking Lake Geneva on April 10, 1947. The majority of attendees were European émigrés from the United States or Britain, including Karl Popper, Ludwig von Mises, and Fritz Machlup. A contingent from the University of Chicago included Milton Friedman and Aaron Director. Henry Hazlitt from *Newsweek* and John Davenport from *Fortune* were there. The assembled individualists could not muster a unanimous vote in favor of the institution of private property, but did so with respect to the principle of individual freedom. They readily agreed that the organization would not publish books or periodicals, engage in political activity, or issue statements, but they rejected Hayek's proposal that they call themselves the Acton-Tocqueville Society after Frank Knight from the University of Chicago objected to naming the group after "two Roman Catholic aristocrats." [11] Ludwig von Mises created a scene during a debate on income distribution by accusing others of harboring Socialist sympathies. Walter Eucken, an economist from Germany, ate his first orange since before the war. When three days of wide-ranging discussion threatened to end without a statement of principles, Lionel Robbins, a veteran of countless committees, managed to draft one that all but Maurice Allais of France felt able to sign. Noting that "freedom of thought and expression, is threatened by the spread of creeds which, claiming the privilege of tolerance when in the position of a minority, seek only to establish a position of power in which they can suppress and obliterate all views but their own," [12] the statement emphasized free enterprise, opposition to historical fatalism, and the obligation of nations as well as individuals to be bound by moral codes and, above all, support for complete intellectual freedom.

As soon as the conference disbanded, Hayek set out for Vienna. The condition of the city and its inhabitants was far worse than anything he had been able to imagine. Under occupation by the four allies for three long years, Vienna was as seedy, demoralized, and dark as it would appear to audiences who saw *The Third Man,* the film noir written by the English novelist Graham Greene, with its immortal line, added by the director and star, Orson Welles: "In Italy for thirty years under the Borgias they had warfare, terror, murder, bloodshed—they produced Michelangelo, Leonardo da Vinci, and the Renaissance. In Switzerland they had brotherly love, five hundred years of democracy, and peace, and what did they produce? The cuckoo clock." [13]

The Russians were still in charge of the eastern suburbs, feared and despised by the Viennese. Hayek protested that the Allies were treating Austria "much worse than Italy or any of the other countries which joined Germany voluntarily." The occupation authorities were applying essentially the same guidelines as in Germany, which meant that virtually all economic activity—except Harry Lime's black market—was banned. Hayek complained, "the Austrians have been prevented from helping themselves to get out of a desperate economic position." [14]

By one of those coincidences that seem to multiply in extraordinary times, Hayek once again ran into his cousin Ludwig Wittgenstein on a train leaving Vienna for Munich. Wittgenstein seemed more morose and angry than ever. He had spent most of his time in the Russian sector, where the Red Army had used the house Wittgenstein had designed and built for one of his sisters as a stable and a garage. Wittgenstein had been a great admirer of the Bolsheviks and had thought seriously of immigrating to Russia in the 1930s.[15] Now, Hayek thought, the philosopher behaved as if he had met the Russians "in the flesh for the first time and that this had shattered all his illusions."

Hayek's visit to Vienna was followed by a tour, arranged by the British Council, of a half dozen German cities. He found Cologne, including its great cathedral, "laid absolutely flat by the war; there didn't seem to be a city left, just great piles of rubble. I climbed through the rubble into an underground big hole to speak." In Darmstadt, he had "my most moving experience as a university lecturer," he wrote to Machlup.

> I didn't have any idea the Germans knew anything about me at that time; and I gave a lecture to an audience so crowded that the students couldn't get in, in an enormous lecture hall. And I discovered then that people were circulating hand-typed copies of *The Road to Serfdom* in German, although it hadn't been published in Germany yet.[16]

Characteristically, Hayek's first thought on returning to London was to organize a drive to collect books published since 1938 that censorship and war had kept out of the hands of Austrian and German scholars. By year's end, the committee had already collected some 2,500 volumes, which were, with great difficulty, eventually shipped to Vienna.

In 1947 the question of how to deal with Germany was still unsettled. Keynes, White, and their respective governments had been embroiled in a bitter debate beginning within weeks of D-day three years before. White was an aggressive advocate of the deindustrialization of Germany, while Keynes argued for economic integration and recovery. Keynes first learned of the Morgenthau Plan from the papers a few weeks later in July 1944. The Versailles Treaty, which he had repeatedly assailed as "Carthaginian" throughout the 1920s and which he blamed for another world war, was punitive. But it was an attempt by the victors to make Germany pay the costs of the war. The Morgenthau Plan was designed to return Germany, a modern economy, to its eighteenth-century preindustrial state. The plan had two strengths, Keynes observed in a letter to the Chancellor of the Exchequer, John Anderson: It was being proposed at a moment of bitter fighting and horrendous casualties when extreme measures—even geno- cidal ones—had become acceptable. It was a plan. The State Department and the War Department had nothing as coherent.

Keynes did not speak out, because he could not afford to alienate Morgenthau or White. He saw that instantly. He eased his conscience by predicting that the plan would never pass Congress. In this he was correct. By the time Eisenhower assumed control of southern Germany in 1945, the Morgenthau Plan had been shelved. But the lack of a positive vision or concrete counterproposal left a vacuum, and Keynes's failure to speak out still had consequences. In the absence of a positive plan, "Morgen- thau principles and Morgenthau men" governed Germany for three full years. As early as June 1945, Austin Robinson, on a fact-finding mission for the Treasury, reported to Keynes that he "felt more worried about the economic system that has completely stopped than about the physical damage." He found "no papers ... no telephones that operate over long distances, little true communication of any sort." Instead "the Germany of the towns is in ruins, with its factories flat, its houses burnt out or bombed out, and its life dead. The Germany of the country[side] still vigorous, the work of the fields proceeding normally ... lacking only incentives to sell to the towns that have so little to offer in exchange for food."[17]

The refusal to permit a resumption of economic activity in Germany had two consequences unforeseen by the American authorities. First, the collapse of the German economy prevented the rest of Europe from recovering. Second, the cost of the occupation to American and British taxpayers soared. The price tag had, according to conservative estimates, multiplied by a factor of three. Robinson warned Keynes that if the Russians "or just possibly the French" extracted too much in reparations, Britain "would have to provide and pay for imports to feed and maintain our zone sufficiently to prevent starvation and disaster." [18] Having witnessed the same phenomenon after World War I, Keynes responded immediately: "For goodness sake, see that we don't have to pay the reparations *this* time." [19]

The United States ultimately adopted the Marshall Plan. With Europe starving and in danger of falling into the Communist camp, the Marshall Plan was a natural heir to Bretton Woods and the commitment of Britain and the United States to create institutions that could help promote growth and stability among the free world's economies. The shift from nationalism to a global perspective in economic policy was thus part of the changing perceptions about security and postwar diplomatic and military strategy. The notion that economic collapse had produced totalitarian regimes created American resolve to restore the economic health of Europe that became more urgent when it became clear in 1947 that Europe was not recovering on its own. Economic revival was in the interest of American business, as well as a necessary condition for European self-defense. Truman's rationale helped win over business leaders to massive government spending on aid and the military in peacetime.

Although Germany got relatively little Marshall aid, Germany's recovery was so strong that it was quickly labeled Wirtschaftswunder or economic miracle. In the three years after the currency reform of 1948, per capita output jumped an average of 15 percent every year. By 1950, despite the destruction of the war and the removal of machinery by Russians, it was 94 percent of prewar level.

What happened? Ludwig Erhard, the finance minister, attributed the Wirtschaftswunder to the introduction of a new currency and lifting of price controls in 1948. "More perhaps than any other economy," he recalled, "the German one experienced the economic and supra-economic consequences of an economic and trading policy subjected to the extremes of nationalism, autarchy and government control. We have learned the lesson." The liberalization "awakened entrepreneurial impulses. The

worker became ready to work, the trader to sell, and the economy to produce." Hitherto there had been a premium on stagnation. Foreign trade
moved languidly in a framework provided by allied instructions. Goods
were lacking; there was a universal cry for supplies; yet the economic impulse was wanting.[20]

For Hayek, Germany's rise from the ashes was both a vindication
of his faith in free markets, free trade, and sound money, and a hopeful
portent that the liberal European civilization he loved was not, after all,
doomed to extinction.

When he got an invitaion to teach at the University of Chicago, he
resigned from the LSE, got a divorce, and married his longtime lover. He
indulged his passion for book collecting and intellectual biography by
writing a charming account of John Stuart Mill's partnership with Harriet
Taylor, and spent his honeymoon retracing Mill's famous pilgrimage from
London to Rome.

Hayek's turn as a darling of American conservatives was short-lived.
He despised most Republican politicians, all cars, and practically everything else about life in America, including the absence of universal health
insurance and government-sponsored pensions. Homesick for Europe
and no longer welcome at the LSE, he finally settled at the University of
Salzburg.

In 1974, the Swedish Academy of Sciences plucked Hayek out of ob-
scurity by awarding him a Nobel Prize for his "penetrating analysis of the
interdependence of economic, social and institutional phenomena." Ironically, he shared it with Swedish socialist Gunnar Myrdal. A few years later,
his *Constitution of Liberty* became the bible of Margaret Thatcher's conservative revival. And in the early 1990s, the collapse of the Soviet Union and
the spread of free market reforms in Eastern Europe and Asia made him a
hero among conservatives around the globe.

Chapter XVI

Instruments of Mastery:
Samuelson Goes to Washington

I don't care who writes the nation's laws—or crafts its advanced
treaties—if I can write its economics textbooks.

—*Paul A. Samuelson*[1]

Paul Anthony Samuelson, the anonymous mind behind the government
report to which President Roosevelt referred in his "radical" State of the
Union message, had whiled away the first months of the war force-feeding
economics to bored engineering students and producing endless calculations for the army at MIT's Radiation Lab.[2] As early as 1940, Lauchlin Currie, FDR's economics aide, had suggested to the president that it was not
too early for the United States to begin planning for the postwar era. The
president agreed, and Currie promptly recruited a new class of freshman
brain trusters to work at the National Resources Planning Board, the nation's first and only planning agency, run by the president's uncle Frederic
A. Delano. Samuelson, a twenty-five-year-old Harvard wunderkind, newly
minted PhD, and MIT assistant professor, soon became the titular head
of a group of twenty or so economists and a bunch of graduate students
from Johns Hopkins University assigned to calculate possible trajectories
for the postwar economy and propose solutions to potential problems.[3]
To reassure his superiors that the new Keynesian economics was no more
subversive than a branch of accounting, Samuelson made a point of wearing a green eyeshade at White House briefings.

On the morning after Labor Day 1944, for the first time in nearly a
year, this foot soldier in the Roosevelt administration's vast wartime army
of university consultants was back in Washington, D.C. Short, wiry, and
crew-cut, Samuelson had come down from Boston by overnight train.
Nattily dressed in a suit and bow tie, he made the rounds of "sweltering
temporaries" that had sprung up all over the capital, buttonholing former
colleagues and students, pumping them for news and gossip.

Samuelson could "smell cuts in war production."[4] Every office he visited was awash in desk calculators, messy piles of green sheets, and stacks
of budget reports. With the end of the war now certain, Washington's attention was shifting from the problems of wartime production to those of
conversion to a peacetime economy. Hundreds of bureaucrats were busy
calculating how much military procurement could be scaled back, how
many GIs could be discharged, how long it would take to convert tank
production lines into car production lines. The first round of reductions
was slated to begin that fall, perhaps not coincidentally during the 1944
presidential election campaign, which pitted the president against Republican Thomas E. Dewey, who held FDR's old job of governor of New York.
As it turned out, the Allied sweep through Europe stalled that fall at the
Battle of the Bulge, and the actual reconversion was postponed until early
1945.[5]

Despite the sultry temperature and oppressive humidity, Samuelson
found the mood in Washington among "experts," as well as Congress,
unexpectedly sanguine. The day before, the New York Times had run a
headline: "Boom After War Almost Certain."[6] He was appalled. The potential problems were staggering: 11 million men and women were in
uniform, and 16 million—almost a third of the labor force—were working
in defense plants. In 1943 the federal government had spent more than
$60 billion—nearly half the nation's annual output—and nearly seven
times as much as in 1940. The more Samuelson looked beyond the war,
the more worried he felt.

His mood coincided with that of other Keynesians who took for
granted that American business could increase production, efficiency, and
per capita income year in and year out but were less certain that businesses

and households could be counted on to spend, rather than save, the profits and wages that all this activity generated. Increasingly, Samuelson leaned toward the view that the tendency of the economy to stagnate was not necessarily a transient illness caused by monetary policy mistakes or external shocks, but rather a chronic disease. David M. Kennedy, the economic historian, observes that the tenor and conclusions of Samuelson's report for the NRPB reflected two sources. One was Keynes's judgment, expressed in his 1940 pamphlet *How to Pay for the War,* of the British economy's poor postwar prospects absent a major and constant infusion of government spending.[7] The other was that of the administration's Keynesian advisors, notably Currie, White, and Alvin Hansen, a Harvard professor and a consultant for the NRPB and the Federal Reserve. It was Hansen who had rallied the conservative department's graduate students and instructors (the "lumpen proletariat," Samuelson liked to say) to the Keynesian banner. If anything, Kennedy points out, Keynes's American disciples were even more pessimistic than their leader. As early as 1938, the year Hansen arrived at Harvard from the Midwest, he published a book, *Full Recovery or Stagnation,* in which he already foresaw a dismal postwar future.

Samuelson, who wrote as fast and breezily as he talked, launched a stellar second career in journalism with a provocative two-part series for the *New Republic* on "the coming economic crisis."[8] His tone was energetic, not fatalistic. Implying that the problem, though dire, was fixable, he urged the same steps as he had in the 1942 NRPB report: slow down demobilization and keep government spending high. The piece radiated confidence that, as New Dealer Chester Bowles once remarked: "We have seen the last of our great depressions for the simple reason that the public [is] wise enough to know that it doesn't have to stand for one."[9]

Samuelson was a child of the Jewish exodus out of Russia to America, the World War I boom in the Midwest, and the go-go twenties. He was born in Gary, Indiana, in 1915, a fact to which he later attributed his lifelong zest for economics and stock market speculation. Gary was not yet an exurb of Chicago. Instead it was a gritty company town of behemoth steel mills and brand-new tenements rising out of the prairie and shrouded in its own special atmosphere of soot, smoke, and money. During World War I, the mills blazed day and night. Steelworkers, mostly immigrants, had the opportunity to work 12/7. When factory hands got sick, they went to the druggist instead of the doctor to avoid losing a day's pay. Frank Samuelson was in the happy position of being one of the town's few pharmacists. A first-generation Jewish immigrant, he spoke Russian and Polish to his customers.

He was also a small-time real estate speculator, as was the typical midwesterner with cash to spare, whether saved or borrowed. The war boom had spilled into the farm economy and sent grain prices into the stratosphere. Farmers, who had never had it so good, borrowed money and poured it into more acreage and machinery. And for several years, they and Frank Samuelson, who invested in property in downtown Gary, prospered. Like Gopher Prairie, the fictional town in Sinclair Lewis's *Main Street*, Gary was full of overachievers like Frank Samuelson and their discontented wives, who despised the town for its ugliness and resented their husbands for marooning them a full day's travel from Chicago. Pretty, vain, and "a great snob," Ella Lipton Samuelson alternately egged on her husband and heaped scorn on him. A woman of uncertain temper and a passionate desire to become a famous hat designer, she longed for daughters. Instead she had three sons whom she farmed out, one after the other, to foster parents not long after they were old enough to walk.

At age seventeen months, the blond, blue-eyed toddler was sent to a farm in Wheeler, Indiana, a crossroads standing amid endless fields of wheat, without electricity, indoor plumbing, a telephone, or an automobile. Later, Samuelson said, "I did experience first hand, in my virtual infancy, the disappearance of the horse economy, the arrival of indoor plumbing and electric lighting. After that radio waves through the air or TV pictures left one blasé." [10] He did not see his mother again until he was ready to start kindergarten.

Maternal abandonment can produce coldness and detachment, but it can also create a longing for attachment and a desire to please. In Samuelson's case, it did a little of both. His foster mother became the first of the many women in his life who adored him: from wives and secretaries to daughters and dogs. Unlike his birth mother, she was ample, warm, kind, and a good cook.

When five-year-old Paul went home again, the armistice had been signed, and the new Federal Reserve was turning off the credit spigot in an effort to reverse wartime inflation. In England and France, the biggest markets for American wheat, central banks were doing the same. In a matter of months, grain prices had fallen by half, the steel mills were standing idle, and banks were failing in droves. "Now, bank failures were not a strange and unfamiliar phenomenon in my part of Indiana," Samuelson recalled. "The farms that were mortgaged up and fully equipped at the peak of the War prosperity were hard hit by the drop in grain prices. And so country banks failed." Inevitably, land prices collapsed, and so did the Samuelsons' financial security.

The economic recovery that began in mid-1921 did little to revive the battered farm economy or the family finances. For four years, Frank Samuelson watched his once-thriving pharmacy business melt away. Finally, in the summer of 1925, lured by delicious visions of warm winters and tropical bounty—oranges at the front door—and tired of his wife's constant scolding, he handed the keys of his drugstore to a new owner. He and Ella got in their car and headed south to Miami, joining tens of thousands of other families in the great Florida land rush. Florida land looked like a sure bet: with 10 percent down, a doubling in price meant a 1,000 percent profit on the original investment. Never mind that the "dream development" was "midway between pine thicket and swamp." [11]

When their parents left Gary, Samuelson, ten, and his twelve-and-one-half-year-old brother, Harold, were back in Wheeler, where they always spent summers. Around Labor Day, they were summoned by their parents. The boys rode from Chicago to Miami in a Pullman. Samuelson recalled that his first sight when he got off the train was not his mother or father but "men in plus-four knickerbockers on the streets buying and selling land." [12]

By mid-1925, the boom had spread all the way north to Jacksonville, a sleepy farm town near the Atlantic Ocean, 350 miles from Miami. It had also attracted an already infamous con man named Charles Ponzi, who sold parcels for $10 that turned out to be sixty-five miles (as the crow flies) from Jacksonville and 1/23 of an acre. In 1926 faith that Florida's streets were lined with gold was beginning to wane, and the influx of new buyers began to tail off. Inevitably, prices did too. Then came two hurricanes, and what looked like a pause in a perpetual upward climb became a plunge. Frank Samuelson lost most of his remaining money and accumulated more of his wife's reproaches. "She didn't hold a lot in," Paul said of his mother, who relished retelling the story of her husband's foolish bets long after his premature death from heart disease at age forty-eight. The nature of the family's economic problem was clear even to a ten-year-old.

Two years later, the Samuelsons were back in the Midwest, settling on Chicago's South Side, then as now a middle-class enclave squeezed between Lake Michigan and an African-American ghetto. The Chicago economy was booming once more. The stench from the stockyards mingled with smog from the steel mills in Gary that came drifting across the lake. Paul entered Hyde Park High School and joined the rest of the country in studying the stock pages daily, often with his high school math teacher.

The cult of F. Scott Fitzgerald, author of *The Great Gatsby*, was at its height. Samuelson wrote stories for the school's literary magazine featuring worldly-wise, cynical youths who fell in and out of love in the time it took them to change their clothes and spit out one-liners like "For the love of Mike, Pat, Pete, and the other seven apostles, shut up!"[13] Living with a mother who was "a screamer," he fantasized about escaping to an eastern college with "a white chapel tower" in a "peaceful green village."[14] By the time Samuelson graduated from high school in 1931 at age sixteen, a Great Depression was settling on Main Streets all over America like a long winter night. Going east for college was no longer in the cards, if it had ever been a realistic possibility. So Samuelson enrolled at the University of Chicago in January 1932, declared a math major, and continued to live at home.

Being trapped in the Midwest had unexpected benefits. Far from the backwater he feared, Chicago was a buzzing hive of intellectual and political activity and a meeting place for economists who wanted Washington to do more to fight the depression. A mix of fiscally conservative midwesterners and Burkean liberals of Central European extraction, the Chicago faculty was alarmed and frustrated by Washington's ineffectual response to the crisis and eager to advise a more activist approach.

Samuelson learned from his freshman tutor that "the world's leading economist," John Maynard Keynes, had spent the previous summer lecturing at the university.[15] His first economics professor was Aaron Director, "a very dry, confident, reactionary economist" and the future brother-in-law of Milton Friedman, who had "quite a big impact" on him. Director's first lecture, on Thomas Malthus's theory of population, got him hooked on economics, he later said. Another professor was Jacob Viner, a Canadian of Romanian extraction with a terrifying reputation as the toughest grader at Chicago. After Roosevelt's inauguration, he became one of Treasury Secretary Henry Morgenthau's closest outside advisors, staffing the Treasury, the Federal Reserve, and the New Deal agencies with dozens of his students. A close friend of Schumpeter and Hayek, Viner became one of the most vocal and influential American critics of Keynes's *General Theory of Employment, Interest, and Money.* He agreed with Keynes on policy and on the need for deficit spending to fight the depression. However, he held that Keynes's theory was not "general" at all but valid only in the short run, and fell apart if applied to longer time frames.

During Samuelson's first month at Chicago, the university hosted a conference at which Irving Fisher, the most famous and simultaneously most notorious American economist, and a bevy of other monetary experts debated how the Hoover administration should fight the depression. Director and Viner both signed Fisher's telegram urging the president to launch an aggressive stimulus plan.

When Samuelson decided three years later that he would make a better economist than mathematician and won a scholarship to graduate school, he chose Harvard over Chicago. The presence of Edward Chamberlin, who had recently published the groundbreaking *The Theory of Monopolistic Competition,* was an attraction, but getting away from home and the fantasy of the "peaceful green village" were far bigger lures. Arriving in Cambridge in the third year of the Roosevelt recovery, Samuelson quickly discovered that Harvard's senior faculty, while politically to the left of Chicago's, was intellectually far more conservative.

A Canadian graduate student who had been attending Keynes's lectures in Cambridge arrived at Harvard during Samuelson's first semester in the fall of 1936. Robert Bryce gave a paper summarizing the ideas in Keynes's as yet unpublished *General Theory*. He emphasized public spending to combat unemployment without fully explaining the underlying theory, leaving Samuelson, who did not regard fiscal activism as a new or uniquely "Keynesian" idea, somewhat puzzled over what the fuss was about. But since the economy was clearly rebounding, he took it for granted that the New Deal was responsible, and he took it on faith that Keynes had a new, rigorous, internally consistent theory explaining how it could be so. "In the end I asked myself why do I refuse a paradigm that enables me to understand the Roosevelt upturn from 1933 'til 1937?"[16]

When Nicholas Kaldor, a Marxo-Keynesian and economic advisor to the Labour Party, visited in 1936, he attended what he thought to be a brilliant talk by a faculty star. "Congratulations, Professor Chamberlin," he prefaced his question to the speaker. The "Professor" turned out to be Samuelson, a first-year graduate student. Samuelson took a class from the mathematician Edwin Bidwell Wilson, Willard Gibbs's last disciple at Yale. He and Schumpeter, who had instantly taken up Samuelson as a protégé, comprised half the students. He took another course with the brilliant Russian émigré and future Nobel laureate Wassily Leontief. Recalled the Japanese economist Tsuru Shigeto, Samuelson's best friend in graduate school, "Leontief, as is well known, was not very eloquent, and he would make a frequent use of a blackboard, drawing a couple of lines which crossed with each other and would start saying: 'You see at this point of intersection. . . .' Thereupon Paul would intervene: 'Yes, that is the point of. . . .' But he cannot finish the sentence, for Leontief would immediately exclaim in approval 'That's right. You see what I mean.' He and Paul both knew each other, but neither of them would reveal it, and the rest of the class had to remain mystified."[17]

The following year, Samuelson became the first economist to be elected to the Society of Fellows, a remarkable Harvard institution inspired by the English university's tradition of high table. It demanded that young scholars from different disciplines suspend work toward their degrees for three years to . . . think. He suddenly found himself in the company of Willard Van Orman Quine, the logician; George Birkhoff, the inventor of lattice theory; Stanislaw Ulam, originator of the Teller-Ulam design for thermonuclear weapons; and other extraordinary mathematical minds.

A heady atmosphere and intellectual thrills were no substitute for a family. Within a year, Samuelson had married a fellow graduate student from Wisconsin, Marion Crawford. By the time the prohibition against finishing his PhD expired in May 1940 on his twenty-fifth birthday, Marion had also finished her PhD, and the young couple had had their first baby.

Like so many young men who came of age in the Great Depression, Samuelson was in a hurry. He horrified his European friends by listening to Beethoven's Ninth Symphony out of order so that he could minimize time wasted in flipping over his seventy-eights. Hoping for a tenure track offer from Harvard, he plunged into his dissertation. Marion did the typing. When he handed it in, it bore the title *Foundations of Economic Analysis*. *Foundations* was inspired partly by Schumpeter's 1931 lament about the crisis in economic theory and bore a family resemblance to Irving Fisher's dissertation. It was an ambitious attack on contemporary economic theory, using "simple arithmetic and logic" to show how much of the theory could be boiled down to simpler, more fundamental propositions. "I felt I was hacking my way through a jungle with a penknife," Samuelson said later. "It was a tangle of contradictions, overlaps, confusions." [18]

Foundations accomplished what Bertrand Russell's *Principia Mathematica* and John von Neumann's *Mathematical Foundations of Quantum Mechanics* sought to achieve—and what, in 1890, Marshall's *Principles of Economics* had achieved. Herbert Stein, a University of Chicago–trained New Dealer, offered the most intuitive explanation of Samuelson's achievement by comparing it to the pre-Fisher, pre-Keynes economics: if people were out of work, you gave them jobs. If people were out of work, you fiddled with something in one corner of the system—say, the money supply or tax rates—on the assumption that it would affect something at the far end of the system: employment. This was the practical implication of the new "economics of the whole," or macroeconomics. This is what was new about Fisher's and Keynes's economics.[19]

The emphasis on links between different parts of the economy and on indirect and secondary effects also explains why the new macroeconomics relied on mathematics. You cannot analyze an integrated system without math. The debate over whether its use to analyze economic problems is a good or bad thing crops up from time to time—as does the debate over the use of computers to prove mathematical theorems. Economists, like engineers, nuclear physicists, and composers, are problem solvers. If they are working on a problem that the old tools aren't quite suited to, they try the new ones. True, the older generation rarely sees the point and often finds it impossible to master new techniques, but to Samuelson's generation that came of age during the Great Depression and World War II, Willard Gibb's point that mathematics is a language seemed perfectly natural. The fear that using mathematics would cause other languages to wither turned out to be overblown. John von Neumann, one of several mathematicians who had a major impact on economics, could translate from German into English in real time and quote verbatim from Dickens. Samuelson's verbal virtuosity was even more pronounced.

It was probably no accident that *Foundations* was a product of the 1930s, an extraordinarily innovative decade. Samuelson, who took his generals at the end of his first year at Harvard, used the three years of his tenure as a Junior Fellow, the academic years 1937 to 1940, to produce the core of *Foundations of Economic Analysis. Foundations* "had no definite moment of conception," he recalled. "Gradually over the period 1936 to 1941, it got itself evolved."[20] When Samuelson defended his dissertation, Schumpeter is said to have turned to Leontief to ask, "Well . . . have *we* passed?" But like so many ideas and inventions of that pregnant era, *Foundations* was kept off the market by World War II. Unlike von Neumann's and Oskar Morgenstern's *Theory of Games and Economic Behavior,* Samuelson's doctoral dissertation had no influential champions or wealthy patrons. Indeed, Harold Burbank, the chairman of the Harvard Economics Department, was so hostile to it—whether because of an aversion to mathematics or Jews is hard to say—that he had the printing plates destroyed and insisted that Samuelson be offered only a temporary lectureship. When *Foundations* finally appeared in print in late 1947, it was all the more warmly received because the war had made the use of new tools and techniques seem natural. Samuelson won the John Bates Clark Medal, the equivalent of the Fields Medal for the best mathematician under age forty. Schumpeter proclaimed *Foundations* a masterpiece and wrote to his former student, "If I read in it in the evening the excitement interferes with my night's rest." [21]

Americans' fears about the postwar economy were rooted in the belief that the war, not the New Deal, was responsible for the economic recovery. Whereas the British were mostly concerned with preventing an outburst of inflation while rewarding the population for its enormous sacrifices, the worry for most Americans was that unemployment would return when Washington slashed military spending and millions of GIs were demobbed.

The National Resources Planning Board, a precursor to the President's Council of Economic Advisors, was charged with planning the economic transition to peace. Everett Hagen, Samuelson's coauthor on the NRPB report, was responsible for producing the administration's consensus forecast. By mid-1944, a sharp split had developed among Washington's economic experts. The New Dealers tended to be optimistic about postwar prospects. Keynesians tended to be pessimistic. Samuelson admitted that a restocking boom was likely at the end of the war, as business built up depleted inventories and replaced worn-out equipment and consumers took similar steps. But he thought that it would be short lived, overwhelmed by the huge military cuts.

Demobilization occurred even faster than Samuelson expected, but the crisis he predicted did not materialize. After a steep but brief recession in 1947, the economy rebounded rapidly. The onset of the Cold War caused the Truman administration to spend hundreds of millions on America's nuclear arsenal, even as spending on conventional ground forces plunged. But what Samuelson had failed to foresee was the magnitude of pent-up demand by consumers, starved for houses, cars, appliances, and other appurtenances of middle-class life and with plenty of savings in the

bank. His embarrassingly wrong prediction slowed the spread of Keynesianism in academe, he always believed. Being disastrously wrong early
in one's career was in some ways a salutary experience for someone who
hated making mistakes and rarely did. It left Samuelson more skeptical of
economic forecasts and more circumspect in the claims he made for policies he favored or opposed.

Demobilization became a bonanza for American colleges, MIT and its
embryo economics department included. The only economic bill of rights
that Congress passed in the wake of FDR's 1944 exhortation was the GI
Bill. But that measure had a large and lasting effect on the postwar economy. In Britain, the Labour government would construct a cradle-to-grave
welfare state to compensate the British people for their wartime sacrifices.
The GI Bill was the American counterpart. The only serious opposition, David Kennedy points out, came from Samuelson's and Friedman's
alma mater, the University of Chicago, and its famous president, Robert
Hutchins, who warned, "Colleges and universities will find themselves
converted into educational hobo jungles." [22] MIT, which had no graduate
program in economics, took a more pragmatic position.

The GI Bill was passed in June 1944, just before demobilization began.
Samuelson was begging to be released from his obligations at the Radiation Lab, which he found tedious, to take up new projects. He considered
but rejected an offer to ghostwrite a history of the Manhattan Project.
Meanwhile, as GIs began streaming into Cambridge, his teaching load
increased exponentially. In April 1945, Ralph Freeman, his department
chairman, proposed that he write an economics textbook for engineers.
"MIT is anxious to have me return to undertake a necessary project that
I alone can do," he wrote to the army, which still claimed his time, adding that "the day is approaching when it will no longer be in the national
interest to convert a good economist into a mediocre mathematician." [23]

All new MIT students were required to take economics, another sign
of changing times. The trouble was, as Freeman confided to Samuelson,
for whom it could hardly have been news, "They all hate it." On the day
after Japan attacked Pearl Harbor, only one professor had been in his office

at the Harvard Economics Department when Basil Dandison, a McGraw-Hill textbook salesman, stopped by. Dandison mentioned to the professor that his company was looking for someone to write an economics textbook and was told about a bright young star who had lately defected to the engineering college at the far end of Cambridge. By the time Japan surrendered, Dandison and the MIT hotshot had struck a deal. "I thought it would do very well," Dandison recalled. The author shrewdly refused an advance and insisted instead on a then unheard-of 15 percent royalty.[24]

Samuelson thought he could knock off the textbook during the summer, provided that the Rad Lab would let him go. But in 1945 he agreed to serve as one of three ghostwriters for Vannevar Bush, an MIT engineer and founder of Raytheon, who headed up a postwar planning group and had been commissioned by FDR to produce a report on research and development, *Science: The Endless Frontier*.[25] *Economics: An Introductory Analysis* wasn't finished until April 1948, although MIT engineering students got previews in mimeographed form.

In *God and Man at Yale: The Superstitions of "Academic Freedom,"* the publishing sensation of 1951, its twenty-five-year-old author, William F. Buckley Jr., leveled a dramatic accusation at his alma mater. "The net influence of Yale economics," he charged, was "thoroughly collectivist," the antithesis of the entrepreneurial values espoused by the university's alumni. As evidence, he cited the textbooks assigned in Economics 10, the introductory course taken on average by one-third of the Yale class.[26] One of the offending texts was Samuelson's *Economics: An Introductory Analysis*. Charging Samuelson with glorifying government and disparaging competition and individual initiative, Buckley was irritated by his "typical glibness . . . and soap opera appeal."[27] He was particularly incensed by the author's suggestion that great fortunes and inheritances were suspect.

The blasphemies in *Economics* were numerous and the bows to traditional wisdom few.[28] Instead of Adam Smith's Invisible Hand, Samuelson invoked the image of a "machine without an effective steering wheel" to describe the private economy.[29] Instead of treating government as a necessary evil, Samuelson called it a modern necessity "where complex economic conditions of life necessitate social coordination and planning,"[30] adding, for emphasis, "No longer is modern man able to believe 'that government governs best which governs least.'"[31] The monetary discipline imposed by the pre–World War I gold standard is breezily dismissed as making "each country a slave rather than the master of its own economic destiny."[32] Samuelson treats budget balancing as a similarly outmoded obsession, assuring students that there is "no technical, financial reason why a nation fanatically addicted to deficit spending should not pursue such a policy for the rest of our lives and even beyond."[33]

Economics was the work of a young man speaking directly to other young men:

> TAKE A GOOD LOOK AT THE MAN ON
> YOUR RIGHT, AND THE MAN ON YOUR LEFT . . .
>
> [T]he first problem of modern economics: the causes of . . . depression; and other of prosperity, full employment and high standards of living. But no less important is the fact—clearly to be read from the history of the 20th century—that the political health of a democracy is tied up in a crucial way with the successful maintenance of stable high employment and living opportunities. It is not too much to say that the widespread creation of dictatorship and the resulting WWII stemmed in no small measure from the world's failure to meet this basic economic problem adequately.[34]

Capturing the Zeitgeist of big government and bottom-up democracy, Samuelson announced portentously, "The capitalistic way of life is on trial."[35] The book's organization reflected a new set of priorities. Samuelson starts by explaining how the national income is produced, distributed, and spent, and how the government's decisions to tax and spend affect the private economy. These are topics "important for understanding the postwar economic world" as well as "topics people find most interesting." Reversing the usual order, he placed the macroeconomy first, with traditional topics such as the theory of the firm and consumer choice left to the book's second half. Cognizant of the new interest in investing created by wartime saving and purchases of government bonds, and of the need to keep the engineers awake, Samuelson included a chapter on personal finance and the stock market.

Essentially, Samuelson integrated the new Keynesian economics with the economic theory inherited from Marshall, while following Marshall's example of inserting his own insights and techniques. In the fourth edition of *Economics*, Samuelson gave his approach the name of "neo-classical synthesis."[36] Marshall and Schumpeter had emphasized productivity growth as the primary driver of gains in living standards. To this Samuelson added "the importance of preventing mass unemployment."[37]

He described the implications of the new theory by invoking *Alice in Wonderland*. In the world of full employment—in other words, a world of scarcity and substitution in which there were no free lunches and getting more of anything meant giving up something else—the old rules, perhaps restated more precisely in the language of mathematics, applied. In the Keynesian world of abundance and less than full employment, impossible things like getting something for nothing became possible. The best example is the "paradox of thrift."[38] In a full-employment economy, if households save a bigger fraction of their incomes, the total amount of savings goes up. In a depression, saving more actually reduces the total pool of savings because cutting spending causes production and incomes—and hence savings—to fall. At less than full employment, "everything goes into reverse." The same applies to governmental thrift.

The Great Depression was a breakdown not of a single market but in coordination among markets, but Samuelson did not coin the term *macroeconomics* that now refers to effective demand by households, businesses, and government; total amount of unemployment; the rate of inflation; and such. If there was one message that Samuelson meant students to take away, it was that monetary policy no longer worked. The Great Depression was proof of that, he asserted: "Today few economists regard Federal Reserve monetary policy as a panacea for controlling the business cycle."[39] The ideas of the Fed seemed as dusty and dated as flapper fashions. The same could be said about Irving Fisher, who had died the year before, or, indeed, of the pre-1933 Keynes.

. . .

Any impression that the runaway success of *Economics* in university class-rooms implied an embrace of the new economists in Washington is incorrect. Despite the gauzy nostalgia in which the 1950s are wrapped today, the decade was notable for three recessions, one of which was severe, and toward the end of the decade relatively high unemployment. Historians have sometimes underestimated the urgency with which Truman, and later Eisenhower, viewed the need to balance the federal budget and, in particular, to cut military spending. They have also sometimes confused Truman's hawkish Cold War rhetoric with actual commitments to back words with resources. But, as Herbert Stein explains, Truman sought major cuts not just in 1945, but in 1946, 1947, and 1948 as well. The Marshall Plan was an exception, not the rule.

How to explain the gap between theory and practice? Fiscal prudence, for one thing. Truman was convinced that a strong defense was predicated on a healthy economy and attributed Allied victory in no small measure to America's ability to fulfill its role as the arsenal of democracy. To Truman, who was a Midwestern fiscal conservative and economic conservative (and, in any case, was dealing with a Republican Congress), the top priority was putting a stop to the run-up in war debt by eliminating the federal government's annual deficit. What's more, as America's complete lack of defense in 1940 shows, there was no tradition in the United States of a large peacetime military. After the defeat of Germany, the pressure for demobilization was irresistible, and Truman's subsequent proposal for a universal peacetime draft was overwhelmingly defeated. So, added to the need to project power globally in order to defend the United States was the need to do it on a shoestring.

The Keynesian revolution didn't capture Washington until the sixties. Of all of Samuelson's students, none was more important than John F. Kennedy, who, shortly before the 1960 presidential election, invited Samuelson to give him an alfresco seminar at his family's home in Hyannis Port on Cape Cod. "I had expected a scrumptious meal," Samuelson later joked. "We had franks and beans."

On the whole, JFK's cold, calculating, cautious temperament appealed to Samuelson. The new president was hard to sway, but he would stick to his guns once he had made up his mind. Despite a large budget deficit, Kennedy called for a huge tax cut to revive the sluggish economy and his abysmal approval ratings. "The worst deficit comes from a recession," he told the nation in a televised address, adding that cutting tax rates for individuals and businesses alike was "the most important step we can take to prevent another recession."

The 1963 Kennedy tax cut, passed after the president's assassination, was a huge success. By 1970, President Richard Nixon was insisting that, "We are all Keynesians now," but the tax cut marked the high-water mark for Keynesian theory of managing the business cycle. Samuelson's view was that Keynesianism was toppled by stagflation—that nasty combination of unemployment, inflation, and stagnating productivity that afflicted the world's richest economies in the 1970s and 1980s—rather than a rival theory. But by the late 1950s and early 1960s, Milton Friedman was already mounting a major assault on the reigning paradigm from the University of Chicago, challenging the notion that the government could pick any combination of unemployment and inflation that it wished by fiddling with the government budget. Reviving the legacy of Irving Fisher, and the theory that the supply of money determines the level of economic output, and reinterpreting the Great Depression as a colossal failure of monetary management, Friedman first convinced young economists and later President Jimmy Carter, who appointed Paul Volcker to tame the inflation monster, that money mattered after all. Neither Friedman nor Samuelson ever returned to government, both confident that they could have a bigger impact by teaching and writing than the staff of a president or the Fed.

Chapter XVII

Grand Illusion:
Robinson in Moscow and Beijing

It's very difficult these days to lecture on economic theory
because now we have both socialist countries and capitalist
countries.

—*Joan Robinson, 1945*[1]

Moscow in April is still frigid, and the snow has not yet melted away, but
daylight lingers until almost nine at night, and old women from the country selling mimosa blossoms suddenly appear on street corners. In the
spring of 1952, not long after Winston Churchill announced that Britain
possessed the atomic bomb, Joan Robinson stared at the golden domes
of the Kremlin, feeling that her heart might burst. The sight was both
intensely familiar and strangely unreal. "I gaze and gaze," she wrote in her
diary, "and wonder if what I see is really there, and if this is really me looking at it."[2]

Later, in the mammoth Hall of Columns, Robinson only half listened
to the bombastic speeches, peace resolutions, and "fraternal" greetings
from the "women of Scotland." Her mind was taken up with impressions
of the new society outside: the farmers' market with its rosy heaps of
radishes and piles of chartreuse lettuces; the sparkling shops with plaster
hams, sausages, and cheeses in their windows (not because, as in England,
the real thing had run out, but to avoid wasting food on a window display); the free day care centers where working mothers dropped off well-
dressed, well-fed children; the clever consignment shop where outgrown

clothes could be recycled ("What a good idea!"); the "'Swedish' standard of public orderliness and cleanliness"—without the joyless Scandinavian atmosphere. What a contrast to dreary, dirty, dilapidated London.[3]

Robinson luxuriated in her hosts' generosity, so lavish that it almost seemed as if money really had been abolished by the first Socialist Great Power. Four hundred and seventy delegates to the Soviet-sponsored International Economic Conference were being treated like royalty.[4] They were housed in a hotel with "sweeping staircases, chandeliers, malachite columns" grand enough for any sultan.[5] Travel from Prague, Czechoslovakia, was gratis. Each delegate was given 1,000 rubles of pocket money to spend on vodka, furs, and caviar in specially stocked stores. A fleet of one hundred limousines, with uniformed chauffeurs, was on perpetual standby, though Robinson gamely insisted on trying out the subway and trams despite the absence of street maps and her lack of Russian.[6] The best seats at the opera and ballet were always available to the delegates. In contrast to English rations of powdered mashed potatoes and sausages that tasted like wet bread, Moscow dinners were "fabulous." Even the homely act of eating was a reminder of what it meant to be a "coming" rather than a "going" Great Power. After one feast, Joan could almost see "the continent stretching around me, as a Victorian diner might have felt the seaways of the world bearing provisions to his table."[7]

"Oriental lavishness" notwithstanding, Robinson stressed that her Russian hosts displayed "Nordic efficiency" in running the conference. Five hundred interpreters, translators, typists, messengers, and other minions, more than one per delegate, were at the visitors' disposal. Joan was confident that "All the interpreters, cars and guides are not to check our movements, but for our convenience." The promise to refrain from overt propaganda was scrupulously observed. (*Time* reported that the Russians had even removed the ubiquitous life-sized portraits of Stalin that normally hung in every public room.)[8] In postracist Moscow, Robinson exulted, "you can freeze off an Oriental bore just as you would an English one."[9] Here was the reality about which the West was in denial.

Instead of dark forebodings about the future of the West, Robinson was filled with Panglossian optimism about the East. The conference was a Socialist Bretton Woods, a United Nations of Socialists. The conference hall was "cleverly outfitted" with simultaneous translation machinery that seemed to embody the delegates' hopes for a unified global economy and global understanding.[10] A rift had developed in the world economy, thanks to a Cold War that Robinson, like most of the other delegates, assumed was instigated by the world's new imperial superpower, the United States. Lord Boyd Orr, head of the twenty-three-member British delegation, spoke for most of the delegates when he called for East and West to "burst the iron curtain by wagons of goods coming from the East bringing a surplus of goods which the West needs, and wagons of goods going through taking the surplus goods from the West that the East needs."[11] Delegate after delegate insisted that once "artificial barriers" such as the new American ban on strategic exports to the Soviet bloc were removed, the current trickle of East-West trade would become a torrent capable of sweeping away economic ills—from unemployment in the British textile industry, to chronic poverty in India. One delegate from the United States was sure that trade agreements would spark a "spiritual chain reaction of the brotherhood of man" and stave off a nuclear holocaust.

After a week in Moscow, Robinson decided that Stalin was no dictator but a solicitous if stern and somewhat distant father. She recorded a story that she found especially touching: An old cook who had been in service with a Moscow family before the war was assigned to factory work in a rural town after the Nazi invasion. When the war was over, her mistress's family got permission to return to Moscow, but the old servant was left behind. "After trying the regular channels in vain," Robinson noted in her diary, "she wrote to Stalin . . . explaining that factory work did not suit her, that all her village had been wiped out and that she had no friend in the world but her old mistress." According to Joan, "She got a permit within three weeks."[12]

Robinson left Moscow more certain than ever that the Cold War was a mistake rooted in American paranoia rather than in Soviet designs. Her *Conference Sketchbook,* published soon after her return to Cambridge, concluded serenely. "I soak through every pore the conviction that the Soviets have not the smallest desire to save our souls, either by word or sword," she wrote. Without alluding explicitly to the imposition of Soviet rule in Eastern Europe, she was convinced that the Soviets were motivated solely by the fear of Western encirclement. "If they could once be really assured that we will let them alone, they would be only too happy to leave us to go to the devil in our own way," she assured readers. "If our local communists think otherwise, they are the more deceived."[13]

Robinson portrayed herself not as a pilgrim in the new Socialist Mecca but as an objective observer, a truth teller. She insisted that she and the other participants in the conference were not "delegates from anyone but, a job lot of individuals" aware of "the importance of telling the exact truth about all we have seen here."[14] She did not necessarily expect to be believed, however. She wrote to Richard Kahn, "We are steeling ourselves to meet all the dirt that will be slung at us when we get back."[15] Instead of dirt, her lectures on Soviet society drew decent-sized crowds in Cambridge when she returned. "But what about the alleged plot of Jewish doctors to kill Joseph Stalin?" one undergraduate with an American accent had ventured. She didn't miss a beat: "And how about your lynchings in the South?"[16]

By then Robinson was well on her way to becoming one of the Communist bloc's *Parade-Intellektuellen,* or trophy intellectuals, a demanding but rewarding role that involved yearly junkets, photo ops with potentates, a Moscow bank account, and a network of friends consisting largely if not exclusively of government apparatchiks, underground Communists, and spies.

Readers of the *Sketchbook* would have been surprised to learn that the wide-eyed narrator who had tumbled Alice-like into a Socialist Wonderland was, in fact, one of the conference organizers. Robinson was one of two British members of the Initiating Committee, although she insisted that she had signed on merely as a favor to "her old friend Oskar Lange," a Polish economist who collaborated with the KGB. The British Foreign Office was convinced that she was "well aware of the origins" of the conference, and other committee members commented on her "extreme views,"[17] which matched those of the other British delegate, Jack Parry, a Communist Party of Great Britain (CPGB) official as well as a businessman. Alec Cairncross, a member of the British delegation in Moscow, reported that the delegates knew that the conference had originally been conceived as "the next installment of the communist peace campaign" and took for granted that Stalin's main motive in hosting the conference was political; namely, "to drive a wedge between the USA and her European allies."[18] The economists in attendance, including Lange, Jurgen Kuczynski, Piero Sraffa, and Charles Madge, were nearly all party members or fellow travelers.

That is not to say that Robinson had any deeper insight than Harry Dexter White into the true drift of Stalin's thinking. For example, she was probably unaware that he had repudiated the entire premise of the conference just weeks before. In remarks distributed to the Central Committee in early February, Stalin had attacked the very notion of peaceful coexistence and economic convergence with the West that was the gospel of the One Worlders like herself. He accused Soviet Communists who predicted a reconstitution of a single global economy of being misled by the flurry of international cooperation during and immediately after the war. The chief legacy of World War II, he warned, was the permanent division of the global economy into "two parallel world markets." Socialist economies and capitalist ones would evolve in isolation from, and opposition to, one another. The "inevitable" outcome would be deepening economic crisis in the West, intensifying imperialist rivalry, and, ultimately, a fratricidal war between the United States and Britain: "The inevitability of wars between capitalist counties remains in force."[19] All this, Stalin assured committee members, was a matter of scientific law.

He was quite sincere, John Gaddis, the American historian of the Cold War, has concluded.[20] Apparently, Stalin was as devout a believer in a secular apocalypse as Marx or Engels a century before. Had they been widely known, the timing of his pronouncements would have put the Russian conference hosts in an awkward position. On the one hand, they had dangled the prospect of huge orders in front of British textile manufacturers and other businessmen to ensure their attendance. On the other hand, Stalin had gone so far as to claim that the Communist bloc would soon be able to dispense with Western imports altogether. If anything, he insisted, the Soviet Union and her allies would soon "feel the necessity of finding an outside market for their surplus products."[21] However, on the eve of the conference, Stalin espoused the more politic view that the "peaceful coexistence of capitalism and communism" was possible, subject to noninterference with the domestic affairs of other countries and other conditions.[22]

If Robinson felt let down in any way by the proceedings, she gave no sign in either her public accounts or her letters to Kahn. In all likelihood, she and the other foreign delegates had not seen Stalin's remarks to the Central Committee, which Stalin kept from the press until releasing an English translation the following October.[23] The trade deals made at the conference added up to very little, particularly when compared to the inflated rhetoric with which they were presented. An economist estimated that the volume of East-West trade implied by the proposals was considerably below prewar levels.

Or perhaps Robinson did suspect the truth. At one point, she wandered into some offices at the commerce ministry where abacuses and desk calculators lay side by side on desks. Possibly it was that anomalous juxtaposition of modern and ancient that drew her attention to another incongruity: namely, that the Soviet economists at the conference had "raised not quoting figures to a fine art."[24]

One of Robinson's biographers, Geoffrey Harcourt, an economist at Cambridge, dates the beginning of her political "conversion" to 1936.[25] For British intellectuals, 1936 is associated less with the Great Depression, which was nearly over in Britain, than with the beginning of the Spanish Civil War. When Germany and Italy intervened on behalf of the Nationalists, and the Soviets on the side of the Republicans, the conflict came to be seen as a proxy war between Fascism and Communism. Stalin's willingness to fight the Fascists in Spain enhanced Soviet prestige, while Britain's and America's refusal to engage in the struggle made them seem, at best, pusillanimous.

But in 1936 Joan was infatuated with her poet in Aleppo, Dr. Ernest Altounyan, and intellectually in thrall with Keynes. It was only in 1939, while she was recovering from her breakdown, that she surprised Schumpeter, a regular and admiring correspondent, by taking up Marx (whom Keynes considered a bore). Her political activity during the war consisted of serving on various Labour Party advisory committees, writing Labour Party pamphlets, and working on reports. These included the *Beveridge Employment Report*, which was drafted by her close friend Nicholas Kaldor, the clever Hungarian LSE lecturer who, like Robinson and Hayek, wound up spending the war in Cambridge. Her assumption that the West was doomed to secular stagnation as well as recurring depressions was shared by Keynesians of all political stripes, but in 1943 she had not yet ruled out that the problem was soluble: "The problem of unemployment overshadows all other post war problems. The economic system under which we are living is on trial. The modern world has seen a great experiment in Socialist planning . . . It remains to be seen if the democratic nations can find a way to plan for peace and prosperity." [26]

Like other Labour Party economists, Robinson advocated a mixture of Socialist planning and Keynesianism demand management through taxes and subsidies. [27] As an advisor to the Trade Unions Congress, she argued for the nationalization of most industries on the grounds that planning required government ownership. [28] Her preferred solution involved government economic planning, government control of investment, and nationalization of key industries, while allowing that "a fringe of small scale private economy might well survive in a controlled economy provided that it did not threaten to encroach too far." [29] All of this was standard for the Labour Party's left wing. "By 1944," one historian commented, "the wartime radicalism had passed its peak, and the proposals advanced by Kaldor and Robinson were noticeably more moderate in tone." [30] When Keynes returned from Washington in December 1945 to announce the terms of the "infamous" American loan—that he had fought so hard to win only to have it attacked furiously by both right and left—Robinson supported him publicly by acknowledging that Britain could not afford to turn down the loan or to alienate the United States.

After Labour swept to power in 1945, Robinson aligned herself with the extreme-left opposition to the leadership. Unlike the Labour government of 1931, the government of Prime Minister Clement Attlee began at once to fulfill its wartime promises to nationalize industry and create a cradle-to-grave welfare state. As unemployment faded away and real wages rose, she became more critical rather than less critical of Labour's leadership, less focused on domestic issues, and increasingly obsessed by American power and the threat of nuclear war. The Labour landslide had not, as expected, resulted in a U-turn away from Churchill's vehement pro-American, anti-Soviet foreign policy. According to the historian Jonathan Schneer, Ernest Bevin, the Labour foreign minister, "did not believe that substantial agreement with Russia about the shape of the postwar world was possible." Bevin's "primary goal, shared by most Conservatives, was to convince the Americans that they must step into the power vacuum in Europe and elsewhere created by Britain's declining strength before the Russians did."[31]

By 1950, Stalin would complain to Harry Pollitt, the British Communist Party boss, that Labour was even more "subservient to the Americans" than the Tories.[32] But his attacks on the Labour Party, which began as soon as the election was over, had the effect of rallying the rank and file around the leadership.[33] The party's left wing resented the allegation that they were no better than Tories while they were waging a furious struggle in Parliament to nationalize heavy industry and create a national health system. Though still leaning toward nonalignment, the left wing was further antagonized by Soviet actions in Bulgaria, Romania, Poland, and East Germany. Already in 1946, the leadership of the Labour Party was convinced that the chief threat to peace came not from the United States but from the Soviet Union.

The hard-core Left opposed the mainstream on the issue of what would today be called human rights. It consisted of no more than a dozen or so activists such as Robinson. The bulk of the Labour Left was far more resolutely anti-Communist than political liberals in the United States. Open Communists such as D. N. Pritt and John Platts-Mills were expelled, and the CPGB's request for affiliation with Labour was denied. As reliably a pro-Soviet figure as Harold Laski—a prominent Marxist political scientist at the LSE who served as Labour Party chairman in 1945—defended the leadership's actions, arguing that Communists "act like a secret battalion of paratroopers within the brigade . . . the secret purpose makes them willing to sacrifice all regard for truth and straight dealing." [34] As far as most of the British Left was concerned, the wartime romance with the Soviet Union was over.

Not for Robinson. Authoritarian by temperament and disdainful of the political compromise that characterizes democracies, she was deterred neither by Stalin's purges at home nor by his fishing in troubled waters abroad—or rather troubling waters so that he could fish. If anything, universal condemnation seemed to heighten his appeal for her. In her mind, the United States was the biggest threat to world peace. "The great question which overshadows everything is whether Russia is planning aggression, for, if not, our whole policy is nonsensical." Accusing the United States of conflating ideological versus military aggression, she argued that "the great boom in America built up on rearmament has gone too far for comfort . . . and yet the prospect of a peaceful détente and a sudden cessation of rearmament expenditure is a menace to their economy . . . The line of least resistance is to keep on with it. That is what seems to me the biggest menace in the present situation." [35]

The Marshall Plan was the bolt from the blue that split the British Left. On June 5, 1947, Secretary of State George C. Marshall gave a speech at Harvard outlining his plan. "The United States should do whatever it is able to assist in the return of normal economic health in the world, without which there can be no political stability and no assured peace." The Marshall Plan eclipsed the IMF, which was "almost inactive," and the World Bank, which was husbanding its resources and refusing to make loans for reconstruction. A 1949 report of the IMF's directors "wrote a poignant epitaph for the wartime hopes of multilateralism," observes Richard Gardner, concluding that, "dependence on bilateral trade and bilateral currencies is far greater than before the war." [36] Within a month, the Soviet foreign minister, Vyacheslav Molotov, publicly rejected the plan at a Paris meeting of Communist bloc countries, calling it an "American plan for the enslavement of Europe."

After the Labour Party welcomed American aid as "an important step toward a united, prosperous Europe," Robinson was quick in her denunciation.[37] On June 25 on the BBC's *London Forum,* she argued that American money would "create a Western anti-communist bloc" and thus increase the chance of war, adding, "I don't think you can say we're going to preserve Western values by accepting dollars and splitting Europe. I think that will imperil Western values."[38] In other words, Britain should reject the offer of American aid, as the Soviets and their Eastern European allies had—a position that put her at odds with virtually the entire British Left, which rallied behind the Labour Party. The only exception was the CPGB, which attacked the Labour government for "selling out to Wall Street."[39]

Robinson's support for Stalin in the 1940s and 1950s was more puzzling—and less conditional—than Beatrice Webb's enthusiasm in the 1930s. It was a bit like her earlier worship of Keynes, whom she seemed to have regarded primarily as an icon.[40] In his 1977 book *The Russia Complex: The British Labour Party and the Soviet Union,* the political scientist Bill Jones estimates that there were no more than twenty or so fellow travelers in the Labour Party in 1946. Robinson's advocacy for the Soviet Union set her apart from Laski—described by George Orwell as "a Socialist by allegiance, and a liberal by temperament"—and most of the Labour Left. It represented a repudiation of her family's traditions, necessarily involving a degree of duplicity as well as complicity in others' deception. "Whereof one cannot speak, thereof one must be silent," Ludwig Wittgenstein had famously concluded in *Tractatus Logico-Philosophicus.* Robinson spoke fearlessly when it came to her opinions yet maintained a cagey silence about the nature of her relationship to the Soviets.

In 1939 she had confessed to Richard Kahn that the "deep rift between my political and tribal loyalty had been a continuous and growing strain all these years."[41] By the time she committed herself to Moscow, her forebodings about Western decline and optimism about the dynamism of the East had become articles of faith, and the strain of divided loyalties greater than ever. Over the summer and fall of 1952, Robinson's elated mood became more fevered and frantic. She wrote that she was discovering great secrets, including the key to her frustrating relationship with Kahn. She developed a conviction that she had discovered a hidden flaw in the foundation of economic theory that would, if people realized its existence, bring down capitalism. By fall, she was no longer sleeping, talking incessantly, obviously delusional. After a three-way consultation among Richard Kahn, Austin Robinson, and Ernest Altounyan, she was once again hospitalized, this time for six months.

Still, she recovered sufficiently to return to Moscow the following spring. Stalin was dead, and Moscow was merely the first stop on an elaborate pilgrimage that took her first to Beijing and then to a string of Russia's third world clients, including Burma, Thailand, Vietnam, Egypt, Lebanon, Syria, and Iraq. She had allowed herself to be appointed vice president of the British Council of International Trade with China, an organization with a board that consisted largely of underground CPGB members and that was suspected of serving as a conduit for CPGB funding. The group's president was Lord Boyd Orr, the food specialist who had headed the British delegation to the Moscow conference and had been a fixture at such events.[42] Possibly it was embarrassment at her role in the so-called Icebreaker Mission that made her hide behind another dignitary almost out of camera range at the "business arrangements signing" ceremony. Milton Friedman, who spent that academic year visiting Cambridge University, was baffled that an economist of Robinson's brilliance "found it possible to rationalize and praise every feature of Russian and Chinese policy."[43]

At forty-nine, Joan Robinson was more formidable than ever, part "magnificent Valkyrie," part houri, part commissar. Imperious, intellectually intimidating, and seductive, she combined Olympian certitude with a fine sarcasm. Although she was not admitted to the British Academy until 1958 and had to wait for Austin to retire in 1965 before being appointed to a university professorship, she stepped into the leadership vacuum that resulted from Keynes's death. She was not the only prominent Keynesian there. But while Sraffa buried himself in collecting and editing the papers of David Ricardo, and Nicholas Kaldor went on to become a political insider in the Labour Party, she defined the agenda. She dominated the men around her.

At a seminar at Oxford run by John Hicks, who later shared the Nobel with the American economist Kenneth Arrow for his work on economic growth, Robinson "kept telling him what he had said," recalled another participant. "He got pinker and pinker and finally said with much stammering, 'I didn't say anything of the sort,' to which she replied that if he didn't say it, that is what he meant to say."[44] Unlike the catholic Keynes, who was anxious not to become too wedded to his own ideas and disliked it when his intellectual offspring became doctrinaire, Joan sought disciples. Her male tutees were either smitten or silenced. One of them recalled,

> Mrs. R. would sit on a hassock, smoking with a long cigarette holder . . . wearing a peignoir, her graying hair pulled into a tight bun in the back, her intelligent eyes set in an expansive brow, focused on me. The scene bore a vague resemblance to Picasso's portrait of Gertrude Stein: the same heavy solidity and presence. But there the resemblance ended. Mrs. R., if not quite a dish, was certainly comely. And the difference between her and Stein was made more evident by a pen sketch which sat on a small table, next to the hassock, of a woman, stark naked, sitting on a hassock, her hands covering her face.[45]

For Robinson, Cambridge, England, became the anti–Cambridge, Massachusetts. Her disdain for mathematics bordered on affectation. She turned down an invitation to become president of the Econometric Society on the grounds that she could not join an editorial committee of a journal she could not read. Arthur Pigou, her old professor, called Joan "a magpie breeding innumerable parrots" and complained that she "propounds the Truth with an enormous T and with such Prussian efficiency that the wretched men become identical sausages without any minds of their own."[46] Michael Straight, whose family owned the *New Republic* and who was recruited by the KGB, called her "the most exciting and brilliant lecturer as far as economics students were concerned."[47]

As a member of the class that had once administered the empire, Joan came of age during British imperialism's terminal decline, and perhaps it was this sense of being on the losing side of history that fueled her determination to side with history's winners. By the time she went to Moscow for the first time, Robinson's new passion was economic growth, and she

was already convinced that she had taken "the wrong turning" intellectually twenty years earlier when she "worked out *The Economics of Imperfect
Competition* on static assumptions."[48] During the Great Depression, she
had striven mightily to answer what she now considered to be the wrong
question. Instead of asking what caused transitory unemployment, she
now realized that she should have focused on what determined the wealth
and poverty of nations. In retrospect, she said, she should have abandoned
"static analysis" and tried instead to "come to terms with Marshall's theory
of development."

The question of long-run growth had originally captured the attention of Keynesians, Robinson included, who were worried about long-run
stagnation in the industrialized West. But several developments caused
them to shift their attention to the "overpopulated, backward countries"—
that is, the former colonies in Asia, Africa, and Latin America.[49] To begin
with, postwar stagnation had not materialized. War-shattered Britain and
Europe rebounded so strongly that by 1950 unemployment had virtually
disappeared and wages were rising rapidly. The Left argued that the arms
race had saved the market economy, but the fact remained that the economic problem was no longer a compelling rationale for Socialism in the
West.

World War II had made decolonization inevitable. Britain's financial
weakness and commitment to building a welfare state at home coincided
with the emergence of indigenous liberation movements. The intensifying
Cold War accelerated the process by improving the third world's bargaining power, and the growing political participation of poor countries in
global organizations, including the United Nations, focused attention on
"underdevelopment" as an economic problem.

The hopeful rhetoric of the Moscow conference sounds absurdly optimistic in retrospect. With one-fifth of the world's population, China had
an average per capita income roughly half that of Africa's in 1952, and a
mere 5 percent of America's. Living standards in India, with 15 percent of
the world's population, were only marginally higher. Had anyone asked
before the war, most economists would have readily conceded that poor
countries could grow rich—eventually. After all, Europe had escaped the

Malthusian trap of universal poverty and life on the edge of starvation by achieving economic growth just 1 or 2 percentage points faster than that of population growth.

But what hope could the European experience offer populous China, India, or the Middle East? Not only was the gulf in material conditions between the populous poor countries and the world's richest countries unimaginably large, but, more disturbingly, the day's poor countries were far poorer than England in the 1840s, before the real wages and living conditions of ordinary Englishmen began their remarkable cumulative rise. "There are today in the plains of India and China men and women, plague-ridden and hungry, living lives little better . . . than those of the cattle that toil with them," T. S. Ashton wrote in 1948. "Such Asiatic standards, and such un-mechanized horrors, are the lot of those who increase their numbers without passing through an Industrial Revolution." At the rate at which Britain, Europe, and America had escaped poverty, it would take China and India another hundred years to reach *that* level.

The pros and cons of central planning and state-run enterprises were not the only issue. There was also the question of international trade and investment. Was integration in the global economy or autarky the fastest path? The answer depended on what one thought had caused underdevelopment in the first place. In Victorian England a century before, Friedrich Engels and Karl Marx had maintained that poverty was a new condition, far worse in Victorian than in Elizabethan England. They blamed the rich. Later, Alfred Marshall, Irving Fisher, Joseph Schumpeter, and John Maynard Keynes, among others, took a different view. They pointed out that poverty had been man's fate long before the emergence of the modern economy. The root cause of low living standards was not a lack of resources or unequal distribution of existing incomes but an inability to use existing resources—land, labor, capital, knowledge—efficiently. Now, in most of the globe, the question was whether the poverty of nations was caused by the Western economic system or by local conditions and institutions inimical to economic growth that Western organization could cure.

Schumpeter had called the triumph of Bolshevism in a precapitalist agrarian economy "nothing but a fluke." In her review of *Capitalism, Socialism, and Democracy,* Robinson acknowledged,

> May be so. But in that case the exception seems rather more important than the rule. Who knows what flukes may accompany the end of the present war? And, even if the Bolshevik fluke remains unique, there cannot be much doubt that the existence of a socialist Great Power will play at least as important a part in the development in other countries (even without any deliberate intervention in their affairs) as the more subtle processes of evolution according to the immanent characteristics of capitalism.

The Soviet victory over Germany, Europe's leading industrial power, in World War II had apparently convinced Robinson that Socialism was a shortcut to industrialization:

> The grand moral of this thirty years of history is not so much for the western industrial countries where the standard of living is already high, as for the undeveloped nations. That communism is destined to supersede capitalism is in the nature of a dogma, but it is a proven fact that the Soviet system shows how the technical achievements of capitalism can be imitated (and in some cases surpassed) by those whom the first industrial revolution kept as hewers of wood and drawers of water.[50]

In 1951 Robinson wrote a brief introduction to a Marxist classic, *The Accumulation of Capital* by Rosa Luxemburg. Luxemburg was the German Communist leader who was murdered in 1919 and was one of the few first-rate minds among Marx's disciples. Today her reputation rests more on her early criticism of the Bolshevik dictatorship than on her economic theory, but in 1951 Robinson was compelled by Luxemburg's argument that the limits to growth—and the source of the inevitable breakdown— of the global market economy were to be found in the third world.

According to Luxemburg, shrinking investment opportunities at home drove entrepreneurs abroad in search of profits and led, inevitably, to rivalries. When these imperialists ran out of fresh territory to exploit—or ran into one another—capitalism had to break down, through either stagnation or war. Robinson acknowledged that Luxemburg's analysis was incomplete in that it identified imperialism as the only means by which capitalism extended its flagging lease on life, omitting any consideration of technological change or rising real wages: "All the same, few would deny that the extension of capitalism into new territories was the mainspring of what an academic economist has called 'the vast secular boom' of the last two hundred years and many academic economists account for the uneasy condition of capitalism in the 20th century largely by 'the closing of the frontier' all over the world." Nonetheless, she concluded, somewhat inaccurately, that Luxemburg's book "shows more prescience than any orthodox contemporary could claim." [51]

Robinson embarked on her own magnum opus on economic growth, for which she intended to borrow Luxemburg's title. [52] A hostile 1949 review of Roy Harrod's classic on economic growth makes it clear what she wanted to accomplish. [53] She castigated Harrod for ignoring conflicts of interest, history, politics, and especially "the distribution of income or measures to increase useful investment." [54] A 1952 article for the *Economic Journal*, written before she left for Moscow, offered a preview of her main argument: growth, she wrote, was the process of accumulating physical capital—roads, office buildings, dams, factories, machinery, and the like. Admittedly, Marx had erred in claiming that free market economies could not grow indefinitely. She would demonstrate that almost none *would*. "Perpetual steady accumulation is not inherently impossible," she wrote, but "the conditions required by the model are unlikely to be found in reality." [55]

Robinson's first visit to China, in 1953, provided for her "the final proof that communism is not a stage beyond capitalism but a substitute for it." [56] She explained later, "Private enterprise has ceased to be the form of organization best suited to take advantage of modern technology." [57] The chief obstacle to growth in poor countries was not a lack of capital or entrepreneurship, she concluded, but interference by the West. North-south trade was a zero-sum game that produced losers as well as winners, and inevitably the poor countries wound up as the losers. She discounted the role of education and innovation. "Only when the advanced countries are satisfied that they need not disturb themselves will they tolerate, and so permit, the drastic social changes required to send the colonial and ex colonial and quasi colonial nations on a hopeful path," adding somewhat irrelevantly that "peaceful coexistence is natural and logical."

While Robinson wrote her book, Richard Kahn hosted what he and she called the "secret seminar." It met every Tuesday during the Michaelmas and Easter terms in Kahn's rooms at King's College and served as a testing site for her work in progress. Visitors were invited to drop in but often found it hard to get airtime. Samuelson described a typical meeting as Robinson's friend "Kaldor talking 75% of the time and Joan talking 75% of the time."[58]

When *The Accumulation of Capital* was published in 1956, the book's "heroic scale" and Robinson's lofty stature guaranteed copious reviews. But although reviewers called the book "monumental" and "important," reaction was muted. Some complained of "few new insights," "no propositions [that] can be empirically tested," "a verbal, graphical exposition" of "long familiar results in linear programming."[59] Others criticized her for not understanding the role of consumers, making logical errors, and ignoring recent research. (This was held to be a typically Cambridge, England, vice, with one reviewer pointing out that Piero Sraffa's *Production of Commodities by Means of Commodities,* written during World War II, contained not a single reference more recent than 1913.) Less charitably, Harry Johnson wrote that his former professor "has proved conclusively to her own satisfaction that Capitalism Cannot Possibly Work."[60] Samuelson compared Robinson's theory to Lenin's rule of three: electricity + Soviets = Communism.[61] Abba Lerner called the book a "pearl," not only for redirecting attention to "the causes of the wealth of nations" but for providing graduate students with a host of "errors and . . . ingenious confusions" on which to flex their muscles.[62] Lawrence Klein, who shared Robinson's political views, dismissed her insights as "ordinary sort of results usually derived in economic theory from some maximizing or minimizing principle."[63]

Robert Solow, a Keynesian at MIT who had published a paper on economic growth that same year—one that would earn him a Nobel Prize in 1987—delivered the coup de grace: "I think there's nothing Keynesian about Joanian economics . . . There's nothing in *The Accumulation of Capital* . . . or any of those papers which strikes me as having a genuine root or inspiration in Keynes."[64]

Solow had not only proposed an elegant theory but produced a stunning empirical result: Nine-tenths of the doubling in output per worker in the United States between 1909 and 1949 was due neither to the accumulation of physical capital nor to improvements in the health or education of the labor force, but rather to technological progress. The implication that an economic environment conducive to innovation mattered more than its stock of factories and machines flatly contradicted Robinson's central premise, not to mention that of the widely imitated Soviet model. Solow, who dismissed Schumpeter, rather unfairly, as a pro-German anti-Semite and an intellectual phony, had supplied most compelling evidence that it was not what a nation had but what a nation did with what it had that determined long-run economic success or failure. This, of course, was pure Schumpeter.

Robert Solow and Kenneth Arrow spent the academic year 1963–64 visiting Cambridge, England, and heard Robinson describe her two months touring Chinese communes. Saying she wanted to counter "the malicious misrepresentation of China in the Western press," she dismissed "the critics [who] were shedding crocodile tears over the 'famine'" and claimed that China's communes were "a method of organizing relief" during the three "bitter years" of flood and drought. Echoing Beatrice and Sidney Webb's glowing reports during the 1932 Ukrainian famine, Robinson called the communes "a brilliant invention" and concluded that "the rationing system worked; the rations were tight, but they were always honored."

We now know that an estimated 15 to 30 million peasants in Henan, Anhui, and Sichuan provinces died between 1958 and 1962—ten times the toll in the 1943 Bengal famine—and that forced collectivization, the disastrous Great Leap Forward, and the refusal of Mao Tse-tung's regime to organize relief, not bad weather, were primarily to blame.

That democracy and well-being go hand in hand is now conventional wisdom. For a long time, it was not. Individual rights were thought, by many intellectuals influenced by the utilitarian tradition, to be a luxury that poor nations simply could not afford. Robinson considered democracy a bit of a fraud and politicians both pusillanimous and deceitful. "The notion of freedom is a slippery one," she wrote during World War II, adding, without the slightest hint of irony, "It is only when there is no serious enemy within or without that full freedom of speech can safely be allowed."[65] She was inclined to dismiss democratic reforms as "premature attempts to pluck the low-lying fruit." That blind spot goes a long way in explaining why Robinson, who visited China frequently in the 1950s and 1960s, "failed utterly to detect the biggest famine in modern history," while others—including Bertrand Russell, Michael Foot, Harold Laski, and Harold Wilson, all of whom were vilified at one time or another as Communist sympathizers or even fellow travelers—saw what was happening and called for international relief.

To be sure, Joan Robinson was hardly the only eminent Western observer to be hoodwinked by Beijing's denials. Lord Boyd Orr, head of the British delegation to the 1952 Moscow economics conference and one of the world's leading experts on food, concluded Mao was ending the "traditional Chinese famine cycle."[66] In fact, the magnitude of the death toll wasn't known outside China until after Mao's death in 1976. But Robinson's willingness to believe a totalitarian regime that forbade free movement, free speech, free press, and free elections was symptomatic of a mind-set—all too common among development economists fifty years ago—that ignored the crucial role of political rights.

Geoffrey Harcourt once remarked that Robinson was "always looking for the next Utopia." Perhaps, but she was also looking for the next Great Leader and, of course, the next worshipful audience. She relished her celebrity, her junkets, the VIP treatment, and bully pulpits. She liked playing the fearless outsider speaking truth to power. Perhaps the Moscow bank account, friendships with Cold War spies, including Solomon Adler, Frank Coe, Donald Wheeler, and Oskar Lange, and the need for veiled allusions and careful elisions gave her a kick as well.

As time went by, Robinson became even more Olympian, imperious, and pessimistic. Her book *Economic Philosophy,* published in 1962, surveys economic ideas since 1700. In his review, George Stigler, Milton Friedman's best friend at the University of Chicago, called Robinson "a superior logician" but accused her of ignoring facts:

> There really isn't a great deal to economics, considered as a logical structure based upon a few indisputable axioms about the world. If one cuts oneself off from two generations of immensely varied and instructive empirical research, and if one thinks economic history had no relevance to economic theory . . . then one is indeed left with a hollow discipline. A logician is a wondrous creature, but he cannot distinguish between the two simple errors: If $A = B$, and $B = C$, then (1) $A = 1.01C$ and (2) $A = 10^{65}C$. An *economist* can.[67]

Chapter XVIII

Tryst with Destiny:
Sen in Calcutta and Cambridge

There haven't been many folk songs written for capitalism, but there have been many composed for social justice.

It is mainly an attempt to see development as a process of expanding the real freedoms that people enjoy. In this approach, expansion of freedom is viewed as both (1) the *primary end* and (2) the *principal means* of development.

—*Amartya Sen*[1]

Joan Robinson wound up her talk at the Delhi School of Economics clutching a copy of Mao's "red book." It was the late 1960s. Her topic was the dismal state of Western economics, but mostly she talked about China and the Cultural Revolution. The audience was in rapture. When the wild applause faded at last, a willowy young man asked a question. His tone implied the mildest and most polite skepticism. Robinson rebuffed him soundly but "with affection."[2] They were, after all, the best of enemies, former professor and favorite student. At Cambridge, she had cultivated students from the third world. One of the most gifted was Amartya Sen, but Sen's interest in human rights and the immediate amelioration of poverty clashed with Robinson's enthusiasm for the Soviet model of industrialization.

Amartya means "destined for immortality." Born into a scholarly and cosmopolitan Hindu family, Amartya Sen grew up amid the horrors of the Bengal famine, communal violence, the collapse of British rule, and partition. As a brilliant student and campus agitator in Calcutta, he overcame a near-lethal bout of cancer, bested one hundred thousand other exam takers, and won admission to the college of Isaac Newton, G. H. Hardy, and the mathematician Srinivasa Ramanujan, Trinity College in Cambridge. Since 1970, Sen has lived mostly in England and America, but his thoughts have never strayed far from India. Drawing on his own experiences, a lifelong study of the disenfranchised, and a deep knowledge of Eastern and Western philosophy, Sen has questioned every facet of contemporary economic thought. Challenging traditional assumptions about what is meant by social welfare and how to measure progress, he has helped restore "an ethical dimension to the discussion of vital economic problems."[3] He is a public intellectual, engaged by issues, from famines and premature female mortality, to multiculturalism and nuclear proliferation. His inspiring journey from impoverished Calcutta in newly independent India to the ivory towers of Cambridge, England, and Cambridge, Massachusetts—and back again—is a triumph of reason, empathy, and a very human determination to overcome incredible odds.

In January 2002, India's Hindu nationalist government of the Bharatiya Janata Party threw a three-day celebration for India's farflung diaspora in Delhi. In a gesture that revealed both how far he had traveled—and how close he had remained to his roots—Sen left that gathering to address an outdoor "hunger hearing" with several hundred peasants and laborers in a chilly dirt field on the far side of town.

One by one, members of the audience went up to the microphone. A scrawny fourteen-year-old from Delhi spoke about going hungry after she lost her dishwashing job. A dark-skinned man from Orissa described how three members of his family had died after a local drought the previous year. Fifty years after independence, a larger fraction of India's population suffered from chronic malnutrition than in any other part of the world, including sub-Saharan Africa. Yet India's government kept food prices high via agricultural price supports and had accumulated the biggest food

stockpile in the world, a third of which was rotting in rat-infested government granaries.

When Sen stood up, shivering in his baggy cords and rumpled jacket,
he spoke less about the "interest of consumers being sacrificed to farmers"
and more about "profoundly lonely deaths." Addressing an audience that
seemed plainly awestruck, he conveyed sympathy and encouragement.
"Without protests like these," he said, "the deaths would be much more.
If there had been something like this, the Bengal famine could have been
prevented." Their willingness to speak out, he told them approvingly, was
"democracy in action."

Sen is Bengali. Like saying that an American is a southerner, that has very
specific connotations. Bengal is a river delta; fish is the mainstay of the
Bengali diet; dhoti, chappals, and panjabi are the traditional garb. All Bengalis, Sen says, are great talkers, as he is. The worst thing about dying, Bengalis like to joke, is the thought that the people will keep talking and that
you won't be able to answer back.

The Bengali word for public intellectual is *bhadralok,* and Bengal has
a long tradition, going back at least two centuries, of learned men with
cosmopolitan outlooks who battled social evils such as untouchability
and suttee. Sen is part of that tradition. His family is from the old part of
Dhaka, an ancient river city 240 kilometers as the crow flies from Calcutta,
now the capital of Muslim Bangladesh. In Jane Austen's day, Dhaka was
"a big, bustling place of first-rate importance," famous the world over for
its fine muslins (called *bafta hawa,* or "woven air").[4] Competition from
Manchester brought decline. By 1900, Dhaka's population had shrunk by
two-thirds, and, according to a contemporary travel guide, "all round the
present city are ruins of good houses, mosques, and temples, smothered in
jungle."[5] Thirty-odd years later, when Sen was born, in 1933, Dhaka had
regained some of its former importance by becoming a regional administrative center for the British Raj.

Sen was born into that class of English-speaking academics and civil
servants who helped run British India. He describes his father, Ashutosh,

as "an adventurous man" who got a PhD in chemistry at London University and fell in love with an English Quaker. After returning home to an arranged marriage, he became head of the agricultural chemistry department at Dacca University. The Sens lived in a typical Dhaka house, fifty or sixty feet long, narrow in the front, "the middle being a courtyard open to the sky," with plenty of room for servants and relatives.[6]

Sen began his education at an English missionary school in 1939. Two years later, as the Japanese advanced toward British India, he was sent to live with his maternal grandparents in Santiniketan, just north of Calcutta, "to keep me safe from the bombs." Santiniketan has special connotations for Bengalis—indeed for all Indians—because of its association with Rabindranath Tagore, the poet. After winning the Nobel Prize for literature in 1913, Tagore used his prize money to expand the Visva Bharati school in Santiniketan, where he tried to apply his ideas about education and his notions of merging Eastern spirituality with Western science. Gandhi visited Santiniketan in 1940, and for years India's nationalist elite, including Prime Minister Jawaharlal Nehru, sent their children to study there.

Sen's maternal grandfather, Kshitimohan Sen, a distinguished Sanskrit scholar, was on the faculty of Visva Bharati. Sen attended classes in Tagore's coeducational school under the eucalyptus trees. His free time was spent mostly with his grandfather. "Everyone found him formidable," Sen recalled. "He woke at four. He knew all the stars. He talked with me about the connections between Greek and Sanskrit. I was the only one of his grandchildren who had a sense of academic vocation. I was going to be the one who carried the mantle."

If Santiniketan was a tranquil oasis, it hardly escaped the upheavals of the time. At the time of his death in 1941, Tagore was deeply disenchanted by the West, professing to see little difference between the Allied and Axis powers. The war accelerated the final break with Britain. After Gandhi launched the "Quit India" movement in 1942, the British arrested sixty thousand Congress Socialist Party supporters, including Amartya Sen's uncle; by end of that year, over one thousand people had been killed in anti-British riots. "My uncle was in preventive detention for a very long time," Sen recalled. "Several other 'uncles' also were jailed, including one who died in prison. I grew up feeling the injustice of this."

The 1943 Bengal famine—the consequence of wartime inflation, censorship, and imperial indifference rather than crop failures—destroyed the last remnant of respect for the British. The new viceroy, Lord Wavell, wrote to Churchill, the "Bengal famine was one of the greatest disasters that has befallen any people under British rule and damage to our reputation here both among Indians and foreigners in India is incalculable."[7] Sen later estimated that 3 million people, mostly poor fishermen and landless laborers, perished from starvation and disease.

At the time, for the boy of ten, the famine meant a steady stream of starving villagers who passed through Santiniketan in a desperate attempt to reach Calcutta. His grandfather allowed him to hand out rice to beggars, "but only as much as would fill a cigarette tin" and only one tin per family. Later, as a university student, he reflected on the fact that only the very poor and members of despised castes had starved, while he and his family—and, indeed, their entire class—remained unaffected. That observation was to inform his theory of famines as man-made, not natural, disasters.

Even more traumatic was the eruption of communal violence on the eve of independence. The idea of a multicultural Indian nation was very much alive in Santiniketan, and, traditionally, Muslims and Hindus achieved a higher degree of assimilation in Bengal than in other parts of India. Yet when religious conflict erupted on the eve of independence, it set neighbor against neighbor in a vast pogrom. Ashutosh Sen, along with the other Hindus on the faculty of Dhaka University, were forced to leave Dhaka in 1945.

On one of his last school holidays in his Dhaka home, Sen witnessed a horrific scene. A Muslim laborer named Kader Mia staggered into the family compound, screaming and covered in blood. Stabbed in the back by some Hindu rioters, he died later that day. "The experience was devastating for me," recalled Sen. Mia told Sen's father, who took him to a hospital, that his wife had pleaded with him to stay home that day. But his family had no food, so he had little choice but to go to the Hindu part of town to seek work. The realization that "extreme poverty can make a person a helpless prey," Sen said, was to inspire his philosophical inquiry into the conflict between necessity and freedom.[8] A more immediate effect, however, was a strong distaste for all forms of religious fanaticism and cultural nationalism.

Presidency College, one of the most elite institutions of higher education in India, looks today much as it did in 1951, when Sen enrolled there, and, for that matter, much as it did in 1817 when British expats and Indian notables founded the Hindu College. Its faded pink stucco façade with peeling green shutters, the black plaques identifying the different rooms, the dim interiors with their ceiling fans and row upon row of long wooden benches, all evoke a long-bygone era. In the years immediately following independence, though, the college was a political hotbed. Sen arrived thinking he would study physics but quickly found economics of greater urgency and interest.

Thanks to the traditions of Indian higher education, Sen was introduced to classical works like Marshall's *Principles of Economics* as well as new work like Hicks's *Value and Capital* and Samuelson's *Foundations*. (Later, at Trinity, he would be disappointed in the relative lack of mathematical sophistication of his Cambridge dons.) His principal passion, however, was politics, and before his first term ended, he was elected as one of the leaders of the Communist-dominated All India Students Federation. He read voraciously, skipped lectures, and spent most of his time debating Marx with his Stalinist friends in the coffeehouse on nearby College Street, a street that then as now was lined with hundreds of booksellers' stalls.

Later he recalled, "[A]s I look back at the fields of academic work in which I have felt most involved throughout my life . . . they were already among the concerns that were agitating me most in my undergraduate days in Calcutta."[9] Those concerns were crystallized by a life-and-death crisis in his second year at Presidency. Just before his nineteenth birthday, Sen felt a pea-sized lump in the roof of his mouth. A street-corner GP dismissed it as a fish bone that had worked its way under the skin. The lump, however, didn't disappear and, in fact, grew larger. After consulting a premed student who lived next door to him at the YMCA, he learned that cancers of the mouth were fairly common among Indian men. A few hours with a borrowed medical textbook convinced Sen that he was suffering from stage two squamous cell carcinoma.

It took months and the intervention of relatives and family friends to arrange a biopsy at Calcutta's Chittaranjan Cancer Hospital. The biopsy confirmed his suspicions. At that time, a diagnosis of oral cancer was a virtual death sentence. Surgery generally only accelerated the spread of the cancer, and, as a result, most sufferers slowly suffocated as their tumors gradually blocked their windpipes. Radiation, the standard treatment in England and the United States since the turn of the century, was still too difficult and costly to be widely available in Calcutta. After reading about radiation in medical journals, Sen was finally able to locate a radiologist willing to treat him. The radiologist urged Sen to let him use a maximum dose, justifying the risk by saying, "I can't repeat it." For Sen, possible death from radiation sickness seemed preferable to certain death by suffocation.

The treatment was unpleasant, if not as awful as its aftermath. A mold was taken. A leaden mask was made. Radium needles were placed inside the mask. Like the hero of the Victor Hugo novel, Sen sat in a tiny hospital room with the mask screwed down "so there would be no movement." The procedure was repeated every day for one week. "I sat there for four hours at a time and read," Sen recalled. "Out of the window, I could see a tree. What a relief it was to see that one green tree."

The dose was massive, some 10,000 rads—four or five times today's standard dose. After he was sent home—his parents now lived in Calcutta—the effects of the radiation appeared: weeping skin, ulcers, bone pain, raw throat, difficulty in swallowing. "My mouth was like putty. I couldn't go to class. I couldn't eat solids. I lived in fear of infection. I couldn't laugh without bleeding. It brought home to me the misery of human life." That misery lasted for nearly six months. And these were only the immediate effects. Over time, radiation destroys bone and tissue, leads to necrosis and fractures, and destroys the teeth.

Cancer was a defining moment. For one thing, learning that you have a devastating illness—especially one that carries a social stigma, that's taboo—isn't just terrifying, it makes you feel polluted, powerless, outcast. The awful things Sen witnessed growing up were shocking, but they were happening to others. This was happening to him. It produced a lasting identification with others who were also hurting, voiceless, deprived.

Overcoming the cancer was also empowering. His mother, Amita, said, "I gave Amartya to God when he was nineteen."[10] But he has said that taking matters into his own hands left him with enormous confidence in his own instincts and initiative. "Psychologically I was in the driver's seat," he recalled. "I was aggressive. I was the one asking whether I would live. What was best? What could I do? I had a sense of victory."

When he returned to his classes, he said, "I came back with a bang," full of fresh purpose. He promptly got a first, won all sorts of prizes, including a debate prize. He applied to Trinity College in Cambridge, where Nehru had studied. He was rejected initially but, some months later, unexpectedly summoned. His father spent half his slender capital to pay for the journey. The airfare on BOAC proved prohibitive, so in September 1953, just before his twentieth birthday, Sen sailed from Bombay to London on the same liner as the Indian women's hockey team.

In Cambridge, new miseries—darkness, cold, awful food, dreadful loneliness—awaited Sen. His teeth, addled by radiation, were a chronic source of pain and embarrassment. The landlady of his rooming house, who had begged the college not to send her "Coloreds," fussed at him about such things as drawing the curtains at night. "You can't see out, but they can see you," she would say, as if he were a stupid child.

At the university, Sen encountered a political minefield, split by rancorous rivalries among Keynes's disciples and critics. Indira Gandhi, who studied in Santiniketan for a year, once remarked that she learned an essential survival skill there; namely, "the ability to live quietly within myself, no matter what was happening outside."[11] Sen, too, got by on inner quiet, eagerly engaging with scholars from different sides of the ideological divide but without giving up his independent way of looking at things.

He did, however, fall under the spell of the brilliant and imperious

Joan Robinson. Newly independent India was divided not just along ethnic lines but also between diametrically opposite visions of the future.
Gandhians envisioned a spiritual and rural India of hand-loom weavers.
Followers of Nehru saw Soviet-style central planning and a landscape
dotted with dams and steel plants. Sen's thesis, *The Choice of Techniques*
(1960), criticized government planning in India by underscoring basic
economic principles. After completing a second BA and finishing his thesis
research, he returned to India, first teaching at Jadvapur University and,
subsequently, at the newly formed Delhi School of Economics.

Had Sen stopped writing in the late 1960s, we would know him, if
at all, as one of a generation of Indian development economists who
favored Nehru's formula of heavy industry, state-run enterprises, and
self-sufficiency—a formula that produced disappointing results and that
has since been disavowed by most economists, including Sen. But beginning around 1970, he shifted his intellectual focus sharply and produced
a series of startling philosophical papers on social welfare that account for
much of his influence today.

This burst of creativity followed a second life crisis. In the space of a
year, he accepted a position at the London School of Economics, his father
died of prostate cancer, and he was forced to confront the possibility that
his own cancer had come back. Once in England, he underwent extensive
reconstructive surgery when it turned out that his symptoms had been
due to the delayed effects of his earlier radiation. After a long and difficult convalescence, he left his wife and two young daughters and fell
passionately in love with Eva Colorni, an Italian economist who was the
daughter of a prominent Socialist philosopher killed by Fascist forces in
World War II. Eva encouraged Sen's new philosophical interests and urged
him to apply his ethical insights to urgent issues like poverty, hunger, and
women's inequality. He and Eva lived together in London from 1973 until
her death from stomach cancer in 1985, and had two children together.

When Sen turned to ethics, Robinson advised her star pupil to "give
up all that rubbish." He ignored her counsel. At Eva's urging, he made
a detailed study of what he saw as a particularly grim consequence of
authoritarian rule, notably famines. "I once weighed nearly 250 children

from two villages in West Bengal to check their nutritional status related to income, sex, etc," he said. "If anyone asked me what I was doing, I would have said, I was doing welfare economics." [12]

Famines like the one in Bengal, Sen argued, occurred despite adequate food supplies when higher prices and joblessness robbed the most vulnerable groups in society of their "entitlements" to food and when the lack of elections and a free press stifled public pressure on the government to intervene. By contrast, Robinson applauded draconian policies such as the Great Leap Forward—and, as Sen later pointed out with some bitterness, "failed utterly to detect the biggest famine in modern history," in which an estimated 15 million to 30 million Chinese perished in the aftermath of forced collectivization. He never broke with her publicly, but by the time Robinson died in 1983, they had not corresponded in years.

In the 1970s and 1980s, Sen proposed a general theory of social welfare that attempted to integrate economists' traditional concern for material well-being with political philosophers' traditional concern with individual rights and justice. Objecting to the utilitarian creed of his fellow economists, which called for judging material progress chiefly by the growth of GDP per head—and citing a long tradition from Aristotle to Friedrich von Hayek and John Rawls—Sen argued that freedom, not opulence per se, was the true measure of a good society, a primary end as well as a principal means of economic development. He wished, as he says in his book on India, to "judge development by the expansion of substantive human freedoms—not just by economic growth ... or technical progress, or social modernization ... [These] have to be appraised ... in terms of their actual effectiveness in enriching the lives and liberties of people—rather than taking them to be valuable in themselves." [13]

Sen asked three separate questions to which he proposed answers: Can society make choices in a way that reflects individual citizen's preferences? Can individual rights be reconciled with economic welfare? And, lastly, what is the measure of a just society?

In the 1930s and 1940s, libertarians worried that the West would trade its commitment to political liberalism for economic security. A generation later, Sen worried that India and other third world nations would sacrifice democracy in the race for economic growth. How, he wondered, could conflicts between social action and individual rights be resolved?

When Sen took up the issue in the late 1960s, two powerful challenges had been laid down to the possibility of reconciling the two. One came from Friedrich von Hayek, who feared that "specialists" and specific interests would impose their own preferences on everyone. By substituting government plans for individual plans, he argued, the authorities were imposing a monolithic set of priorities on individuals who would prefer to make their own trade-offs among diverse alternatives.

The other, even more daunting, challenge came from a wholly unexpected quarter: a highly theoretical tract, *Social Choice and Individual Values*, published in 1951 by a politically moderate American economist, Kenneth Arrow. Sen first encountered Arrow's impossibility theorem at Presidency College. The theorem appeared to be a logically unassailable proof that no system of voting could produce results that reflected the preferences of individual citizens. Except when there was complete consensus, all voting procedures yielded outcomes that were, in some sense, undemocratic. Most of Sen's college friends were Stalinists. While Sen shared their enthusiasm for equality, he "worried about political authoritarianism." Was Arrow's theorem a rationale for dictatorship?

Since Arrow's result could not be challenged directly, Sen chose to probe Arrow's seemingly innocuous assumptions—the conditions any democratic procedure had to meet. In *Collective Choice and Social Welfare*, published in 1970, he argued that one of Arrow's axioms—which ruled out comparisons between different citizens' well-being—was not, in fact, essential, and was indeed arbitrary. If such comparisons were allowed, Sen suggested, the impossibility result no longer held. Sen, and researchers inspired by him, went on to pinpoint the conditions that would enable decision-making rules consistent with individual rights to work. In fact, Sen's "comparative metrics of well-being" launched his pursuit of yardsticks that could prod democratic governments to adopt social reforms, and launched a long-running debate over the best ways to define and measure poverty.

Is there a conflict between individual rights and economic welfare? Sen proceeded to mount a much broader attack on utilitarianism, inspired, in part, by John Rawls's magisterial 1971 *A Theory of Justice*, widely seen as a philosophical justification for the modern welfare state. Utilitarians, including most economists, believe that society needs only to take account of the welfare of its citizens. Rights enter their thinking, if at all, only indirectly, as contributors to happiness or satisfaction. In a twist on Jeremy Bentham's rule "the greatest good for the greatest number," Rawls's "difference principle" states that a just society should maximize the welfare of the worst-off group. This, of course, is a very utilitarian idea. But Rawls's primary focus is on individual rights, which take precedence over material well-being, and which economists have traditionally ignored.

In another 1970 journal article, "The Impossibility of the Paretian Liberal," Sen made an urgent case for paying attention to rights as well as welfare, pointing out a potentially serious conflict between the two.[14] Most economists accept a criterion for economic welfare far less demanding than those proposed by Bentham or Rawls. The optimal state, argued the nineteenth-century Italian economist Vilfredo Pareto, is one in which it is no longer possible to make anyone better off without making someone else worse off. In other words, it is a society in which all conflict-free opportunities for improving overall utility have been exploited.

But Sen showed that even this seemingly innocuous standard can run afoul of individual rights. When many people define their own welfare in terms of restricting the freedom of others—Muslim clerics are happier if schooling is prohibited for girls, Catholic nuns feel better if abortion is illegal, parents like the idea of outlawing recreational drugs—free choice can conflict with Pareto optimality.

Suppose, to use an updated version of Sen's original example, "Prude" values the freedom to practice his own religion, but not as much as he would a ban on pornography. "Lewd" values the freedom to read pornography, but not as much as he would a ban on religion. If the government outlawed both pornography and religion, both would be happier—but also less free.

Economics hasn't necessarily come to grips with Sen's message, but economists are now more apt to reflect on what's left out of the equation when they use GDP to measure material gains. In particular, they have become more circumspect in equating GDP with well-being. Sen argues that GDP leaves out individuals' opportunities that may be more important to them than their income, a serious shortcoming. To be sure, one could argue (as does Eric Maskin, a Nobel laureate in economics) that while rights and welfare may sometimes conflict, in general, rights can be seen as a way of protecting welfare. The right to read what you want—as opposed to having people tell you what you can read—usually results in higher incomes, for example. Still, given how polarizing such conflicts are in many societies, it was remarkably prescient of Sen to have pointed it out three decades ago.

In expanding his assault on utilitarianism, Sen argued that growth alone is an inadequate measure of welfare because it doesn't reveal how well or badly deprived individuals are doing, and that utility—based on people's current preferences and satisfaction—is similarly misleading, because deprived individuals often tailor their aspirations to their impoverished circumstances. To get around these and other difficulties, he proposed a new way of thinking about the goals of development. He called it "the capabilities approach."

What creates welfare aren't goods per se, but the activity for which they are acquired, he argued. I value my car for increasing my mobility, for example. You might value your education for giving you the chance to participate in discussions like ours. According to Sen's view, income is significant because of the opportunities it creates. But the actual opportunities (or capabilities, as Sen calls them) also depend on a number of other factors—not just preferences that might be constrained by deprivation—such as life spans, health, and literacy. These factors should also be considered when measuring welfare. He constructed alternative welfare indicators, such as the UN's Human Development Index, in this spirit.

Paralleling his approach to welfare measurement, Sen maintains that individuals' capabilities constitute the principal dimension in which societies should strive for greater equality—though he stops short of saying which capabilities and what degree of equality. He admits, however, that one problem with this definition of justice is that individuals make decisions—whether to work hard or to complete an education—that determine their capabilities at a later stage.

How does postcolonial India measure up in Sen's view? His book with Jean Drèze, *India: Development and Participation,* begins by quoting Nehru's stirring speech at the hour of independence: "Long years ago we made a tryst with destiny, and now the time comes when we shall redeem our pledge." Nehru pledged among other things, "the ending of poverty and ignorance and disease and inequality of opportunity." [15] For Sen, "the ambitious goals . . . remain largely unaccomplished." A student once asked Sen why he hadn't changed the "content" of his thoughts since the 1950s. "Because," said Sen, "the surrounding environment hasn't changed. I'll probably die saying the same things."

To be sure, he points out, much has changed in the third world. Life expectancy has expanded from forty-six to sixty-five, and real per capita income has more than tripled. Many once-poor countries now have more in common with rich ones than with the ones they left behind.

Yet, Sen says, the 1 billion citizens of the world's biggest democracy are still among the world's most deprived. Extreme deprivation, he points out, is now concentrated in just two regions of the world: South Asia and sub-Saharan Africa. Life expectancy is higher in India than in Africa because India has escaped large-scale famine and avoided civil war. But in terms of illiteracy, chronic malnutrition, and economic and social inequality, India does as badly or worse than sub-Saharan Africa, especially with respect to the condition of women.

India and China were comparably poor in the 1940s. Today, however, China's life expectancy is seventy-three versus India's sixty-four. Infant mortality is less than half of India's, seventeen deaths per one thousand births versus fifty. Nutritional yardsticks also show that China is far ahead in eliminating chronic malnourishment. Literacy rates for adolescents are well over 90 percent in China—with no gap between girls and boys—versus much lower, and far more divergent, rates in India. [16] Of course, India's citizens enjoy democratic rights—including a free press—that China's more prosperous citizens can still only dream of. The challenge for Sen and other economists advising India is how to nudge its economy along China's path of globalization without sacrificing the democracy of which Sen and India are so proud.

Robert Solow, who won a Nobel for his theory of economic growth, once called Sen the "conscience of our profession." For a long time, however, Sen's approach to economics was decidedly suspect on both left and right. At Cambridge, Calcutta, and Delhi in the 1950s and 1960s, when Soviet-style planning was in vogue, Sen was persona non grata on the Left. In the 1980s and 1990s, when free markets were once again the rage, the then chairman of the Nobel Prize committee confidently predicted, "Sen will never get the prize." Sen won the Nobel in 1998 "for his contributions to welfare economics."

But times have changed. These days, when he travels to Asia, Sen is treated more like Gandhi than like a professor of economics as he travels about with police escorts. In Santiniketan in January 2002, crowds lined the streets to watch him come and go, and young girls at Visva-Bharati dropped to the ground to touch his feet (something he brusquely discourages). Determined, like the poet who named him, to use his Nobel Prize to draw attention to issues he cares about, he has donated half of his $1 million winnings to establish two foundations, one in West Bengal, the other in Bangladesh, to promote the spread of elementary education in rural areas.

As India's Soviet-style, autarkic, and bureaucratic economy became increasingly dysfunctional, while Japan and the Asian tigers achieved modern standards of living, Sen moved away from the view that Western aid and better terms of trade were the keys to third world growth and closer to the Schumpeterian perspective that local conditions are decisive and that nations do, ultimately, control their own destinies. He embraced deregulation and opening the Indian economy to foreign trade and investment while insisting on government intervention on behalf of the poor, especially in health, education, and nutrition. The argument ended when Mao suspended the Cultural Revolution and Deng Xiaoping introduced economic freedoms. China's remarkable leap into modernity left the Soviet Union in the dust and fatally discredited the Soviet economic model.

Epilogue

Imagining the Future

Most journeys start in the imagination. The grand pursuit to make mankind the master of its circumstances is no exception.

The eighteenth-century founders of economics had a vision of economic organization in which voluntary cooperation would replace coercion. But they assumed that nine out of ten human beings were sentenced by God or nature to lives of grinding poverty and toil. Two thousand years of history convinced them that the bulk of humanity had as much chance of escaping its fate as prisoners of a penal colony surrounded by a vast ocean had of escaping theirs.

Dickens, Mayhew, and Marshall came to economics in Victorian London during a revolution in productivity and living standards. They were animated by a brighter, more hopeful vision. To them, the ocean looked more like a moat. They could imagine humanity on the far side, advancing a step at a time toward an ever-receding horizon. These economic thinkers were driven not only by intellectual curiosity and a hunger for theory but also by the desire to put mankind in the saddle. They were searching for instruments of mastery: ideas that could be used to foster societies characterized by individual freedom and abundance instead of moral and material collapse.

Economic intelligence, they learned, was far more critical to success than territory, population, natural resources, or even technological leadership. Ideas matter. Indeed, as Keynes famously put it during the Great Depression, "the world is ruled by little else."[1] Like Marshall, Keynes thought of economics as an engine of analysis that could separate the wheat of experience from the chaff, and he was convinced that economic ideas had done more to transform the world than the steam engine. Economic truths might be less permanent than mathematical truths, but economic theory was essential for learning what worked, what didn't, what mattered, what did not. Inflation could lift output in the short run but not the long run. Gains in productivity are the primary driver of wages and living standards. Education and a safety net could reduce poverty without producing economic stagnation. A stable currency was necessary for economic stability, a healthy financial system is essential for innovation. As Robert Solow observed, "The questions keep changing and the answers to even old questions keep changing as society evolves. That doesn't mean we don't know quite a bit that is useful, at any given moment."[2]

Economic calamities—financial panics, hyperinflations, depressions, social conflicts and wars—have always triggered crises of confidence, but they have not come close to wiping out the cumulative gains in average living standards. The Great Depression put economics as well as the modern decentralized economy on trial. World War II ended on a note of gloom and self-doubt with Keynesian economists anticipating a twilight age of stagnation and disciples of Hayek fearing the triumph of socialism in the West. Instead, growth rebounded and living standards shot up. Governments achieved some success in managing their economies. Since World War II, history has been dominated by the escape of more and more of the world's population from abject poverty. Defeated Germany and Japan rose phoenixlike from the ashes in the 1950s and 1960s. China launched its remarkable growth spurt around 1970. More recently, India has emerged from decades of stagnation.

Reality has mostly outstripped imagination. Even Schumpeter could not have imagined that the world's population would be six times greater but ten times more affluent. Or that the fraction of the earth's citizens who lived in abject poverty would dwindle by five-sixths. Or that the average Chinese lives at least as well today, if not better, than the average English-man did in 1950. Only Fisher would not have been surprised to learn that the average lifespan has risen to two and one-half times that of 1820 and continues to edge higher. Remarkably, even the Great Recession of 2008 to 2009, the most severe economic crisis since the 1930s, did not reverse the prior gains in productivity and income. Life expectancy kept going up. The world financial system did not collapse. There was no second great depression.

Madmen in authority from Hitler to Stalin and Mao have repeatedly tried—still try—to ignore or even suppress economic truths. But the more nations escape poverty and make their own economic destinies, the less compelling the rationalizations of dictators become. Rather than overtaking the West, the Soviet Union collapsed in 1990.

There is no going back. Nobody debates any longer whether we should or shouldn't control our economic circumstances, only how. Asked about their fondest hope for the future, protestors in Cairo named economic improvement. The men and women on the streets of Tunisia, Syria, and other Middle Eastern nations in 2011 represent the latest wave of citizens to imagine an economic future characterized by growth, stability, and a business climate favorable to entrepreneurship. Once such a future can be imagined, returning to the nightmare of the past seems increasingly impossible.

Acknowledgments

I've accumulated a staggering number of debts while researching and writing this book.

The largest are owed to three individuals without whom *Grand Pursuit* would never have been started, sustained, or completed: My editor, Alice Mayhew, who showed me, patiently and with extraordinary dedication, how to turn economics, history, and biography into a story; my agent, Kathy Robbins, who launched the whole enterprise with her customary elan; and my eldest daughter, Clara O'Brien, who helped bring the project to its conclusion.

Many individuals and institutions generously supported my research. Topping the list are Amartya Sen, Emma Rothschild, Eric Maskin, Philip Griffiths, Alan Krueger, Orley Aschenfelter, and Eric Wanner. I am grateful to the Institute for Advanced Study, Russell Sage Foundation, Churchill and Kings Colleges at the University of Cambridge, the Yaddo Foundation, and the MacDowell Colony for stimulating and productive visits.

At Columbia, I got some of my best ideas from the extraordinary Bruce C. N. Greenwald. My journalism colleague Jim Stewart was a constant source of support and sage advice. And I can't thank my teaching partner, Ed McKelvey, enough for giving his all to our students for the past two years, to my benefit as well as theirs.

I had the incredible good fortune to work with a remarkable team at Simon & Schuster. Special thanks to Jonathan Karp, Richard Rhorer, Roger Labrie, Rachel Bergmann, Irene Kheradi, Gina DiMascia, John Wahler,

Nancy Inglis, Jackie Seow, Ruth Lee-Mui, Tracey Guest, Danielle Lynn, Rachelle Andujar, and the imperturbable Phil Metcalf.

For granting me interviews and sources, I am grateful to William Barber; Peter Singer; Harold James; Bruce Caldwell; Meghnad Desai; Marina Whitman; Peter Dougherty; Geoffrey Harcourt; Prue Kerr; Frances Stewart; Francis Cairncross; Barbara Jeffrey; Dutta Jayasri; Avinash Dixit; Lawrence Hayek; Luigi Pasinetti; Bill Gibson; Laurie Kahn-Leavitt; Jim Mirlees; Hans Jörg Hennecke; Hans Jörg Klausinger; Nils Eric-Sahlin; Geoffrey Heal; the family of Margaret Paul; Harold Kuhn; Hugh Mellor; Peter Passell; Edmund Phelps; Jagdish Bhagwati; Andrew Scull; Ruth and Carl Kaysen; Peter Boettke; Guido Hulsmann; William Barnett; Vernon Smith; Peter Temin; Elizabeth Darling; Robert Skidelsky; Andrew Scull; John Whitaker; Ray Monk; Amartya Sen; Paul Samuelson, his wife, Risha, and longtime assistant, Janice Murray; Robert and Anita Summers; Robert and Bobbie Solow; Milton and Rose Friedman; and Kenneth Arrow.

Ruth Tenenbaum waged a ruthless but always good-natured campaign against errors and omissions. Alexandra Saunders, Louise Story, Jonathan Hull, Barry Harbaugh, Melanie Hollands, Rachel Elbaum, Catherine Viette, and Tori Finkle provided helpful research assistance at various points. I am especially thankful to Bill Gibson for spotting logical and other lapses in the final galleys.

Most of the research for this book was done in archives and libraries, and I would especially like to thank the staffs of the following for their kindness and expert guidance: Marshall Library, University of Cambridge, Trinity College Archive, Kings College Archive, City of Cambridge Archive, Harvard University Archive, London School of Economics Archive, MIT Archive, Hoover Institution Archive. My gratitude extends, naturally, to the creators of Google Books, J-Stor, Lexis-Nexis, the Marx-Engels Archive, and numerous online archives and libraries that have revolutionized historical research.

The last word is, as always, for my children, Clara, Lily, and Jack, and my dear friends. They know that it's all about the journey . . . and who travels with you.

Notes

NOTES ON SOURCES

In researching *Grand Pursuit* I consulted and read hundreds of inspiring and informative works of biography, history, and economics. But the following are the books on which I especially relied for facts, ideas, and understanding:

Preface: Claire Tomalin, *Jane Austen: A Life*, (New York: Knopf, 1997); Gregory Clark, *A Farewell to Alms: A Brief Economic History of Modern Britain* (Princeton: Princeton University Press, 2009); Bradford DeLong, unpublished economic history of the twentieth century; Harold Perkin, *The Origins of Modern British Society* (London: Routledge, 1990); Angus Maddison, *The World Economy: A Millennial Perspective* (Paris: OECD Publishing, 2006) and *The World Economy: Historical Statistics* (Paris: OECD Publishing, 2006); Mark Blaug, *Economic Theory in Retrospect* (Cambridge: Cambridge University Press, 1983); T. W. Hutchison, *A Review of Economic Doctrines 1870–1939* (London: Clarendon Press, 1966); W. W. Rostow, *Theorists of Economic Growth from David Hume to the Present* (Oxford: Oxford University Press, 1992); Niall Ferguson, *Cash Nexus* (New York: Basic Books, 2001).

Act I Prologue: Kitson Clark, "Hunger and Politics in 1842" (*Journal of Modern History*, 24, no. 4 (December, 1953); James P. Henderson, " 'Political Economy Is a Mere Skeleton Unless . . .': What Can Social Economists Learn from Charles Dickens" (*Review of Social Economy*, 58, no. 2 (June, 2000); Michael Slater, *Charles Dickens* (New Haven: Yale University Press, 2009).

Chapter I: David McLellan, *Karl Marx: Interviews and Recollections* (New York: Barnes & Noble, 1981); Gustav Mayer, *Friedrich Engels: A Biography* (Berlin: H. Fertig, 1969); Steve Marcus, *Engels, Manchester and the Working Class* (New York: Norton, 1974); Gertrude Himmelfarb, *The Idea of Poverty: England in the Early Industrial Age* (New York: Alfred A. Knopf, 1984) and *Poverty and Compas-*

sion: The Moral Imagination of the Late Victorians (New York: Random House, 1991); David McLellan, *Karl Marx: His Life and Thought* (London: Macmillan, 1973); Isaiah Berlin, *Karl Marx: His Life and Environment* (London: Thornton Butterworth, 1939); Francis Wheen, *Karl Marx: A Life* (New York: W. W. Norton & Co., 1999); Dirk Struik, *Birth of the Communist Manifesto* (New York: International Publishers, 1986); Anne Humphereys, *Travels into the Poor Man's Country: The Work of Henry Mayhew* (Athens: University of Georgia Press, 1977); Francis Sheppard, *London 1808-1870: The Infernal Wen* (London: Seeker and Warburg, 1971); Asa Briggs, *Victorian Cities* (Berkeley: University of California Press, 1993); Gareth Stedman Jones, *Outcast London* (London: Penguin Books, 1982).

Chapter II: Mary Paley Marshall, *What I Remember* (Cambridge: Cambridge University Press, 1947); J. M. Keynes, "Alfred Marshall 1842-1924," in Arthur Pigou, ed., *Memorials of Alfred Marshall* (London: MacMillan, 1925); Gertrude Himmelfarb, *Poverty and Compassion: The Moral Imagination of the Late Victorians* (New York: Alfred A. Knopf, 1991); Peter Groenewegen, *A Soaring Eagle: Alfred Marshall 1842-1924* (London: E. Elgar, 1995); John Whitaker, *Early Economic Writings of Alfred Marshall, Vols. 1-2* (London: The Royal Economic Society, 1975); John Whitaker, *The Correspondence of Alfred Marshall, Vols. 1-3* (Cambridge: Cambridge University Press, 1996); Tizziano Raffaeli, Eugenio F. Biagini, Rita McWilliams Tullberg, eds., *Alfred Marshall's Lectures to Women: Some Economic Questions Directly Connected to the Welfare of the Laborer* (Aldershott, UK: Edward Elgar Publishing Company, 1995).

Chapter III: Barbara Caine, *Destined to Be Wives: The Sisters of Beatrice Webb* (Oxford: Clarendon Press, 1986); Carole Seymour Jones, *Beatrice Webb: Woman of Conflict* (Chicago: Ivan R. Dee, 1992); Royden Harrison, *The Life and Times of Sidney and Beatrice Webb: The Formative Years, 1858-1903* (London: Palgrave, 1999); Kitty Muggeridge and Ruth Adam, *Beatrice Webb: A Life, 1858-1943* (New York: Alfred A. Knopf, 1968); Margaret Cole, *Beatrice Webb* (New York: Harcourt Brace, 1946); Michael Holroyd, *Bernard Shaw* (London: Chatto and Windus, 1997); William Manchester, *The Last Lion: Winston Spencer Churchill: Visions of Glory, 1874-1932* (New York: Little Brown, 1983); Gertrude Himmelfarb, *Poverty and Compassion: The Moral Imagination of the Late Victorians* (New York: Random House, 1991); Elie Halevy, *A History of the English People in the Nineteenth Century, Vol. 6, The Rule of Democracy (1905-1914)*, (London: Ernest Benn Ltd., 1952); Jeanne and Norman MacKenzie, *The Diary of Beatrice Webb*, vols. 1-4 (London: Virago, 1984); Norman MacKenzie, *The Letters of Sidney and Beatrice Webb*, vols. 1-3 (Cambridge: Cambridge University Press, 2008).

Chapter IV: Muriel Rukeyser, *Willard Gibbs* (New York: Doubleday, Doran and Co., 1942), William J. Barber, ed., *The Works of Irving Fisher*, vols. 1-17 (London: Pickering and Chatto, 1997); Irving Norton Fisher, *My Father: Irving Fisher* (New York: Comet Press, 1956); Muriel Rukeyser, *Willard Gibbs: American Genius*

(New York: Doubleday, Doran and Co., 1942); Robert Loring Allen, *Irving Fisher: A Biography* (Cambridge: Blackwell Publishers, 1993); Richard Hofstadter, *The Age of Reform: From Bryan to FDR and Social Darwinism in American Thought* (New York: George Braziller, Inc., 1969); Jeremy Atack and Peter Passell, *A New Economic View of American History* (New York: W. W. Norton, 1994); Perry Mehrling, "Love and Death: The Wealth of Irving Fisher," *Research in the History of Economic Thought and Methodology*, Warren J. Samuels and Jeff E. Biddle, eds. (Amsterdam: Elsevier Science, 2001; New York: Harcourt Brace Jovanovich, 1992), 47–61.

Chapter V: Seymour Harris, *Joseph Schumpeter: Social Scientist* (Cambridge, Mass.: Harvard University Press, 1951); Wolfgang F. Stolper, *Joseph Alois Schumpeter: The Public Life of a Private Man* (Princeton: Princeton University Press, 1994); Robert Loring Allen, *Opening Doors: The Life and Works of Joseph Schumpeter*, vol. I (New Brunswick: Transaction Publishers, 1991); Richard Swedberg, *Joseph A. Schumpeter: His Life and Work* (Cambridge, UK: Polity Press, 1991); Thomas K. McCraw, *Prophet of Innovation: Joseph Schumpeter and Creative Destruction* (Cambridge, Mass.: Harvard University Press, 2007); Charles A. Gulik, *Austria from Habsburg to Hitler*, vol. I (Berkeley: University of California Press, 1948); David F. Good, *The Economic Rise of the Hapsburg Empire 1750–1914* (Berkeley: University of California Press, 1990); Joseph Schumpeter, *History of Economic Analysis* (Cambridge, Mass.: Harvard University Press, 1954).

Act II Prologue: Charles John Holmes, *Self and Partners (Mostly Self): Being the Reminiscences of C. J. Holmes* (London: Macmillan, 1936); Anne Emberton, "Keynes and the Degas Sale," *History Today*, December 31, 1995; Ray Monk, *Ludwig Wittgenstein: The Duty of Genius* (New York: Penguin Books, 1991); Ray Monk, *Bertrand Russell: The Spirit of Solitude 1872–1921*, vol. I (New York: Simon & Schuster, 1996); Hugh Mellor, *Frank Ramsey: Better Than the Stars* (London: BBC, 1994); Henry Andrews Cotton, with a Foreword by Adolf Meyer, *The Defective, Delinquent and Insane: The Relation of Focal Infections to Their Causation, Treatment and Prevention, Lectures delivered at Princeton University, January 11, 13, 14, 15, 1921* (Princeton: Princeton University Press, 1922).

Chapter VI: Eduard Marz, *Joseph A. Schumpeter: Forscher, Lehrer und Politiker*, Munchen: R. Oldenbourg, 1983); Eduard Marz, "Joseph Schumpeter as Minister of Finance in X Helmut Frisch, in *Schumpeterian Economics* (New York: Praeger, 1981); F. L. Carsten, *The First Austrian Republic* (Aldershot, UK: Wildwood House, 1986); F. L. Carsten, *Revolution in Central Europe: 1918–1919*, Aldershot, UK: Wildwood House, 1988); David Fales Strong, *Austria (October 1918–March 1919)* (New York: CUP, 1939); Norbert Schausberger, *Der Griff nach Oesterreich: Der Anschluss* (Wien, Muenchen: Jugend und Volk, 1988); Otto Bauer, *The Austrian Revolution*, (London: Parsons, 1925); Eduard Marz, *Austrian Banking and Financial Policy: Creditanstalt at a Turning Point, 1913–1923* (New York: St. Martin's Press, 1984); Christine Klusacek and Kurt Stimmer, *Dokumentation zur Oesterreichische Zeitgeschichte 1918–1928* (Wien und Muenchen: Jugend und Volk, 1984); Joseph A. Schumpeter, *Aufsatze zur Wirtschaftspolitik*, Wolfgang F. Stolper and Christian Seidl, eds. (Tuebingen: JCB Mohr, 1985); Joseph A. Schumpeter, *Politische Reden*, Seidl and Stolper, eds. (Tubingen: JCB Mohr, 1992).

Chapter VII: D. E. Moggridge, ed., *The Collected Writings of John Maynard Keynes*, vols. 1–30 (London: Macmillan, 1971–1989); Paul Mantoux, *The Carthaginian Peace or The Economic Consequences of Mr. Keynes* (Oxford: Oxford University Press, 1946); Robert Skidelsky, *John Maynard Keynes, Vol. 1, Hopes Betrayed* (New York: Viking, 1986); Donald E. Moggridge, *Maynard Keynes: An Economist's Biography* (London: Routledge, 2009); Margaret MacMillan, *Paris 1919: Six Months That Changed the World* (New York: Random House, 2002).

Chapter VIII: Peter Gay, *Freud: A Life of Our Time* (New York: W. W. Norton, 1988); F. L. Carsten, *The First Austrian Republic* (Aldershot: Wildwood House, 1986); Otto Bauer, *The Austrian Revolution* (London: Parsons, 1925); Eduard Marz, *Austrian Banking and Financial Policy: Creditanstalt at a Turning Point, 1913–1923* (New York: St. Martin's Press, 1984).

Chapter IX: Robert Skidelsky, *John Maynard Keynes, Vol. 2: The Economist as Savior 1920–1937* (London: Macmillan, 1992); D. E. Moggridge, *Maynard Keynes: An Economist's Biography* (London, Routledge, 1992); Irving Norton Fisher, *My Father: Irving Fisher* (New York: Comet Press, 1956), Robert Loring Allen, *Irving Fisher: A Biography* (Cambridge: Blackwell Publishers, 1993); Milton Friedman, *Money Mischief: Episodes in Monetary History* (New York: Harcourt Jovanovich Brace, 1992).

Chapter X: Robert Skidelsky, *John Maynard Keynes, Vol. II: The Economist as Savior 1920–1937* (London: Macmillan, 1992); D. E. Moggridge, *Maynard Keynes: An Economist's Biography* (London, Routledge, 1992); Irving Norton Fisher, *My Father: Irving Fisher* (New York: Comet Press, 1956), Robert Loring Allen, *Irving Fisher: A Biography* (Cambridge: Blackwell Publishers, 1993); Milton Friedman, *Money Mischief: Episodes in Monetary History* (New York: Harcourt Brace Jovanovich, 1992).

Chapter XI: Nahid Aslanbeigui and Guy Oakes, *The Provocative Joan Robinson: The Making of a Cambridge Economist* (Durham, N.C.: Duke University Press, 2009); Marjorie Shepherd Turner, *Joan Robinson and the Americans* (Armonk, N.Y.: ME Sharpe, 1989).

Chapter XII: Robert Skidelsky, *John Maynard Keynes, Vol. 3: Fighting for Freedom, 1937–1946* (New York: Viking, 2001); David Kennedy, *Freedom from Fear: The American People and in Depression and War* (Oxford: Oxford University Press, 1999); Milton Friedman and Rose Friedman, *Two Lucky People* (Chicago: University of Chicago Press, 1998), Herbert Stein, *Presidential Economics: The Making of Economic Policy from Roosevelt to Clinton* (Washington, D.C.: American Enterprise Institute, 1994); Stephen Kresge and W. W. Bartley III, eds., *The Collected Works of F. A. Hayek*, vols. 1–17 (Chicago: University of Chicago Press, 1989).

Chapter XIII: Seymour Harris, *Joseph Schumpeter: Social Scientist* (Cambridge, Mass.: Harvard University Press, 1951); Wolfgang F. Stolper, *Joseph Alois Schumpeter: The Public Life of a Private Man* (Princeton: Princeton University Press, 1994); Robert Loring Allen, *Opening Doors: The Life and Works of Joseph Schumpeter*, vol. I (New Brunswick: Transaction Publishers, 1991); Richard Swedberg, *Joseph A. Schumpeter: His Life and Work* (Cambridge: Polity Press, 1991); Thomas K. McCraw, *Prophet of Innovation: Joseph Schumpeter and Creative Destruction* (Cambridge, Mass.: Harvard University Press, 2007).

Act III Prologue: James McGregor Burns, *Roosevelt: The Soldier of Freedom, 1940–1945* (New York: Harcourt Brace Jovanovich, 1970).

Chapter XIV: Robert Skidelsky, *John Maynard Keynes, Vol. 3: Fighting for Freedom* (New York, Viking, 2000).

Chapter XV: Alan Ebenstein, *Hayek's Journey* (Chicago: University of Chicago Press, 2005); Hans Jorg Hennecke, *Friedrich von Hayek* (Hamburg: Junius Verlag GmbH, 2010); Werner Erhard, *Germany's Comeback in the World Market* (New York: Macmillan, 1954).

Chapter XVI: Richard Reeves, *President Kennedy* (New York: Simon & Schuster, 1993); Herbert Stein, *Presidential Economics: The Making of Economic Policy from Roosevelt to Clinton* (Washington, D.C: American Enterprise Institute, 1994).

Chapter XVII: John Lewis Gaddis, *The Cold War: A New History* (New York: Alfred A. Knopf, 2009); Marjorie Shepherd Turner, *Joan Robinson and the Americans* (Armonk, N.Y.: ME Sharpe, 1989).

Chapter XVIII: Amartya Sen, *Development as Freedom* (New York: Alfred A. Knopf, 1999); Amartya Sen, *The Idea of Justice* (Cambridge, Mass.: Harvard University Press, 2009).

PREFACE: THE NINE PARTS OF MANKIND

1. John Kenneth Galbraith, *The Affluent Society* (Boston: Houghton Mifflin, 1958).
2. Edmund Burke, "A Vindication of Natural Society Or, a View of the Miseries and Evil Arising to Mankind from Every Species of Artificial Society, In a Letter to Lord **** by a Late Noble Writer, 1756," *Writings and Speeches* (New York: Little Brown and Co., 1901), 59.
3. Patrick Colquhoun, *A Treatise on the Wealth, Power, and Resources of the British Empire* (London: Jay Mawman, 1814(1812)), 49.

4. James Heldman, "How Wealthy is Mr. Darcy—Really? Pound and Dollars in the World of *Pride and Prejudice*," *Persuasions* (Jane Austen Society), 38–39.

5. Author's calculation based on data from Colquhoun, *Wealth, Power, and Resources*; Harold Perkin, *The Origins of Modern British Society* (London: Routledge), 20–21; and Roderick Floud and Paul Johnson, *Cambridge Economic History of Modern Britain* (Cambridge: Cambridge University Press, 2004), 92.

6. Jane Austen to Cassandra Austen, *Jane Austen's Letters*, Deirdre le Fay, ed. (Oxford: Oxford University Press, 1995) and Anonymous, *How to Keep House! Or Comfort and Elegance on 150 to 200 a Year* (London: James Bollaert, 1835, 14th edition).

7. Claire Tomalin, *Jane Austen, A Life* (New York: Knopf, 1997).

8. Burke, *Vindication*, 59.

9. Gregory Clark, *A Farewell to Alms: A Brief Economic History of the World* (Princeton: Princeton University Press, 2009).

10. James Edward Austen Leigh, *A Memoir of Jane Austen* (London: Richard Bentley & Son, 1871), 13.

11. Clark, *A Farewell to Alms*.

12. Robert Giffen, *Notes on the Progress of the Working Classes (1883) and Further Notes on the Progress of the Working Classes, Essays in Finance* (London: Putnam & Sons, 1886), 419.

13. Burke, *Vindication*, 60.

14. Tomalin, *Jane Austen*, 96.

15. Patrick Colquhoun, *A Treatise on Indigence* (London: J. Hatchard, 1806).

16. Leigh, *A Memoir of Jane Austen*, 13.

17. Giffen, 379.

18. Alfred Marshall, *The Present Position of Economics: An Inaugural Lecture* (1885), 57.

19. John Maynard Keynes, "Economic Possibilities for our Grandchildren," *Essays in Persuasion* (London: Macmillan, 1931), 344.

20. John Maynard Keynes, Toast on the occasion of his retirement from the editorship of *The Economic Journal*, 1945, quoted in Roy Harrod, *The Life of John Maynard Keynes* (London: Harcourt Brace, 1951), 193–94.

ACT I: PROLOGUE: MR. SENTIMENT VERSUS SCROOGE

1. G. Kitson Clark, "Hunger and Politics in 1842," *Journal of Modern History*, 24, no. 4 (December 1953), 355–74.

2. Thomas Carlyle, *Past and Present* (London: Chapman and Hall, 1843), 26.

3. Charles Dickens, *Daily News* (London), January 21, 1846.

4. Asa Briggs, ed., *Chartist Studies* (London: Macmillan, 1959).

5. Carlyle, *Past and Present*, 335.

6. Thomas Carlyle to John A. Carlyle, Chelsea, London, March 17, 1840. The Carlyle Letters Online, 2007, http://carlyleletters.org (accessed January 2, 2011).

7. John Stuart Mill to John Robertson, London, July 12, 1837 in *The Earlier Letters of John Stuart Mill*, vol. 1, *1812–1848*, ed. Francis E. Mineka (University of Toronto Press, 1963), 343 (paraphrasing Carlyle's description of Camille Desmoulins in *The French Revolution: A History*, [1837]).

8. Quoted in Michael Slater, *Charles Dickens: A Life Defined by Writing* (New Haven, Conn.: Yale University Press, 2009), 143.

9. Thomas Carlyle, "Occasional Discourse on the Negro Question," *Fraser's Magazine for Town and Country* 40 (February 1849), 672.

10. Edmund Burke, *A Vindication of Natural Society: or, a View of the Miseries and Evils Arising to Mankind from Every Species of Artificial Society* (1756), Frank N. Pagano, ed. (Indianapolis: Liberty Fund, Inc., 1982), 87.

11. Thomas Robert Malthus, *An Essay on the Principle of Population, as It Affects the Future Improvement of Society with Remarks on the Speculations of Mr. Godwin, M. Condorcet, and Other Writers* (London: J. Johnson, 1798), 30.

12. Ibid., 139.

13. Ibid., 31.

14. Leviticus 19:18, Romans 13:9.

15. Charles Dickens, *Oliver Twist*, vol. 1 (London: Richard Bentley, 1838), 25.

16. Nicholas Bakalar, "In Reality, Oliver's Diet Wasn't Truly Dickensian," *New York Times*, December 29, 2008.

17. Charles Dickens, *American Notes for General Circulation*, vol. 2 (London: Chapman and Hall, 1842), 304.

18. Charles Dickens to Dr. Southwood Smith, March 10, 1843, in *The Letters of Charles Dickens*, vol. 3, *1842–1843*, eds. Madeline House, Graham Storey, Kathleen Mary Tillotson, Angus Eanon, Nina Burgis (Oxford: Oxford University Press, 2002), 461.

19. James P. Henderson, "'Political Economy is a Mere Skeleton Unless . . .': What Can Social Economists Learn from Charles Dickens?," *Review of Social Economy*, 58, no. 2 (June 2000): 141–51.

20. Charles Dickens, *A Christmas Carol; in Prose: Being a Ghost Story of Christmas* (London: Chapman Hall, 1843).

21. Henderson, "Political Economy," 146.

22. Dickens, *A Christmas Carol*, 96.

23. Thomas Malthus, *An Essay on the Principle of Population: Or, a View of Its Past and Present Effects on Human Happiness: With an Inquiry Into Our Prospects Respecting the Future Removal or Mitigation of the Evils Which It Occasions*, 2nd ed. (London: J. Johnson, 1803), 532.

24. Dickens, *A Christmas Carol*, 94.

25. Michael Slater, introduction and notes to Charles Dickens, *A Christmas Carol and Other Christmas Writings* (London: Penguin, 2003), xi.

26. Anthony Trollope, *The Warden* (London: Longman, Brown, Green, and Longmans, 1855), chap. 15.

27. Charles Dickens, "The Bemoaned Past," *All the Year Round: A Weekly Journal, With Which is Incorporated Household Words*, no. 161 (May 24, 1862).

28. Sir Robert Peel to Sir James Graham, August 1842, quoted in Clark, "Hunger and Politics in 1842."

29. Charles Dickens, "On Strike," *Household Words; A Weekly Journal* no. 203 (February 11, 1854).

30. Ibid.

31. Joseph A. Schumpeter, *The Economics and Sociology of Capitalism,* ed. Richard Swedberg (Princeton: Princeton University Press, 1991), 290. Schumpeter coined this phrase to describe Alfred Marshall's view that economics "is not a body of concrete truth, but an engine for the discovery of concrete truth." Alfred Marshall, *The Present Position of Economics: An Inaugural Lecture* (London: Macmillan and Co., 1885), 25.

32. John Maynard Keynes, introduction to *Cambridge Economic Handbooks,* I (London: Nesbit and Co. and Cambridge: Cambridge University Press, 1921).

I: PERFECTLY NEW: ENGELS AND MARX IN THE AGE OF MIRACLES

1. Walter Bagehot, *Lombard Street: A Description of the Money Market* (New York: Scribner, Armstrong & Co., 1873), 20.

2. Friedrich Engels to Karl Marx, November 19, 1844, Marxists Internet Archive, www.marxists.org/archive/marx/works/1844/letters/44_11_19.htm.

3. Ibid.

4. Friedrich Engels to Arnold Ruge, June 15, 1844, quoted in Steven Marcus, *Engels, Manchester and the Working Class* (New York: Random House, 1976), 82.

5. Friedrich Engels, writing as "X," four-part series on political and economic conditions in England, *Rheinische Zeitung,* December 8, 9, 10, and 25, 1842.

6. Edwin Chadwick, *Report on the Sanitary Condition of the Labouring Population of Great Britain* (1842).

7. Friedrich Engels, *Rheinische Zeitung,* December 8, 1842.

8. Charles Dickens, *Nicholas Nickleby,* chap. 43.

9. Friedrich Engels, *The Condition of the Working Class in England in 1844, With a Preface Written in 1892,* trans. Florence Kelley Wischnewetzky (London: Swan Sonnenschein & Co., 1892).

10. Quoted in David McLellan, *Friedrich Engels* (New York: The Viking Press, 1977), 22.

11. Friedrich Engels, "Outlines of a Critique of Political Economy," *Deutsch-Französiche Jahrbücher* 1, no. 1 (February 1844).

12. Karl Marx, preface to *A Contribution to the Critique of Political Economy* (1859) in Karl Marx and Friedrich Engels, *Selected Works* (Moscow: Foreign Languages Publishing House, 1951).

13. Karl Heinzen, *Erlebtes* [Experiences], vol. 2 (Boston: 1864), 423–24.

14. Isaiah Berlin, *Karl Marx: His Life and Environment,* London: Thompson Butterworth, 1939), 26.

15. George Bernard Shaw, "The Webbs," in Sidney and Beatrice Webb, *The Truth About Soviet Russia* London: Longmans Green, (1942).

16. Arnold Ruge to Ludwig Feuerbach, May 15, 1844, in Arnold Ruge, *Briefwechsel und Tagebuchblatter aus den Jahren 1825–1880* [Correspondence and Diaries from the Years 1825–1880], vol. 1 (Berlin: Weidmannsche Buchhandlung, 1886), 342–49.

17. Karl Marx to Arnold Ruge, July 9, 1842, in ed., *Marx/Engels Collected Works,* vol. 1, 398–91.

18. Karl Marx, "A Contribution to the Critique of Hegel's Philosophy of Right," *Deutsch-Französische Jahrbücher* 1, no. 1 (February 1844).

19. Karl Marx to Arnold Ruge, September 1843; *Deutsch-Französische Jahrbücher* 1, no. 1 (1844), www.marxists.org/archive/marx/works/1843/letters/43_09-alt. htm.

20. Gertrude Himmelfarb, *The Idea of Poverty: England in the Early Industrial Age* (New York: Alfred A. Knopf, 1984), 278.

21. Friedrich Engels to Karl Marx, November 19, 1844, in *Der Briefwechsel Zwischen F. Engels und K. Marx,* vol. 1 (Stuttgart, 1913), Marxist Internet Archive, www.marxists.org/archive/marx/works/1844/letters/44_11_19.htm.

22. Friedrich Engels to Karl Marx, January 20, 1845, in *Der Briefwechsel Zwischen F. Engels und K. Marx,* vol. 1 (Stuttgart, 1913), Marxist Internet Archive, www.marxists.org/archive/marx/works/1845/letters/45_01_20.htm.

23. Engels, *Condition of the Working Class in England,* 296.

24. Friedrich Engels to Karl Marx, Paris, January 20, 1845. Marxist Internet Archive, http://www.marxists.org/archive/marx/works/1845/letters/45_01_20 .htm (accessed March 15, 2011).

25. Karl Marx, Preface to *Das Kapital* (1867), Friedrich Engels, ed., trans. S. Moore and E. Aveling (New York: Charles H. Kerr & Company, 1906), 14.

26. Henry Mayhew, letter 47, *The Morning Chronicle,* April 11, 1850. *The Morning Chronicle Survey of Labour and the Poor: The Metropolitan Districts,* vol. 4 (Sussex or London: Caliban Books, 1981), 97.

27. Asa Briggs, *Victorian Cities* (Berkeley: University of California Press, 1993), 311.

28. William Lucas Sargant, "On the Vital Statistics of Birmingham and Seven Other Large Towns," *Journal of the Statistical Society of London* 29, no. 1 (March 1866): 92–111.

29. Roy Porter, *London: A Social History* (Cambridge, Mass.: Harvard University Press, 1998), 187.

30. Engels, *Condition of the Working Class in England,* 23.

31. Charles Dickens, *Dombey and Son* (London: Bradbury and Evans, 1846–1848).

32. Niall Ferguson, *The House of Rothschild*, vol. 1 (New York: Penguin Books, 2000), 401.

33. Bagehot, *Lombard Street*, 4.

34. Ferguson, *The House of Rothschild*, vol. 12, 65.

35. Peter Geoffrey Hall, *The Industries of London* (London: Hutchison, 1962), 21.

36. Francis Sheppard, *London 1808–1870: The Infernal Wen* (London: Secker and Warburg, 1971), 158–59.

37. George Dodd, *Dodd's Curiosities of Industry* (Henry Lea's Publications, 1858), 158.

38. Hall, *The Industries of London*, 6.

39. Henry Mayhew, *The Daily Chronicle*, October 19, 1849, in *The Unknown Mayhew: Selections from the* Daily Chronicle *1849–1850* (London: Penguin Books 1884), 13.

40. John Maynard Keynes, *The Economic Consequences of the Peace* (London: Macmillan, 1919), 9.

41. Henry James, *Essays in London and Elsewhere* (New York: Harper and Brothers, 1893), 19.

42. George Augustus Sala, *Twice Around the Clock; or the Hours of the Day and Night in London* (London: Richard Marsh, 1862), 157.

43. Henry Mayhew and John Binney, *The Criminal Prisons of London and Scenes of Prison Life* (London: Griffin, Bohn and Co., 1862), 28.

44. *The Economist*, May 19, 1866.

45. Harold Perkin, *The Origins of Modern English Society 1780–1880* (London: Routledge and Kegan Paul, 1969), 91. Sala, *Twice Around the Clock*, 157.

46. Mayhew and Binny, *The Criminal Prisons of London*, 28.

47. Ibid., 32.

48. Henry James, "London," *Century Illustrated Magazine*, December 1888, 228.

49. Charles Dickens, *Bleak House* (London: Chapman and Hall, 1853), 1.

50. Friedrich Engels to Karl Marx, Paris, November 23–24, 1847. Marxist Internet Archive, http://www.marxists.org/archive/marx/works/1844/letters/44_11_19.htm (accessed March 14, 2011). Friedrich Engels, "Introduction to English Edition of *The Communist Manifesto*," (1888), in Karl Marx and Friedrich Engels, *The Communist Manifesto*, Gareth Stedman Jones, ed. (London: Penguin Books, 2002).

51. David McLellan, *Karl Marx: His Life and Thought* (London: Macmillan, 1973), 169.

52. Friedrich Lessner, quoted in David McLellan, ed., *Karl Marx: Interviews and Recollections* (London: Barnes & Noble, 1981), 45.

53. *The Rules of the Communist League*, adopted by the Second Congress of the Communist League in December 1847, in Karl Marx and Friedrich Engels, *The Communist Manifesto* (London: Lawrence & Wishart, 1930).

54. Friedrich Engels, "The Book of Revelation" (1883), in *Marx and Engels on Religion* (Moscow: Foreign Languages Publishing House, 1957), 204.

55. Karl Marx, preface to *The Poverty of Philosophy* (1847), trans. H. Quelch (Chicago: Carles H. Kerr & Company, 1920).

56. Anonymous [Robert Chambers], *Vestiges of the Natural History of Creation* (London: John Churchill, 1844).

57. Marx and Engels, *Communist Manifesto,* 223.

58. Friedrich Engels, "The English Constitution," *Vorwaerts!,* no. 75 (September 1844).

59. Angus Maddison, *Statistics on World Population, GDP and Per Capita GDP, 1–2008 AD,* www.ggdc.net/maddison/.

60. Marx and Engels, *Communist Manifesto,* 224.

61. Gregory Clark, *A Farewell to Alms: A Brief Economic History of the World* (Princeton, N.J.: Princeton University Press, 2007); Roderick Floud and Bernard Harris, "Health, Height and Welfare: Britain 1700–1800," in *Health and Welfare During Industrialization,* eds. Richard H. Steckel and Roderick Floud (Chicago: University of Chicago Press, 1997), 91–126.

62. Charles H. Feinstein, "Pessimism Perpetuated: Real Wages and the Standard of Living in Britain During and After the Industrial Revolution," *Journal of Economic History* vol. 58, no. 3 (September 1998), 630.

63. Thomas Carlyle, *Past and Present* (London: Chapman and Hall, 1843), 4.

64. Arnold Toynbee, *Lectures on the Industrial Revolution of the Eighteenth Century in England* (London: Rivingtons, 1884), 84.

65. John Stuart Mill, *The Subjection of Women* (London: Longmans, Green, Reader, and Dyer, 1869), 29–30.

66. John Stuart Mill, *Principles of Political Economy,* vol. 2 (London: John W. Parker, 1848), 312.

67. Marx and Engels, *Communist Manifesto,* 233, 258.

68. McLellan, *Karl Marx,* 35.

69. Charles Dickens, "Perfidious Patmos," in *Household Words; A Weekly Journal* 7, no. 155 (March 12, 1853).

70. *Times* (London), October 26, 1849.

71. Anne Humpherys, *Travels into the Poor Man's Country: The Work of Henry Mayhew* (Athens: University of Georgia Press, 1977), 203.

72. Henry Mayhew, "A Visit to the Cholera Districts of Bermondsey," *The Morning Chronicle,* September 24, 1849.

73. E. P. Thompson and Eileen Yeo, eds., *The Unknown Mayhew* (London: The Merlin Press Ltd., 2009), 102–3.

74. Quoted in Humpherys, *Travels,* 31.

75. Charles Dickens, *Oliver Twist* (London: Richard Bentley, 1838), 252.

76. Gareth Stedman Jones, *Outcast London: A Study in the Relationship Between Classes in Victorian Society* (New York: Penguin Books, 1984).

77. Henry Mayhew, letter 11, *The Morning Chronicle,* November 23, 1849.

78. Ibid.

79. Ibid., letter 15, December 7, 1849.

80. Henry Mayhew, "Needlewomen Forced into Prostitution," letter 8, *The Morning Chronicle,* November 13, 1849.

81. Thomas Carlyle, "The Present Time," *Latter Day Pamphlets,* issue 9 (February 1, 1850).

82. Douglas Jerrold to Mary Cowden Clarke, February 1850.

83. Henry Mayhew, *London Labour and the London Poor,* no. 40, September 13, 1851.

84. John Stuart Mill, "The Claims of Labor," *Edinburgh Review,* April 1845.

85. Quoted in James Anthony Froude, *Thomas Carlyle: A History of the First Forty Years of His Life (1795–1835)* (Montana: Kessinger Publishing, 2006), 298.

86. Ibid., 282.

87. Thomas Carlyle, "Chartism," *Latter Day Pamphlets,* London, December 1839.

88. John Stuart Mill to Macvey Napier, November 9, 1844.

89. H. G. Wells, "Men Like Gods," *Hearst's International* 42, no. 6 (December 1922); David Ricardo, *On the Principles of Political Economy and Taxation* (London: John Murray, 1817).

90. Mill, *Principles of Political Economy,* vol. 3, ch. 1.

91. Thomas Carlyle, "Occasional Discourses on the Negro Question," *Fraser's Magazine,* 1849.

92. *Archiv für die Geschichte des Sozialismus und der Arbeiterbewegung* [Archive for the History of Socialism and the Workers' Movement] (1922), 56ff 10, quoted in McLellan, *Karl Marx,* 268–69.

93. Karl Marx to Joseph Weydemeyer, London, August 2, 1851, in Saul K. Padover, ed., *The Letters of Karl Marx* (Englewood Cliffs, N.J.: Prentice-Hall, 1979), 72–73.

94. John Tallis, *Tallis's History and Description of the Crystal Palace, and the Exhibition of the World's Industry in 1851* London and New York: John Tallis and Co., 1852), quoted in Jeffrey A. Auerbach, *The Great Exhibition of 1851,* (1999).

95. "The Revolutionary Movement," *Neue Rheinische Zeitung,* no. 184, January 1, 1850.

96. Ibid.

97. Karl Marx and Friedrich Engels, *Neue Rheinische Zeitung,* May–October, 1850.

98. Marx and Engels, *Communist Manifesto,* chap. 1.

99. Karl Marx to Ludwig Kugelmann, December 28, 1862.

100. Ibid.

101. Marx and Engels, *Communist Manifesto,* chap. 1.

102. Marx, *Das Kapital*, 671.
103. John Stuart Mill, *Essays on Some Unsettled Questions of Political Economy* (London, 1844), 94.
104. Mark Blaug, *Economic Theory in Retrospect* (Cambridge, UK: Cambridge University Press, 1997).
105. Marx, *Das Kapital*, 711.
106. Robert Giffen, "The Recent Rate of Material Progress in England," *Opening Address to the Economic Science and Statistics Section of the British Association* (London: George Bell and Sons, 1887), 3.
107. R. Dudley Baxter, *National Income, the United Kingdom* (London: Macmillan, 1868), B1.
108. E. J. Hobsbawm, "The Standard of Living During the Industrial Revolution: A Discussion," *Economic History Review*, New Series, vol. 16, no. 1 (1963), 119–34.
109. Charles H. Feinstein, "Pessimism Perpetuated: Real Wages and the Standard of Living in Britain During and After the Industrial Revolution," *Journal of Economic History* 58, no. 3, 625–58.
110. Gareth Stedman Jones, introduction to Marx and Engels, *Communist Manifesto*.
111. Marx, *Das Kapital*, 264–65, note 3.
112. Egon Erwin Kisch, *Karl Marx in Karlsbad* (Weimar, Germany: Aufbau Verlag, 1968); Saul Kussiel Padover, *Karl Marx: An Intimate Biography* (New York: McGraw-Hill, 1978).
113. Karl Marx to Friedrich Engels, July 22, 1859. Reviews appeared in *Das Volk*, no. 14, August 6, 1859, and no. 16, August 20, 1859.
114. Berlin, *Karl Marx*, 13.
115. Ibid.
116. Karl Marx, "The Right of Inheritance," August 2 and 3, 1869, endorsed by the General Council on August 3, 1869. Marxist Internet Archive, www.marxists.org/archive/marx/iwma/documents/1869/inheritance-report.htm.
117. Karl Marx to Eleanor Marx, quoted in McLellan, *Karl Marx*, 334.
118. Karl Marx to Ludwig Kugelmann, December 28, 1862.
119. Fyodor Dostoyevsky, *Winter Notes on Summer Impressions* (Illinois: Northwestern University Press, 1988).
120. Author's calculation.
121. *The Bankers Magazine*, vol. 26 (1886), 639; *Illustrated London News*, May 19, 1866; *Times* (London), May 12, 1866.
122. *New York Times*, May 26, 1866.
123. Sidney Pollard and Paul Robertson, *The British Shipbuilding Industry, 1870–1914* (Cambridge, Mass: Harvard University Press, 1999), 77–79.
124. Marx, *Das Kapital*, 733–34.

125. J. H. Clapham, *An Economic History of Modern Britain,* vol. 3, *Machines and National Rivalries (1887–1914) with an Epilogue (1914–1929)* (Cambridge: Cambridge University Press, 1932), 117.

126. Karl Marx to Friedrich Engels, April 6, 1866.

127. Friedrich Engels to Karl Marx, May 1, 1866.

128. Karl Marx to Friedrich Engels, July 7, 1866.

129. Marx, *Das Kapital,* 715.

130. William Gladstone, "Budget Speech of 1863, House of Commons," *Times* (London), April 16, 1863.

131. Honore de Balzac, *The Unknown Masterpiece* (1845), www.gutenberg.org/files/23060/23060-h/23060-h.htm.

132. John Maynard Keynes, *Essays in Persuasion* (W. W. Norton and Co., 1963), 300.

II: MUST THERE BE A PROLETARIAT? MARSHALL'S PATRON SAINT

1. Ralph Waldo Emerson, "Ode, Inscribed to William H. Channing," in *Poems* (London: Chapman Bros., 1847).

2. Alfred Marshall, "Speech to the Cambridge University Senate," in John K. Whitaker, ed., *The Correspondence of Alfred Marshall,* vol. 3, *Towards the Close, 1903–1924* (Cambridge: Cambridge University Press, 1996), 399.

3. *Morning Star,* quoted in Karl Marx, *Das Kapital* (1867), Modern Library edition. 734; W.D.B, "Distress in Poplar," letter to the editor, *The Times* (London), January 12, 1867; "Able-Bodied Poor Breaking Stones for Roads, Bethnel Green London," *Illustrated London News,* February 15 (or 16?); "The Distress at the East End: A Soup Kitchen in Ratcliff Highway," *Illustrated London News,* February 16, 1867; "The Distress at the East End: A Soup Kitchen in Ratcliff Highway," *Illustrated London News,* February 16, 1867.

4. Sara Horrell and Jane Humphries, "Old Questions, New Data, and Alternative Perspectives: Families' Living Standards in the Industrial Revolution," *Journal of Economic History* 52, no. 4 (December 1992): 849–80.

5. Florence Nightingale to Charles Bracebridge, January 1867, in Lynn McDonald, ed., *The Collected Works of Florence Nightingale,* vol. 6, *Florence Nightingale on Public Health Care* (Ontario: Wilfred Laurier University Press, 2002).

6. Francis Sheppard, *London: 1808–1870* (London: Secker & Warburg, 1971), 340.

7. *Times* (London) May 6, 1867.

8. Robert Giffen, "Proceedings of the Statistical Society," *Journal of the Statistical Society of London* 30, no. 4 (December 1867), 564–65.

9. Henry Fawcett, *Pauperism: Its Causes and Remedies* (London: Macmillan, 1871), 1–2.

10. Edward Denison, *A Brief Record: Being Selections from Letters and Other Writings of Edward Denison,* ed. Sir Bryan Baldwin Leighton (London: E. Barrett and Sons, 1871), 46.

11. Alfred Marshall, in John Maynard Keynes, "Alfred Marshall, 1842–1924," in Arthur Pigou, ed., *Memorials of Alfred Marshall* (London: Macmillan, 1925), 358.

12. Alfred Marshall, "Lecture Outlines," in Tiziano Raffaelli, Eugenio F. Biagini, Rita McWilliams Tullberg, eds., *Alfred Marshall's Lectures to Women: Some Economic Questions Directly Connected to the Welfare of the Laborer* (Aldershott, UK: Edward Elgar Publishing Company, 1995), 141.

13. Ronald H. Coase, "Alfred Marshall's Mother and Father," and "Alfred Marshall's Family and Ancestry," in *Essays on Economics and Economists* (Chicago: University of Chicago Press), 1994.

14. Charles Dickens, *Great Expectations* (London: Chapman and Hall, 1861).

15. *The Times* (London), October 8, 1859.

16. Anthony Trollope, *The Vicar of Bullhampton* (London: Bradbury and Evans, 1870).

17. K. Theodore Hoppen, *The Mid-Victorian Generation 1846–1886* (Oxford, UK: Clarendon Press, 1998), 40.

18. Anthony Trollope, *The Warden* (London: Longman, Brown, Green, and Longmans, 1855), 289.

19. Peter D. Groenewegen, *A Soaring Eagle: Alfred Marshall: 1842–1924* (London: E. Elgar, 1995), 51.

20. David McLellan, *Karl Marx: His Life and Thought* (New York: Harper and Row, 1974).

21. William Dudley Baxter, *National Income: The United Kingdom* (London: Macmillan, 1868), *Global Prices and Income History Website,* http://gpih.ucdavis.edu.

22. Groenewegen, *A Soaring Eagle,* 107.

23. John Maynard Keynes, "Alfred Marshall," in *Essays in Biography* (New York: W. W. Norton, 1951), 126.

24. Mary Paley Marshall, quoted in Keynes, "Alfred Marshall, 1842–1924," 37.

25. Ibid.

26. Groenewegen, *A Soaring Eagle,* 62.

27. Leslie Stephen, *Sketches from Cambridge by a Don* (London: Macmillan and Co., 1865), 37–38.

28. Alfred Marshall to James Ward, in John King Whitaker, ed., *The Correspondence of Alfred Marshall,* vol. 2, *At the Summit, 1891–1902* (Cambridge: Cambridge University Press, 1996), 441.

29. Mary Paley Marshall, quoted in Keynes, "Alfred Marshall, 1842–1924," 37.

30. Alfred Marshall, "Speech to Promote a Memorial for Henry Sidgwick," in Whitaker, ed., *Correspondence,* vol. 2, 441.

31. Groenewegen, *A Soaring Eagle*, 3.

32. Alfred Marshall, preface to *Money, Credit and Commerce* (London: Macmillan, 1923).

33. Beatrice Webb, *My Apprenticeship* (London: Macmillan, 1926).

34. Alfred Marshall to James Ward, September 23, 1900, in Whitaker, ed., *Correspondence*, vol. 2.

35. Gertrude Himmelfarb, "The Politics of Democracy: The English Reform Act of 1867," *Journal of British Studies* 6, no. 1 (November 1966): 97.

36. Henry James, preface to *The Princess Casamassima* (New York: Charles Scribner's Sons, 1908 [1886]), vi.

37. Keynes, "Alfred Marshall, 1842–1924," 37.

38. Marshall to Ward, September 23, 1900.

39. Henry Sidgwick, *Principles of Political Economy* (London: Macmillan and Co., 1883), 4.

40. John E. Cairnes, *The Character and Logical Method of Political Economy; Being a Course of Lectures Delivered in the Hilary Term, 1857* (London: Longmans, Brown, Green, Longmans and Roberts, 1857), 38.

41. John Ruskin, *Unto This Last: Four Essays in the First Principles of Political Economy* (London: Smith Elder, 1862).

42. Gertrude Himmelfarb, *The Idea of Poverty: England in the Early Industrial Age* (New York: Alfred A. Knopf, 1984).

43. Leslie Stephen, *The Life of Henry Fawcett* (London: Smith, Elder and Co., 1886), 222.

44. Ruskin, *Unto This Last*, 20.

45. J. E. Cairnes, *Some Leading Principles of Political Economy* (London: University College London, 1874), 291.

46. John Stuart Mill, *Principles of Political Economy* (London: Longmans, Green and Co., 1885), 220.

47. Francis Bowen, *The Principles of Political Economy Applied to the Condition, the Resources, and the Institutions of the American People* (Boston: Little, Brown and Co., 1859) 197.

48. Millicent Garrett Fawcett, *Political Economy for Beginners* (London: Macmillan, 1906), 100.

49. John Francis Bray, *Labour's Wrongs and Labour's Remedy, or the Age of Might and the Age of Right* (Leeds, UK: David Green Briggate, 1839).

50. Alfred Marshall, *Alfred Marshall's Lectures to Women, Some Economic Questions Directly Connected to the Welfare of the Labourer* (Aldershot, UK: Edward Elgar, 1995), lecture 5, 119.

51. Ibid., 156.

52. Ibid., quotes from April and May 1873 notes by Mary Paley, 47, 53, and 54.

53. Joseph Schumpeter, *The History of Economic Thought* (Cambridge, Mass.: Harvard University Press, 1954), 290.

54. Arnold Toynbee, *Lectures on the Industrial Revolution of the Eighteenth Century in England* (London: Rivingtons, 1884) 175.

55. Marshall, *Lectures to Women.* May 9, 1873.

56. Ibid.

57. Mary Paley Marshall, *What I Remember,* 9.

58. Winnie Seebohm in Martha Vicinus, *Independent Women: Work and Community for Single Women 1850–1920* (Chicago: University of Chicago Press, 1985), 151.

59. W. S. Gilbert and Arthur Sullivan, *Princess Ida,* 1884.

60. Mary Paley Marshall, *What I Remember,* 16.

61. George Eliot, *The Mill on the Floss* (London: William Blackwood and Sons, 1860).

62. Mary Paley Marshall, *What I Remember,* 20–21.

63. Lord Ernle, *English Farming Past and Present,* 3d ed. (London: Longmans, Green and Co., 1922), 407.

64. *The Cambridge Chronicle,* April 11, 1874.

65. Alf Peacock, "Revolt of the Fields in East Anglia," *Our History* (London: Communist Party of Britain, 1968).

66. *Times* (London), April 13, 1874.

67. George Eliot, *Middlemarch* (Edinburgh: William Blackwood and Son, 1874).

68. *The Cambridge Chronicle,* April 25, 1874, and May 8, 1874.

69. *The Cambridge Independent Press,* May 16, 1874.

70. Alfred Marshall, "Beehive Articles," 1874; in R. Harrison, "Two Early Articles by Alfred Marshall," *Economic Journal* 73 (September 1963): 422–30.

71. Alfred Marshall, quoted in *The Cambridge Independent Press,* May 16, 1874.

72. Alfred Marshall to Rebecca Marshall, Niagara Falls, July 10, 1875, in John K. Whitaker, ed., *The Correspondence of Alfred Marshall, Economist, vol. 1, Climbing, 1868–1890* (Cambridge: Cambridge University Press, 1996), 68–70.

73. Ibid., Alfred Marshall to Rebecca Marshall, Springfield, Mass., June 12, 1875.

74. Ibid.

75. Ibid., Alfred Marshall to Rebecca Marshall, Boston, June 20, 1875, 54.

76. Ibid., Alfred Marshall to Rebecca Marshall, Cleveland, July 18, 1875, 71.

77. Alfred Marshall, "Some Features of American Industry," November 17, 1875, lecture to the Cambridge Moral Sciences Club, in John K. Whitaker, ed., *The Early Economic Writings of Alfred Marshall, 1867–1890,* vol. 2 (London: The Royal Economic Society, 1975), 369.

78. Alfred Marshall to Rebecca Marshall, Cleveland, July 18, 1875, in Whitaker, *Correspondence,* vol. 1, 72.

79. Keynes, "Alfred Marshall: 1842–1924," *Essays in Biography* (New York: W. W. Norton and Co., 1951), 142.

80. John K. Whitaker, "The Evolution of Alfred Marshall's Economic Thought and Writings Over the Years," in Whitaker, *Early Economic Writings*, 57.

81. Alfred Marshall, "Some Features of American Industry," in Whitaker, *Early Economic Writings*, 354.

82. *Reminiscences of America in 1869 by Two Englishmen* (London: Sampson, Low and Son and Marston, 1870).

83. Mary Paley Marshall, *What I Remember*.

84. Marshall, "Some Features of American Industry," 357.

85. Alfred Marshall to Rebecca Marshall, Lowell, Mass., and Cambridge, Mass., June 22, 1875, in Whitaker, *Correspondence*, vol. 1, 58.

86. *Reminiscences of America*, 86.

87. Samuel Bowles, *The Pacific Railroad—Open: How to Go, What to See* (Boston: Fields, Osgood and Co., 1869).

88. Marshall, "Some Features of American Industry," 357.

89. Alfred Marshall to Rebecca Marshall, Springfield, Mass., June 12, 1875, in Whitaker, *Correspondence*, vol. 1, 44.

90. *Reminiscences of America*, 242.

91. Marshall, "Some Features of American Industry," 359.

92. Alfred Marshall, *Principles of Economics* (London: Macmillan, 1890).

93. Marshall, "Some Features of American Industry," 353.

94. Alfred Marshall to Rebecca Marshall, Cleveland, July 18, 1875, in Whitaker, *Correspondence*, vol. 1, 71.

95. Ibid., June 5, 1875.

96. Marshall, "Some Features of American Industry," 372.

97. Karl Marx, *Das Kapital* (1887), Friedrich Engels, ed., trans. S. Moore and E. Aveling (New York: Charles H. Kerr & Company, 1906), 709.

98. Marshall, "Some Features of American Industry," 375.

99. Alfred Marshall to Rebecca Marshall, June 5, 1875, in Whitaker, *Correspondence*, vol. 1, 36.

100. Mary Paley Marshall, *What I Remember*, 19.

101. Phyllis Rose, *Parallel Lives: Five Victorian Marriages* (New York: Alfred A. Knopf, 1983).

102. Mary Paley Marshall, *What I Remember*, 23.

103. Alfred Marshall, testimony, December 1880, Governmental Committee on Intermediate and Higher Education in Wales and Monmouthshire, quoted in J. K. Whitaker, "Marshall: The Years 1877 to 1885," in *History of Political Economy* 4, no. 1 (Spring 1972): 6.

104. Mary Paley Marshall, *What I Remember*, 24.

105. Marion Fry Pease, "Some Reminiscences of University College, Bristol" (University of Bristol Library, Special Collections, 1942).

106. John Maynard Keynes, "Mary Paley Marshall," in *Essays in Biography*.

107. Marshall, in Whitaker, *Early Economic Writings*, 355.

108. Alfred Marshall, "The Present Position of Economics," in Whitaker, ed., *Early Economic Writings,* 51.

109. Marshall, *Principles of Economics,* 1.

110. Mill, *Principles of Political Economy,* vol. 2.

111. Mary Paley Marshall, unpublished notes, Marshall Archive, Cambridge University.

112. Charles Dickens, *Hard Times,* 1854, chap. 5.

113. Marx, *Das Kapital,* 462.

114. Alfred Marshall, in Whitaker, *Correspondence,* vol. 1, 59.

115. Alfred and Mary Marshall, *The Economics of Industry* (London MacMillan, 1879).

116. Mary Paley Marshall, *What I Remember,* 24.

117. Edwin Cannan, "Alfred Marshall, 1842–1924," *Economica* 4 (November 1924): 257–61.

118. Alfred Marshall to Macmillan, June 1878, in Whitaker, *Correspondence,* vol. 1, 97.

119. Henry George, *Progress and Poverty* (New York: Appleton, 1879).

120. *Jackson's Oxford Journal,* March 15, 1884. An account of the meeting is reprinted in an appendix to George Stigler, "Three Lectures on Progress and Poverty by Alfred Marshall," *Journal of Law and Economics* 12, no. (April 1969): 184–226.

121. Ibid., 186.

122. Ibid., 188.

123. Ibid., 208.

124. Ibid.

125. Ibid.

III: MISS POTTER'S PROFESSION:
WEBB AND THE HOUSEKEEPING STATE

1. George Eliot, *Middlemarch* (Edinburgh: William Blackwood and Son, 1874).

2. Daniel Pool, *What Jane Austen Ate and Charles Dickens Knew . . .* (New York: Simon & Schuster, 1993), 50–56.

3. Beatrice Webb, *My Apprenticeship* (London: Longmans, Green and Co., 1926), 48.

4. Michelle Jean Hoppe, "The London Season," *Literary Liaisons,* accessed March 14, 2011, www.literary-liaisons.com/article024.html.

5. Norman and Jeanne MacKenzie, eds., *The Diary of Beatrice Webb,* vol. 1, *1873–1892: "Glitter Around and Darkness Within"* (Cambridge, Mass.: Harvard University Press, 1982), 90 (July 15, 1883).

6. Ibid., 75 (February 22, 1883).

7. Ibid., 76 (February 26, 1883).

8. Ibid., 74 (January 2, 1883).

9. Beatrice Webb, *My Apprenticeship*, 157.

10. Henry James, preface to *The Portrait of a Lady* (New York: Charles Scribner's Sons, 1908).

11. Margaret Harkness to Beatrice Potter, n.d., 2/2/2 Papers of Beatrice and Sidney Webb, Passfield Archive, British Library of Political and Economic Science, London School of Economics and Political Science.

12. Henry James, *The Portrait of a Lady*, vol. 1 (London: Macmillan and Co., 1881), 193.

13. MacKenzie, *Diary of Beatrice Webb*, vol. 1, 80 (March 31, 1883).

14. Ibid., 54 (July 24, 1882).

15. Eliot, *Middlemarch*, 61.

16. Barbara Caine, *Destined to Be Wives: The Sisters of Beatrice Webb* (Oxford, UK: Clarendon Press, 1986), 12.

17. Webb, *My Apprenticeship*, 39.

18. Ibid., 42.

19. MacKenzie, *Diary of Beatrice Webb*, vol. 1, 4.

20. Norman and Jean MacKenzie, eds., *The Diary of Beatrice Webb*, vol. 2, *1892–1905: All the Good Things of Life* (Cambridge, Mass.: Harvard University Press, 1983), 132 (n.d. [March 1883]).

21. Herbert Spencer, *An Autobiography*, vol. 1 (New York: D. Appleton and Co., 1904), 298.

22. Webb, *My Apprenticeship*, 10.

23. MacKenzie, *Diary of Beatrice Webb*, vol. 2.

24. Spencer, *An Autobiography*, vol. 1, 298.

25. Ibid.

26. Webb, *My Apprenticeship*, 10.

27. MacKenzie, *Diary of Beatrice Webb*, vol. 1, 112 (April 8, 1884).

28. Ibid., 16 (March 6, 1874).

29. Webb, *My Apprenticeship*, 25 (emphasis added).

30. Kitty Muggeridge and Ruth Adam, *Beatrice Webb: A Life, 1858–1943* (New York: Alfred A. Knopf, 1968).

31. Ibid.

32. MacKenzie, *Diary of Beatrice Webb*, vol. 1, 19 (September 27, 1874).

33. Webb, *My Apprenticeship*, 56, 106, 112; MacKenzie, *Diary of Beatrice Webb*, vol. 1, 74 (January 2, 1883).

34. Webb, *My Apprenticeship*, 112–13.

35. MacKenzie, *Diary of Beatrice Webb*, vol. 1, 77 (March 1, 1883).

36. Ibid.

37. Ibid., 81 (March 31, 1883).

38. Ibid., 88 (May 24, 1883).

39. Ibid., 79 (March 24, 1883).

40. Helen Dandy Bosanquet, *Social Work in London, 1869–1912: A History of the Charity Organization Society* (New York: E. P. Dutton, 1914), 95.

41. MacKenzie, *Diary of Beatrice Webb,* vol. 1, 85 (May 18, 1883).

42. Ibid., 89 (July 7, 1883).

43. Ibid., 81 (March 31, 1883).

44. J. L. Garvin, *The Life of Joseph Chamberlain,* vol. 1 (London: Macmillan, 1932), 202.

45. MacKenzie, *Diary of Beatrice Webb,* vol. 1, 90–91 (July 15, 1883).

46. Ibid., 88 (June 3, 1883).

47. Ibid., 89 (June 27, 1883).

48. Ibid., 91 (July 15, 1883).

49. Ibid., 111 (March 16, 1884).

50. Ibid., 95 (September 22, 1883).

51. Ibid., 94 (September 26, 1883).

52. Ibid.

53. "The Bitter Cry of Outcast London," *The Pall Mall Gazette,* October 16, 1883 (issue 5808), 11.

54. Andrew Mearns, *The Bitter Cry of Outcast London: An Inquiry into the Condition of the Abject Poor* (London: James Clarke and Co., 1883), 5, 7; Earl Grey Pamphlets Collection (1883), Durham University Library, www.jstor.org/stable/60237726 (accessed January 13, 2011); Gertrude Himmelfarb, *Poverty and Compassion: The Moral Imagination of the Late Victorians* (New York: Alfred A. Knopf, 1991).

55. MacKenzie, *Diary of Beatrice Webb,* vol. 1, 137 (August 22, 1885).

56. Webb, *My Apprenticeship,* 150.

57. Ibid., 152.

58. MacKenzie, *Diary of Beatrice Webb,* vol. 1, 101 (December 31, 1883).

59. Ibid.

60. Ibid., 100 (December 27, 1883).

61. Ibid., 102–3 (January 12, 1884).

62. Ibid.

63. Webb, *My Apprenticeship,* 23.

64. Terence Ball, "Marx and Darwin: A Reconsideration," *Political Theory* 7, no. 4 (November 1979), 469–83.

65. Herbert Spencer, *The Man Versus the State* (London: Williams and Norgate, 1884), vii.

66. Arnold Toynbee, "Progress and Poverty: A Criticism of Mr. Henry George— Mr. George in England," London, January 18, 1883, in *Lectures on the Industrial Revolution of the 18th Century in England: Popular Addresses, Notes and Fragments by the Late Arnold Toynbee,* 6th ed. (London: Longmans, Green, and Co., 1902), 318.

67. MacKenzie, *Diary of Beatrice Webb,* vol. 1, 91 (July 15, 1883).

68. Beatrice Webb to Anna Swanwick, London, 1884 (not sent), in MacKenzie, ed., *The Letters of Sidney and Beatrice Webb*, vol. 1 (Cambridge: Cambridge University Press, 1978), 23.

69. MacKenzie, *Diary of Beatrice Webb*, vol. 1, 115 (April 22, 1884).

70. Webb, *My Apprenticeship*, 138.

71. MacKenzie, *Diary of Beatrice Webb*, vol. 1, 105–12 (March 16, 1884).

72. Joseph Chamberlain, "Work for the New Parliament," Birmingham, UK, January 5, 1885, in *Speeches of the Right Honorable Joseph Chamberlain, M.P.*, Henry W. Lucy, ed. (London: George Routledge and Sons, 1885), 104.

73. MacKenzie, *Diary of Beatrice Webb*, vol. 1, 117 (May 9, 1884).

74. Ibid., 119 (July 28, 1884).

75. Ibid. (August 1, 1884).

76. Webb, *My Apprenticeship*, 272.

77. MacKenzie, *Diary of Beatrice Webb*, vol. 1, 145 (December 19, 1885).

78. Ibid., 153 (January 1, 1886).

79. Ibid., 154 (February 11, 1886).

80. "London Under Mob Rule," *New York Times*, February 8, 1886.

81. Ibid.

82. Ibid.

83. "London's Recent Rioting," *New York Times*, February 10, 1886.

84. "The Rioting in the West-End," *Times* (London), February 10, 1886.

85. Queen Victoria to William Ewart Gladstone, Windsor Castle, February 11, 1886, in *The Letters of Queen Victoria; Third Series: A Selection of Her Majesty's Correspondence and Journal Between the Years 1886 and 1901*, vol. 1, George Earle Buckle, ed. (New York: Longmans, Green and Co., 1932), 52.

86. Ibid.

87. Margaret Harkness [John Law], *Out of Work* (London: Swan Schonnenschein, 1888).

88. Webb, *My Apprenticeship*, 273.

89. MacKenzie, *Diary of Beatrice Webb*, vol. 1, 154.

90. Beatrice Webb, "A Lady's View of the Unemployed at the East," *Pall Mall Gazette*, February 18, 1886.

91. Joseph Chamberlain to Beatrice Potter, February 25, 1886, 2/1/2 Passfield Archive.

92. Joseph Chamberlain to Beatrice Potter, February 28, 1886, 2/1/2 Passfield Archive.

93. Ibid.

94. Beatrice Potter to Joseph Chamberlain, Bournemouth, n.d. [March 1886], in *Letters*, ed. MacKenzie, vol. 1, 53–54.

95. Joseph Chamberlain to Beatrice Potter, March 5, 1886, 2/1/2 Passfield Archive.

96. Royden Harrison, *The Life and Times of Sidney and Beatrice Webb: The Formative Years, 1858–1903* (London: Palgrave, 1999), 125.

97. MacKenzie, *Diary of Beatrice Webb,* vol. 1, 160 (April 4, 1886).

98. Webb, *My Apprenticeship,* 212.

99. MacKenzie, *Diary of Beatrice Webb,* vol. 1, 164 (April 18, 1886).

100. Charles Booth, "The Inhabitants of Tower Hamlets (School Board Division), Their Condition and Occupations," Royal Statistical Society, London, May 17, 1887, in *Journal of the Royal Statistical Society,* vol. 50 (London: Edward Stanford, 1887), 326–91.

101. MacKenzie, *Diary of Beatrice Webb,* vol. 1, 164 (April 17, 1886).

102. Ibid., 173 (July 2, 1886).

103. Ibid.

104. Ibid., 174 (July 18, 1886).

105. Ibid., 213 (n.d.).

106. Webb, *My Apprenticeship,* 300.

107. MacKenzie, *Diary of Beatrice Webb,* vol. 1, 241 (April 11, 1888).

108. Beatrice Potter, "Pages from a Work-Girl's Diary," *The Nineteenth Century: A Monthly Review* 24, issue 139 (September 1888): 301–14.

109. MacKenzie, *Diary of Beatrice Webb,* 249 (April 13, 1888).

110. "The Peers and the Sweaters," *Pall Mall Gazette,* May 12, 1888.

111. MacKenzie, *Diary of Beatrice Webb,* vol. 1; 261 (September 14, 1888).

112. Ibid., 264 (November 8, 1888).

113. Ibid., 269 (December 29, 1888).

114. Ibid., 250 (April 26, 1888).

115. Ibid., 274 (March 8, 1889).

116. Webb, *My Apprenticeship,* 341.

117. Review of *Labour and Life of the People,* ed. Charles Booth, *The Times* (London), April 15, 1889, 9.

118. Webb, *My Apprenticeship,* 374.

119. MacKenzie, *Diary of Beatrice Webb,* vol. 1, 321 (February 1, 1890).

120. Ibid., 328 (March 29, 1890).

121. Ibid., 321 (February 1, 1890).

122. Ibid., 310 (November 26, 1889); Beatrice Potter to Sidney Webb [December 7, 1890], in *Letters,* vol. 1, ed. MacKenzie, 239.

123. *Letters,* vol. 1, ed. MacKenzie, 70.

124. Webb, *My Apprenticeship,* 390.

125. Muggeridge and Adam, *Beatrice Webb: A Life,* 123.

126. MacKenzie, *Diary of Beatrice Webb,* vol. 1, 184 (October 31, 1886).

127. Ibid., 324 (February 14, 1890).

128. Sidney and Beatrice Webb, *The History of Trade Unionism* (London: Green and Co., 1907), 400.

129. G. M. Trevelyan, *British History in the Nineteenth Century (1782–1901)* (London: Longmans, Green and Co., 1922), 403.

130. Sidney Webb to Edward Pease, London, in *Letters*, vol. 1, ed. MacKenzie, 101; Sidney Webb, *Socialism in England* (London: American Economic Association, 1889), 11, 20.

131. Sidney Webb, "Historic," in *Fabian Essays in Socialism*, ed. G. Bernard Shaw, 30–61 (London: The Fabian Society, 1889), 38.

132. MacKenzie, *Diary of Beatrice Webb*, vol., 322 (February 1, 1890).

133. William Harcourt, Speech to the House of Commons, August 11, 1887. *Parliamentary Debates*, 3rd series, vol. 319.

134. MacKenzie, *Diary of Beatrice Webb*, 330 (April 26, 1890).

135. Beatrice Potter to Sidney Webb, Gloucestershire, May 2, 1890, in *Letters*, vol. 1, ed. MacKenzie, 133.

136. Ibid., Sidney Webb to Beatrice Potter, April 6, 1891, 269.

137. MacKenzie, *Diary of Beatrice Webb*, vol. 1, 354.

138. Beatrice Potter to Sidney Webb, Gloucestershire, in *Letters*, vol. 1, ed. MacKenzie, 281.

139. MacKenzie, *Diary of Beatrice Webb*, vol. 1, 357 (June 20, 1891).

140. Friedrich August Hayek, review of *Our Partnership* by Beatrice Webb, eds. Barbara Drake and Margaret I. Cole (London: Longmans, Green and Co., 1948); *Economica*, New Series 15, no. 59 (August 1948): 227–30.

141. MacKenzie, *Diary of Beatrice Webb*, vol. 1, 371 (July 23, 1892).

142. Ibid., vol. 2, 37 (September 17, 1893).

143. Michael Holroyd, *Bernard Shaw: The One-Volume Definitive Edition* (London: Chatto and Windus, 1997), 164.

144. George Bernard Shaw to Archibald Henderson, June 30, 1904, in Archibald Henderson, *George Bernard Shaw: His Life and Works* (Cincinnati: Stewart and Kidd Company, 1911), 287.

145. George Bernard Shaw, preface to *Mrs. Warren's Profession: A Play in Four Acts* (London: Constable, 1907), xvii.

146. George Bernard Shaw to the editor of the *Daily Chronicle*, April 30, 1898, in *Bernard Shaw: Collected Letters, 1874–1897* (New York: Dodd, Meade and Company, 1965), 404.

147. H. G. Wells, *The New Machiavelli* (New York: Duffield and Co., 1910), 194–95.

148. James A. Smith, *The Idea Brokers: Think Tanks and the Rise of the New Policy Elite* (New York: Free Press, 1991), xiii.

149. Wells, *The New Machiavelli*, 199.

150. Ibid., 197.

151. A. G. Gardiner, *The Pillars of Society* (London: James Nisbet, 1913); Wells, *The New Machiavelli*, 195.

152. Wells, *The New Machiavelli*, 194.

153. Ibid., 190.

154. MacKenzie, *Diary of Beatrice Webb,* vol. 2, 262 (November 28, 1902), 325 (June 8, 1904).

155. Gardiner, *The Pillars of Society,* 204, 206.

156. Wells, *The New Machiavelli,* 196.

157. Richard Henry Tawney, *The Webbs in Perspective: The Webb Memorial Lecture Delivered 9 December 1952* (London: The Athlone Press, 1953), 4.

158. MacKenzie, *Diary of Beatrice Webb,* vol. 3, 69 (March 22, 1907).

159. Wells, *The New Machiavelli,* 196.

160. Wells, *The New Machiavelli,* 191.

161. MacKenzie, *Diary of Beatrice Webb,* 287 (July 8, 1903).

162. Ibid., 321 (May 2, 1904), 326–27 (June 10, 1904).

163. Elie Halevy, *A History of the English People in the Nineteenth Century,* vol. 6, *The Rule of Democracy (1905–1914),* 2nd ed. (London: Ernest Benn Limited, 1952), 267.

164. Edward Marsh, *A Number of People: A Book of Reminiscences* (New York: Harper and Brothers, 1939), 163; Winston Churchill and Henry William Massingham, introduction to *Liberalism and the Social Problem: A Collection of Early Speeches as a Member of Parliament* (London: Hodder and Stoughton, 1909).

165. Winston S. Churchill, "H. G. Wells," in *The Collected Essays of Sir Winston Churchill,* vol. 3, *Churchill and People,* ed. Michael Wolff (London: Library of Imperial History, 1976), 52–53.

166. Marsh, *A Number of People,* 150.

167. William Manchester, *The Last Lion: Winston Spencer Churchill, Visions of Glory (1874–1932)* (Boston: Little, Brown and Company, 1983), 403.

168. Peter de Mendelssohn, *The Age of Churchill,* vol. 1, *Heritage and Adventure, 1874–1911* (New York: Alfred A. Knopf, 1961), 365.

169. *Never Give In! The Best of Winston Churchill's Speeches,* Winston S. Churchill, ed. (New York: Hyperion, 2003), 25.

170. Beatrice Webb, *Our Partnership,* eds. Barbara Drake and Margaret I. Cole (London: Longmans, Green and Co., 1948), 149.

171. Sidney and Beatrice Webb, *Industrial Democracy,* vol. 2 (London: Longmans, Green, and Co., 1897), 767.

172. Sidney and Beatrice Webb, *The Prevention of Destitution* (London: Longmans, Green and Co., 1911), 1.

173. Ibid., 17, 97.

174. Ibid., 5.

175. Ibid., 90.

176. Ibid., 285.

177. MacKenzie, *Diary of Beatrice Webb,* vol. 3, 95 (July 27, 1908).

178. Webb, *Our Partnership,* 481–82.

179. George Bernard Shaw, "Review of the Minority Report," quoted in Holroyd, *Bernard Shaw*, 398.

180. Webb, *Our Partnership*, 481–92.

181. MacKenzie, *Diary of Beatrice Webb*, February 10, 1908.

182. Ibid., October 16, 1908.

183. Ibid., April 18/20, 1908.

184. John Grigg, *Lloyd George: The People's Champion, 1902–1911* (London: Eyre Methuen, 1978), 100.

185. Charles Frederick Gurney Masterman to Lucy Blanche Masterman, February 1908.

186. Roy Jenkins, *Churchill: A Biography* (London: Hill and Wang, 2001), 143–44.

187. Winston S. Churchill to H. H. Asquith, March 14, 1908, quoted in Martin Gilbert, *Churchill: A Life* (New York: Henry Holt and Company, 1991), 193.

188. Churchill to Asquith, December 29, 1908.

189. MacKenzie, *Diary of Beatrice Webb*, vol. 3, 100 (October 16, 1908), 118 (June 18, 1909).

190. Ibid., June 18, 1909.

191. Ibid., vol. 3, 90 (March 11, 1908).

192. Manchester, *The Last Lion*, 371.

193. Himmelfarb, *Poverty and Compassion*, 378.

194. Baron William Henry Beveridge, *Power and Influence* (London: Hodder and Stoughton, 1953), 86.

IV: CROSS OF GOLD: FISHER AND THE MONEY ILLUSION

1. David A. Shannon, ed., *Beatrice Webb's American Diary* (Madison: The University of Wisconsin Press, 1963), 72. Remark made by Professor H. Morse Stephens to Beatrice Webb during a tour of Cornell University in May 1898.

2. Norman and Jeanne MacKenzie, eds., *The Diary of Beatrice Webb*, vol. 2, *1892–1905: All the Good Things of Life* (Cambridge, Mass.: Harvard University Press, 1983), 137.

3. Beatrice Webb, *Our Partnership* (London: Longmans, Green and Co., 1948), 146.

4. Niall Ferguson, *Empire: The Rise and Demise of the British World Order* (New York: Basic Books, 2004), 242.

5. See, for example, the following articles in *The Manchester Guardian:* "An American Invasion," June 21, 1871 (rumors of Susan B. Anthony's trip to Ireland with the American Woman's Rights League); "From Our London Correspondent," October 21, 1890 (American girls invade the market for Britain's eligible noblemen); "Cycling Notes," October 29, 1894 (American-made bicycles threaten to dominate the British market); "By-ways of Manchester Life, XI. An American Invasion," April 9, 1898 (American firm builds a grain elevator on the Manchester Ship Canal).

6. Frederick Arthur McKenzie, *The American Invaders: Their Plans, Tactics and Progress* (London: Grant Richards, 1902), 142–43.

7. William Ewart Gladstone, *Gleanings of Past Years,* vol. I, *1843–78: The Throne and the Prince Consort; The Cabinet and the Constitution* (London: John Murray, 1879), 206.

8. Angus Maddison, *The World Economy: A Millennial Perspective* (Paris: OECD, 2001), 265.

9. Ferguson, *Empire,* 242.

10. Dudley Baines, *Migration in a Mature Economy: Emigration and Internal Migration in England* (Cambridge: Cambridge University Press, 2003), 63, table 3.3.

11. William Ewart Gladstone, "Free Trade" in Gladstone et. al., *Both Sides of the Tariff Question by the World's Leading Men* (New York: Alonzo Peniston, 1890), 44.

12. Jeremy Atack and Peter Passell, *A New Economic View of American History from Colonial Times to 1940* (New York: W. W. Norton, 1994), 468.

13. Shannon, *American Diary,* 27 (April 12, 1898).

14. Ibid., 136 (July 2–7, 1898).

15. Ibid., 137–50 (July 2–7 and July 10, 1898).

16. Ibid., 89, 90–91 (May 24, 1898), and 92–93 (May 29, 1898).

17. Beatrice Webb to Catherine Courtney, Chicago, May 29, 1898, in Norman McKenzie, ed., *The Letters of Sidney and Beatrice Webb,* vol. 2, *Partnership: 1892–1912* (Cambridge: Cambridge University Press, 1978).

18. Norman and Jean MacKenzie, eds., *The Diary of Beatrice Webb,* vol. 2, *1892–1905: All the Good Things of Life* (Cambridge, Mass.: Harvard University Press, 1983), 159 (May 16, 1889); Charles Philip Trevelyan to Beatrice Webb, Chicago, April 19, 1898, quoted in Shannon, *American Diary,* 88, note 4.

19. Shannon, *American Diary,* 60 (April 29, 1898), 10 (April 1, 1898), May 24, 1898, and 68 (May 7, 1898).

20. Milton Friedman, *Money Mischief: Episodes in Monetary History* (New York: Harcourt Brace Jovanovich 1992), 37.

21. Henry James, *The Ambassadors* (New York: Harper and Brothers Publishers, 1903), 257.

22. Alfred Marshall to Rebecca Marshall, St. Louis, August 22, 1875, in John K. Whitaker, ed., *The Correspondence of Alfred Marshall, Economist,* vol, 1, *Climbing, 1868–1890* (Cambridge: Cambridge University Press, 1996), 73.

23. Henry Seidel Canby, *Alma Mater: The Gothic Age of the American College* (New York: Farrar Reinhart, 1936), 71, 32.

24. Irving Norton Fisher, *My Father: Irving Fisher* (New York: Comet Press, 1956), 21, 26–27, 29–30, 33.

25. Muriel Rukeyser, *Willard Gibbs: American Genius* (New York: Doubleday, Doran and Co., 1942), 158.

26. Edward Bellamy, *Looking Backward: 2000–1887* (London: George Routlege and Sons, 1887).

27. Rukeyser, *Willard Gibbs,* 146.

28. Ibid., 231.

29. Paul A. Samuelson, "Economic Theory and Mathematics—An Appraisal," in Joseph E. Stiglitz, ed., *The Collected Scientific Papers of Paul A. Samuelson,* vol. 2 (Cambridge, Mass.: The M.I.T. Press, 1966), 1751.

30. Irving Fisher to William G. Eliot, Jr., Berlin, N.J., May 29, 1886, in Irving Norton Fisher, *My Father,* 25–26.

31. Irving Fisher to Will Eliot, Fisher to Eliot, Jr., Pittsfield, Mass., July 25, 1886, in Irving Norton Fisher, *My Father,* 26.

32. Arthur Twining Hadley, *Economics: An Account of the Relations Between Private Property and Public Welfare* (New York: G.P. Putnam's Sons, 1896), iv.

33. Richard Hofstadter, *Social Darwinism in American Thought* (New York: George Braziller, Inc., 1959), 8.

34. Albert Galloway Keller, introduction to *War and Other Essays by William Graham Sumner,* Keller, ed. (New Haven, Conn.: Yale University Press, 1911), xx, xxiv; Hofstadter, *Social Darwinism,* 51.

35. Fisher to Eliot, Peace Dale, R.I., September 1892, in Irving Norton Fisher, *My Father,* 52.

36. William James to Thomas W. Ward, Berlin, n.d. [November 1867], in Henry James, ed., *The Letters of William James,* vol. 1 (Boston: Atlanta Monthly Press, 1920), 118.

37. Irving Fisher, "Mathematical Investigations in the Theory of Value and Prices (April 27, 1892)," in William J. Barber, ed., *The Works of Irving Fisher,* vol. 1 (London: Pickering and Chatto, 1997), 162.

38. Ibid., 68.

39. Ibid., 145.

40. Ibid., 4.

41. Francis Ysidro Edgeworth, review of "Mathematical Investigations in the Theory of Value and Prices" by Irving Fisher, *Economic Journal,* vol. 3, no. 9 (March 1893), 112.

42. Alfred Marshall, *Principles of Economics,* 3rd ed. (London: Macmillan, 1895), 450, 148 (note 1).

43. Barbara W. Tuchman, *The Proud Tower: A Portrait of the World Before the War, 1890–1914* (New York: Macmillan and Co., 1966).

44. *Narragansett Times,* June 23, 1893, quoted in Irving Norton Fisher, *My Father,* 60.

45. *New York Times* wedding announcement, June 18, 1893.

46. Daniel T. Rogers, *Atlantic Crossings: Social Politics in a Progressive Age* (Cambridge, Mass.: Harvard University Press, 1998).

47. Irving Fisher to Ella Wescott Fisher.

48. Fisher, Jr. *My Father,* 69.

49. Douglas Steeples and David O. Whitten, *Democracy in Desperation: The Depression of 1893* (New York: Greenwood, 1998).

50. Reverend T. De Witt Talmage, sermon delivered in Washington on September 27, 1896, quoted in William Jennings Bryan, *The First Battle: A Story of the Campaign of 1896* (Chicago: W. B. Conkey Company, 1896), 474.

51. Albro Martin, *James J. Hill and the Opening of the Northwest* (Minneapolis: Minnesota Historical Society Press, 1975), 428.

52. Bryant, *The First Battle,* 439.

53. Paxton Hibben and Charles A. Beard, *The Peerless Leader: William Jennings Bryan* (Whitefish, MT: Kessinger Publishing, 2004), 189.

54. Bryan, *The First Battle,* 485–86.

55. Ibid.

56. Ibid.

57. "Bryan's Backers Are Shy," *New York Times,* September 27, 1896; Canby, *Alma Matter,* 27; Martin L. Fausold, *James W. Wadsworth, Jr.: The Gentleman from New York* (Syracuse, N.Y.: Syracuse University Press, 1975), 17.

58. "Yale Would Not Listen," *New York Times,* September 25, 1896, 15.

59. Fisher to Eliot, summer 1895, quoted in Irving Norton Fisher, *My Father,* 71.

60. Fisher to Eliot, July 29, 1895, quoted in Barber, *Works of Irving Fisher,* 10.

61. Fisher to Eliot, summer 1895, quoted in Irving Norton Fisher, *My Father,* 71.

62. Fisher to Eliot, New Haven, November 1865, quoted in Irving Norton Fisher, *My Father,* 71.

63. William Graham Sumner, *The Absurd Effort to Make the World Over,* in Keller, *War, and Other Essays,* 195–210.

64. Fisher to Eliot, summer 1895, quoted in Irving Norton Fisher, *My Father,* 71.

65. Irving Fisher, "The Mechanics of Bimetallism," *Economic Journal,* 4 (September 1894), 527–36; Irving Norton Fisher, *My Father,* 187.

66. Harold James, *The End of Globalization: Lessons from the Great Depression* (Cambridge, Mass.: Harvard University Press, 2001), 24–25.

67. Walter Bagehot, *Lombard Street: A Description of the Money Market* (New York: Scribner, Armstrong, 1873), 123.

68. Fisher, *Mathematical Investigations,* in Barber, *Works of Irving Fisher,* 147.

69. Katherine Ott, *Fevered Lives: Tuberculosis in American Culture Since 1870* (Cambridge, Mass.: Harvard University Press, 1996), 113.

70. Ibid.,79.

71. Irving Fisher, May 1901, "Self Control," a talk given at the Thacher School in Ojai, California, a high school founded by William L. Thacher.

72. Fisher to Eliot, Saranac, December 11, 1898, in Irving Norton Fisher, *My Father,* 75.

73. Fisher to Margaret Hazard Fisher, Battle Creek, Michigan, December 31, 1904, in ibid., 108.

74. Irving Fisher, "Memorial Relating to the Conservation of Human Life," S. Doc. No. 493, at 7–8 (1912).

75. Irving Fisher, "Why Has the Doctrine of Laissez Faire Been Abandoned?" Address at the Fifty-fifth Annual Meeting of the American Association for the Advancement of Science, New Orleans, December 1905-January 1906.

76. Perry Mehrling, "Love and Death: The Wealth of Irving Fisher," in Warren J. Samuels and Jeff E. Biddle, eds., *Research in the History of Economic Thought and Methodology*, vol. 19 (New York: Elsevier Science BV, 2001), 47–61.

77. Fisher, "Why Has the Doctrine of Laissez Faire Been Abandoned?"

78. Ibid.

79. Ibid.

80. Ibid.

81. Fisher to Bert, Peace Dale, Rhode Island, January 1, 1903, in Irving Norton Fisher, *My Father*, 84–85.

82. Irving Fisher, *The Rate of Interest: Its Nature, Determination and Relation to Economic Phenomena* (New York: The Macmillan Company, 1907), 326.

83. Ibid., 327.

84. Ibid., 288.

85. Ibid.

V: CREATIVE DESTRUCTION:
SCHUMPETER AND ECONOMIC EVOLUTION

1. Rosa Luxemburg, *The Accumulation of Capital* (1913) (London: Routledge and Keegan Paul, 1951), 458.

2. National Bureau of Economic Research, UK Bank Rate, www.nber.org/databases/macrohistory/rectdata/13/m13013.data.

3. Felix Somary, *Erinnerungen aus Meinem Leben* [Memories from My Life] (Zurich: Manesse Verlag, 1959).

4. Oszkár Jászi, *The Dissolution of the Habsburg Monarchy* (Chicago: University of Chicago Press, 1929), 210.

5. Carl Schorske, *Fin de Siècle Vienna* (New York: Knopf, 1979).

6. Erich Streissler, "Schumpeter's Vienna and the Role of Credit in Innovation," in H. Frisch, ed., *Schumpeterian Economics* (New York: Praeger, 1981), 60.

7. Joseph Roth, *The Radetzky March*, trans. Geoffrey Dunlop (New York: Viking, 1933), 212.

8. "Opening of the International Exhibition of Electricity at Vienna," *Manufacturer and Builder*, vol. 15, no. 9 (September 1883), 214–15; "An Electric Exhibition," *New York Times*, August 12, 1883.

9. Quoted in Roman Sandgruber, "The Electrical Century: The Beginnings of Electricity Supply in Vienna," trans. Richard Hockaday, in Mikulas Teich and Roy Porter, eds., *Fin de Siècle and Its Legacy* (Cambridge, UK: Cambridge University Press, 1990), 42.

10. Richard L. Rubenstein, *The Age of Triage: Fear and Hope in an Overcrowded World* (Boston: Beacon Press, 1983), 8; Raymond James Sontag, *Germany and England: Background of Conflict, 1848–1894* (New York: Russell & Russell, 1964), 146.

11. David F. Good, *The Economic Rise of the Habsburg Empire, 1750–1914* (Berkeley: University of California Press, 1984), 256.

12. Gottfried Haberler, *Quarterly Journal of Economics,* vol. 64, no. 3 (August 1950), 338.

13. Arthur Smithies, "Memorial: Joseph Alois Schumpeter, 1883–1950," *American Economic Review,* vol. 40, no. 4 (September 1950), 628–48.

14. Marcel Proust, *Swann's Way,* trans. C. K. Scott Moncrieff (London: Chatto and Windus, 1922), 73.

15. Joseph A. Schumpeter, "Preface to the Japanese Edition of *The Theory of Economic Development,*" in Schumpeter, *Essays on Entrepreneurs, Innovations, Business Cycles, and the Evolution of Capitalism,* Richard Clemence, ed. (New York: Transaction Publishers, 1951), 166.

16. Alfred Marshall, *Principles of Economics,* vol. 1, 5th ed. (London: Macmillan, 1907), xxix, 820.

17. Joseph A. Schumpeter, "Review of *Essays in Biography* by J. M. Keynes," *Economic Journal* 43, no. 172 (December 1933), 652–57.

18. "Wills and Bequests," *Times* (London), January 12, 1933.

19. Richard Swedberg, "Appendix II: Schumpeter's Novel Ships in Fog (a Fragment)," in *Schumpeter, a Biography* (Princeton, N.J.: Princeton University Press, 1991), 207.

20. W. W. Rostow, *Theorists of Economic Growth from David Hume to the Present,* 234–35.

21. Anthony Trollope, *The Bertrams* (London: Chapman and Hall, 1859), 465.

22. Rosa Luxemburg, *The Accumulation of Capital (1913)* (London: Routledge and Keegan Paul, 1951), 434.

23. Quoted in Alexander D. Noyes, "A Year After the Panic of 1907," *Quarterly Journal of Economics* 23 (February 1909); 185–212.

24. "The Progress of the World," *American Monthly Review of Reviews,* vol. 35, no. 1 (January 1907).

25. Evelyn Baring Cromer, *The Situation in Egypt: Address Delivered to the Eighty Club on December 15th, 1908 by the Earl of Cromer* (London: Macmillan, 1908), 9.

26. William Jennings Bryan, "The Government of Egypt Beyond Definition," in *The Old World and Its Ways* (St. Louis: Thompson, 1907), 323.

27. "Railroad Up Cheops," *Los Angeles Times,* February 12, 1907, II.

28. Quoted in Noyes, "A Year After the Panic," 202.

29. "Cotton Crops and Gold in Egypt," *New York Times,* January 5, 1908, AFR 28.

30. Harry Boyle to Lord Rennell, April 21, 1907, in Clara Boyle, *A Servant of the Empire: A Memoir of Harry Boyle with a Preface by the Earl of Cromer* (London: Methuen, 1938), 107.

31. "Egyptian Finance," *New York Times,* December 8, 1907, 54.

32. Noyes, "A Year After the Panic," 202–3.

33. Ibid., 194.

34. Desmond Stewart, "Herzl's Journeys in Palestine and Egypt," *Journal of Palestine Studies* vol. 3, no. 3 (spring, 1974), 18–38.

35. Wassily Leontief, "Joseph A. Schumpeter," *Econometrica,* vol. 8, no. 2 (April 1950).

36. Quoted in Trevor Mostyn, *Egypt's Belle Époque, 1869–1952: Cairo and the Age of the Hedonists* (London: Quartet Books, 1989), 154.

37. Douglas Sladen, quoted in Max Rodenbeck, *Cairo: The City Victorious* (New York: Alfred A. Knopf, 1999), 138.

38. Joseph A. Schumpeter, *Das Wesen und Hauptinhalt der Theoretischen Nationalekonomie* (Altenburg: Stefan Geibel, 1908), 621, trans. by Bruce McDaniel as *The Nature and Essence of Economic Theory* (New Brunswick, N.J.: Transaction Publishers, 2010), x.

39. Ibid., 621.

40. Smithies, "Memorial," 629.

41. Joseph A. Schumpeter, *The Theory of Economic Development: An Inquiry Into Profits, Capital, Credit, Interest and the Business Cycle,* (1911) trans. Redvers Opie (New York: Transaction Publishers, 2004), 91.

42. *The Norton Anthology of English Literature,* vol. 2, *The Age of Victoria* (New York: Norton, 2000).

43. Joseph Schumpeter, *History of Economic Analysis* (Cambridge, Mass.: Harvard University Press, 1952), 571.

44. Alfred Marshall, "The Social Possibilities of Economic Chivalry," *Economic Journal* 17, no. 5 (March 1907); 7–29.

45. Angus Maddison, "GDP per Capita in 1990 International Geary-Khamis Dollars," *The World Economy: Historical Statistics* (Paris: OECD Publishing, 2003).

46. Jeffrey Williamson, "Real Wages and Relative Factor Prices in the Third World Before 1940: What Do They Tell Us About the Sources of Growth?" October 1998, Conference on Growth in the 19th and 20th Century: A Quantitative Economic History, December 14–15, 1998, Valencia, Spain, 37, table 2, www.economics.harvard.edu/pub/HIER/1998/1855.pdf; Michael D. Bordo, Alan M. Taylor, Jeffrey G. Williamson, *Globalization in Historical Perspective* (Chicago: University of Chicago Press, 2005), 285.

47. Joseph A. Schumpeter, *Capitalism, Socialism and Democracy,* 87.

48. Karl Marx and Friedrich Engels, *The Communist Manifesto* (1848), trans. Samuel Moore, introduction and notes by Gareth Stedman Jones (London: Penguin Books, 1967), 222.

49. Marshall, *Principles*.

50. Schumpeter, *Theory of Economic Development*, 95.

51. Beatrice Webb, *My Apprenticeship* (1926) (Longmans, Green, 1950), 380.

52. Schumpeter, *Capitalism, Socialism and Democracy*, 132.

53. Schumpeter, *Theory of Economic Development*, 85.

54. Schumpeter, *Capitalism, Socialism and Democracy*, 132.

55. Friedrich von Wieser, *The Theory of Social Economics* (New York: Augustus M. Kelly, 1927 and 1967).

56. Joseph A. Schumpeter, "The Communist Manifesto in Sociology and Economics," *Journal of Political Economy* (June 1949), 199–212.

57. Ibid.

58. David Landes, *Bankers and Pashas: International Finance and Imperialism in Egypt* (Cambridge, Mass.: Harvard University Press, 1980), 57.

59. Joseph A. Schumpeter to David Pottinger, June 4, 1934, in Swedberg, *Schumpeter*, 219.

60. Edwin A. Seligman, Professor of Economics at Columbia, to Nicholas Murray Butler, President of the University, October 22, 1913, quoted in Robert Loring Allen, *Opening Doors: The Life and Work of Joseph Schumpeter* (New Brunswick: Transaction Publishers, 1991), 130.

ACT II: PROLOGUE: WAR OF THE WORLDS

1. Irving Fisher, "The Need for Health Insurance," *American Labor Legislation Review* 7 (1917): 10.

2. Norman and Jeanne MacKenzie, eds., *The Diary of Beatrice Webb* vol. 3, *1905–1924: The Power to Alter Things* (Cambridge, Mass.: Harvard University Press, 1984), 204.

3. Ibid., August 5, 1914.

4. Ibid., November 4, 1918.

5. George Bernard Shaw, "Common Sense About the War," 1914.

6. Bertrand Russell, quoted in Niall Ferguson, *The Pity of War* (New York: Basic Books, 1999), 318.

7. Robert Skidelsky, John Maynard Keynes: *Hopes Betrayed*, vol. I (New York: Viking, 1986).

8. John Maynard Keynes to Neville Chamberlain.

9. Richard Shone with Duncan Grant, "The Picture Collector," in Milo Keynes, *Essays on John Maynard Keynes* (Cambridge: Cambridge University Press, 1975), 283.

10. Charles John Holmes, *Self & Partners (Mostly Self): Being the Reminiscences of C. J. Holmes* (London: Macmillan, 1936); Anne Emberton, "Keynes and the Degas Sale," *History Today*, December 31, 1995.

11. John Maynard Keynes to Florence Keynes.

12. Vanessa Bell to Roger Fry.

13. Sigmund Freud, in Peter Gay, *Sigmund Freud: A Life of Our Time* (New York: W.W. Norton, 1988).

14. Friedrich Hayek, "Remembering My Cousin Ludwig Wittgenstein (1889–1951)," *Encounter,* August 19, 1977, 20–21, and Ray Monk, *Ludwig Wittgenstein: The Duty of Genius* (New York: Penguin Books, 1991)

15. Hayek, "Remembering My Cousin," 20.

16. D. H. Mellor, "Better than Stars: Portrait of Frank Ramsey," BBC; D. H. Mellor (1995), "Cambridge Philosophers, vol. I: F. P. Ramsey," *Philosophy 70* (1995), 259.

17. "National Society to Conserve Life," *New York Times,* December 30, 1913; Irving Fisher and Eugene Lyman Fisk, Preface to *How to Live: Rules for Healthful Living Based on Modern Science,* 2nd ed. (New York: Funk & Wagnalls Company, 1915).

18. Henry Andrews Cotton, *The Defective, Delinquent, and Insane: The Relation of Focal Infections to Their Causation, Treatment, and Prevention, by Henry A. Cotton, lectures delivered at Princeton University, January 11, 13, 14, 15, 1921,* with a foreword by Adolf Meyer (Princeton, N.J.: Princeton University Press, 1922).

19. Bette M. Epstein, New Jersey State Archives, to author.

20. Irving Fisher, *American Labor Legislation Review,* p. 10.

21. MacKenzie, *Diary of Beatrice Webb,* vol. 3, 324 (November 17, 1918).

22. Ibid., 318 (November 11, 1918).

23. Ray Monk, *Bertrand Russell: The Spirit of Solitude 1872–1921, Vol. I* (New York: Simon & Schuster, 1996).

VI: THE LAST DAYS OF MANKIND: SCHUMPETER IN VIENNA

1. Joseph A. Schumpeter, *Politische Reden* [Political Speeches], Wolfgang F. Stolper and Christian Seidl, eds. (Tubingen: J.C.B. Mohr, 1992).

2. Francis Oppenheimer, *The Stranger Within: Autobiographical Pages* (London: Faber, 1960).

3. Norman and Jeanne MacKenzie, eds., *The Diary of Beatrice Webb,* vol. 3, *1905–1924* (Cambridge, Mass.: Harvard University Press, 1982–84), November 11, 1918.

4. Sigmund Freud, quoted in Peter Gay, *Freud: A Life of Our Time* (New York: W. W. Norton and Co., 1988), 382.

5. F. L. Carsten, *Revolution in Central Europe: 1918–1919* (Aldershot, UK: Wildwood House, 1988), 41.

6. Karl Kraus, *The Last Days of Mankind: A Tragedy in Five Acts* (New York: Unger, 2000).

7. Edmund von Glaise-Horstenau, "The Armistice of Villa Giusti 1918," in *The Collapse of the Austro-Hungarian Empire* (London: J. M. Dent and Sons, 1930).

8. Sigmund Freud, quoted in Gay, *Freud.*

9. F. O. Lindley, British high commissioner, quoted in Carsten, *Revolution in Central Europe,* 11–12.

10. Friedrich Wieser, "The Fight Against Famine in Austria," in *Fight the Famine Council, International Economic Conference* (London: Swarthmore Press, 1920), 53.

11. *The Memoirs of Herbert Hoover,* vol. 1, *Years of Adventure 1874–1920* (New York: Macmillan, 1951), 392.

12. Ibid.

13. Stefan Zweig, *The World of Yesterday: An Autobiography* (Lincoln: University of Nebraska Press, 1984), 289.

14. Ludwig von Mises, "The Austro-Hungarian Empire," *Encyclopedia Britannica,* 1921.

15. Quoted in Gay, *Freud,* 378.

16. Felix Salten, *Florian, the Emperor's Horse* (New York: Aires Scribner Sons, 1934).

17. "Austria Willing to Pawn Anything," *New York Times,* January 22, 1920.

18. Carsten, *Revolution in Central Europe,* 37.

19. Joseph Schumpeter, *Die Arbeiter Zeitung,* November 22, 1919, in *Dokumentation zur Oesterreichischen Zeitgeschichte, 1918–1928.* [Documentation of Austrian History, 1918–1928], eds. Christine Klusacek, Kurt Stimmer (Vienna: Jugend und Volk, 1984).

20. Sir T. Montgomery-Cuninghame, *Dusty Measure* (London: John Murray, 1939), 309.

21. SHB to ASB, December 30, 1918, quoted in William Beveridge, *The Power and Influence,* 153.

22. Karl Kautsky, *The Social Revolution and On the Morrow of the Social Revolution* (London: Twentieth Century Press, 1907), part 2, 1.

23. Felix Somary, *Erinnerungen aus Meinem Leben* [Memories from My Life] (Zurich: Manesse Verlag, 1955), 171.

24. Eduard Bernstein

25. Otto Bauer, *The Austrian Revolution* (London: Parsons, 1925).

26. Albert Einstein to Hedwig and Max Born, January 15, 1919, *Albert Einstein, Collected Papers,* vol. 4.

27. Joseph Schumpeter, quoted in Eduard Marz, *Joseph A. Schumpeter: Forscher, Lehrer und Politiker* [Researcher, Teacher, and Politician] (Munchen: R. Oldenbourg, 1983).

28. Somary, *Erinnerungen,* 172.

29. Karl Corino, *Robert Musil* (Hamburg: Rowolt, 2003), 598.

30. Friedrich von Wieser, *Tagebuch* Gertrud Enderle-Burcel, Staatsarchiv Wien Nachlass Wieser in the Haus-, Hof- und Staatsarchiv Extracts in Seidl, *Politische Reden,* 10–12.

31. Wolfgang F. Stolper, *Joseph Alois Schumpeter: The Public Life of a Private Man* (Princeton, N.J.: Princeton University Press, 1994), 123.

32. Karl Kraus, *Die Fackel*, April 1919.

33. Joseph Schumpeter, *Politische Reden*.

34. Otto Bauer, *The Austrian Revolution* (London: Parsons, 1925).

35. Gabor Betony, *Britain and Central Europe 1918–1933* (Oxford, UK: Clarendon Press, 1999), 10.

36. Joseph Schumpeter, "The Sociology of Imperialism," in Richard Sweds, *The Economics and Sociology of Capitalism* (Princeton: Princeton University Press, 1991), 156–57.

37. Joseph Schumpeter, *Politische Reden*.

38. Joseph Schumpeter, *Politische Reden*.

39. David Lloyd George, "Fontainebleau Memorandum," March 25, 1919, www.fullbooks.com/Peaceless-Europe2.html.

40. Winston Churchill, House of Commons, May 29, 1919, http://www.winstonchurchill.org; Randolph Spencer Churchill and Martin Gilbert, *Winston S. Churchill*, vol. 4, *The Stricken World* (New York: Houghton Mifflin, 1966), 308.

41. Bauer, *The Austrian Revolution*, 106.

42. *The Memoirs of Herbert Hoover*, vol. 1, *Years of Adventure 1874–1920* (New York: Macmillan, 1951); Bauer, *The Austrian Revolution*, 103.

43. Hans Loewenfeld-Russ, *Im Kampf Gegen den Hunger* [In the Fight Against Hunger] (Munich: R. Oldenburg, 1986).

44. T. Montgomery-Cuninghame, *Dusty Measure*, (London: John Murray, 1939).

45. Ellis Ashmead-Bartlett, *The Tragedy of Central Europe* (London: Thornton Butterworth, 1924), 159.

46. Ibid.

47. Friedrich von Wieser, *Tagebuch*, Gertrud Enderle-Burcel, Staatsarchiv Wien Nachlass Wieser in the Haus-, Hof- und Staatsarchiv Extracts in Seidl, *Politische Reden*, pp. 10–12.

48. Eduard Marz, *Austrian Banking and Financial Policy: Creditanstalt at a Turning Point, 1913–1923* (New York: St. Martin's Press, 1984), 333.

49. "Entretien avec le Docteur Schumpeter," De notre envoye spécial, Vienne, Mai, *Le Temps*, June 2, 1919, translated and quoted in W. F. Stolper, *Joseph Alois Schumpter, The Public Life of a Private Man* (Princeton: Princeton University Press, 1994), 219.

50. Bauer, *The Austrian Revolution*, 110.

51. Ibid., 257.

52. Schumpeter, *Politische Reden*.

53. Ibid.

54. Francis Oppenheimer to John Maynard Keynes, May 18, 1919, Kings College Archive.

55. Francis Oppenheimer, *The Stranger Within: Autobiographical Pages* (London: Faber, 1960), 369.

56. Bauer, *The Austrian Revolution*.

57. Ibid.

58. Joseph Schumpeter, *Neue Freie Presse*, June 24, 1919, in Klusacek et al., eds., *Dokumentation*.

59. Joseph Schumpeter, *Neue Freie Presse*, June 28, 1919. "Es ist nicht leicht ein Volk zu vernichten. Im allgemeinen ist es sogar unmöglich. Hier haben wir aber einen der seltenen Fälle for uns, wo es möglich ist."

60. Friedrich Wieser, "The Fight Against Famine in Austria," in *Fight the Famine Council, International Economic Conference* (London: Swarthmore Press, 1920), 53.

61. Quoted in Stolper, *Joseph Alois Schumpeter*.

62. Richard Kola, *Rückblick ins Gestrige: Erlebtes und Empfundenes* [Looking Back to Yesterday: Experiences and Perceptions] (Vienna: Rikola, 1922).

63. Schumpeter, *Politische Reden*.

64. Somary, *Erinnerungen*.

65. Richard Swedberg, *Joseph A. Schumpeter, His Life and Work* (Cambridge, UK: Polity Press, 1991).

66. Ibid., 144–45.

67. Friedrich von Wieser, *Tagebuch*, November 19, 1919: "Es scheint, dass Schumpeter in der Meinung aller Parteien und aller gebildeten Menschen völlig abgewirtschaftet hat. Wie mir Kelsen erzählte, auch unsere jüngeren Nationalökonomen, die ihn als ihren Führer betrachteten, sind von ihm abgekommen und geben ihn wissenschaftlich auf, es sei nichts mehr von ihm zu erwarten."

68. Eduard Marz, "Joseph Schumpeter as Minister of Finance," in Helmut Frisch, ed., *Schumpeterian Economics* (New York: Praeger, 1981).

VII: EUROPE IS DYING: KEYNES AT VERSAILLES

1. Frances Oppenheimer, *The Stranger Within: Autobiographical Pages* (London: Faber, 1960), 374.

2. Lord William Beveridge, *Power and Influence* (New York: Beechhurst Press, 1955), 149–50.

3. David Lloyd George to Woodrow Wilson, April 1919.

4. John Maynard Keynes to Vanessa Bell, March 16, 1919, Keynes Papers, King's College Archive.

5. Harold Nicolson, *Peacemaking 1919: Being Reminiscences of the Paris Peace Conference* (Boston: Houghton Mifflin, 1933), 44.

6. Ibid., 275–76.

7. David Lindsay, *The Crawford Papers: The Journals of David Lindsay, Twenty-seventh Earl of Crawford and Tenth Earl of Balcarres (1871–1940), During the Years 1892 to 1940*, April 9, 1919.

8. Robert Skidelsky, *John Maynard Keynes*, vol. 1, *Hopes Betrayed* (New York: Viking, 1986), 304.

9. John Maynard Keynes, "My Early Beliefs," September 9, 1938, in *Essays in Biography* (London: MacMillan St. Martin's Press for the Royal Economic Society, 1972), 436.

10. John Maynard Keynes to Lytton Strachey, November 23, 1905, quoted in Skidelsky, *Keynes,* vol. 1, 166.

11. John Maynard Keynes to Lytton Strachey, November 15, 1905, Skidelsky, 165.

12. "A Key for the Prurient: Keynes's Loves, 1901–15," Donald E. Moggridge, *Maynard Keynes: An Economist's Biography* (London: Routledge, 1992), annex 1.

13. C. R. Fay, "The Undergraduate," in Milo Keynes, ed., *Essays on John Maynard Keynes* (Cambridge: Cambridge University Press, 1975), 36.

14. Lionel Robbins, *Autobiography of an Economist* (London: Macmillan, 1971).

15. Winston Churchill to Clementine Churchill, *Speaking for Themselves: The Personal Letters of Winston and Clementine Churchill,* ed. Mary Soames, (London and New York: Doubleday, 1998).

16. Elizabeth Johnson, "Keynes' Attitude Toward Compulsory Military Service," *Economic Journal* 70, no. 277 (March 1960): 160–65.

17. David Lloyd George, *Memoirs of the Peace Conference,* vol. 1 (New Haven, Conn.: Yale University Press, 1939), 302.

18. William Shakespeare, *A Midsummer Night's Dream* (New York: Palgrave, 2010).

19. Lloyd George, *Memoirs of the Peace Conference,* vol. 1, 302.

20. John Maynard Keynes to Florence Keynes, quoted in Skidelsky, *Keynes,* vol. 1, *Hopes Betrayed,* 353.

21. John Maynard Keynes to Florence Keynes, Keynes Papers, King's College Archive.

22. Quoted in Macmillan, *Paris 1919,* 60.

23. John Maynard Keynes, "Dr. Melchior: A Defeated Enemy," in *Essays in Biography,* 210.

24. Max Warburg, "Aus Meinem Aufzeichnungen" [From My Records], quoted in *Collected Writings of John Maynard Keynes,* vol. 16, *Activities 1914–1919, The Treasury and Versailles* (Cambridge: Cambridge University Press), 417.

25. Keynes, "Dr. Melchior," 214.

26. Ibid., 216.

27. Ibid., 218.

28. Ibid., 221.

29. Ibid., 223.

30. George Allerdice Riddell, *Lord Riddell's Intimate Diary of the Peace Conference and After, 1918–1923* (New York: Reynal & Hitchcock, 1924), 30.

31. Keynes, "Dr. Melchior," 231.

32. Thomas W. Lamont, "The Final Reparations Settlement," *Foreign Affairs*, 1930.

33. Nicolson, *Peacemaking 1919*, 86.

34. Peter Rowland, *David Lloyd George* (London: Macmillan, 1975), 485–86.

35. Nicolson, *Peacemaking 1919*, 78.

36. Skidelsky, 367.

37. Jan Smuts, quoted in Skidelsky, *Keynes*, vol. 1, *Hopes Betrayed*, 373.

38. *The Memoirs of Herbert Hoover*, vol. 1, *Years of Adventure 1874–1920* (New York: Macmillan, 1951), 461–62.

39. John Maynard Keynes to Florence Keynes, in Skidelsky, *Keynes*, vol. 1, *Hopes Betrayed*, 371.

40. John Maynard Keynes to Florence Keynes, Keynes Papers, King's College Archive.

41. John Maynard Keynes, *The Economic Consequences of the Peace* (London: Macmillan and Co., 1920), 233 (note 1).

42. John Maynard Keynes to Duncan Grant, May 14, 1919.

43. Rowland, *David Lloyd George*, 480.

44. John Maynard Keynes to Austin Chamberlain, June 5, 1919.

45. Alec Cairncross, "Austin Robinson," *Economic Journal* 104 (July, 1994): 903–15.

46. Ibid.

47. Jan Smuts, quoted in Skidelsky, *Keynes*, vol. 1, 373.

48. John Maynard Keynes, *The Economic Consequences of the Peace* (London: Macmillan, 1920).

49. Henry Wickham Steed, "A Critic of the Peace," "The Candid Friend at Versailles," "Comfort for Germany," *John Maynard Keynes: Critical Responses*, ed. Charles Robert McCons (London: Taylor and Francis, 1998), 51–60.

50. Quoted in Niall Ferguson, *Paper and Iron* (Cambridge: Cambridge University Press, 1995), 206.

51. Keynes, "Dr. Melchior," 234.

52. Keynes, *The Economic Consequences of the Peace*, 39.

53. Lytton Strachey to John Maynard Keynes, quoted in Michael Holroyd, *Lytton Strachey* (London: Heineman, 1978), 374.

54. Austin Chamberlain to Ida Chamberlain.

55. A. J. P. Taylor, *The Origins of the Second World War* (London: Penguin Books, 1964), 26.

56. Paul Mantoux, *The Carthaginian Peace or the Economic Consequences of Mr. Keynes* (Oxford: Oxford University Press, 1946).

57. Wickham Steed, "A Critic of the Peace," "The Candid Friend at Versailles," "Comfort for Germany," *John Maynard Keynes: Critical Responses* (Charles Robert McCann, ed. (London: Taylor & Francis, 1998), 51–60.

58. Thorstein Veblen, "Review of J. M. Keynes' *The Economic Consequences of the Peace*," *Political Science Quarterly* 35 (1920): 467–72.

59. "Europe a Year Later," *New York Times*, May 16, 1920.

60. "Solution of Europe's Disorder, as Seen by Baruch," *New York Times*, April 20, 1920.

61. Joseph A. Schumpeter, *History of Economic Analysis* (London: Allen & Unwin, 1954), 39.

VIII: THE JOYLESS STREET: SCHUMPETER AND HAYEK IN VIENNA

1. Joseph A. Schumpeter, *The Theory of Economic Development* (Oxford: Oxford University Press, 1961), 215.

2. Ludwig von Mises, "The Austro-Hungarian Empire," *Encyclopedia Britannica*, 1921.

3. Schober, quoted in F. L. Carsten, *The First Austrian Republic* (Aldershot, UK: Wildwood House, 1986), 41.

4. Ibid., 45.

5. Peter Gay, *Freud: A Life of Our Time* (New York: W. W. Norton and Co., 1988), 386.

6. Ibid., 382.

7. Anna Eisenmenger, *Blockade: The Diary of an Austrian Middle-Class Woman, 1914–1924* (London: Constable Publishers, 1932), 149.

8. Pierre Hamp, *La Peine des Hommes: Les Chercheurs D'Or* [The Pain of Men: The Seekers of Gold], 1920.

9. Quoted in Carsten, *The First Austrian Republic*, 13.

10. Charles A. Gulik, *Austria from Habsburg to Hitler*, vol. 1 (Berkeley: University of California Press, 1948), 248.

11. Eisenmenger, *Blockade*, 149.

12. Ibid.

13. C. A. Macartney, *The Social Revolution in Austria* (Cambridge, UK: Cambridge University Press, 1926), 215.

14. Alois Mosser and Alice Teichova, "Investment Behavior of Joint Stock Companies," in *The Role of Banks in the Interwar Economy*, Harold James, Hekan Lindgren, Alice Teichova, eds. (Cambridge, UK: Cambridge University Press, 2002), 127.

15. Quoted in Richard Swedberg, *Joseph A. Schumpeter: His Life and Work* (Cambridge, UK: Polity Press, 1991), 68.

16. Quoted in Wolfgang F. Stolper, *Joseph Alois Schumpeter: The Public Life of a Private Man* (Princeton, N.J.: Princeton University Press, 1994), 3.

17. Charles A. Gulik, *Austria from Hapsburg to Hitler*, vol. 1 (Berkeley: University of California Press, 1948), 251.

18. Fritz Machlup, *Tribute to Mises, 1881–1973* (Chislehurst, UK: Quadrangle, 1974).

19. "Ships in Fog," a fragment of a novel Schumpeter started in the 1930s, in Swedberg, *Joseph A. Schumpeter,* appendix 2.

20. Thomas K. McCraw, *Prophet of Innovation: Joseph Schumpeter and Creative Destruction* (Cambridge, Mass.: Harvard University Press, 2007), 140.

21. Quoted in Robert Loring Allen, *Opening Doors: The Life and Work of Joseph A. Schumpeter,* vol. 1, *Europe* (New Brunswick, N.J., and London: Transaction Publishers, 1991), 274.

22. Israel Kirzner, "Austrian Economics," lecture at Foundation for Economic Education, July 26, 2004.

23. Joseph A. Schumpeter, *Business Cycles: A Theoretical, Historical and Statistical Analysis of the Capitalist Process* (New York: McGraw-Hill Company, 1939).

24. Joseph A. Schumpeter, *The Theory of Economic Development: An Inquiry into Profits, Capital, Credit, Interest and the Business Cycle* (New Brunswick, N.J.: Transaction Publishers, 1934).

25. Ibid.

26. Ibid., 245.

27. Joseph A. Schumpeter, *Essays on Entrepreneurs, Innovations, Business Cycles, and The Evolution of Capitalism,* ed. Richard Clemence (New York: Transaction Publishers, 1951), 71–72.

28. Friedrich A. Hayek, *Hayek on Hayek: An Autobiographical Dialogue,* ed. Stephen Kresge (Chicago: University of Chicago Press, 1984).

29. Fritz Machlup to Barbara Chernow, June 12, 1978.

30. Gulik, *Austria from Hapsburg to Hitler,* vol. 1, 134–35.

31. Max Weber, "Der Sozialismus" (1918), in *Gesammelte Aufsätze zur Soziologie, Economy and Society*

32. Otto Bauer, "Der Weg zum Sozialismus" [The Way to Socialism], 1921, serialized in *Arbeiter Zeitung,* January 1919.

33. *Hayek on Hayek,* 54–59.

34. *Monatsberichte,* April and October 1929, pp. 69 and 182. Text and translation provided to the author by Hansjorg Klausinger.

IX: IMMATERIAL DEVICES OF THE MIND:
KEYNES AND FISHER IN THE 1920s

1. Irving Fisher, et al, *Report on National Vitality Bulletin 30 of the Committee of One Hundred on Public Health* (Washington, D.C.: Government Printing Office, 1908), 1.

2. Irving Fisher, "Unstable Dollar and the So-called Business Cycle," *Journal of the American Statistical Association,* vol. 20, no. 150 (June, 1925), 179–202.

3. John Maynard Keynes, quoted in Robert Skidelsky, *John Maynard Keynes,* vol. 2, *The Economist as Savior, 1920–1937* (London: Macmillan, 1992).

4. Ibid.

5. Peter Clarke, *Keynes; The Rise, Fall, and Return of the 20th Century's Most Influential Economist* (New York: Bloomsbury, 2009).

6. John Maynard Keynes, "Alternative Theories of the Rate of Interest," *Economic Journal* 47 (June 1937).

7. John Maynard Keynes, "How Far Are Bankers at Fault for Depressions?," 1913, quoted in Angel N. Rugina, "A Monetary and Economic Dialogue with Lord Keynes," *International Journal of Social Economics* 28, vol. 1, No. 2, 200, www.emeraldinsight.com/journals.htm?articleid=1453937&show=html.

8. John Maynard Keynes, *Tract on Monetary Reform,* 1923.

9. Ibid.

10. Quoted in D. E. Moggridge, *Keynes: An Economists' Biography* (London: Routledge, 1992), 429.

11. John Maynard Keynes, *A Short View of Russia* (London: Hogarth Press, 1925).

12. Ibid.

13. Ibid.

14. Norman and Jean MacKenzie, eds., *The Diary of Beatrice Webb,* vol. 4, *1924–1943: The Wheel of Life* (Cambridge, Mass.: Harvard University Press, 1985) (August 9, 1926).

15. John Maynard Keynes, "My Visit to Berlin," *Collected Writings of John Maynard Keynes,* vol. 10, 383–84; "Das Ende des Laissez-Faire, Ideen zur Verbindung von Privat und Gemeinwirtschaft" [The End of Laissez-Faire: Ideas for Combining the Private and Public Economy], *Zeitschrift für die Gesamte Staatswissenschaft* 82 (1927): 190–91. A review of a lecture given by Keynes in Berlin. In papers: October 1925–June 1926 correspondence, autograph manuscript "My Visit to Berlin," June 23, "The General Strike," June 24, given to Berlin University; Conditions in Germany; Keynes at Melchior's apartment in Berlin for dinner, 1926 visit; source: Felix Somary, *Erinnerungen Aus Meinem Leben,* (Zurich: 1926), 199.

16. *The Letters of Virginia Woolf,* vol. 3.

17. John Maynard Keynes addressing the National Liberal Federation, March 27, 1928, quoted in Robert Skidelsky, *John Maynard Keynes,* vol. 2, *The Economist as Savior, 1920–1937* (London: Macmillan, 1992), 297.

18. Skidelsky, *Keynes,* vol. 2, *The Economist as Savior,* 231.

19. Ibid., 232.

20. Charles Loch Mowat, *Britain Between the Wars, 1918–1940* (London: Methuen and Co., 1956), 262.

21. Skidelsky, *Keynes,* vol. 2, *The Economist as Savior,* 258.

22. John Maynard Keynes to H. G. Wells, January 18, 1928.

23. Mowat, *Britain Between the Wars,* 349.

24. Skidelsky, *Keynes,* vol. 2, *The Economist as Savior,* 302.

25. Irving Norton Fisher, *My Father: Irving Fisher* (New York: Comet Press, 1956), 171.

26. Alan Milward, *War, Economy and Society, 1939–1945* (Berkeley: University of California Press, 1979), 17.

27. Angus Maddison, "Statistics of World Population, GDP, per Capita GDP, 1–2008 AD," www.ggdc.net/maddison/.

28. Joseph Schumpeter, "The Decade of the Twenties," *American Economic Review, 1946* and "Business Cycle Dates," National Bureau of Economic Research.

29. Geoffrey Keynes, quoted in D. E. Moggridge, *Maynard Keynes: An Economist's Biography* (London: Routledge, 1992), 103.

30. Irving Norton Fisher, *My Father: Irving Fisher,* 200.

31. Ibid., 232.

32. Ibid., 117–18.

33. Irving Fisher, address to the American Public Health Association, October 23, 1926.

34. Irving Fisher et al., *Report on National Vitality,* bulletin 30 of the Committee of One Hundred on Public Health (Washington, D.C.: GPO, 1908), 1.

35. Irving Fisher, *Stabilizing the Dollar* (New York: Macmillan, 1920), 75.

36. Irving Fisher, *The Purchasing Power of Money: Its Determination and Relation to Credit Interest and Crises* (New York: Macmillan, 1912).

37. Irving Fisher, "Our Unstable Dollar and the So-Called Business Cycle," *Journal of the American Statistical Association* (June 1925): 181.

38. John Maynard Keynes, "Opening remarks: The Galton Lecture," *Eugenics Review,* vol. 38, no. 1 (1946), 39–40.

39. See Robert W. Dimand, "Economists and 'the Other' Before 1912," *The American Journal of Economics and Sociology,* July 2005, http://findarticles.com/p/articles/mi_m0254/is_3_64/ai_n15337798/?tag=content;col1, and *New International Year Book* (New York: Dodd Meade & Co., 1913).

40. Irving Fisher, "Lecture on The Irving Fisher Foundation," *Collected Works,* vol. I (1997), 35.

41. Ibid.

42. Irving Fisher, "Our Unstable Dollar and the So-Called Business Cycle," 197.

43. Irving Fisher, "Depressions and Money Problems," April 4, 1941.

44. Irving Fisher, "I Discovered the Phillips Curve: 'A statistical relation between unemployment and price changes'" *Journal of Political Economy* 81, no 2; 496–502, reprinted from *International Labour Review,* 1926.

45. Irving Fisher, *New York Times,* September 2, 1923.

46. Irving Fisher, "The Unstable Dollar and the So-called Bisiness Cycle" (1925). 179–202.

47. Irving Fisher, "A Statistical Relation Between Unemployment and Price Changes" (1926), 496–502.

48. Ibid.
49. Irving Fisher, *Battle Creek Sanitarium News,* 25, 7, July 1925.
50. Irving Norton Fisher, *My Father: Irving Fisher,* 57.
51. Ibid., 192, from autobiographical appendix in *Stable Money, A History of the Movement.*
52. Jeremy Siegel, *Stocks for the Long Run* (New York: McGraw-Hill, 2008).
53. Irving Norton Fisher, *My Father: Irving Fisher,* 264.
54. *Recent Economic Changes in the United States* (Chicago: National Bureau of Economic Research, 1929), xii.
55. "Fisher Sees Stocks Permanently High," *New York Times,* October 16, 1929.

X: MAGNETO TROUBLE: KEYNES AND FISHER IN THE GREAT DEPRESSION

1. Arnold J. Toynbee, *Journal of International Affairs,* 1931, 1.
2. David Fettig, "Something Unanticipated Happened," in *The Region* (Minneapolis: Federal Reserve Bank of Minneapolis, 2000).
3. John Maynard Keynes to F. C. Scott, August 15, 1934.
4. John Maynard Keynes, "A British View of the Wall Street Slump," *New York Evening Post,* October 25, 1929.
5. Charles A. Selden, "Big British Labor Gains; Third of Vote Counted; Tory Control Seems Lost," *New York Times,* May 31, 1929, 1.
6. Winston Churchill, "Disposal of Surplus," *Hansard 1803–2005,* April 15, 1929, Commons Sitting, Orders of the Day, www.hansard.millbanksystems .com/commons/1929/apr/15/disposal-of-surplus.
7. Lionel Robbins, *Autobiography of an Economist* (London: Macmillan, 1971), 151.
8. John Maynard Keynes to Lydia Keynes, 1929.
9. Joseph J. Thorndike, "Tax Cuts, Confidence, and Presidential Leadership," September 8, 2008, www.taxhistory.org/thp/readings.nsf/.
10. John Maynard Keynes, "The Great Slump of 1930," *The Nation & Athenæum,* December 20, 1930, and December 27, 1930, www.gutenberg.ca/ ebooks/keynes-slump/keynes-slump-00-h.html.
11. John Maynard Keynes, *The General Theory,* book 6, chapter 22, section 3 (London: Macmillan, 1936), 322.
12. Keynes, "The Great Slump," *Nation.*
13. Ibid.
14. Godfrey Harold Hardy, "Mathematical Proof," in Raymond George Ayoub, *Musings of the Masters: An Anthology of Mathematical Reflections* (New York: American Mathematical Association, 2004), 59.
15. Keynes, *The Great Slump of 1930.*
16. Robert Skidelsky, *John Maynard Keynes,* vol. 2, *The Economist as Savior, 1920–1937* (London: Macmillan, 1992), 333.
17. *Minority Report,* 35, 507n, 657–59, 660, 661, 662.

18. Skidelsky, *Keynes,* vol. 2, *The Economist as Savior,* 32.
19. Sir John Anderson to Ramsay MacDonald, July 31, 1930.
20. October 20, 1930.
21. Ross McKibbin, "The Economic Policy of the Second Labour Government, 1929–1931," *Past and Present* 65 (1975); 95–123.
22. Skidelsky, *Keynes,* vol. 2, *The Economist as Savior,* 524.
23. Irving Fisher, September 2, 1929, quoted in Kathryn M. Dominguez, Ray C. Fair, Matthew D. Shapiro, "Forecasting the Depression: Harvard Versus Yale," *American Economic Review* 78, no. 4 (September 1988); 607.
24. "Fisher Sees Stocks Permanently High," *New York Times,* October 16, 1929, 8.
25. Irving Fisher, January 6, 1930, *Collected Works,* ed. Robert Barber, vol. 14, 4.
26. Harvard Economic Society, *Weekly Letter,* vols. 8 and 9 (Cambridge, Mass.: Harvard University Press, 1929), quoted in Dominguez et al., "Forecasting the Depression," 606.
27. Irving Fisher, The *Stock Market Crash and After* (New York: Macmillan, 1930).
28. Milton Friedman and Anna Jacobson Schwartz, *A Monetary History of the United States, 1867–1960* (Princeton, N.J.: Princeton University Press, 1971).
29. "Scores Coolidge in Market Slump," *New York Times,* January 12, 1930.
30. Robert W. Dimond, "Irving Fisher's Monetary Macroeconomics," in *The Economics of Irving Fisher* (London: Elgar, 1999).
31. Irving Norton Fisher, *My Father, Irving Fisher,* 263.
32. "Harvard Group Sees Debt Plan Benefits: Believes Moratorium Will Balance Exchanges and Remove Pressure on Commodities," *Wall Street Journal,* July 17, 1931, 20; "The 1929 Speculation and Today's Troubles: Controversy as to How Far the 'Great Boom' Caused the Great Depression," *New York Times,* January 1, 1932, 33.
33. Irving Fisher, "The Stock Market Panic in 1929," *Proceedings of the American Statistical Association,* 1930.
34. June 22–23, 1931, quoted in Skidelsky, *Keynes,* 391.
35. John Maynard Keynes, typewritten notes, King's College Archive.
36. John Maynard Keynes, discussion leader, typewritten notes, King's College Archive.
37. Bank of England rate of discount, 1836–1939, National Bureau of Economic Research Macro Data Base, www.nber.org/databases/macrohistory/rectdata/13/m13013.dat.
38. Irving Fisher to Ramsay MacDonald, December 1931.
39. Vanessa Bell Skidelsky, *Keynes,* vol. 2, *The Economist as Savior,* 430.
40. Irving Fisher to Henry Stimson, November 11, 1932, quoted in Fisher, 273.
41. Lauchlin Bernard Currie, *Memorandum Prepared by L. B. Currie, P. T. Ellsworth, and H. D. White* (Cambridge, Mass., 1932), reprinted in *History of Political Economy* 34, no. 3 (Fall 2002): 533–52.

42. Irving Fisher to Margaret Fisher, quoted in Irving Norton Fisher, *My Father: Irving Fisher,* 267.

43. Walter Lippmann, *Interpretations 1933–1935* (New York: Macmillan, 1936), 15.

44. K. M. Dominguez, R. C. Fair, and M. D. Shapiro, "Forecasting the Great Depression: Harvard Versus Yale," *American Economic Review,* 78 (September, 1988), 595–612.

45. David Fettig, "Something Unanticipated Happened," (Minneapolis Fed, 2000).

46. Irving Fisher, *Booms and Depressions: Some First Principles* (New York: Adelphi, 1932).

47. Irving Fisher, "Cancellation of War Debts," Southwest Foreign Trade Conference Address, July 2, 1931, quoted in Giovanni Pavanelli, "The Great Depression in Irving Fisher's Thought," *Fifth Annual Conference of the European Society for the History of Economic Thought,* February 2001.

48. Irving Fisher, *The Depression: Causes and Cures* (Miami: Committee of One Hundred, March 1, 1932).

49. "Economists Urge Release of Gold," *New York Times,* October 28, 1931.

50. *New York Times,* December 9, 1931.

51. Irving Fisher, *Booms and Depressions,* viii.

52. R. G. Tugwell, *Brains Trust* (New York: Viking, 1964), 97.

53. Kennedy, *Freedom from Fear,* 113.

54. Tugwell, 98.

55. Franklin Delano Roosevelt, *Oglethorpe University Commencement Speech,* May 22, 1932, http://georgiainfo.galileo.usg.edu/FDRspeeches.htm.

56. Franklin Delano Roosevelt, *Address to Commonwealth Club,* September 23, 1932, San Francisco, in *Great Speeches* (New York: Courier Dover, 1999).

57. Kennedy, *Freedom from Fear,* 123.

58. John Maynard Keynes, *The Means to Prosperity* (London: Macmillan, 1933).

59. Irving Fisher, George Warren of Cornell, and John Commons of the University of Wisconsin to Franklin Roosevelt, February 25, 1933.

60. *The New York Times,* December 31, 1933.

61. Irving Fisher to Irving Norton Fisher, August 15, 1933.

62. Irving Fisher to Margaret Hazard Fisher, quoted in Irving Norton Fisher, *My Father, Irving Fisher.*

63. Skidelsky, *Keynes,* vol. 3, 506.

64. Ibid.

65. *The New York Times,* May 29, 1933.

66. D. E. Moggridge, *Maynard Keynes: An Economists' Biography* (London: Routledge, 1992), 584.

67. Irving Fisher to Howe (FDR's secretary), May 18, 1934.

68. Irving Fisher to Margaret Hazard Fisher, June 7, 1934.

69. John Maynard Keynes, *American Economic Review,* 1933.

70. John Maynard Keynes, *Lecture Notes*

71. Quoted in Skidelsky, *Keynes*, 503.

72. John Maynard Keynes to George Bernard Shaw, January 1, 1935.

73. Marriner S. Eccles, *Fortune*, April 1937, reproduced in *The Lessons of Monetary Experience: Essays in Honor of Irving Fisher Presented to Him on the Occasion of His 70th Birthday* (New York: Farrar and Rhinehart, 1937), 6.

74. Friedrich Hayek, Austrian Institute of Economic Research Report, February 1929.

75. Friedrich A. Hayek, interview. *Gold and Silver Newsletter* (Newport Beach, Calif.: Monex International, June, 1976).

76. Lionel Robbins, *The Great Depression*, 1934.

77. Ibid.

78. Robbins, *Autobiography of an Economist*, 154.

79. Skidelsky, *Keynes*, vol. 2, *The Economist as Savior*, 469.

80. Beatrice Webb, quoted in José Harris, *William Beveridge: A Biography* (Oxford: Clarendon Press, 1977), 330.

81. Fritz Machlup to Barbara Chernow, June 12, 1978.

82. John Maynard Keynes "The Pure Theory of Money: A Reply to Dr. Hayek," *Econometrica*, vol. 11 (November, 1931), 387–97.

83. Alan Ebenstein, *Friedrich Hayek: A Biography* (New York: Palgrave, 2001), 81.

84. Erich Schneider, *Joseph A. Schumpeter: Leben und Werk eines grossen Sozialekonomenen* [Life and Work of a Great Social Scientist]

85. Harold James, *The German Slump: Politics and Economics, 1924–1936* (Oxford: Clarendon Press, 1986), 6.

86. Joseph Schumpeter, "The Present World Depression: A Tentative Diagnosis," in American Economic Association, *Proceedings*, March 31, 1931.

87. Joseph Dorfman, *The Economic Mind in America*, vol. 4, 168.

88. Joseph Schumpeter, to Rev. Harry Emerson Fosdick at Riverside Church, April 19, 1933.

89. Douglas V. Brown, *The Economics of the Recovery Program* (New York: McGraw-Hill, 1934), reprinted in Joseph Schumpeter, *Essays: On Entrepreneurs, Innovations, Business Cycles, and the Evolution of Capitalism* (New York: Transaction Publishers, 1989).

90. Joseph Schumpeter, review of Keynes's *General Theory of Employment, Interest and Money*, *Journal of the American Statistical Association* (December 1936), 791–95.

XI: EXPERIMENTS: WEBB AND ROBINSON IN THE 1930s

1. Walter Duranty, *New York Times*, July 20, 1931, 1.

2. Beatrice Webb to Arthur Salter, April 12, 1932, Norman and Jeanne MacKenzie, eds., *The Letters of Sidney and Beatrice Webb* (Cambridge, Mass.: Harvard University Press, 1978).

3. Norman and Jean MacKenzie, eds., *The Diary of Beatrice Webb,* vol. 4, *1924–1943: The Wheel of Life* (Cambridge, Mass.: Harvard University Press, 1985), September 23, 1931, and October 10, 1931.

4. Ibid.

5. Ibid., 272.

6. Ibid., May 14, 1932.

7. Ibid.

8. Ibid., September 2, 1931.

9. Ibid.

10. Ibid.

11. Walter Duranty, *New York Times,* November 13, 1932, 1.

12. MacKenzie, *Diary of Beatrice Webb,* vol. 4, 299–301, 315, 328 (March 29, 1933; March 30, 1933; October 21, 1933; February 22, 1934).

13. Beatrice and Sidney Webb, *Soviet Communism: A New Civilization* (London: Longmans, Green and Co., 1935), 265.

14. Bertrand Russell, *Autobiography* (London: George Allen and Unwin, 1967), 74–75.

15. Robert Conquest, *Reflections on a Ravaged Century* (New York: W. W. Norton and Co., 2001), 148.

16. John Maynard Keynes, *Collected Writings,* vol. 23, *Activities 1940–1943* (London: Macmillan, 1979), 5.

17. Malcolm Muggeridge, *Chronicles of Wasted Time,* vol. 1, *The Green Stick* (New York: William Morrow, 1973), 207.

18. MacKenzie, *Diary of Beatrice Webb,* vol. 4, 371 (June 19, 1936).

19. John Maynard Keynes to Kingsley Martin, 1937, in *The Collected Writings of John Maynard Keynes,* vol. 28, *Social, Political and Literary Writings* (London: Macmillan, 1928), 72.

20. John Maynard Keynes, quoted in Muggeridge, *Chronicles,* 469.

21. John Maynard Keynes, "Democracy and Efficiency," *New Statesman and Nation,* January 28, 1939.

22. Ibid.

23. Rita McWilliams Tullberg, "Alfred Marshall and Evangelicalism," in Claudio Sardoni, Peter Kriesler, Geoffrey Colin Harcourt, eds., *Keynes, Post-Keynesianism and Political Economy* (London: Psychology Press, 1999), 82.

24. Austin Robinson to Joan Robinson, Robinson Papers, Kings College Archive.

25. Major General Sir Edward Speers, "Forward," in Sir Frederick Maurice and Nancy Maurice, *The Maurice Case* (London: Archon Books, 1972), 95–96.

26. Quoted in Marjorie Shepherd Turner, *Joan Robinson and the Americans* (New York: M. E. Sharpe, 1989), 13.

27. Margaret Gardiner, *A Scatter of Memories* (London: Free Association Books, 1988), 65.

28. Interview with Geoffrey Harcourt, Jesus College, University of Cambridge, 2000.

29. Joan Robinson to Richard Kahn, n.d., November 1930.

30. Joan Robinson to Stevie Smith

31. Ibid.

32. Austin Robinson to Joan Robinson, n.d., April 1926.

33. *Diary of Beatrice Webb.*

34. Dorothy Garratt to Joan Robinson, January 26, 1932.

35. Joan Robinson to Richard Kahn, March 1931.

36. Ibid.

37. Nahid Aslanbeigui and Guy Oakes, *The Provocative Joan Robinson: The Making of a Cambridge Economist* (Durham, N.C.: Duke University Press, 2009).

38. James Meade, quoted in George R. Feiwell, *Joan Robinson and Modern Economic Theory* (New York: New York University Press, 1989), 917.

39. Ibid., 916.

40. Aslanbeigui and Oakes, *The Provocative Joan Robinson.*

41. Joan Robinson to Austin Robinson, October 11, 1932.

42. Joan Robinson to Richard Kahn, Michaelmas term, 1932; Joan Robinson to Austin Robinson, October 11, 1932; Richard Kahn to Joan Robinson.

43. Joan Robinson to Richard Kahn, March 2, 1933.

44. Joan Robinson, introduction to *The Theory of Employment* (London: Macmillan, 1969), xi.

45. Richard Kahn to Joan Robinson, March 1933.

46. Joseph Schumpeter, "Review of Joan Robinson's Theory of Imperfect Competition," *Journal of Political Economy,* 1934.

47. Dorothy Garratt to Joan Robinson, May 25, 1934.

48. Joan Robinson to Richard Kahn, September 5, 1934.

49. John Maynard Keynes to Richard Kahn, February 19, 1938.

50. Andrew Boyle, *Climate of Treason* (London: Hutchinson, 1979), 63, 453 (note 4).

51. Geoffrey Harcourt, "Joan Robinson," *Economic Journal.*

52. Joan Robinson, "Review of *The Nature of the Capitalist Crisis* by John Strachey," *Economic Journal* 46, no. 182 (June 1936): 298–302.

53. Joan Robinson, "Review of Britain Without Capitalists," *Economic Journal* (December 1936).

54. Taqui Altounyan, *Chimes from a Wooden Bell* (London: I. B. Taurus and Co., 1990) and *In Aleppo Once* (London: John Murray, 1969).

55. Ernest Altounyan to Joan Robinson, May 30, 1936.

56. Agatha Christie, *Murder on the Orient Express* (New York: Collins, 1934), 17.

57. Quoted in Altounyan, *Chimes from a Wooden Bell.*

58. Interview with Frank Hahn, Churchill College, University of Cambridge, 2000.

XII: THE ECONOMISTS' WAR:
KEYNES AND FRIEDMAN AT THE TREASURY

1. John Maynard Keynes, *How to Pay for the War* (London: Macmillan, 1940), 17.
2. Friederich von Hayek to Fritz Machlup, October 1940.
3. Robert Skidelsky, *John Maynard Keynes,* vol. 3, *Fighting for Freedom, 1937–1946* (New York: Viking, 2001), 51.
4. Friedrich Hayek to Fritz Machlup, March 19, 1934 (Machlup Papers, box 43, folder 15).
5. John Maynard Keynes, "Paying for the War I: The Control of Consumption," *Times* (London), November 14, 1939, 9, and "Paying for the War II: Compulsory Savings," *Times* (London), November 15, 1939, 9.
6. Skidelsky, *Keynes,* vol. 3, *Fighting for Freedom,* 142.
7. John Maynard Keynes to F. A. Hayek, guoted in Skidelsky, ibid., 56.
8. John Maynard Keynes to J. T. Sheppard, August 14, 1940.
9. Skidelsky, *Keynes,* vol. 3, 179.
10. Winston Churchill to Clementine Churchill, July 18, 1914, in Mary Soames, *Winston and Clementine: The Personal Letters of the Churchills* (New York: Houghton Mifflin Harcourt, 2001), 96.
11. John Maynard Keynes to Russell Leffingwell, July 1, 1942.
12. John Maynard Keynes to P. A. S. Hadley, September 10, 1941.
13. "Wheeler Doubts President Will Order Convoys," *Chicago Daily Tribune,* May 10, 1941.
14. Sir John Wheeler Bennet, *New York Times,* November 24, 1940, 7.
15. Alan Milward, *War, Economy and Society, 1939–1945* (Berkeley: University of California Press, 1979), 49.
16. Gerhard L. Weinberg, *A World at Arms: A Global History of World War II* (Cambridge: Cambridge University Press, 2005); David Kennedy, *Freedom from Fear: The American People in Depression and War* (Oxford: Oxford University Press, 1999), 446.
17. Winston Churchill to Franklin D. Roosevelt, December 7, 1940, Great Britain Diplomatic Files.
18. Franklin D. Roosevelt, press conference, White House, December 17, 1940, http://docs.fdrlibrary.marist.edu/ODLLPc2.html.
19. Ibid.
20. Franklin Roosevelt, "Fireside Chat" radio address, White House, December 29, 1940, http://docs.fdrlibrary.marist.edu/122940.html.
21. Winston S. Churchill to Franklin D. Roosevelt, December 31, 1940, in Martin Gilbert, ed., *The Churchill War Papers* (New York: W. W. Norton and Co., 2000), 3:11.
22. Winston S. Churchill to Sir Kingsley Wood, March 20, 1941, in Gilbert, *The Churchill War Papers,* 3:372.

23. Franklin D. Roosevelt, campaign address, Boston, October 30, 1940, www .presidency.ucsb.edu.

24. Franklin D. Roosevelt, conversation in the Oval Office with unidentified aides, October 4, 1940, White House Office Transcripts, 48–61:1, Franklin D. Roosevelt Presidential Library and Museum, Hyde Park, New York, http:// docs.fdrlibrary.marist.edu:8000/transcr7.html.

25. Weinberg, *A World at Arms,* 240.

26. John Maynard Keynes, quoted in Skidelsky, *Keynes,* vol. 3, *Fighting for Freedom,* 102.

27. Paul A. Samuelson in *The Coming of Keynesianism,* 170.

28. Ibid.

29. Quoted in Skidelsky, *Keynes,* vol. 3, *Fighting for Freedom,* 116.

30. John Kenneth Galbraith, *A Life in Our Times,*

31. F. Scott Fitzgerald, *This Side of Paradise* (New York, 1920).

32. Milton Friedman and Rose Friedman, *Two Lucky People* (Chicago: University of Chicago Press, 1998).

33. Ibid.

34. Ibid.

35. Ibid.

36. Herbert Stein, *Presidential Economics: The Making of Economic Policy from Roosevelt to Clinton* (Washington, D.C.: American Enterprise Institute, 1994).

37. Friedman and Friedman, *Two Lucky People.*

38. Ibid., 107

39. Galbraith, *A Life in Our Times,* 163.

40. Ibid. Galbraith was assistant, then deputy, chief of the Price Division. Richard Gilbert, George Stigler, Walter Salant, and Herbert Stein belonged to OPA's economics staff.

41. Quoted in ibid., 133. The General Maximum Price Regulation of 1942 went into effect on April 28.

42. Friedman and Friedman, *Two Lucky People,* 113. See also Milton Friedman and Walter Salant, *American Economic Review* 32 (June 1942); 308–20; Milton Friedman, "The Spendings Tax as a Wartime Fiscal Measure," *American Economic Review* (March 1943); 50–62.

43. Friedman and Friedman, *Two Lucky People.*

44. Ibid., 113

45. Ibid.

46. Withholding was first imposed on 1943 income, but the Ruml Plan, the subject of the 1942 debate, called for it to be imposed on 1942 income. The Revenue Act of 1942 passed on October 21, 1942; the Current Tax Payment Act of 1943, on June 9, 1943.

47. Friedman and Friedman, *Two Lucky People.*
48. Ibid., 116.
49. Isaiah Berlin, March 3, 1942, *Washington Dispatches,* 25.
50. Ibid.
51. Herbert Stein, *Presidential Economics,* 68.

XIII: EXILE: SCHUMPETER AND HAYEK IN WORLD WAR II

1. Friedrich Hayek, *The Road to Serfdom* (Chicago: University of Chicago Press, 1944).
2. Joseph Schumpeter, *Capitalism, Socialism and Democracy* (New York: Harper and Co., 1942).
3. Ibid.
4. Joseph Schumpeter to Irving Fisher, February 18, 1946.
5. Joseph Schumpeter, Diary, October 30, 1942.
6. John Hicks, "The Hayek Story," in *Critical Essays in Monetary Theory* (Oxford, UK: Oxford University Press, 1967).
7. Friedrich Hayek to Fritz Machlup, January 1935.
8. Friedrich Hayek to Fritz Machlup, May 1, 1936.
9. Friedrich Hayek to Fritz Machlup.
10. Friedrich Hayek to Lord Macmillan, September 9, 1939.
11. Friedrich Hayek to Fritz Machlup, December 14, 1940.
12. Friedrich Hayek to Fritz Machlup, June 21, 1940.
13. Friedrich Hayek to Alvin Johnson, August 8, 1940.
14. Friedrich Hayek to Alfred Schutz, September 26, 1943.
15. Friedrich Hayek to Fritz Machlup.
16. Friedrich Hayek to Fritz Machlup, June 21, 1940.
17. Friedrich Hayek to Herbert Furth, January 27, 1941.
18. Friedrich Hayek to Fritz Machlup, January 2, 1941.
19. Friedrich Hayek to Fritz Machlup.
20. Friedrich Hayek to Fritz Machlup, July 31, 1941.
21. Friedrich Hayek, *The Road to Serfdom.*
22. Ibid.
23. Ibid., 135.
24. Friedrich Hayek, "The Road to Serfdom: Address Before the Economic Club of Detroit, April 23, 1945," typescript, Hoover Institution.
25. Quoted in Fritz Machlup to Friedrich Hayek, January 21, 1943.
26. Ordway Tead to Fritz Machlup, September 25, 1943.

ACT III: PROLOGUE: NOTHING TO FEAR

1. James MacGregor Burns, *Roosevelt: The Soldier of Freedom, 1940–1945* (New York: Harcourt Brace Jovanovich, 1970), 424.

2. Franklin Delano Roosevelt, "Economic Bill of Rights," State of the Union Address, January 11, 1944, transcript, Franklin D. Roosevelt Presidential Library and Museum, Hyde Park, New York, http://www.fdrlibrary.marist.edu/archives/stateoftheunion.html.

3. Ibid.

4. James McGregor Burns, *Roosevelt: The Soldier of Freedom,* vol. 2 (New York: Harcourt Brace Jovanovich, 1970), 426.

5. John Maynard Keynes to Sir J. Anderson, August 10, 1944, quoted in Robert Jacob Alexander Sidelsky, *John Maynard Keynes,* vol. 3, Fighting for Freedom (New York: Viking Press, 2001), 360.

6. Gunnar Myrdal, "Is American Business Deluding Itself?," *Atlantic Monthly* (November 1944), 51–58.

7. Roosevelt, State of the Union Address, January 11, 1944.

8. Ibid.

9. Alvin H. Hansen, "The Postwar Economy," in Seymour E. Harris, ed., *Postwar Economic Problems* (New York: McGraw-Hill Book Company, 1943), 12.

10. Paul A. Samuelson, "Full Employment After the War, in Harris, *Postwar Economic Problems,* 27, 52.

11. Joseph A. Schumpeter, "Capitalism in the Postwar World," in Harris, *Postwar Economic Problems,* 120–21.

12. Ibid.

13. Roosevelt, State of the Union Address, January 11, 1944.

14. Myrdal, "Is American Business Deluding Itself?"

15. George Orwell, *Nineteen Eighty-Four* (London: Penguin Classics, 2009), 231.

16. Roosevelt, State of the Union Address, January 11, 1944.

17. John Lewis Gaddis, *The Cold War: A New History* (New York: Penguin, 2006), 14.

18. John Maynard Keynes, *The General Theory* (1936; repr. London: MacMillan & Co., 1954), 383–84.

XIV: PAST AND FUTURE: KEYNES AT BRETTON WOODS

1. FDR, Message to Delegates at Bretton Woods, July 1944.

2. John Maynard Keynes to Florence Keynes, June 28, 1944.

3. Robert Skidelsky, *John Maynard Keynes,* vol. 3, *Fighting for Freedom 1937–1946* (New York: Viking, 2000), 343.

4. John Maynard Keynes to Friedrich Hayek, July 1944.

5. John Maynard Keynes, "My Early Beliefs," in *Essays in Biography.*

6. Lionel Robbins, *Autobiography of an Economist* (London: Macmillan, 1976).

7. John Maynard Keynes to Friedrich Hayek, July 1944.

8. Lydia Keynes quoted in Liaquat Ahmed, *Lords of Finance: The Bankers Who Broke the World* (New York: Penguin, 2009).

9. Cordell Hull, *The Memoirs of Cordell Hull* (New York: Macmillan, 1948), 1:81.

10. Papers of Harry Dexter White, Princeton University Archive.

11. Skidelsky, *Keynes,* vol. 3, *Fighting for Freedom,* 348.

12. Ibid.

13. Ibid.

XV: THE ROAD FROM SERFDOM: HAYEK AND THE GERMAN MIRACLE

1. George Orwell, review of *The Road to Serfdom* (1944).

2. Isaiah Berlin, March 31, 1945, *Washington Despatches, 1941–1945: Weekly Political Reports from the British Embassy* (Chicago: University of Chicago Press, 1981).

3. Berlin, *Despatches,* May 6, 1945.

4. Berlin, *Despatches,* June 10, 1945.

5. Friedrich Hayek to Fritz Machlup, and *Message to Congress on the Concentration of Economic Power,* April 29, 1938.

6. Marquis Childs. "Washington Calling: Hayek's 'Free Trade,'" *Washington Post,* June 6, 1945, http://www.proquest.com.ezproxy.cul.columbia.edu/ (accessed February 10, 2011).

7. George Kennan, *Memoirs 1925–1950* (New York: Atlantic Monthly Press, 1967), 292.

8. Friedrich Hayek to Lydia Keynes, April 21, 1946.

9. Harry S. Truman, March 12, 1947, transcript of the Truman Doctrine (1947), http://www.ourdocuments.gov/; Robert A. Pollard, *Economic Security and the Origins of the Cold War, 1945–1950* (New York: Columbia University Press, 1985), 123, http://questia.com.

10. Friedrich Hayek, "Opening address to a conference at Mont Pelerin," 1947, P. G. Klein, ed., *The Collected Works of F. A. Hayek, Volume IV: The Fortunes of Liberalism,* (Chicago, Ill.: University of Chicago Press, 1992), 238.

11. Friedrich A. Hayek, *Nobel Prize Winning Economist Friedrich A. von Hayek* (Los Angeles: University of California at Los Angeles Oral History Program, 1983), http://www.archive.org/stream/nobelprizewinnin00haye#page/n11/mode/2up.

12. *Statement of Aims,* Mont Pelerin Society, https://www.montpelerin.org/montpelerin/mpsGoals.html.

13. Orson Welles's contribution to *The Third Man,* 1949 in Robert Andrews, *The Columbia Dictionary of Quotations,* (New York: Columbia University Press, 1993), 888.

14. Quoted in Kurt R. Leube, "Hayek in War and Peace," *Hoover Digest,* no. 1, 2006.

15. Ray Monk, *Wittgenstein: The Duty of Genius* (New York: Penguin Books), 518.

16. Friedrich Hayek, *Hayek on Hayek: An Autobiographical Dialog,* Stephen Kresge, ed. (Chicago: University of Chicago Press, 1994), 105–6.

17. Austin Robinson, *First Sight of Postwar Germany, May–June, 1945* (Cambridge: The Canteloupe Press, 1986).

18. Ibid.

19. John Maynard Keynes to Austin Robinson, June, 1945.

20. Ludwig Erhard, *Germany's Comeback in the World Market* (New York: Macmillan, 1954).

XVI: INSTRUMENTS OF MASTERY: SAMUELSON GOES TO WASHINGTON

1. Quoted in Philip Saunders and William Walstead, *The Principles of Economics Course* (New York: McGraw-Hill, 1990), ix.

2. Paul A. Samuelson, *The Samuelson Sampler* (Glen Ridge, N.J.: Thomas Horton & Co., 1973), vii.

3. Paul A. Samuelson with Everett Hagen, "Studies in Wartime Planning for Continuing Full Employment" (Washington, D.C.: National Resources Planning Board, 1944); Paul A. Samuelson et al., *After the War 1918–1920* (Washington, D.C.: National Resources Planning Board, 1943); and Paul A. Samuelson et al, (Washington, D.C.: National Resources Planning Board, 1942).

4. Paul Samuelson, Godkin Lecture I.

5. Alan Millward, *War, Economy and Society, 1939–1945* (Berkeley: University of California Press, 1980).

6. Will Lissner, *New York Times,* September 3, 1944, 23.

7. Paul Samuelson, "Unemployment Ahead and the Coming Economic Crisis," *New Republic,* September, 1944.

8. Quoted in Polenberg, 94.

9. Interview, Paul Samuelson.

10. Paul A. Samuelson and William Nordhaus, *Economics: The Original 1948 Edition,* 573.

11. Robert Summers, father of Lawrence Summers. He and Harold Samuelson, Paul Samuelson's older brother, changed their names to "Summers" in an attempt to avoid anti-Semitism.

12. Florence Wieman, South Chicago, *The Scroll,* May, 1930.

13. Paul A. Samuelson, "Reflections on the Great Depression," typescript.

14. Ibid., p. 58.

15. Paul A. Samuelson, "How Foundations Came To Be," *Journal of Economic Literature* (1998), 1376.

16. Tsuru Shigeto, "Reminiscences of Our 'Sacred Decade of Twenties,'" *The American Economist* (Fall 2007).

17. Samuelson, "Reflections on the Great Depression."

18. Herbert Stein, *Presidential Economics.*

19. Paul A. Samuelson, interview.

20. Joseph Schumpeter to Paul A. Samuelson, November 3, 1947.

21. Robert Maynard Hutchins, quoted in David Kennedy, *Freedom from Fear: The American People in Depression and War* (Oxford, UK: Oxford University Press, 2001).

22. Paul A. Samuelson to F. Wheeler Loomis, director, M.I.T. Radiation Laboratory, April 26, 1945.

23. Kenneth Elzinga, "The Eleven Principles of Economics," *Southern Economic Review* (April 1992).

24. Stanley Fisher, interview with Paul A. Samuelson, typescript transcript.

25. William F. Buckley, *God and Man at Yale* (Washington, D.C.: Regnery Gateway, 1951).

26. Ibid., 49.

27. Ibid., 60.

28. Ibid., 81.

29. Paul A. Samuelson, *Economics* (New York: McGraw-Hill, 1948), 412.

30. Ibid., 434.

31. Ibid., 152.

32. Ibid., 380.

33. Ibid., 433.

34. Ibid., 3.

35. Ibid., 584.

36. Paul A. Samuelson, *Economics,* 4th ed. (New York: McGraw-Hill), 209–210.

37. Samuelson, *Economics,* 1st ed., 607.

38. Ibid., 271.

39. Ibid.

XVII: GRAND ILLUSION: ROBINSON IN MOSCOW AND BEIJING

1. Joan Robinson, lecture, Cambridge University, quoted in Harry G. Johnson, *On Economics and Society* (Chicago: University of Chicago Press, 1975), 110.

2. Joan Robinson, *Conference Sketch Book, Moscow, April 1952* (Cambridge: W. Heffer and Sons, 1952), 19.

3. Ibid., 6, 21, 23–24.

4. Alec Cairncross, "The Moscow Economic Conference," *Soviet Studies* 4, no. 2 (October 1952), 114.

5. Robinson, *Conference Sketch Book,* 5.

6. Robinson, *Conference Sketch Book,* 7–8; Cairncross, "The Moscow Economic Conference," 119.

7. Robinson, *Conference Sketch Book,* 23.

8. "Russia: Two Faces West," *Time,* April 14, 1952.

9. Robinson, *Conference Sketch Book,* 11.

10. Committee for the Promotion of International Trade, *International Economic Conference in Moscow April 3–12, 1952* (Moscow, 1952); Oleg Hoeff-

ding, "East-West Trade Possibilities: An Appraisal of the Moscow Economic Conference," *American Slavic and East European Review,* 1953; Richard B. Day, *Cold War Capitalism: The View from Moscow, 1945–1975* (Armonk, NY: M. E. Sharpe, 1995), 79.

11. Committee for the Promotion of International Trade, *International Economic Conference,* 85.

12. Robinson, *Conference Sketch Book,* 28.

13. Ibid.

14. Ibid., 3, 5.

15. Joan Robinson to Richard Kahn, April 4, 1952, Papers of Richard Ferdinand Kahn, RFK/13/90/5, King's College, University of Cambridge.

16. Paul Samuelson, "Remembering Joan," in G. R. Feiwell, ed., *Joan Robinson and Modern Economic Theory* (London: Macmillan, 1989), 135.

17. Paul Preston, Michael Partridge, and Piers Ludlow, "British Documents on Foreign Affairs: Reports and Papers from the Foreign Office Confidential Print" (Lexis Nexis, 2006).

18. Cairncross, "The Moscow Economic Conference," 113, 118.

19. *Economic Problems of Socialism in the U.S.S.R.* (New York: International Publishers, 1952), 26, 30. Stalin's "Remarks on Economic Questions in Connection with Discussion of November 1951" were distributed around February 7, 1952, to Central Committee members working on Stalin's textbook on Soviet economic theory. "Remarks" was published later that year as *Economic Problems.*

20. John Lewis Gaddis, *We Now Know: Rethinking Cold War History* (New York: Oxford University Press USA, 1997), 195.

21. Stalin, *Economic Problems of Socialism,* 27.

22. Richard B. Day, *Cold War Capitalism: The View from Moscow, 1945–1975* (Armonk, NY: M. E. Sharpe, 1995), 76.

23. Ethan Pollock, "Conversations with Stalin on Questions of Political Economy," July 2001, Working Paper No. 33, Cold War International History Project, Woodrow Wilson International Center for Scholars, http://www.wilsoncenter.org/topics/pubs/ACFB07.pdf.

24. Robinson, *Conference Sketch Book.*

25. Geoffrey Colin Harcourt, "Some Reflections on Joan Robinson's Changes of Mind and Their Relationship to Post-Keynesianism and the Economics Profession," in *Capitalism, Socialism and Post-Keynesianism: Selected Essays of George Harcourt* (Cheltenham, UK: Edward Elgar, 1995), 111.

26. Joan Robinson, *The Problem of Full Employment: An Outline for Study Circles* (London: Workers Educational Association, 1943).

27. Stephen Brooke, "Revisionists and Fundamentalists: The Labour Party and Economic Policy During the Second World War," *Historical Journal* (March 1989), 158.

28. Elizabeth Durbin, *New Jerusalems: The Labour Party and the Economics of Democratic Socialism* (London: Routledge and Keegan Paul, 1985), 164.

29. Quoted in C. W. Guillebaud, "Review of Joan Robinson, *Private Enterprise or Public Control: Handbook for Discussion Groups,*" *Economica* 10, no. 39 (August 1943), 265.

30. J. E. King, "Planning for Abundance: Joan Robinson and Nicholas Kaldor, 1942–1945," in European Society for the History of Economic Thought, *Political Events and Economic Ideas* (London: Elgar), 307.

31. Jonathan Schneer, "Hopes Deferred or Shattered: The British Labour Left and the Third Force Movement, 1945–1949," *Journal of Modern History* (June 1984), 197.

32. Joseph Stalin, *Meeting Between Comrades Stalin and H. Pollitt 31st May 1950,* transcript, Russian State Archive of Social and Political History, 4.

33. Eric Shaw, *Discipline and Discord in the Labour Party* (Manchester, UK: University of Manchester Press, 1988).

34. Harold Laski, *The Secret Battalion,* a 1946 pamphlet defending the Labour Party's rejection of the Communist Party of Great Britain's application for affiliation.

35. Joan Robinson, "Preparation for War," *Cambridge Today,* October 1951, reprinted in *Monthly Review,* no 2 (1951), 194–95.

36. Richard Gardner, *Sterling Dollar Diplomacy: Anglo-American Collaboration in the Reconstruction of Multilateral Trade* (London: Clarendon, 1956), 298.

37. Schneer, "Hopes Deferred or Shattered."

38. Joan Robinson, BBC, *London Forum,* June 25, 1947, quoted, ibid., 221.

39. "Why the CP Says Reject the Marshall Plan," July 5, 1947, quoted in Keith Laybourn, *Marxism in Britain: Dissent, Decline and Re-emergence, 1945–c.2000* (New York: Taylor and Francis, 2006), 35.

40. Robert Solow, quoted in Marjorie Shepherd Turner, *Joan Robinson and the Americans* (Armonk, NY: M. E. Sharpe, 1989), 143.

41. Joan Robinson to Richard Kahn, King's College Archive.

42. Christopher Andrew, *Defend the Realm: The Authorized History of MI5* (New York: Alfred A. Knopf, 2009), 400; Marjorie S. Turner, *Joan Robinson and the Americans,* 86; Percy Timberlake, *The 48 Group: The Story of the Icebreakers in China* (London: 48 Group Club, 1994).

43. Milton Friedman and Rose Friedman, *Two Lucky People: Memoirs* (Chicago: University of Chicago Press, 1998), 245–46.

44. Robert Clower, quoted in Turner, *Joan Robinson and the Americans,* 133.

45. Alvin L. Marty, "A Reminiscence of Joan Robinson," *American Economic Association Newsletter,* (October 1991), 5–8.

46. Arthur Pigou to John Maynard Keynes, June 1940, King's College Archive.

47. Michael Straight, quoted in Turner, *Joan Robinson and the Americans,* 56.

48. Brian Loasby, "Joan Robinson's Wrong Turning," in Ingrid H. Rima, ed., *The Joan Robinson Legacy* (London: M. E. Sharpe, 1991), 34.

49. Joan Robinson, "Mr. Harrod's Dynamics," *Economic Journal* (March 1949), 81.

50. Joan Robinson, "Review of Joseph Schumpeter, *Capitalism, Socialism and Democracy,*" *Economic Journal*, 1943.

51. Sidney Hook, "Review of Rosa Luxemburg, *The Accumulation of Capital, with a Preface by Joan Robinson,*" 1951,

52. Joan Robinson, *The Accumulation of Capital* (London: MacMillan, 1956).

53. Roy Forbes Harrod, *Towards a Dynamic Economics* (London: Macmillan, 1948).

54. Robinson, "Mr. Harrod's Dynamics," 85.

55. Joan Robinson, "Model of an Expanding Economy," *Economic Journal* (March 1952).

56. Joan Robinson, *Letters from a Visitor to China* (Cambridge: Students' Bookshop, 1954), 8.

57. Joan Robinson, "Has Capitalism Changed?" *Monthly Review,* 1961.

58. Samuelson, "Remembering Joan," 121–43.

59. Stanislaw H. Wellisz, review, *Review of Economics and Statistics* 40, no. 1 (February 1958): 87–88.

60. Elizabeth S. Johnson and Harry G. Johnson, *The Legacy of Keynes* (Oxford: Basil Blackwell, 1978).

61. Samuelson, "Remembering Joan."

62. Abba Lerner, "*The Accumulation of Capital,*" *American Economic Review* (September 1957): 693, 699.

63. L. R. Klein, "*The Accumulation of Capital* by Joan Robinson," *Econometrica* 26, no. 4 (October 1958), 622, 624.

64. Robert Solow, "Technical Change and the Aggregate Production Function," *Review of Economics and Statistics* 39, no. 3 (August 1957); 320; and Robert Solow, quoted in Turner, *Joan Robinson,* 143.

65. Joan Robinson, *Private Enterprise or Public Control* (London: English University Press Ltd.), 13–14.

66. Quoted in Jason Becker, *Hungry Ghosts: Mao's Secret Famine* (London: Macmillan, 1998), 292.

67. George J. Stigler, review of *Economic Philosophy* by Joan Robinson, *The Journal of Political Economy* 71, no. 2 (April 1963), 192–93 (emphasis added).

XVIII: TRYST WITH DESTINY: SEN IN CALCUTTA AND CAMBRIDGE

1. Amartya Sen, *Development as Freedom* (New York: Alfred A. Knopf, 1999), 36.

2. Sankar Ray, "The Third World Apologist Finally Strikes," *Calcutta Online,* October 15, 1998, http://www.nd.edu/~kmukhopa/cal300/sen/art1014m.htm.

3. The Royal Swedish Academy of Sciences, "The Prize in Economics 1998—Press Release," news release, October 14, 1998, http://nobelprize.org/nobel_prizes/economics/laureates/1998/press.html.

4. John B. Seely, *The Road Book of India* (London: J. M. Richardson and G. B. Whittaker, 1825), 12: "Dacca . . . is celebrated for the manufacture of the finest and most beautiful muslins." Muslin was a favorite topic of Jane Austen's letters to her sister Cassandra. In *Northanger Abbey* (1818), a potential suitor wows a chaperone with the "prodigious bargain" he got on a gown for his sister made of "true Indian muslin."

5. William Sproston Caine, *Picturesque India: A Handbook for European Travellers* (London: George Routledge and Sons Limited, 1891), 367.

6. Amartya Sen, interview by the author. Except where otherwise noted, quotes of Mr. Sen are from discussions and interviews with the author.

7. Archibald Percivel Wavell to Winston Churchill, telegram, February 1944, in Penderel Moon, ed., *Wavell: The Viceroy's Journal* (Oxford University Press, 1973), 54.

8. Amartya Sen, "Autobiography," http://nobelprize.org/nobel_prizes/economics/laureates/1998/sen-autobio.html.

9. Ibid.

10. Amita Sen, interview by the author.

11. Indira Gandi, *Selected Speeches and Writings of Indira Gandi*, vol. 5, *January 1, 1982–October 30, 1984* (Delhi: Publications Division, Ministry of Information and Broadcasting, Government of India, 1986), 457.

12. Arjo Klamer, "A Conversation with Amartya Sen," *Journal of Economic Perspectives* 3, no. 1 (Winter 1989), 148.

13. Jean Drèze and Amartya Sen, *India, Development and Politics* (Oxford University Press, 2002), 3.

14. Amartya Sen, "The Impossibility of a Paretian Liberal," *Journal of Political Economy* 78 (1970): 152–57.

15. Drèze and Sen, *India, Development and Politics*, 2.

16. World Bank World Development Indicators (accessed April 13, 2011), http://data.worldbank.org/indicators.

EPILOGUE: IMAGINING THE FUTURE

1. John Maynard Keynes, *The General Theory of Employment, Interest and Money* (New York: Harcourt, Brace, 1936), 383.

2. Robert Solow, "Faith, Hope and Clarity" in David Colander and Alfred William Coats, eds., *The Spread of Economic Ideas* (Cambridge, Mass.: Cambridge University Press, 1993), 37.

Index

V

Photo Credits